T5-AFD-224

A SHORT COURSE IN ORGANIC CHEMISTRY

A SHORT COURSE IN ORGANIC CHEMISTRY

EDWARD E. BURGOYNE, Professor of Chemistry, Arizona State University

McGraw-Hill Book Company • New York • St. Louis • San Francisco
Auckland • Bogotá • Düsseldorf • Johannesburg • London • Madrid
Mexico • Montreal • New Delhi • Panama • Paris • São Paulo
Singapore • Sydney • Tokyo • Toronto

A SHORT COURSE IN ORGANIC CHEMISTRY

1234567890 VHVH 7832109

This book was set in Times Roman by Progressive Typographers.
The editors were Donald C. Jackson, Sue Barnes, and Sibyl Golden; the designer was Hermann Strohbach; the production supervisor was Leroy A. Young. The photo editor was Libby Forsyth. The drawings were done by J & R Services, Inc.
Von Hoffmann Press, Inc., was printer and binder.

Library of Congress Cataloging in Publication Data

Burgoyne, Edward E

A short course in organic chemistry.

Includes index.

1. Chemistry, Organic. I. Title.

QD251.2.B83 547 78-17079

ISBN 0-07-009171-4

TO MARY IDA

CONTENTS

PREFACE

This text has been written for a one-semester or a one-quarter course to introduce students of nursing, biology, agriculture, forestry, and home economics to the principles of structure and reactions in organic chemistry and to the properties of organic compounds related to the various fields of study and life in general. At least a semester of general chemistry is assumed to have been previously completed; in it the principles of atomic and molecular theory, bonding, the periodic table, states of matter, solutions, acids and bases, and ionic equilibrium will have been studied. Some of those subjects are reviewed briefly in Chapter 1, but the discussion there is not adequate for an initial presentation.

The principles of structure and isomerism of molecules, functional groups, and nomenclature are introduced in Chapter 2. That should be considered only an introduction, and the student should not be expected to fully master those subjects at this time. The principles of Chapter 2 will be expanded, with learning in detail expected, as the chapters on the various functional groups are considered. The subject of organic chemistry cannot be arranged in one unique order of topics that can be followed as in going through a tunnel. The subject matter spreads out more like the surface of a sphere and has many related reactions and ideas. One could begin the study at several different places in the body of knowledge and spread out from there. One successful textbook used in the past began with the alcohols rather than with the alkanes. Since even the reactions and the preparation of the alkanes and alkenes cannot be discussed without some reference to alkyl halides and alcohols, it is useful to begin with a little knowledge of functional groups and nomenclature.

A brief introduction to the use of spectra is included in Chapter 2. It is intended to make the student aware of the use of spectra in analysis and structure determination, and not to make him or her proficient in using them. Some instructors will wish to omit the material altogether; others may wish to supplement it.

Although covering such a large field in one semester requires brevity, the presentation must be modern, as comprehensive as the fields of the students require, and as rational and palatable as possible within such a framework. Consequently, the simpler concepts of molecular orbitals, resonance, stereoisomerism, and reaction mechanisms are included. My experience has been that students can master those concepts and that they do appreciate them as an aid in understanding the diverse properties and reactions of organic compounds rather than relying entirely on rote memorization of a mass of descriptive organic chemistry. Each instructor will make a decision as to depth of understanding and retention he or she will require for the reaction mechanisms given. In a course of this type, however, the chief value of the mechanisms is to facilitate understanding and learning of the important reactions; the mechanisms themselves have little importance to the students.

Some synthetic reactions are given not only as examples of reactions of organic compounds but also as a way of giving the student an appreciation of the sources of materials. Synthetic chemistry is involved in some of the problems as a drill to familiarize the student with important compounds, reactions, and concepts. However, the major emphasis is not on synthesis; it is on an understanding of properties through study of structure and reactions.

The chapters on fats, carbohydrates, proteins, enzymes, and metabolism apply the principles of organic chemistry to the more complicated compounds involved in biochemical reactions. In order for the contents to be covered in one semester, this introduction to biochemistry is necessarily brief. Some students will want to take an additional semester course in elementary biochemistry.

I am grateful to the following persons for their helpful comments upon reading the manuscript: Professor David L. Adams, North Shore Community College; Professor Jon Michael Bellama, University of Maryland; Professor Jack E. Leonard, Texas A&M University; Professor Lawrence T. Scott, University of Nevada at Reno; Professor Walter S. Trahanovsky, Iowa State University. I acknowledge with gratitude my colleagues at Arizona State University who read portions of the original manuscript and offered suggestions and encouragement. They are: Professor William J. Burke, Professor Wayne W. Luchsinger, and Professor Tom R. Thomson. I thank my son Edward R. Burgoyne for taking and finishing the photos of the molecular models from which the drawings were made. I wish to thank Sue Barnes for helping to develop this manuscript and Libby Forsyth for selecting the photographs.

Edward E. Burgoyne

TO THE STUDENT

Organic chemistry has become a very complex, highly organized, and structured science, and reading a textbook or lecture notes over and over is not a very efficient method for learning it. Like physics and calculus, it can be learned thoroughly and in depth only by working problems. Before each lecture you should rapidly read the corresponding textual material as an introduction and to learn where the topics are discussed, so that you can later look the topics up again and consider them more carefully as you attempt the problems. Following the lecture, you should work out as many of the problems in the chapter and at the end of the chapter as you can. The work will go slowly at first because you will probably find it necessary to go back and consult the textual material and your lecture notes. Gradually, however, as your expertise increases, the work will go faster, and effective problem-oriented learning will take place.

There is emphasis on application of principles as well as mere understanding. How well would a football team play if it had heard the coach lecture on the plays in the locker room but had never tried the plays in scrimmage? How can a student demonstrate a knowledge of organic chemistry in the laboratory or on an examination if he or she has done nothing more than attend lectures and read the text? *About two-thirds of your study time should be devoted to working problems,* whether they are specifically assigned or not. You will find that working the problems can be interesting and fun. You will have a feeling of accomplishment when the problems are completed. Working on the problems with one or two other students may be mutually stimulating, and may make the task more interesting and efficient. But remember that although the old idea that two heads may be better than one has some merit, each of you should pull your own load.

Edward E. Burgoyne

INTRODUCTION TO ORGANIC CHEMISTRY

At about the time when the American colonists were declaring and fighting for their independence from England, chemistry was developing in Europe. *Chemistry* may be defined as the science that deals with the composition, structure, and properties of matter not only as matter may now exist but also as matter may change with respect to those aspects of its nature. *Organic chemistry* is the chemistry of the compounds of carbon. It is one of the most important branches of chemistry. Since carbon constitutes by weight less than 0.1% of the crust of the earth, it may seem strange that its compounds are so important. However, all things that are living, or that have lived, contain the element carbon. Carbohydrates, fats, and proteins are carbon compounds. An atom of carbon, because of its unique electronic arrangement, has the ability to unite with other atoms of carbon to form chains and rings of carbon atoms of great diversity. The number of known compounds of carbon (now about 3 to 4 million) is greater than the number of compounds of all of the rest of the elements put together. We shall find that carbon compounds are important in almost every phase of our existence.

Why the name "organic chemistry"?

We might now ask why the chemistry of the compounds of carbon is called organic chemistry. Berzelius first used the term *organic* in 1808 to apply to all compounds occurring in, or deriving from, living organisms or their remains. That was in accord with the vital force theory, which held that a living, or vital, force was essential to the production of all such compounds. The theory was discredited and finally overthrown. In 1828 Wöhler produced urea (a component of urine) by heating ammonium cyanate, and in 1845 Kolbe synthesized a derivative of acetic acid (a component of vinegar) from inorganic or nonliving source materials. In 1848 Gmelin noted that the only essential difference between organic and inorganic compounds is that organic compounds always contain carbon. Today, although our major sources of organic compounds are still the fossil remains of living things in the form of petroleum and coal, we have the knowledge and technology to produce a great many organic compounds from simple carbon molecules and ions considered a part of the nonliving or inorganic world.

How organic chemistry has developed

Some organic compounds were actually prepared and studied when the science of chemistry was just beginning. In about 1776 the Swedish apothecary, Scheele (who discovered oxygen independently of Priestley), prepared oxalic acid by the action of nitric acid on sugar. Since oxalic acid had previously been isolated from rhubarb, his was an early example of conversion of one organic compound into another. The complexity and challenge of organic compounds was recognized by Wöhler in 1840 when he made the remark, "Organic chemistry gives me the impression of a primeval tropical forest, full of the most remarkable things."

It was not until the discovery of Kekulé and Couper, in 1858, that carbon has a valence of 4 and can unite with itself indefinitely that organic chemistry began to develop as an important branch of the science of chemistry. The concept of carbon uniting with itself, called *catenation,* was necessary for the foundation of the structural theory of organic chemistry. (Structural theory is concerned with how atoms are arranged to form molecules.) Also at about that time, organic chemistry received a most important economic stimulus. In England, in 1857, young William Henry Perkin accidentally discovered the first synthetic dye. He obtained a patent on his discovery and had the dye in production about a year later. It was a purple color called mauve, and it was so fantastically successful that it gave its name to a whole decade. More than that, it was an important factor in stimulating the beginning of the chemical industry in England and also in Germany. The practice of chemistry became a profitable profession which attracted investment, workers, and attention. The result was a relatively rapid growth of chemistry not only as an art and a science but also as a business.

The period when organic chemistry was beginning to develop in Europe coincided with the Civil War in the United States. After the war, population growth and the industrial revolution greatly increased consumer demand and products in the United States. Later, the two World Wars demonstrated the need for the United States to be independent of European chemicals. The result was the growth of the science of chemistry and the chemical industry in the United States to its present position of world prominence. Today huge chemical industries manufacture dyes, synthetic fibers, plastics, synthetic rubbers, fertilizers, insecticides, pharmaceuticals, cosmetics, explosives, fuels, lubricants, and all kinds of surface coatings. Most of those materials involve organic chemistry, and over half of all chemists are employed in the organic fields. You are now going to have the opportunity to learn something of this important and interesting subject.

1

BASIC CONCEPTS OF CHEMISTRY: A REVIEW

Organic chemistry is the study of the 3 to 4 million carbon-containing compounds known. Although the large number and the kinds of reactions characteristic of organic compounds make the subject complex, some of the complexity is more apparent than real. There are indeed many, many more organic compounds than inorganic compounds; and many organic molecules are extremely large and complex. But the basic principles governing the formation and reaction of compounds are the same as those you have met previously in general chemistry.

In this chapter we shall review the most important of the basic principles and take a preliminary look at the application of those principles in organic chemistry. We shall consider chemical bonds, both covalent and ionic; intermolecular attractive forces; the nature of chemical reactions, including mechanisms and energy changes; and the nature of acids and bases and the application of acid-base theories to organic chemistry.

The concepts considered in this chapter will be of continuing importance as you progress through the course. If you find this brief review insufficient to help you recall them clearly, you may wish to study them in greater depth in a textbook of general chemistry.

1-1 Chemical bonds

How are the atoms comprising organic compounds held together? By the same types of chemical bonds you encountered in general chemistry. They are the covalent bonds and the ionic bonds. Let us review them briefly.

Covalent bonds

The most common type of chemical bond is the *covalent bond,* which consists of a pair of shared electrons of opposite spin. Thus the hydrocarbon methane, CH_4, is held together by four equivalent covalent bonds linking a single carbon atom with four hydrogen atoms. It may be represented by the formula

$$\begin{array}{c} H \\ H:\ddot{C}:H \\ \ddot{H} \end{array}$$

where the dots represent the outer, or valence, electrons which are paired up and shared. Methane may also be represented by the somewhat simpler formula

$$\begin{array}{c} H \\ | \\ H-C-H \\ | \\ H \end{array}$$

where the connecting lines represent the pairs of shared electrons.

You will note that the electron arrangement for methane effectively gives each hydrogen atom two electrons to complete its outer (and only) electron shell, while the carbon atom effectively completes its outer electron shell with eight electrons. Those electron configurations are possessed respectively by a helium atom and a neon atom—arrangements that have special stability and minimum energy.

Bond energy. Bond formation is usually an energy-releasing process. In the methane example, each hydrogen atom can be considered as contributing one electron to a bonding pair of electrons and the carbon atom as contributing its four valence electrons—one to each bonding pair—with attendant release of energy. The following equations illustrate the release of energy in bond formation for a mole of hydrogen and for a mole of methane.

$$2H\cdot \longrightarrow H:H + \underbrace{104 \text{ kilocalories per mole (kcal/mol)}}_{\text{Bond energy}} \tag{1-1}$$

[1]

$$4H\cdot + \cdot\dot{\underset{\cdot}{C}}\cdot \longrightarrow \begin{array}{c} H \\ H:\ddot{C}:H \\ \ddot{H} \end{array} + \underbrace{4(99 \text{ kcal/mol}) = 396 \text{ kcal/mol}}_{\text{Bond energy}} \tag{1-2}$$

Pairing of electrons. The concept of pairing of electrons is an important one. Each electron can be considered as spinning either clockwise or counterclockwise. Since each electron has a unit negative charge of electricity, its motion results in a magnetic field with poles located at the ends

[1] A note on units: In 1960, international agreement established the standard system of measurement called *Le Système International d'Unités*, abbreviated SI. The SI units are gradually being adopted in chemistry, but many older units continue to appear in the literature and in everyday use. The *calorie* and its multiple the *kilocalorie* are such units. The kilocalorie is retained in this text. You should know, however, that the SI unit for energy is the *joule* and that 1 kilocalorie (kcal) = 4.184 kilojoules (kJ).

of the axis of spin of the electron. Electrons that have the same direction of spin will possess magnetic fields of the same orientation. They will thus repel each other with a magnetic force, as well as with the electrostatic repulsion caused by possessing like charges. Such electrons could not pair up. However, electrons of opposite direction of spin will have opposed magnetic fields which will exert a magnetic force of attraction sufficient to counteract the inherent electrostatic repulsion. That makes it possible for two electrons of opposite spin to pair up and to occupy the same atomic orbital of an atom (unshared pair), or the same molecular orbital of a molecule (bonding pair). This is an application of Pauli's exclusion principle that no more than two electrons may occupy the same orbital and that two electrons occupying the same orbital must have opposite spins. Spins are of only two types, clockwise or counterclockwise. If more than two electrons (three, for example) were to occupy a given orbital, two of them would have to have the same spin. But electrons of the same spin would repel each other, so only two electrons of opposite spin could remain in the orbital. Thus the covalent bond consists of a pair of electrons of opposite spin shared by two atoms.

PROBLEM 1-1 By using black dots and colored dots to represent the valence electrons, show electronic structures for (a) hydrogen peroxide, H_2O_2, (b) chlorine, Cl_2, (c) nitric acid, HNO_3 ($HONO_2$).

Solution (a) H:Ö:Ö:H (b) :C̈l:C̈l: (c) H:Ö:N(:Ö:)(:Ö:)

Coordinate covalent bond

Note that in (c), nitric acid, the nitrogen atom contributes *both* electrons of the bonding pair to one of the oxygen atoms instead of each atom contributing one electron to the pair. This type of bond is termed a *coordinate covalent bond*.

Geometry of single covalent bonds of carbon. Compounds in which one or more of the hydrogen atoms of methane, CH_4, are replaced by some other atom or group of atoms can be formed; they are called *substituted methanes*. When one hydrogen is replaced by another element Y, only one product—one monosubstituted methane, CH_3Y—is formed. This indicates that all of the hydrogen atoms of the methane molecule are equivalent. If a disubstituted methane, CH_2YZ, is prepared, again only one product is formed. This indicates that the molecule cannot be planar. If it were planar, we would find two distinct compounds produced:

$$\begin{array}{ccc} & H & \\ & | & \\ H- & C & -Y \\ & | & \\ & Z & \end{array} \quad \text{and} \quad \begin{array}{ccc} & Y & \\ & | & \\ H- & C & -H \\ & | & \\ & Z & \end{array}$$

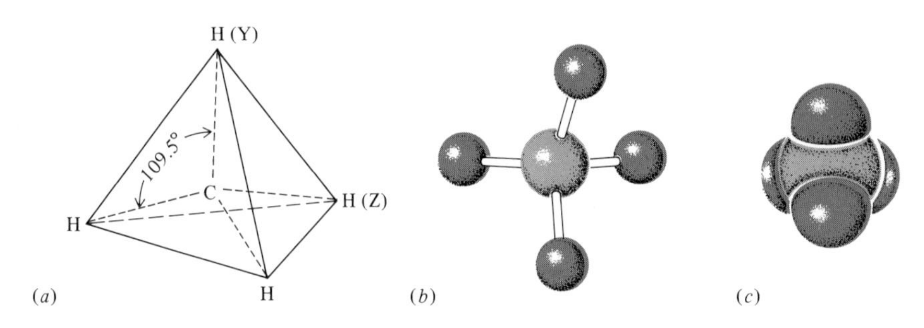

FIGURE 1-1
(a) The tetrahedral structure of methane. (b) Ball-and-stick and (c) space-filling models of methane.

Consideration of those facts led to the idea that methane and its derivatives must have bond angles of approximately 109°28′, the bond angle which would maximize use of the space about the carbon atom. If the ends of the bonds are connected, a tetrahedron is formed, so the bond angle is referred to as the *tetrahedral bond angle*. Since 28′ is almost half a degree, the angle is sometimes expressed approximately as 109.5°. See Fig. 1-1.

Thus methane is a three-dimensional molecule, and the structural formula

```
     H
     |
 H—C—H
     |
     H
```

is to be considered a planar projection of the actual structure. The bond angle, then, is not really 90° as might be implied from the planar projection; it is actually 109°28′. Since all four hydrogen atoms are equivalent in position relative to each other as well as to the central carbon atom, there is only one possible CH_2YZ.

Modern orbital theory helps explain the tetrahedral nature of a *saturated carbon atom,* i.e., carbon with four single bonds. The arrangement of the six electrons of a carbon atom in the ground state is shown in Table 1-1. The arrows represent the direction of electron spin. The arrangement would suggest that carbon would have a valence of 2, since it has only two electrons that are unpaired. However, the fact is that carbon has a valence of 4.

Bond formation is generally an energy-releasing process, and atoms tend to maximize bond formation. The empty $2p$ orbital is of only slightly higher energy than the $2s$ orbital. The release of energy from additional bond formation is more than enough to justify the concept that one of the electrons in the $2s$ orbital is promoted to the empty $2p$ orbital. In that

TABLE 1-1
Carbon atom in the ground state

Electron shell	K	L			
Orbitals	$1s$	$2s$	$2p_x$	$2p_y$	$2p_z$
Electrons	⇅	⇅	↑	↑	

TABLE 1-2
Carbon atom in excited state

Electron shell	**K**	**L**			
Orbitals	$1s$	$2s$	$2p_x$	$2p_y$	$2p_z$
Electrons	⇅	↑	↑	↑	↑

slightly excited state there are then four electrons with which pairing can occur, so the valence is 4. See Table 1-2.

As you may recall, p orbitals are directional and are at an angle of 90° to each other. Since the 90° angle is *not* observed and all the bonds are the same in length, it is suggested that the $2s$ orbital and the three $2p$ orbitals mix and adjust their shapes to provide four equivalent *hybrid orbitals*. The hybrid orbitals have one part s character and three parts p character, and so they are designated as *sp^3 hybrid orbitals*. The best geometry for maximizing the use of space for overlapping of the four sp^3 hybrid orbitals to produce the molecular bonding orbitals is tetrahedral geometry. Thus, sp^3 hybridization always results in bond angles that approximate the tetrahedral bond angle of 109.5°. See Fig. 1-2.

Geometry of single covalent bonds of nitrogen and oxygen. Other common molecules with possible sp^3 hybrid orbitals are ammonia and water:

Ammonia	Water
$\ddot{N}$ bonded to H, H, H; H–N–H angle 107°	$\ddot{O}$: bonded to H, H; H–O–H angle 105°

In the case of ammonia an unshared pair of electrons occupies one corner of the tetrahedron (Fig. 1-3). In the case of water, there are two pairs of unshared electrons (Fig. 1-4). Unshared pairs of electrons, which are not concentrated between two nuclei, actually take up more space than do shared electrons. The result is a compression of the angle between the bonds holding the attached atoms. That nitrogen and oxygen atoms are

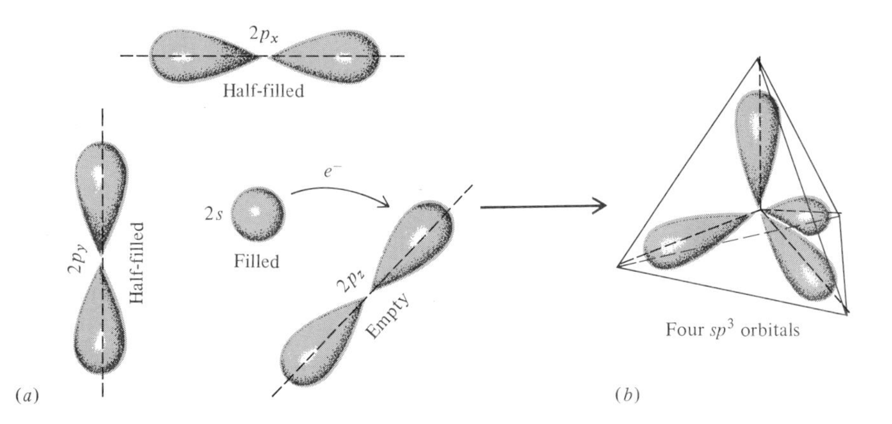

FIGURE 1-2
(*a*) Three p orbitals and one s orbital hybridize to produce (*b*) the four sp^3 hybrid orbitals of a carbon atom. The angle between orbitals is 109.5°.

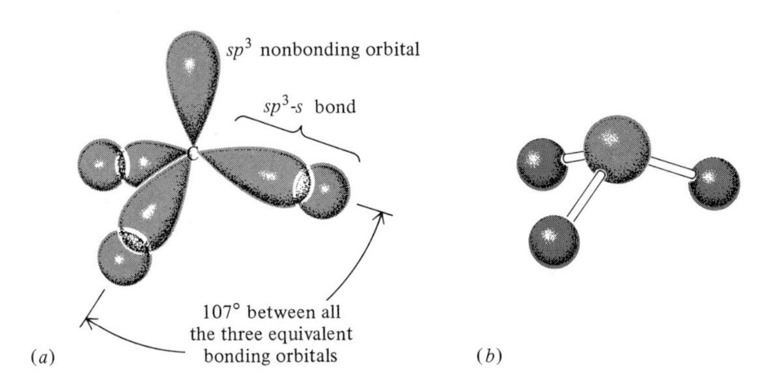

FIGURE 1-3
The ammonia (NH_3) molecule: (*a*) orbitals and (*b*) ball-and-stick model.

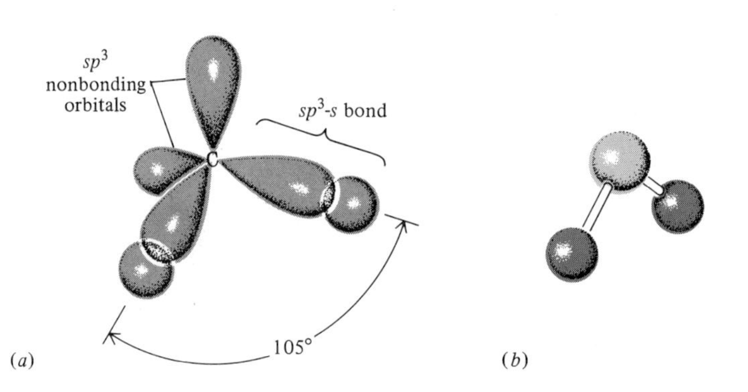

FIGURE 1-4
The water (H_2O) molecule: (*a*) orbitals and (*b*) ball-and-stick model.

similar in size and structure to carbon atoms explains why they are often found substituted for carbon atoms in organic molecules.

PROBLEM 1-2 (a) What is the geometry of the bonds about the carbon atom in methanol, CH_3OH? (b) What is the geometry of the bonds and unshared pairs of electrons about the oxygen atom in methanol? (*Hint:* Methanol is similar in structure to water, but it has a CH_3 group substituted for one of the hydrogen atoms.)

Solution to (a) The geometry of the bonds about the *carbon atom* in methanol, CH_3OH, is tetrahedral, just as it is in methane:

```
      OH
      |   109.5°
      C
    / | \
   H  H  H
```

Polar covalent bonds. Unless a pair of electrons is shared by the same kind of atoms carrying the same substituents (attached atoms or groups), if any, the electron pair will not be shared equally; it will be closer to the atom having the greater *electronegativity,* or attraction for electrons. The result is a *polar bond* with one end relatively negative and the other end relatively positive. The bonds in H—H or Cl—Cl are nonpolar covalent, but the bonds in

$$H{-}Cl \qquad O{=}C\begin{matrix} \diagup H \\ \diagdown H \end{matrix} \qquad \text{and} \qquad H{-}\overset{\overset{\displaystyle H}{|}}{\underset{\underset{\displaystyle H}{|}}{C}}{-}Cl$$

are polar covalent. The resulting imbalance of electron density results in a polar molecule—one that is more negative on one end and more positive on the other end. That is depicted by representing partial charges ($\delta+$ or $\delta-$) at each end as follows:

$$^{\delta+}H{-}Cl^{\delta-} \qquad ^{\delta+}H{-}\overset{\overset{\displaystyle H}{|}}{\underset{\underset{\displaystyle H}{|}}{C}}{-}Cl^{\delta-}$$

It may also be represented by arrows over the bonds, with the head of the arrow pointing to the atom or region of greater concentration of negative charge and the tail of the arrow indicating the atom or region of greater concentration of positive charge:

$$H \overset{+\!\!\rightarrow}{-} Cl \qquad H{-}\overset{\overset{\displaystyle H}{|}}{\underset{\underset{\displaystyle H}{|}}{C}} \overset{+\!\!\rightarrow}{-} Cl$$

That such polar molecules attract each other is due to the electrostatic effect of the partial positive charge of the end of one molecule on the partial negative charge of the end of another molecule. The effect is termed *dipole-dipole attraction,* and such molecules must have dipole moments greater than zero.

Dipole moment of a molecule. The *dipole moment* of a molecule is a measure of the tendency of the molecule to orient itself in an electrostatic field. Mathematically it is the product of the magnitude of the charge at the ends of the molecule expressed in coulombs, and the distance between the charges in meters to the -10th power.[1] For example, a negative charge of 1.6×10^{-19} coulomb and a positive charge of 1.6×10^{-19} coulomb separated by a distance of 1.0×10^{-10} meter (m) constitute a dipole with a moment calculated as follows:

$$\begin{aligned} \text{Dipole moment} &= \text{charge} \times \text{distance} \\ &= (1.6 \times 10^{-19}\ \text{C})(1.0 \times 10^{-10}\ \text{m}) \\ &= 1.6 \times 10^{-29}\ \text{C} \cdot \text{m} \\ &= 4.8\ \text{Debye units} \end{aligned}$$

[1] In older books and references, you will often find the distance between the charges given in angstroms (Å) and the charge given in electrostatic units (esu). The angstrom, a unit which is *not* used in SI, is equivalent to 1.0×10^{-10} m. The electrostatic unit is equivalent to 3.33×10^{-20} coulomb (C), the approved SI unit.

TABLE 1-3
Some examples of dipole moments

Common name	Formula	Dipole moment, Debye units
Hydrogen	H_2	0
Carbon dioxide	$O{=}C{=}O$	0
Water	H_2O	1.84
Ammonia	NH_3	1.46
Methanol	CH_3OH	1.70
Methane	CH_4	0
Methyl chloride	CH_3Cl	1.86
Carbon tetrachloride	CCl_4	0
Acetone	$CH_3C({=}O)CH_3$	2.88

Dipole moments are generally expressed in Debye units. A Debye unit equals $(3.33 \times 10^{-20}\ C)(1.0 \times 10^{-10}\ m)$, or 3.33×10^{-30} coulomb-meter $(C \cdot m)$.

However, not all compounds which contain polar covalent bonds have polar molecules with dipole moments. A compound such as carbon tetrachloride,

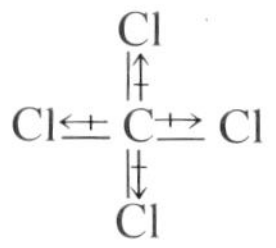

Cl
Cl ⇷ C ⇸ Cl
Cl

has four equivalent polar covalent bonds. The partial negative charges are symmetrically distributed, so no end of the molecule has a charge different from any other end. Another example of a nonpolar molecule having polar bonds is BF_3, which also has a dipole moment of zero.

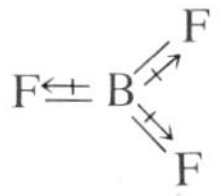

F ⇷ B (F)(F)

However, NH_3, ammonia, does have a dipole moment. The unshared pair of electrons on the nitrogen is accommodated in an approximate tetrahedral shape which results in a polar molecule.

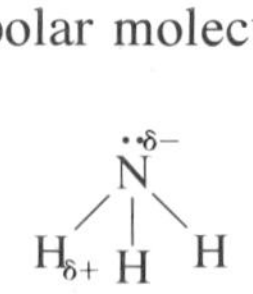

$\ddot{N}^{\delta-}$
$H^{\delta+}$ H H

Table 1-3 gives the dipole moments of some familiar compounds.

PROBLEM 1-3 Which of the following compounds would have polar molecules, and which would have nonpolar molecules? (a) CH_3NH_2, (b) $CH_3{-}CH_3$, (c) $CH_3{-}O{-}CH_3$.

Solution to (a) CH_3NH_2 is a derivative of NH_3 in which one of the hydrogen atoms has been substituted by the CH_3 group, so its structure is like that of ammonia, NH_3. The unshared pair of electrons on the nitrogen is accommodated in the approximate tetrahedral shape of the molecule. The molecule, like ammonia, NH_3, is polar.

$$\overset{\cdot\cdot\ \delta-}{N}(H_{\delta+})(H)(CH_3)$$

Double and triple covalent bonds. Atoms of a compound may share two pairs of electrons to form a double covalent bond, which is usually simply called a *double bond* (Fig. 1-5). Some examples are

$$O::C::O \quad \text{or} \quad O{=}C{=}O$$

Carbon dioxide

$$H_2C::CH_2 \quad \text{or} \quad H_2C{=}CH_2$$

Ethylene

Atoms of some compounds may share three pairs of electrons to form a *triple bond*. Examples are

$$H:C\vdots\vdots C:H \quad \text{or} \quad H{-}C{\equiv}C{-}H$$

Acetylene

$$H:C\vdots\vdots N: \quad \text{or} \quad H{-}C{\equiv}N$$

Hydrogen cyanide

$$C_6H_5:\overset{+}{N}\vdots\vdots N:\ :\ddot{\underset{\cdot\cdot}{Cl}}:^- \quad \text{or} \quad C_6H_5{-}\overset{+}{N}{\equiv}N\ :\ddot{\underset{\cdot\cdot}{Cl}}:^-$$

Benzenediazonium chloride

The geometry of compounds having double and triple covalent bonds will be discussed when such compounds are considered in detail in later chapters.

Ionic bonds In general chemistry you encountered simple ions such as the sodium ion, in which a sodium atom has lost an electron to acquire a positive charge,

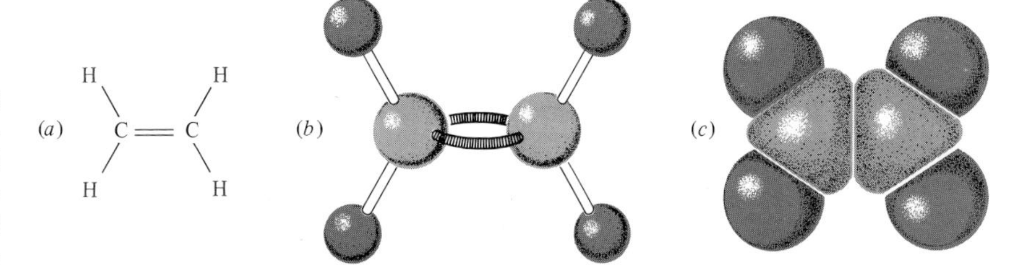

FIGURE 1-5
Ethylene, a compound containing a double bond: (*a*) structural formula, (*b*) ball-and-spring model, and (*c*) space-filling model.

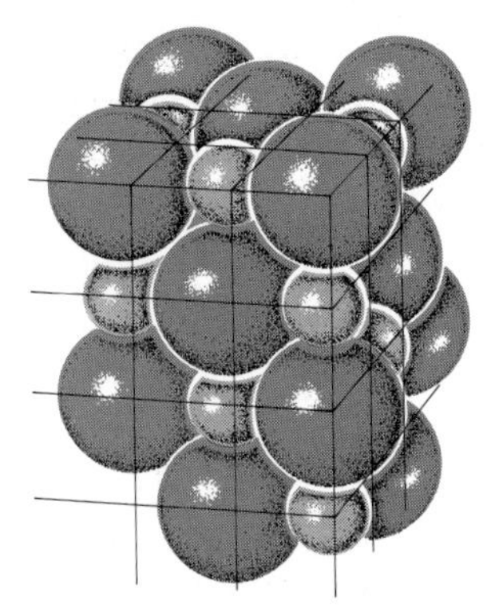

FIGURE 1-6
The structure of solid sodium chloride (NaCl).

Na^+, and the chloride ion, in which a chlorine atom has gained an electron to acquire a negative charge, Cl^-. Solid crystalline sodium chloride is considered to be an ordered array of positive sodium ions and negative chloride ions held together by electrostatic attraction (Fig. 1-6). The attraction of ions of opposite charge for each other constitutes the *ionic bond*. The bond is very strong; most ionic compounds are solids with relatively high melting and boiling points. You also encountered complex ions such as the sulfate ion, SO_4^{2-}; the hydroxide ion, OH^-; and the ammonium ion, NH_4^+. Organic compounds may also consist of groups of atoms which, by being covalently bonded together, have acquired an electric charge; there is either a net gain or a net loss of an electron or several electrons to produce respectively an anion or a cation. Some organic compounds, that is, are ionic compounds. The compounds are held together in the crystalline state by ionic bonds resulting from the electrostatic attraction of ions of opposite charge. Some examples are the organic salts sodium acetate, $Na^+C_2H_3O_2^-$, and dimethylammonium chloride, $(CH_3)_2NH_2^+Cl^-$, and the organic base, tetramethylammonium hydroxide, $(CH_3)_4N^+OH^-$. There are also a great many weak organic acids and bases which produce low concentrations of ions in a solvent like water. You may already be familiar with such examples as carbonic acid, acetic acid, citric acid, and ascorbic acid (vitamin C). Organic analogs of ammonia such as methylamine, CH_3NH_2, are weak bases like ammonia and produce low concentrations of ions in water.

$$\underset{\text{Acetic acid}}{CH_3CO_2H} + HOH \rightleftharpoons \underset{\substack{\text{Hydronium acetate}\\ \text{(Aqueous acetic acid)}}}{H_3O^+ + CH_3CO_2^-} \tag{1-4}$$

$$\underset{\text{Methylamine}}{CH_3NH_2} + HOH \rightleftharpoons \underset{\substack{\text{Methylammonium}\\ \text{hydroxide}}}{CH_3NH_3^+ + OH^-} \tag{1-5}$$

PROBLEM 1-4 For each of the following compounds identify single covalent bonds, any double covalent bonds, any triple covalent bonds, and any ionic bonds. (a) HCO_2Na, (b) $NaCN$, (c) HC_2Na, (d) K_2CO_3.

Solution to (a) This compound, commonly known as sodium formate, contains two single covalent bonds, one double covalent bond, and one ionic bond.

$$H{-}C({=}O){-}O^- \quad Na^+$$

1-2 Intermolecular attractive forces

Several types of intermolecular attractive forces serve to hold molecules together in the liquid or solid state. They are much weaker than the covalent and ionic bonds between the atoms of molecules and the ions of ionic compounds.

Dipole-dipole attractive forces The dipole-dipole intermolecular attractive forces have already been mentioned in the discussion of polar bonds. All molecules which have dipole moments demonstrate dipole-dipole attraction. The greater the dipole moment the greater the dipole-dipole attraction of the more positive end of one molecule for the more negative end of another molecule. The carbonyl group, $\gt C{=}O$, imparts a rather large dipole moment to a molecule. On the other hand, the $-\overset{|}{\underset{|}{C}}-O-\overset{|}{\underset{|}{C}}-$ group imparts only a very small dipole moment. The effect of dipole-dipole attraction is to help hold the molecules to each other and thus lower the vapor pressure and increase the boiling point.

PROBLEM 1-5 By considering dipole-dipole attraction, indicate which of the following pairs would have the highest boiling point.

(a) CH_3-O-CH_3 or $CH_3-\overset{\overset{O}{\|}}{C}-H$

(b) $CH_3-\overset{\overset{O}{\|}}{C}-CH_3$ or $CH_3\underset{\underset{CH_3}{|}}{C}HCH_3$

Solution to (a) Acetaldehyde, $CH_3-\overset{\overset{O}{\|}}{C}-H$, should have a higher boiling point than CH_3-O-CH_3, dimethyl ether. The $\gt C{=}O$ group gives a molecule a larger dipole than does the $-\overset{|}{\underset{|}{C}}-O-\overset{|}{\underset{|}{C}}-$ group. A check of a handbook indicates boiling points of 21°C for acetaldehyde and −23.65°C for dimethyl ether.

Hydrogen bonds So-called hydrogen bonding is an especially strong type of dipole-dipole attraction between molecules in which hydrogen is covalently bonded to an atom of one of the strongly electronegative elements, fluorine, oxygen, or nitrogen. Since the hydrogen is thus strongly electron-deficient (partially positive), hydrogen bonding is also referred to as *proton bonding*. The order of the strength of the bonding is

$$-H{---}F- \; > \; -H{---}O\lt \; > \; -H{---}N\lt$$

The result is an intermolecular attraction of the partially positive hydrogen of one molecule to the electronegative atom of an adjacent molecule. It may be depicted as shown in Fig. 1-7.

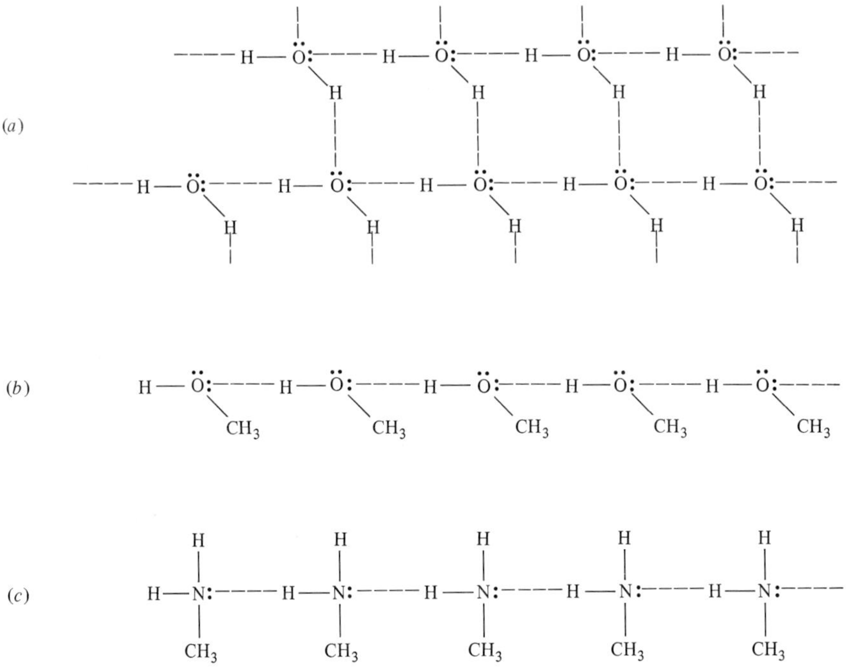

FIGURE 1-7
Hydrogen bonding in (*a*) water, (*b*) methanol, and (*c*) methylamine. The bonding, shown here in two dimensions, actually occurs in three dimensions.

(*a*)

(*b*)

(*c*)

Because of the additional intermolecular attraction, the compounds of Fig. 1-7 will have relatively lower vapor pressures and higher boiling points than will compounds of comparable size without hydrogen bonding. The energy of hydrogen bonding is about 5 kcal/mol, and it is stronger than ordinary dipole-dipole attraction.

PROBLEM 1-6 By taking into consideration any hydrogen bonding or dipole-dipole attraction due to polar covalent bonds, number the following compounds in order of decreasing boiling point. That is, designate the compound with the highest boiling point as number 1.

(a) ___ $CH_3C(=O)-CH_3$, ___ $CH_3CH_2-O-CH_3$, ___ $CH_3CH_2CH_2OH$, ___ $CH_3CH_2CH_2CH_3$

(b) ___ $CH_3CH(CH_3)-CH(=O)$, ___ $CH_3CH(CH_3)-CH_2CH_3$, ___ $CH_3CH(CH_3)-O-CH_3$, ___ $CH_3CH(CH_3)-CH_2OH$

Solution to (a) All these compounds have about the same molecular weight. However, only $CH_3CH_2CH_2OH$ is capable of hydrogen bonding, the strongest type of intermolecular attractive force, so it will have the highest boiling point.

Next comes $CH_3C(=O)—CH_3$ with its large dipole moment.

__2__ $CH_3C(=O)—CH_3$, __3__ $CH_3CH_2—O—CH_3$, __1__ $CH_3CH_2CH_2OH$, __4__ $CH_3CH_2CH_2CH_3$

London forces

London forces operate even on molecules that are ordinarily nonpolar. They are considered to be due to *instantaneous induced dipoles*. As one molecule approaches another, mutual repulsion between the electron fields of the molecules results in a partial shift or *polarization* of electrons. A slight attraction is then possible between the partially positive end or side of one molecule and the partially negative end or side of another molecule. In general, the larger the molecule the greater the number and magnitude of such forces can be. Consequently, in a family of similar compounds, vapor pressure decreases and boiling point increases with increase in size of the molecule. However, the shape of the molecule also has an effect, so London forces are greater when effective surface area and closeness of approach can be maximized. Consequently, those forces will be greater in a long, continuous-chain compound than they will be in a branched-chain compound of the same number and kind of atoms. Consider the three forms of pentane, C_5H_{12} (see also Fig. 1-8):

$CH_3CH_2CH_2CH_2CH_3$	$CH_3CH(CH_3)CH_2CH_3$	$CH_3—C(CH_3)_2—CH_3$
n-Pentane bp 36°C	Isopentane bp 28°C	Neopentane bp 9°C

London forces are weaker than ordinary dipole-dipole attractive forces, so a dipole-dipole attraction will result in a higher boiling point, and hy-

FIGURE 1-8 Space-filling models of (*a*) *n*-pentane, (*b*) isopentane, and (*c*) neopentane.

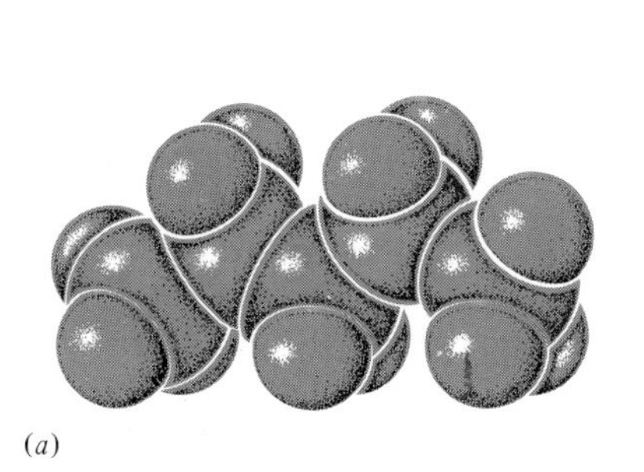

(*a*)

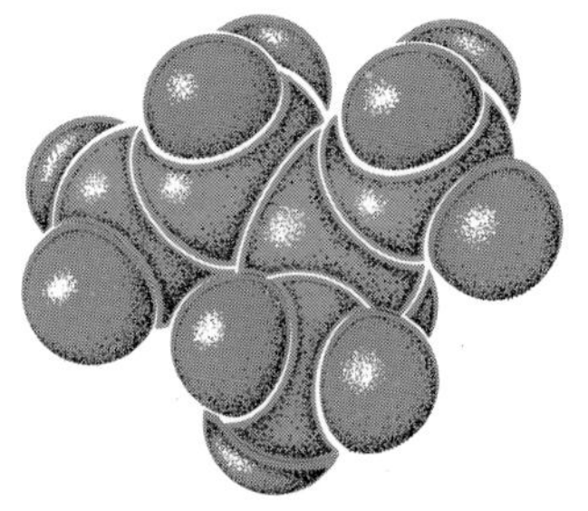

(*b*)

(*c*)

drogen bonding will result in a still higher boiling point for molecules of similar molecular weight.

PROBLEM 1-7 Number the following compounds in order of decreasing boiling point; that is, designate the compound with the highest boiling point as number 1.

(a) ___ $CH_3CH(CH_3)-CH_3$, ___ $CH_3CH(NH_2)-CH_3$, ___ $CH_3CH_2CH_2CH_3$,

___ $CH_3CH(OH)-CH_3$

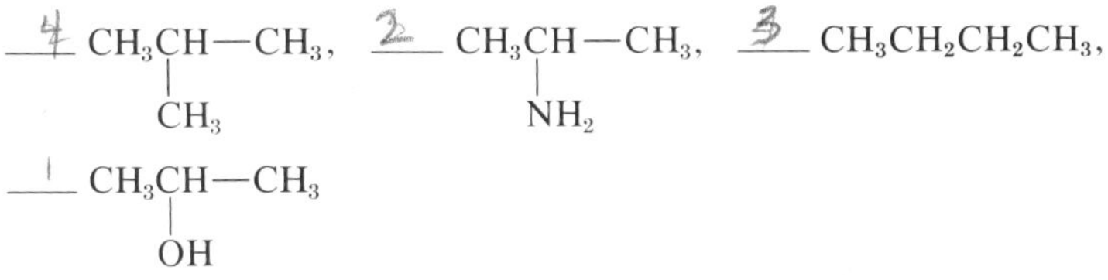

(b) ___ $CH_3-C(CH_3)_2-CH_3$, ___ $CH_3-C(CH_3)_2-NH_2$, ___ $CH_3-C(CH_3)_2-OH$,

___ $CH_3-C(CH_3)(CH_2CH_3)-H$

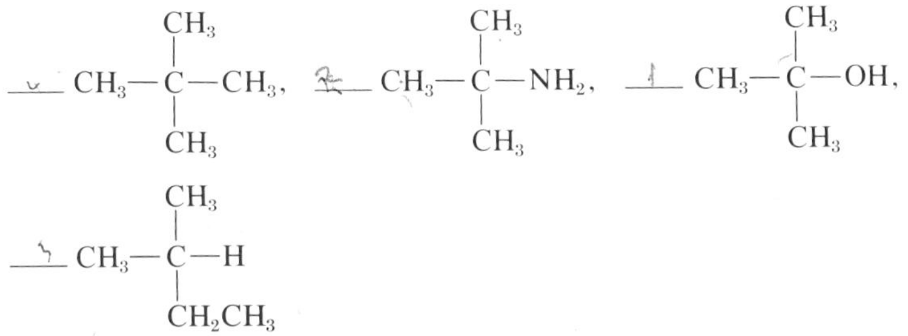

Solution to (a) 4 $CH_3CH(CH_3)-CH_3$, 2 $CH_3CH(NH_2)-CH_3$, 3 $CH_3CH_2CH_2CH_3$,

1 $CH_3CH(OH)-CH_3$

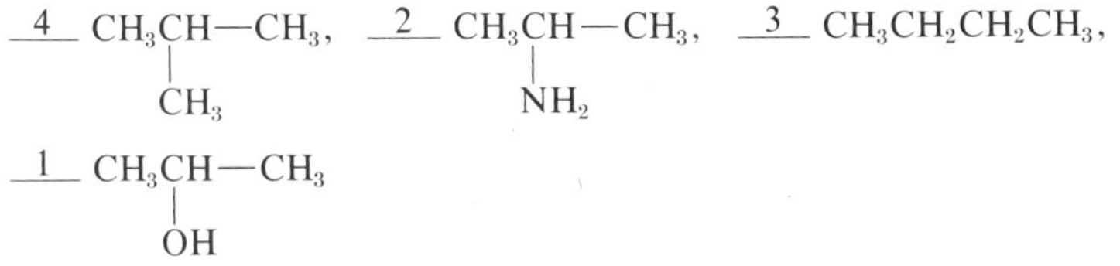

Number 1 has hydrogen attached to oxygen, so it has strong hydrogen bonding and the highest boiling point.

Number 2 has hydrogen attached to nitrogen, so it has hydrogen bonding and the next highest boiling point.

Numbers 3 and 4 are hydrocarbons with only London forces, but number 4 is the most highly branched. That lessens its London forces, so it has the lowest boiling point.

1-3 Nature of chemical reactions

All chemical changes or reactions result in a change in composition and/or structure of the molecules involved. Such reactions require the breaking of existing chemical bonds and the forming of new ones.

Reaction intermediates

Since the chemical bond consists of a pair of shared electrons, the bond may be (a) broken *homolytically* to produce reaction intermediates having unpaired electrons, called *free radicals,* or (b) broken *heterolytically* to produce reaction intermediates with electric charges, called *ions*. The free radicals or ions which are so produced are very reactive intermediates which then combine in some way to form a product different from the starting materials. That may be illustrated as follows:

HOMOLYTIC BONDING PAIR CLEAVAGE.

$$\mathrm{A{:}B} \longrightarrow \underset{\text{Free radicals}}{\mathrm{A\cdot + \cdot B}} \tag{1-6}$$

HETEROLYTIC BONDING PAIR CLEAVAGE.

$$\mathrm{A{:}B} \longrightarrow \underset{\text{Anion A}}{\mathrm{A{:}^-}} + \underset{\text{Cation B}}{\mathrm{B^+}} \tag{1-7}$$

or

$$\mathrm{A{:}B} \longrightarrow \underset{\text{Cation A}}{\mathrm{A^+}} + \underset{\text{Anion B}}{\mathrm{{:}B^-}} \tag{1-8}$$

These concepts, applied to organic reactions, describe reaction intermediates with an unpaired electron on a *carbon* atom or with a negative charge or a positive charge on a *carbon* atom. For example, consider the rupture of one of the C—H bonds in methane. The possible intermediates are

```
     H                H               H
     |                |               |
   H—C·             H—C+            H—C:−
     |                |               |
     H                H               H
```

Methyl free radical — Methyl cation (a carbonium ion) — Methyl anion (a carbanion)

Such reaction intermediates are generally very short-lived.

Reaction mechanisms

The detailed pathway involving all of the intermediates by which a chemical reaction occurs, called the *reaction mechanism,* helps to explain why and how the reaction takes place. However, once a mechanism has been presented, it is not necessary to illustrate it each time the equation for the reaction is written. That is, the formulas for the reaction intermediates, electron shifts, and so on, need not be included unless called for. When reaction mechanisms are presented, you will find them helpful in understanding the reactions, so that learning the reactions becomes less a matter of rote memorization.

Energy changes in chemical reactions

Reactions are always accompanied by energy changes. The energy changes may involve changes in heat content and/or molecular organization. Heat, light, and electricity are forms of energy which may be involved in chemical reactions.

Free-energy change. Once a chemical reaction has begun, it will proceed spontaneously if the change in free energy, ΔG, is negative, i.e., if there is a loss of free energy to the system. Such reactions are said to be *exergonic*. If the free-energy change is positive, then the reaction will not proceed spontaneously; it is said to be *endergonic*. The *change in free energy* is determined by a combination of *enthalpy*, or heat content, and *entropy*. The equation is

$$\Delta G^\circ = \Delta H - T\,\Delta S^\circ \tag{1-9}$$

where $\Delta G'$ is the standard free-energy change, ΔH is the enthalpy change (more precisely ΔH°), ΔS° is the standard entropy change, and T is the temperature, in kelvins.[1]

Effect of change in entropy. *Entropy* corresponds roughly to the randomness, chaos, or disorder of the system. To the extent that $T\,\Delta S^\circ$ contributes to ΔG°, equilibrium tends to shift toward the side in which fewer restrictions are placed on the positions of atoms and molecules. (Clausius, in 1865, stated, "Die Entropie der Welt strebt einem Maximum zu." Randomness or disorder of the world tends to increase to a maximum.) Since the entropy change is multiplied by the temperature, its effect upon the free-energy change increases with increasing temperature. Thus, in general, synthesis of molecules through formation of bonds between atoms with energy release is favored at lower temperatures, but decomposition reactions into a greater number of particles (increase in randomness) is favored at higher temperatures. A negative free-energy change may be caused either by a relatively large negative enthalpy change, ΔH, or by a large positive increase in $T\,\Delta S^\circ$.

PROBLEM 1-8 Is ΔS positive or negative for the following processes? (a) Freezing a liquid, (b) vaporizing a liquid, (c) expanding a gas, (d) convening a chemistry class, (e) tearing down a building.

Solution to (a) For the freezing of a liquid ΔS is negative. In the solid state, the molecules are essentially fixed in place in the crystalline structure, although they still possess vibrational motion. The arrangement of the molecules is much more ordered and much less random than in the liquid state.

PROBLEM 1-9 In the equilibrium reaction

$$\underset{\text{Benzene}}{C_6H_6} + 3H_2 \overset{\text{Pt}}{\rightleftharpoons} \underset{\text{Cyclohexane}}{C_6H_{12}}$$

a high temperature should favor displacement of the equilibrium in favor of production of which compound, benzene or cyclohexane? Why?

[1] The unit *kelvin* (K) is the SI unit for thermodynamic or absolute temperature. As one of the base units of the SI system, it is not capitalized. Further, the word "degrees" and the degree sign are *not* used with the name or the symbol (0°C = 273.1 K).

Change of enthalpy. In attempting to understand the effect of structure on position of equilibrium, differences in relative stabilities of reactants and products may be estimated. However, what is estimated in that way is not differences in free-energy changes; it is differences in changes of potential energy or heat content known as *enthalpy change,* ΔH. Often the differences in ΔH are proportional to differences in $\Delta G°$, so predictions made by this approach are generally valid. In other words, an *exothermic reaction* (one in which the value of ΔH is negative) can be assumed to be an exergonic one also and hence to proceed spontaneously if the product of entropy change and temperature can be ignored. This assumption is made in the progress-of-reaction diagrams to be discussed next. Enthalpy tends to a minimum for spontaneous processes.

Progress-of-reaction diagrams. One or more of the steps in a chemical reaction may be represented by a *progress-of-reaction diagram* to portray the energy requirements and changes as the reaction proceeds. The reaction is considered to take place along the x axis, and potential energy changes are represented in the direction of the y axis. Figure 1-9 is an example of the diagram for an *endothermic reaction* ($+\Delta H$), and Fig. 1-10 is an example of the diagram for an *exothermic reaction* ($-\Delta H$). The highest point on the curve represents the energy of the activated state. In Fig. 1-9 the energy level of the products is higher than that of the starting materials, so the reaction is *endothermic*. The methyl radical formed in the reaction may then react with bromine (Fig. 1-10) and the reaction is exothermic.

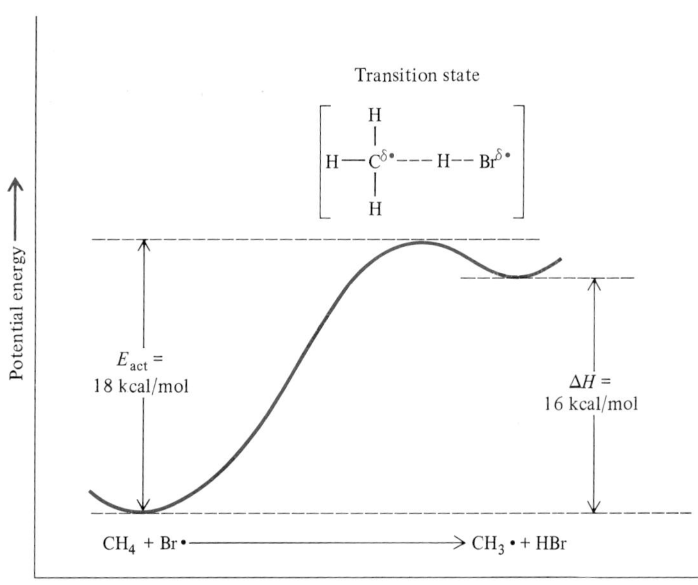

FIGURE 1-9
Progress-of-reaction diagram for an endothermic reaction.

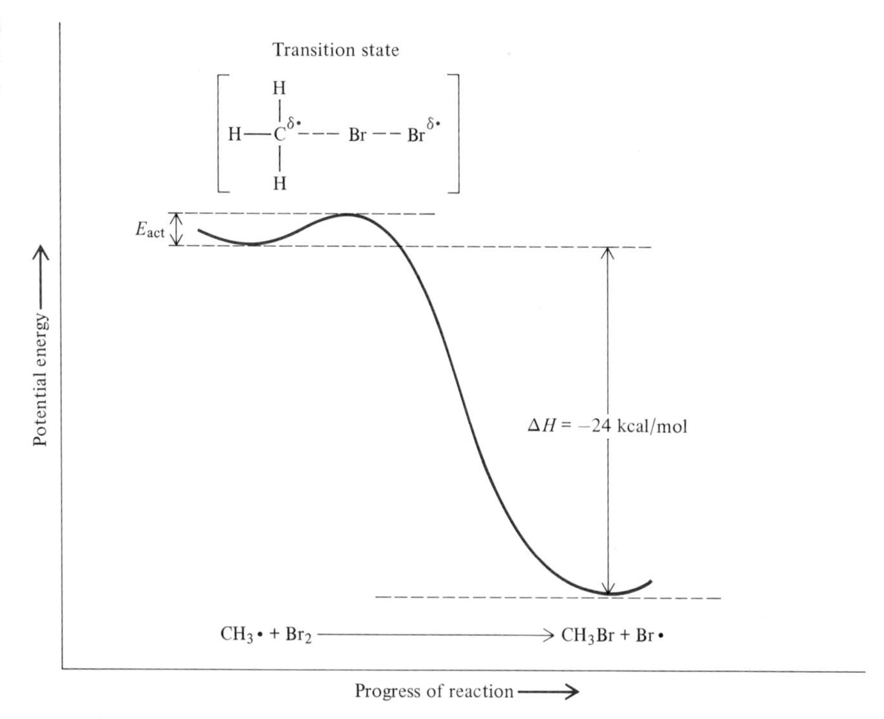

FIGURE 1-10
Progress-of-reaction diagram for an exothermic reaction.

Activated state and energy of activation. The *activation energy,* E_{act}, is the difference in energy of the initial state and the activated state. It is the minimum amount of energy the reaction particles must have in order to react. The *activated state* is the *transition state* in which bonds are being broken and other bonds are being made as the reactants change into products. The activation energy must always be at least equal to and generally greater than ΔH in order for the transition state to be achieved and the reaction to proceed. The activation energy is sometimes called an energy hump, and it is compared to a mountain one may have to cross in traveling from one valley to another. The higher the activation energy the fewer will be the molecules having that amount of energy to react, so the slower the reaction will be. *Catalysts* are substances that can lower the activation energy and so speed up the reaction.

1-4 Acids and bases in organic chemistry

As was mentioned in our review of ionic bonds, there is a large number of organic acids and bases; also, many organic reactions are catalyzed by acids or bases. Consequently, the concepts of acids and bases are very important to organic chemistry. Since a great many organic compounds are not soluble in water, the Arrhenius concepts of an acid as a substance which will furnish hydrogen ions in water solution and a base as a substance which will furnish hydroxide ions in water solution are of limited application. We need to review some more extensive acid-base theories.

Brønsted-Lowry acid-base theory

Much more generally useful than the Arrhenius theory is the Brønsted-Lowry theory, which, by defining an *acid* as a *proton donor* and a *base* as a *proton acceptor,* is applicable to both aqueous and nonaqueous systems. Acid-base reactions may be considered to take place in such a way that the base forms a *conjugate acid* (by gaining a proton) and the acid forms a *conjugate base* (by losing a proton). The reactions may be considered reversible. The position of equilibrium will favor the formation of the relatively weaker acid and the relatively weaker base. Here are some examples:

$$\underset{\text{Stronger acid}}{HCl} + \underset{\text{Stronger base}}{HOH} \rightleftharpoons \underset{\text{Weaker acid}}{H_3O^+} + \underset{\text{Weaker base}}{Cl^-} \tag{1-10}$$

$$\underset{\text{Stronger acid}}{HCl} + \underset{\text{Stronger base}}{CH_3OH} \rightleftharpoons \underset{\text{Weaker acid}}{CH_3OH_2^+} + \underset{\text{Weaker base}}{Cl^-} \tag{1-11}$$

$$\underset{\text{Stronger base}}{Na^{+\,-}OC_2H_5} + \underset{\text{Stronger acid}}{HOH} \rightleftharpoons \underset{\text{Weaker base}}{Na^{+\,-}OH} + \underset{\text{Weaker acid}}{HOC_2H_5} \tag{1-12}$$

$$\underset{\text{Weaker base}}{NH_3} + \underset{\text{Weaker acid}}{HOH} \rightleftharpoons \underset{\text{Stronger acid}}{HNH_3^+} + \underset{\text{Stronger base}}{OH^-} \tag{1-13}$$

Note that in the second reaction CH_3OH, methanol, reacts similarly to the way water reacts in the first reaction. Note also that in the third reaction $OC_2H_5^-$, the ethoxide ion, is an even stronger base than OH^-. In the fourth reaction the weaker acid and weaker base are found on the left because of displacement of the equilibrium to the left as the reaction is written.

Addition of a proton to a base forms the conjugate acid; removal of a proton from an acid forms the conjugate base. For example, the conjugate acid of the base water is H_3O^+, and the conjugate base of the acid water is OH^-.

PROBLEM 1-10 Write the formula of the conjugate base for each of the following acids and the formula of the conjugate acid for each of the bases given.

Acid	**Conjugate base**	**Base**	**Conjugate acid**
(a) HCl	____________	(d) OH^-	____________
(b) H_3O^+	____________	(e) NH_2^-	____________
(c) NH_4^+	____________	(f) OCH_3^-	____________

Solution to (a) The conjugate base of hydrochloric acid, HCl, is the chloride ion, Cl^-.

Lewis acid-base theory

The Lewis acid-base theory extends the concepts of acids and bases to compounds which act as acids but do not transfer protons. An *acid* is defined as an *electron-pair acceptor,* and a *base* is defined as an *electron-pair donor*. Electron-deficient atoms in compounds such as BF_3, $AlCl_3$, and CH_3^+ can serve as Lewis acids, i.e., nonproton acids. The following equation represents the reaction of the catalyst $AlCl_3$ in a Friedel-Crafts reaction, which will be considered in more detail in Chap. 5.

$$\underset{\text{Acid}}{AlCl_3} + \underset{\text{Base}}{C_2H_5Cl} \rightleftharpoons \underset{\text{Base}}{AlCl_4^-}\ \underset{\text{Acid}}{{}^+C_2H_5} \qquad (1\text{-}14)$$

Aluminum chloride is electron-deficient, and so it can serve as a Lewis acid. The ethyl cation, ${}^+C_2H_5$, has a carbon atom with only six electrons about it instead of the more stable eight, so it also may serve as a Lewis acid.

PROBLEM 1-11 In the following equation for the production of fluoboric acid from boron trifluoride and hydrogen fluoride, label the reactants as Lewis acid or base.

$$BF_3 + HF \rightleftharpoons H^+\ BF_4^-$$

pH and hydrogen-ion concentration

The concentration of dilute aqueous solutions of hydrogen ion is commonly expressed as pH. The term is used in the discussions of amino acids and proteins, enzymes, and metabolism. The pH is the negative logarithm of the hydrogen-ion concentration, in moles per liter, which is written $[H^+]$. Then,

$$pH = -\log [H^+] \qquad (1\text{-}15)$$

or more exactly in aqueous solutions,

$$pH = -\log [H_3O^+] \qquad (1\text{-}16)$$

Logarithms are exponents, and the common logarithm of a number (abbreviated *log*) is the power to which the number 10 must be raised in order to produce that number.

$$n = 10^{\log n} \quad \text{for any number } n \qquad (1\text{-}17)$$

For example, the log of 10 is 1, since $10^1 = 10$, and the log of 100 is 2, since $10^2 = 100$. The log of 50 would be in between that of 10 and 100, and it may be found with the aid of a table or a suitable calculator to be 1.6990; that is, $10^{1.6990} = 50$. The log of 0.1 is -1, since $10^{-1} = 0.1$; the log of 0.01 is -2, since $10^{-2} = 0.01$. The log of 0.05 is found with the aid of a table or calculator to be -1.3010; that is, $10^{-1.3010} = 0.05$.

Study Table 1-4 to better understand the relation of concentration to pH.

pH as related to aqueous solutions

Water is slightly ionized,

$$2H_2O \rightleftharpoons H_3O^+ + OH^- \qquad (1\text{-}18)$$

TABLE 1-4
Relation of concentration to pH

$[H_3O^+]$, mol/liter	log $[H_3O^+]$	pH (−log $[H_3O^+]$)	
1.0 = 10^0	0	0	Acidic
0.1 = 10^{-1}	−1	1	
0.01 = 10^{-2}	−2	2	
0.001 = 10^{-3}	−3	3	
0.0001 = 10^{-4}	−4	4	
0.00001 = 10^{-5}	−5	5	
0.000001 = 10^{-6}	−6	6	
0.0000001 = 10^{-7}	−7	7	Neutral
10^{-8}	−8	8	Basic
10^{-9}	−9	9	
10^{-10}	−10	10	
10^{-11}	−11	11	
10^{-12}	−12	12	
10^{-13}	−13	13	
10^{-14}	−14	14	

and the usual equilibrium constant expression would be

$$K = \frac{[H_3O^+][OH^-]}{[H_2O]^2} \qquad (1\text{-}19)$$

where K is the equilibrium constant. Because water is so slightly ionized, the concentration of water remains almost constant at 1000 grams per liter (g/liter) or,

$$\frac{(1000\text{ g})(1\text{ mol})}{(1\text{ liter})(18\text{ g})} = 55.5\text{ mol/liter} \qquad (1\text{-}20)$$

When 55.5 mol/liter is substituted for $[H_2O]$, the equilibrium constant expression becomes

$$[H_3O^+][OH^-] = 55.5^2K = K_w = 10^{-14} \qquad (1\text{-}21)$$

where K_w is the ion product constant for water. In pure water the concentration of hydrogen ions must equal the concentration of hydroxide ions. It has been experimentally determined that, in pure water,

$$[H_3O^+] = [OH^-] = 10^{-7}\text{ mol/liter} \qquad (1\text{-}22)$$

Such a solution is neutral, and its pH is 7. Solutions with a hydrogen-ion concentration greater than 10^{-7} have a pH less than 7 and are acidic; solutions with a hydrogen-ion concentration less than 10^{-7} have a pH greater than 7 and are basic.

Since

$$[H_3O^+][OH^-] = 10^{-14} \qquad (1\text{-}21)$$

$$\log [H_3O^+] + \log [OH^-] = \log 10^{-14} = -14 \qquad (1\text{-}23)$$

and

$$\text{pH} + \text{pOH} = 14 \quad \text{by definition} \qquad (1\text{-}24)$$

PROBLEM 1-12 What is the pH of a water solution of each of the following? (a) 0.10 M HCl, (b) 0.01 M HCl, (c) 0.15 M HCl, (d) 0.10 M NaOH.

Solution to (a) $[H_3O^+] = 0.10\ M = 1.0 \times 10^{-1}\ M$; $\log [H_3O^+] = -1$; pH = 1

Summary

The *covalent bond* consists of a pair of shared electrons; it is the type of bond that predominates in organic compounds. Generally, one electron is donated by each bonded atom; if both electrons come from only one atom of the bonded pair, the bond is called a *coordinate covalent bond.* When bonds are broken, energy is required; when bonds are formed, energy is generally released.

Single covalent bonds about a neutral carbon atom have a *tetrahedral geometry* and *bond angle.* Also, single covalent bonds about oxygen and nitrogen atoms have an approximately tetrahedral bond angle.

Atoms of differing electronegativity are bonded together by *polar covalent bonds;* the result is a *polarized bond.* A *dipole moment* may be calculated for polar bonds; and if the vector sum of all dipole moments is not zero, the molecule may have a net dipole moment and be a polar molecule.

Double covalent bonds consist of two pairs of shared electrons, and *triple covalent bonds* consist of three pairs of shared electrons.

Ionic bonds are formed by the electrostatic attraction between oppositely charged ions, and such ionic bonds are found in many organic compounds.

The important *intermolecular attractive forces,* in order of decreasing strength, are *hydrogen bonds, dipole-dipole attractive forces,* and *London forces.*

Chemical reactions may take place by the formation of *reaction intermediates. Homolytic bond cleavage* produces an intermediate with an unpaired electron called a *free radical; heterolytic bond cleavage* produces electrically charged intermediates, *cations* and *anions.* The *reaction mechanism* is the detailed pathway involving all of the intermediates by which a chemical reaction occurs.

The *free-energy change* for a reaction is the sum of a change in enthalpy and a change in entropy. If the free-energy change is negative, the reaction will proceed spontaneously once it has begun; it will be *exergonic. Entropy* is roughly a measure of randomness and tends to a maximum. The term is $T\,\Delta S$, so entropy is more important at higher temperatures. Change of *enthalpy* or heat content, ΔH, will be negative for exothermic reactions and positive for endothermic reactions.

Progress-of-reaction diagrams represent enthalpy change and energy-of-activation requirements graphically. The *activated state* is a transition state involved in the breaking and making of bonds. The energy required to attain the activated state is the minimum amount of energy required for the reaction to take place, the *activation energy.* It is the difference in energy of the initial and activated states.

Acids and *bases* are important as types of organic compounds and as *catalysts* for organic reactions. The *Brønsted-Lowry acid-base theory defines* an acid as a *proton donor* and a base as a *proton acceptor*. The *Lewis acid-base theory* defines an acid as an *electron-pair acceptor* and a base as an *electron-pair donor*. Defined as $-\log [H_3O^+]$, *pH* is a common method for expressing hydrogen-ion concentration of dilute aqueous solutions.

Key terms

Acid: a proton donor in the Brønsted-Lowry theory and an electron-pair acceptor in the Lewis theory.

Activated state: a transition state in which bonds are being broken and new bonds are formed in a chemical reaction. The energy required to reach the activated state is the activation energy.

Activation energy: the energy which a molecule must acquire to form the activated state and for the reaction to proceed.

Base: a proton acceptor in the Brønsted-Lowry theory and an electron-pair donor in the Lewis theory.

Catalyst: a substance that can lower the activation energy and speed up the reaction without being permanently altered or consumed.

Coordinate covalent bond: a covalent bond in which one atom contributes both electrons of the bonding pair.

Covalent bond: a chemical bond consisting of a pair of shared electrons of opposite spin.

Dipole-dipole attraction: the attraction of the partially positive end or side of one polar molecule for the partially negative end or side of another polar molecule.

Dipole moment: the tendency of a molecule to orient itself in an electrostatic field. A polar molecule has a dipole moment.

Double bond: a covalent bond in which two pairs of electrons are shared.

Electronegativity: the attraction for bonding electrons.

Endergonic reaction: a reaction in which the free-energy change is positive; such a reaction is not spontaneous.

Enthalpy: the potential energy or heat content of a molecule; equilibrium in a reaction tends toward a minimizing of enthalpy.

Entropy: a measure of randomness, chaos, or disorder in a system; equilibrium in a reaction tends toward a maximizing of entropy.

Exergonic reaction: a reaction in which the free-energy change is negative; such a reaction is spontaneous.

Free-energy change: the sum of the change in enthalpy and the change in entropy that results from a chemical reaction.

Free radical: a reaction intermediate having an unpaired electron.

Heterolytic bond cleavage: the cleavage of a covalent bond in which ions are produced.

Homolytic bond cleavage: the cleavage of a covalent bond in which a reaction intermediate having an unpaired electron—a free radical—is produced.

Hydrogen bond: an especially strong type of dipole-dipole attraction between molecules in which hydrogen is covalently bonded to fluorine, oxygen, or nitrogen.

Instantaneous induced dipole: a dipole induced in nonpolar molecules by the mutual repulsion of the electron fields as the molecules approach each other.

Intermolecular attractive force: the force which holds molecules together in the liquid or solid state; it is much weaker than the covalent or ionic bond.

Ion: a charged particle. It is called **cation** if positively charged and **anion** if negatively charged.

Ionic bond: a bond formed by the mutual attraction of ions of opposite charge.

London force: the attraction between molecules caused by an instantaneous induced dipole.

pH: the negative logarithm of the hydrogen-ion concentration, in moles per liter.

Polar covalent bond: a covalent bond in which the bonding electron pair is closer to one atom than the other; one end is relatively negative and the other is relatively positive.

Progress-of-reaction diagram: a diagram that represents the energy requirements and changes as a chemical reaction proceeds.

Reaction intermediate: a short-lived species produced by bond cleavage in a chemical reaction; the species then combine to form a new product.

Reaction mechanism: the detailed pathway, involving all intermediates, by which a chemical reaction occurs.

Saturated carbon atom: a carbon atom with four single bonds.

***sp^3* hybrid orbital:** a molecular orbital formed by the mixing or hybridizing of one 2*s* orbital and three 2*p* orbitals.

Tetrahedral geometry: the geometry of a methane molecule and many other molecules which are bonded by sp^3 hybrid orbitals; the bond angles are approximately 109.5°.

Triple bond: a covalent bond in which three pairs of electrons are shared.

Additional problems

1-13 Write the electron-dot structural formula for each of the following compounds:

(a) ethane, C_2H_6
(b) ethene, C_2H_4
(c) sulfuric acid, H_2SO_4
(d) nitric acid, HNO_3
(e) carbonic acid, H_2CO_3
(f) water, H_2O
(g) methanol, CH_3OH
(h) sodium hydrogen sulfite, $NaHSO_3$

1-14 Write the electron-dot structural formula for each of the following species:

(a) ammonium nitrate, NH_4NO_3
(b) sodium cyanide, NaCN
(c) ethyl free radical, $\cdot C_2H_5$
(d) ethyl cation, $C_2H_5^+$
(e) ethyl anion, $C_2H_5^-$
(f) nitrogen, N_2
(g) bromine, Br_2
(h) carbon dioxide, CO_2

1-15 Taking into consideration any hydrogen bonding or dipole-dipole attraction due to polar covalent bonds, number the following compounds in order of decreasing boiling point; that is, the compound with the highest boiling point will be designated number 1.

____ CH_3CH_2OH	____ $CH_3CH_2CH_3$
____ $CH_3—O—CH_3$	____ $CH_3CH_2NH_2$

1-16 For the following compounds identify single covalent bonds, any double covalent bonds, any triple covalent bonds, and any ionic bonds.

(a) $(CH_3)_4NOH$
(b) C_2H_4
(c) CaC_2
(d) $NaOCH_3$
(e) $(CH_3COO)_2Ca$

1-17 Write the equations for the following acid-base reactions; label the Brønsted-Lowry acids and bases involved; and indicate each acid and base as stronger or weaker. (*Hint:* Omit nonparticipating ions.)

(a) sodium bicarbonate and hydrochloric acid (H_3O^+ Cl^-)
(b) methylamine (CH_3NH_2) and water
(c) ammonia and hydrogen chloride
(d) sodium amide (Na^+ NH_2^-) and water to give NaOH and NH_3

1-18 Write the formula of the conjugate base for each of the following acids and the formula of the conjugate acid for each of the bases given.

Acid	**Conjugate base**	**Base**	**Conjugate acid**
(a) HCO_3^-	____	(d) $^-OC(=O)—CH_3$	____
(b) $CH_3C{\equiv}C—H$	____	(e) HSO_4^-	____
(c) H_2SO_4	____	(f) CH_3NH_2	____

1-19 Identify the following reactants as Lewis acids or as Lewis bases.

(a) $CH_3CH_2^+ + Cl^- \longrightarrow CH_3CH_2Cl$
(b) $BF_3 + NH_3 \longrightarrow BF_3NH_3$

1-20 Indicate whether the following pH values are acidic or basic: (a) 6.7, (b) 0.5, (c) 7.1, (d) 0.0, (e) 14.1.

1-21 For the following substances whose pH is given, calculate their hydrogen-ion concentrations. (Items d to f will require antilogs from a table or a calculator.)

	pH		**pH**
(a) gastric juice	2.0	(e) cow's milk	6.5
(b) vinegar	3.0	(f) blood	7.4
(c) orange juice	4.0	(g) seawater	8.0
(d) beer	4.5		

2

STRUCTURE, NOMENCLATURE, AND IDENTIFICATION OF ORGANIC COMPOUNDS

In Chap. 1 we reviewed some of the basic chemical principles which are applicable to organic chemistry also. But organic compounds are so numerous, so varied and complex in structure, and so pervasive in our world that they create special problems for the student. Why are there so many of them? How are their atoms arranged to give such diversity? How can we keep track of them so we know when we are talking about the same one? And how can we even identify specific ones?

In this chapter we shall look at some of the ways organic chemists deal with those questions. First we shall discuss the general principles of organic structure—the ways in which atoms can be arranged to create such a great variety of compounds. Then we shall introduce ways of naming compounds. Finally, we shall look at some of the techniques organic chemists use in the laboratory to identify specific compounds.

2-1 The structures of organic compounds

Of all the elements, carbon best possesses the ability to combine with itself to form long chains, both continuous and branched, and also closed-chain or cyclic compounds. Open-chain compounds are called *acyclic* (noncyclic) *compounds,* and closed-chain compounds are called *cyclic compounds*.

Isomerism

Isomers are compounds which have the same molecular formula (the same number and kinds of atoms) but differ in some way in their structures (the way in which the atoms are arranged in the molecule). Since, as the number of carbon atoms increases, the number of possible arrange-

TABLE 2-1
Number of isomeric alkanes*

Molecular formula	Number of possible isomers	Molecular formula	Number of possible isomers
CH_4	1	C_7H_{16}	9
C_2H_6	1	C_8H_{18}	18
C_3H_8	1	C_9H_{20}	35
C_4H_{10}	2	$C_{10}H_{22}$	75
C_5H_{12}	3	$C_{20}H_{42}$	366,319
C_6H_{14}	5		

* *Alkanes* are open-chain compounds with the general molecular formula C_nH_{2n+2}, where n represents the number of carbon atoms. They will be discussed in detail in Chap. 3.

ments of a given number and kind of atoms also greatly increases, the number of isomeric organic compounds can be very great. See Table 2-1.

Many chemists classify isomers as either structural isomers or stereoisomers. *Structural isomers* include chain or skeletal isomers, position isomers, and functional group isomers. They will be discussed next, and then the types of stereoisomers will be considered. *Stereoisomers* also differ in structure, but the difference is a much more subtle one. It is a difference in spatial arrangement of the constituent atoms. Stereoisomers include geometric, or cis-trans, isomers, configurational, or optical, isomers, and conformational isomers. Some chemists do not regard conformations as isomers, since they usually cannot be separated and isolated as other isomers can be. However, conformations do exist as an equilibrium mixture for most compounds and do help determine physical and chemical properties, so we need to be aware of their existence.

Chain or skeletal isomerism. *Chain* or *skeletal isomerism* exists when two or more compounds have the same molecular formula but differ in the branching of the skeletal chain of carbon atoms. For example, the two isomers models of which are shown in Fig. 2-1 correspond to the molecular formula C_4H_{10}. They have the structural formulas

```
                                              H
                                              |
    H   H   H   H                     H  H—C—H  H
    |   |   |   |                     |       |       |
H—C—C—C—C—H        and        H—C———C———C—H
    |   |   |   |                     |       |       |
    H   H   H   H                     H       H       H
```

n-Butane, bp −0.5°C
(*n* stands for "normal")

Isobutane, bp −10.2°C

These are different compounds with different physical and chemical properties. Since both isomers have the molecular formula C_4H_{10}, that formula is not adequate to express the difference between them. The structural formulas shown above are required to express the *structural difference*. Since such structural formulas take a long time to write, condensed structural formulas which require less time to write may be used.

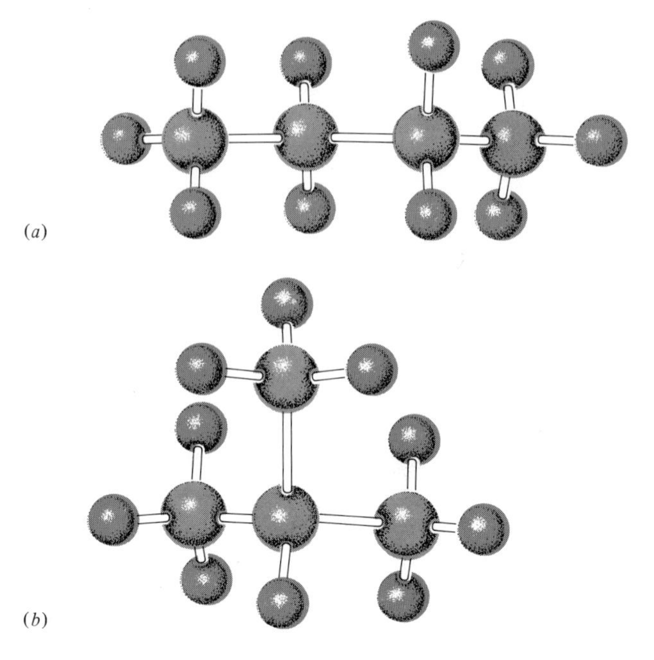

FIGURE 2-1
Models of (*a*) *n*-butane and (*b*) isobutane (a branched-chain isomer of *n*-butane).

For example, the butane isomers can be written

	n-Butane	Isobutane
	$CH_3CH_2CH_2CH_3$	$CH_3CH(CH_3)CH_3$ (drawn with CH_3 above the central CH: $CH_3\overset{|}{C}HCH_3$)
or	$CH_3(CH_2)_2CH_3$	$CH_3CH(CH_3)CH_3$

Another example, but one involving a cyclic structure, is C_6H_{12}:

Cyclohexane (ring of six CH_2 groups: H_2C, H_2C, CH_2, CH_2, with $C H_2$ at top and bottom) and Methylcyclopentane (ring of H_2C, CH_2, H_2C, $C H_2$ at top, and $CHCH_3$)

or in more abbreviated form:

(hexagon) and (pentagon)—CH_3

Position isomerism. *Position isomerism* exists when two or more compounds have the same molecular formula and the same skeletal chain of carbon atoms but differ in the position or location of a substituent atom or

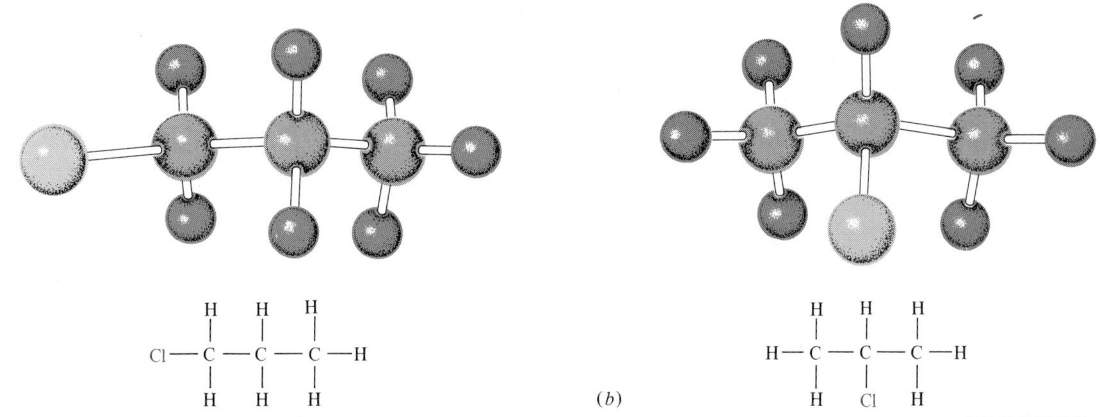

FIGURE 2-2 **(*a*) 1-Chloropropane, $CH_3CH_2CH_2Cl$ and (*b*) 2-chloropropane, $CH_3CHClCH_3$, are position isomers.**

group on the chain. The monochloropentanes are an example.

$$\overset{5}{C}H_3\overset{4}{C}H_2\overset{3}{C}H_2\overset{2}{C}H_2\overset{1}{C}H_2Cl \qquad \overset{5}{C}H_3\overset{4}{C}H_2\overset{3}{C}H_2\overset{2}{C}H(Cl)\overset{1}{C}H_3 \qquad \overset{1}{C}H_3\overset{2}{C}H_2\overset{3}{C}H(Cl)\overset{4}{C}H_2\overset{5}{C}H_3$$

1-Chloropentane 2-Chloropentane 3-Chloropentane

Note the use of the locator numbers, 1, 2, and 3, in the names to indicate the position of the substituent chlorine atom on the carbon chain. The names are the ones assigned by the rules of the International Union of Pure and Applied Chemistry (IUPAC). Here the applicable rule is to begin numbering the carbon atoms at the end of the chain which will give the lowest number for the carbon atom holding the substituent. The IUPAC system of naming compounds will be considered in more detail in Sec. 2-2 (see also the chloropropanes shown in Fig. 2-2).

An example of position isomerism involving a cyclic structure is given by the dichlorocyclohexanes.

1,2-Dichlorocyclohexane: ring C1 (H)(Cl), C2 CH—Cl, C3 CH_2, C4 CH_2, C5 H_2C, C6 H_2C

1,3-Dichlorocyclohexane: ring C1 (H)(Cl), C2 CH_2, C3 CH—Cl, C4 CH_2, C5 H_2C, C6 H_2C

1,4-Dichlorocyclohexane: ring C1 (H)(Cl), C2 CH_2, C3 CH_2, C4 (H)(Cl), C5 H_2C, C6 H_2C

1,2-Dichlorocyclohexane 1,3-Dichlorocyclohexane 1,4-Dichlorocyclohexane

PROBLEM 2-1 If any of the following compounds are position isomers of each other, identify them by check-marking their formulas.

(a) ____ $ClCH_2CH_2Cl$, ____ $Cl{-}CH(Br){-}CH_3$, ____ $BrCH_2CH_2Br$,

____ $ClCH_2CH_2Br$, ____ $ClCH_2CH(Cl){-}Br$

(b) ___ $Cl{-}CH(Cl){-}CH_2CH_3$, ___ $CH_3CCl_2{-}CH_3$, ___ $Cl{-}CH(CH_3){-}CH_3$,

___ $ClCH_2CH(Cl){-}CH_3$, ___ (cyclopropane ring: H_2C, H_2C, $C(H)(Cl)$)

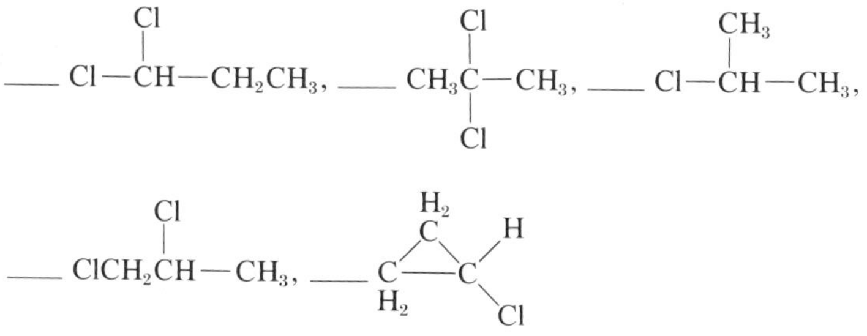

Solution to (a) ___ $ClCH_2CH_2Cl$, √ $Cl{-}CH(Br){-}CH_3$, ___ $BrCH_2CH_2Br$,

√ $ClCH_2CH_2Br$, ___ $ClCH_2CH(Cl){-}Br$

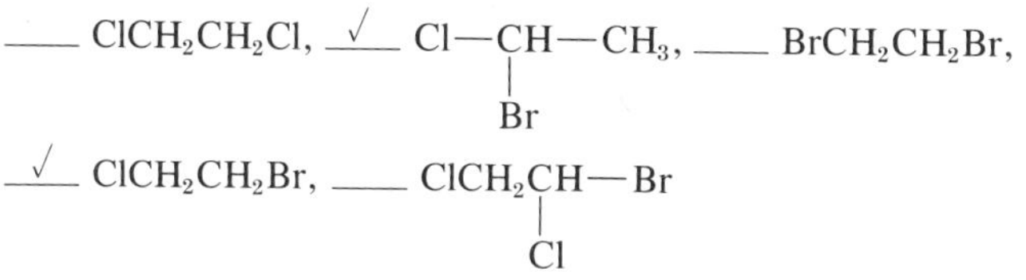

Functional group isomerism. *Functional group isomerism* exists when two or more compounds have the same molecular formula but contain different functional groups. The carbon chains and rings constitute the skeletal framework to which various reactive groups called *functional groups* may be attached. Functional groups will be discussed in more detail on pages 39 to 42. The following are some examples of functional group isomers.

ISOMERS OF C_2H_6O (Fig. 2-3).

CH_3CH_2OH Ethanol (Ethyl alcohol) and CH_3OCH_3 Methoxymethane (Methyl ether)

ISOMERS OF C_3H_6O (Fig. 2-4).

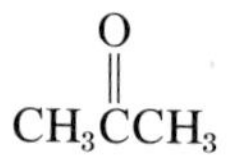

$CH_3C(=O)CH_3$ Propanone (Acetone or dimethyl ketone) and $CH_3CH_2CH(=O)$ Propanal (Propionaldehyde)

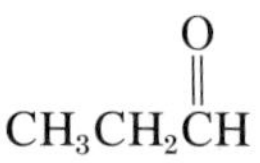

ISOMERS OF $C_3H_6O_2$ (Fig. 2-5).

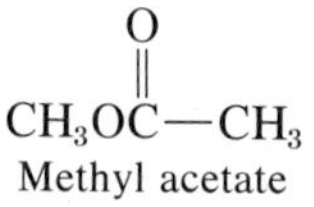

$CH_3OC(=O){-}CH_3$ Methyl acetate and $HO{-}C(=O){-}CH_2CH_3$ Propanoic acid (Propionic acid)

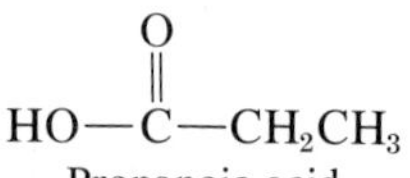

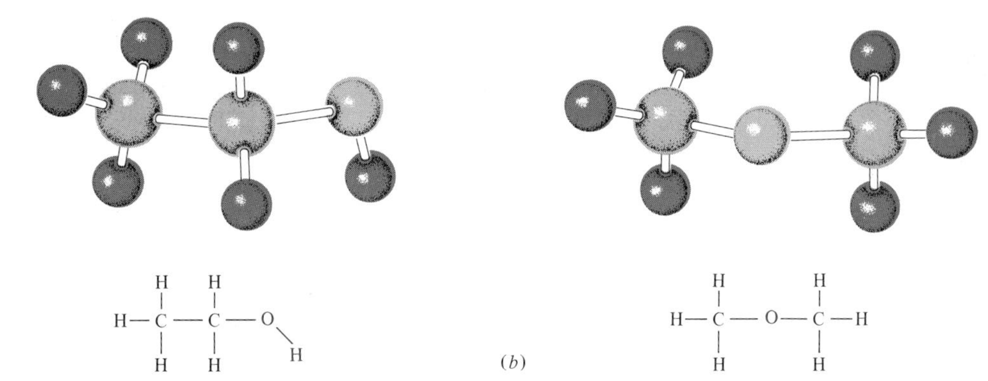

FIGURE 2-3 (*a*) Ethanol (ethyl alcohol) and (*b*) methoxymethane (methyl ether) are functional group isomers of formula C_2H_6O.

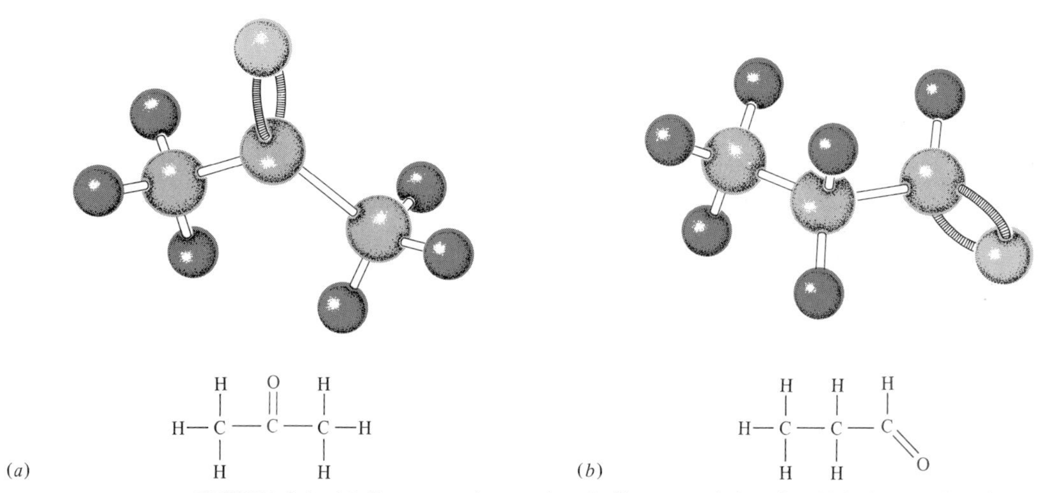

FIGURE 2-4 (*a*) Propanone (acetone) and (*b*) propanal (propionaldehyde) are functional group isomers of formula C_3H_6O.

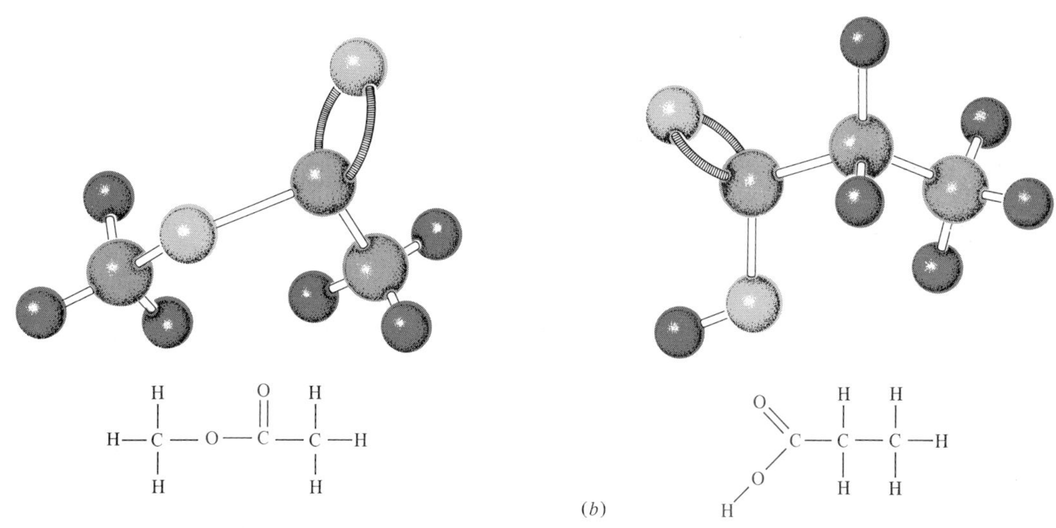

FIGURE 2-5 (*a*) Methyl acetate and (*b*) propanoic acid (propionic acid) are functional group isomers of formula $C_3H_6O_2$.

PROBLEM 2-2 Indicate which members of the following group of compounds are functional group isomers of each other.

```
(a) CH2OH      (b) H—C=O         (c)    CH2OH     (d)      O
     |                |                   |                 ||
    C=O           H—C—OH            H—C—OH                C—OH
     |                |                   |                 |
    CH2OH            CH2OH               CH2OH          H—C—OH
                                                            |
                                                           CH3
```

Solution Compound (a) has the molecular formula $C_3H_6O_3$. It contains two —OH groups and one $\gt C{=}O$ group.

Compound (b) has the molecular formula $C_3H_6O_3$. It contains one —HC=O group and two —OH groups.

Compound (c) has the molecular formula $C_3H_8O_3$. It contains three —OH groups.

Compound (d) has the molecular formula $C_3H_6O_3$. It contains one

```
  O
  ||
—C—OH
```

group and one —OH group.

Compounds (a), (b), and (d) are functional group isomers of each other.

PROBLEM 2-3 Indicate which of the following compounds are functional group isomers of each other.

```
                                       O
                                       ||
(a) CH3O—CH2O—CH3        (b) CH3CH2C—OH        (c) CH3CH—CH2
                                                          |    |
                                                          OH   OH

        O
        ||
(d) CH3C—CH2OH
```

Geometric isomerism. It is possible to have two or more isomeric compounds in which the type of carbon skeleton is the same and the same functional groups are attached to the corresponding carbon atoms in their respective chains, but which differ in structure because of the spatial arrangement of their groups. Such isomers are called *stereoisomers*.

One type of stereoisomerism is called *geometric isomerism*. Generally, two carbon atoms can rotate about a single bond joining them. However, such rotation is restricted when the single bond is part of a closed chain (cyclic structure). Rotation is restricted about a double bond also. Geometric isomerism becomes possible when there is substitution on the chain at a position where rotation is restricted. For example, there are two isomeric 1,2-dichlorocyclopropanes (Fig. 2-6):

```
        H2                        H2
  H    3C     H             H    3C     Cl
    \  / \  /                 \  / \  /
     C1—C2                     C1—C2
    /      \                  /      \
  Cl        Cl              Cl        H
```

cis-1,2-Dichlorocyclopropane   *trans*-1,2-Dichlorocyclopropane

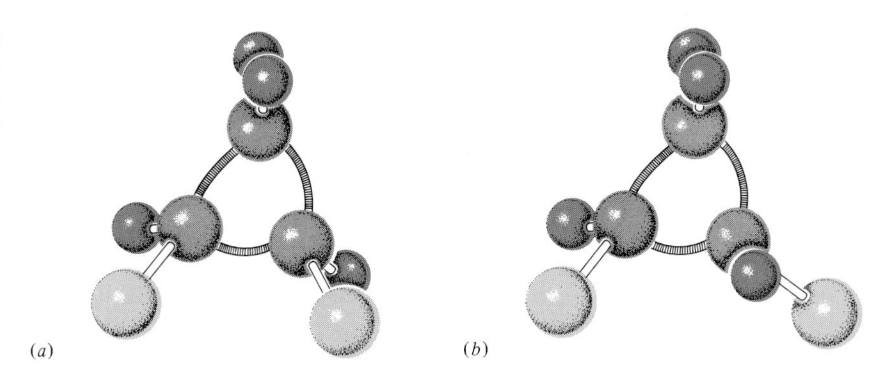

FIGURE 2-6 **Geometric isomers: (*a*) *cis*-1,2-dichlorocyclopropane and (*b*) *trans*-1,2-dichlorocyclopropane.**

Also, there are two isomeric 2-butenes (Fig. 2-7):

```
   1          4            1
 H3C         CH3         H3C          H
    \       /               \        /
     C == C                  C == C
    /  2  3 \               /  2  3 \
   H         H             H         CH3
                                      4
  cis-2-Butene             trans-2-Butene
```

The geometric isomer with the substituents on the same side of the ring or the double bond is referred to as *cis*. The isomer with the substituents on opposite sides of the ring or the double bond is referred to as *trans*. Note that the locator number in the name of the butenes refers to the position of the double bond between the carbon atoms in positions 2 and 3.

Observe that a compound like the following could not have a geometric isomer.

```
   1          4
 H3C         CH3
    \       /
     C == C
    /  2  3 \
 H3C         H
 2-Methyl-2-butene
```

The atoms or groups attached to the doubly bonded carbon atoms of geometric isomers must differ from each other respectively on *each* end. It

FIGURE 2-7 **(*a*) *cis*-2-Butene and (*b*) *trans*-2-butene are geometric isomers.**

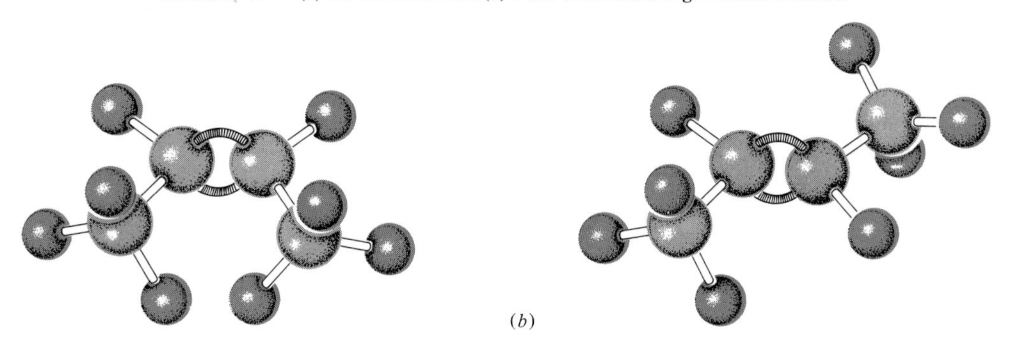

might also be noticed now that each of the dichlorocyclohexanes, whose formulas were given earlier, could exist as cis-trans, or geometric, isomers.

PROBLEM 2-4 Label the following geometric isomers as either cis or trans.

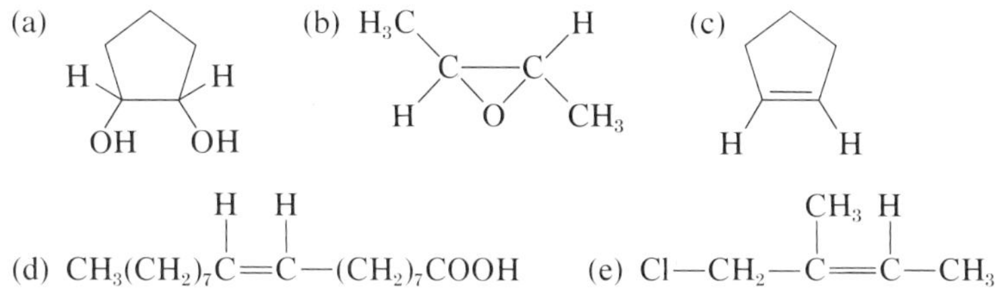

(a) [cyclopentane ring with H and OH on each of two adjacent carbons; both OH groups below the ring]

(b) [epoxide: H_3C and H on one carbon, H and CH_3 on the other; the two CH_3 groups on opposite sides]

(c) [cyclopentene with an H on each double-bond carbon]

(d) $CH_3(CH_2)_7C(H){=}C(H){-}(CH_2)_7COOH$ (both H drawn above the double bond)

(e) $Cl{-}CH_2{-}C(CH_3){=}C(H){-}CH_3$ (CH_3 and H drawn above the double bond)

Solution to (a) This isomer is cis. The two —OH groups are on the same side of the ring. Note for (e): This type is considered cis if the carbon groups making up the main carbon chain are on the same side of the double bond and trans if they are on opposite sides. For example, the following geometric isomer is trans:

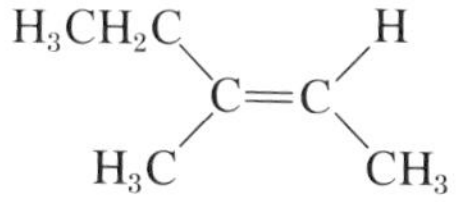

$$\begin{matrix} H_3CH_2C & & H \\ & C{=}C & \\ H_3C & & CH_3 \end{matrix}$$

Configurational isomerism. *Configurational,* or *optical, isomerism* is a type of stereoisomerism which results from a compound having *chiral molecules* (from the Greek *cheir,* ''hand''). If two molecules are identical, they are superimposable one upon the other. But a chiral molecule may have a mirror-image stereoisomer (called an *enantiomer*) which is not superimposable and hence is not identical. Mirror-image stereoisomers differ in the way one's hands differ. This very subtle structural difference is both important and interesting. Further study of configurational isomers will be postponed until Chap. 11.

Conformational isomerism. If you examine a model of ethane, C_2H_6, you will observe that the model suggests that free rotation is possible about the carbon-carbon bond. Such rotation actually is possible, and different forms of the molecule are thereby produced. They are illustrated in Fig. 2-8. These structural forms, called *conformations,* may be represented by *Newman projection formulas,* in which one looks down the axis of the carbon-carbon bond. Examples for ethane are given in Fig. 2-9. The relatively close proximity of the substituents in the eclipsed conformation, with attendant repulsion between the substituents, contributes toward making this the least stable of the possible conformations. A molecule does not remain in this conformation; it merely passes through it in changing from one staggered conformation to another staggered conformation.

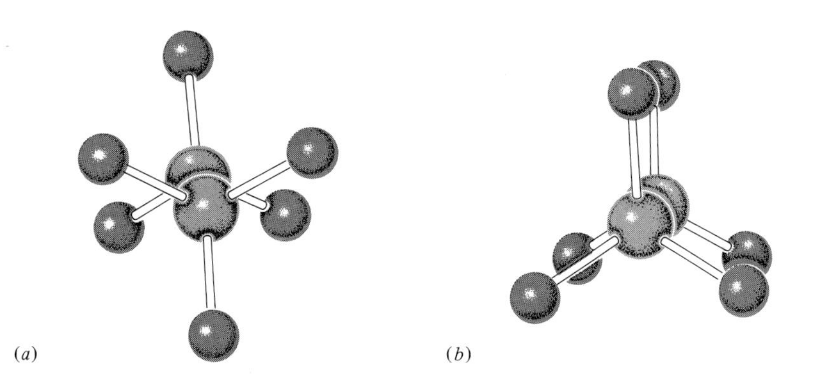

FIGURE 2-8
Models of *(a)* the staggered and *(b)* the eclipsed conformations of ethane.

The staggered conformations are the more stable forms, and the eclipsed conformations may be regarded as transition states.

Three staggered conformations for butane are possible; they are shown in Fig. 2-10. Such staggered conformations represent energy minima in the rotation process; they are referred to as *conformational isomers* or *conformers*. The anti conformation of butane is the most stable one because it allows the most room for the methyl, CH_3—, groups. However, butane actually consists of an equilibrium mixture of all three conformers, since the three are so readily interconvertible by rotation about the center carbon-carbon bond. *Because the conformers cannot be isolated as separate compounds, except in some special cases of very restricted rotation, some chemists do not regard them as isomers at all.*

FIGURE 2-9
Newman projection formulas of conformations of ethane: *(a)* staggered conformation, *(b)* an almost eclipsed conformation, and *(c)* one of an infinite number of skew conformations.

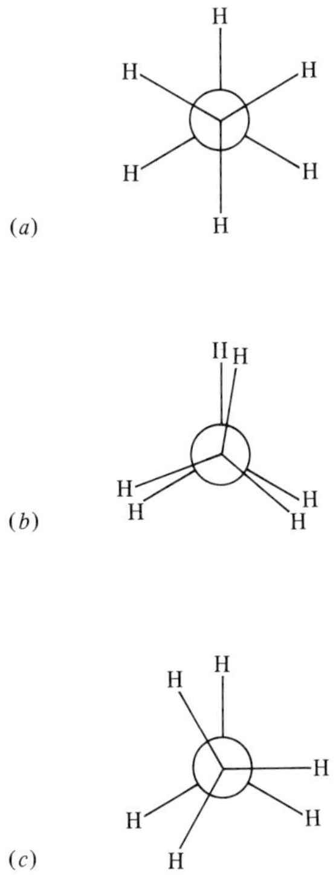

The longer the chain, the greater the number of carbon-carbon bonds about which rotation can occur, and so the greater the number of conformations in which the molecule may occur. In Fig. 2-11 is shown a model of propane in an all-staggered conformation such that none of the hydrogen atoms are eclipsing one another. The model is so oriented that a planar projection of it would appear as the structural formula is commonly written,

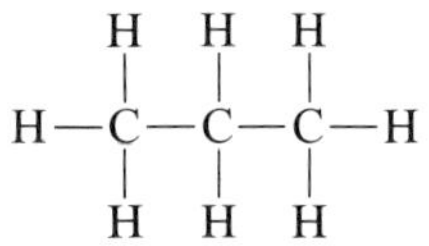

$$\begin{array}{ccccccc} & & H & & H & & H & & \\ & & | & & | & & | & & \\ H & — & C & — & C & — & C & — & H \\ & & | & & | & & | & & \\ & & H & & H & & H & & \end{array}$$

CONFORMATIONAL ISOMERISM OF RING COMPOUNDS. Partial rotation about single carbon-carbon bonds is even possible in rings of six carbon atoms or more and to a lesser extent in smaller rings. Examples are cyclobutane (four carbon atoms) and cyclopentane (five carbon atoms), which twist slightly to become a little anti planar so that their hydrogens are not eclipsing each other and have a little more space. Cyclohexane can exist in boat, twist, and chair conformations; see Figs. 2-12 and 2-13.

The chair conformation, which has all of its hydrogen atoms staggered and none eclipsed, is the most stable form, and so it is the form in which most of the cyclohexane molecules will be found at any given time. Half

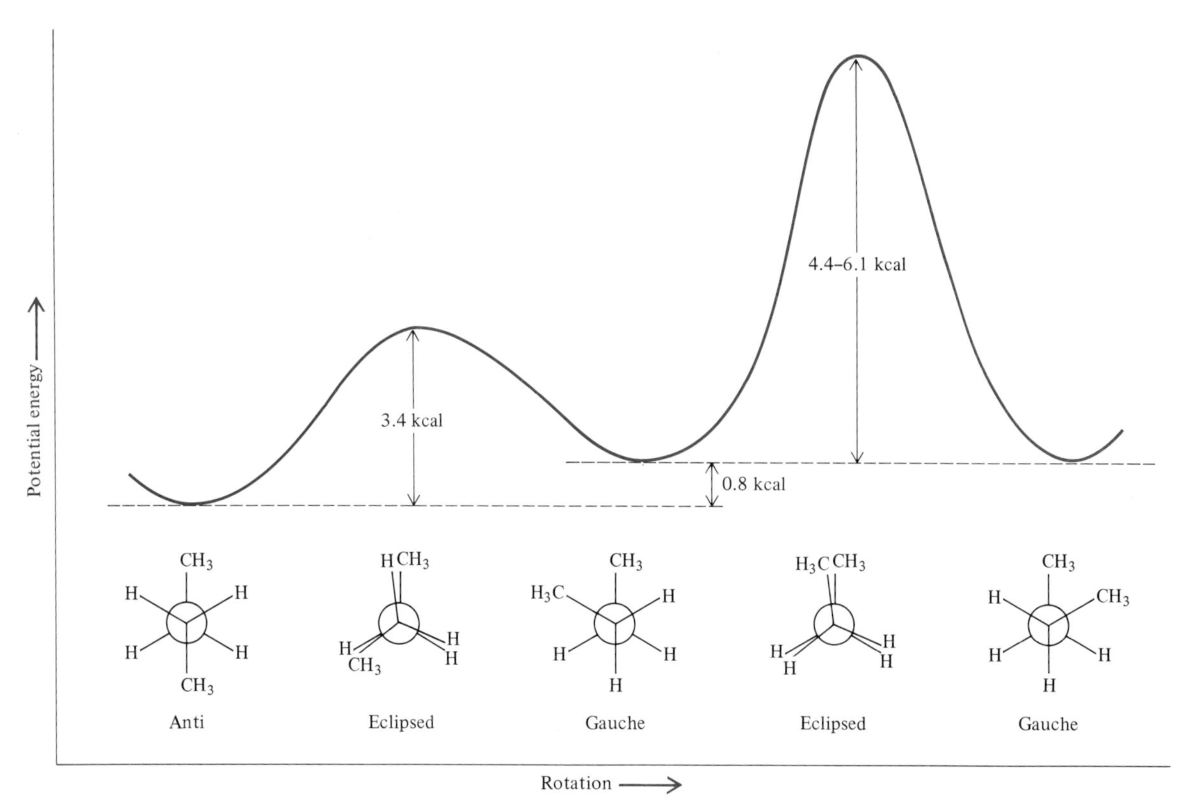

FIGURE 2-10 Potential-energy changes associated with conformations of *n*-butane produced by rotation about the C-2–C-3 bond.

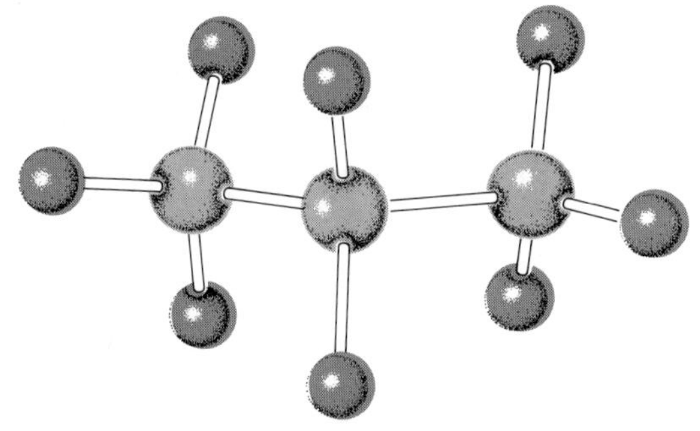

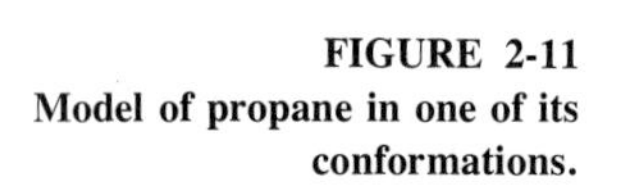
FIGURE 2-11
Model of propane in one of its conformations.

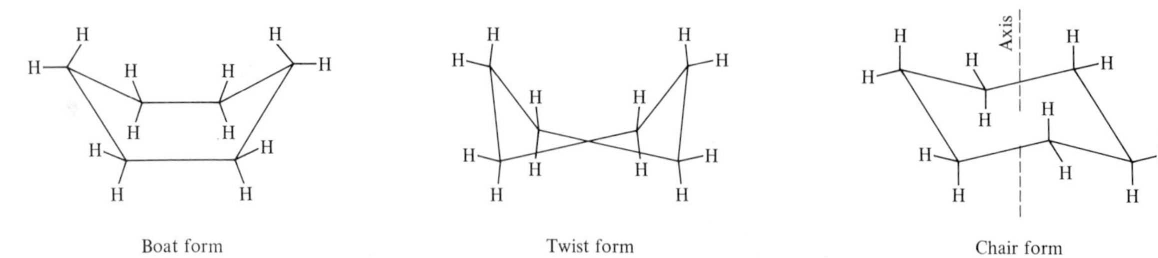

FIGURE 2-12 Structural formulas of three conformations of cyclohexane.

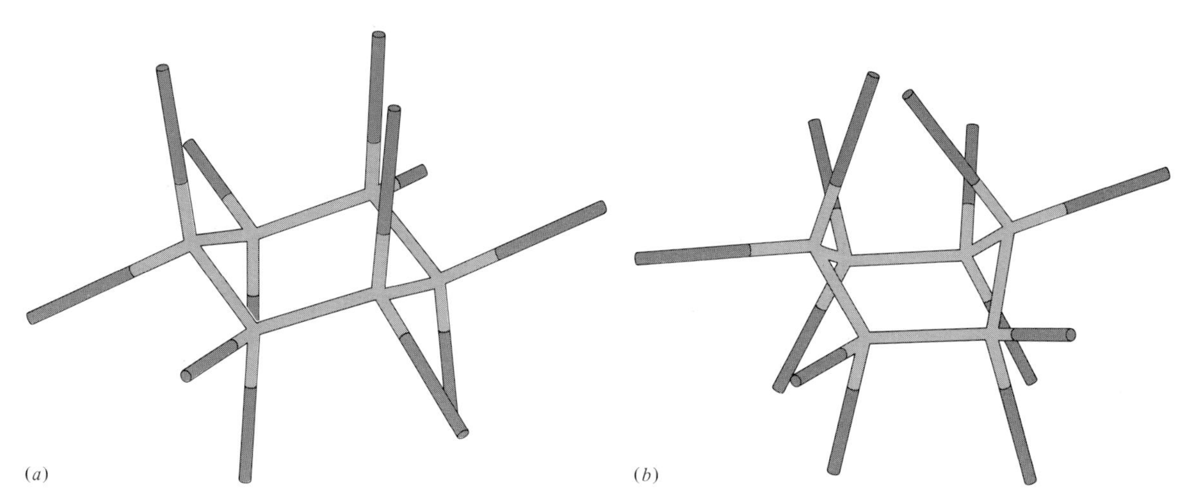

FIGURE 2-13 Framework Molecular Models of (*a*) the chair conformation and (*b*) the boat conformation of cyclohexane. These models show the bonding but not the space occupied by the orbitals.

of the hydrogen atoms of the chair form of cyclohexane are attached to *axial bonds*—bonds parallel to the imaginary axis passing through the center of the molecule. The other half of the hydrogen atoms are attached to *equatorial bonds*. Equatorial bonds are usual for any substituents larger than hydrogen because more space is available.

Functional groups and homologous series

The compounds of organic chemistry may be considered as *functional group derivatives* of the alkanes, or saturated hydrocarbons, in which functional groups have been substituted for one or more hydrogen atoms. The *alkanes* are straight or branched-chain hydrocarbons with a general molecular formula of C_nH_{2n+2}, where n equals the number of carbon atoms. Table 2-2 is a partial listing of some of the smaller alkane molecules. Examples of functional groups are X (halogen), $—NH_2$, —OH, —COOH, =O, and $—NO_2$.

Table 2-3 lists some important functional groups with which it would be well for you to become acquainted. The functional group is shown in color in the structural formula of each example. You need not commit the functional groups to memory at this time. Additional groups will be introduced as the course progresses.

Organic chemistry is readily organized by putting compounds into families according to a common functional group; each such family forms a *homologous series*. The members of a homologous series, called *homologs,* differ from each other only by some multiple of $—CH_2—$ groups, called *methylene groups*. Such a classification greatly facilitates the study of organic compounds. All homologs will have similar chemical and physical properties because they have the same functional group.

Table 2-4 gives portions of several homologous series. Each column

TABLE 2-2
Examples of alkanes, C_nH_{2n+2}

Formula	IUPAC name	Common name
CH_4	Methane	Methane
CH_3CH_3	Ethane	Ethane
$CH_3CH_2CH_3$	Propane	Propane
$CH_3CH_2CH_2CH_3$	Butane	*n*-Butane
CH_3 | CH_3CHCH_3	2-Methylpropane	Isobutane
$CH_3(CH_2)_3CH_3$	Pentane	*n*-Pentane
CH_3 | $CH_3CHCH_2CH_3$	2-Methylbutane	Isopentane
CH_3 | CH_3CHCH_3 | CH_3	2,2-Dimethylpropane	Neopentane
$CH_3(CH_2)_4CH_3$	Hexane	*n*-Hexane
$CH_3(CH_2)_5CH_3$	Heptane	*n*-Heptane
$CH_3(CH_2)_6CH_3$	Octane	*n*-Octane

TABLE 2-3
Some important functional groups

Formula of example compound	Name of functional group	IUPAC name of example compound
H_3C CH_3 $C{=}C$ H H	Alkenyl or vinyl group	Propene
H | $CH_3CH_2{-}C{-}OH$ | H	Alcohol (hydroxyl group)	1-Propanol
H | $CH_3CH_2{-}C{=}OH$	Aldehyde (carbonyl group)	Propanal
O ‖ $H_3C{-}C{-}CH_3$	Ketone (carbonyl group)	Propanone
O ‖ $CH_3CH_2{-}C{-}OH$	Carboxyl group	Propanoic acid
O ‖ $CH_3{-}C{-}O{-}CH_3$	Ester	Methyl acetate
$H_2N{-}CH_2CH_2CH_3$	Amine (primary)	1-Aminopropane
$CH_3{-}C{\equiv}C{-}H$	Alkynyl or acetylenic group	Propyne
O ‖ $CH_3CH_2{-}C{-}Cl$	Acyl chloride	Propanoyl chloride
O ‖ $CH_3CH_2{-}C{-}NH_2$	Amide	Propanamide
$CH_3CH_2{-}C{\equiv}N$	Nitrile	Propanonitrile
H H | | $CH_3{-}C{-}O{-}C{-}CH_3$ | | H H	Ether	Ethyl ether

TABLE 2-4 Oxidative and structural relations of the simple alkanes and their families*

Alkynes, C_nH_{2n-2} (acetylenes)	Alkenes, C_nH_{2n} (olefins)	Alkanes, C_nH_{2n+2} (paraffins)	Alkanols, R—OH (alcohols)	Alkanals, RCHO (aldehydes)	Alkanones, RC(=O)R (ketones)	Alkanoic acids, RCOOH (carboxylic acids)
. . .	. . .	CH_4 Methane	CH_3OH Methanol† (Methyl alcohol)	$HC(H)=O$ Methanal (Formaldehyde)	. . .	$HC(=O)OH$ Methanoic acid (Formic acid)
$HC{\equiv}CH$ Ethyne (Acetylene)	$H_2C{=}CH_2$ Ethene (Ethylene)	CH_3CH_3 Ethane	CH_3CH_2OH Ethanol (Ethyl alcohol)	$CH_3C(H){=}O$ Ethanal (Acetaldehyde)	. . .	CH_3COOH Ethanoic acid (Acetic acid)
$CH_3C{\equiv}CH$ Propyne (Methylacetylene)	$CH_3CH{=}CH_2$ Propene (Propylene)	$CH_3CH_2CH_3$ Propane	$CH_3CH_2CH_2OH$ 1-Propanol (*n*-Propyl alcohol	$CH_3CH_2C(H){=}O$ Propanal (Propionaldehyde)	. . .	CH_3CH_2COOH Propanoic acid (Propionic acid)
			$CH_3CHOHCH_3$ 2-Propanol (Isopropyl alcohol)	. . .	CH_3COCH_3 Propanone (Acetone, dimethyl ketone)	
$CH_3CH_2C{\equiv}CH$ 1-Butyne (Ethylacetylene)	$CH_3CH_2CH{=}CH_2$ 1-Butene (α-Butylene)	$CH_3CH_2CH_2CH_3$ Butane (*n*-Butane)	$CH_3CH_2CH_2CH_2OH$ 1-Butanol (*n*-Butyl alcohol)	$CH_3CH_2CH_2C(H){=}O$ Butanal (*n*-Butyraldehyde)	. . .	$CH_3CH_2CH_2COOH$ Butanoic acid (*n*-Butyric acid)
$CH_3C{\equiv}CCH_3$ 2-Butyne (Dimethylacetylene)	$CH_3CH{=}CHCH_3$ 2-Butene (β-Butylene)		$CH_3CH_2CHOHCH_3$ 2-Butanol (*sec*-Butyl alcohol)		$CH_3CH_2COCH_3$ Butanone (Methyl ethyl ketone)	
. . .	$CH_3C(CH_3){=}CH_2$ 2-Methylpropene (Isobutylene)	$CH_3CH(CH_3)CH_3$ 2-Methylpropane (Isobutane)	$CH_3CH(CH_3)CH_2OH$ 2-Methyl-1-propanol (Isobutyl alcohol)	$CH_3CH(CH_3)C(H){=}O$ 2-Methylpropanal (Isobutyraldehyde)	. . .	$CH_3CH(CH_3)COOH$ 2-Methylpropanoic acid (Isobutyric acid)
			$CH_3{-}C(CH_3)(OH){-}CH_3$ 2-Methyl-2-propanol (*tert*-Butyl alcohol)			

* The compounds to both the left and the right of the alkanes are in increasingly higher oxidation states.
† The first name given is the IUPAC name: the second name, given in parentheses, is the common name.

represents a homologous series, and each row shows compounds derived from the parent alkane (given in the third column). You should study this table now for an overview of the relations shown. Do not, however, attempt to master the material at this time. In subsequent chapters, you will study these homologous series in detail and will then find Table 2-4 a valuable reference.

PROBLEM 2-5 A molecule may have more than one functional group. Identify *all* of the functional groups in each of the following compounds by using the information in Table 2-3.

(a) 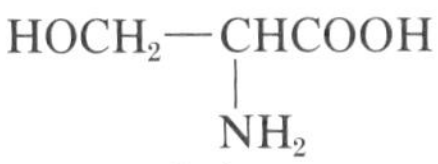 $HOCH_2—CH(NH_2)COOH$

Serine

(b) 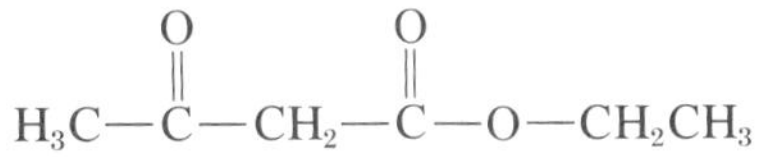 $H_3C—C(=O)—CH_2—C(=O)—O—CH_2CH_3$

Ethyl acetoacetate

(c) 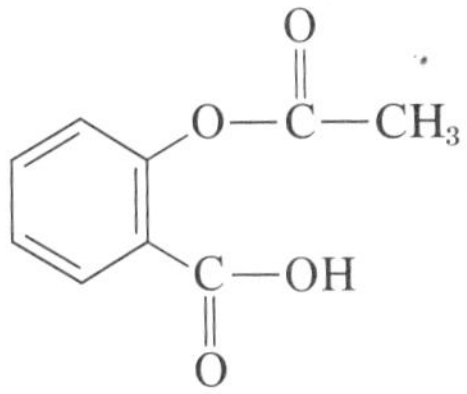 $C_6H_4(—O—C(=O)—CH_3)(—C(=O)—OH)$

Aspirin

(d) 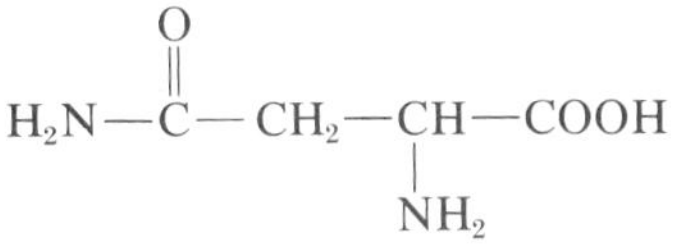 $H_2N—C(=O)—CH_2—CH(NH_2)—COOH$

Asparagine

Solution to (a) Serine contains one alcohol or hydroxyl group, one carboxyl group, and one amine group. The functional groups are shown in black in the following formula:

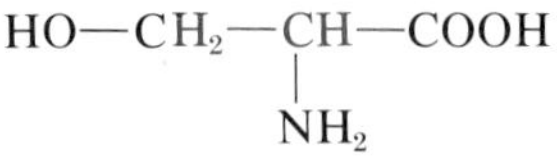

$$HO—CH_2—CH(NH_2)—COOH$$

2-2 Nomenclature

As organic compounds were discovered during the early years of chemistry, they were given common or trivial names such as formic acid, lactic acid, and glycerin. Later, however, when new organic compounds began to be synthesized, it became apparent that they would be exceedingly numerous and that a systematic way of naming them was needed. Accordingly, in 1892, an international meeting was held in Geneva, Switzerland, to devise a standard system of nomenclature. The system agreed upon was called the *Geneva system,* and the international assembly was eventually formalized as the International Union of Pure and Applied Chemistry. Now the system of nomenclature developed and endorsed by that body is known as the *IUPAC system.* The basic rules of the IUPAC system for naming compounds include the following:

1 The root or stem of the name is derived from the name of the alkane (C_nH_{2n+2}) corresponding to the *parent chain*—the *longest continuous carbon chain containing the functional group if there is one.*

2 Groups attached to the parent chain are called *substituents*. They are designated by adding prefixes or suffixes to the stem of the name.
3 The location of each substituent is indicated by a *locator number,* which is the number of the carbon atom to which the substituent is attached. The parent chain containing the functional group of *highest nomenclature priority* should be so numbered that the functional group has the lowest possible locator number. The locator numbers are set off by hyphens from the names of substituents.
4 If a substituent appears more than once in a formula, it is given a locator number for each appearance, and the numbers are set off from each other by commas. Also, the total number of identical substituents must be indicated by a prefix such as *di-, tri-,* or *tetra-*.
5 The substituents are identified by name and arranged in alphabetical order. Prefixes such as *di-* and *tri-* are usually ignored in alphabetical ordering.

In practice, you will probably find that the most perplexing questions of nomenclature involve rule 3. If a compound has multiple substituents, you have to determine which substituent has the highest nomenclature priority so you will know in what direction to number the carbon atoms of the parent chain. You may also have to determine which substituent should be indicated by a suffix and which by a prefix. Table 2-5 gives the nomenclature priority for functional groups. The highest priority is at the top of the table, and the lowest is at the bottom.

Generally, the first functional group to be named in the compound is identified by a suffix. Table 2-5 does not give prefixes for the first several functional groups, because such prefixes occur only in the names of extremely complex compounds. You will note that two suffixes are given for several of the functional groups. When the carbon atom of the functional group is counted as part of the chain, the first suffix is used; when it is not counted as part of the chain, the second suffix is used. Note also the symbol R in the eleventh entry of the first column; it denotes a carbon branch or group of unspecified structure. The last seven nomenclature priorities in the table apply, strictly, only to open-chain compounds.

Carbon branches or groups attached to the parent carbon chain are given names related to the alkanes with corresponding numbers of carbons. The names of the most commonly used carbon groups are given in Table 2-6. Carbon groups which are derived from the alkanes by loss of one hydrogen atom are called *alkyl groups*. They are named by dropping the *-ane* from the name of the parent alkane and adding the suffix *-yl* as in the following example:

```
     H                          H
     |                          |
  H—C—H    or    CH4         H—C—     or    —CH3
     |                          |
     H                          H
       Methane                     Methyl
```

TABLE 2-5
Functional group nomenclature priority (highest to lowest)*

	Denoted by	
Functional group	**Prefix**	**Suffix**
$-\mathrm{C(=O)}-\mathrm{OH}$		-oic acid -carboxylic acid
$-\mathrm{S(=O)}-\mathrm{O}-\mathrm{OH}$		-sulfonic acid
$-\mathrm{C(=O)}-\mathrm{Cl}$		-oyl chloride -carbonyl chloride
$-\mathrm{C(=O)}-\mathrm{NH_2}$		-amide -carboxamide
$-\mathrm{C(H)}{=}\mathrm{O}$		-al -carbaldehyde
$-\mathrm{C}-\mathrm{C(=O)}-\mathrm{C}-$	Oxo-	-one
$-\mathrm{C}{\equiv}\mathrm{N}$	Cyano-	-nitrile -carbonitrile
$-\mathrm{C}-\mathrm{OH}$	Hydroxy-	-ol
$-\mathrm{SH}$		-thiol
$-\mathrm{N}-$	Amino-	-amine
$-\mathrm{O}-\mathrm{R}$, $-\mathrm{O}-\mathrm{C_6H_5}$	Alkoxy- or phenoxy-	ether
$>\mathrm{C}{=}\mathrm{C}<$		-ene
$-\mathrm{C}{\equiv}\mathrm{C}-$		-yne
$-\mathrm{C}-\mathrm{C}-$		-ane
$-\mathrm{F}$	Fluoro-	
$-\mathrm{Cl}$	Chloro-	
$-\mathrm{Br}$	Bromo-	
$-\mathrm{I}$	Iodo-	
$-\mathrm{CH_3}$	Methyl-	
$-\mathrm{C_6H_5}$	Phenyl-	
$-\mathrm{NO_2}$	Nitro-	

* Some prefixes or suffixes have been omitted because they are so seldom used.

You will note in Table 2-6 that both the IUPAC name and the older common name are given for some groups. We shall generally use the IUPAC names in this course, but you may encounter the older names in other books or on reagent bottles in the laboratory.

TABLE 2-6 Commonly used carbon groups

Formula	IUPAC name	Common name	Classification of alkyl
$CH_3—$	Methyl	Methyl	
$CH_3CH_2—$ ($C_2H_5—$)	Ethyl	Ethyl	Primary
$CH_3CH_2CH_2—$	Propyl	*n*-Propyl*	Primary
$(CH_3)_2CH—$	1-Methylethyl	Isopropyl	Secondary
$CH_3CH_2CH_2CH_2—$	Butyl	*n*-Butyl	Primary
$CH_3CH_2CH(CH_3)—$	1-Methylpropyl	*sec*-Butyl (*s*-Butyl)	Secondary
$(CH_3)_2CHCH_2—$	2-Methylpropyl	Isobutyl	Primary
$(CH_3)_3C—$	1,1-Dimethylethyl	*tert*-Butyl (*t*-Butyl)	Tertiary
$CH_2{=}CH—$	Ethenyl	Vinyl	
$CH_2{=}CHCH_2—$	Allyl	Allyl	
$C_6H_5—$	Phenyl	Phenyl	
$C_6H_5CH_2—$	Benzyl	Benzyl	

* The *n* stands for normal, and it is so read.

Table 2-6 also indicates whether a group is classified as primary, secondary, or tertiary. A *primary* carbon atom is bonded to one other carbon atom; a *secondary* carbon atom is bonded to two other carbon atoms; a *tertiary* carbon atom is bonded to three other carbon atoms.

$$CH_3—CH_2—CH(CH_3)—CH_3$$

Primary (CH_3 at left); Secondary (CH_2); Tertiary (CH); Primary (both CH_3 on the CH)

Likewise, then, an alkyl group is primary if the carbon atom at the point of attachment is joined to only one other carbon atom; it is secondary if the carbon at the point of attachment is joined to two other carbon atoms; and it is tertiary if the carbon at the point of attachment is joined to three other carbon atoms. These terms are useful in classification. Secondary (abbreviated *sec* or *s*) and tertiary (abbreviated *tert* or *t*) are sometimes used in the naming of compounds or groups.

The following examples should be studied. You should obtain practice in nomenclature by working out the problems in the text and at the end of the chapter.

$$\overset{1}{C}H_3—\overset{2}{C}(CH_3)_2—\overset{3}{C}H_2\overset{4}{C}H(CH_3)—\overset{5}{C}H_3$$

2,2,4-Trimethylpentane

$$\overset{6}{C}H_3\overset{5}{C}H_2\overset{4}{C}H_2\overset{3}{C}H_2\overset{2}{C}H(CH_2CH_3)\overset{1}{C}H_2OH$$

2-Ethyl-1-hexanol

$$\overset{6}{C}H_3\overset{5}{C}H(CH_3)\overset{4}{C}H_2\overset{3}{C}H(OH)\overset{2}{C}H_2\overset{1}{C}H_3$$

5-Methyl-3-hexanol

Note that the —OH group is indicated by the suffix *-ol*. In the second example, its position on the chain is given by the number 1; in the third example, its position is given by the number 3.

PROBLEM 2-6 Give the IUPAC name for each of the following:

$$\text{(a)}\ \begin{array}{c} CH_3 \\ | \\ CH_3C{-}CH_2OH \\ | \\ CH_3 \end{array} \qquad \text{(b)}\ \begin{array}{ccccc} CH_3 & - & CH & - & CH_2 \\ & & | & & | \\ & & OH & & Cl \end{array} \qquad \text{(c)}\ CH_3CH_2CH_2CO_2H$$

Solution to (a) This compound is correctly named 2,2-dimethyl-1-propanol.

$$\begin{array}{ccccc} & & CH_3 & & \\ & & |^{2} & & \\ \overset{3}{CH_3} & - & C & - & \overset{1}{CH_2OH} \\ & & | & & \\ & & CH_3 & & \end{array}$$

PROBLEM 2-7 Give the structural formulas for (a) 2,2,3,3-tetramethylbutane, (b) 2-ethyl-1-hexanol, (c) 3-pentanone.

Solution to (a)

$$\begin{array}{ccccccccccc} & & H & & CH_3 & & CH_3 & & H & & \\ & & |^{1} & & |^{2} & & |^{3} & & |^{4} & & \\ H & - & C & - & C & - & C & - & C & - & H \\ & & | & & | & & | & & | & & \\ & & H & & CH_3 & & CH_3 & & H & & \end{array}$$

PROBLEM 2-8 Give the IUPAC name for each of the following formulas:

$$\text{(a)}\ CH_3CH_2CH_3 \qquad \text{(b)}\ \begin{array}{c} CH_3 \\ | \\ CH_3CHCH_3 \end{array} \qquad \text{(c)}\ \begin{array}{ccccc} & & CH_3 & & \\ & & | & & \\ CH_3 & - & C & - & CH_2CH_3 \\ & & | & & \\ & & CH_3 & & \end{array}$$

$$\text{(d)}\ \begin{array}{ccccccc} & & CH_3 & & CH_3 & & \\ & & | & & | & & \\ CH_3 & - & C & - & C & - & CH_3 \\ & & | & & | & & \\ & & CH_3 & & H & & \end{array} \qquad \text{(e)}\ \begin{array}{ccccccc} & & CH_3 & & CH_3 & & \\ & & | & & | & & \\ CH_3 & - & C & - & C & - & CH_3 \\ & & | & & | & & \\ & & CH_3 & & H & & \end{array}$$

Solution to (a) This compound is correctly named propane. It is a parent alkane.

2-3 Analysis

In the investigation of natural products, compounds are isolated and must be identified. Likewise, in carrying out a synthesis (preparation) of a desired compound or an exploratory or test reaction, unexpected compounds are sometimes isolated. Analysis and identification become important and desirable procedures.

Empirical-formula determination

The organic nature of a compound may be revealed by complete oxidation of the compound to carbon dioxide and water, called carbon-hydrogen analysis. Compounds which are not readily combustible may be heated with copper(II) oxide[1] to effect combustion. The presence of carbon dioxide is determined by conducting the effluent gas through limewater (saturated calcium hydroxide solution). If a precipitate of calcium carbonate is formed, carbon dioxide is present. The equation for the reaction is

$$CO_2(g) + Ca(OH)_2(aq) \longrightarrow CaCO_3(s) + H_2O \qquad (2\text{-}1)$$

To make the procedure quantitative, a sample of the unknown is carefully weighed and the water and carbon dioxide evolved from the combustion are absorbed and weighed in absorption tubes of known weight. The percents of carbon and hydrogen can then be calculated. Suitable quantitative combustion methods are available for the determination of sulfur, nitrogen, and halogen also. When the percentage composition is known, the *empirical formula* (the simplest atomic ratio formula) can be calculated. (You may wish to review a general chemistry text on this subject for more detail.) Problem 2-9 is calculated for you as an example.

PROBLEM 2-9 On combustion, a sample weighing 120 milligrams (mg) of a compound found in the vanilla bean was found to produce 278.1 mg of CO_2 and 57.2 mg of H_2O. Calculate the percent composition and empirical formula of the compound.

Solution To set up the proper proportions to solve this kind of problem, you must be clear on the relations involved. The equation for the combustion of carbon is

$$C + O_2(g) \longrightarrow CO_2(g) \qquad (2\text{-}2)$$

$$1 \text{ mol C atoms} \longrightarrow 1 \text{ mol } CO_2 \text{ molecules} \qquad (2\text{-}3)$$

The equation for the combustion of hydrogen is

$$H_2 + \tfrac{1}{2}O_2(g) \longrightarrow H_2O(g) \qquad (2\text{-}4)$$

$$2 \text{ mol H atoms} \longrightarrow 1 \text{ mol } H_2O \text{ molecules} \qquad (2\text{-}5)$$

The following equation yields the percent by weight of carbon in the sample:

$$\frac{\text{Wt } CO_2 \text{ produced}}{\text{Wt sample}} \times \frac{1 \text{ mol } CO_2}{44.0 \text{ g } CO_2} \times \frac{1 \text{ mol C atoms}}{1 \text{ mol } CO_2} \times \frac{12.0 \text{ g C atoms}}{1 \text{ mol C atoms}} \times 100 = \%\text{ C} \qquad (2\text{-}6)$$

[1] In the system of nomenclature recommended by IUPAC, the Cu^{2+} ion is named the copper(II) ion and the Cu^{+} ion is named the copper(I) ion. In older systems of nomenclature, those ions were known respectively as the cupric and cuprous ions.

The following equation yields the percent by weight of hydrogen in the sample:

$$\frac{\text{Wt H}_2\text{O produced}}{\text{Wt sample}} \times \frac{1\text{ mol H}_2\text{O}}{18.0\text{ g H}_2\text{O}} \times \frac{2\text{ mol H atoms}}{1\text{ mol H}_2\text{O}} \times \frac{1.0\text{ g H atoms}}{1\text{ mol H atoms}} \times 100 = \%\text{ H} \tag{2-7}$$

The percent of oxygen is calculated by subtracting the values for carbon and hydrogen from 100%:

$$\%\text{ O} = 100\% - (\%\text{ C} + \%\text{ H}) \tag{2-8}$$

By substituting the data for the problem into Eqs. (2-6) and (2-7), we obtain

$$\%\text{ C} = \frac{(0.2781\text{ g CO}_2)(12.0\text{ g C})(100)}{(0.1200\text{ g sample})(44.0\text{ g CO}_2)} = 63.2\%\text{ C} \tag{2-9}$$

$$\%\text{ H} = \frac{(0.0572\text{ g H}_2\text{O})(2)(1.0\text{ g H})(100)}{(0.1200\text{ g sample})(18.0\text{ g H}_2\text{O})} = 5.3\%\text{ H} \tag{2-10}$$

To obtain the percent of oxygen, we now substitute those values into Eq. (2-8):

$$\%\text{ O} = 100\% - (63.2\% + 5.3\%) = 31.5\%\text{ O} \tag{2-11}$$

The simplest atomic ratio is found by dividing the percent of each constituent by its atomic weight:

$$\text{C}_{63.2/12.0}\text{H}_{5.3/1.0}\text{O}_{31.5/16.0} = \text{C}_{5.27}\text{H}_{5.3}\text{O}_{1.97} \tag{2-12}$$

As a first step in reducing these ratios to simple whole numbers, we can divide each by the smallest of the three, 1.97:

$$\text{C}_{5.27/1.97}\text{H}_{5.3/1.97}\text{O}_{1.97/1.97} = \text{C}_{2.67}\text{H}_{2.7}\text{O}_{1.00} \tag{2-13}$$

We may now observe that 2.67 is $2\frac{2}{3}$ or $\frac{8}{3}$, so we can get a ratio of whole numbers by multiplying through by the number 3. Doing so gives us $C_8H_8O_3$ as the empirical formula.

PROBLEM 2-10 If a 105.0-mg sample of a compound produces 154.0 mg CO_2 and 63.3 mg H_2O, what are the percent composition and empirical formula of the compound?

Molecular-formula determination

An experimentally determined molecular weight will enable one to know what multiple of the empirical formula should be used to calculate the molecular formula. Some of the methods that can be used, which you may have encountered in general chemistry, are measurement of vapor density, freezing-point depression, and boiling-point rise. The following problem illustrates how the molecular formula may be determined from the empirical formula if even an approximate molecular weight is known.

PROBLEM 2-11 A compound with an empirical formula of CH_2 was found to have an experimental molecular weight of about 100. What is its molecular formula?

Solution The molecular formula is a multiple of the empirical formula that for this case may be represented by $(CH_2)_n$, where n is an unknown integer. The integer n may be found by choosing the integer closest to the quotient of the experimental molecular weight divided by the empirical formula weight. Thus, $\frac{100}{14} = 7.1$, or $n = 7$. The molecular formula is then $(CH_2)_7$ or C_7H_{14}.

PROBLEM 2-12 A molecular-weight determination was made, by the freezing-point-depression method, for the compound discussed in Prob. 2-10. The method gave an experimental molecular weight of 178 ± 2. By using that information and the empirical formula you calculated in Prob. 2-10, determine the molecular formula.

Molecular weight from mass spectra

The most accurate molecular weight may be obtained by analyzing a sample with a high-resolution mass spectrometer. A small sample is injected into the evacuated chamber of a mass spectrometer. There it is vaporized and subjected to bombardment by high-energy electrons. The result is that electrons are knocked out of the molecules and bonds are broken to fragment some of the molecules and produce a variety of positive ions. The instrument separates the positive ions on the basis of their differences in mass-to-charge ratio. The ion with the greatest mass-to-charge ratio has a mass essentially the same as that of the parent compound, so it is called the parent or molecular ion. The mass of the parent ion is a close approximation of the molecular weight. Identification of many of the fragment ions gives useful insight into the probable structure of the original compound.

2-4 Structure determination

Once the molecular formula of a compound is known, structure of the compound must be determined. Presence of functional groups may be demonstrated by a great variety of special test reactions, some of which will be discussed as particular functional groups are studied in detail. Information concerning not only functional groups but also the nature of the carbon skeletal framework may be obtained from infrared, ultraviolet, mass, nuclear magnetic resonance, and x-ray spectra.

Use of electromagnetic absorption spectra

Electromagnetic radiation has wavelike properties and a wide spectrum of types related to wavelengths (see Fig. 2-14). Wavelength of radiation is

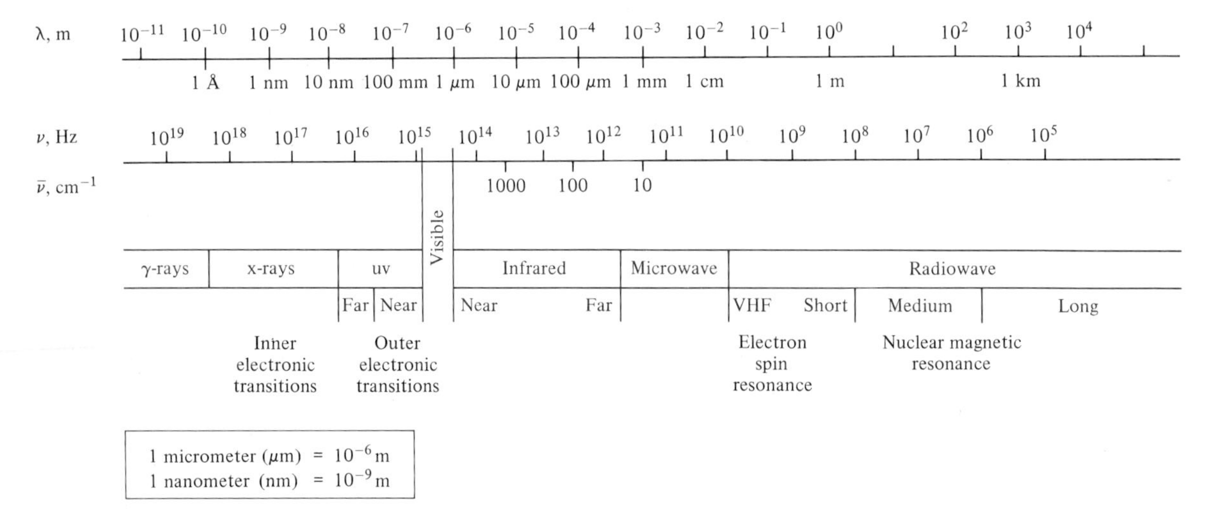

FIGURE 2-14 The electromagnetic spectrum.

inversely proportional to frequency, since the velocity of such radiation (light) is a product of wavelength and frequency.

$$c = \lambda \nu \tag{2-14}$$

$$\lambda = \frac{c}{\nu} \tag{2-15}$$

where c = velocity of light = 3.0×10^8 meters per second (m/s)
$\lambda \approx$ wavelength, in m/cycle
ν = frequency, in cycles/s or hertz (Hz)

Electromagnetic radiation also has characteristics of small units of energy called *photons*. The energy E of a photon is given by the following relation:

$$E = h\nu \quad \text{or} \quad E = \frac{hc}{\lambda} \tag{2-16}$$

where h is Planck's constant, 6.63×10^{-34} joule per second (J/s). Thus the energy of electromagnetic radiation is proportional to frequency and inversely proportional to wavelength.

When organic compounds are exposed to radiation, they absorb part of the radiation. By plotting the amount of radiation absorbed against the wavelength of incident radiation, an absorption spectrum for the compound is obtained (see Fig. 2-15). Both the amount and wavelength of absorption observed depend on the nature of the compound being irradiated. Since certain kinds of absorptions are typical of specific structural features, an absorption spectrum for an unknown organic compound may reveal structural details.

According to the quantum theory, molecules can possess only certain amounts of energy referred to as *energy levels*. The structure of the molecule determines exactly which energy levels are available. If the molecule is irradiated with exactly the same amount of energy as the difference between its present energy level and a higher energy level available to it, it can absorb the radiant energy and be excited to the higher energy level.

The types of energy levels which give rise to spectra that are easily measured in the laboratory are electronic and vibrational. *Electronic energy levels* are associated with molecular orbitals. Excitation of electrons to higher-energy-level orbitals results in absorption of energy in the regions of visible and ultraviolet light. Ultraviolet spectra are especially valuable for identifying *conjugated systems,* which are composed of alternating double and single bonds. *Vibrational energy levels* are related to several vibrational frequencies available to bonded atoms within the constraints of the molecular bond. The higher-energy vibrations can be considered as stretching and bending of the bonds. The energy difference between adjacent vibrational levels corresponds to wavelengths in the infrared region of the electromagnetic spectrum.

Infrared spectra

Organic compounds typically absorb radiation in the 2- to 15-micrometer (μm) region of the infrared spectrum (1 μm = 10^{-6} m). That corresponds to 5000 to 667 cm^{-1} (reciprocal centimeters, or wave numbers), a scale more often used. The wave number $\bar{\nu}$ is found by the equation:

$$\bar{\nu} = \frac{1}{\lambda} \tag{2-17}$$

where λ is in centimeters per cycle (cm/cycle). The wave number represents the number of wave cycles per centimeter. If $\lambda = 2.0\ \mu m$, or 2.0×10^{-6} m, or 2.0×10^{-4} cm, then,

$$\bar{\nu} = \frac{1}{2.0 \times 10^{-4}\ cm} = 0.50 \times 10^{4}\ cm^{-1} = 5000\ cm^{-1} \tag{2-18}$$

Some of the absorption bands, especially those resulting from bending vibrations, are found in the so-called fingerprint region of the spectrum, 10 to 15 μm, or 1000 to 667 cm^{-1}. The fingerprint region is so characteristic of a compound that positive identification of an unknown compound may be made by superimposing the spectrum of the unknown on that of a known compound.

In addition to its being a means of identification, the infrared spectrum reveals much molecular structure detail. The frequency range at which a functional group shows an infrared absorption band does not vary greatly from one compound to the next. The O—H stretching frequency common to alcohols and phenols is found in the range 3200 to 3700 cm^{-1}. The stretching frequency of the carbonyl group, $-\overset{|}{C}{=}O$, is 1715 to 1725 cm^{-1} for aldehydes and ketones.

The frequencies of the more reliable infrared absorption bands characteristic of the common functional groups are given in Table 2-7. By

TABLE 2-7
Some infrared absorption bands*

Bond	Compound type in which located	Frequency range, cm^{-1}
C—H	Alkanes	2850–2960 1350–1470
C—H	Alkenes	3020–3080 (medium) 675–1000
C—H	Aromatic rings	3000–3100 (medium) 675–870
C—H	Alkynes	3300
C═C	Alkenes	1640–1680 (variable)
C≡C	Alkynes	2100–2260 (variable)
C—C	Aromatic rings	1500–1600 (variable)
C—O	Alcohols, ethers, esters, carboxylic acids	1080–1300
C═O	Aldehydes, ketones, esters, carboxylic acids	1690–1760
O—H	Alcohols, phenols	3200–3640 (variable)
O—H	Carboxylic acids	2500–3000 (broad)
N—H	Amines	3300–3500 (medium)
C—N	Amines	1180–1360
C≡N	Nitriles	2210–2260
—NO_2	Nitro compounds	1515–1560 1345–1385

* All absorption bands are strong unless otherwise indicated.

inspecting the spectrum of an unknown compound to locate significant absorption bands and then identifying the bands by comparison with values listed in a table, an experimenter can determine which functional groups are present in a compound. Interpretation of infrared spectra is often complicated, however, because absorption bands sometimes overlap or are shifted by interaction between groups. Even if an absorption band differs by as much as 20 to 30 cm^{-1} from the value listed, it may still represent the group in question.

Infrared absorption spectra are usually recorded with wavelength increasing to the right along the horizontal coordinate and transmittance increasing upward from 0 to 100% along the vertical coordinate. Radiation which is not transmitted is absorbed.

$$\text{Percent absorption} = 100\% - \text{percent transmittance} \qquad (2\text{-}19)$$

The absorption peaks (or maxima) are thus actually the valleys portrayed on the usual recorder chart, such as the one shown in Fig. 2-15.

Nuclear magnetic resonance spectra involve radiation in the shortwave radio region. They are exceptionally useful tools for determining the structure of an organic compound.

PROBLEM 2-13 Examine the spectrum of glycine shown in Fig. 2-15. Could the compound have an —OH group? an —NH_2 group? a >C═O group? Refer to Table 2-7 for the infrared absorption bands for those functional groups. Look up the structural formula of glycine and check your decisions.

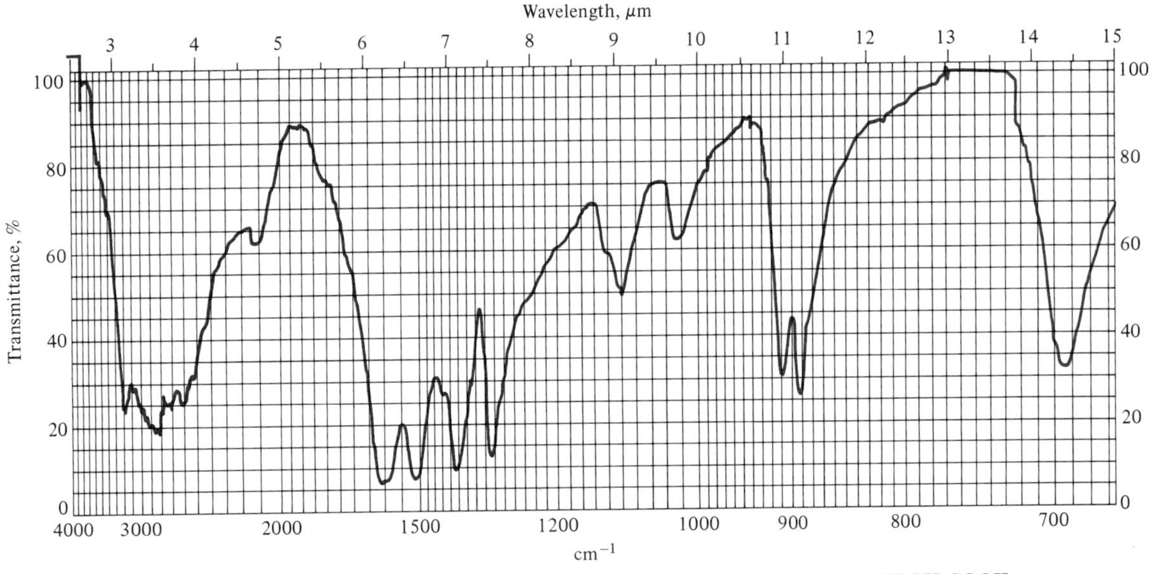

FIGURE 2-15 An infrared absorption spectrum of glycine, NH_2CH_2COOH.

Summary

Open-chain compounds are called *acyclic compounds* and *closed-chain compounds* are called *cyclic compounds*.

Isomers are compounds which have the same *molecular formula* but differ in some way in their structures. *Structural formulas* are required to depict them. *Chain,* or *skeletal, isomers* differ in the branching of the skeletal chain of carbon atoms. *Position isomers* have identical skeletal chains of carbon atoms, but they differ in the position of a substituent on the chains. *Functional group isomers* have different *functional groups,* which are reactive groups attached to the skeletal framework.

Stereoisomers have identical skeletal chains of carbon atoms, and they have the same functional groups attached to corresponding carbon atoms in their respective chains. They differ, however, in the *spatial arrangement* of their groups. Stereoisomers include geometric isomers, configurational isomers, and conformational isomers. *Geometric,* or *cis-trans, isomers* are stereoisomers in which a different spatial arrangement of substituent parts is maintained because rotation about a carbon-carbon bond is restricted. If substituents are on the same side of the double bond or ring, the isomer is said to be *cis;* if they are on opposite sides, it is said to be *trans*.

Configurational isomers are stereoisomers which have *chiral* (handed) *molecules* (to be studied in Chap. 11). *Conformational isomers* are stereoisomers that result from the different forms a molecule can take from rotation about single carbon-carbon bonds. Although some conformations are more stable than others, all conformations are so readily interconvertible that an equilibrium mixture is formed and from it the individual conformations are usually not separable. *Staggered conformations* are the most stable and *eclipsed conformations* are the least stable.

The compounds of organic chemistry can be considered as *functional group derivatives* of the *alkanes,* or *saturated hydrocarbons,* in which functional groups have been substituted for one or more hydrogen atoms. The study of organic chemistry can be organized about functional groups. A family of compounds the members of which differ from each other only by some multiple of —CH_2— groups and which contain a common functional group is called a *homologous series*. The members of such a group, called *homologs,* have similar chemical and physical properties but with a gradation of differences.

The chief system of *nomenclature* (naming) is the *IUPAC system*. The rules given in Sec. 2-2 should be noted and then learned and applied as the various families of functional groups are studied in later chapters.

Alkyl groups are carbon groups derived from the alkanes by loss of a hydrogen atom. They are named by dropping the *-ane* from the name of the parent alkane and adding the suffix *-yl*. Those listed in Table 2-6 should be learned. A *primary carbon atom* is bonded to one other carbon atom; a *secondary* carbon atom is bonded to two other carbon atoms; a *tertiary* carbon atom is bonded to three other carbon atoms.

Quantitative analysis of an organic compound makes possible the computation of *percent composition,* from which an *empirical formula* can be calculated. If an approximate molecular weight is determined, the *molecular formula* can be calculated from the empirical formula. The *mass spectrum* of an organic compound gives an accurate molecular weight and also insight into the possible structure of the molecule.

After the molecular formula of an unknown compound has been found, the molecular structure must be determined in order to really understand the compound. *Functional group tests* and *electromagnetic spectra* aid in the determination of structure, and the spectra are especially useful. The *wavelength* of electromagnetic radiation is equal to the *velocity of light* divided by the *frequency*. The *energy* of electromagnetic radiation is proportional to frequency and inversely proportional to wavelength. When organic compounds are exposed to electromagnetic radiation, they *absorb* part of the radiation. The *absorption spectrum* produced consists of a plot of the radiation transmitted versus the wavelength or frequency of the incident radiation. *Ultraviolet, infrared,* and *nuclear magnetic resonance spectra* are especially useful to the organic chemist.

Key terms

Absorption spectra: (1) spectra produced by the absorption of electromagnetic radiation by organic compounds and (2) plots of the amount of radiation transmitted versus the wavelength of incipient radiation. Spectra are extremely useful in determining the structural features of a molecule.

Acyclic compound: a long-chain compound of carbon, either continuous or branched, in which the chain is open.

Alkane: an open-chain compound with the general molecular formula C_nH_{2n+2}.

Alkyl group: a carbon group derived from an alkane by the loss of one hydrogen atom.

Axial bond: a bond parallel to the imaginary axis passing through the center of a cyclohexane molecule.

Chain isomers: isomers which differ in the branching of the skeletal chain of carbon atoms; also known as **skeletal isomers.**

Chiral molecules: molecules which are "handed"; they differ as your right hand differs from your left.

Cis isomers: geometric isomers in which the substituents are on the same side of a double bond or ring.

Configurational isomers: isomers which result from a compound having chiral molecules; such isomers are mirror images of each other.

Conformational isomers: differing forms of a molecule resulting from rotation about carbon-carbon single bonds; they exist in an equilibrium mixture and cannot be separated from each other.

Conformers: staggered conformations which represent energy minima in the rotation about a single carbon-carbon bond.

Conjugated system: a system composed of alternating double and single bonds.

Cyclic compound: a compound of carbon in which the chain is closed to form a ring.

Eclipsed conformation: a conformation in which the substituent groups are as close as possible to each other; the least stable conformation.

Electromagnetic spectrum: the entire spectrum of the radiation arising from vibrating electric and magnetic fields in space, including visible light. (A spectrum is an array of entities ordered in accordance with the magnitudes of a common physical property.)

Electronic energy levels: energy levels associated with molecular orbitals; they give rise to absorption of energy in the regions of visible and ultraviolet radiation.

Empirical formula: a formula giving the simplest atomic ratio of a compound; the molecular formula is often a multiple of the empirical formula.

Energy levels: the specific amounts of energy which molecules can possess; the structure of a molecule determines which energy levels are available.

Energy of electromagnetic radiation: the energy of a photon of a specific frequency or wavelength given by the equation $E = h\nu$ or $E = hc/\lambda$.

Equatorial bond: a bond almost coplanar with the imaginary equator of a cyclohexane molecule and at approximately a tetrahedral angle to axial bonds.

Frequency: the number of waves of electromagnetic radiation passing a given point per second; frequency ν, in cycles per second, is equal to the velocity of light c divided by the wavelength λ.

Functional group: a reactive group of atoms such as —X, —OH, —COOH, and —NH_2; to be discussed in detail in subsequent chapters.

Functional group isomers: isomers which contain different functional groups.

Functional group test: a special test reaction to determine the presence of a particular functional group.

Geometric isomers: isomers which result from substitution on a carbon chain at a position where rotation about the carbon-carbon bond is restricted; they include cis and trans isomers.

Homologous series: a family of compounds containing a common functional group the members of which differ from each other only by some multiple of —CH_2— (methylene) groups.

Homologs: members of a homologous series.

Infrared spectra: absorption spectra resulting from exposure of compounds to infrared radiation.

Isomers: compounds which have the same molecular formula but differ in some way in the arrangement of atoms.

Mass spectrum: a spectrum of charged particles arranged in accordance with the masses or mass-to-charge ratios of the particles.

Molecular formula: a formula indicating the kinds of atoms and the number of each kind that occur in a molecule; it is calculated from experimental determination of the empirical formula and the molecular weight.

Newman projection formulas: formulas in which one looks down the axis of a carbon-carbon bond about which rotation occurs; used to depict conformational isomers.

Nomenclature: a system of naming; the IUPAC system of nomenclature is the one we shall generally use.

Nuclear magnetic resonance spectra: absorption spectra resulting from exposure of compounds to radiation in the shortwave radio region.

Percent composition: the percent by weight of each of the elements in a compound.

Position isomers: isomers which differ in the position or location of a substituent atom or group on the chain.

Primary carbon atom or group: a carbon atom that is bonded to one other carbon atom; an alkyl group in which the carbon atom at the point of attachment is bonded to only one other carbon atom.

Secondary carbon atom or group: a carbon atom that is bonded to two other carbon atoms; an alkyl group in which the carbon atom at the point of attachment is bonded to two other carbon atoms.

Skeletal isomers: isomers which differ in the branching of the skeletal chain of carbon atoms; also known as **chain isomers.**

Staggered conformation: a conformation in which the substituent groups are as far as possible from each other; the most stable conformation.

Stereoisomers: isomers in which the bonding of the atoms is the same but the spatial arrangements of the atoms differ; they include geometric isomers, configurational isomers, and conformational isomers.

Structural formula: a formula which shows the arrangement of the atoms in a molecule.

Structural isomers: isomers in which the bonding arrangements of atoms differ; they include chain or skeletal isomers, position isomers, and functional group isomers.

Tertiary carbon atom or group: a carbon atom that is bonded to three other carbon atoms; an alkyl group in which the carbon atom at the point of attachment is bonded to three other carbon atoms.

Trans isomers: geometric isomers in which the substituents are on opposite sides of a double bond or ring.

Ultraviolet spectra: absorption spectra resulting from the exposure of compounds to ultraviolet radiation.

Velocity of light: 3×10^8 m/s.

Vibrational energy levels: energy levels associated with bending and stretching of molecular bonds; they give rise to absorption of energy in the region of infrared radiation.

Wavelength: the length of waves of electromagnetic radiation; wavelength λ = velocity of light c divided by frequency ν, in cycles per second.

Additional problems

2-14 Draw the structural formulas for all of the isomeric hexanes, C_6H_{14}. Give the IUPAC name for each one.

2-15 The following list contains the formulas of six pairs of isomers. Indicate which ones are isomers of each other, and tell what kind of isomers they are.

(a) $CH_3CHOHCH_3$

(b) $CH_3CHOHCH_2CH_3$

(c) $CH_3CH_2OCH_2CH_3$

(d) $CH_3CH_2CH_2OH$

(e) $CH_2{=}CHCH_2CH_2CH_3$

(f) $ClCH_2{-}C(CH_3){=}C(H){-}CH_3$

(g) O

(h) OH OH

(i)

(j) H—C=O

(k) $CH_3{-}C(CH_2Cl){=}C(H){-}CH_3$

(l) HO OH

2-16 Give the IUPAC name for each of the following compounds.

(a) 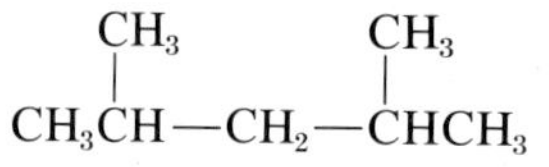 $$\begin{array}{ccccc} & CH_3 & & CH_3 & \\ & | & & | & \\ CH_3 & CH & -CH_2- & CH & CH_3 \end{array}$$

(b) $CH_3CH_2CH_2OH$

(c) 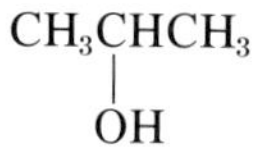 $$\begin{array}{ccc} CH_3 & CH & CH_3 \\ & | & \\ & OH & \end{array}$$

(d) 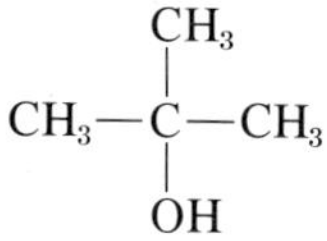 $$\begin{array}{ccc} & CH_3 & \\ & | & \\ CH_3- & C & -CH_3 \\ & | & \\ & OH & \end{array}$$

(e) 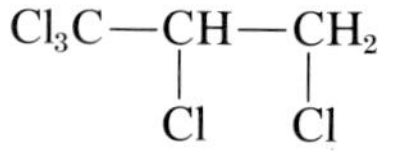 $$\begin{array}{ccccc} Cl_3C & - & CH & - & CH_2 \\ & & | & & | \\ & & Cl & & Cl \end{array}$$

(f) 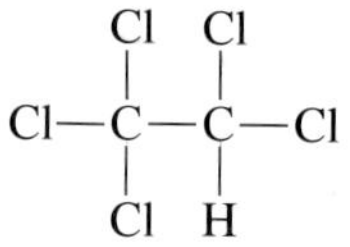 $$\begin{array}{ccccccc} & & Cl & & Cl & & \\ & & | & & | & & \\ Cl & - & C & - & C & - & Cl \\ & & | & & | & & \\ & & Cl & & H & & \end{array}$$

2-17 Draw a structural formula for each of the following:

(a) 2,2,3-trimethylbutane
(b) 1,1,1,2,2-pentachloroethane
(c) *cis*-1,2-dichlorocyclopentane
(d) *trans*-1,4-dimethylcyclohexane
(e) 2-methyl-1-propanal

2-18 What is the IUPAC name for each of the following compounds?

(a) 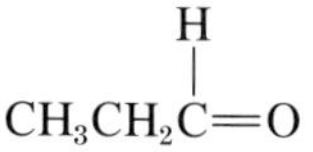 $$\begin{array}{cc} & H \\ & | \\ CH_3CH_2 & C{=}O \end{array}$$

(b) $$\begin{array}{ccccc} & CH_3 & & H & \\ & | & & | & \\ CH_3- & CH & - & C & {=}O \end{array}$$

(c) $$\begin{array}{ccc} & O & \\ & \| & \\ CH_3CH_2 & C & -CH_3 \end{array}$$

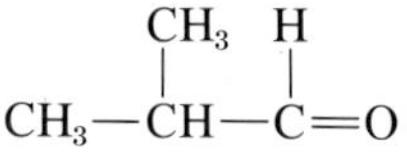

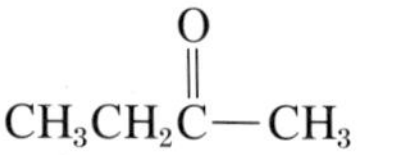

(d) 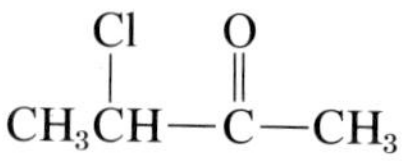 $$\begin{array}{ccccc} & Cl & & O & \\ & | & & \| & \\ CH_3 & CH & - & C & -CH_3 \end{array}$$

(e) CH_3CH_2SH

(f) 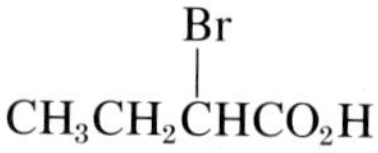 $$\begin{array}{ccc} & Br & \\ & | & \\ CH_3CH_2 & CH & CO_2H \end{array}$$

2-19 Draw a structural formula for each of the following compounds.

(a) 4-isobutyl-2-octyne
(b) 1-penten-4-yne
(c) 2-aminopropanoic acid
(d) *trans*-2-pentene
(e) 2-pentanone

2-20 Give the IUPAC name for each of the following:

(a) $CH_3CH_2CH_2CH_2NH_2$

(b) $H_2N{-}CH_2CH_2CH_2CH_2{-}NH_2$

(c) 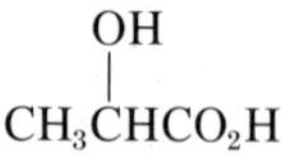 $$\begin{array}{ccc} & OH & \\ & | & \\ CH_3 & CH & CO_2H \end{array}$$

(d) $N{\equiv}C{-}CH_2CO_2H$

(e) 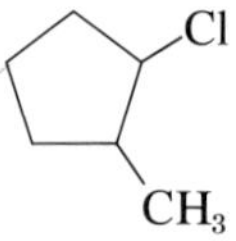 cyclopentane ring bearing Cl and CH_3 on adjacent carbons

(f) 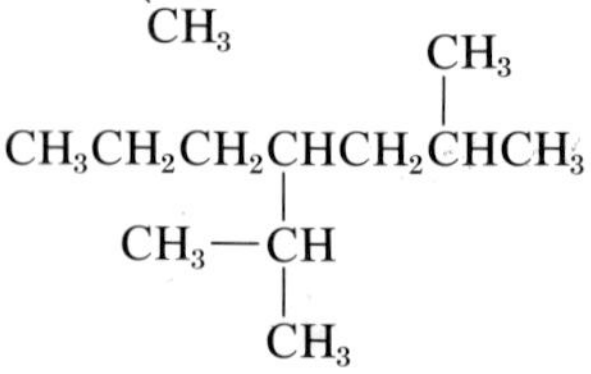 $$\begin{array}{ccccc} & & & & CH_3 \\ & & & & | \\ CH_3CH_2CH_2 & CH & CH_2 & CH & CH_3 \\ & | & & & \\ CH_3- & CH & & & \\ & | & & & \\ & CH_3 & & & \end{array}$$

2-21 Give the IUPAC name for each of the following compounds. Use *cis* or *trans* when appropriate.

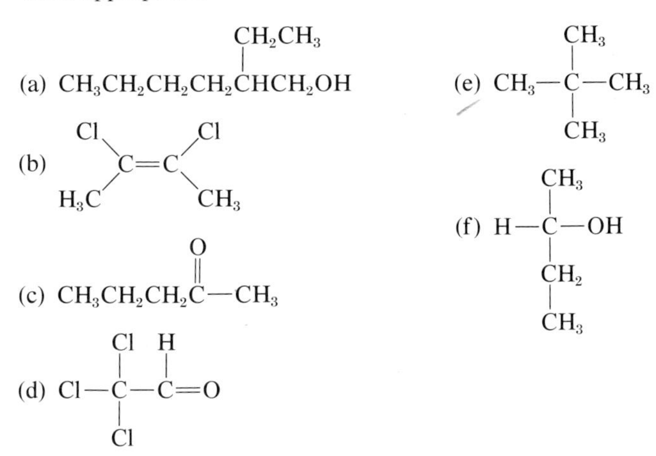

2-22 Indicate which of the following compounds could have cis-trans isomers.

(a) 2-methyl-2-butene
(b) 1-chloro-2-methyl-2-butene
(c) 1,1-dichlorocyclopropane
(d) 1,2-dichlorocyclopropane
(e) 1,1,2-trichloroethane
(f) 1,2-dichlorocyclopentane

2-23 A sugar isolated from a nucleic acid has the following percentage composition: C, 40.0%, and H, 6.7%. The molecular weight is about 147 ± 3. Determine the empirical formula and then the molecular formula.

2-24 An acidic growth-promoting compound is found to have 61.3% C, 5.1% H, and 10.2% N. The molecular weight is about 140 ± 3. Determine the empirical and molecular formulas.

2-25 Cyclohexanone can be made by oxidation of cyclohexanol:

OH $\xrightarrow{(O)}$ O

How can infrared spectra be used to ascertain whether the conversion is complete?

3

SATURATED HYDROCARBONS

The *saturated hydrocarbons* are so named because they will not react with hydrogen. The other types of hydrocarbons, which are referred to as *unsaturated* and which have a higher oxidation state, will react with hydrogen (see Table 2-4). The saturated hydrocarbons consist of the *alkanes* and the *cycloalkanes*. They are important as solvents, fuels, and raw materials, and they also constitute the skeletal framework of other organic compounds.

The alkanes have the general formula C_nH_{2n+2}, where n is the number of carbon atoms. They were formerly called *paraffins,* which means *having little affinity,* and the name was given them because they are generally unreactive. They not only are rather unreactive to acids, bases, and oxidizing agents; they also do not react with reducing agents because they are already in a highly reduced state.

The cycloalkanes have the general formula C_nH_{2n}, and they also are rather unreactive. The reactivity of cyclopropane is exceptional.

3-1 Structure of the saturated hydrocarbons

Each carbon atom of a saturated hydrocarbon has four single covalent bonds consisting of sp^3 hybrid orbitals bonding it to either hydrogen atoms or to other carbon atoms. The bond angles are the tetrahedral bond angle of 109°28′. Figure 3-1 depicts ethane. Rotation about the carbon-carbon covalent bonds is possible, and it produces a variety of conformations which are not isolable. The conformations exist as an equilibrium mixture to produce the observed properties of each alkane. Chain or skeletal isomers are possible in increasing numbers as the number of carbon atoms in the molecule increases.

Table 3-1 lists the names, formulas, melting points, and boiling points of the first 15 straight-chain (normal) alkanes. Note that the first four members of this homologous series are gases under ordinary conditions. The alkanes are insoluble in water, and all of them are less dense than water. Their densities range from about 0.63 gram per milliliter (g/ml) for pentane to 0.77 g/ml for pentadecane.

FIGURE 3-1
The bonding and geometry in ethane: (*a*) structural formula; (*b*) representation of molecular orbitals.

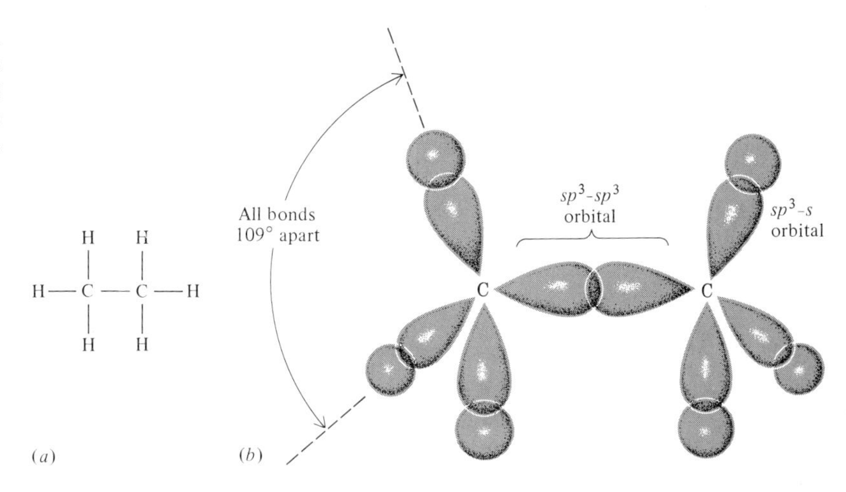

Observe in the list of names in Table 3-1 that, except for the first four members of the series, the first part of each name is a prefix, derived from Greek or Latin, that indicates the number of carbon atoms in the molecule. You are already familiar with the use of some of the prefixes in words like "pentagon" and "hexagon."

Structure of methane and other alkanes

Methane is the alkane of lowest molecular weight; it consists of only one carbon atom and four hydrogen atoms. The tetrahedral structure of methane and its derivatives was discussed in Chap. 1. Figure 3-2 depicts two different models of a molecule of methane; (*a*) is a space-filling model and (*b*) is a ball-and-stick model. Both models show that the molecule is not planar but tetrahedral. The space-filling model represents the overlap-

TABLE 3-1
The straight-chain alkanes

Molecular formula	Structural formula	IUPAC name	mp, °C	bp, °C
CH_4	CH_4	Methane	−183	−161
C_2H_6	CH_3CH_3	Ethane	−172	−88
C_3H_8	$CH_3CH_2CH_3$	Propane	−188	−42
C_4H_{10}	$CH_3(CH_2)_2CH_3$	Butane	−135	−0.6
C_5H_{12}	$CH_3(CH_2)_3CH_3$	Pentane	−130	36
C_6H_{14}	$CH_3(CH_2)_4CH_3$	Hexane	−95	69
C_7H_{16}	$CH_3(CH_2)_5CH_3$	Heptane	−91	98
C_8H_{18}	$CH_3(CH_2)_6CH_3$	Octane	−57	125
C_9H_{20}	$CH_3(CH_2)_7CH_3$	Nonane	−54	150
$C_{10}H_{22}$	$CH_3(CH_2)_8CH_3$	Decane	−30	174
$C_{11}H_{24}$	$CH_3(CH_2)_9CH_3$	Undecane	−26	196
$C_{12}H_{26}$	$CH_3(CH_2)_{10}CH_3$	Dodecane	−10	216
$C_{13}H_{28}$	$CH_3(CH_2)_{11}CH_3$	Tridecane	−5.5	234
$C_{14}H_{30}$	$CH_3(CH_2)_{12}CH_3$	Tetradecane	6	253
$C_{15}H_{32}$	$CH_3(CH_2)_{13}CH_3$	Pentadecane	10	270

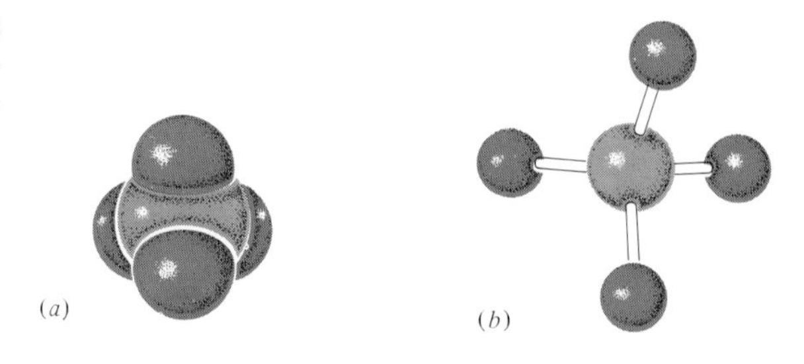

FIGURE 3-2
(*a*) Space-filling and (*b*) ball-and-stick models of methane.

ping of the sp^3 hybrid orbitals of the carbon atom with the *s* orbitals of the four hydrogen atoms.

Figure 2-11 shows a model of propane in its most stable conformation—one in which all the hydrogen atoms on adjacent carbon atoms are staggered and none are eclipsed. A planar projection of that conformation, as shown in Fig. 2-11, results in our usual structural formula for propane, shown as formula (*a*) in Fig. 3-3. Formulas (*b*), (*c*), and (*d*) in Fig. 3-3 are among other projections which could represent propane. Thus (*a*) is really only *one* of the projections of *one* of the conformations of a normal alkane that appears to be straight, but the alkanes are usually most conveniently represented on paper in that way. The branched-chain alkanes also can be represented in several ways.

PROBLEM 3-1 With the aid of a model, convince yourself that all the following formulas represent 2-methylbutane, the common name of which is isopentane.

```
   CH3                    CH3
    |                      |
CH3CHCH2CH3        CH3CH2CHCH3        CH3CHCH2CH3
                                         |
                                        CH3

                   CH3
                    |
CH3CH2CHCH3        CHCH2CH3           CH3CHCH3
        |           |                    |
       CH3         CH3                  CH2CH3

  CH3  CH3         CH3                     CH3
   |    |           |                       |
CH3CH—CH2          CHCH3              CH2—CH
                    |                  |     |
                   CH2                CH3   CH3
                    |
                   CH3
```

FIGURE 3-3 Planar projections of propane.

```
(a)   H  H  H        (b)        H        (c)   H              (d)    H   H
      |  |  |                   |              |                      \ /
   H—C—C—C—H            H H—C—H        H—C—H H             H     C     H
      |  |  |            |      |              |    |               \   / \   /
      H  H  H         H—C——C—H         H—C——C—H             C     C
                         |      |              |    |               / \   / \
                         H      H              H    H              H   H H   H
```

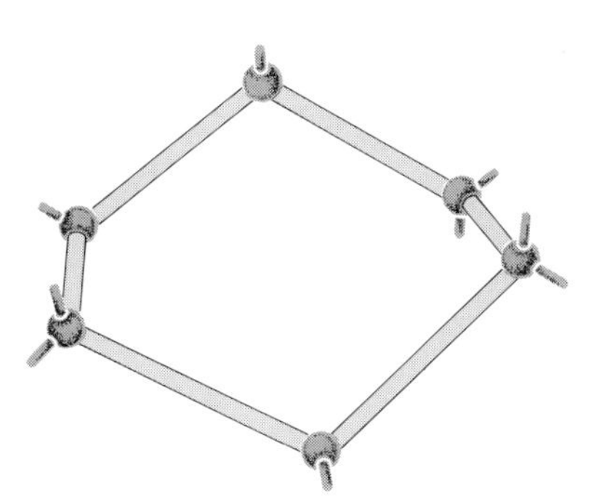

FIGURE 3-4
A model of cyclohexane showing the overlap of the sp^3 orbitals and the nonplanar ring.

Structure of cycloalkanes

The atomic orbitals on carbon not only in the alkanes but also in the cycloalkanes are sp^3 hybrids. However, only in rings of six or more atoms is complete head-on "overlap" of the sp^3 orbitals with 109° bond angles possible, and only then because not all of the carbon atoms must be in the same plane (Fig. 3-4). The most stable nonplanar conformation for cyclohexane is the so-called chair form, in which the tetrahedral bond angle is approximately maintained and all the hydrogen atoms on adjacent carbon atoms are staggered. The chair conformation is conventionally represented by the structural formula in Fig. 3-5. Refer to Fig. 2-13 for a Framework Molecular Model of the chair conformation of cyclohexane. Framework Molecular Models represent the bonding, bond angles, and approximate bond lengths; they do not depict the space occupied by the orbitals, or clouds of electrons.

The types of atomic orbital overlap are shown in Fig. 3-6. Examples of head-on overlap to produce sigma (σ) molecular orbitals are shown in Fig. 3-6*a*, *b*, and *c*. Maximum overlap of this type can occur only if the overlap is coaxial with respect to the axes of the orbitals being overlapped. A relatively strong bond results.

Sideways overlap is shown in Fig. 3-6*d*. This type of overlap is parallel with the axes of the orbitals. The molecular orbital so produced is called a pi (π) orbital. Such a bond is weaker than the sigma type.

Figure 3-6*e* shows orbital overlap which is mostly head-on, but not coaxial, so it has some sideways overlap. This type of bond is weaker than a full sigma bond but stronger than a pi bond. It is the type of carbon-carbon bond in cycloalkanes of fewer than six carbon atoms. In cyclopentane there is only a trace of sideways overlap; in cyclobutane more sideways overlap is required; and still more is required in cyclopropane. The carbon-carbon bonds in cyclopropane are almost as weak as a

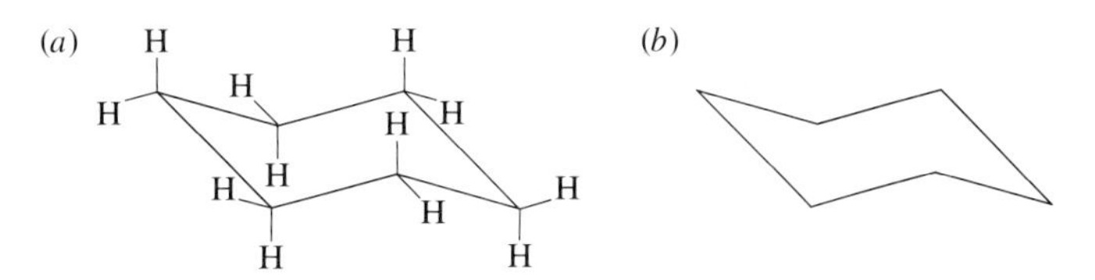

FIGURE 3-5
(*a*) Conventional structural formula for the chair conformation of cyclohexane. (*b*) Simplified structural formula.

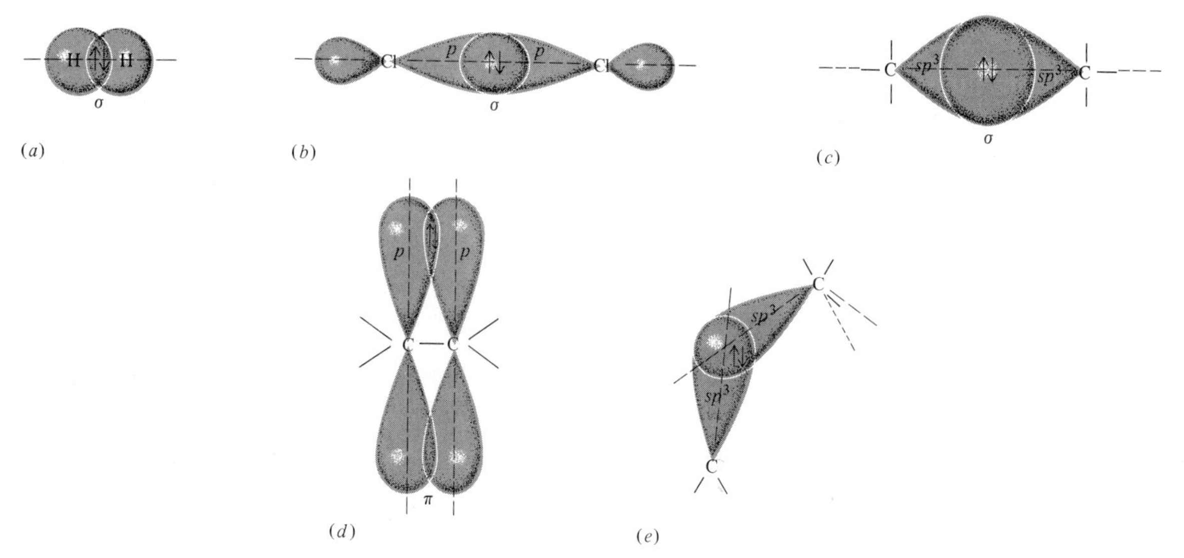

FIGURE 3-6 Types of orbital overlap: (*a*) head-on overlap (overlap along the axes) of the *s* orbitals on two hydrogen atoms to produce a sigma molecular orbital; (*b*) head-on overlap (overlap along the axes) of the *p* orbitals on two chlorine atoms to produce a sigma molecular orbital; (*c*) head-on overlap (overlap along the axes) of sp^3 hybrid orbitals on two carbon atoms to form a sigma molecular orbital; (*d*) sideways or lateral overlap (parallel to axes) of *p* orbitals on two carbon atoms to form a pi molecular orbital; (*e*) partial sideways overlap of sp^3 hybrid orbitals on two carbon atoms, as in cyclopropane. The resulting molecular orbital is essentially a sigma molecular orbital, but with some pi character because of the partial sideways overlap. (The axes of the overlapped orbitals are neither coaxial nor parallel, but intersect.)

carbon-carbon pi bond. See the representations in Fig. 3-7 of (*a*) cyclopentane, (*b*) cyclobutane, and (*c*) cyclopropane. The carbon-carbon bonds of both cyclopropane and cyclobutane may be broken, with resultant ring opening, by some reagents.

In forming cyclopentane, all of the carbon atoms must lie in almost the

FIGURE 3-7 Overlap of sp^3 orbitals on carbon in (*a*) cyclopentane, (*b*) cyclobutane, and (*c*) cyclopropane.

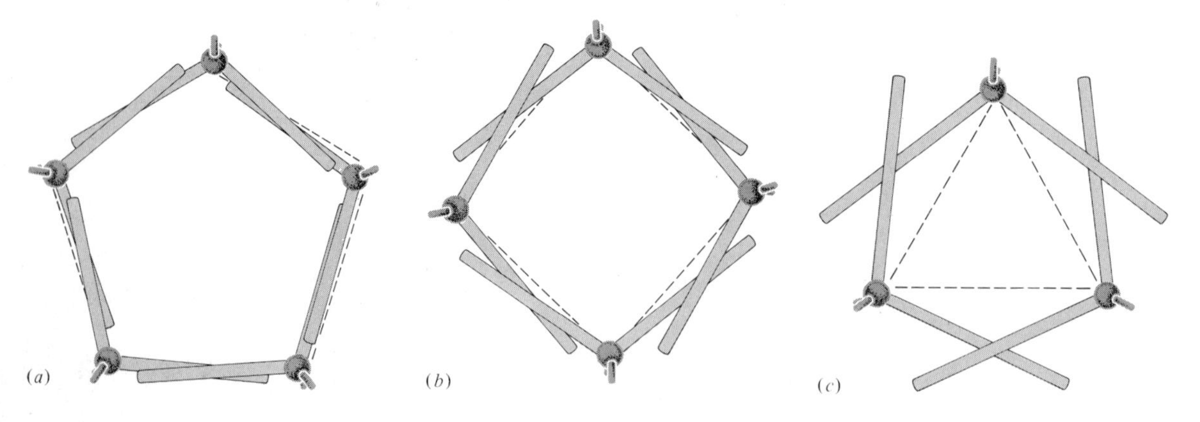

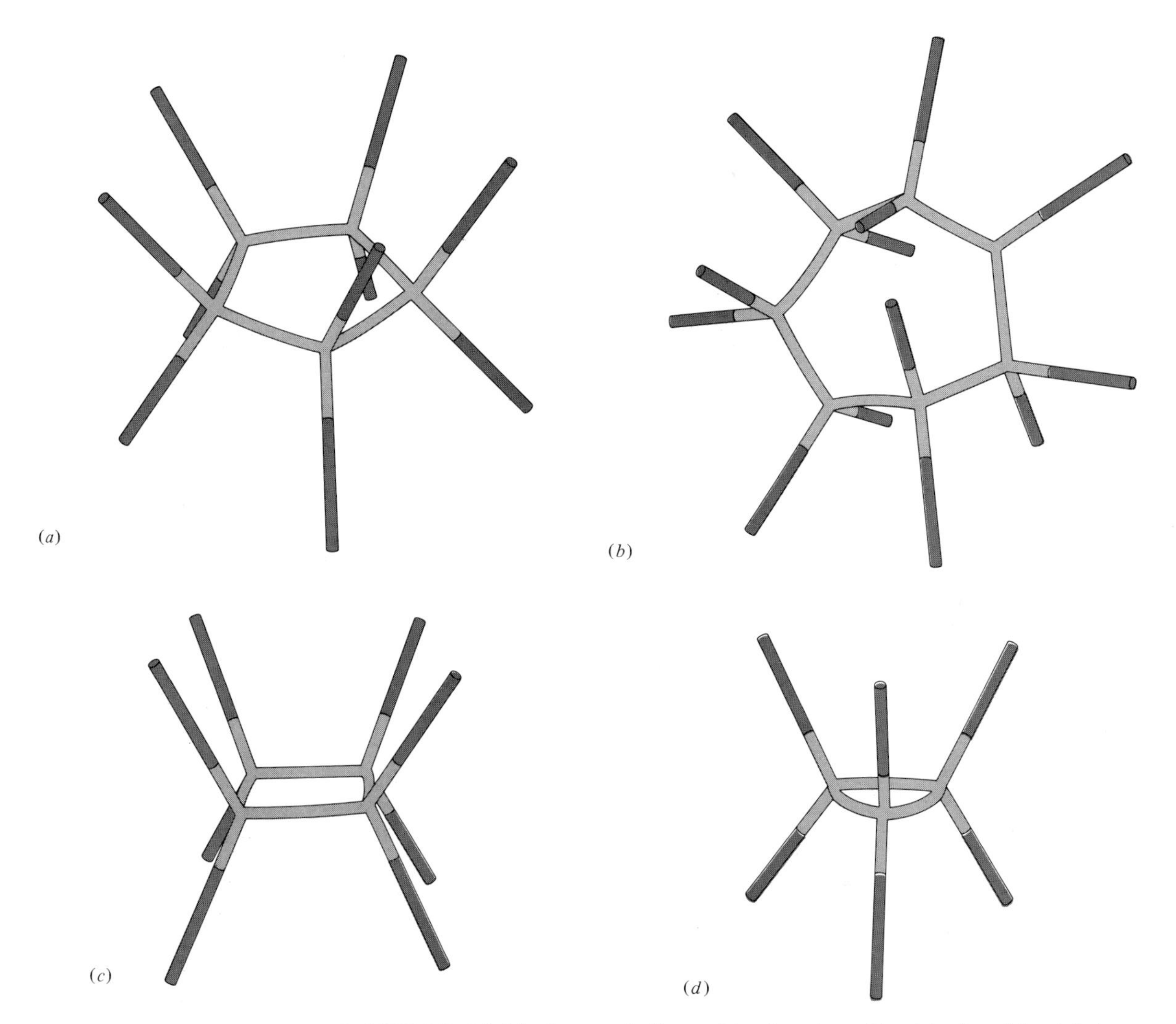

FIGURE 3-8 Models of some cycloalkanes: (*a*) cyclopentane, (*b*) cycloheptane, (*c*) cyclobutane, and (*d*) cyclopropane.

same plane if head-on overlap is to be maximized. The carbon atoms in cyclopentane tend to form almost a regular, planar pentagon (108° angles). The bond angles are a little less than 109.5°, the tetrahedral bond angle. If the structure were not planar, the bond angles would be reduced still more. That would result in less head-on overlap and more of the weaker sideways overlap. Nevertheless, the cyclopentane ring is not quite planar because the repulsion of its hydrogens, all of which would be eclipsing each other if the ring were planar, is relieved by a slight twisting out of the plane. See Fig. 3-8*a*.

In the case of cyclobutane and cyclopropane, the sp^3 hybrid orbitals of the carbon atoms must be overlapped sideways to a still greater extent to

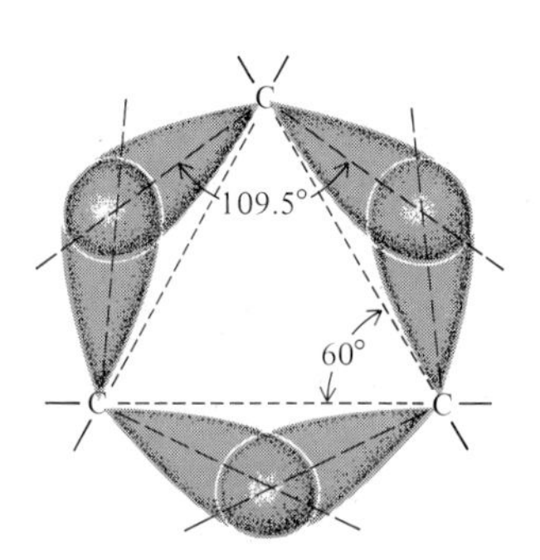

FIGURE 3-9
sp^3 Orbital overlap for cyclopropane. The interorbital bond angle is 109.5°, and the internuclear angle is 60°.

accommodate internuclear angles of 90° for cyclobutane and 60° for cyclopropane. Figure 3-9 represents sp^3 orbital overlap for cyclopropane. To avoid a complete eclipsing conformation, the cyclobutane ring also is not quite planar.

The formulas, names, melting points, and boiling points of some cycloalkanes are given in Table 3-2.

3-2 Nomenclature of the saturated hydrocarbons

The IUPAC system is applied to the alkanes as follows:

1 The name of the *longest continuous-chain alkane* is used as the *parent name* for all *aliphatic* compounds (alkanes and their derivatives). Table 3-1 lists the parent names. Commit them to memory.

TABLE 3-2
Cycloalkanes

Molecular formula	Structural formula	Simplified structural formula	IUPAC name	mp, °C	bp, °C
C_3H_6	H_2 C H_2C—CH_2	△	Cyclopropane	−126	−34
C_4H_8	H_2C—CH_2 | | H_2C—CH_2	□	Cyclobutane	−50	13
C_5H_{10}	H_2 C H_2C CH_2 H_2C—CH_2	⬠	Cyclopentane	−93	50
C_6H_{12}	H_2 C H_2C CH_2 H_2C CH_2 C H_2	⬡	Cyclohexane	6	81
C_7H_{14}	H_2C—CH_2 H_2C CH_2 H_2C CH_2 C H_2	(heptagon)	Cycloheptane	−12	118

2 A branched-chain alkane is named as a *derivative* of its *parent alkane* (longest continuous chain of carbon atoms).
3 The *positions of branches* on the parent alkane are indicated by *locator numbers*.
4 The *numbering* of the longest continuous carbon chain is such that the locator numbers are *as low as possible*.
5 If the same branch occurs *two or more times,* a suitable prefix (for example, *di-, tri-, tetra-,* or *penta-*) is used to designate the number of the like branches, and appropriate locator numbers are used to give the positions of the branches.
6 If two or more different alkyl groups are attached to a parent alkane as branches, their naming sequence follows the *alphabetical order of the substituent* (alkyl) *groups,* regardless of the number of each present and without consideration of the prefixes *di-, tri-,* and so on.

For example, the IUPAC name of $CH_3CH_2CH_2CH_3$ is simply butane; not *n*-butane, which is the common or trivial name. The IUPAC name of $(CH_3)_2CHCH_3$ is 2-methylpropane; the older trivial name is isobutane. The IUPAC name for neohexane, $CH_3C(CH_3)_2CH_2CH_3$, is 2,2-dimethylbutane.

PROBLEM 3-2 Give the IUPAC name for each of the following alkanes.

(a) $\begin{array}{lllll} & CH_3 & & CH_3 & \\ & | & & | & \\ CH_3 & CH & -CH_2 & CH & CH_3 \end{array}$

(b) $\begin{array}{llll} & CH_3 & CH_3 & \\ & | & | & \\ CH_3 & CH- & CH & CH_2CH_3 \end{array}$

(c) $\begin{array}{lll} & CH_3 & \\ & | & \\ CH_3 & C & -CH_2CH_2CH_3 \\ & | & \\ & CH_3 & \end{array}$

(d) $\begin{array}{llll} & & CH_3 & \\ & & | & \\ CH_3CH_2 & CH & CH & CH_3 \\ & | & & \\ & CH_2CH_3 & & \end{array}$

(e) $\begin{array}{lllll} & CH_3 & CH_3 & CH_3 & \\ & | & | & | & \\ CH_3 & CH- & C- & CH & CH_3 \\ & & | & & \\ & & CH_3 & & \end{array}$

(f) $\begin{array}{lll} & H & \\ & | & \\ CH_3 & C & -CH_3 \\ & | & \\ CH_3 & CH & CH_2CH_2CH_3 \end{array}$

Solution to (a) The IUPAC name for this alkane is 2,4-dimethylpentane.

$$\begin{array}{ccccccccc} & & CH_3 & & & & CH_3 & & \\ 1 & & 2| & & 3 & & 4| & & 5 \\ CH_3 & - & CH & - & CH_2 & - & CH & - & CH_3 \end{array}$$

PROBLEM 3-3 Draw structural formula for each of the following compounds.

(a) 2,2,3-trimethylbutane
(b) 3-methylheptane
(c) 3,3-diethylpentane
(d) 3,4-diethylhexane
(e) 3-ethyl-2-methylpentane
(f) 2,3-dimethylbutane

Solution to (a) The correct structural formula for 2,2,3-trimethylbutane is

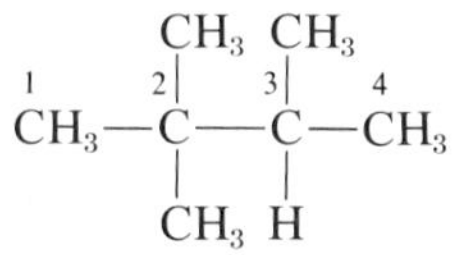

```
        CH3 CH3
  1    2|   3|   4
CH3—C——C—CH3
        |    |
        CH3  H
```

The IUPAC system is applied to cycloalkanes in much the same way as to alkanes:

1 The *parent name* of the cycloalkane is derived by adding the prefix *cyclo-* to the name of the *linear alkane* having the *same number* of carbon atoms. See Table 3-2.
2 For a *substituted cycloalkane* each substituent is given the *lowest possible locator number*. That means the numbering must begin with a carbon atom bearing a substituent.
3 If the same branch or group occurs more than once, then an appropriate prefix (for example, *di-*, *tri-*, *tetra-*) must be used.

Two examples of multiply substituted cycloalkanes are 1,1,3-trimethylcyclohexane and *cis*-1-ethyl-2-methylcyclopentane:

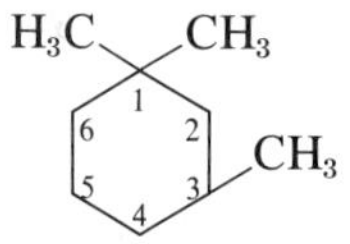

H_3C CH_3 CH_3 — 1,1,3-Trimethylcyclohexane

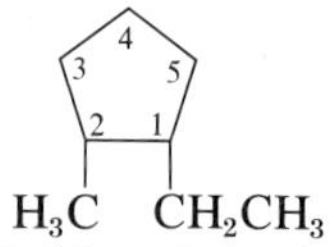

H_3C CH_2CH_3 — *cis*-1-Ethyl-2-methylcyclopentane

PROBLEM 3-4 Give the IUPAC name for each of the following cycloalkanes. Indicate cis or trans when it is appropriate.

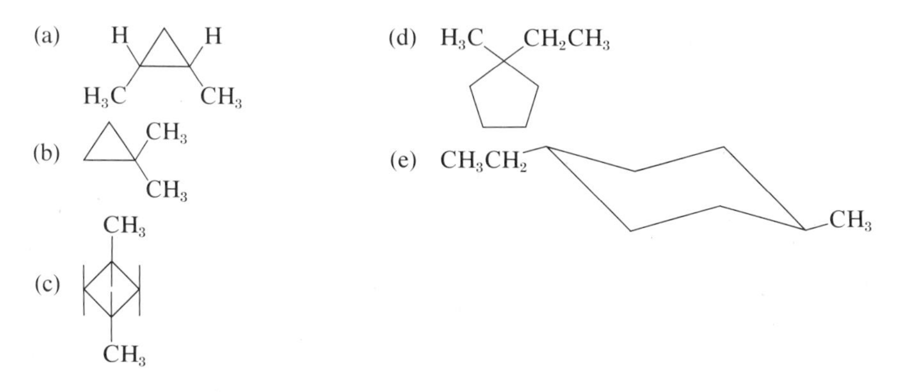

Solution to (a) The IUPAC name for this compound is *cis*-1,2-dimethylcyclopropane.

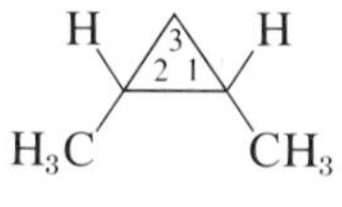

H H H_3C CH_3

PROBLEM 3-5 Draw the structural formula for each of the following:

(a) methylcyclopentane
(b) *cis*-1,3-dimethylcyclopentane
(c) 1-ethyl-1-methylcyclobutane
(d) *trans*-1-(1-methylethyl)-4-methylcyclohexane
(e) all *cis*-1,2,3-trimethylcyclopropane

Solution to (a) The correct structural formula for methylcyclopentane is shown below. Note that the locator number 1 can be omitted from the IUPAC name when the location of the substituent group is unambiguous.

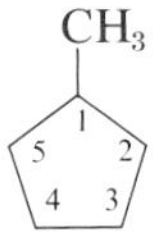

CH_3

1 2 3 4 5

3-3 Sources and preparations of alkanes

The alkanes occur abundantly in various natural deposits in the earth, generally as complex mixtures. Laboratory methods are available to synthesize pure alkanes when they are required.

Alkanes from natural gas and petroleum

Petroleum is the chief source of many acyclic as well as some cyclic alkanes. Indeed, as petroleum becomes more scarce, it may be necessary to reserve it for raw materials for industry rather than burn so much of it for heat and energy production. *Natural gas* consists of about 80% methane, 10% ethane, and 10% of a mixture of the other relatively low boiling alkanes. The alkanes which are obtained from petroleum may be separated to some extent by *fractional distillation*. It is difficult to get them pure in this way. It is especially difficult to so obtain the higher-molecular-weight homologs, which have more isomers and relatively smaller differences in boiling points.

Methane in marsh gas

Methane is also a product of the decay of plant and marine organisms which have become buried in swamps and marshes. Methane from such sources is referred to as *marsh gas*. It is not an important source of methane, however. Technology is now being developed to produce methane and other low-molecular-weight alkanes from organic garbage and trash.

Other sources of alkanes

The technology has now been developed to produce saturated hydrocarbons from coal, oil shale, and the reaction of carbon monoxide with hydrogen (*Fischer-Tropsch process*). As petroleum becomes less readily available and its price rises still higher, those processes will become economically feasible and competitive.

Laboratory methods for preparation of alkanes

There are several laboratory methods for synthesizing particular saturated hydrocarbons when they are required. Here are some of them.

REDUCTION OF AN ALKENE TO PRODUCE AN ALKANE. This is a general reaction for all alkenes. A nickel, palladium, or platinum catalyst is required. Note, in this and other general reactions, that R represents a hydrogen atom or an alkyl group. The general reaction for the reduction of an alkene is

$$R_2C{=}CR_2 + H_2 \xrightarrow{Ni} R{-}\underset{R}{\overset{H}{C}}{-}\underset{R}{\overset{H}{C}}{-}R \qquad (3\text{-}1)$$

An example of this reaction is the reduction of 4-methyl-2-pentene to yield 2-methylpentane:

$$\underset{\text{4-Methyl-2-pentene}}{CH_3\overset{CH_3}{\overset{|}{C}}H{-}CH{=}CHCH_3} + H_2 \xrightarrow{Ni} \underset{\text{2-Methylpentane}}{CH_3\overset{CH_3}{\overset{|}{C}}H{-}CH_2CH_2CH_3} \qquad (3\text{-}2)$$

FUSION OF THE SODIUM SALT OF AN ORGANIC ACID WITH SODIUM HYDROXIDE. In the following general equation, Δ (delta) represents heat applied:

$$R{-}C\begin{matrix}{/\!\!/}O \\ \backslash O^-Na^+\end{matrix} + NaOH \xrightarrow{\Delta} RH + Na_2CO_3 \qquad (3\text{-}3)$$

An example of this reaction is the production of methane from sodium acetate:

$$\underset{\text{Sodium acetate}}{CH_3C\begin{matrix}{/\!\!/}O \\ \backslash O^-Na^+\end{matrix}} + NaOH \xrightarrow{\Delta} \underset{\text{Methane}}{CH_3H} + Na_2CO_3 \qquad (3\text{-}4)$$

REDUCTION OF AN ALKYL HALIDE. One method uses an active metal and an acid. In the general equation, X represents a halogen atom:

$$RX + Zn + 2H^+ \longrightarrow RH + Zn^{2+} + HX \qquad (3\text{-}5)$$

A specific example of this type of reaction is

$$\underset{\text{2-Bromopropane}}{CH_3CHBrCH_3} \xrightarrow{Zn/HCl,\ CH_3COOH} \underset{\text{Propane}}{CH_3CH_2CH_3} + HBr \qquad (3\text{-}6)$$

An elegant laboratory method for reduction of an alkyl halide is to convert the halide into an alkyl magnesium halide, called a *Grignard reagent,* and then hydrolyze the Grignard reagent with water. If heavy water (D_2O) is used, a deuterium-substituted alkane can be obtained. Deuterated alkanes have been used to identify specific absorption lines in infrared spectra by comparison of their spectra with the spectra of the corresponding undeuterated alkanes. Equation (3-7) shows the preparation of a Grignard reagent from 2-bromopropane. Equation (3-8) shows the hydrolysis of the Grignard reagent with heavy water to yield 2-deuteropropane.

$$CH_3CHBrCH_3 + Mg \xrightarrow{\text{dry ether}} CH_3\overset{\overset{\displaystyle MgBr}{|}}{C}HCH_3 \tag{3-7}$$

$$CH_3\overset{\overset{\displaystyle MgBr}{|}}{C}HCH_3 + D_2O \longrightarrow \underset{\text{2-Deuteropropane}}{CH_3\overset{\overset{\displaystyle D}{|}}{C}HCH_3} + Mg\begin{matrix} \diagup Br \\ \diagdown OD \end{matrix} \tag{3-8}$$

PROBLEM 3-6 Write equations showing the synthesis of the following alkanes from the starting material indicated.

(a) cyclohexane from cyclohexene

(b) undecane from the sodium salt of dodecanoic acid, $CH_3(CH_2)_{10}COOH$

(c) cyclopropane from $\triangleright\!-\overset{\overset{\displaystyle O}{\|}}{C}O^-Na^+$

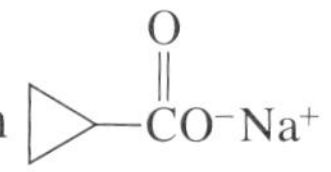

(d) 2,2,4-trimethylpentane from 2,4,4-trimethyl-2-pentene

(e) 1-deuteropropane from 1-bromopropane

Solution to (a)

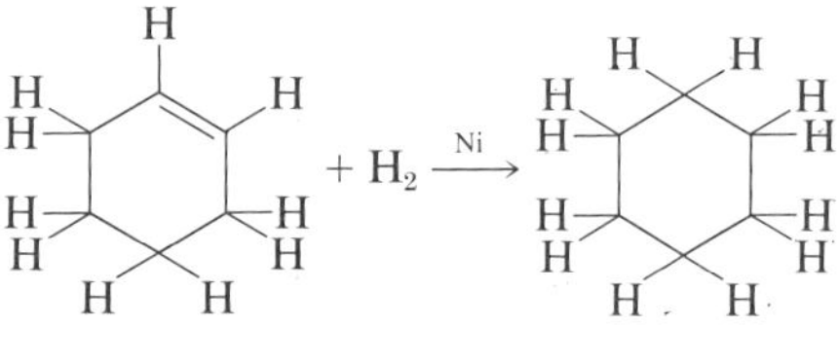

$$\text{(cyclohexene, } C_6H_{10}\text{)} + H_2 \xrightarrow{Ni} \text{(cyclohexane, } C_6H_{12}\text{)}$$

3-4 General reactions of the alkanes

The chief reactions of the alkanes are types of oxidation—specifically, chlorination, bromination, and combustion with oxygen. All three reactions proceed spontaneously, with the production of heat, once they have been initiated. They involve free-radical intermediates.

Combustion of alkanes

The combustion of alkanes is chiefly of value for reason of the heat produced for home, industry, and power production. The amount of heat produced per mole increases as the size of the alkane molecule increases. Approximately 156 kcal of energy per mole is produced for each additional carbon atom.

$$CH_4 + 2\,O_2 \longrightarrow CO_2 + 2H_2O + 210\ \text{kcal/mol} \qquad (3\text{-}9)$$

$$CH_3CH_2CH_2CH_2CH_3 + 8\,O_2 \longrightarrow 5CO_2 + 6H_2O + 833\ \text{kcal/mol} \qquad (3\text{-}10)$$

A lesser amount of oxygen results in incomplete combustion and the formation of soot, or carbon black, and the toxic gas carbon monoxide.

$$CH_4 + \tfrac{3}{2}\,O_2 \longrightarrow CO + 2H_2O + 143\ \text{kcal/mol} \qquad (3\text{-}11)$$

$$CH_4 + O_2 \longrightarrow \underset{\text{Carbon black}}{C} + 2H_2O + 118\ \text{kcal/mol} \qquad (3\text{-}12)$$

Halogenation of alkanes

The halogenation of alkanes is an interesting reaction, but in most cases it is of limited value for synthesis because of the mixture of products obtained. Multiple substitution may occur. For example, chlorination of methane produces a mixture of chloromethane, dichloromethane, trichloromethane, and tetrachloromethane.

$$CH_4 + Cl_2 \xrightarrow{\text{light}} CH_3Cl + HCl \qquad (3\text{-}13)$$

$$CH_3Cl + Cl_2 \xrightarrow{\text{light}} CH_2Cl_2 + HCl \qquad (3\text{-}14)$$

$$CH_2Cl_2 + Cl_2 \xrightarrow{\text{light}} CHCl_3 + HCl \qquad (3\text{-}15)$$

$$CHCl_3 + Cl_2 \xrightarrow{\text{light}} CCl_4 + HCl \qquad (3\text{-}16)$$

The yield of monosubstituted product may be increased by using an excess of the alkane. The excess alkane competes with the monohalide product in reacting with halogen.

$$\underset{\text{Excess methane}}{CH_4} + Cl_2 \xrightarrow{\text{light}} CH_3Cl + HCl \qquad (3\text{-}13)$$

Cycloalkanes react similarly.

$$\underset{\text{Cyclopentane in excess}}{C_5H_{10}} + Cl_2 \xrightarrow{\text{light}} \underset{\text{Chlorocyclopentane}}{C_5H_9Cl} + HCl \qquad (3\text{-}17)$$

Halogenation of the alkanes is initiated by light or by temperatures near 300°C. The order of reactivity of the halogens with alkanes is $F_2 > Cl_2 > Br_2 > I_2$. However, the fluorination reaction is too violent to be practical, and iodine actually does not react at all.

Mechanism of halogenation. A free radical is the reaction intermediate in the halogenation of alkanes. The reaction seems to be a *chain reaction*—a multiple-step reaction which regenerates one of its needed reactants so that it is self-propagating. The mechanism of halogenation is considered to include the following steps for methane:

Step 1 $$Cl_2 \xrightarrow{\text{heat or light}} 2Cl\cdot \quad \text{(atomic chlorine)} \tag{3-18}$$

Step 2 $$Cl\cdot + CH_4 \longrightarrow CH_3\cdot + HCl \tag{3-19}$$

Step 3 $$CH_3\cdot + Cl_2 \longrightarrow CH_3Cl + Cl\cdot \tag{3-20}$$

Step 1 is an endothermic reaction, but it is required only to get the reaction started. It is called the *chain-initiating reaction*. Steps 2 and 3 are called the *chain-propagating reactions*. Table 3-3 gives thermodynamic data for the halogenation of methane by each of the halogens. On examination of the table, step 2 is seen to be the most difficult reaction. It is slightly endothermic for chlorination, very endothermic for bromination, and still more endothermic for iodination—so much so that iodination does not occur.

It is worthwhile to examine each of the reaction steps with the data in Table 3-3. Step 1 involves dissociation of molecules into atoms and is of course endothermic, but it is only the chain-initiating reaction. Relatively few molecules must be dissociated to get the chain reaction started, after which the reaction generates its own halogen atoms. There are no great differences in ΔH for the various halogens for this reaction.

Step 2, abstraction of a hydrogen atom from methane by a halogen atom, shows great differences in ΔH. The variation is from the strongly exothermic -32 kcal/mol for the very active fluorine to the highly endothermic $+33$ kcal/mol for the unreactive iodine. This is therefore the step which accounts for the difference in reactivity of the halogens.

Step 3, attack of methyl radicals on halogen, is exothermic for all four halogens, and the value of ΔH is about the same for Cl, Br, and I.

The progress-of-reaction diagrams and transition states for steps 2 and 3 of the bromination of methane were given as examples in Figs. 1-9 and 1-10. You may wish to refer to them now.

PROBLEM 3-7 By using the data given in Table 3-3 and a value of 4 kcal/mol for the activation energy E_{act}, construct progress-of-reaction diagrams for steps 2 and 3 of the chlorination of methane. Compare the diagrams with the

TABLE 3-3 Thermodynamic data for halogenation of methane

Reaction step	ΔH, kcal/mol			
	F	Cl	Br	I
1 $X_2 \longrightarrow 2X\cdot$	+38	+58	+46	+36
2 $X\cdot + CH_4 \longrightarrow CH_3\cdot + HX$	−32	+1	+16	+33
3 $CH_3\cdot + X_2 \longrightarrow CH_3X + X\cdot$	−70	−26	−24	−20

steps for bromination in Figs. 1-9 and 1-10. Does it now seem more reasonable that the chlorination reaction is many times faster than the bromination reaction?

In addition to the three reaction steps for the mechanism of halogenation of methane which have already been given, there is a fourth reaction step. It consists of *chain-terminating reactions*. The chain reaction may be terminated before either the alkane or the halogen has reacted completely. The termination occurs because of the combination of the reaction-intermediate free radicals with each other rather than with alkane or halogen. For the chlorination of methane the chain-terminating reactions are

$$2Cl\cdot \longrightarrow Cl_2 \tag{3-21}$$

$$CH_3\cdot + Cl\cdot \longrightarrow CH_3Cl \tag{3-22}$$

$$2CH_3\cdot \longrightarrow CH_3CH_3 \tag{3-23}$$

Because of such reactions, rupture of more than one halogen molecule is actually required to initiate the reaction. Several chains must be initiated for the reaction to go to completion.

The mechanism for the halogenation of other alkanes is considered to be the same as that of methane.

Selectivity in halogenation. Monochlorination of propane produces both 1- and 2-chloropropane (Fig. 3-10):

$$CH_3CH_2CH_3 + Cl_2 \xrightarrow{\text{light}} \left\{ \begin{array}{l} CH_3CH_2CH_2Cl \\ \overset{\displaystyle Cl}{\overset{|}{CH_3CHCH_3}} \end{array} \right. + HCl \tag{3-24}$$

FIGURE 3-10 The products of monochlorination of propane are (*a*) 1-chloropropane and (*b*) 2-chloropropane.

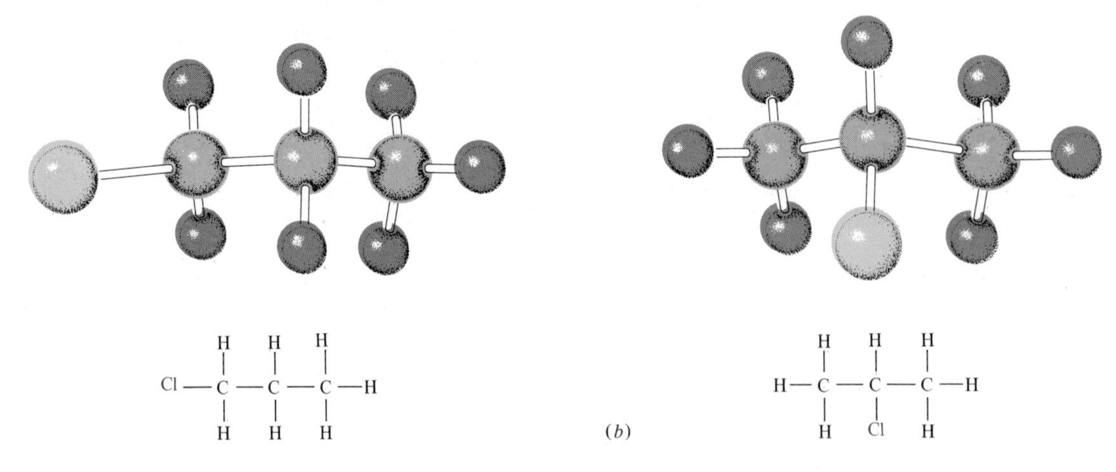

Since there are six primary hydrogens to only two secondary hydrogens, the chance of collision of a chlorine atom with a primary hydrogen atom versus collision with a secondary hydrogen atom should be 6:2, or 3 times as great. If the activity of a primary hydrogen atom is the same as the activity of a secondary hydrogen atom, there should then be 3 times as much 1-chloropropane produced as 2-chloropropane. Actually, the two isomers are produced in about the same amount. That brings us to the conclusion that the secondary hydrogen atoms must be about 3 times as reactive as the primary ones in order to compensate for their lesser number. Similar experiments reveal that tertiary hydrogen atoms are still more reactive; toward a chlorine atom they are about 5 times as reactive as primary hydrogen atoms. For example, monochlorination of 2-methylpropane produces 2-chloro-2-methylpropane and 1-chloro-2-methylpropane[1] in the ratio of about 1:2, although there is only one tertiary hydrogen atom to nine primary hydrogen atoms.

$$CH_3-\underset{\displaystyle H}{\overset{\displaystyle CH_3}{C}}-CH_3 + Cl_2 \xrightarrow{\text{light}} \begin{cases} CH_3-\underset{\displaystyle Cl}{\overset{\displaystyle CH_3}{C}}-CH_3 & \text{2-Chloro-2-methylpropane (1 part produced)} \\ CH_3-\underset{\displaystyle H}{\overset{\displaystyle CH_3}{C}}-CH_2Cl & \text{1-Chloro-2-methylpropane (2 parts produced)} \end{cases} + HCl \quad (3\text{-}25)$$

The reactions just presented indicate that a hydrogen on a tertiary carbon atom is more readily substituted by a chlorine atom than is a hydrogen on a secondary carbon atom, and the latter, in turn, is more readily substituted than is a hydrogen on a primary carbon atom. A similar order is found for substitution of hydrogen by bromine. The critical reaction step for each of these substitutions involves formation of an alkyl free radical and a molecule of hydrogen halide by loss of a hydrogen atom from the alkane to the halogen atom.

$$R-H + X\cdot \longrightarrow R\cdot + H-X \quad (3\text{-}26)$$

It has been found from thermodynamic data that the energy required to break the carbon-hydrogen bond to produce an alkyl free radical is least for a tertiary carbon-hydrogen bond and greatest for a methyl carbon-hydrogen bond. The energy required for the formation of sec-

[1] In older systems of nomenclature, 2-methylpropane is known as isobutane. 2-Chloro-2-methylpropane is called *tert*-butyl chloride, and 1-chloro-2-methylpropane is called isobutyl chloride. You may still encounter the older names.

ondary and primary radicals lies in between. Relative to the alkane from which each radical is formed, that means a tertiary free radical is more stable than a secondary free radical. The latter, in turn, is more stable than a primary free radical, which is more stable than a methyl free radical. If 3° = tertiary, 2° = secondary, and 1° = primary, then the order of stability is $3° > 2° > 1° > CH_3\cdot$.

3-5 Ring-opening reactions of cycloalkanes

You will recall that, in our earlier consideration of the structure of the cycloalkanes, we noted that the bonding orbitals in cyclopropane and cyclobutane are not as strong as those in the higher homologs. Basically, that is a result of the kind of overlap of orbitals required by the geometry of the smaller rings. The relative weakness of the bonding orbitals means that the bonds in cyclopropane and cyclobutane are vulnerable to attack by certain reagents. Cyclopropane and, to some extent, cyclobutane undergo special *ring-opening reactions*. Some examples are as follows:

$$\triangle + H_2 \xrightarrow{\text{Ni, 80°C}} CH_3CH_2CH_3 \tag{3-27}$$

$$\square + H_2 \xrightarrow{\text{Ni, 100°C}} CH_3CH_2CH_2CH_3 \tag{3-28}$$

$$\triangle + \underset{\text{(Brown)}}{Br_2} \xrightarrow{CCl_4} \underset{\text{(Colorless)}}{BrCH_2CH_2CH_2Br} \tag{3-29}$$

$$\triangle + HBr \longrightarrow CH_3CH_2CH_2Br \tag{3-30}$$

Cyclopropane also undergoes a ring-opening reaction with hydriodic acid, HI, and with concentrated sulfuric acid analogous to that with HBr:

$$\triangle + H_2SO_4 \xrightarrow{\Delta} \underset{\text{Propyl hydrogen sulfate}}{CH_3CH_2CH_2OSO_2OH} \tag{3-31}$$

Cyclopentane and the higher homologs do not undergo the ring-opening reaction with hydrogen in the presence of nickel and heat. Cyclobutane and the higher homologs do not undergo the reaction with bromine in carbon tetrachloride[1] solution. Although $KMnO_4$ reacts with alkenes, it does not react with cyclopropane, so it can be used to distinguish cyclopropane from its isomer, propene.

3-6 Chemistry of petroleum

Petroleum is considered to be of biorganic origin. It consists of a complex mixture of hydrocarbons, mostly alkanes with some cycloalkanes. That from some sources even includes small amounts of aromatic hydrocarbons (benzene and its derivatives, to be discussed in Chap. 5). It can

[1] The IUPAC name for CCl_4 is tetrachloromethane. However, the common name, carbon tetrachloride, is so familiar that we shall use it frequently.

TABLE 3-4
Fractions distilled from petroleum

Carbon content	Use of fraction	Boiling point, °C
C_3–C_4	Bottled gas	−45–0
C_5–C_{12}	Natural gasoline	25–200
C_{11}–C_{15}	Kerosene, jet engine fuel	175–275
C_{14}–C_{25}	Fuel oil, diesel oil	300–400
C_{25}–C_{35}	Lubricating oils, greases, paraffin	400–500

be distilled to yield fractions with the boiling points and uses shown in Table 3-4.

Development of gasoline

When the first petroleum well was drilled and developed in Pennsylvania in 1859, the most important distillation fraction of the petroleum was kerosene. It was used as a fuel for lamps. With the development of the internal-combustion engine, gasoline became the fraction in greatest demand. Since the yield of natural or straight-run gasoline obtained by fractional distillation is only about 20% of a barrel of crude oil (petroleum), processes were developed for converting other fractions to hydrocarbons boiling in the range of gasoline. Today about 50% of a barrel of crude oil may be converted to gasoline by those processes.

Cracking of petroleum. One of the processes for converting more of a barrel of crude oil to gasoline is *cracking,* in which the longer-chain alkanes of the higher-boiling fractions are broken down into shorter-chain alkanes and alkenes. For example, the 12-carbon straight-chain alkane, dodecane, upon cracking yields alkanes and alkenes of from 5 to 10 carbon atoms boiling in the range of gasoline (25 to 200°C). It also yields some by-product alkanes and alkenes such as ethylene, propylene, and butylenes.

Both *thermal cracking* and *catalytic cracking* are employed. Temperatures of 400 to 600°C are required for thermal cracking. In catalytic cracking, petroleum fractions are either passed over a *fixed-bed catalyst* or mixed with a *fluidized catalyst* at temperatures of about 500°C (Fig. 3-11). The catalysts used consist primarily of aluminum silicates.

Some carbon-hydrogen bonds as well as carbon-carbon bonds are broken in cracking. That produces quite a lot of by-product hydrogen, which may be used in further refining the petroleum or in the synthesis of ammonia and methanol. Alkenes of two to four carbon atoms also are by-products of cracking. They can be used to produce hydrocarbons with longer chains that boil in the range of gasoline. They can also be used by the chemical industry as raw materials for the production of all sorts of chemicals, plastics, and pharmaceuticals.

Octane rating of gasoline. The gasoline prepared by industrial processes gives better engine performance than straight-run gasoline. The engine

(*a*)

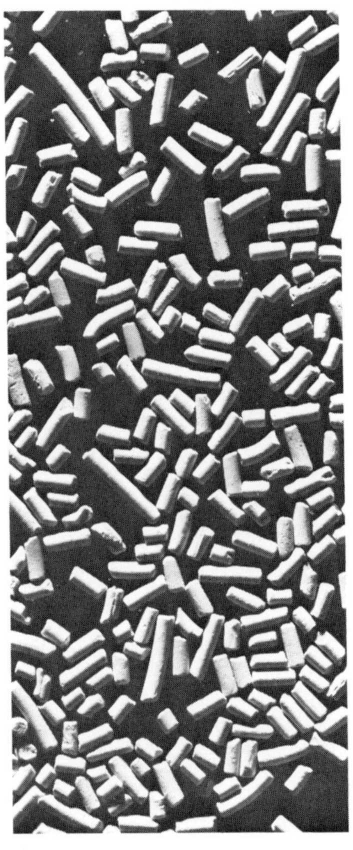

(*b*)

FIGURE 3-11
(*a*) A catalytic cracking unit at Exxon USA's Baytown, Texas refinery; this plant can refine 640,000 barrels of crude oil per day. (*An Exxon photo.*) (*b*) Photomicrograph of catalyst particles. (*Exxon Research and Engineering Co.*)

performance of gasoline for high-compression internal-combustion engines (compression ratios of 9: or 10: 1) is rated by the *octane number*. When the octane scale was developed, *n*-heptane gave the worst engine performance and so was given a rating of zero. Isooctane, as the branched-chain alkane 2,2,4-trimethylpentane was called by petroleum technologists, gave the best performance and was given a rating of 100. Subsequently, hydrocarbons with both worse and better performance have been found, so the octane scale has had to be extended. For example, *n*-octane has a rating of − 17, and 2,2,3-trimethylbutane (triptane) has a rating of 116. A gasoline rated 90 octane would have the same performance as a mixture of 90% 2,2,4-trimethylpentane and 10% *n*-heptane.

The higher the compression ratio of an engine, the higher the octane number of the fuel must be if the engine is to perform without the inefficient ignition effects called knock. In general, branched-chain alkanes have higher octane ratings than the straight-chain alkanes have. Since cracking produces some branched-chain hydrocarbons, gasoline from the cracking process has a higher octane rating than the product of fractional distillation. See Table 3-5 for some examples of octane ratings.

TABLE 3-5 Octane ratings of some hydrocarbons

Number of carbon atoms	Straight-chain alkanes	Octane rating	Branched-chain or cyclic hydrocarbons*	Octane rating
3	Propane	112	. . .	
4	Butane	94	. . .	
5	Pentane	62	2-Methylbutane (Isopentane)	93
6	Hexane	25	Methylcyclopentane	91
7	Heptane	0	Methylcyclohexane	75
7	. . .	. . .	Methylbenzene (Toluene)	103
7	. . .	. . .	2,2,3-Trimethylbutane (Triptane)	116
8	Octane	−19	2,2,4-Trimethylpentane (Isooctane)	100
9	. . .	. . .	(1-Methylethyl)benzene (Cumene or isopropyl benzene)	113

* The first name given is, in each case, the IUPAC name. An older, common name which is still in use is given in parentheses below the IUPAC name.

Isomerization of alkanes. It was found that, in the presence of certain catalysts, straight-chain alkanes can be isomerized to branched forms (Fig. 3-12). An example of the reaction is

$$\underset{\text{Pentane}}{CH_3CH_2CH_2CH_2CH_3} \xrightarrow{AlCl_3} \underset{\substack{\text{2-Methylbutane} \\ \text{(Isopentane)}}}{CH_3\overset{\overset{\displaystyle CH_3}{|}}{C}HCH_2CH_3} \qquad (3\text{-}32)$$

As a result of this reaction, the octane rating is improved appreciably. Pentane has an octane number of 62, whereas 2-methylbutane has an octane number of 90. *Catalytic isomerization reactions* such as Eq. (3-32) are one of the ways of improving gasoline quality.

Production of gasoline components by alkylation. *Alkylation* is a reaction of an alkene with an alkane in the presence of an acid catalyst. It is a very

FIGURE 3-12 Catalytic isomerization of (*a*) pentane produces (*b*) isopentane.

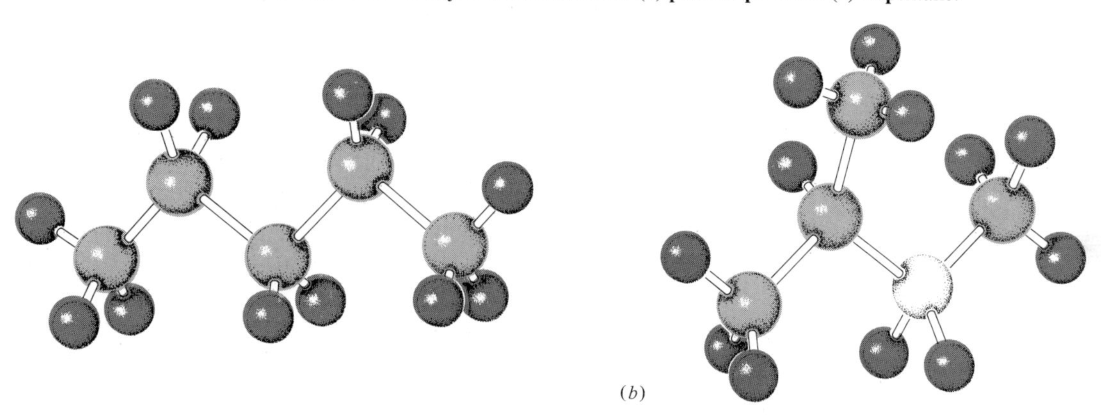

useful method of producing highly branched alkanes of good octane number from some of the by-products of cracking and from the low-boiling petroleum fraction. An example is

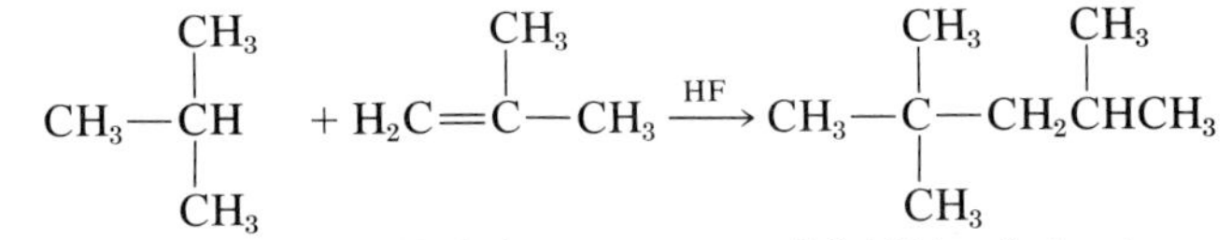

$$\underset{\substack{\text{2-Methylpropane}\\ \text{(Isobutane)}}}{CH_3-\underset{CH_3}{\overset{CH_3}{\overset{|}{\underset{|}{C}}}}H} + \underset{\substack{\text{2-Methylpropene}\\ \text{(Isobutylene)}}}{H_2C=\overset{CH_3}{\overset{|}{C}}-CH_3} \xrightarrow{HF} \underset{\substack{\text{2,2,4-Trimethylpentane}\\ \text{(isooctane)}}}{CH_3-\underset{CH_3}{\overset{CH_3}{\overset{|}{\underset{|}{C}}}}-CH_2\overset{CH_3}{\overset{|}{C}}HCH_3} \tag{3-33}$$

You will recall that 2,2,4-trimethylpentane has an octane number of 100.

Reforming to produce high-octane hydrocarbons. Cyclic and even acyclic alkanes of six to eight carbon atoms can be converted to aromatic hydrocarbons (discussed in Chap. 5) by heating them in the presence of a hydrogenation-dehydrogenation catalyst at about 400°C. Platinum seems to be the best catalyst, and so the process is sometimes called *platforming*. However, the metals nickel and palladium, which are in the same periodic table group as platinum, also are possible catalysts for both dehydrogenation and hydrogenation. (Remember that a catalyst merely hastens the attainment of equilibrium and alters the rates of both the forward and the reverse reactions.) An example of the platforming reaction is

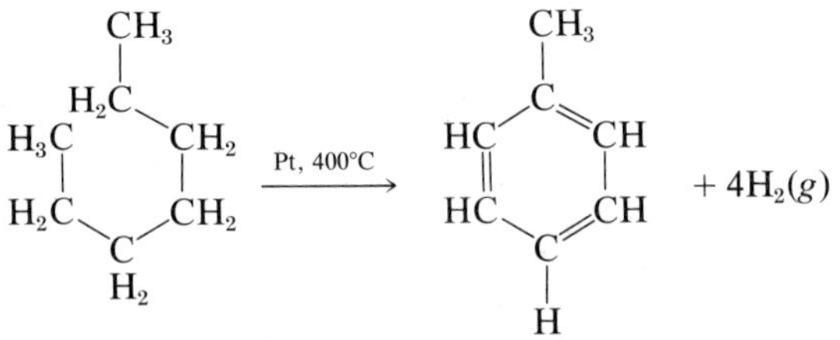

$$\underset{\substack{\text{Heptane}\\ \text{Octane number 0}}}{\text{(methylcyclohexane ring: } CH_3\text{–}H_2C,\ H_3C,\ CH_2,\ H_2C,\ CH_2,\ C H_2)} \xrightarrow{\text{Pt, 400°C}} \underset{\substack{\text{Toluene}\\ \text{Octane number 103}}}{\text{(ring: } CH_3\text{–}C,\ HC,\ CH,\ HC,\ CH,\ C\text{–}H)} + 4H_2(g) \tag{3-34}$$

This reaction is an important source of aromatic hydrocarbons as well as another method for improving the octane number of gasoline.

Gasoline additives The most common examples of additives used to raise the octane number of gasoline are tetraethyl lead and tetramethyl lead. Tetraethyl lead is most commonly prepared by the following reaction:

$$4C_2H_5Cl + \underset{\substack{\text{(Sodium-lead}\\ \text{alloy)}}}{4Pb{\cdot}Na} \longrightarrow (C_2H_5)_4Pb + 3Pb + 4NaCl \tag{3-35}$$

Addition of up to 3 ml of tetraethyl lead per gallon of gasoline raises the octane number by up to 15 units. The deposition of lead oxide in the engine is prevented by also adding some 1,2-dibromoethane and 1,2-dichloroethane. The lead is then expelled in the exhaust gases as lead halides, which are volatile at engine temperatures. The tetraethyl or tetramethyl lead approach to octane improvement has the disadvantage of contaminating the environment with toxic lead compounds. The contamination is a health hazard when a large number of motor vehicles exhaust their combustion products in relatively stagnant air.

$$(CH_3CH_2)_4Pb + BrCH_2CH_2Br + 16\ O_2 \longrightarrow PbBr_2 + 10CO_2 + 12H_2O \qquad (3\text{-}36)$$

Leaded gasoline will some day be eliminated entirely, because cars now being produced use unleaded gasoline. A reason other than direct environmental protection is that lead compounds poison the catalysts in the antipollution devices now required on nearly all cars.

Tetraethyl lead has a catalytic effect on the combustion of the gasoline. Although the meaning usually taken is that a catalyst speeds up the rate of the reaction, in this case tetraethyl lead functions as a negative catalyst and slows the rate of combustion down. Engine knock is the result of preignition combustion, too rapid a combustion (detonation), or both. The result is the "ping" in the engine called engine knock and its accompanying loss of power. The tetraethyl lead dissociates into free ethyl radicals which cause combustion to begin and proceed at a slower rate.

A satisfactory octane rating is achieved for unleaded gasolines by including a much higher proportion of highly branched alkanes and such aromatic hydrocarbons as benzene, toluene, xylenes, and cumene. That puts a bigger load on the refinery operations of isomerization, catalytic cracking, and reforming and increases the cost per gallon.

Summary

The *saturated hydrocarbons* are so called because they are already in a highly reduced state and so do not generally react with hydrogen. They consist of the *alkanes* (C_nH_{2n+2}) and the *cycloalkanes* (C_nH_{2n}). *Chain* or *skeletal isomers* are possible in increasing numbers for homologs with a greater number of carbon atoms than propane. Methane, ethane, propane, and the isomeric butanes are *gases* under ordinary conditions; the higher homologs are *liquids*. The alkanes are insoluble in water and are less dense than water.

The carbon atoms of alkanes have four *single covalent bonds* utilizing *sp^3 hybrid* orbitals with bond angles of 109.5°. Conformations of ethane and its higher homologs are formed by *rotation* about carbon-carbon *single* bonds.

The ring structure of cyclohexane is *nonplanar* to accommodate the tetrahedral interorbital bond angle of 109.5°. The rings of cyclopentane and cyclobutane are *almost planar,* but they are twisted out of plane slightly to relieve the eclipsing of their hydrogen atoms. The ring bonds of cyclobutane and cyclopropane are weaker than those of larger rings and are subject to some *ring-opening reactions*.

A *straight-chain* or *normal alkane* is named by its longest continuous chain. A *branched-chain alkane* is named as a *derivative* of its *parent alkane,* the alkane with the same number of carbon atoms as in its longest continuous chain. The positions of the branches are indicated by *locator numbers,* which are so assigned to the continuous-chain carbon atoms that they are *as low as possible.* A suitable *prefix* is used to designate the number of like branches.

The chief source of alkanes is petroleum, but other sources are coal, oil shale, and a high-pressure reaction of carbon monoxide and hydrogen. In the laboratory, alkanes may be produced by (a) *reduction of an alkene with hydrogen,* (b) *fusion of the sodium salt of an organic acid with sodium hydroxide,* and (c) *reduction of an alkyl halide.* The reduction of an alkyl halide may be accomplished by reacting the halide with magnesium to produce a *Grignard reagent* (RMgX), which is then hydrolyzed to the alkane.

The chief reactions of the alkanes involve *oxidation*—with oxygen as in *combustion* or with halogen as in *substitution of halogen for hydrogen.* The chief product of combustion of alkanes is *heat* for homes and industries and for power production.

The *halogenation of alkanes* generally results in *multiple substitution.* The *order of reactivity* of the halogens with the alkanes is $F_2 > Cl_2 > Br_2 > I_2$. The fluorination reaction is generally too violent to be controlled, and iodine does not react. The halogenation *chain reaction* proceeds by a *free-radical mechanism* after being initiated by light or by temperatures near 300°C. *Tertiary hydrogen atoms* are more easily removed from alkanes to produce a *free radical* than are *secondary hydrogen atoms,* the latter more easily than *primary hydrogen atoms,* and primary atoms more easily than *methyl hydrogen atoms.*

Cyclopropane undergoes *ring-opening reactions* with reagents like H_2, Br_2, and HBr to produce derivatives of propane. Cyclobutane opens only with H_2 at 200°C.

Petroleum is an oily residue from the remains of ancient living things. When *fractionally distilled,* it may be separated into bottled gas, natural gasoline, kerosene and jet fuel, fuel oil and diesel fuel, and lubricating oils and greases. Since the gasoline so obtained constitutes only about 20% of the total products, *cracking* and *alkylation* procedures are utilized to increase the total yield of gasoline to about 50%. Cracking of petroleum is a high-temperature reaction carried out with or without *catalysts.* In it, higher-molecular-weight hydrocarbons are broken down into molecules of lower molecular weight, many of which boil in the gasoline range (25 to 200°C).

The engine performance of gasoline for high-compression internal-combustion engines is rated by *octane number.* Branched-chain alkanes have higher octane numbers than straight-chain alkanes. Thus performance can be increased by *isomerizing straight-chain alkanes* by passing them over an $AlCl_3$ catalyst.

Alkylation consists in reaction of an alkene with an alkane over an acid catalyst to produce a branched-chain alkane boiling in the gasoline range. *Reforming* is the process of converting alkanes and cycloalkanes to *aromatic hydrocarbons* by passing them over a platinum catalyst at about 400°C. Aromatic hydrocarbons have a high octane rating.

Additives are used in gasoline to improve its performance. The chief additive is *tetraethyl lead,* which is very effective in raising the octane rating. It does, however, contaminate the atmosphere to some degree with toxic lead compounds.

Key terms

Additive: a compound added to gasoline to raise the octane number.

Alkane: an open-chain, saturated hydrocarbon with the general molecular formula C_nH_{2n+2}.

Alkene: a hydrocarbon containing a double bond; to be discussed in Chap. 4.

Alkylation: reaction of an alkene with an alkane in the presence of an acid catalyst to produce a branched alkane with higher octane number.

Aromatic hydrocarbon: benzene or one of its derivatives; to be discussed in Chap. 5.

Ball-and-stick model: a model employing balls for the atoms and sticks for the bonds holding the atoms together in a molecule. Bond lengths are not represented to scale, and bond angles are only approximately correct, but the model does show which atoms are bonded together.

Catalytic cracking: a cracking method in which petroleum fractions are passed over a fixed-bed catalyst or mixed with a fluidized catalyst.

Catalytic isomerization: the reaction of straight-chain alkanes in the presence of certain catalysts to produce branched alkanes of higher octane number.

Chain reaction: a multiple-step reaction which regenerates one of its needed reactants; it is self-propagating.

Chain-initiating reaction: the initial reaction needed to begin a chain reaction.

Chain-propagating reactions: the intermediate reactions in a chain reaction; they keep the reaction going.

Chain-terminating reactions: reactions of reaction intermediate free radicals with each other; they tend to stop a chain reaction before it goes to completion.

Cracking: a process, used in the refining of petroleum, in which longer-chain alkanes are broken into shorter-chain alkanes and alkenes.

Cycloalkane: a saturated hydrocarbon consisting of a closed chain or ring of carbon atoms; it has the general molecular formula C_nH_{2n}.

Engine knock: a "ping" in an internal-combustion engine caused by preignition combustion, too rapid combustion, or both; it results in a loss of power.

Fischer-Tropsch process: the reaction of carbon monoxide with hydrogen to produce hydrocarbons.

Fractional distillation: a distillation process yielding fractions with different boiling point ranges. It is used in the refining of petroleum.

Framework Molecular Models: models which represent the bonding, bond angles, and approximate bond lengths in molecules; they do not depict the space occupied by the orbitals.

Gasoline: a mixture of saturated hydrocarbons, ranging from 5 to 12 carbon atoms, boiling in the range of 25 to 200°C.

Grignard reagent: an alkyl magnesium halide, RMgX, used as an intermediate in the laboratory reduction of an alkyl halide to an alkane.

Halogenation: substitution of one or more halogen atoms for one or more hydrogen atoms in an alkane.

Hydrogenation-dehydrogenation catalyst: a catalyst for either the addition of molecular hydrogen to a compound or the removal of molecular hydrogen from a compound; the most common such catalysts are Ni, Pd, or Pt.

Leaded gasoline: gasoline to which tetraethyl lead or tetramethyl lead has been added to raise its octane number and improve its performance.

Octane number: a rating of the engine performance of gasoline for high-compression internal-combustion engines.

Oxidation: a reaction in which a reactant undergoes an increase in oxidation state by a complete or a partial loss of electrons; loss of hydrogen by a compound is a common type of electron loss (oxidation) in organic chemistry.

Paraffins: a name formerly given to the alkanes; it means "having little affinity."

Petroleum: an oily residue from the remains of ancient living things; it is the chief source of many alkanes and some alkenes.

Pi molecular orbital: a type of molecular orbital formed by sideways overlapping (parallel to orbital axes) of atomic orbitals; the most common type is formed by such overlap of *p* orbitals.

Platforming: a name sometimes given to reforming, since platinum is the best catalyst for the reaction.

Primary free radical: a free radical in which the unpaired electron is on a carbon atom bonded to one other carbon atom. It is less stable than a secondary or tertiary free radical but more stable than a methyl free radical.

Reduction: a reaction in which a reactant undergoes a decrease in oxidation state by a gain or partial gain of electrons. Gain of hydrogen by a compound is a common type of electron gain (reduction) in organic chemistry.

Reforming: converting cyclic and some acyclic alkanes to aromatic hydrocarbons by heating them in the presence of a hydrogenation-dehydrogenation catalyst. Sometimes called **platforming.**

Ring-opening reactions: reactions of cyclopropane and cyclobutane with certain reagents to yield propane, butane, or their derivatives.

Saturated hydrocarbons: hydrocarbons in a highly reduced state, which will not react with hydrogen. They consist of the alkanes and the cycloalkanes.

Secondary free radical: a free radical in which the unpaired electron is on a carbon atom bonded to two other carbon atoms. It is less stable than a tertiary free radical, but more stable than a primary free radical.

Sigma molecular orbital: a type of molecular orbital formed by head-on overlapping (coaxial) of atomic orbitals of any type. It is ellipsoidal in shape.

Space-filling model: a model which represents the overlapping of molecular orbitals (electron clouds) in space, as well as the geometry of the molecule.

Tertiary free radical: a free radical in which the unpaired electron is on a carbon atom bonded to three other carbon atoms. It is more stable than a secondary or primary free radical.

Thermal cracking: a cracking method in which petroleum fractions are heated to about 600°C.

Unsaturated hydrocarbons: hydrocarbons which will react with hydrogen; to be studied in Chaps. 4 and 5.

Additional problems

3-8 The following are structural formulas for the nine isomeric heptanes. Give the IUPAC name of each compound.

(a) $CH_3CH_2CH_2CH_2CH_2CH_2CH_3$

(b) $CH_3\overset{\overset{CH_3}{|}}{C}HCH_2CH_2CH_2CH_3$

(c) $CH_3CH_2\overset{\overset{CH_3}{|}}{C}HCH_2CH_2CH_3$

(d) $CH_3\underset{\underset{CH_3}{|}}{\overset{\overset{CH_3}{|}}{C}}{-}CH_2CH_2CH_3$

(e) $CH_3\overset{\overset{CH_3}{|}}{C}HCH_2\overset{\overset{CH_3}{|}}{C}HCH_3$

(f) $CH_3\overset{\overset{CH_3}{|}}{C}H\underset{\underset{CH_3}{|}}{C}HCH_2CH_3$

(g) $CH_3CH_2\underset{\underset{CH_3}{|}}{\overset{\overset{CH_3}{|}}{C}}{-}CH_2CH_3$

(h) $CH_3\underset{\underset{CH_3}{|}}{\overset{\overset{CH_3}{|}}{C}}{-}\overset{\overset{CH_3}{|}}{C}HCH_3$

(i) $CH_3CH_2\overset{\overset{\overset{\overset{CH_2}{|}}{CH_3}}{|}}{C}HCH_2CH_3$

3-9 Write the structural formulas for all of the isomeric cycloalkanes and alkylcycloalkanes of molecular formula C_6H_6 and name them. Besides cyclohexane itself, there will be alkylcyclopentane, mono- and dialkylcyclobutanes, and mono-, di-, and trialkylcyclopropanes. Remember cis and trans isomers. There are a total of 15.

3-10 Write the structural formulas for the following compounds:

(a) 2,2,3-trimethyl-1-butanol
(b) 2-chlorocyclopentanol (the locator number 1 for the —OH is understood in this name)
(c) 1,2,3-trichloropropane
(d) cyclopentylcyclohexane

3-11 Complete the following equations for the preparation of alkanes.

(a) $CH_3CH(CH_3)CH_2COO^-Na^+ + NaOH \xrightarrow{\Delta}$

(b) $(CH_3)_3CCH{=}C(CH_3)_2 + H_2 \xrightarrow{Ni}$
(c) $(CH_3)_3CMgBr + D_2O \longrightarrow$

3-12 Complete the following equations for reactions of the alkanes.

(a) $(CH_3)_3CCH(CH_3)_2 + O_2 \xrightarrow{ignite}$

(b) $CH_3CH_2CH_2CH_3 + Cl_2 \xrightarrow{light}$

(c) $(CH_3)_2CHCH_3 + Br_2 \xrightarrow{light}$

3-13 Write the equations showing the mechanism for the bromination of methane. (It is analogous to that for the chlorination.)

3-14 Give the structural formulas for the two monochlorinated products of chlorination of butane. What percent of the total monochlorinated product would be expected for each of the two position isomers? (Remember that the secondary hydrogen atoms are about three times as reactive as the primary ones.)

3-15 Complete the equations for the following reactions:

(a) $\triangle + HI \longrightarrow$

(b) $CH_3(CH_2)_6CH_3 \xrightarrow{Pt,\ 400°C}$

(c) $CH_3(CH_2)_4CH_3 \xrightarrow{AlCl_3}$

3-16 What simple chemical reaction would enable you to readily distinguish a dimethylcyclopropane from methylcyclobutane? (You would like a reaction which gives a readily discernible change such as a change in color, state, or solubility.)

3-17 Which five of the isomeric heptanes in Prob. 3-8 would you predict would have the best octane ratings?

UNSATURATED HYDROCARBONS

The *unsaturated hydrocarbons* include all of the acyclic and cyclic compounds with one or more double bonds, one or more triple bonds, or both. This chapter will deal primarily with those with one double bond (*alkenes*), those with two double bonds (*dienes*), and those with a triple bond (*alkynes*). A special type of unsaturated hydrocarbon (*aromatic*) will be considered in Chap. 5.

4-1 Alkenes

The *alkenes* are unsaturated hydrocarbons with a carbon-carbon double bond. The acyclic alkenes have the general formula C_nH_{2n}, and the cyclic alkenes have the general formula C_nH_{2n-2}. The chief commercial source of the simple alkenes is the petroleum industry, in which the alkenes are obtained as by-products from the cracking of petroleum. However, many of our more complicated natural products also contain carbon-carbon double bonds; examples are vitamin A and cholesterol. The *functional group* of an alkene is the *carbon-carbon double bond*. The simplest alkene containing that functional group is $CH_2{=}CH_2$. Its IUPAC name is *ethene*, but its older common name, *ethylene*, is still in common use. Figure 4-1 depicts two different types of models of ethene.

The carbon-carbon double bond

In the carbon-carbon double-bond functional group each carbon atom is bonded to three other atoms. Three of the available bonding electrons of each carbon atom would be required (Table 4-1). It would be expected

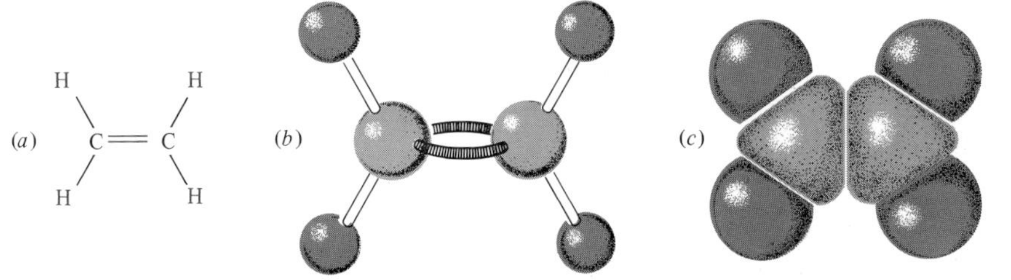

FIGURE 4-1
Ethene (ethylene): (*a*) structural formula; (*b*) ball-and-spring model; (*c*) space-filling model.

TABLE 4-1
Carbon atom in a double bond

Electron shell	K	L			
Orbitals	1*s*	2*s*	$2p_x$	$2p_y$	$2p_z$
Electrons	⇅	↑	↑	↑	↑
		Hybridize to produce three sp^2 hybrid orbitals			

that the 2*s* orbital electron (being of lower energy) would be used along with two of the 2*p* orbital electrons. Those three orbitals mix or *hybridize* to produce three almost equivalent *sp^2 hybrid orbitals*. The hybrid orbitals then pair and share electrons with hydrogen or other atoms to form three *sigma-type molecular orbitals*. The best geometry for utilization of space for maximum overlap to produce the strongest sigma bonds is *planar trigonal* with 120° bond angles (Fig. 4-2). Each carbon atom still has an electron in a 2*p* orbital not yet considered. If the two 2*p* orbitals are rotated into coplanarity, they overlap laterally (sideways) to produce the type of molecular orbital called a *pi* (π) *orbital,* which consists of a pair of shared and paired *p* electrons. The pi bond constitutes the second bond of the carbon-carbon double bond. The double bond then consists of a sigma bond and a pi bond (Fig. 4-3). The pi bond with lateral overlap is the weaker and more easily broken of the two bonds.

The distance between doubly bonded carbons is 1.34×10^{-10} m, compared with 1.54×10^{-10} m between the singly bonded carbons of an alkane. The sigma bond connecting alkenyl carbons is composed of sp^2 orbitals, so it has more *s* character and is shorter than the sigma bond of an alkyl carbon atom, which is composed of sp^3 hybrid orbitals.

The pi electrons in the pi orbital project in a plane on either side of the plane of the rest of the molecule; that is, the two planes are perpendicular to each other. That makes it relatively easy for the pi bond to be ruptured, and the electrons can then be used in making stronger bonds to other

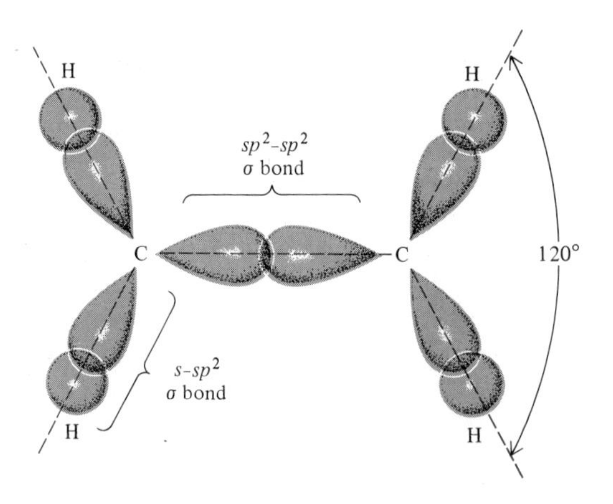

FIGURE 4-2
The sigma bonds in ethene (ethylene). Each carbon atom has an additional electron in a 2*p* orbital.

atoms. That explains why the chief reaction of the alkenes is *electrophilic addition* to the double bond. *Electrophiles* are *electron-deficient atoms and groups*. They are sometimes described as "electron-seeking" or "electron-loving." Electrophiles can break the pi bond and use the electrons to bond themselves to the carbon atoms of the former double bond with sigma bonds. Since rotation of doubly bonded carbon atoms is not possible without rupturing the pi bond, we can now understand the restricted rotation about doubly bonded carbon atoms which makes cis-trans isomerism possible.

Nomenclature

In the older system of nomenclature, the first part of the name of an alkene was the same as that of the alkane with the same number of carbon atoms. The suffix *-ane* was changed to *-ylene* to form the common name. Thus C_3H_6 was called propylene and C_4H_8 was called butylene.

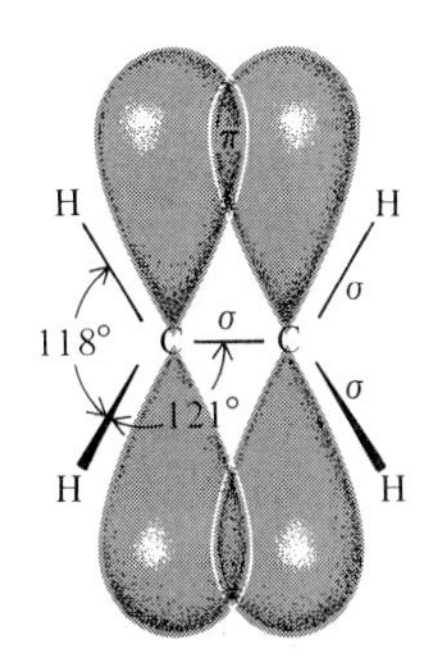

FIGURE 4-3
The sigma bonds and the pi bond in ethene (ethylene).

In the IUPAC system, the longest continuous chain containing the carbon-carbon double bond is the parent alkane. Its name is used for the stem of the name, to which the suffix *-ene* is added. Locator numbers designate on which carbon atom the double bond commences, beginning at the end which gives the lowest number for the location of the double bond. Note carefully the differences in these two examples:

$$\overset{1}{CH_2}{=}\overset{2}{C}(CH_2CH_3)\overset{3}{C}H_2\overset{4}{C}H_3$$

2-Ethyl-1-butene

$$CH_3{-}\overset{3}{C}(\overset{4}{C}H_2\overset{5}{C}H_3){=}\overset{2}{C}H{-}\overset{1}{C}H_3$$

3-Methyl-2-pentene

In Table 4-2 you will note that there are four isomeric butenes (C_4H_8). Two of them are geometric isomers: *cis*-2-butene, with the segments of the chain on the same side of the double bond, and *trans*-2-butene, with

TABLE 4-2
Names and properties of some alkenes

Molecular formula	Structural formula	IUPAC name	Melting point, °C	Boiling point, °C
C_2H_4	$CH_2{=}CH_2$	Ethene	−169	−104
C_3H_6	$CH_3CH{=}CH_2$	Propene	−185	−47
C_4H_8	$CH_3CH_2CH{=}CH_2$	1-Butene	−130	−65
C_4H_8	$CH_3C(H){=}C(H){-}CH_3$ (both H on the same side)	*cis*-2-Butene	−139	+4
C_4H_8	$(H)(H_3C)C{=}C(CH_3)(H)$ (H and CH_3 on opposite sides)	*trans*-2-Butene	−106	+1
C_4H_8	$(H_3C)_2C{=}CH_2$	2-Methylpropene	−141	−69
C_5H_{10}	$CH_3CH_2CH_2CH{=}CH_2$	1-Pentene	−138	32
C_6H_{10}	(cyclohexene ring)	Cyclohexene	−104	83

the segments of the chain on opposite sides of the double bond. The geometric isomers are not readily interconvertible because of the restricted rotation about the carbon-carbon double bond. Which of the two would you expect to have a dipole moment? Which of the two would you expect to be thermodynamically the more stable? (The more stable one generally has the higher melting point.)

Cycloalkenes are possible, but the configuration must be cis about the double bond unless there are eight or more carbon atoms in the ring. If there are fewer than eight ring atoms, the trans configuration results in excessive bond angle strain. You may wish to construct models of *cis*- and *trans*-cyclooctene to demonstrate that they can be constructed without undue strain. Then attempt to construct models of *trans*-cycloheptene and *trans*-cyclohexene. Because of this restriction, cis and trans designations for the cycloalkenes are often omitted. Is the cyclohexene shown in Table 4-2 cis or trans?

For cyclic alkenes, numbering begins with a doubly bonded carbon atom and goes *through* the double bond. The number 1 may simply be understood rather than mentioned. For example,

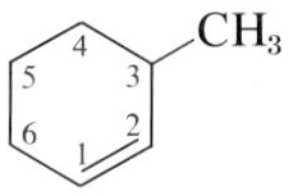

CH_3 (ring positions 1, 2, 3, 4, 5, 6)

3-Methylcyclohexene

PROBLEM 4-1 Give the IUPAC names for the compounds represented by the following structural formulas.

(a) $CH_3C(CH_3)_2CH_2\overset{CH_3}{\overset{|}{C}}{=}CH_2$

(b) $CH_3CH_2CH_2\overset{CH_3}{\overset{|}{C}}{=}CH_2$

(c) $CH_3CH_2CH_2CH_2\overset{CH_2}{\overset{\|}{C}}CH_2CH_3$

(d) 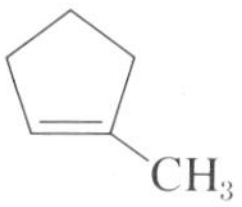 CH_3 (methylcyclopentene ring)

Solution to (a) The correct name for the compound is 2,4,4-trimethyl-1-pentene.

$$\overset{5}{C}H_3\overset{4}{\underset{CH_3}{\overset{CH_3}{C}}}{-}\overset{3}{C}H_2{-}\overset{2}{\overset{CH_3}{C}}{=}\overset{1}{C}H_2$$

PROBLEM 4-2 Give the structural formulas for the compounds with the following names: (a) 4-methylcyclopentene, (b) 2,4,4-trimethyl-2-pentene, (c) 2-ethyl-1-pentene.

Solution to (a)

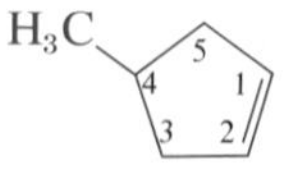

H_3C (ring positions 1, 2, 3, 4, 5)

Preparation of alkenes

Alkenes are generally prepared by *elimination reactions* in which two groups on adjacent carbon atoms of a saturated chain are removed. The result is the formation of a double bond between the two carbon atoms. The most commonly employed elimination reactions are dehydrohalogenation of alkyl halides and dehydration of alcohols.

Dehydrohalogenation of alkyl halides. In the dehydrohalogenation process, a hydrogen atom and a halogen atom are removed from adjacent carbon atoms in the presence of heat and alcoholic KOH solution. The general reaction is

$$-\underset{|}{\overset{\overset{\displaystyle X}{|}}{C}}-\underset{\underset{\displaystyle H}{|}}{\overset{|}{C}}- + KOH \xrightarrow{\Delta,\ C_2H_5OH} -\underset{|}{C}=\overset{|}{C}- + KX + HOH \qquad (4\text{-}1)$$

The reaction mechanism is shown in Eq. (4-2). The curved arrows in color represent the transfer of electrons to make or to break bonds.

$$-\underset{|}{\overset{\overset{\displaystyle Cl}{|}}{C}}-\underset{\underset{\displaystyle H \quad OH^- \; K^+}{|}}{\overset{|}{C}}- \xrightarrow{\Delta,\ C_2H_5OH} -\underset{|}{C}=\overset{|}{C}- + KCl + H_2O \qquad (4\text{-}2)$$

The mechanism shown for the general reaction is called a *concerted mechanism,* which means that bonds are broken and formed simultaneously. As the strong base OH^- abstracts the proton (H^+, hydrogen as a cation), the sp^3 orbitals convert to sp^2 orbitals and the pair of electrons which were bonding the hydrogen to carbon form a pi bond with an adjacent carbon atom. At the same time, the chlorine atom breaks loose from carbon to become a chloride ion. The departure of the chloride ion is thought to be assisted by coordination of the ion with the solvent alcohol (a process known as *solvation*).

An example of dehydrohalogenation is the reaction of 2-chlorobutane to produce 2-butene as the major product and 1-butene as the minor product:

$$H-\underset{\underset{\displaystyle H}{|}}{\overset{\overset{\displaystyle H}{|}}{C}}-\underset{\underset{\displaystyle H}{|}}{\overset{\overset{\displaystyle Cl}{|}}{C}}-\underset{\underset{\displaystyle H}{|}}{\overset{\overset{\displaystyle H}{|}}{C}}-\underset{\underset{\displaystyle H}{|}}{\overset{\overset{\displaystyle H}{|}}{C}}-H + KOH \xrightarrow{\Delta,\ C_2H_5OH} \begin{cases} CH_3CH=CHCH_3 & \text{2-Butene} \\ CH_2=CHCH_2CH_3 & \text{1-Butene} \end{cases} \qquad (4\text{-}3)$$

In this example it should be noted that a proton on an adjacent carbon atom either to the left or to the right of the halogen-bearing carbon atom

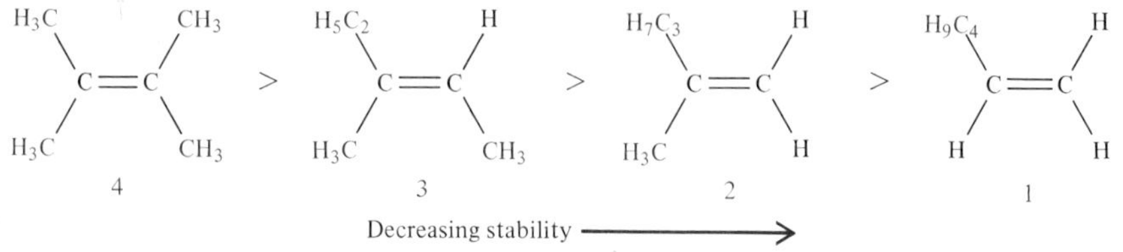

FIGURE 4-4 Relative stability of isomeric alkylated alkenes of formula C_6H_{12}.

$$(H_3C)_2C{=}C(CH_3)_2 > (H_5C_2)(H_3C)C{=}C(H)(CH_3) > (H_7C_3)(H_3C)C{=}CH_2 > (H_9C_4)HC{=}CH_2$$

4, 3, 2, 1

Decreasing stability →

can be removed. The product formed in the greatest amount follows the rule formulated in 1875 by Alexander Saytzeff at the University of Kazan in Russia. The *Saytzeff rule* states that the major product will be the one that has *the most alkyl groups attached to the resultant carbon-carbon double bond.* The rule parallels the order of thermodynamic stability of the alkenes; that is, the alkene with the most alkyl groups attached to the carbon-carbon double bond is the most stable.

Because of the relative stability of the resultant alkenes, tertiary halides dehydrohalogenate more readily than secondary halides, which dehydrohalogenate more readily than primary halides. Figure 4-4 shows four isomeric alkylated alkenes (C_6H_{12}) in order of decreasing stability. They have four, three, two, and one alkyl groups.

Dehydration of alcohols. When alcohols are heated with a catalytic amount of a strong acid like H_2SO_4 or H_3PO_4 at about 200°C, or passed over Al_2O_3 at a still higher temperature, water is eliminated and a carbon-carbon double bond is formed. The general reaction is

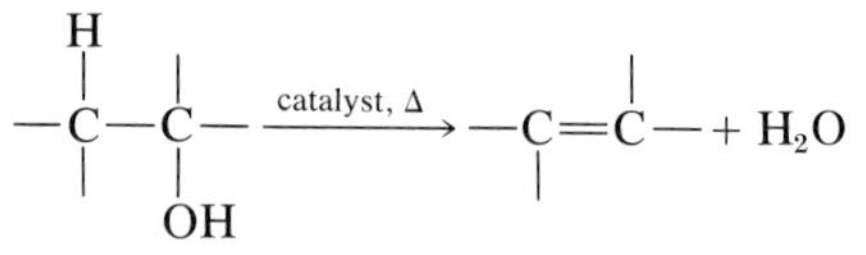

$$-\underset{|}{\overset{H}{\overset{|}{C}}}-\underset{OH}{\underset{|}{\overset{|}{C}}}- \xrightarrow{\text{catalyst, }\Delta} -\underset{|}{C}=\overset{|}{C}- + H_2O \qquad (4\text{-}4)$$

The mechanism is as follows:

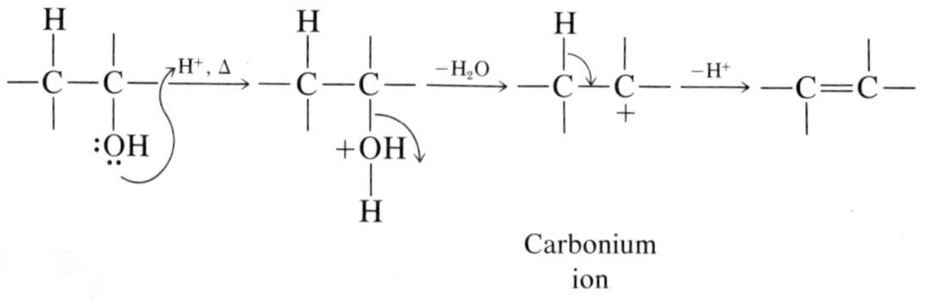

$$-\underset{|}{\overset{H}{\overset{|}{C}}}-\underset{:\ddot{O}H}{\underset{|}{\overset{|}{C}}}- \xrightarrow{H^+,\ \Delta} -\underset{|}{\overset{H}{\overset{|}{C}}}-\underset{\underset{H}{|}}{\underset{+OH}{\underset{|}{\overset{|}{C}}}}- \xrightarrow{-H_2O} -\underset{|}{\overset{H}{\overset{|}{C}}}-\underset{+}{\overset{|}{C}}- \xrightarrow{-H^+} -\underset{|}{C}=\overset{|}{C}- \qquad (4\text{-}5)$$

Carbonium ion

Note that a *carbonium ion* (an alkyl cation) is the reaction intermediate. The direction and rate of the reaction again follow the Saytzeff rule. Tertiary alcohols dehydrate more readily than secondary alcohols, which dehydrate more readily than primary alcohols. The mechanism for the dehydration of secondary and tertiary alcohols is as shown in Eq. (4-5). An example of this reaction for a secondary alcohol is the dehydration of

2-butanol to yield 2-butene as the major product and 1-butene as the minor product.

$$CH_3CH_2\underset{\underset{:\ddot{O}H}{|}}{C}HCH_3 \xrightarrow{H^+,\ \Delta} CH_3CH_2\underset{\underset{\underset{H}{|}}{+\ddot{O}H}}{C}HCH_3 \xrightarrow{-H_2O}$$

2-Butanol
(*sec*-Butyl alcohol)

$$\overset{3}{C}H_3\overset{2}{C}H_2-\overset{1}{\underset{+}{C}}H-CH_3 \longrightarrow \begin{cases} CH_3CH{=}CHCH_3 & \text{2-Butene} \\ CH_3CH_2CH{=}CH_2 & \text{1-Butene} \end{cases} + H^+ \qquad (4\text{-}6)$$

1-Methylpropyl cation
(*sec*-Butyl cation)

The dehydration of primary alcohols may take place partly by the mechanism of Eq. (4-6) and partly by a concerted elimination reaction similar to the dehydrohalogenation mechanism shown in Eq. (4-2). The general mechanism for a primary alcohol is

$$\underset{|}{\overset{H}{\overset{|}{-C}}}-\underset{\underset{:\ddot{O}H}{|}}{\overset{|}{C}}-H \xrightarrow{H^+} \underset{|}{\overset{H}{\overset{|}{-C}}}-\underset{\underset{\underset{H}{|}}{+\ddot{O}H}}{\overset{|}{C}}- \xrightarrow{\Delta} -\underset{|}{C}{=}\overset{H}{\overset{|}{C}}- + H_3O^+ \qquad (4\text{-}7)$$

An example is the dehydration of 1-propanol to yield propene:

$$CH_3-\underset{\underset{H}{|}}{\overset{H}{\overset{|}{C}}}-\underset{\underset{:\ddot{O}H}{|}}{\overset{H}{\overset{|}{C}}}-H \xrightarrow{H^+} CH_3-\underset{\underset{H}{|}}{\overset{H}{\overset{|}{C}}}-\underset{\underset{\underset{H}{|}}{+\ddot{O}H}}{\overset{H}{\overset{|}{C}}}-H \xrightarrow{\Delta} CH_3-\underset{\underset{H}{|}}{C}{=}\overset{H}{\overset{|}{C}}-H + H_3O^+ \qquad (4\text{-}8)$$

1-Propanol — Propene

PROBLEM 4-3 Write equations for the preparation of alkenes from the following starting materials. Indicate the conditions required. If there is more than one product, indicate which alkene is the major product.

(a) $CH_3CH_2CH_2Br$

(b) $CH_3CHBrCH_3$

(c) $CH_3\underset{\underset{OH}{|}}{\overset{CH_3}{\overset{|}{C}}}-CH_3$

(d) Cyclohexane ring–OH

Solution to (a)

$$\underset{\text{1-Bromopropane}}{CH_3-\underset{\substack{|\\H}}{\overset{\substack{H\\|}}{C}}-\underset{\substack{|\\Br}}{\overset{\substack{H\\|}}{C}}-H} + KOH \xrightarrow{\Delta,\ C_2H_5OH} \underset{\text{Propene}}{CH_3CH{=}CH_2} + KBr + HOH$$

Electrophilic addition reactions of alkenes

As was mentioned earlier, the pi bond is the reactive part of an alkene. *Electrophiles,* which are electron-deficient atoms or groups such as H^+ or X^+, can break the pi bond. They use the electrons to bond themselves to the carbon atoms of the former double bond with new sigma bonds. The reagent may be either a symmetrical one like X_2 or an unsymmetrical one like HX. Using an unsymmetrical reagent with a general symbol AB as an example, *electrophilic addition* to an alkene may be represented as follows.

$$>C{=}C< + A{-}B \xrightarrow{\text{addition}} -\overset{\substack{A\\|}}{\underset{|}{C}}-\overset{|}{\underset{\substack{|\\B}}{C}}- \qquad (4\text{-}9)$$

Figure 4-5 shows an orbital representation of this general reaction.

Alkenes readily add hydrogen, halogen, hydrohalogen, ozone, and even water with the help of an acid catalyst. Addition of hydrogen has already been mentioned as one method of preparing an alkane from an alkene.

Addition of halogen. The general reaction for the addition of halogen to an alkene is

$$>C{=}C< + X_2 \longrightarrow -\overset{|}{\underset{\substack{|\\X}}{C}}-\overset{|}{\underset{\substack{|\\X}}{C}}- \qquad (4\text{-}10)$$

FIGURE 4-5 An orbital representation of electrophilic addition to an alkene.

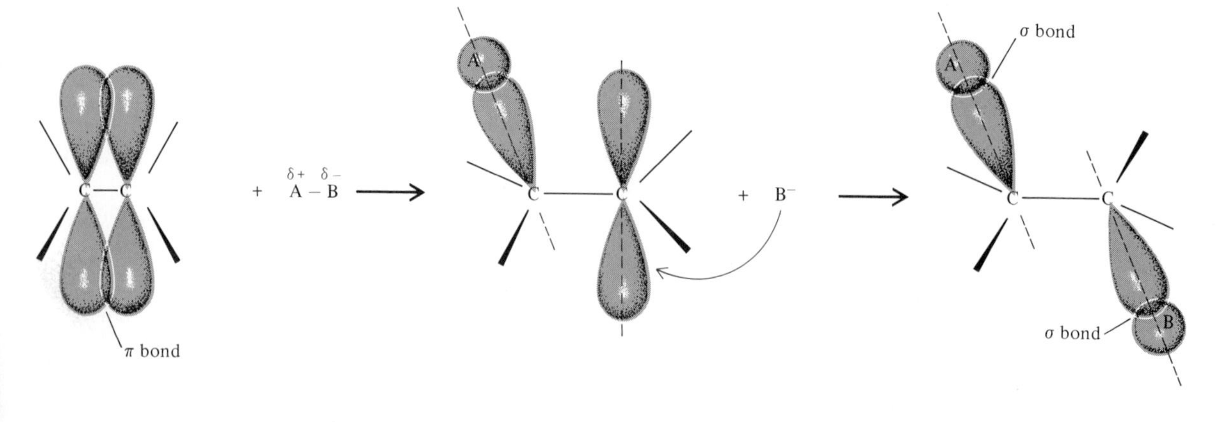

A somewhat oversimplified description of the mechanism is given in Eq. (4-11). Remember that the colored curved arrows represent the transfer of electrons to make or break bonds.

$$>C{=}C< + \overset{\delta+}{X}{-}\overset{\delta-}{X} \longrightarrow -\overset{|}{\underset{+}{C}}-\overset{|}{\underset{\underset{X}{|}}{C}}- + X^- \longrightarrow -\overset{|}{\underset{\underset{X}{|}}{C}}-\overset{|}{\underset{\underset{X}{|}}{C}}- \qquad (4\text{-}11)$$

Note that the normally nonpolar halogen molecule acquires an *induced polarization* in the proximity of the negative pi electron cloud of the alkene. That culminates in breaking of the halogen-halogen bond and formation of a carbon-halogen bond with what were the pi electrons. The resulting carbonium ion is then attracted to and reacts with the negative halogen ion. An example of the reaction is the addition of bromine to propene to yield 1,2-dibromopropane:

$$CH_3CH{=}CH_2 + Br_2 \longrightarrow CH_3{-}\underset{+}{C}HCH_2Br + Br^- \longrightarrow CH_3\underset{\underset{Br}{|}}{C}HCH_2Br \qquad (4\text{-}12)$$

The reality of a carbonium-ion intermediate has been demonstrated by the addition of some negative ion, such as chloride ion or nitrate ion, to compete with the bromide ion for the carbonium ion. Incorporation of the competitive ions in products would seem to indicate that a carbonium ion must indeed have been present, since solutions of such ions would not react with the alkene itself.

$$CH_3{-}\underset{+}{C}HCH_2Br \begin{cases} \xrightarrow{Cl^-} CH_3\underset{\underset{Cl}{|}}{C}HCH_2Br \\ \xrightarrow{ONO_2^-} CH_3\underset{\underset{ONO_2}{|}}{C}HCH_2Br \end{cases} \qquad (4\text{-}13)$$

If water is present, it will react with the intermediate carbonium ion to form a product called a *halohydrin,* a halogen-substituted alcohol:

$$\underset{\text{Carbonium-ion intermediate}}{\underset{+}{C}HCH_2Br} + HOH \longrightarrow CH_3\underset{\underset{\underset{\underset{H}{|}}{+OH}}{|}}{C}HCH_2Br \longrightarrow \underset{\text{1-Bromo-2-propanol (Propylene bromohydrin)}}{CH_3\underset{\underset{OH}{|}}{C}HCH_2Br} + H^+ \qquad (4\text{-}14)$$

Chlorine adds similarly to bromine. Iodine is not reactive enough for practical results, and fluorine is too reactive.

Addition of hydrohalogen. The general equation for the addition of hydrohalogen to an alkene is

$$>C{=}C< + HX \longrightarrow -\underset{|}{\overset{\overset{H}{|}}{C}}-\underset{\underset{X}{|}}{\overset{|}{C}}- \qquad (4\text{-}15)$$

The reaction mechanism is similar to that for addition of a halogen:

$$\rangle C{=}C\langle + H{-}X \longrightarrow -\overset{H}{\underset{|}{\overset{|}{C}}}-\underset{+}{\overset{|}{C}}- + X^- \longrightarrow -\overset{H}{\underset{|}{\overset{|}{C}}}-\underset{X}{\underset{|}{\overset{|}{C}}}- \qquad (4\text{-}16)$$

A specific example of the addition of hydrohalogen is the reaction of propene with hydrogen chloride:

$$CH_3CH{=}CH_2 + \overset{\delta+}{H}-\overset{\delta-}{Cl} \longrightarrow \begin{cases} CH_3-\underset{+}{C}HCH_3 + Cl^- \longrightarrow CH_3\underset{Cl}{\underset{|}{C}}HCH_3 \text{ (2-Chloropropane)} \\ CH_3CH_2CH_2^+ + Cl^- \longrightarrow CH_3CH_2CH_2Cl \text{ (1-Chloropropane, crossed out)} \end{cases} \qquad (4\text{-}17)$$

The major product is 2-chloropropane. Little, if any, 1-chloropropane is actually formed.

MARKOVNIKOV'S RULE. The direction that the addition of HX takes in forming the major product follows the *rule* established by Vladimir Markovnikov, University of Kazan, in 1869. One way to express *Markovnikov's rule* is to say that usually *the hydrogen* of the reagent *adds to the carbon of the double bond which is already bonded to the larger number of hydrogen atoms*. Markovnikov addition is explained by the order of relative stability of the intermediate carbonium ions. The order of stability is that, relative to their parent alkanes, tertiary carbonium ions are more stable than secondary carbonium ions, which in turn are more stable than primary carbonium ions. The order of stability of the carbonium ions can be expressed as $3° > 2° > 1° > CH_3^+$. It should be noted that the order of stability of the carbonium ions is the same as that of carbon free radicals.

The observed product is, then, the product resulting from the direction of addition which produces the most stable intermediate carbonium ion. In the example of the addition of HCl to propene, the intermediate 2-propyl cation, which is a secondary carbonium ion, is more stable than the 1-propyl cation, which is a primary carbonium ion. Hence the product is 2-chloropropane.

The observed order of stability of carbonium ions is explained in part by the slightly positive *inductive effect* of alkyl groups, which results in some *delocalization of charge* with attendant stabilization. Hence the greater the number of alkyl groups attached to a carbonium ion the more stable the ion. If R represents an alkyl group, then

$$\underset{\text{Tertiary}}{R\rightleftarrows\overset{R}{\underset{R}{C}}+} > \underset{\text{Secondary}}{R\rightleftarrows\overset{R}{\underset{H}{C}}+} > \underset{\text{Primary}}{R\rightleftarrows\overset{H}{\underset{H}{C}}+} > \underset{\text{Methyl}}{H-\overset{H}{\underset{H}{C}}+}$$

The colored arrows represent the direction of shifting of electrons toward the positive carbon atom.

A *positive inductive effect* is a type of *electron release* in which there is a shift in electron cloud density *away from* a given atom. A *negative inductive effect* is a type of *electron gain* in which electron cloud density about a given atom is *increased*. (Electronegative atoms such as halogen, nitrogen, and oxygen will usually have negative inductive effects.) It is a fact of physics that delocalizing an electric charge by spreading it over a larger surface area stabilizes it, whereas concentrating it in a small area, like a point, makes it unstable and less likely to be retained.

ANTI-MARKOVNIKOV ADDITION OF HBr. In the presence of an organic peroxide catalyst, the addition of hydrogen bromide to alkenes differs from that of the other hydrogen halides. Under that condition, hydrogen bromide can be made to add in a manner *contrary* to that expected from Markovnikov's rule. The reaction mechanism was elucidated in 1933 by M. S. Kharasch and F. W. Mayo at the University of Chicago. The reaction intermediate is a free radical rather than a carbonium ion; the mechanism is a chain reaction similar to the halogenation of an alkane.

Step 1 $$\underset{\text{Peroxide}}{R{-}O{-}O{-}R} \longrightarrow 2RO\cdot \qquad (4\text{-}18)$$

Step 2 $$RO\cdot + HBr \longrightarrow ROH + Br\cdot \qquad (4\text{-}19)$$

Step 3 $$Br\cdot + CH_3CH{=}CH_2 \longrightarrow CH_3\dot{C}HCH_2Br \qquad (4\text{-}20)$$

Step 4 $$CH_3\dot{C}HCH_2Br + HBr \longrightarrow CH_3CH_2CH_2Br + Br\cdot \qquad (4\text{-}21)$$

In step 3 the direction of addition of the Br· atom is such as to give the most stable intermediate free radical. In this case, the secondary free radical, $CH_3\dot{C}HCH_2Br$, is formed. The other possibility, the primary free radical, $CH_3CH(Br)\dot{C}H_2$, is not formed. (Remember from Sec. 3-4 that the order of stability of free radicals is 3° > 2° > 1°.) The mechanism is somewhat similar to that for the halogenation of an alkane, but the halogen atom attacks the pi electron bond, which is easier to rupture than the C—H sigma bond.

Addition of sulfuric acid. Alkenes readily add cold concentrated sulfuric acid at room temperatures or below to form water-soluble alkyl hydrogen sulfate esters. The addition follows Markovnikov's rule. The general reaction is

$$>C{=}C< + H_2SO_4 \longrightarrow -\underset{|}{\overset{H}{\overset{|}{C}}}-\underset{OSO_2OH}{\underset{|}{\overset{|}{C}}}- \qquad (4\text{-}22)$$

The more alkyl groups there are attached to the carbon-carbon double bond, the more reactive is the alkene toward the addition of sulfuric acid. The order of reactivity is related to the stability of the intermediate carbonium ions (3° > 2° > 1°). It applies also to the addition of the other electrophilic reagents.

An example of the addition of sulfuric acid is

$$\underset{\substack{\text{Propene}\\\text{(Propylene)}}}{CH_3CH{=}CH_2} + HOSO_2OH \longrightarrow \underset{\substack{\text{1-Methylethyl hydrogen sulfate}\\\text{(Isopropyl hydrogen sulfate)}}}{CH_3\underset{\displaystyle OSO_2OH}{\underset{|}{C}}HCH_3} \qquad (4\text{-}23)$$

Warming an alkyl hydrogen sulfate ester in aqueous solution hydrolyzes the ester to an alcohol and liberates sulfuric acid:

$$CH_3\underset{\displaystyle OSO_2OH}{\underset{|}{C}}HCH_3 + 2HOH \longrightarrow \underset{\substack{\text{2-Propanol}\\\text{(Isopropyl alcohol)}}}{CH_3\underset{\displaystyle OH}{\underset{|}{C}}HCH_3} + H_3O^+\ OSO_2OH^- \qquad (4\text{-}24)$$

Addition of water. In the presence of an acid catalyst, an alkene will add water:

$$\gt C{=}C\lt + HOH \xrightarrow{H^+} -\overset{\displaystyle H}{\overset{|}{\underset{|}{C}}}-\overset{|}{\underset{\underset{\displaystyle OH}{|}}{C}}- \qquad (4\text{-}25)$$

This addition follows Markovnikov's rule. Thus, except for ethanol (ethyl alcohol) from ethene (ethylene), the alcohols produced in this way are secondary or tertiary alcohols. The mechanism is

$$\underset{\text{Propene}}{CH_3CH{=}CH_2} + H^+ \longrightarrow \underset{\substack{\text{1-Methylethyl cation}\\\text{(a secondary carbonium ion)}}}{CH_3-\underset{+}{C}HCH_3} \qquad (4\text{-}26)$$

$$CH_3-\underset{+}{C}HCH_3 + HOH \longrightarrow CH_3-\underset{\underset{\displaystyle H}{\underset{|}{{}^+OH}}}{\underset{|}{C}}HCH_3 \xrightarrow{H_2O} \underset{\substack{\text{2-Propanol}\\\text{(Isopropyl alcohol)}}}{CH_3\underset{\displaystyle OH}{\underset{|}{C}}HCH_3} + H_3O^+ \qquad (4\text{-}27)$$

This is an important industrial process for manufacturing secondary and tertiary alcohols.

Other reactions of alkenes

Allylic substitution by chlorine or bromine. Although propene adds chlorine at room temperature to give 1,2-dichloropropane,

$$CH_3CH{=}CH_2 + Cl_2 \longrightarrow CH_3\underset{\displaystyle Cl}{\underset{|}{C}}H-\underset{\displaystyle Cl}{\underset{|}{C}}H_2 \qquad (4\text{-}28)$$

at high temperatures (about 600°C) substitution of chlorine for hydrogen takes place. It occurs by a free-radical mechanism and produces 3-chloro-1-propene, commonly called allyl chloride.

$$\underset{\substack{\text{Propene}\\ \text{(Propylene)}}}{CH_3CH{=}CH_2} + Cl_2 \xrightarrow{600^\circ C} \underset{\substack{\text{3-Chloro-1-propene}\\ \text{(Allyl chloride)}}}{ClCH_2CH{=}CH_2} + HCl \qquad (4\text{-}29)$$

As you may recall from Table 2-6, the $—CH_2CH{=}CH_2$ group is commonly called the *allyl* group. In this group, or in an alkene containing it, the *allylic hydrogen atoms* are those on the third carbon atom when numbering is as follows:

$$\overset{3}{C}H_3—\overset{2}{C}H{=}\overset{1}{C}H_2 \qquad \text{or} \qquad R—\overset{3}{C}H_2—\overset{2}{C}H{=}\overset{1}{C}H_2$$

In allylic substitution, allylic hydrogen atoms are replaced by halogen.

Bromine reacts similarly with propene, or with any other alkene containing allylic hydrogen atoms.

THE ALLYLIC FREE RADICAL. The mechanism for allylic substitution involves the formation of an intermediate *allylic free radical,* $\cdot CH_2—CH{=}CH_2$. We need to understand the nature of this free radical before we consider the mechanism for allylic substitution.

Allylic free radicals are even more stable and more easily formed than tertiary (and therefore secondary) free radicals. That is because the unpaired electron in the free radical is delocalized by *resonance.* Resonance exists whenever two or more electronic arrangements for a given species (molecule, ion, or radical) are possible. The same arrangement of atoms is maintained in all the various electronic arrangements; only the electrons are moved. In general, the greater the number of such possible arrangements, the greater the resonance and the greater the stability of the species. The actual electronic arrangement is considered to be a *resonance hybrid* of all of the possibilities. A double-headed arrow between two electronic forms denotes that the forms are contributors to a state of resonance.

$$H—\underset{\cdot}{\overset{H}{\overset{|}{C}}}—\overset{H}{\overset{|}{C}}{=}\overset{H}{\overset{|}{C}}—H \xleftrightarrow[\text{Resonance}]{} H—\overset{H}{\overset{|}{C}}{=}\overset{H}{\overset{|}{C}}—\underset{\cdot}{\overset{H}{\overset{|}{C}}}—H \qquad (4\text{-}30)$$

This is *not* a representation of equilibrium between two different forms.

The resonance hybrid has properties derived from all of its contributing electronic arrangements. It is a definite entity of itself, just as the mule is a hybrid of the horse and the donkey. The mule is not fluctuating between being a horse or a donkey; it is a real animal with some of the properties of both the horse and the donkey.

In the formula for the resonance hybrid,

$$\mathrm{H-\overset{\overset{H}{|}}{C}_1 \overset{\cdot}{\underbrace{=\!\!\!-\,\overset{\overset{H}{|}}{C}_2 =\!\!\!-}} \overset{\overset{H}{|}}{C}_3 - H}$$

Resonance hybrid for the allylic free radical

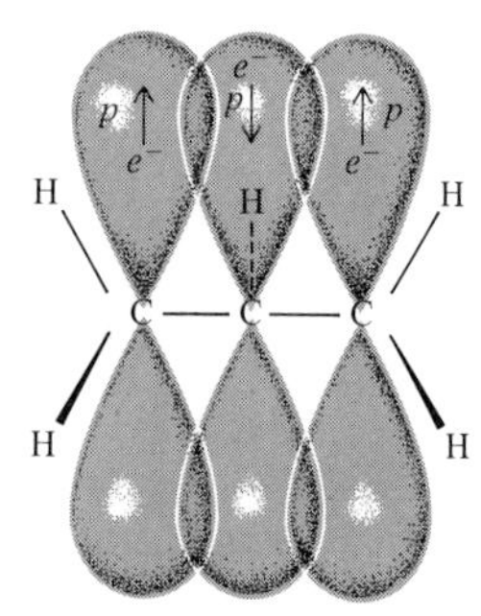

FIGURE 4-6 Molecular orbital representation of the allylic free radical.

the dashed lines for bonds signify that the bonds are intermediate between double and single bonds. The unpaired electron is represented as being delocalized over both carbon atoms 1 and 3. Resonance is an important concept which will be encountered frequently.

The concept of the allylic free radical as a resonance hybrid is clarified somewhat by the molecular orbital representation (Fig. 4-6). The three carbon atoms and five hydrogen atoms are bonded together with sigma bonds utilizing sp^2 hybrid orbitals on the carbon atoms. Each carbon atom then has a remaining electron in a p orbital at right angles to the plane of the molecule, so that the p orbitals can overlap sideways with each other. The p electrons of the carbon atoms on either end can be considered to have their spins paired with the spin of the p electron on the center carbon atom. Thus the unpaired electron can be considered to be delocalized.

MECHANISM FOR ALLYLIC SUBSTITUTION. The mechanism for the allylic substitution of chlorine in propene is shown in Eqs. (4-31) through (4-33). In step 2, Eq. (4-32), two competing reactions are possible. The chlorine atom as a type of free radical can add to the pi bond of the carbon-carbon double bond (step 2a). At the rather high temperature of 600°C, however, the reaction is reversible to a considerable extent (denoted by the double arrows in the equation). The competing reaction, removal of an allylic hydrogen atom by the chlorine atom to produce HCl and an allylic free radical (step 2b), is favored.

Step 1

$$\underset{\text{Molecular chlorine}}{Cl_2} \xrightarrow{\Delta} \underset{\text{Atomic chlorine}}{2Cl\cdot} \tag{4-31}$$

Step 2a

$$CH_3CH{=}CH_2 + Cl\cdot \overset{600°C}{\rightleftharpoons} CH_3{-}\underset{\underset{Cl}{|}}{C}H\dot{C}H_2 \tag{4-32a}$$

Step 2b

$$CH_3CH{=}CH_2 + Cl\cdot \xrightarrow{600°C}$$

$$\underbrace{\dot{C}H_2CH{=}CH_2 \longleftrightarrow CH_2{=}CH\dot{C}H_2}_{} + HCl \tag{4-32b}$$

$$\mathrm{H-\overset{\overset{H}{|}}{C} \overset{\cdot}{\underbrace{=\!\!\!-\,\overset{\overset{H}{|}}{C} =\!\!\!-}} \overset{\overset{H}{|}}{C} - H}$$

Allylic free radical (resonance hybrid)

Step 3a

$$CH_3-\underset{\displaystyle Cl}{\underset{|}{\dot{C}}}HCH_2 \overset{Cl_2}{\rightleftharpoons} CH_3-\underset{\displaystyle Cl}{\underset{|}{C}}H-\underset{\displaystyle Cl}{\underset{|}{C}}H_2 + Cl\cdot \quad (4\text{-}33a)$$

Step 3b

$$\dot{C}H_2CH=CH_2 \longleftrightarrow CH_2=CH\dot{C}H_2 \xrightarrow{Cl_2} \underset{\displaystyle Cl}{\underset{|}{C}}H_2CH=CH_2 + Cl\cdot \quad (4\text{-}33b)$$

Allyl chloride

As noted above, at the temperature of the reaction, steps 2a and 3a are reversible and the competing reaction, step 2b, is strongly favored because step 3b is not reversible. Thus the reaction proceeds through steps 1, 2b, and 3b.

Allylic substitution is an important step in the industrial synthesis of the trihydric alcohol commonly known as glycerol or glycerin, CH_2OH-$CHOHCH_2OH$.

$$CH_3CH=CH_2 + Cl_2 \xrightarrow{600°C} ClCH_2CH=CH_2 + HCl \quad (4\text{-}29)$$

Allyl chloride

$$\underset{\text{Allyl chloride}}{ClCH_2CH=CH_2} \xrightarrow{Na_2CO_3/H_2O} \underset{\text{Allyl alcohol}}{HOCH_2CH=CH_2} \quad (4\text{-}34)$$

$$\underset{\text{Allyl alcohol}}{HOCH_2CH=CH_2} \xrightarrow{Cl_2/H_2O} HOCH_2\underset{\displaystyle OH}{\underset{|}{C}}HCH_2Cl \xrightarrow{Na_2CO_3/H_2O} \underset{\displaystyle OH}{\underset{|}{C}}H_2-\underset{\displaystyle OH}{\underset{|}{C}}H-\underset{\displaystyle OH}{\underset{|}{C}}H_2 \quad (4\text{-}35)$$

Glycerol (Glycerin)

The common names, allyl alcohol and glycerol or glycerin, are still widely used. The IUPAC names for the alcohols are 2-propen-1-ol and 1,2,3-propanetriol, respectively.

Oxidation of alkenes. The double bond is also a vulnerable functional group for strong oxidizing agents such as ozone and potassium permanganate. The reactions are complex and may rupture the molecule. Examples of the reaction of ozone with alkenes are

$$\underset{\text{1-Butene}}{CH_3CH_2CH=CH_2} + \underset{\text{Ozone}}{O_3} \longrightarrow \underset{\text{An ozonide}}{CH_3CH_2C(H)\langle O, O{-}O \rangle CH_2} \xrightarrow{Zn/H_2O}$$

$$\underset{\text{Propanal}}{CH_3CH_2\overset{\displaystyle H}{\overset{|}{C}}=O} + \underset{\text{Methanal}}{O=\overset{\displaystyle H}{\overset{|}{C}}H} \quad (4\text{-}36)$$

$$CH_3CH{=}CHCH_3 + O_3 \longrightarrow CH_3\overset{H}{C}\underset{O-O}{\overset{O}{\diagup\ \diagdown}}\overset{H}{C}CH_3 \xrightarrow{Zn/H_2O} 2CH_3\overset{H}{\overset{|}{C}}{=}O \qquad (4\text{-}37)$$

Butene — Ozone — An ozonide — Ethanal

In the older system of nomenclature, methanal was called formaldehyde, ethanal was called acetaldehyde, and propanal was called propionaldehyde. Those names are still used very widely.

Rubber has many double bonds, so atmospheric ozone will cause rubber to crack and deteriorate after a time. The effect was used as an indication of ozone in Los Angeles smog at one time.

Cold, dilute aqueous $KMnO_4$ with an alkene produces some dihydric alcohol, called a diol in the IUPAC system or a glycol in the older system of nomenclature. The glycol itself may be oxidized further, so the reaction is of limited preparative use. But it does have use as a qualitative test for a carbon-carbon double bond. Alkenes will essentially decolorize cold, dilute aqueous $KMnO_4$ solutions. The reaction is called the *Baeyer test*. Alone it is not conclusive, because other easily oxidized compounds also will react. In a positive test, the bright, dark purple color of the reagent fades and disappears as the $KMnO_4$ is reduced. The brown precipitate of MnO_2 which is formed is almost insignificant when a dilute solution of $KMnO_4$ is used.

$$4H_2O + 3CH_2{=}CH_2 + 2KMnO_4 \longrightarrow 3\underset{OH}{\underset{|}{C}}H_2{-}\underset{OH}{\underset{|}{C}}H_2 + 2MnO_2(s) + 2KOH \qquad (4\text{-}38)$$

Since the Baeyer test for an alkene is fallible, it is supplemented by a test using bromine dissolved in carbon tetrachloride. An alkene will add bromine to form a colorless alkyl halide, so that the reddish-brown color of the reagent disappears. Although alkanes also react with bromine in the presence of light, they react much more slowly and with the production of HBr.

Like the alkanes, the alkenes are very flammable and readily undergo combustion in the presence of oxygen.

$$\underset{\text{1-Hexene}}{CH_3CH_2CH_2CH_2CH{=}CH_2} + 9O_2 \longrightarrow 6CO_2 + 6H_2O \qquad (4\text{-}39)$$

PROBLEM 4-4 Write the structural formulas for the products of the reaction of 2-methyl-2-butene with each of the following. (a) HCl, (b) H_2SO_4, (c) Br_2/CCl_4, (d) Cl_2/H_2O, (e) HBr/ROOR, (f) H_2O/H^+ catalyst, (g) $Br_2/NaCl(aq)$.

Solution to (a)

$$CH_3{-}\underset{Cl}{\underset{|}{\overset{CH_3}{\overset{|}{C}}}}{-}\underset{H}{\underset{|}{\overset{H}{\overset{|}{C}}}}{-}CH_3$$

This addition follows Markovnikov's rule.

PROBLEM 4-5 Write the equations for the following reactions.

(a) 1-methylcyclohexene + HBr
(b) 1-methylcyclohexene + HBr/ROOR
(c) 1-bromobutane + KOH, alcohol, heat
(d) 1-butene + HBr
(e) isobutylene (2-methylpropene) + H_2O/H^+ catalyst
(f) 3-hexene + Cl_2, 600°C
(g) combustion of cyclohexene
(h) cyclohexene + Cl_2, 600°C

Solution to (a)

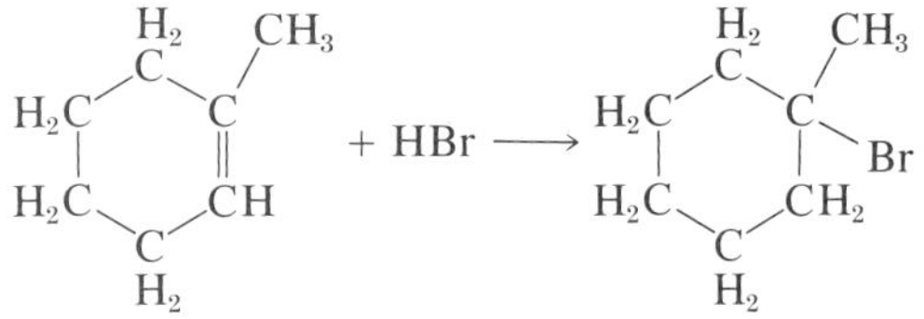

1-methylcyclohexene (ring: H_2C, $C(H_2)$, $C(CH_3)$=CH, $C(H_2)$, H_2C) + HBr ⟶ 1-bromo-1-methylcyclohexane (ring: H_2C, $C(H_2)$, C(CH_3)(Br), CH_2, $C(H_2)$, H_2C)

PROBLEM 4-6 What *simple chemical test* or reagent will *readily* distinguish between members of each of the following pairs of compounds?

(a) $CH_3CH_2CH_3$ and $CH_3CH{=}CH_2$
(b) $CH_3CH{=}CH_2$ and cyclopropane
(c) $CH_3CH_2CH_2OH$ and $CH_2{=}CHCH_2OH$

Solution to (a) Cold, dilute $KMnO_4$ solution will distinguish between propane and propene. That is the Baeyer test. The $KMnO_4$ solution will not react with propane, but it will react with propene according to the following reaction:

$$3CH_3CH{=}CH_2 + 4H_2O + 2KMnO_4 \longrightarrow 3CH_3\underset{\displaystyle OH}{\underset{|}{CH}}{-}\underset{\displaystyle OH}{\underset{|}{CH_2}} + 2MnO_2(s) + 2KOH$$

The observed change will be the disappearance of the dark purple color of the $KMnO_4$.

4-2 Dienes

Dienes are compounds having two carbon-carbon double bonds. There are three types of dienes:

$CH_2{=}C{=}CHCH_2CH_3$ 1,2-Pentadiene	Cumulated double bonds
$CH_2{=}CHCH{=}CHCH_3$ 1,3-Pentadiene	Conjugated double bonds
$CH_2{=}CHCH_2CH{=}CH_2$ 1,4-Pentadiene	Isolated double bonds

Compounds that contain cumulated double bonds are known, but they are not very common. Compounds with isolated double bonds react like compounds with only one double bond, except that more reagent is used. However, compounds with conjugated double bonds react somewhat differently. All of the above pentadienes will add 2 mol of hydrogen to produce pentane. However, the heat of hydrogenation is least for 1,3-pentadiene, which indicates it to be the most stable isomer thermodynamically.

1,4 Addition reactions of conjugated dienes

If the conjugated diene, 1,3-butadiene, is reacted with 1 mol of HBr, two products are obtained.

$$CH_2{=}CH{-}CH{=}CH_2 + HBr \xrightarrow[40°C]{1,4\ \text{addition}} \underset{\displaystyle H}{\underset{|}{CH_2}}{-}CH{=}CH{-}\underset{\displaystyle Br}{\underset{|}{CH_2}} \quad \text{Major product (80\%)} \tag{4-40}$$

$$CH_2{=}CH{-}CH{=}CH_2 + HBr \xrightarrow[40°C]{1,2\ \text{addition}} \underset{\displaystyle H}{\underset{|}{CH_2}}{-}\underset{\displaystyle Br}{\underset{|}{CH}}{-}CH{=}CH_2 \quad \text{Minor product (20\%)}$$

The type of addition giving the major product is referred to as *1,4 addition* to indicate the carbon atoms to which the reagent bonds. The direction of the initial addition of the proton from HBr is such as to give the most stable intermediate carbonium ion (secondary rather than primary).

$$CH_2{=}CH{-}CH{=}CH_2 + HBr \longrightarrow \underset{\displaystyle H}{\underset{|}{CH_2}}{-}\underset{+}{CH}{-}CH{=}CH_2 + Br^- \tag{4-41}$$

Reaction of this carbonium ion with the bromide ion produces the observed product of 1,2 addition (the minor product, but a significant one). However, note that the intermediate carbonium ion still retains one double bond with its pi electrons. The pi electrons could shift to produce another electronic form.

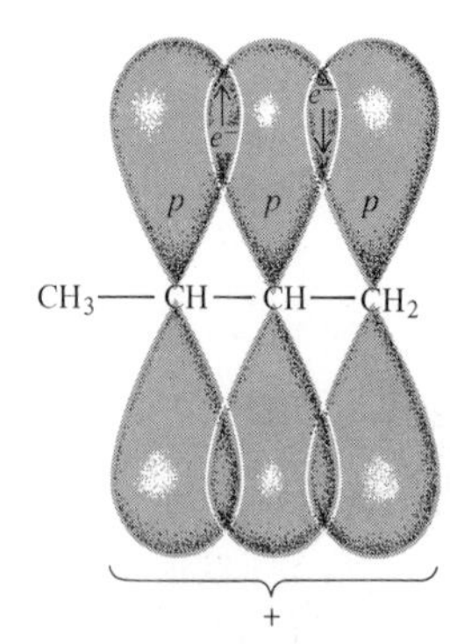

FIGURE 4-7
Molecular orbital representation of the allylic carbonium ion.

$$CH_3{-}\underset{+}{CH}{-}CH{=}CH_2 \longleftrightarrow CH_3{-}CH{=}CH{-}\underset{+}{CH_2} \tag{4-42}$$

Allylic carbonium ion

Note the use of the double-headed, single arrow to denote resonance. This carbonium ion is considered a resonance hybrid of the two electronic forms, and *not* to be rapidly changing from one to the other.

The resonance hybrid can be represented by the single structure

$$CH_3{-}\underbrace{CH{\overset{\cdot\cdot}{-}}CH{\overset{\cdot\cdot}{-}}CH_2}_{+}$$

Allylic carbonium ion

The dashed lines signify that the bonds are in between double and single bonds. All three carbon atoms involved have some pi-type overlap of their p orbitals; thus there is one pair of shared electrons for the three overlapped p orbitals. The positive charge is delocalized over both the second and the fourth carbon atoms. (See Fig. 4-7.) The result is an especially stable type of carbonium ion called an *allylic carbonium ion*. Although both the second and the fourth carbon atoms bear some positive charge, the product of 1,4 addition is most stable and so predominates. This is an application of the Saytzeff rule that the major product will be the one with the most alkyl groups attached to the carbon-carbon double bond.

PROBLEM 4-7 Write the equations for the reaction of 1 mol of each of the following reagents at about 40°C with 1,3-cyclohexadiene, [1,3-cyclohexadiene structure]. (a) Br_2, (b) HBr, (c) H_2/Ni.

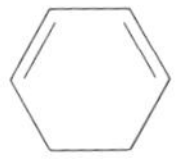

Solution to (a) The predominant reaction is 1,4 addition:

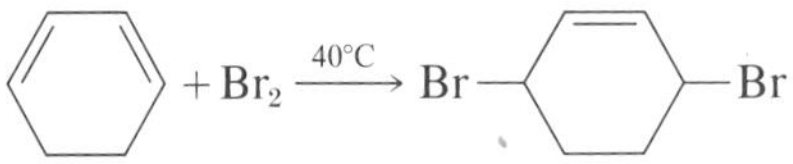

$$\text{1,3-cyclohexadiene} + Br_2 \xrightarrow{40°C} \text{Br–(cyclohexene)–Br (3,6-dibromocyclohexene)}$$

PROBLEM 4-8 Predict the chief product of the following reaction.

$$\text{3-bromocyclohexene} + KOH \xrightarrow{\Delta,\ C_2H_5OH}$$

4-3 Addition polymerization of alkenes

All compounds that contain a carbon-carbon double bond can conceivably be made to utilize their pi electrons to bond together (*polymerize*) a great number of the individual molecules (*monomers*) to form a very long molecule (*polymer*) composed of the repeating units. Examples of familiar polymers are given in Table 4-3.

Polymerization may be initiated by (a) a free radical, (b) a cation, or (c) an anion.

$$\text{(a)}\quad R\cdot + \rangle C{=}C\langle \longrightarrow R{-}\overset{|}{\underset{|}{C}}{-}\overset{|}{\underset{|}{C}}\cdot \xrightarrow{\rangle C=C\langle} R{-}\overset{|}{\underset{|}{C}}{-}\overset{|}{\underset{|}{C}}{-}\overset{|}{\underset{|}{C}}{-}\overset{|}{\underset{|}{C}}\cdot \xrightarrow{\rangle C=C\langle} \text{etc.} \qquad (4\text{-}43)$$

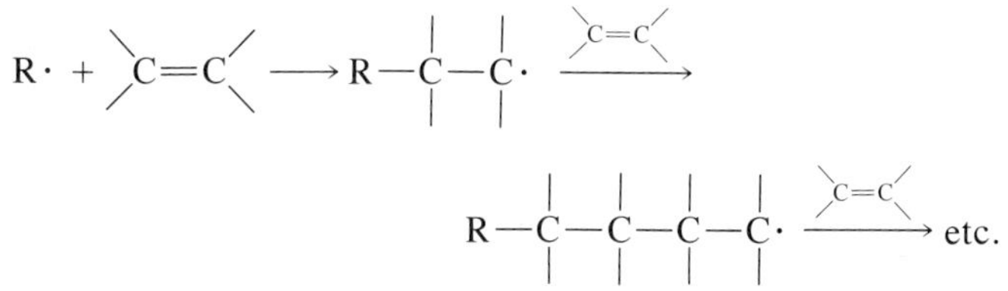

$$\text{(b)}\quad Y^+ + \rangle C{=}C\langle \longrightarrow Y{-}\overset{|}{\underset{|}{C}}{-}\overset{|}{\underset{|}{C}}+ \xrightarrow{\rangle C=C\langle} Y{-}\overset{|}{\underset{|}{C}}{-}\overset{|}{\underset{|}{C}}{-}\overset{|}{\underset{|}{C}}{-}\overset{|}{\underset{|}{C}}+ \xrightarrow{\rangle C=C\langle} \text{etc.} \qquad (4\text{-}44)$$

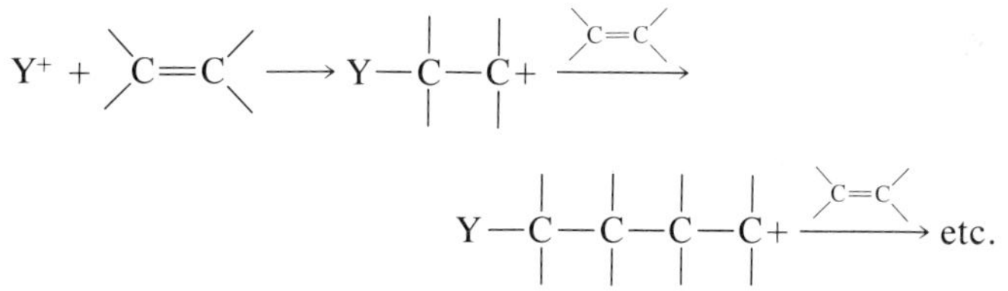

$$\text{(c)}\quad Z:^- + \;\rangle C{=}C\langle \;\longrightarrow Z-\overset{|}{\underset{|}{C}}-\overset{|}{\underset{|}{C}}:^- \xrightarrow{\;\rangle C=C\langle\;}$$

$$Z-\overset{|}{\underset{|}{C}}-\overset{|}{\underset{|}{C}}-\overset{|}{\underset{|}{C}}-\overset{|}{\underset{|}{C}}:^- \xrightarrow{\;\rangle C=C\langle\;} \text{etc.} \qquad (4\text{-}45)$$

Initiation of polymerization by a free radical

Production of polystyrene is an example of initiation by a free radical. The mechanism, in which R· is a free-radical initiator such as that obtained by

TABLE 4-3
Some examples of polymers of alkenes

Monomer	Polymer	Common name or trade name	Uses
$CH_2{=}CH_2$	$\left[\begin{array}{cc} H & H \\ -C- & C- \\ H & H \end{array}\right]_n$	Polyethylene	Plastic pipe, sheeting, and containers
$(H)(C_6H_5)C{=}CH_2$	$\left[\begin{array}{cc} H & H \\ -C- & C- \\ C_6H_5 & H \end{array}\right]_n$	Polystyrene	Plastic wall tiles; Styrofoam for many uses
$CH_3\overset{CH_3}{\overset{\mid}{C}}{=}CH_2$	$\left[\begin{array}{cc} H_3C & H \\ -C- & C- \\ H_3C & H \end{array}\right]_n$	Polyisobutylene	Inner tubes for tires requiring them
$CH_3CH{=}CH_2$	$\left[\begin{array}{cc} H_3C & H \\ -C- & C- \\ H & H \end{array}\right]_n$	Polypropylene, Herculon	Outdoor carpets, water pipe
$F_2C{=}CF_2$	$\left[\begin{array}{cc} F & F \\ -C- & C- \\ F & F \end{array}\right]_n$	Teflon	Lining kitchen and laboratory ware
$CH_2{=}\overset{H}{\overset{\mid}{C}}-CN$	$\left[\begin{array}{cc} H & H \\ -C- & C- \\ H & CN \end{array}\right]_n$	Orlon, Acrilan	Fiber for carpets and clothing
$CH_2{=}C(CH_3)-C({=}O)-OCH_3$	$\left[\begin{array}{cc} H & CH_3 \\ -C- & C- \\ H & C{=}O \\ & OCH_3 \end{array}\right]_n$	Lucite, Plexiglas	Plastic substitute for glass
$CH_2{=}CHCl$	$\left[\begin{array}{cc} H & H \\ -C- & C- \\ H & Cl \end{array}\right]_n$	Polyvinyl chloride (PVC)	Garden hoses, water pipe, shoe soles

the decomposition of an organic peroxide, is as follows:

$$R\cdot + \underset{\text{Styrene}}{(H)(C_6H_5)C{=}CH_2} \longrightarrow R{-}CH(C_6H_5){-}\dot{C}H_2 \xrightarrow{CH(C_6H_5){=}CH_2} \underset{\text{Dimer of styrene}}{R{-}CH(C_6H_5){-}CH_2{-}CH(C_6H_5){-}\dot{C}H_2} \xrightarrow{CH(C_6H_5){=}CH_2} \text{etc.} \quad (4\text{-}46)$$

The overall reaction is

$$n\underset{\text{Styrene}}{\left(\begin{matrix}H \\ C_6H_5\end{matrix}\Big> C{=}CH_2\right)} \xrightarrow{R\cdot} \underset{\text{Polystyrene}}{\left[\begin{matrix} & H & & H & \\ - & C & - & C & - \\ & C_6H_5 & & H & \end{matrix}\right]_n} \quad (4\text{-}47)$$

Polystyrene is a plastic used for making wall tiles and Styrofoam is used for insulation, packing, and art work.

Initiation of polymerization by a cation

The use of anhydrous hydrogen fluoride as a catalyst for the polymerization of 2-methylpropene (isobutylene) is an example of initiation by a cation, H^+.

$$n\underset{\text{Isobutylene}}{\left(\begin{matrix} CH_3 \\ | \\ CH_3C{=}CH_2\end{matrix}\right)} \xrightarrow{HF} \underset{\text{Polyisobutylene}}{\left[\begin{matrix} & CH_3 & & H & \\ - & C & - & C & - \\ & CH_3 & & H & \end{matrix}\right]_n} \quad (4\text{-}48)$$

Initiation of polymerization by an anion

An example of the initiation of polymerization by an anion is the use of a metal alkyl such as triethyl aluminum to initiate the formation of polyethylene from ethylene (IUPAC: ethene). The triethyl aluminum has a metal-carbon bond with enough ionic character to be considered to contain an ethyl *carbanion*.

$$(CH_3CH_2^-)_3Al^{3+}$$
Ethyl carbanion

Catalysts of this type were developed and used by Karl Ziegler of Germany and Giulio Natta of Italy, for which they received a Nobel prize in chemistry (1963).

$$\underset{\text{Ethylene}}{nCH_2{=}CH_2} \xrightarrow{(C_2H_5)_3Al} \underset{\text{Polyethylene}}{\left[\begin{matrix} & H & & H & \\ - & C & - & C & - \\ & H & & H & \end{matrix}\right]_n} \quad (4\text{-}49)$$

Polyethylene is a tough, chemically resistant plastic used to make pipe, garbage cans, buckets, laboratory ware, plastic sheeting, and so on.

Polymerization of dienes

Conjugated dienes can polymerize by either 1,2 or 1,4 addition of the monomer. The 1,4 addition is the most common.

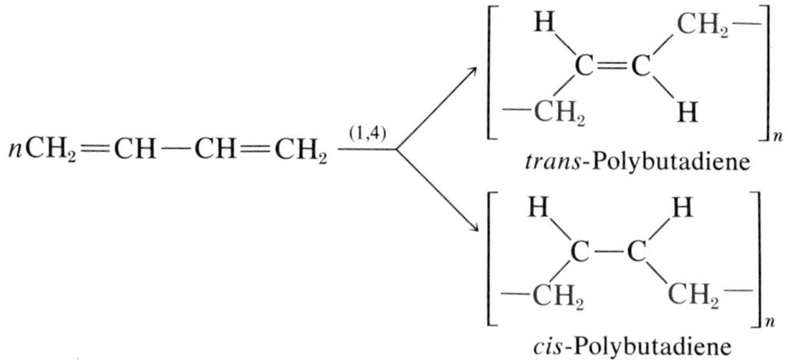

$$nCH_2{=}CH{-}CH{=}CH_2 \xrightarrow{(1,4)} \left[\begin{array}{c} H \quad\quad CH_2{-} \\ C{=}C \\ {-}CH_2 \quad\quad H \end{array} \right]_n \quad \text{or} \quad \left[\begin{array}{c} H \quad\quad H \\ C{-}C \\ {-}CH_2 \quad\quad CH_2{-} \end{array} \right]_n \tag{4-50}$$

trans-Polybutadiene

cis-Polybutadiene

The mechanism for the polymerization of conjugated dienes is as follows for 1,2 addition:

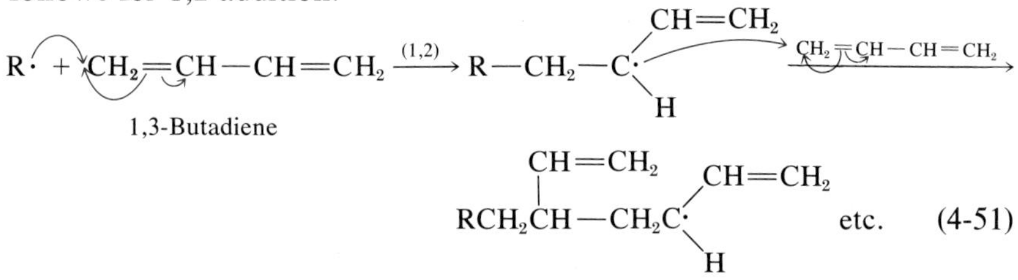

$$R\cdot + CH_2{=}CH{-}CH{=}CH_2 \xrightarrow{(1,2)} R{-}CH_2{-}\overset{CH=CH_2}{\underset{H}{C\cdot}} \xrightarrow{CH_2=CH-CH=CH_2}$$

1,3-Butadiene

$$RCH_2\overset{CH=CH_2}{CH}{-}CH_2\overset{CH=CH_2}{\underset{H}{C\cdot}} \quad \text{etc.} \tag{4-51}$$

For 1,4 addition, the mechanism is

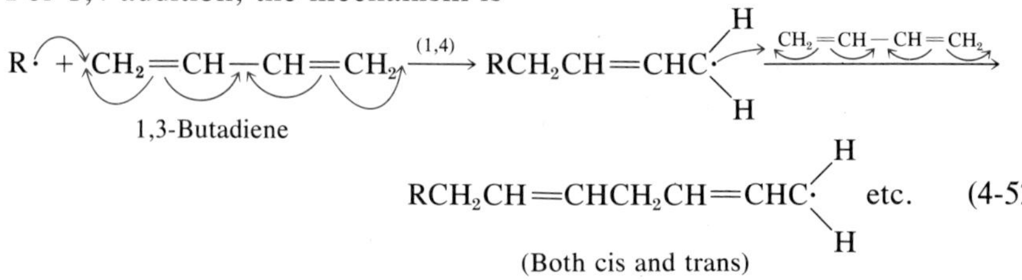

$$R\cdot + CH_2{=}CH{-}CH{=}CH_2 \xrightarrow{(1,4)} RCH_2CH{=}CH\overset{H}{\underset{H}{C\cdot}} \xrightarrow{CH_2=CH-CH=CH_2}$$

1,3-Butadiene

$$RCH_2CH{=}CHCH_2CH{=}CH\overset{H}{\underset{H}{C\cdot}} \quad \text{etc.} \tag{4-52}$$

(Both cis and trans)

Note that 1,2 polymerization results in a polymer with vinyl group ($-CH{=}CH_2$) side chains, whereas 1,4 polymerization results in an unsaturated, continuous-chain polymer.

Natural rubber is essentially the all cis isomer of polyisoprene. (Isoprene is 2-methyl-1,3-butadiene.) It can now be made synthetically by using a special catalyst.

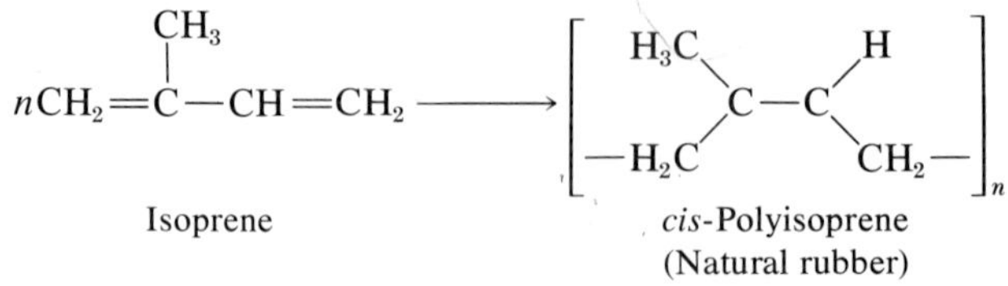

$$nCH_2{=}\overset{CH_3}{C}{-}CH{=}CH_2 \longrightarrow \left[\begin{array}{c} H_3C \quad\quad H \\ C{-}C \\ {-}H_2C \quad\quad CH_2{-} \end{array} \right]_n \tag{4-53}$$

Isoprene

cis-Polyisoprene
(Natural rubber)

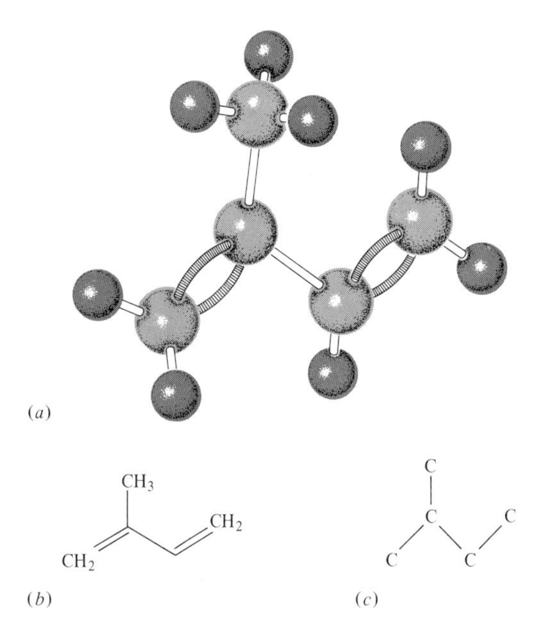

FIGURE 4-8
(*a*) Ball-and-spring model and (*b*) simplified structural formula of isoprene. (*c*) The isoprene unit of structure.

4-4 Terpenes

A large class of natural products derive from combinations of isoprene units (Fig. 4-8). A *dimer* (two units) of, or a derivative of a dimer of, isoprene is called a *terpene*. A *tetramer,* or a derivative of a tetramer, is called a *diterpene*. Rubber is a polyterpene. The *isoprene rule* is that the most likely structure for an unknown natural product will be that which utilizes the greatest number of *skeletal* isoprene units.

The terpenes may be cyclic as well as acyclic. Camphor is an example of a cyclic terpene. The two isoprene skeletal units are shaded in the formula for camphor (Fig. 4-9*a*). Look only for the basic five-carbon skeletal unit, double bonds and substituents do not affect the terpene designation.

Vitamin A is an example of a diterpene, i.e., it consists of four units of isoprene. In its structural formula (Fig. 4-9*b*), the isoprene units have been shaded.

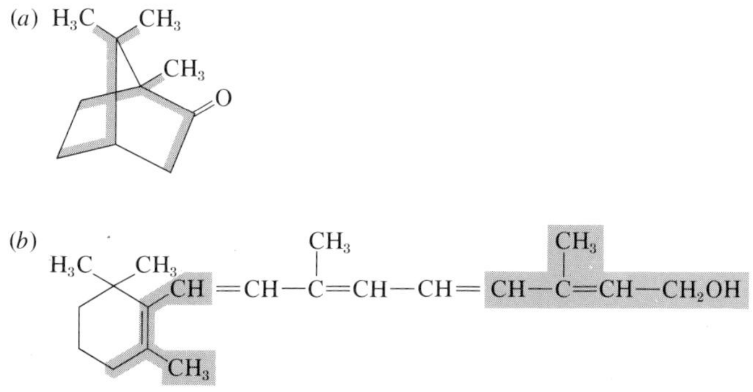

(*a*) H_3C CH_3 CH_3 O

(*b*) H_3C CH_3 CH_3 CH_3

$CH{=}CH{-}\overset{|}{C}{=}CH{-}CH{=}CH{-}\overset{|}{C}{=}CH{-}CH_2OH$

CH_3

FIGURE 4-9
(*a*) Camphor, a cyclic terpene, and (*b*) vitamin A, a diterpene.

PROBLEM 4-9 Write the structural formula representing the polymer of each of the following monomers.

(a) $\mathrm{ClHC{=}CHCl}$ (Cl and H on each carbon, Cl groups trans)

(b) $CH_2{=}C(CH_3)_2$

(c) $CH_2{=}CHCOOCH_3$

(d) $CH_2{=}C(H){-}C(CH_3){=}CH_2$

The all-trans 1,4 polymer

Solution to (a)

$$\left[\begin{array}{c}\mathrm{Cl}\quad\mathrm{Cl}\\ -\mathrm{C}-\mathrm{C}-\\ \mathrm{H}\quad\mathrm{H}\end{array}\right]_n$$

4-5 Alkynes

The *alkynes* are hydrocarbons with a carbon-carbon triple bond; they have the general formula C_nH_{2n-2}. Since there must be at least two carbons for the carbon-carbon triple bond, the first homolog of the series is C_2H_2, acetylene, which has the IUPAC name *ethyne*.

Structure of the alkynes

The carbon atoms of the triple bond are bonded to each other and to a hydrogen atom. Formation of the sigma bonds for this arrangement would utilize only two of the bonding electrons of each carbon atom: the 2*s* electron and one of the 2*p* electrons. The orbitals of the two electrons would be hybridized to produce two *sp* hybrid orbitals (Table 4-4).

The best geometry to allow maximum overlap for production of the three sigma bonds would be linear with bond angles of 180°. Each carbon atom would still have two electrons left in their usual 2*p* orbitals oriented 90° to each other. The 2*p* orbitals of each carbon atom may then overlap laterally, or sideways, to produce two pi molecular orbitals at right angles to each other. Thus the triple bond between the carbon atoms is composed of one sigma bond and two pi bonds (Fig. 4-10). The distance between the carbon atoms is shortened to 1.2×10^{-10} m because of the

TABLE 4-4 Carbon atom in a triple bond

Electron shell	**K**			**L**	
Orbitals	1*s*	2*s*	$2p_x$	$2p_y$	$2p_z$
Electrons	⇅	↑	↑	↑	↑
		Hybridize to produce two *sp* hybrid orbitals (2*s* + $2p_x$)			

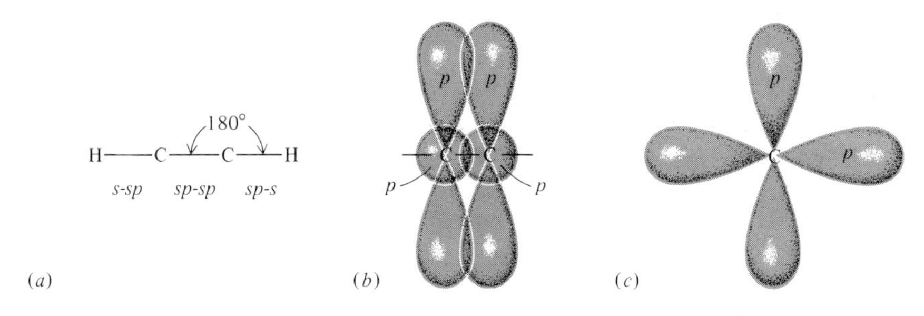

FIGURE 4-10
The triple bond in ethyne (acetylene): (*a*) sigma bonds; (*b*) side view of pi bonds; (*c*) end view of pi bonds.

larger amount of *s* character of the *sp* orbitals as compared with sp^2 orbitals for double bonds and sp^3 orbitals for *single carbon bonds.*

Figure 4-11 depicts models of acetylene. The Framework Molecular Model shows the two pi-bond–type molecular orbitals at right angles to each other. The space-filling model depicts the two pi bonds as having overlapped and diffused into each other to form a cylindrical space concentric about the sigma bond.

Cycloalkynes are uncommon. A cycloalkyne must have at least eight carbon atoms in order to accommodate the linear structure of the triply bonded carbon atoms and the carbon atoms immediately attached to them without too much bond angle strain. Consequently, stable cycloalkynes with fewer than seven carbon atoms are not known.

Nomenclature

The alkynes are commonly named as derivatives of acetylene. $CH_3C{\equiv}CH$ is methylacetylene; $CH_3C{\equiv}C{-}CH_3$ is dimethylacetylene; and $CH_3CH_2C{\equiv}CH$ is ethylacetylene. The IUPAC names are derived by changing the suffix *-ane* of the parent alkane to *-yne:* $HC{\equiv}CH$ is ethyne; $CH_3C{\equiv}CH$ is propyne; $CH_3C{\equiv}C{-}CH_3$ is 2-butyne; and $CH_3CH_2C{\equiv}CH$ is 1-butyne. Note the use of locator numbers to indicate the position of the triple bond in the chain.

PROBLEM 4-10 Give the IUPAC name for each of the following alkynes.

(a) $CH_3CH{-}C{\equiv}CH$ (with CH_3 attached to the CH)   (b) $CH_3CH_2C{\equiv}C{-}CH_2CH_3$

(c) $HC{\equiv}C{-}CH_2CH_2CH_3$

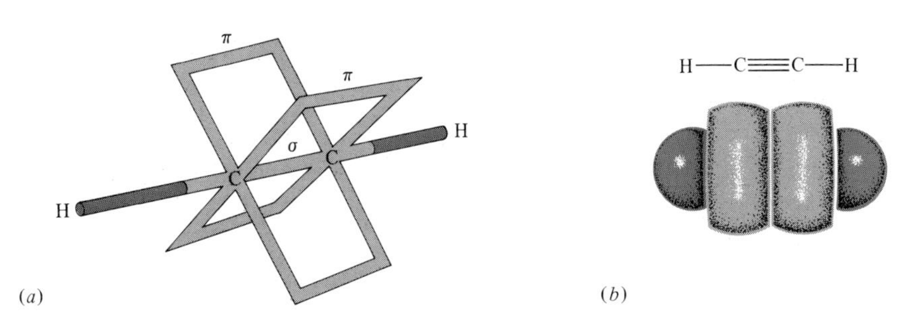

FIGURE 4-11
(*a*) Framework Molecular Model and (*b*) space-filling model of acetylene.

Solution to (a) The IUPAC name for this compound is 3-methyl-1-butyne.

$$\overset{4}{C}H_3-\overset{3}{C}H(CH_3)-\overset{2}{C}\equiv\overset{1}{C}H$$

PROBLEM 4-11 Write the structural formulas for the following alkynes. (a) 2-pentyne, (b) 1-hexyne, (c) 2-hexyne, (d) 4-methyl-2-pentyne.

Solution to (a)

$$\overset{5}{C}H_3-\overset{4}{C}H_2-\overset{3}{C}\equiv\overset{2}{C}-\overset{1}{C}H_3$$

The common name for 2-pentyne is ethylmethylacetylene.

Properties and uses of alkynes

Like the alkanes and the alkenes, alkynes containing up through four carbon atoms are gases and the C_5 homologs and above are liquids. The higher ratio of carbon to hydrogen results in formation of some carbon during combustion, so the alkynes burn with a luminous smoky flame. See Table 4-5 for the names, formulas, and properties of some simple alkynes.

Acetylene was burned in miners' lamps before electric lamps were developed. Because it produces a relatively high temperature when burned with pure oxygen, it is used in oxyacetylene torches for cutting and welding metals. Acetylene is an endothermic compound, which means that it can decompose into carbon and hydrogen with liberation of heat. For that reason, liquid and solid acetylene, and gaseous acetylene at a pressure above about two atmospheres, are violently explosive when subjected to shock, spark, or sudden heating even in the absence of oxygen. Since acetylene has a tendency to explode under pressure, it is dissolved in acetone under low pressure for storage and dispensing from tanks. It is an important intermediate for production of some synthetic fibers and some types of rubber.

Preparation of alkynes

Acetylene is prepared industrially (a) from limestone and coke or (b) from methane.

(a)

$$\underset{\text{Limestone}}{CaCO_3} \xrightarrow{\Delta} \underset{\text{Lime}}{CaO} + CO_2 \quad (4\text{-}54)$$

$$\underset{\text{Lime}}{CaO} + 3C \xrightarrow{2000°C} \underset{\text{Calcium carbide}}{CaC_2} + CO \quad (4\text{-}55)$$

$$\underset{\text{Calcium carbide}}{CaC_2} + 2H_2O \longrightarrow \underset{\text{Acetylene}}{C_2H_2} + Ca(OH)_2 \quad (4\text{-}56)$$

(b)

$$\underset{\text{Methane}}{6CH_4} + 6O_2 \xrightarrow{600°C} \underset{\text{Acetylene}}{2C_2H_2} + 2CO + 10H_2O \quad (4\text{-}57)$$

TABLE 4-5
Some simple alkynes

Structural formula	IUPAC name	Common name	mp, °C	bp, °C
$HC{\equiv}CH$	Ethyne	Acetylene	−81.8	−83.6
$CH_3C{\equiv}CH$	Propyne	Methylacetylene	−101.5	−23.2
$CH_3CH_2C{\equiv}CH$	1-Butyne	Ethylacetylene	−122.5	8.6
$CH_3C{\equiv}C{-}CH_3$	2-Butyne	Dimethylacetylene	−32.3	27.2
$CH_3CH_2CH_2C{\equiv}CH$	1-Pentyne	*n*-Propylacetylene	−90	39.3
$CH_2{=}CH{-}C{\equiv}CH$	1-Buten-3-yne	Vinylacetylene		5.0

Preparation of alkynes by dehydrohalogenation of dihalides. A general method for the preparation of alkynes is the dehydrohalogenation of either vicinal or geminal dihalides. The word *vicinal,* which means "neighboring," signifies that the two substituents are on adjacent carbon atoms. *Geminal* signifies that the two substituents are on the same carbon atom.

An example of dehydrohalogenation of a vicinal dihalide is the preparation of propyne from 1,2-dibromopropane.

$$\underset{\text{A vicinal dihalide}}{CH_3{-}\overset{H}{\underset{Br}{C}}{-}\overset{H}{\underset{Br}{C}}H} + KOH \xrightarrow{\Delta,\ C_2H_5OH} CH_3\overset{H}{C}{=}\underset{Br}{C}H + KBr + H_2O \qquad (4\text{-}58)$$

$$CH_3{-}\overset{H}{C}{=}\underset{Br}{C}H + \underset{\text{Sodium amide}}{NaNH_2} \xrightarrow{\Delta} \underset{\text{Propyne}}{CH_3C{\equiv}C{-}H} + NaBr + NH_3 \qquad (4\text{-}59)$$

The product of elimination of the first mole of HBr is a substituted vinyl bromide, 1-bromopropene. (You will recall that the vinyl group is $CH_2{=}CH{-}$.) The substituted vinyl bromide is less reactive than an alkyl bromide, so the stronger base $NaNH_2$ (sodium amide) is used to remove the second mole of HBr.

An example of dehydrohalogenation of a *gem*-dihalide is the preparation of propyne from 2,2-dibromopropane.

$$\underset{\text{A gem-dihalide}}{CH_3{-}\overset{Br}{\underset{Br}{C}}{-}CH_3} + KOH \xrightarrow{\Delta,\ C_2H_5OH} CH_3\underset{Br}{C}{=}CH_2 + KBr + H_2O \qquad (4\text{-}60)$$

$$CH_3{-}\underset{Br}{C}{=}CH_2 + NaNH_2 \xrightarrow{\Delta} CH_3C{\equiv}C{-}H + NaBr + NH_3 \qquad (4\text{-}61)$$

Initial use of $NaNH_2$ with the dihalide will take the dihalide directly to the alkyne, and so this method is preferred.

Synthesis of alkynes from other alkynes with terminal triple bonds. The higher homologs of acetylene may be prepared in the manner of the following examples.

$$\underset{\text{Acetylene}}{2H{-}C{\equiv}C{-}H} + 2Na \xrightarrow{NH_3(l)} \underset{\text{Sodium acetylide}}{2H{-}C{\equiv}C{:}^{-}Na^{+}} + H_2 \qquad (4\text{-}62)$$

$$H{-}C{\equiv}C{:}^{-}Na^{+} + \underset{\text{Bromoethane}}{CH_3CH_2Br} \longrightarrow \underset{\text{1-Butyne}}{H{-}C{\equiv}C{-}CH_2CH_3} + NaBr \qquad (4\text{-}63)$$

The second acetylenic hydrogen may also be replaced by an alkyl group of choice by repeating the process.

$$2H{-}C{\equiv}C{-}CH_2CH_3 + 2Na \xrightarrow{NH_3(l)} 2Na^{+-}{:}C{\equiv}C{-}CH_2CH_3 + H_2 \qquad (4\text{-}64)$$

$$Na^{+-}{:}C{\equiv}C{-}CH_2CH_3 + CH_3I \longrightarrow \underset{\text{2-Pentyne}}{CH_3{-}C{\equiv}C{-}CH_2CH_3} + NaI \qquad (4\text{-}65)$$

When the alkyl group desired is CH_3—, liquid CH_3I, rather than gaseous CH_3Br, is used in the reaction. The liquid is more convenient to use.

The reaction is limited to the use of primary halides because sodium acetylide is a strong base and it dehydrohalogenates the more reactive secondary and tertiary halides to produce alkenes.

PROBLEM 4-12 Write equations showing how each of the following alkynes could be synthesized beginning with the compound indicated and any necessary inorganic reagents. *More than one reaction will be necessary.*

(a) propyne beginning with 2-bromopropane
(b) 2-hexyne beginning with 1-bromopropane
(c) 1-pentyne beginning with coke, limestone, water, and 1-bromopropane

Solution to (a)

$$\underset{\text{2-Bromopropane}}{CH_3\underset{\displaystyle Br}{\underset{|}{C}}HCH_3} + KOH \xrightarrow{\Delta,\ C_6H_5OH} \underset{\text{Propene}}{CH_3CH{=}CH_2} + KBr + H_2O$$

$$\underset{\text{Propene}}{CH_3CH{=}CH_2} + Br_2 \longrightarrow \underset{\text{1,2-Dibromopropane}}{CH_3\underset{\displaystyle Br}{\underset{|}{C}}HCH_2Br}$$

$$\underset{\text{1,2-Dibromopropane}}{CH_3\underset{\displaystyle Br}{\underset{|}{C}}HCH_2Br} + KOH \xrightarrow{\Delta,\ C_2H_5OH} \underset{\text{1-Bromopropene}}{CH_3CH{=}CHBr} + KBr + H_2O$$

$$\underset{\text{1-Bromopropene}}{CH_3CH{=}CHBr} + NaNH_2 \xrightarrow{\Delta} \underset{\text{Propyne}}{CH_3C{\equiv}CH} + NaBr + NH_3$$

Reactions of alkynes

The reactions of the alkynes are of two general types: (a) electrophilic addition to the triple bond and (b) heterolytic cleavage of the triply bonded carbon-hydrogen bond ($\equiv$C—H) to form an acetylide ion and release a proton.

Electrophilic addition to the triple bond. For reasons not yet known, the alkynes are less reactive than the alkenes in addition of electrophilic reagents. They generally require a catalyst. The catalysts are usually the mercury(II) or copper(I) salts of appropriate acids.

Either one pi bond of the triple bond may be broken by the addition of one mole of an electrophilic reagent to produce a vinyl compound

$$-C\equiv C- + HA \longrightarrow -\underset{\displaystyle A}{\underset{|}{C}}=\overset{\displaystyle H}{\overset{|}{C}}- \tag{4-66}$$

or both pi bonds may be broken by the reaction and addition of *two moles* of an electrophilic reagent to produce a substituted alkane,

$$-C\equiv C- + 2HA \longrightarrow -\underset{\displaystyle A}{\underset{|}{\overset{\displaystyle A}{\overset{|}{C}}}}-\underset{\displaystyle H}{\underset{|}{\overset{\displaystyle H}{\overset{|}{C}}}}- \tag{4-67}$$

The following equations are examples of the addition of *one mole* of reagent to the triple bond.

$$\underset{\text{Ethyne (Acetylene)}}{H-C\equiv C-H} + HCN \xrightarrow{CuCl/NH_4Cl} \underset{\text{Propenenitrile (Acrylonitrile or vinyl cyanide)}}{H-\overset{\displaystyle H}{\overset{|}{C}}=\overset{\displaystyle H}{\overset{|}{C}}-CN} \tag{4-68}$$

$$H-C\equiv C-H + HOH \xrightarrow{HgSO_4/H_2SO_4} \left[\underset{\text{Vinyl alcohol (unstable)}}{H\overset{\displaystyle H}{\overset{|}{C}}=\overset{\displaystyle H}{\overset{|}{C}}-OH}\right] \longrightarrow \underset{\text{Ethanal (Acetaldehyde)}}{H\underset{\displaystyle H}{\underset{|}{\overset{\displaystyle H}{\overset{|}{C}}}}-\overset{\displaystyle H}{\overset{|}{C}}=O} \tag{4-69}$$

$$\underset{\text{Propyne}}{CH_3C\equiv C-H} + HOH \xrightarrow{HgSO_4/H_2SO_4} \left[CH_3-\underset{\displaystyle OH}{\underset{|}{C}}=\overset{\displaystyle H}{\overset{|}{C}}H\right] \longrightarrow$$

$$\underset{\text{Propanone (Acetone)}}{CH_3-\underset{\displaystyle O}{\underset{\|}{C}}-\underset{\displaystyle H}{\underset{|}{\overset{\displaystyle H}{\overset{|}{C}}}}-H} \tag{4-70}$$

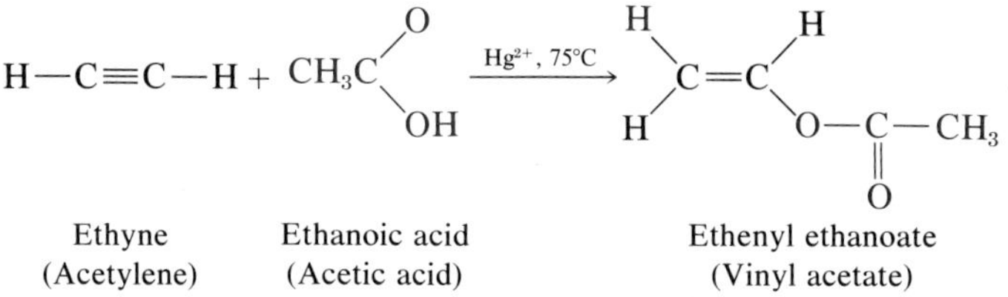

$$H{-}C{\equiv}C{-}H + CH_3C(=O)OH \xrightarrow{Hg^{2+},\ 75°C} H_2C{=}CH{-}O{-}C(=O){-}CH_3 \qquad (4\text{-}71)$$

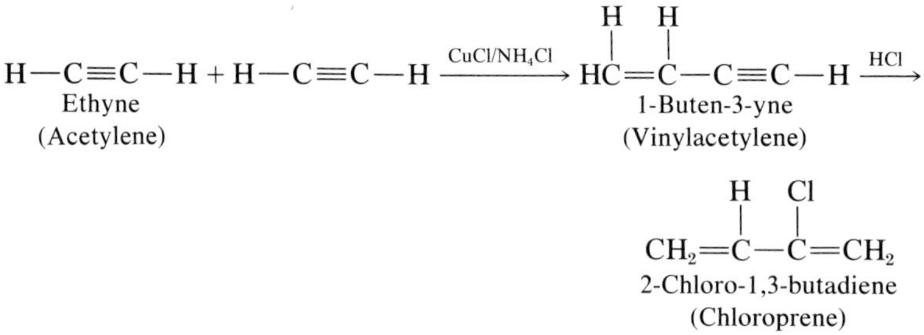

Ethyne (Acetylene) — Ethanoic acid (Acetic acid) — Ethenyl ethanoate (Vinyl acetate)

$$H{-}C{\equiv}C{-}H + H{-}C{\equiv}C{-}H \xrightarrow{CuCl/NH_4Cl} HC(H){=}C(H){-}C{\equiv}C{-}H \xrightarrow{HCl} CH_2{=}C(H){-}C(Cl){=}CH_2 \qquad (4\text{-}72)$$

Ethyne (Acetylene) — 1-Buten-3-yne (Vinylacetylene) — 2-Chloro-1,3-butadiene (Chloroprene)

The product of the last reaction, chloroprene, undergoes a 1,4 addition type of polymerization to produce a synthetic rubber called *neoprene*. Neoprene is resistant to grease and oil, and so it is used for rubber components likely to come into contact with petroleum lubricants. It is also much used as a material for soles for shoes. The polymerization reaction is

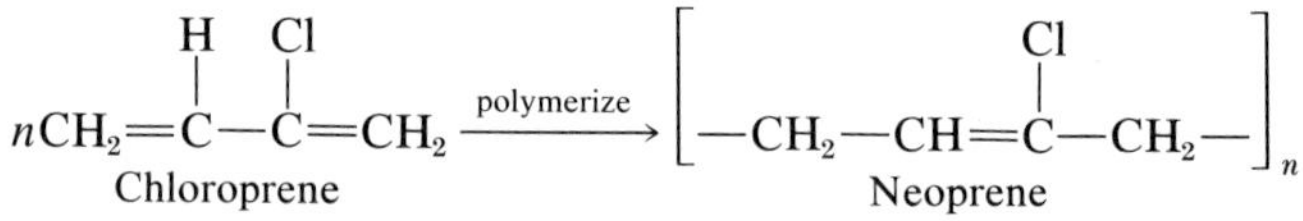

$$nCH_2{=}C(H){-}C(Cl){=}CH_2 \xrightarrow{\text{polymerize}} \left[{-}CH_2{-}CH{=}C(Cl){-}CH_2{-}\right]_n \qquad (4\text{-}73)$$

Chloroprene — Neoprene

The following reactions are examples of the addition of *two moles* of reagent to the triple bond.

$$CH_3{-}C{\equiv}CH + 2H_2 \xrightarrow{Ni} CH_3CH_2CH_3 \qquad (4\text{-}74)$$

Propyne

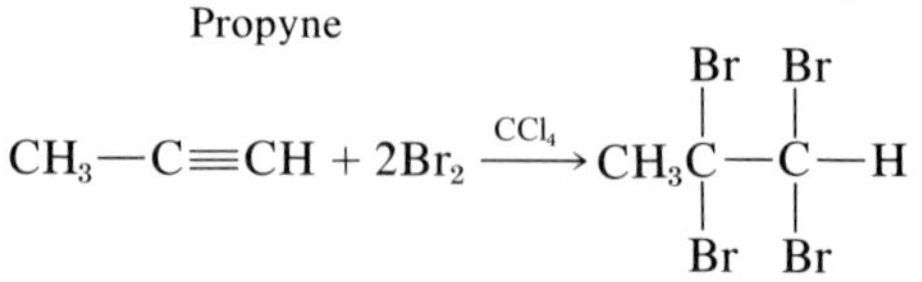

$$CH_3{-}C{\equiv}CH + 2Br_2 \xrightarrow{CCl_4} CH_3CBr_2{-}CBr_2{-}H \qquad (4\text{-}75)$$

$$CH_3{-}C{\equiv}CH + 2HBr \longrightarrow CH_3CBr_2{-}CH_2{-}H \qquad (4\text{-}76)$$

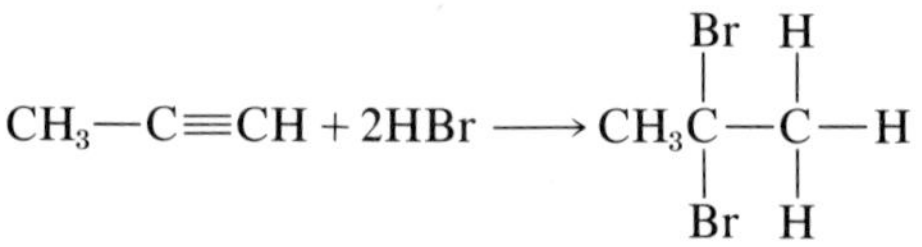

Note that Markovnikov's rule is followed in Eq. (4-76).

STEREOSPECIFIC REDUCTION OF ALKYNES TO YIELD CIS OR TRANS ALKENES. Alkynes, upon partial reduction, yield alkenes. Properly substituted alkynes may yield either cis or trans alkenes depending upon the reducing agent used. Catalytic reduction with molecular hydrogen produces cis alkenes; chemical reduction with sodium or lithium in liquid ammonia produces trans alkenes.

$$CH_3{-}C{\equiv}C{-}CH_3 + H_2 \xrightarrow{\text{Pd-C}} \underset{H}{\overset{H_3C}{}}\!\!\!\!\!\!>C{=}C<\!\!\!\!\!\!\underset{H}{\overset{CH_3}{}} \tag{4-77}$$

2-Butyne — *cis*-2-Butene

$$CH_3{-}C{\equiv}C{-}CH_3 \xrightarrow{\text{Na/NH}_3(l)} \underset{H}{\overset{H_3C}{}}\!\!\!\!\!\!>C{=}C<\!\!\!\!\!\!\underset{CH_3}{\overset{H}{}} \tag{4-78}$$

2-Butyne — *trans*-2-Butene

Replacement of acetylenic hydrogen. The hydrogen atom bonded to triply bonded carbon may be cleaved heterolytically to produce an *acetylide ion* with release of a proton.

$$RC{\equiv}C{-}H \longrightarrow RC{\equiv}C^{-} + H^+ \tag{4-79}$$

As this reaction indicates, the acetylenic hydrogen is very slightly acidic. Its acidity relative to water, ammonia, and ethene is

$$H\ddot{\underset{\cdot\cdot}{O}}H > H{-}C{\equiv}CH > H{-}\ddot{N}H_2 > H{-}CH{=}CH_2$$

Relative acidity

$$^{-}{:}\ddot{\underset{\cdot\cdot}{O}}H < {}^{-}{:}C{\equiv}CH < {}^{-}{:}\ddot{N}H_2 < {}^{-}{:}CH{=}CH_2$$

Relative basicity

Acetylene will react with the strong base sodium amide ($NaNH_2$) to form sodium acetylide. This reaction occurs because acetylene is a stronger acid than ammonia.

$$\underset{\text{Stronger acid}}{HC{\equiv}CH} + \underset{\text{Stronger base}}{{}^{-}{:}\ddot{N}H_2} \rightleftharpoons \underset{\text{Weaker base}}{{}^{-}{:}C{\equiv}CH} + \underset{\text{Weaker acid}}{\ddot{N}H_3} \tag{4-80}$$

Acetylene also reacts with active metals such as sodium, potassium, and calcium to produce gaseous hydrogen.

$$2H{-}C{\equiv}C{-}H + 2Na \xrightarrow{NH_3(l)} 2H{-}C{\equiv}C{:}^{-}Na^+ + H_2(g) \tag{4-81}$$

$$H{-}C{\equiv}C{-}H + 2Cu(NH_3)_2Cl \longrightarrow Cu{-}C{\equiv}C{-}Cu(s) + 2NH_4Cl + 2NH_3 \tag{4-82}$$

$$H{-}C{\equiv}C{-}H + 2Ag(NH_3)_2NO_3 \longrightarrow Ag{-}C{\equiv}C{-}Ag(s) + 2NH_4NO_3 + 2NH_3 \tag{4-83}$$

The ammoniacal copper(I) chloride reagent is made by mixing copper(I) chloride with ammonium hydroxide. The ammoniacal silver nitrate is made by mixing silver nitrate with excess ammonium hydroxide. The heavy-metal acetylides of copper and silver precipitate out. They are extremely sensitive to shock when they become dry, so that they may explode.

The preceding reactions may be used to differentiate terminal alkynes from alkenes, since the ethylenic hydrogens are not replaceable in such reactions. However, it should be noted that only *terminal alkynes* having acetylenic hydrogen atoms (hydrogen atoms attached to a triply bonded carbon) should be expected to react. Since both alkenes and alkynes will decolorize bromine in carbon tetrachloride and also change cold, dilute potassium permanganate, they cannot be differentiated by use of these reagents.

PROBLEM 4-13 Write equations for the reaction of propyne with each of the following reagents. (a) 2 mol HCl, (b) 1 mol Cl_2, (c) ammoniacal CuCl.

Solution to (a)

$$CH_3{-}C{\equiv}CH + 2HCl \longrightarrow CH_3{-}\overset{\overset{\displaystyle Cl}{|}}{\underset{\underset{\displaystyle Cl}{|}}{C}}{-}\overset{\overset{\displaystyle H}{|}}{\underset{\underset{\displaystyle H}{|}}{C}}{-}H$$

PROBLEM 4-14 Write equations for the following conversions. (More than one reaction is required.)

(a) 2,3-dibromopentane into pure *trans*-2-pentene
(b) bromoethane into pure *cis*-3-hexene

Solution to (a)

$$CH_3{-}\overset{\overset{\displaystyle Br}{|}}{\underset{\underset{\displaystyle H}{|}}{C}}{-}\overset{\overset{\displaystyle Br}{|}}{\underset{\underset{\displaystyle H}{|}}{C}}{-}CH_2CH_3 + 2NaNH_2 \xrightarrow{\Delta} CH_3{-}C{\equiv}C{-}CH_2CH_3 + 2NaBr + 2NH_3$$

$$CH_3{-}C{\equiv}C{-}CH_2CH_3 \xrightarrow{Na/NH_3(l)} \begin{matrix} H_3C & & H \\ & C{=}C & \\ H & & CH_2CH_3 \end{matrix}$$

PROBLEM 4-15 What reaction could be used to remove small amounts of acetylene from ethylene?

Summary

Unsaturated hydrocarbons include all of the acyclic and cyclic compounds with one or more *double* and/or *triple carbon-carbon bonds*. The unsaturated hydrocarbons include the *alkenes* (one carbon-carbon double bond), the *dienes* (two carbon-carbon double bonds), the *alkynes* (a carbon-carbon triple bond), and the *aromatic hydrocarbons* (Chap. 5).

The *functional group* of an alkene is its carbon-carbon *double bond*. Its general formula is C_nH_{2n}. The chief commercial source of the alkenes is the cracking of petroleum, but double bonds occur in many natural products.

The sigma bond of the carbon-carbon double bond utilizes *sp^2 hybrid orbitals,* and the geometry is *planar trigonal* with bond angles of approximately 120°. The remaining electrons of each carbon atom are in *p* orbitals which overlap sideways to produce a *pi molecular orbital* at right angles to the plane of the molecule. The carbon-carbon double bond consists of a sigma bond and a pi bond; the pi bond is the weaker. The distance between the carbon atoms is shorter (1.34×10^{-10} m) than for the usual carbon-carbon single bond.

The first part of the name of an alkene is the same as that of the corresponding alkane, but the suffix *-ane* is changed to *-ylene* for the common name or to *-ene* for the IUPAC name. Locator numbers are employed in the IUPAC system to designate on which carbon atom the double bond commences, beginning at the end resulting in the lowest number for the double bond. For cycloalkenes, numbering begins with a doubly bonded carbon atom, and the number 1 may be implied rather than stated.

Alkenes are generally prepared by *elimination reactions,* in which two groups on adjacent carbon atoms of a saturated chain are removed, with a double bond between the two carbon atoms as a result. Two such reactions are *dehydrohalogenation of alkyl halides* and *dehydration of alcohols.* In those reactions to produce alkenes, the *Saytzeff rule* is followed; the *major product* will be the one that has the *most alkyl groups attached to the resultant carbon-carbon double bond.* Tertiary alkyl halides or tertiary alcohols react faster than the corresponding secondary ones, which in turn react faster than the primary ones.

The chief reaction of the alkenes is *electrophilic addition* to the double bond in which *electron-deficient groups* (*electrophiles*) break the pi bond and bond to the carbon atoms of the former double bond. These addition reactions include *addition of halogen,* X_2; addition of *halogen and water,* HOX; addition of *hydrohalogen,* HX; addition of *sulfuric acid;* and addition of *water.* All those reactions follow *Markovnikov's rule* that the *hydrogen* of the reagent *adds to the carbon of the double bond which is already bonded to the most hydrogen atoms.* Markovnikov's rule is explained by the fact that addition takes place in such a way as to form the most stable *intermediate carbonium ion.* The order of stability is $3° > 2° > 1°$. The relative stability of the types of carbonium ions is explained in part by the concept of *delocalization of the positive charge* of the ion by *positive inductive effects* of the alkyl groups attached to the carbonium ion.

Another reaction of the alkenes is *anti-Markovnikov addition of HBr* in the presence of an organic *peroxide catalyst.* The mechanism involves a *free-radical intermediate* rather than a carbonium ion. Addition takes place in such a way as to favor production of the most stable free radical.

At high temperatures, *allylic substitution* of alkenes by halogens, rather than addition, occurs. The reaction proceeds by a free-radical mechanism through an *intermediate resonance-stabilized allylic radical. Resonance* exists whenever two or more electronic arrangements for a molecule are possible.

The alkenes can be oxidized by ozone and by $KMnO_4$. The permanganate reaction is used as a test reaction (the *Baeyer test*) for the carbon-carbon double bond.

Dienes have *two carbon-carbon double bonds* which may be arranged into three types: *cumulative* (double bonds adjacent), *conjugated* (double, single, double), and *isolated* (double bonds separated by more than one single bond). *Conjugated dienes* are the most important because they undergo *1,4 addition reactions*. The product of 1,4 addition is more stable than that of 1,2 addition; this is an application of the Saytzeff rule. The intermediate is a *resonance-stabilized carbonium ion*.

Alkenes undergo *addition polymerization* which may be initiated by a *free radical,* a *cation,* or an *anion*. *Polymerization* is the connecting together of many units of a starting molecule (*monomer*) to form a very long molecule (*polymer*). Some of these polymers constitute useful fibers and plastics, e.g., Orlon, Teflon, Lucite, and Herculon.

Conjugated dienes can polymerize by 1,4 addition. *Isoprene* is 2-methyl-1,3-butadiene, and natural rubber is the all-cis 1,4 polymer of isoprene. *Terpenes* are natural compounds which are combinations of *skeletal units of isoprene*. A terpene has *two units* of isoprene, and a *diterpene* has *four units* of isoprene. The *isoprene rule* is that the most likely structure for an unknown natural product will be the structure that utilizes *the greatest number of skeletal isoprene units*.

Alkynes are hydrocarbons with a carbon-carbon *triple bond* and the general formula C_nH_{2n-2}. The first member of the series is ethyne (acetylene), C_2H_2 or $H—C{\equiv}C—H$.

The sigma bond of the carbon-carbon triple bond utilizes *sp hybrid orbitals,* and the geometry is *linear* with 180° bond angles. The two remaining valence electrons of each carbon are in *p* orbitals at right angles to each other. They are overlapped sideways to form two *molecular pi orbitals* at right angles to each other and merging to produce an almost cylindrical cloud about the sigma bond as an axis. The distance between carbon atoms is shortened to 1.2×10^{-10} m because of the large amount of *s* character in the *sp* hybrid orbitals.

The alkynes are commonly named as derivatives of acetylene. The IUPAC names are derived by changing the suffix *-ane* of the parent alkane to *-yne*. Locator numbers indicate the position of the triple bond in the chain.

The first four homologs of the alkynes are gases. The alkynes burn with a smoky, luminous flame. Acetylene burns with oxygen to produce a relatively high temperature, so it is used in oxyacetylene torches.

Acetylene is prepared industrially from limestone and coke or from methane. Alkynes in general are prepared by *dehydrohalogenation* of *vicinal* or *geminal dihalides* by using the strong base sodium amide, $NaNH_2$.

Alkynes undergo *addition reactions* with hydrogen, halogen, hydrohalogen, water, and hydrogen cyanide; Markovnikov's rule is followed when it is applicable. Alkynes with *terminal triple bonds* (having *acety-*

lenic hydrogen) react with sodium metal to give off hydrogen. Such alkynes also react with heavy-metal salts to give *heavy-metal acetylides* as precipitates. *Stereospecific reduction* of alkynes may be accomplished to yield either cis or trans alkenes.

Key terms

Acetylene: the first homolog of the alkyne series, C_2H_2; its IUPAC name is ethyne.

Acetylenic hydrogen: a hydrogen atom attached to a triply bonded carbon atom.

Acetylide ion: an anion, $R—C \equiv C:^-$, formed by removal of a proton from a terminal alkyne.

1,4 Addition: reaction of a conjugated diene with one mole of reagent. The reagent bonds to carbon atoms 1 and 4, and one double bond, between carbon atoms 2 and 3, remains.

Addition polymerization: bonding together of individual molecules originally containing double bonds to form very long molecules composed of repeating units.

Alkenes: unsaturated hydrocarbons with one double bond; acyclic alkenes have the general formula C_nH_{2n}; cyclic alkenes have the general formula C_nH_{2n-2}.

Alkenyl carbon atoms: carbon atoms joined by a double bond.

Alkynes: unsaturated hydrocarbons with one triple bond; the general formula is C_nH_{2n-2}.

Allylic carbonium ion: the resonance hybrid ion represented by the structure $\underbrace{—CH \overset{\cdots}{=} CH \overset{\cdots}{=} CH_2}_{+}$; an especially stable ion.

Allylic hydrogen atoms: those on the third carbon of an allyl group,

$$—\overset{3}{C}H_2\overset{2}{C}H=\overset{1}{C}H_2$$

Allylic substitution: replacement of allylic hydrogen atoms by halogen; it occurs at high temperatures (600°C).

Anti-Markovnikov addition: addition of hydrogen bromide to an alkene in the presence of an organic peroxide catalyst; it occurs in a manner contrary to that expected from Markovnikov's rule.

Baeyer test: the reaction of cold, dilute $KMnO_4$ solution with an alkene or alkyne in which the solution is decolorized; it is a test for presence of a double or triple bond.

Carbanion: an alkyl anion, i.e., an alkyl group bearing a negative charge.

Carbonium ion: an alkyl cation, i.e., an alkyl group bearing a positive charge.

Concerted mechanism: a reaction mechanism in which bonds are broken and formed simultaneously.

Conjugated double bonds: double bonds separated by one single bond.

Cumulative double bonds: adjacent double bonds.

Dehydration of alcohol: the reaction in which a hydrogen atom and a hydroxide group are eliminated from adjacent carbon atoms of an alcohol to yield water and a carbon-carbon double bond.

Dehydrohalogenation: an elimination reaction in which a hydrogen atom and a halogen atom are removed from adjacent carbon atoms of an alkyl halide; it produces a double bond between the carbon atoms.

Delocalization of charge: a shift in electron cloud density so that the cloud is spread over a larger area; it leads to increased stability of the species.

Diene: an unsaturated hydrocarbon with two double bonds.

Electronegative atoms: atoms of elements having a relatively high attraction for the shared electrons in a covalent bond. Atoms of the halogens, nitrogen, and oxygen are good examples.

Electrophile: an electron-deficient atom or group of atoms.

Electrophilic addition: reactions in which electrophiles break the pi bond of a carbon-carbon double bond and use the electrons to bond themselves to the carbon atoms of the former double bond.

Elimination reaction: a reaction in which two groups on adjacent carbon atoms of a saturated chain are removed; the result is formation of a double bond between the two carbon atoms.

Geminal dihalide: a dihalide in which the halogen atoms are on the same carbon atom.

Halohydrin: a halogen-substituted alcohol.

Isolated double bonds: double bonds separated by two or more single bonds.

Isoprene: the diene 2-methyl-1,3-butadiene; the monomer of natural rubber.

Isoprene rule: the prediction that the most likely structure of an unknown natural product will be that structure that utilizes the greatest number of skeletal isoprene units.

Linear: the geometry of the triple bond; bond angles are 180°.

Markovnikov's rule: in the addition of hydrohalogen to a double bond, the hydrogen of the reagent adds to the carbon of the double bond which is already bonded to the larger number of hydrogen atoms.

Monomers: individual molecules which bond together to form a polymer.

Negative inductive effect: a type of electron withdrawal in which the electron cloud density about a given atom is increased at the expense of adjacent atoms.

Pi orbital or bond: a molecular orbital consisting of a pair of shared and paired *p* electrons. It constitutes the second bond of the carbon-carbon double bond, and it is formed by sideways overlap of two atomic orbitals.

Planar trigonal: the geometry of the carbon-carbon double bond. Bond angles are approximately 120°, the geometry resulting from sp^2 orbitals.

Polymer: a very long molecule composed of repeating units.

Positive inductive effect: a type of electron release in which there is a shift in electron cloud density away from a given atom.

Resonance: delocalization of charge or of an unpaired electron. It occurs whenever two or more electronic arrangements for a given species are possible.

Resonance hybrid: the actual electronic arrangement that exists when two or more electronic arrangements are possible; it is a hybrid of all the possibilities.

Resonance-stabilized ion or radical: an ion or free radical which is a resonance hybrid; the delocalization of charge or of the unpaired electron confers added stability.

Saytzeff rule: the major product in an elimination reaction will be the product that has the most alkyl groups attached to the resultant double bond.

Sigma bond: an ellipsoidal bonding orbital of a molecule formed by coaxial overlap of any two atomic orbitals; the main type of covalent bond and generally stronger than a pi bond.

Solvation: coordination of an ion with solvent molecules.

***sp* hybrid orbital:** an orbital resulting from the hybridizing of the orbitals of an *s* electron and a *p* electron.

sp^2 hybrid orbital: an orbital resulting from the hybridizing of an *s* orbital with two *p* orbitals.

Stereospecific reduction: in this chapter, reduction of an alkyne to yield either cis or trans alkenes.

Terminal alkyne: an alkyne in which a triply bonded carbon atom is at one end of the chain.

Terpenes: dimers, or derivatives of dimers, of isoprene; diterpenes contain four isoprene units.

Unsaturated hydrocarbons: all of the acyclic and cyclic hydrocarbons with one or more double bonds, one or more triple bonds, or both.

Vicinal dihalide: a dihalide in which the halogen atoms are on adjacent carbon atoms.

Additional problems

4-16 Give the IUPAC names for the compounds represented by the following structural formulas.

(a) $CH_3-\overset{H}{C}=\underset{H}{C}-CH_2CH_3$

(b) $CH_3CH_2\overset{H}{C}=\overset{H}{C}-CH_3$

(c) $CH_3\underset{CH_3}{\overset{CH_3}{C}}-\underset{H}{C}=CH_2$

(d) $CH_3\overset{CH_3}{C}=\underset{H}{C}-CH_3$

(e) $CH_3\overset{CH_3}{CH}-\underset{CH_3}{C}=CH_2$

(f) $CH_3C(CH_3)_2\overset{H}{C}=\overset{CH_3}{C}-CH_3$

(g) (cyclopentene ring) CH_3

(h) H_3C (cyclopentene ring) CH_3

4-17 Give the structural formulas for the compounds with the following names.

(a) 1,2-dimethylcyclopentene
(b) *trans*-3-hexene
(c) *cis*-2,3,4,5-tetramethyl-3-hexene

4-18 Write equations for the preparation of alkenes from the following starting materials. Indicate the conditions required. If there is more than one product, indicate which alkene is the major product.

(a) $CH_3CH_2CH_2CH_2Br$

(b) 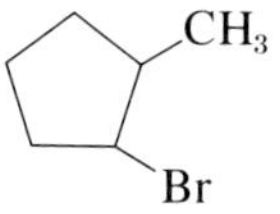 cyclopentane bearing CH_3 and Br on adjacent carbons

(c) $CH_3CH(CH_3)-C(CH_3)(OH)-CH_3$

(d) 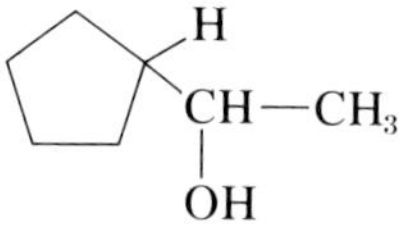 cyclopentane bearing H and $CH(OH)-CH_3$ on the same carbon

4-19 Write equations for the following reactions. If there is more than one product, indicate which is the major product.

(a) $CH_3CH_2C(CH_3)(Br)-CH_3 + KOH \xrightarrow{C_2H_5OH}$

(b) 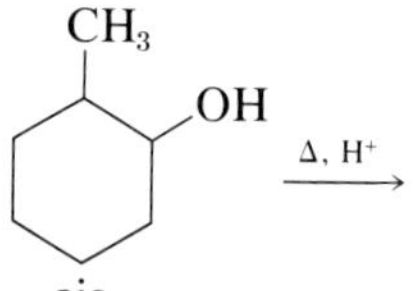 cis-2-methylcyclohexanol (cyclohexane with CH_3 and OH on adjacent carbons, cis) $\xrightarrow{\Delta,\ H^+}$

4-20 Write equations for the following reactions. If there is more than one product, indicate which is the major product.

(a) $CH_3CH{=}CH_2 + Cl_2 \xrightarrow{CCl_4\ \text{solvent}}$

(b) $CH_3CH{=}CH_2 + Cl_2/H_2O \longrightarrow$

(c) $CH_3C(CH_3){=}CH_2 + HCl \longrightarrow$

(d) $CH_3C(CH_3){=}CH_2 + HBr \xrightarrow{\text{peroxide}}$

(e) $CH_2{=}CH_2 + HOH \xrightarrow{H^+}$

(f) 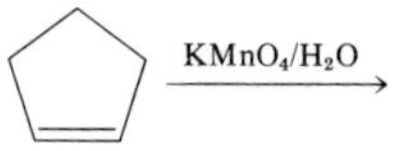 cyclopentene $\xrightarrow{KMnO_4/H_2O}$

(g) $CH_3CH_2CH{=}CH_2 + H_2SO_4 \longrightarrow$

4-21 Write equations for the following reactions. If there is more than one product, indicate which is the major product.

(a) $CH_2{=}C(CH_3)-CH{=}CH_2 + \underset{(1\ \text{mol})}{HBr} \xrightarrow{40°C}$

(b) $n\text{CH}_2\text{=CH—O—}\overset{\overset{\text{O}}{\|}}{\text{C}}\text{—CH}_3 \xrightarrow{\text{peroxide catalyst}}$

(c) $\text{HC}\equiv\text{CH} + \text{CH}_3\overset{\overset{\text{O}}{\|}}{\underset{\underset{\text{OH}}{|}}{\text{C}}} \xrightarrow{\text{Hg}^{2+},\ 75°\text{C}}$

(d) $CH_3C\equiv C—CH_3 + 2HBr \longrightarrow$

(e) $CH_3C\equiv C—CH_3 \xrightarrow{Na/NH_3(l)}$

(f) $CH_3C\equiv CH + Ag(NH_3)_2NO_3 \longrightarrow$

4-22 Name a *single chemical test* or reagent which could be used to distinguish between each of the following pairs. Indicate which member of the pair gives the positive test or greater reaction.

(a) $CH_3CH_2CH_2CH_2CH_3$ and $CH_3CH_2CH_2CH{=}CH_2$
(b) $CH_3CH_2CH_2C\equiv CH$ and $CH_3CH_2CH_2CH{=}CH_2$
(c) $CH_3C\equiv C—CH_3$ and $CH_3CH_2C\equiv C—H$
(d) $CH_2{=}CH—CH{=}CH_2$ and $CH_3CH_2C\equiv CH$

4-23 Write equations showing how each of the following conversions could be effected. (More than one reaction will be required in each case.)

(a) $CH_3CH_2CH_3$ to $\text{CH}_3\underset{\underset{\text{D}}{|}}{\text{C}}\text{HCH}_3$

(b) 1-bromobutane to 2-bromobutane
(c) *trans*-2-butene to *cis*-2-butene
(d) 2-bromopropane to 1-bromopropane
(e) 1,2-dibromopropane to 2,2-dibromopropane
(f) ethene to 3-hexyne
(g) chlorocyclohexane to 2-chlorocyclohexanol

4-24 (a) Tetrahydrocannabinol, which is the active principle of marijuana, has the molecular formula $C_{21}H_{30}O_2$. When it is treated at 150°C with hydrogen under pressure over a nickel or platinum catalyst, 4 mol of hydrogen is absorbed.

$$C_{21}H_{30}O_2 + 4H_2 \longrightarrow C_{21}H_{38}O_2$$

How many double bonds and how many rings does tetrahydrocannabinol have? (*Hint:* The corresponding saturated acyclic compound, or alkane, would have the molecular formula $C_{21}H_{44}$. Remember that there would then be two fewer hydrogen atoms for every double bond or for every ring.)
(b) Testosterone, a male sex hormone of formula $C_{19}H_{28}O_2$, upon hydrogenation yields a compound, $C_{19}H_{32}O_2$. How many double bonds and how many rings does testosterone have?
(c) Vitamin D_2, calciferol, $C_{28}H_{44}O$, upon hydrogenation absorbs hydrogen to give $C_{28}H_{52}O$. How many double bonds and how many rings does vitamin D_2 have?

5
AROMATIC HYDROCARBONS

Aromatic hydrocarbons are unsaturated compounds of a special type. They were originally called aromatic because of the aroma or odor that many of them have. Not all are odorous, however, and a great many compounds with a fragrant odor are not of the aromatic type.

5-1 Benzene

One of the simplest and earliest known compounds with the special structure characteristic of aromatic compounds was the hydrocarbon *benzene*. Michael Faraday discovered benzene in illuminating gas in 1825. The same compound was produced in Germany by Mitscherlich in 1834 by heating benzoic acid from gum benzoin with lime. Mitscherlich established the molecular formula as C_6H_6. The name *pheno* was proposed for C_6H_6 because Faraday had discovered the compound in illuminating gas and *pheno* derives from a Greek word referring to light. Although the name pheno was adopted, in the form of *phenyl,* for the C_6H_5— group, the name *benzene* was adopted for the compound itself because of the relation with benzoic acid and gum benzoin.

Properties of benzene

Benzene is a volatile liquid, and its vapor is quite toxic in high concentrations. It is also carcinogenic (causes cancer), and so it should be used only with good ventilation. The alkylbenzenes (e.g., toluene) are only a little less toxic, so precautions should be taken that their vapors also are not inhaled.

The formula C_6H_6 indicates a high degree of unsaturation, and benzene does burn with a very smoky, luminous flame like that of burning alkynes and dienes. However, benzene decolorizes neither bromine in carbon tetrachloride solution nor cold dilute $KMnO_4$ solution, indicating that it does not react as do the alkenes and alkynes. In general, it does not undergo the addition reactions characteristic of alkenes and alkynes; instead, it exhibits a stability akin to that of the alkanes. Indeed, benzene is stable even to boiling $KMnO_4$ solution. It does, however, react with bromine in the presence of iron as a catalyst,

$$C_6H_6 + Br_2 \xrightarrow{Fe} C_6H_5Br + HBr \tag{5-1}$$

but the reaction produces HBr. That indicates a substitution rather than an addition reaction. Furthermore, only a single bromobenzene is produced; no isomers of bromobenzene have ever been found. That indicates all six carbon atoms to be equivalent in structure. Three isomeric dibromobenzenes are produced by further reaction with bromine:

$$C_6H_5Br + Br_2 \xrightarrow{Fe} \underset{\text{(Three isomers)}}{C_6H_4Br_2} + HBr$$

The Kekulé structural formula

Reconciling the properties of benzene with a rational structural formula was a great problem for the chemists of the middle of the nineteenth century. It was not until 40 years after Faraday's discovery of benzene that August Kekulé, in 1865, proposed a structure which could account for most of the known chemical properties of benzene. He conceived of a planar, cyclic arrangement of six carbon atoms with alternate single and double bonds and one hydrogen atom attached to each carbon atom.

or

The structure proposed by Kekulé would have six equivalent hydrogen atoms, so only one monosubstituted benzene would be expected. That is what was found. For example,

Bromobenzene   Nitrobenzene

See also Fig. 5-1. The three isomeric disubstituted benzene derivatives

FIGURE 5-1
Space-filling model of bromobenzene. No isomers of bromobenzene have ever been found.

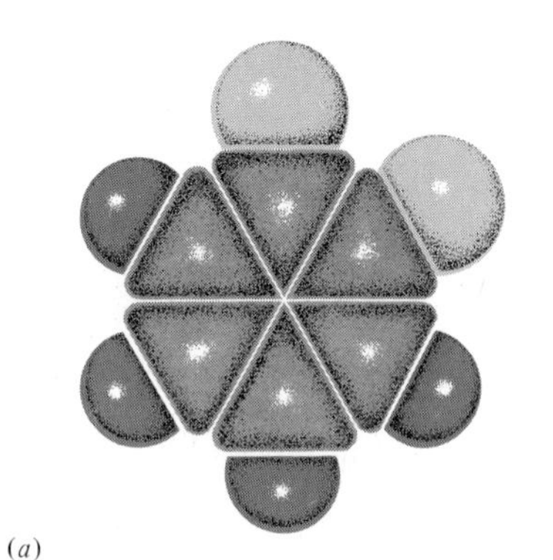

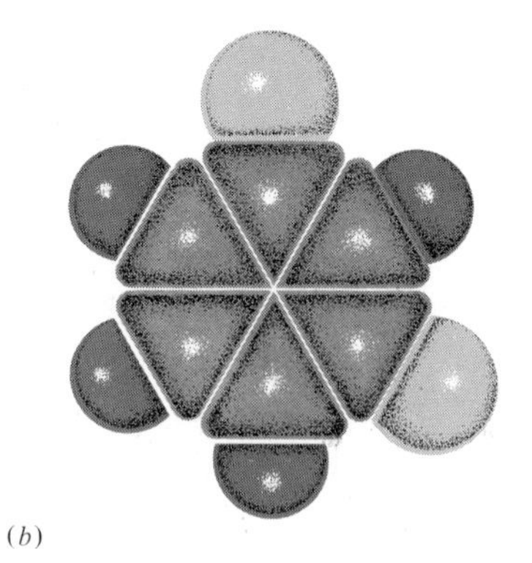

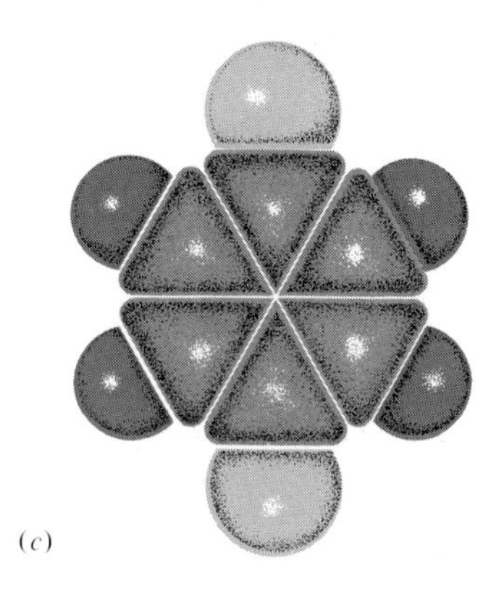

FIGURE 5-2
Space-filling models of (*a*) 1,2-dibromobenzene, (*b*) 1,3-dibromobenzene, and (*c*) 1,4-dibromobenzene.

which were usually found could then be represented by these structures (see also Fig. 5-2):

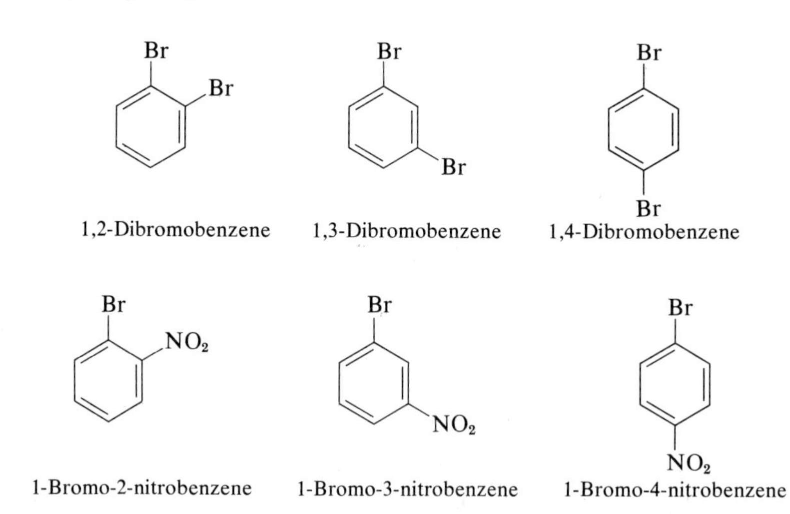

The concept of a cyclic molecule was an important advance. It was the key to the structure of benzene, the first cyclic organic compound to be recognized as such. Other chemists immediately recognized the correctness of the cyclic structure. They were not entirely satisfied with Kekulé's solution, however, so they proposed cyclic structures of their own. The Kekulé structure indicates the possibility of *two* isomeric 1,2-dibromobenzenes, and only one has ever been found. In one of the isomers the carbon atoms bearing the bromine atoms would be separated by a double bond, and in the other they would be separated by a single bond:

Br Br

Br Br

To explain the discrepancy, Kekulé proposed that the two forms of benzene, and of benzene derivatives, are in a state of equilibrium, an equilibrium so rapid in attainment as to prevent isolation of the separate molecules.

(5-3)

We now know the proposal to be wrong; *no such equilibrium exists.* However, of the several formulas proposed at that time, only Kekulé's formula was to survive. In 1965 a centennial celebration honored Kekulé for his achievement.

Benzene as a resonance hybrid

There were difficulties with Kekulé's formula and its double bonds, because, without catalysts or in the dark, benzene does not undergo the addition reactions typical of carbon-carbon double bonds. Furthermore, when it became possible to measure bond lengths with x-ray diffraction, it was found that the usual carbon-carbon single bond has a bond length of 1.54×10^{-10} m, whereas the usual carbon-carbon double bond has a bond length of 1.33×10^{-10} m. However, all the carbon-carbon bonds in benzene were found to be of the *same* length—1.39×10^{-10} m. That lies between the lengths of single and double bonds. The benzene molecule was also found to be planar with 120° bond angles.

Those observations can be explained and accounted for by the theory of resonance. Benzene is actually a *resonance hybrid* of the two Kekulé forms. It is not oscillating between the two forms in equilibrium; it is a real hybrid with properties of both forms. (Analogously, the liger found in the Salt Lake City Zoo is not oscillating between being a lion and a tiger; it is a hybrid with some of the properties of both species.) Benzene, then, is a resonance hybrid of these two forms:

(5-4)

The resonance represents a *delocalization of benzene's pi electrons.* The greater the possibilities of delocalizing the pi electrons, the greater is the stability of the resonance hybrid. Benzene is 36 kcal more stable than a real 1,3,5-cyclohexatriene would be if it existed. Let us examine Fig. 5-3. Observe that the heat of hydrogenation of cyclohexene to cyclohexane is 28.6 kcal/mol. That is the value for any isolated carbon-carbon double bond.

$$\text{Cyclohexene} + H_2 \xrightarrow{\text{Ni}} \text{Cyclohexane} + 28.6\ \text{kcal/mol} \quad (5\text{-}5)$$

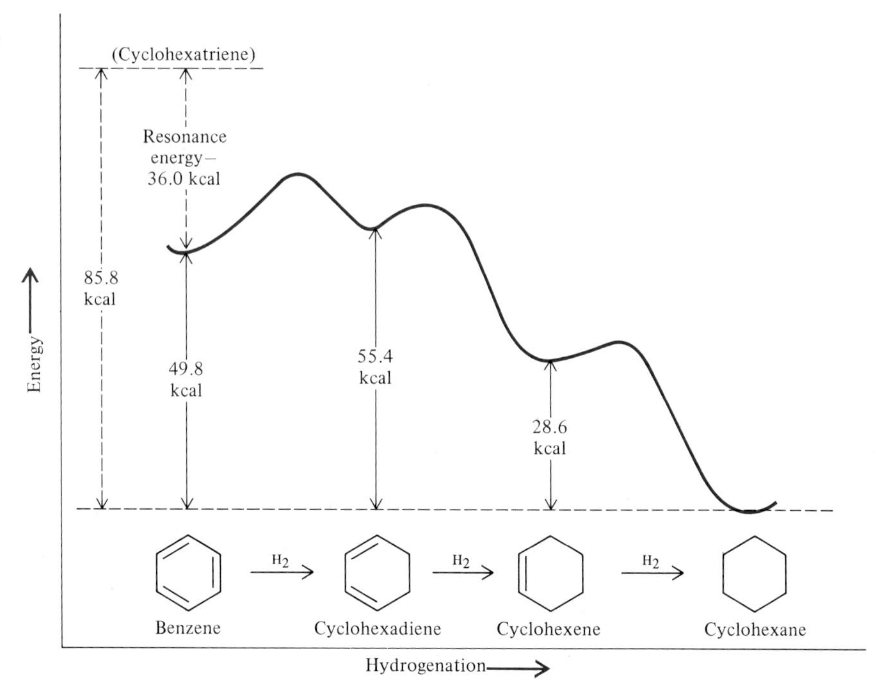

FIGURE 5-3
Heat of hydrogenation of benzene.

The heat of hydrogenation of cyclohexadiene to cyclohexane is 55.4 kcal/mol, about twice 28.6 kcal/mol, as would be expected.

$$\text{Cyclohexadiene} + 2H_2 \xrightarrow{\text{Ni}} \text{Cyclohexane} + 55.4\ \text{kcal/mol} \qquad (5\text{-}6)$$

However, the heat of hydrogenation of benzene to cyclohexane is only 49.8 kcal/mol compared with the 3 times 28.6, or 85.8, kcal/mol that would be expected for the hypothetical cyclohexatriene.

$$\text{Benzene} + 3H_2 \xrightarrow{\text{Ni, 100°C}} \text{Cyclohexane} + 49.8\ \text{kcal/mol} \qquad (5\text{-}7)$$

The difference between 85.8 kcal/mol and the observed 49.8 kcal/mol, or 36.0 kcal/mol, is the *resonance energy;* it is a consequence of the *aromaticity* of the molecule resulting from the special delocalization of the pi electrons. Aromaticity will be discussed further in the next sections of the chapter.

The heat of hydrogenation of 49.8 kcal/mol for benzene compared with 55.4 kcal/mol for cyclohexadiene indicates that benzene is even more stable than cyclohexadiene. That means that hydrogenation of benzene to

FIGURE 5-4
The bonding in benzene.

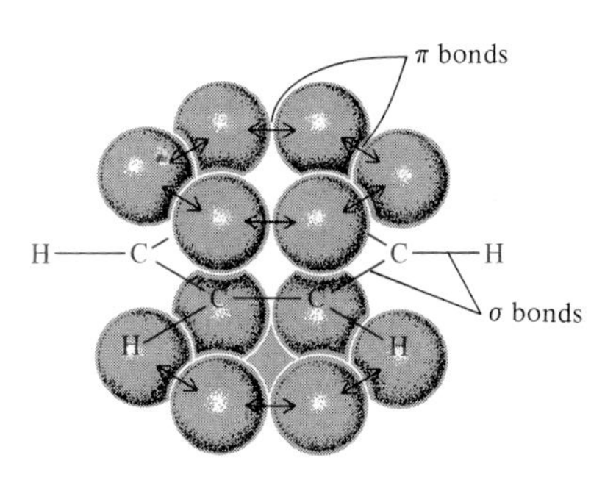

cyclohexadiene is an endothermic reaction, and it explains the high activation energy required for the hydrogenation of benzene. Whereas alkenes may be hydrogenated at even below room temperature, benzene requires a temperature of at least 100°C to be hydrogenated. Once begun, the hydrogenation proceeds completely to cyclohexane, since the intermediate cyclohexadiene and cyclohexene are more easily hydrogenated than benzene is.

The electronic structure of benzene

Each carbon atom in benzene is bonded to two adjacent carbon atoms and to a hydrogen atom. Thus the hybridization of the orbitals on carbon is sp^2 with 120° bond angles (similarly to the alkenes), and therefore the entire molecule is *planar*. Each carbon atom has one electron in a remaining $2p$ orbital. The $2p$ orbitals laterally overlap the $2p$ orbitals on adjacent carbon atoms to form a *circular pi orbital*. Since all the bonds are of the same length, each of the $2p$ orbitals must be overlapped with its neighbor on one side just as much as it is overlapped with its neighbor on the other side. One phase of the circular pi orbital is above the plane of the ring, and the other phase lies below the plane of the ring. One might think of benzene as resembling a hamburger with the carbon and hydrogen atoms, connected with their sigma bonds, being like the meat patty and the circular pi orbital resembling the two halves of the bun. See Figs. 5-4 and 5-5.

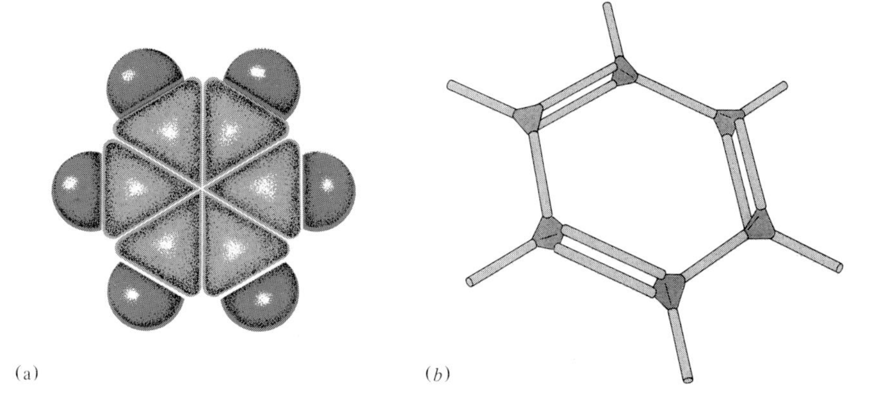

FIGURE 5-5
(*a*) Space-filling model and (*b*) Rinco model of the benzene molecule.

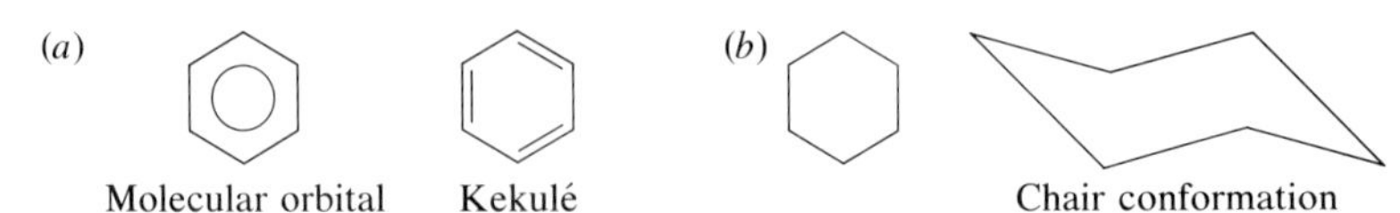

FIGURE 5-6 **Structural formulas used conventionally to represent (*a*) benzene and (*b*) cyclohexane.**

The actual uniformity of bonds and electron distribution has caused many chemists to reject the Kekulé formula in favor of representing benzene by a hexagon with a circle in the center to represent the unique pi bonding. We shall generally follow that practice. Whenever the Kekulé formula is used, however, it should be with the understanding that it is not really like cyclohexatriene. Inserting the alternate double bonds is simply a way to indicate the degree of unsaturation and to distinguish benzene from cyclohexane. (See Fig. 5-6.)

5-2 Nomenclature of benzene derivatives

The *monosubstituted derivatives* of benzene are systematically named as derivatives of benzene. Examples are

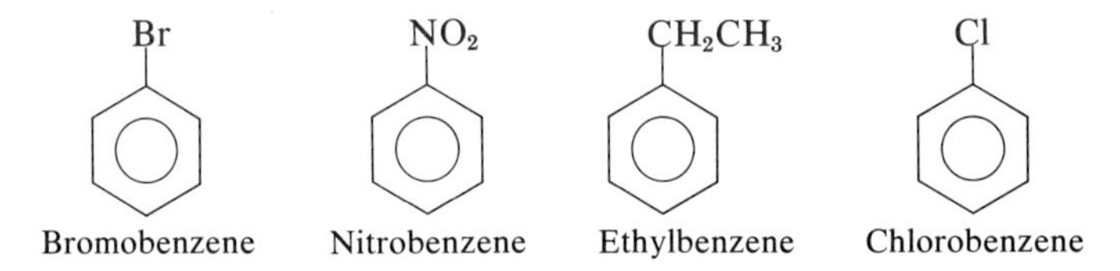

However, some of the monosubstituted derivatives of benzene have their own special names which are still in use.

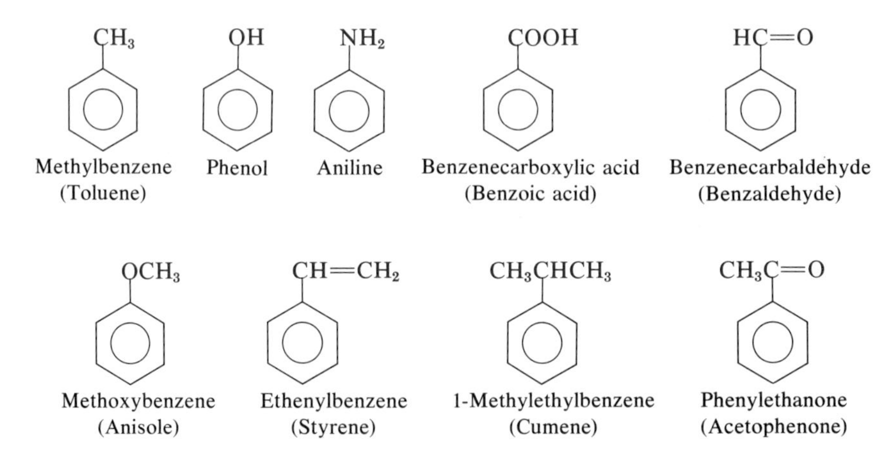

The *disubstituted* derivatives of benzene are systematically named as derivatives of benzene. However, three isomers are always possible for each given disubstituted benzene. In the IUPAC system, they are designated by locator numbers indicating the positions of the substituents. In the older system of nomenclature, they were designated by the prefixes *ortho-* (for 1,2-), *meta-* (for 1,3-), and *para-* (for 1,4-). The prefixes were abbreviated as *o*-, *m*-, and *p*- when written. The use of the prefixes is still widespread, so you should be familiar with them. Some examples are the following:

1,2-Dibromobenzene (*o*-Dibromobenzene) 1,3-Dibromobenzene (*m*-Dibromobenzene)

1,4-Dibromobenzene (*p*-Dibromobenzene) 1-Bromo-4-nitrobenzene (*p*-Bromonitrobenzene)

A few of the *disubstituted* derivatives of benzene have their own special names which are still in use. They include

1,2-Dimethylbenzene (*o*-Xylene) 3-Methylphenol (*m*-Cresol) 3-Methylaniline (*m*-Toluidine)

1,2-Benzenedicarboxylic acid (Phthalic acid)

In addition to *o*-xylene, there are the meta and para forms. Cresol and toluidine have ortho and para isomers as well as the meta forms shown.

The monosubstituted derivatives of toluene, phenol, aniline, and benzoic acid are often named as such. For example,

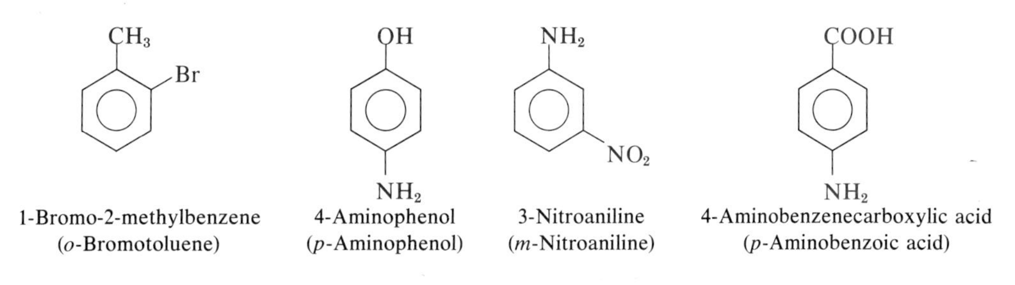

1-Bromo-2-methylbenzene (*o*-Bromotoluene) 4-Aminophenol (*p*-Aminophenol) 3-Nitroaniline (*m*-Nitroaniline) 4-Aminobenzenecarboxylic acid (*p*-Aminobenzoic acid)

If there are three or more substituents, then locator numbers *must* be used, beginning with the suffix group as number one. The 1 is often omitted in the name. (See Table 2-5 for group priorities.) Examples are

2,4-Dinitrophenol	3,5-Dinitrobenzenecarboxylic acid (3,5-Dinitrobenzoic acid)	1-Bromo-3-methyl-2-nitrobenzene (3-Bromo-2-nitrotoluene)

The benzene group, C_6H_5—, , is called the *phenyl* group. The name is used in designating compounds in which more than one benzene ring is attached to an alkane chain or a more important functional group is attached to the same chain as a benzene ring.

Diphenylmethane

2-Phenylethanol ($C_6H_5CH_2CH_2OH$)

A *benzyl* group is also still in use, but it is $C_6H_5CH_2$—, and *not* C_6H_5—, phenyl.

Chloromethylbenzene (Benzyl chloride)

Phenylmethanol (Benzyl alcohol)

1-Bromo-4-bromomethylbenzene (*p*-Bromobenzyl bromide)

PROBLEM 5-1 Name the following compounds:

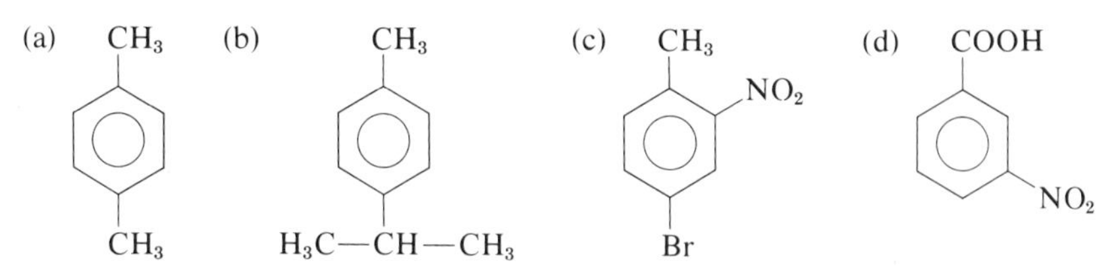

Solution to (a) The IUPAC name of this compound is 1,4-dimethylbenzene. The compound is also commonly known as *p*-xylene.

PROBLEM 5-2 Write structural formulas for the following compounds:

(a) triphenylchloromethane
(b) 3-methylphenol
(c) 2,4,6-trinitrophenol (picric acid)
(d) *m*-bromonitrobenzene

(e) *m*-xylene
(f) *p*-toluenesulfonic acid
(g) 3-methylaniline
(h) 1,3,5-trinitrobenzene
(i) *p*-methylbenzyl chloride
(j) 1,2-diphenylethane

Solution to (a)

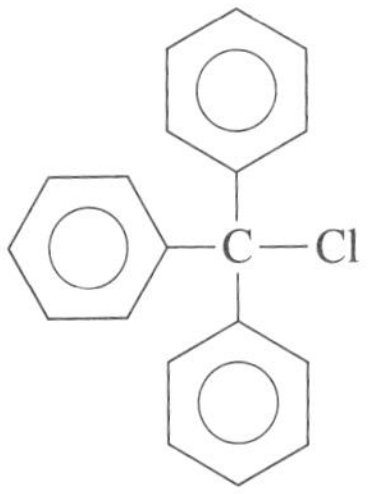

$$(C_6H_5)_3C—Cl$$

5-3 Sources of aromatic hydrocarbons

Coal tar was once the chief source of benzene, toluene, and the xylenes (dimethylbenzenes). It is the volatile material distilled off when coal is heated in ovens to produce coke. Today benzene, toluene, and the xylenes are produced in even larger amounts from petroleum. Passing a six-carbon distillation fraction over a platinum dehydrogenation catalyst at 400°C produces benzene. Toluene is produced from a seven-carbon fraction, and xylenes from an eight-carbon fraction. For example, toluene may be produced as follows:

$$\underset{\text{Heptane}}{CH_3(CH_2)_5CH_3} \xrightarrow{\text{Pt, 400°C}} \underset{\text{Toluene}}{C_6H_5CH_3} + 4H_2 \qquad (5\text{-}8)$$

The reaction is referred to as *reforming* or *platforming* (Chap. 3).

The Friedel-Crafts reaction is an important laboratory and industrial method of preparing alkylbenzenes (called *arenes*). It will be considered in Sec. 5-4.

5-4 Reactions of aromatic hydrocarbons

Electrophilic substitution

Aromatic electrophilic substitution is one of the chief reactions of benzene and other aromatic compounds. The electrophilic (electron-seeking) reagent attacks the electrons of the circular pi orbital and temporarily bonds to a ring carbon atom to form a resonance-stabilized intermediate carbonium ion. (Y represents an electrophilic reagent.)

Step 1

$$\underset{\text{Benzene}}{C_6H_5—H} + Y^+ \longrightarrow \underset{\text{A carbonium ion}}{\overset{+}{C_6H_5}\!\begin{smallmatrix}H\\Y\end{smallmatrix}} \qquad \text{(slow)} \qquad (5\text{-}9)$$

The intermediate carbonium ion is a resonance hybrid of the following electronic forms:

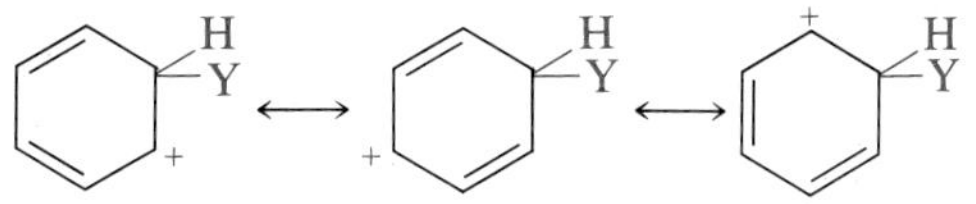

$$\text{(cyclohexadienyl cation, H and Y on sp}^3\text{ carbon, + ortho)} \longleftrightarrow \text{(+ para)} \longleftrightarrow \text{(+ ortho)} \qquad (5\text{-}10)$$

The resonance hybrid may be depicted by the following formula:

A resonance hybrid

The relatively slow, rate-determining step is then followed by a fast reaction in which the hydrogen atom attached to the same carbon atom as the entering group is abstracted by a base.

Step 2

$$\overset{+}{C_6H_5}\!\left<{}^{H}_{Y}\right. + B\!:\longrightarrow HB^+ + C_6H_5Y \quad \text{(fast)} \qquad (5\text{-}11)$$

This step restores the more stable aromatic character of the ring.

Nitration. The general reaction for nitration of benzene is as follows:

$$C_6H_6 + HONO_2 \xrightarrow{H_2SO_4} C_6H_5NO_2 + H_2O \qquad (5\text{-}12)$$

Benzene, Nitric acid, Nitrobenzene

The actual electrophilic reagent is $^+NO_2$ produced by the reaction

$$HONO_2 + 2HOSO_2OH \longrightarrow H_3O^+ + {}^+NO_2 + 2\,{}^-OSO_2OH \qquad (5\text{-}13)$$

Then,

$$C_6H_6 + {}^+NO_2 \longrightarrow \overset{+}{C_6H_5}\!\left<{}^{H}_{NO_2}\right. \xrightarrow{HSO_4^-} C_6H_5NO_2 + H_2SO_4 \qquad (5\text{-}14)$$

Nitrobenzene is easily reduced to aniline, an important intermediate in the dye industry. Nitrobenzene is a toxic yellow liquid with an odor of bitter almonds.

Sulfonation. Benzenesulfonic acid is prepared by the reaction of benzene with fuming sulfuric acid, $H_2SO_4 \cdot SO_3$.

$$\underset{\text{Benzene}}{C_6H_6} + \underset{\substack{\text{Fuming}\\\text{sulfuric acid}\\\text{(Pyrosulfuric acid)}}}{H_2SO_4 \cdot SO_3} \longrightarrow \underset{\substack{\text{Benzenesulfonic}\\\text{acid}}}{C_6H_5SO_2OH} + H_2SO_4 \qquad (5\text{-}15)$$

In sulfonation, the electrophilic reagent is sulfur trioxide, SO_3. When fuming sulfuric acid is not used, the SO_3 is produced by the reaction

$$2HOSO_2OH \longrightarrow H_3O^+ + SO_3 + {}^-OSO_2OH \qquad (5\text{-}16)$$

Although H^+ or H_3O^+ from HNO_3 or H_2SO_4 could conceivably react as an electrophilic reagent, and likely does, a productive reaction does not result because hydrogen is just being substituted for hydrogen.

Halogenation. The following is the general equation for the halogenation of benzene. X represents Cl or Br; the product obtained is chlorobenzene or bromobenzene.

$$\underset{\text{Benzene}}{C_6H_6} + X_2 \xrightarrow{\text{Fe}} C_6H_5X + HX \tag{5-17}$$

The actual electrophilic reagent is X^+. It is produced by a two-step reaction of the halogen with the iron catalyst.

$$2Fe + 3X_2 \longrightarrow 2FeX_3 \tag{5-18}$$

$$FeX_3 + X_2 \longrightarrow FeX_4^-X^+ \tag{5-19}$$

The iron for the catalyst can be in the form of powdered iron or scrap iron. In the laboratory a small iron nail works quite well. Chlorobenzene is an intermediate in the Dow process for conversion of benzene to phenol (Chap. 7).

Friedel-Crafts alkylation. The benzene ring and other aromatic rings may be *alkylated* with alkyl halides, alkenes, or alcohols with the aid of an acid catalyst. The process is called *Friedel-Crafts alkylation,* and the following reactions are examples of it:

$$\underset{\text{Benzene}}{C_6H_6} + \underset{\text{Chloroethane}}{ClCH_2CH_3} \xrightarrow{AlCl_3} \underset{\text{Ethylbenzene}}{C_6H_5CH_2CH_3} + HCl \tag{5-20}$$

$$\underset{\text{Benzene}}{C_6H_6} + \underset{\text{Ethene}}{CH_2{=}CH_2} \xrightarrow{HF} \underset{\text{Ethylbenzene}}{C_6H_5CH_2CH_3} \tag{5-21}$$

$$\underset{\text{Benzene}}{C_6H_6} + \underset{\text{Ethanol}}{HOCH_2CH_3} \xrightarrow{H_3PO_4} \underset{\text{Ethylbenzene}}{C_6H_5CH_2CH_3} + H_2O \tag{5-22}$$

The electrophilic reagent in Friedel-Crafts alkylation is a carbonium ion. In the three examples given, it is the ethyl cation, $CH_3CH_2^+$. The Lewis acid $AlCl_3$ is used as the catalyst with alkyl chlorides. It produces the ethyl cation as follows:

$$CH_3CH_2Cl + AlCl_3 \longrightarrow CH_3CH_2^+ \ {}^-AlCl_4 \tag{5-23}$$

When 1-chloropropane is used as the alkyl halide, the chief product is primarily (1-methylethyl)benzene rather than propylbenzene. That is because the intermediate propyl cation (a primary carbonium ion) rearranges to the more stable 1-methylethyl cation (a secondary carbonium ion) before reacting with the aromatic ring.

$$CH_3CH_2CH_2Cl + AlCl_3 \longrightarrow CH_3CH_2CH_2^+ \ {}^-AlCl_4 \longrightarrow CH_3\underset{+}{C}HCH_3 \ {}^-AlCl_4 \tag{5-24}$$

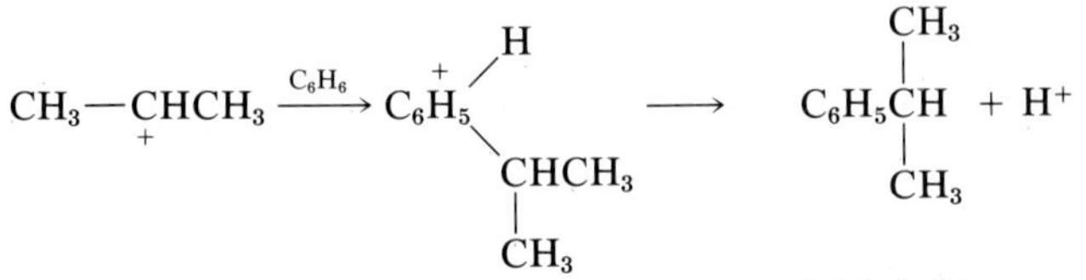

$$CH_3-\underset{+}{C}HCH_3 \xrightarrow{C_6H_6} \overset{+}{C_6H_5}\langle^{H}_{CHCH_3(CH_3)} \longrightarrow C_6H_5CH(CH_3)_2 + H^+ \qquad (5\text{-}25)$$

(1-Methyl)benzene

[In the older system of nomenclature, (1-methylethyl)benzene was called isopropylbenzene. It also has the common name of cumene.]

PROBLEM 5-3 By using the actual electrophilic reagent as given in the text, write the two-step mechanism for (a) bromination, (b) sulfonation, and (c) alkylation using ethene and HF as a catalyst.

Solution to (a)

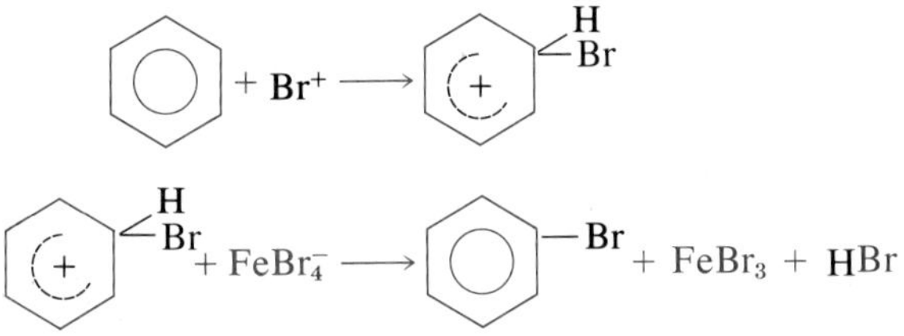

$$C_6H_6 + Br^+ \longrightarrow [C_6H_6Br]^+ \text{ (H, Br on ring carbon)}$$

$$[C_6H_6Br]^+ + FeBr_4^- \longrightarrow C_6H_5-Br + FeBr_3 + HBr$$

Friedel-Crafts acylation. Just as Friedel-Crafts alkylation is a reaction for substituting an alkyl group, R—, for hydrogen on the benzene ring, so Friedel-Crafts *acylation* is a reaction for substituting an acyl group, $R-\overset{O}{\overset{\|}{C}}-$, for hydrogen on the benzene ring. An acyl chloride, $R-\overset{O}{\overset{\|}{C}}-Cl$, may be used with $AlCl_3$ as the catalyst. (Carboxylic acid anhydrides, $R-\overset{O}{\overset{\|}{C}}-O-\overset{O}{\overset{\|}{C}}-R$, to be discussed in Sec. 9-6, may also be used to supply the acyl group.) The reaction intermediate is an *acyl cation,* instead of an alkyl cation as in alkylation. The acyl cation is resonance-stabilized, and so it has no tendency to rearrange as primary alkyl cations do.

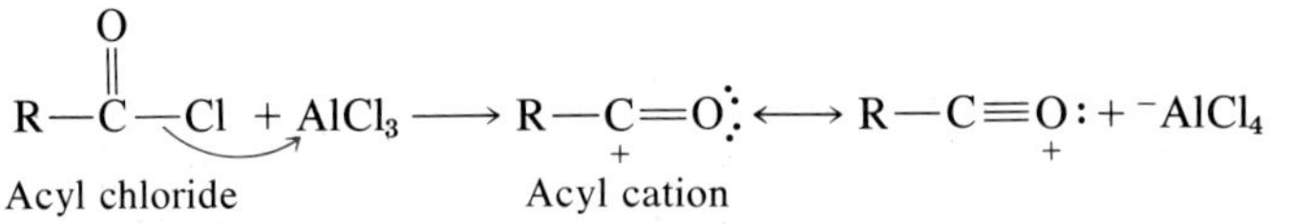

$$\underset{\text{Acyl chloride}}{R-\overset{O}{\overset{\|}{C}}-Cl} + AlCl_3 \longrightarrow \underset{\text{Acyl cation}}{R-\overset{+}{C}=\ddot{O}: \longleftrightarrow R-C\equiv\overset{+}{O}:} + {}^-AlCl_4 \qquad (5\text{-}26)$$

The electronic form $R-C\equiv\overset{+}{O}:$ is especially stable because all the atoms have a complete octet of electrons. The acyl cation then serves as an electrophilic reagent to substitute for hydrogen on the benzene ring. The mechanism is as follows:

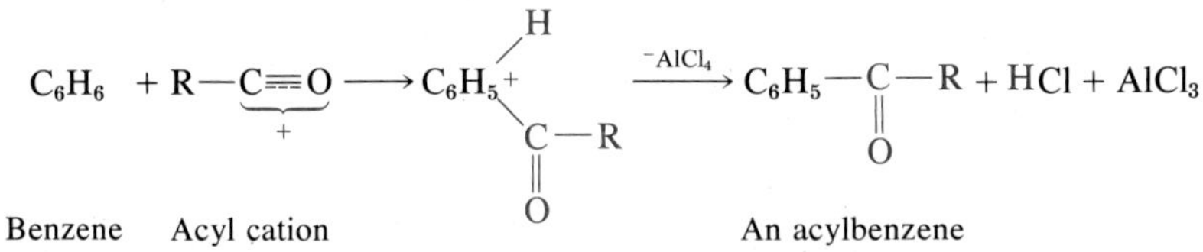

$$\underset{\text{Benzene}}{C_6H_6} + \underset{\text{Acyl cation}}{R-\underbrace{C\equiv O}_{+}} \longrightarrow \overset{+}{C_6H_5}\langle^{H}_{\overset{\|}{C}(O)-R} \xrightarrow{^-AlCl_4} \underset{\text{An acylbenzene}}{C_6H_5-\underset{\underset{O}{\|}}{C}-R} + HCl + AlCl_3 \qquad (5\text{-}27)$$

Acylation of benzene with propanoyl chloride is an example.

$$\underset{\text{Benzene}}{C_6H_6} + \underset{\text{Propanoyl chloride}}{Cl{-}\overset{\overset{\displaystyle O}{\|}}{C}{-}CH_2CH_3} \xrightarrow{AlCl_3} \underset{\text{1-Phenyl-1-propanone}}{C_6H_5\overset{\overset{\displaystyle O}{\|}}{C}{-}CH_2CH_3} + HCl \tag{5-28}$$

The resulting acylbenzene is actually an alkyl phenyl ketone (Chap. 8), which can be reduced to an alkylbenzene by using a special reaction called the *Clemmensen reduction*. The use of amalgamated zinc, Zn(Hg), instead of zinc causes the reduction of the ketone to proceed to the formation of an alkylbenzene. The Clemmensen reduction of 1-phenyl-1-propanone is an example.

$$\underset{\text{1-Phenyl-1-propanone}}{C_6H_5{-}\overset{\overset{\displaystyle O}{\|}}{C}{-}CH_2CH_3} \xrightarrow[\Delta]{Zn(Hg)/HCl} \underset{\text{Propylbenzene}}{C_6H_5CH_2CH_2CH_3} \tag{5-29}$$

The combination of the Friedel-Crafts acylation and the Clemmensen reduction is better than direct Friedel-Crafts alkylation for production of alkylbenzenes, because straight-chain alkylbenzenes can be produced without rearrangement complications and polysubstitution (next section) is not a problem.

Multiple substitution in the benzene ring

The rate of reaction and the orientation of a second substituent on the benzene ring depend on the electron-attracting ability of the first substituent. As we have seen, the mechanism for aromatic electrophilic substitution involves formation of an intermediate resonance-stabilized carbonium ion. If Y^+ represents an electrophilic reagent, the mechanism is

$$C_6H_6 + Y^+ \longrightarrow [C_6H_6Y]^+ \text{ (ring with H and Y on one carbon, positive charge delocalized)} \longrightarrow C_6H_5Y + H^+ \tag{5-30}$$

Electron-releasing substituents, G→, further delocalize the positive charge of the intermediate carbonium ion. They thereby increase the ease of formation of the ion and activate the ring for further substitution.

$$G \rightarrow [C_6H_5(H)(Y)]^+$$

Electron-withdrawing substituents, G←, tend to increase the relative magnitude of the positive charge on the intermediate carbonium ion. They thereby make formation of the ion more difficult and *so deactivate* the ring for further substitution. Electron release or withdrawal may involve *induction, resonance*, or both.

Electronegative atoms such as halogens, nitrogen, oxygen, and sulfur have a *negative* (electron-withdrawing) *inductive effect* ($-$I effect). Alkyl groups have a *positive* (electron-releasing) *inductive effect* ($+$I effect).

An alkylbenzene may be substituted with the substituent ortho, meta, or para to the alkyl group. However, ortho and para substitutions are favored because, in each case, an especially stable electronic form of the intermediate carbonium ion is produced—one in which electron release is right at the site of the positive charge. For ortho substitution, the representation of the carbonium ion is

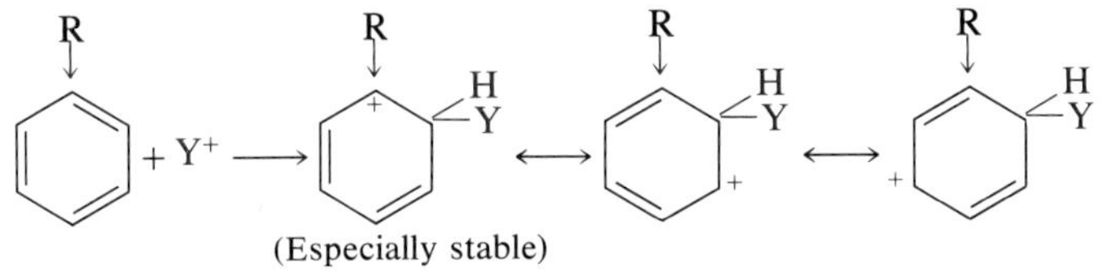

(Especially stable) (5-31)

For para substitution, the representation of the carbonium ion is

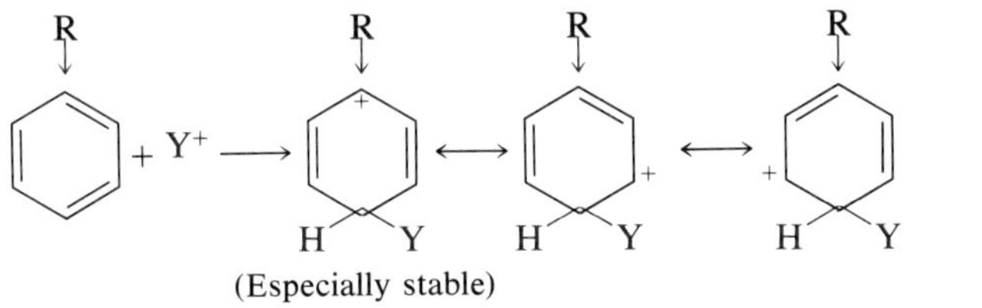

(Especially stable) (5-32)

For meta substitution, the representation of the carbonium ion is

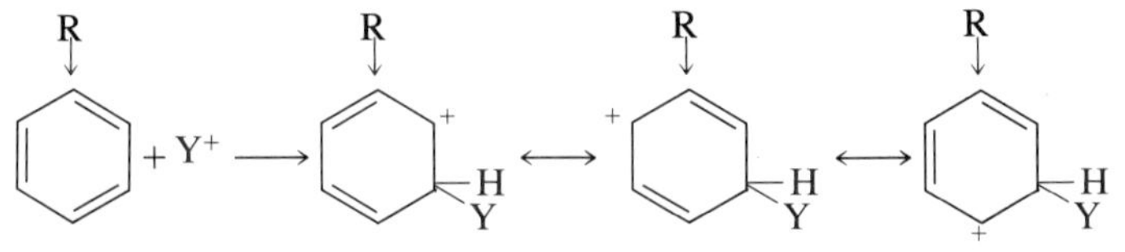

(5-33)

A group on the benzene ring with a negative inductive effect tends to direct a second substituent to the meta position because ortho or para substitution produces an intermediate carbonium ion with an especially unstable electronic form. (The same electronic forms for ortho and para substitution, which are especially stable for electron release, are especially unstable for electron withdrawal.) Any substitution which does take place therefore goes to the meta position.

The resonance effect is generally stronger than the inductive effect. It may act to reinforce the inductive effect; but if it is opposite in sign, it will be in opposition to and may overwhelm the inductive effect.

If there is conjugated pi bonding involving the substituent and the ring, then resonance may occur and cause a more pronounced electron movement than is possible with a simple inductive effect. Substituents with an unshared pair of electrons are able to shift the electrons to a pi orbital in conjugation with the ring (pi overlap with the ring). That is why such groups as

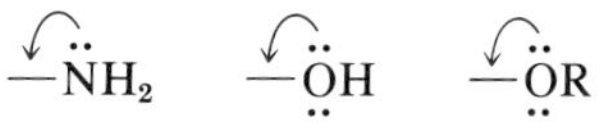

$$-\ddot{N}H_2 \qquad -\ddot{\underset{..}{O}}H \qquad -\ddot{\underset{..}{O}}R$$

exhibit a strong *positive* (electron-releasing) *resonance effect* (+R effect), which *activates* the ring. (That is the case even though nitrogen and oxygen, being electronegative elements, have negative inductive effects with a tendency to deactivate the ring.) When a substituent has a positive resonance effect, ortho and para substitution occur more readily than meta substitution because an especially stable electronic form of the intermediate carbonium ion is produced. The especially stable form has the positive charge residing on the group with the +R effect, and each of the atoms has an octet of electrons. The amino group, $-\ddot{N}H_2$, in aniline has a strong +R effect, and its ring substitution is given as an example. For ortho substitution, the representation of the carbonium ion is

$:NH_2$ + Y^+ ⟶ $:NH_2$ (H, Y) ⟷ $:NH_2$ (H, Y) ⟶

$:NH_2$ (H, Y) ⟷ $^+NH_2$ (H, Y)

(Especially stable)

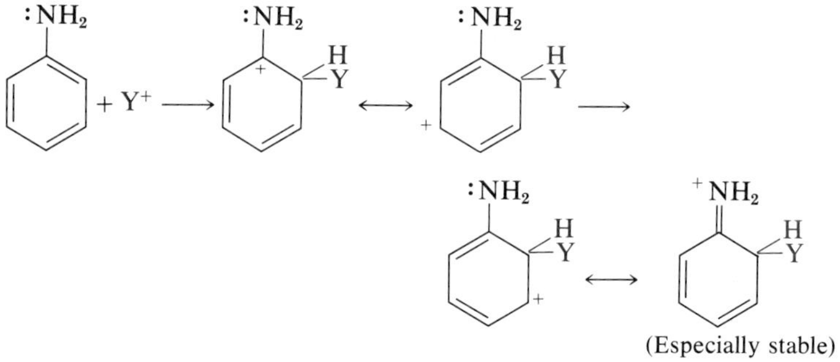

(5-34)

For para substitution, the representation of the carbonium ion is

$:NH_2$ + Y^+ ⟶ $:NH_2$ (H, Y) ⟷ $:NH_2$ (H, Y) ⟷

$:NH_2$ (H, Y) ⟷ $^+NH_2$ (H, Y)

(Especially stable)

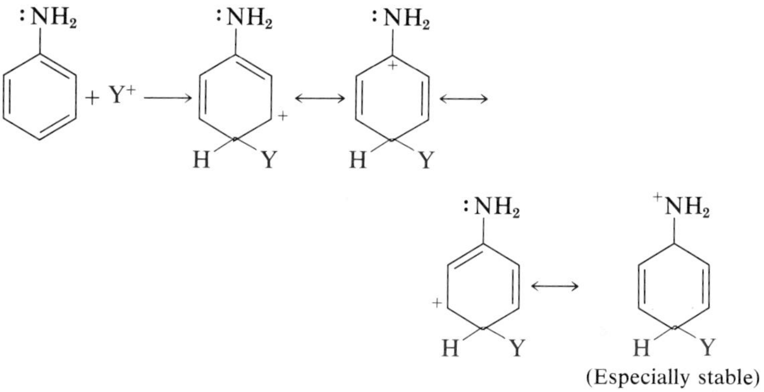

(5-35)

The especially stable electronic form of the intermediate carbonium ion is not possible for meta substitution because of the limitation of eight bonding electrons about a carbon atom. (You may wish to demonstrate that to yourself by attempting to draw such a form.)

Substituents composed of electronegative atoms bonded together with some pi bonding (multiple bonds) exhibit a strong *negative resonance effect* (−R effect). For example,

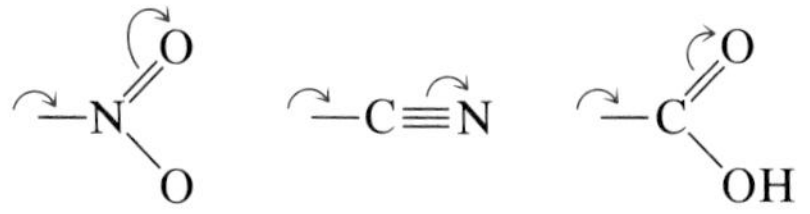

$$-NO_2 \qquad -C\equiv N \qquad -C(=O)OH$$

Such electron withdrawal depletes the electron density more at the ortho and para positions than it does at the meta position, so what substitution does occur favors the meta position. The −R effect of the nitro group, $-NO_2$, in nitrobenzene is shown as an example. Note that the ortho and para positions can be considered partially positive compared with the meta position. (The nitro group is exercising a strong negative inductive effect at the same time.)

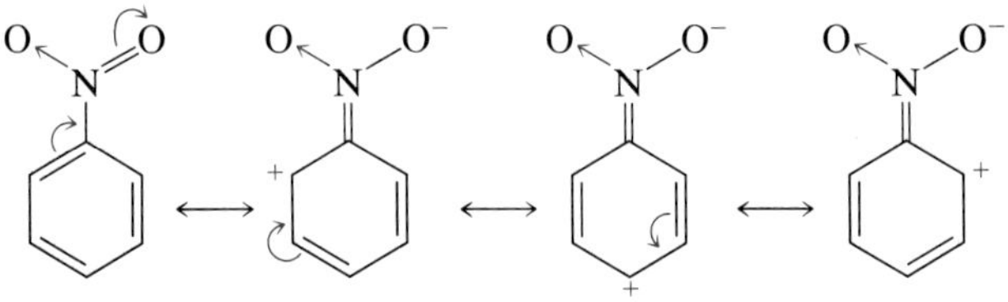

(5-36)

Forms contributing to the resonance hybrid of nitrobenzene

Halogen atoms also have unshared pairs of electrons, but their positive resonance effect is insufficient to overcome the deactivating effect of their very strong negative inductive effect. Although it is overall deactivating, halogen is still ortho,para-directing because of its positive resonance effect. It is the only deactivating group which is not meta-directing (see Table 5-1). The effects may be represented in simplified form as shown in

TABLE 5-1
Effect of substituents on further ring substitution

Activating; ortho,para-directing
Strongly $-NR_2$, $-NHR$, $-NH_2$ } (+R, −I)
Moderately $-O^-$, $-OH$, $-OR$ } (+R, −I)
Weakly $-R$, C_6H_5- (+I)
Deactivating; meta-directing (−R, −I)
$-NH_3^+$, $-NO_2$
$-NO$, $-SO_2OH$, $-SO_2Cl$
$-C\equiv N$, $-CH{=}O$, $-C(=O)-R$
$-C(=O)OH$, $-C(=O)OR$
Weakly deactivating, ortho,para-directing (large −I, small +R)
$-F$, $-Cl$, $-Br$, $-I$

Fig. 5-7, where R represents the resonance effect, I represents the inductive effect, the curved arrows represent electron shifts, and the partial charges represent relative densities of the electron cloud.

Any ortho,para-activating group is stronger in orienting effect than even the strongest of the deactivating groups. Some ortho,para-directing groups are stronger than other ortho,para-directing groups (see Table 5-1). The following reactions illustrate the principles which have been discussed. In the first reaction, bromine on the ring directs the entering nitro group to the ortho and para positions.

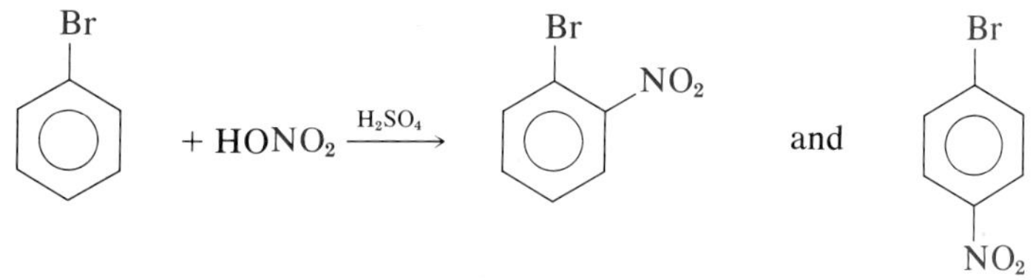

Bromobenzene + $HONO_2$ $\xrightarrow{H_2SO_4}$ *o*-Bromonitrobenzene and *p*-Bromonitrobenzene (5-37)

In the next reaction, the nitro group is deactivating and meta-directing, so *m*-bromonitrobenzene is formed.

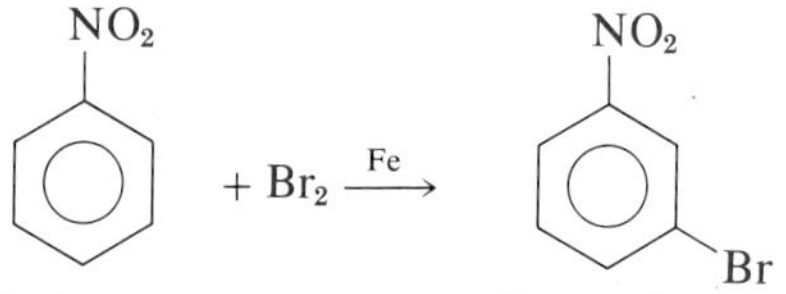

Nitrobenzene + Br_2 $\xrightarrow{Fe}$ *m*-Bromonitrobenzene (5-38)

In the following reaction, both the methyl group and the methoxy group are ortho,para-directing in a competitive way. Since the methoxy group has the stronger effect (Table 5-1), the bromine atom is substituted ortho to the methoxy group even though that positions it meta to the methyl group. Iron is not required for formation of a catalyst in this reaction because the methoxy group is such a strong activating group.

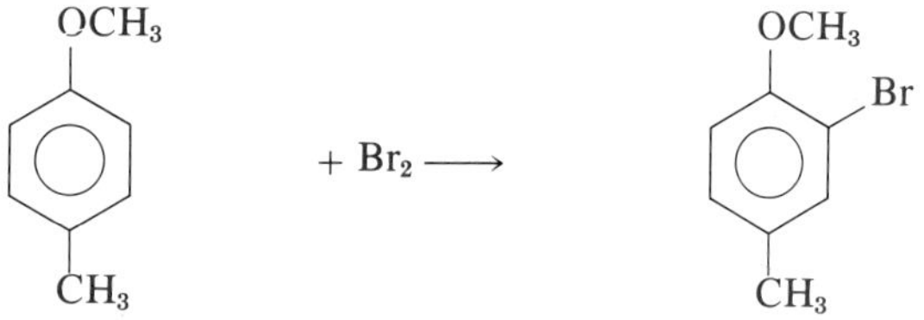

1-Methoxy-4-methylbenzene (*p*-Methylanisole) + Br_2 $\longrightarrow$ 1-Bromo-2-methoxy-5-methylbenzene (2-Bromo-4-methylanisole) (5-39)

In the last of the above examples, the ortho,para-directing methoxy group is in competition with the meta-directing nitro group. The stronger effect of the methoxy group prevails to produce substitution ortho and

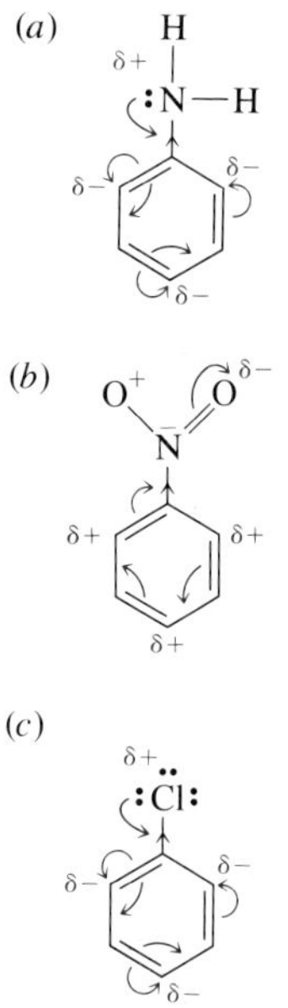

FIGURE 5-7
(*a*) The —NH_2 group has a relatively large +R effect and a −I effect; it is activating and ortho,para-directing. (*b*) The —NO_2 group has a relatively large −R effect and a −I effect; it is deactivating and meta-directing. (*c*) The —X group has a relatively small +R effect and a large −I effect; it is deactivating but ortho,para-directing.

para to itself even though that also produces an orientation of the entering bromine atom ortho and para to the meta-directing nitro group.

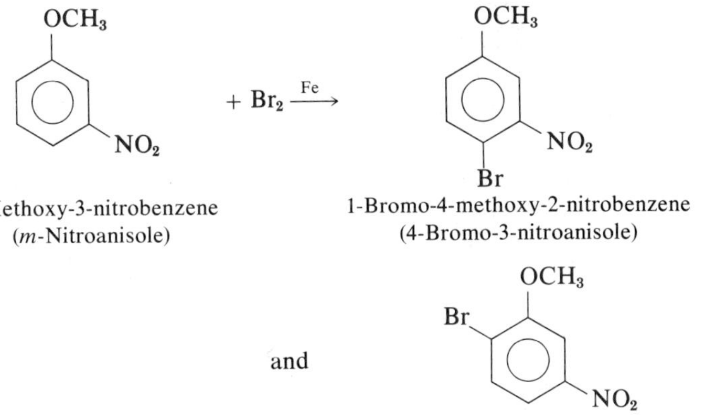

$$C_6H_4(OCH_3)(NO_2) + Br_2 \xrightarrow{Fe} C_6H_3(OCH_3)(NO_2)Br$$

1-Methoxy-3-nitrobenzene (*m*-Nitroanisole)

1-Bromo-4-methoxy-2-nitrobenzene (4-Bromo-3-nitroanisole)

and

1-Bromo-2-methoxy-4-nitrobenzene (2-Bromo-5-nitroanisole) (5-40)

Since alkyl groups are activating groups, alkylbenzenes are more reactive than benzene in all electrophilic substitution reactions including the Friedel-Crafts alkylation reaction. For that reason the Friedel-Crafts reaction is a poor one for making methylbenzene (toluene) from benzene and chloromethane. The following equations illustrate the mixture of polymethylbenzenes which result.

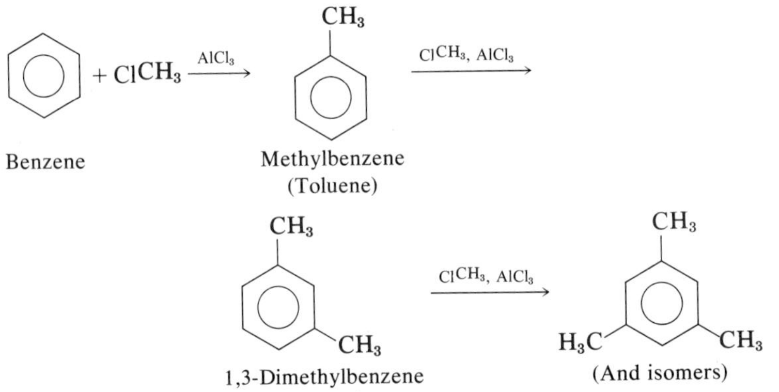

$$C_6H_6 + ClCH_3 \xrightarrow{AlCl_3} C_6H_5CH_3 \xrightarrow{ClCH_3,\ AlCl_3}$$

Benzene; Methylbenzene (Toluene)

$$C_6H_4(CH_3)_2 \xrightarrow{ClCH_3,\ AlCl_3} C_6H_3(CH_3)_3 \quad (5\text{-}41)$$

1,3-Dimethylbenzene (*m*-Xylene; also get *o* and *p* isomers); (And isomers)

On the other hand, Friedel-Crafts *acylation* does not result in a mixture of polysubstituted products because the acyl group, $R{-}\overset{O}{\overset{\|}{C}}{-}$, is a deactivating group (Table 5-1). The product of the initial acylation reaction, $C_6H_5\overset{O}{\overset{\|}{C}}{-}R$, is *less* reactive than benzene, so more benzene is acylated instead of further acylation of product taking place. However, *the Friedel-Crafts reaction will not take place if the ring carries any substituent that is more deactivating than halogen,* e.g., any meta-directing group.

PROBLEM 5-4 (a) Write the equations for the preparation of butylbenzene free of isomers. (b) Write the equations for the preparation of polysubstituted butylbenzenes.

Oxidation of alkyl side chains of benzene

Although the benzene ring itself is generally resistant to oxidation by hot potassium permanganate solution, organic substituents are subject to extensive oxidation. Alkyl groups attached to the benzene ring are oxidized to carboxyl groups, —COOH. Some examples are

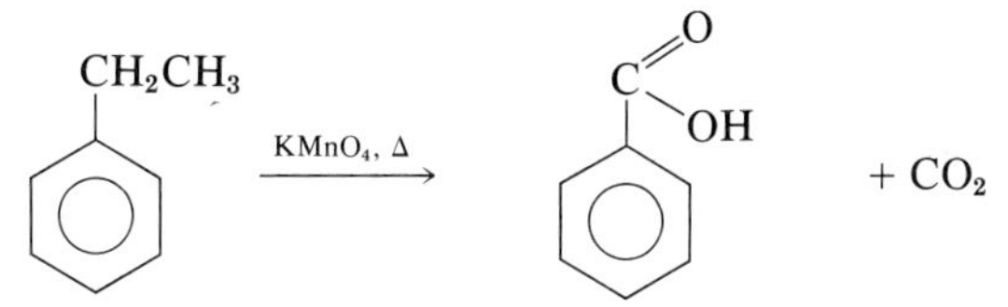

$$C_6H_5CH_2CH_3 \xrightarrow{KMnO_4,\ \Delta} C_6H_5COOH + CO_2 \qquad (5\text{-}42)$$

Ethylbenzene → Benzenecarboxylic acid (Benzoic acid)

$$CH_3C_6H_4OCH_3 \xrightarrow{KMnO_4,\ \Delta} HOOCC_6H_4OCH_3 \qquad (5\text{-}43)$$

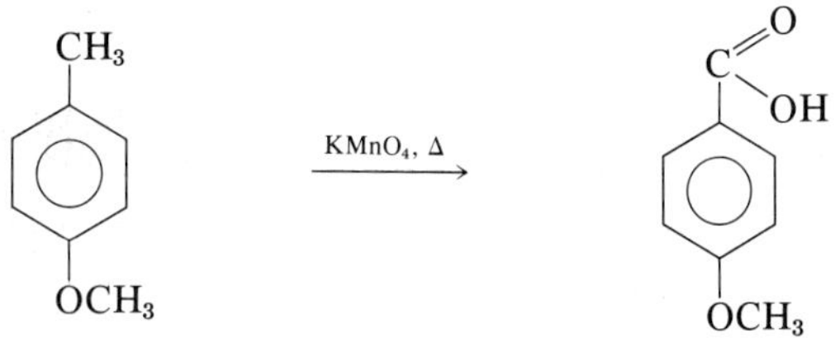

1-Methoxy-4-methylbenzene (p-Methylanisole) → (4-Methoxybenzenecarboxylic acid (p-Methoxybenzoic acid)

$$(CH_3)_2CHC_6H_4CH_3 \xrightarrow{KMnO_4,\ \Delta} HOOCC_6H_4COOH + 2CO_2 \qquad (5\text{-}44)$$

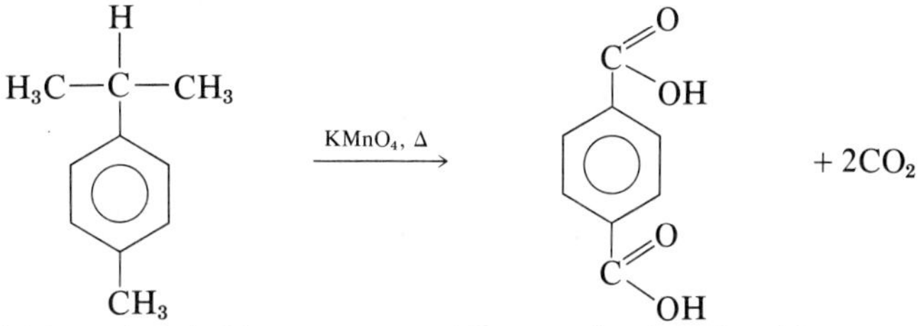

1-Methyl-4-(1-methylethyl)benzene (p-Isopropyltoluene) → 1,4-Benzenedicarboxylic acid (Terephthalic acid)

PROBLEM 5-5 Write equations for the reaction of each of the following compounds with bromine. Use iron to generate the catalyst.

(a) bromobenzene
(b) toluene
(c) nitrobenzene
(d) phenol (C_6H_5OH)
(e) benzenesulfonic acid
(f) benzoic acid

Solution to (a)

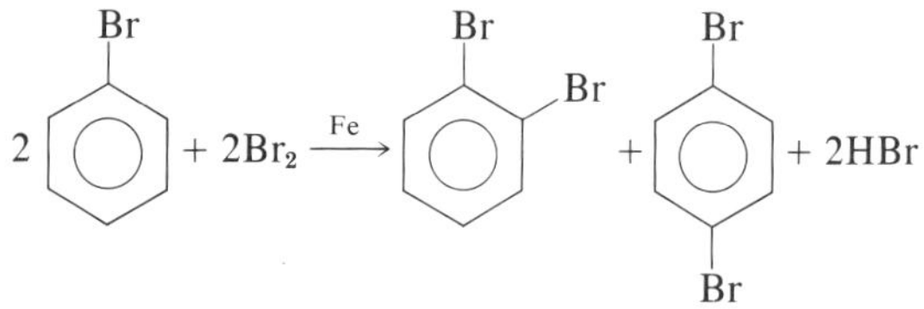

$$2\,C_6H_5Br + 2Br_2 \xrightarrow{Fe} o\text{-}C_6H_4Br_2 + p\text{-}C_6H_4Br_2 + 2HBr$$

PROBLEM 5-6 Write equations to show how the following conversions could be effected. More than one equation may be required for each conversion.

(a) toluene to *m*-bromobenzoic acid
(b) toluene to *p*-toluenesulfonic acid
(c) benzene to *p*-nitroethylbenzene
(d) toluene to *p*-nitrobenzoic acid
(e) benzene to *p*-bromoethylbenzene

Solution to (a)

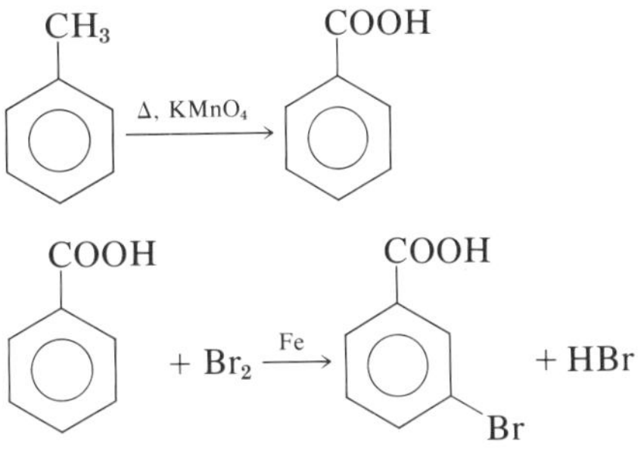

$$C_6H_5CH_3 \xrightarrow{\Delta,\ KMnO_4} C_6H_5COOH$$

$$C_6H_5COOH + Br_2 \xrightarrow{Fe} m\text{-}BrC_6H_4COOH + HBr$$

PROBLEM 5-7 Write the equation for the oxidation of each of the following with hot aqueous $KMnO_4$.

(a) butylbenzene
(b) (1,1-dimethylethyl)benzene
(c) 1,4-dimethylbenzene
(d) *o*-xylene
(e) 1,3,5-trimethylbenzene
(f) $H_3CO-C_6H_3(CH_3)-OCH_3$

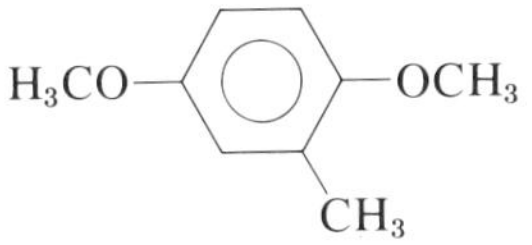

Solution to (a)

$$C_6H_5CH_2CH_2CH_2CH_3 \xrightarrow{\Delta,\ KMnO_4} C_6H_5COOH$$

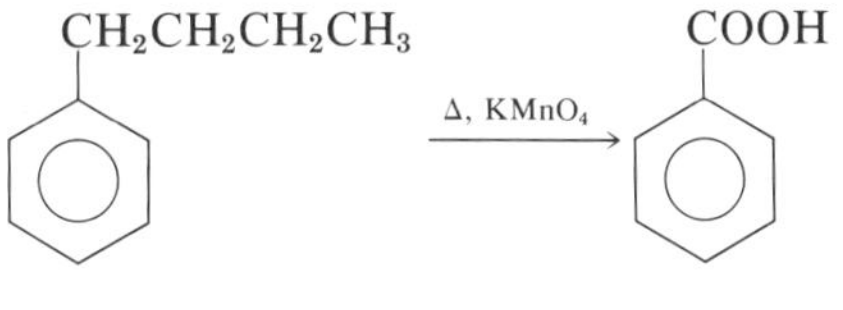

5-5 Condensed benzene ring systems

Another type of aromatic hydrocarbon is the *condensed benzene ring system*. Compounds of that type are called *aromatic polycyclic hydrocarbons,* and their chief source is coal tar. They consist of two or more benzene rings condensed together.

Naphthalene

Naphthalene is a solid, mp 80°C, consisting of two benzene rings. Structurally it is a resonance hybrid whose chief electronic forms are given in Eq. (5-45). Note that, in the older system of nomenclature, carbon atoms

1 and 2 were identified as alpha (α) and beta (β).

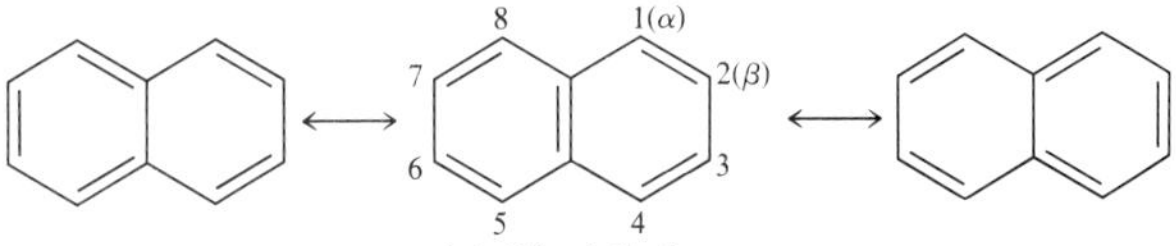

8 1(α) 7 2(β) 6 3 5 4

(a) Naphthalene

(5-45)

The fact that naphthalene is a resonance hybrid is sometimes represented by the structural formula

8 1(α) 7 2(β) 6 3 5 4

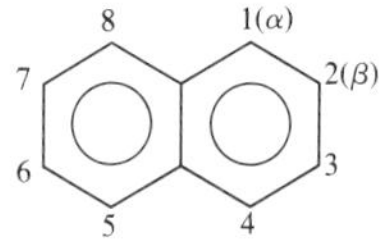

(b) Naphthalene

In this case, however, the structural formula (b) with the circles is not as satisfactory as it is for benzene, because the bonds are *not* all the same length and the distribution of pi electrons is *not* uniform over all of the bonds. The bonds actually do have some of the double-bond and single-bond character shown in formula (a). Consequently, the structure for naphthalene is quite often represented by formula (a). That particular electronic form is more stable and contributes more to the structure of the resonance hybrid than does the electronic structure shown to either its left or right. So, (a) is a better representation.

Representation of the structure of condensed benzene ring systems is governed by the *Fries rule,* which states that *the most stable electronic form and the greatest contributor to the resonance hybrid of such a system is the form which contains the greatest number of complete benzenoid rings.* Note that the central (a) form of naphthalene has two benzenoid rings, whereas the other forms, shown to the left and to the right of (a), have only one benzenoid ring each.

Electrophilic substitution in naphthalene takes place most readily in the 1, or α, position.

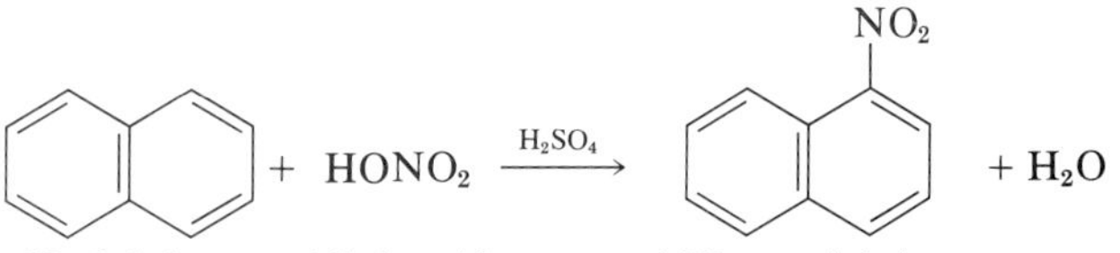

$$\text{Naphthalene} + HONO_2 \xrightarrow{H_2SO_4} \text{1-Nitronaphthalene} + H_2O \qquad (5\text{-}46)$$

Naphthalene — Nitric acid — 1-Nitronaphthalene (α-Nitronaphthalene)

Additional substitution of naphthalene also is possible. If one ring contains an activating group, that ring will be substituted by an additional substituent. If one ring contains a deactivating group, then additional substitution will occur on the other ring.

Since naphthalene has a resonance energy of 61.0 kcal/mol, less than twice the 36 kcal of resonance energy for benzene, the resonance energy per ring is less than for benzene, and naphthalene is considered less aromatic than benzene. It is rather easily hydrogenated to tetrahydronaphthalene, at which point it becomes a benzene derivative. It then requires a higher temperature for further hydrogenation to decahydronaphthalene (decalin).

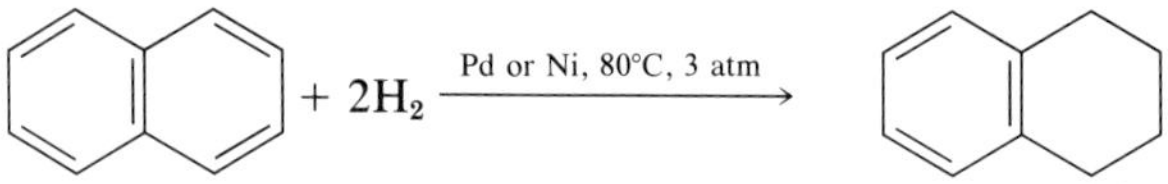

$$\text{Naphthalene} + 2H_2 \xrightarrow{\text{Pd or Ni, 80°C, 3 atm}} \text{Tetrahydronaphthalene (Tetralin)} \quad (5\text{-}47)$$

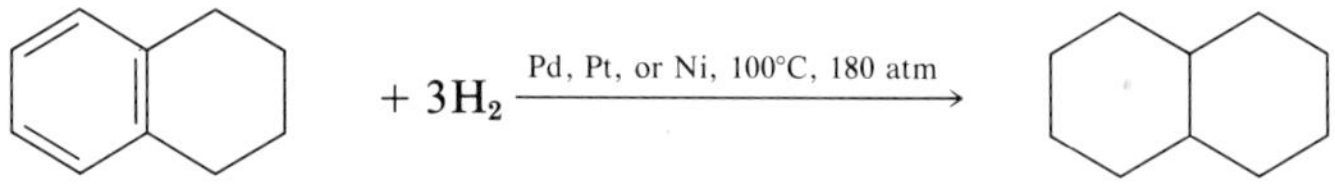

$$\text{Tetrahydronaphthalene (Tetralin)} + 3H_2 \xrightarrow{\text{Pd, Pt, or Ni, 100°C, 180 atm}} \text{Decahydronaphthalene (Decalin)} \quad (5\text{-}48)$$

Naphthalene is the most abundant single constituent of coal tar; the average yield is about 11% of the weight of tar. It is important as a source of phthalic acid (1,2-benzenedicarboxylic acid), which is used in the preparation of dyes and also polyester resins for paints. Tetralin and decalin are utilized in motor fuels, lubricants, and solvents.

Other condensed benzene ring hydrocarbons

A number of condensed benzene ring hydrocarbons are obtained from coal tar. They probably do not exist in the coal itself, but they are formed because of the intense heating required to distill out the coal tar. Examples of such compounds are

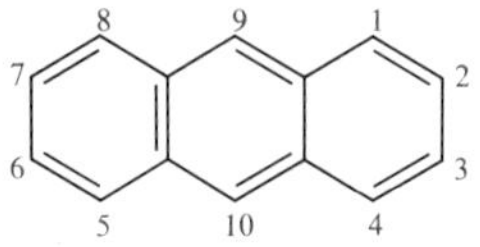

Anthracene
mp 218°C

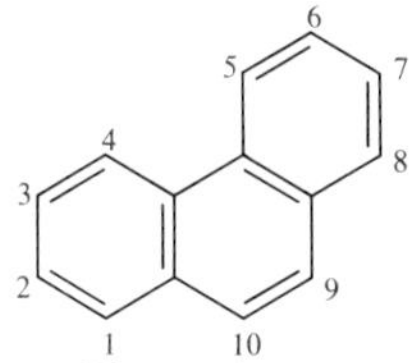

Phenanthrene
mp 101°C

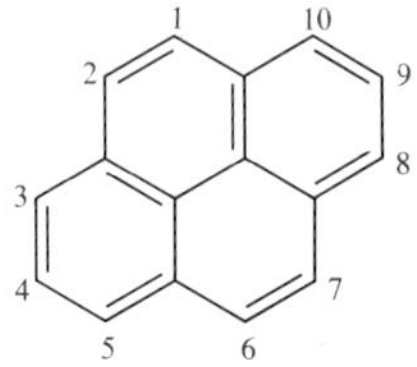

Pyrene
mp 156°C

A number of the aromatic polycyclic hydrocarbons have the ability to cause cancer when they contact the skin; that is, they are *carcinogens*. One of them, 3-4-benzpyrene, is found to the extent of 1.5% in coal tar. Another is 1,2,5,6-dibenzanthracene.

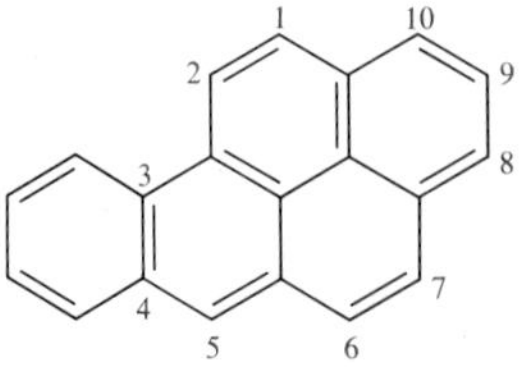

3,4-Benzpyrene

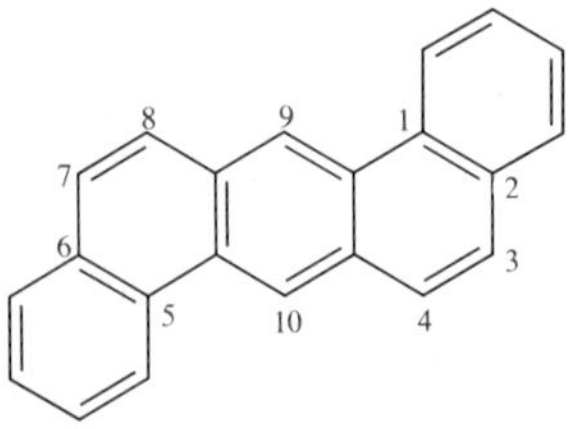

1,2,5,6-Dibenzanthracene

Compounds of that type are formed and found in the smoke of any organic compound that is burned at minimum temperatures (about 600°C); higher temperatures result in oxidation to carbon dioxide and water. The compounds have been isolated from cigarette smoke, smoke from burning paper, smoke in polluted air, and smoked fish, and they are likely to be found to some extent in all smoke. That suggests that any smoke from burning organic compounds is a health hazard and should be avoided whenever possible.

PROBLEM 5-8 Write the equations for the reaction of naphthalene with each of the following reagents: (a) bromine, (b) $Cl-\overset{\overset{O}{\|}}{C}-CH_3$, $AlCl_3$ catalyst, CS_2 solvent, (c) 2 mol of HNO_3.

Solution to (a)

$$\text{Naphthalene} + Br_2 \xrightarrow{Fe} \text{1-Bromonaphthalene} + HBr$$

Naphthalene

1-Bromonaphthalene
(α-Bromonaphthalene)

5-6 Aromaticity

Perhaps you have been wondering whether other cyclic systems with pi electrons might also be considered aromatic and be similar to benzene and its derivatives in reactions and stability. Some of those other systems are indeed aromatic, but not all of them. We shall now examine the special type of unsaturation called *aromaticity* and consider some other aromatic systems. First, in order to be aromatic, the ring must be *essentially planar* to acquire sufficiently strong pi-type overlap around the ring. Second, the *p orbitals* must be *in phase* with respect to their wave functions, as described by the quantum theory.

Hückel's rule

The requirement that the *p* orbitals be in phase is governed by *Hückel's rule* (proposed in 1931 by Erich Hückel, Institute for Theoretical Physics, Stuttgart, Germany), which states that *for a planar ring system to be aromatic, the pi electron system of the ring must consist of* (4n + 2) *pi electrons, where* n = *0, 1, 2, 3, ...* Benzene would then be a six-pi-electron system, where $n = 1$.

The cyclopropenyl cation ($C_3H_3^+$) and the cyclobutenyl dication ($C_4H_4^{2+}$) are two-ring systems which have been found to be aromatic in confirmation of the Hückel rule. They are two-pi-electron systems, where $n = 0$. They are represented by the following formulas:

Cyclopropenyl cation Cyclobutenyl dication

Examples of six-pi-electron systems, where $n = 1$, are the cyclopentadienyl anion and the cycloheptatrienyl cation:

or   or

Cyclopentadienyl anion   Cycloheptatrienyl cation

Note that it is possible to have a six-pi-electron system with only five atoms in the ring if one of the atoms is able to furnish *two* pi electrons. Also, it is possible to have rings of more than six atoms but with only six pi electrons; an example is the cycloheptatrienyl cation.

The six-pi-electron systems include a great many of the important *heterocyclic systems*. All of the *monocyclic* (one-ring) *systems* in the following section are six-pi-electron systems. A nitrogen, oxygen, or sulfur atom contributes two of the six pi electrons in the five-atom rings. Of course, benzene is the most familiar six-pi-electron system.

Aromatic heterocyclic systems

Heterocyclic compounds contain one or more atoms of elements other than carbon in the ring. Those which conform to the Hückel rule and can be almost planar are aromatic. They are important ring systems and will be encountered later in the study of alkaloids, amino acids, and nucleic acids. The most common heteroatoms are those of oxygen, sulfur, and nitrogen. The structural formulas for some of the more important ones are given here.

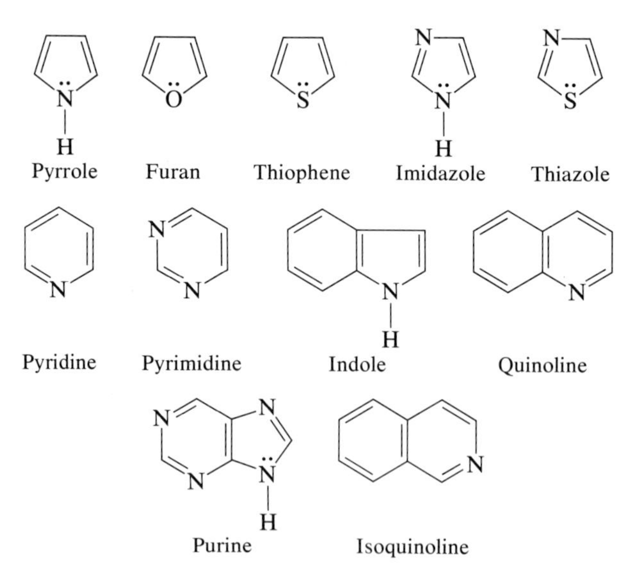

PROBLEM 5-9 Which of the following compounds conform to Hückel's rule and could be aromatic ring systems?

(a) cyclooctatetraene
(b) cyclooctatetraenyl dianion (The dianion will have two additional electrons which could occupy a pi orbital.)
(c) 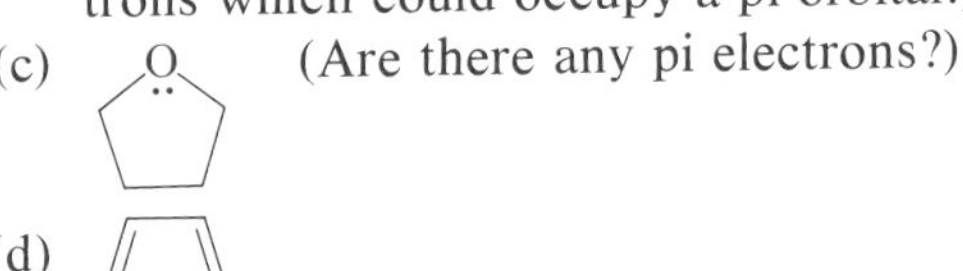

(d) 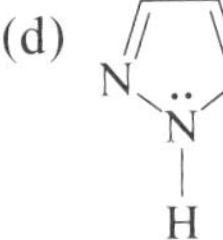

Solution to (a) The name "cyclooctatetraene" indicates an eight-carbon cyclic hydrocarbon with four double bonds (*tetraene*), assumed to be conjugated.

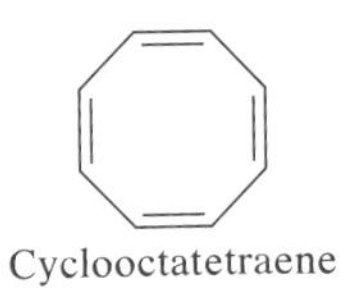

Cyclooctatetraene

There are two pi electrons for each double bond for a total of eight. The Hückel rule states that the number of pi electrons for a stable aromatic monocyclic system is equal to $4n + 2$, where $n = 0, 1, 2, 3, \ldots$. Such a system could then contain 2, 6, 10, 14, 18, . . . pi electrons. Therefore, eight pi electrons do not conform to the Hückel rule, and cyclooctatetraene is not an aromatic system.

Summary

Aromatic compounds are highly unsaturated cyclic compounds of a special type. To be aromatic, the ring must be essentially *planar* to permit *strong pi bonding* all around the ring.

Benzene, C_6H_6, and its derivatives are among the best known aromatic compounds. Benzene does not undergo the electrophilic *addition* reactions typical of alkenes, but it does undergo *electrophilic substitution reactions*. It forms only a single *monosubstituted* benzene from a given reagent, which indicates that all six carbon atoms are equivalent. There are three isomeric *disubstituted* benzenes.

In 1865 Kekulé proposed for benzene a planar, cyclic structure of six carbon atoms with alternating single and double bonds and one hydrogen atom attached to each carbon atom. Benzene is a *resonance hybrid* of the two chief electronic structures. It is more stable than the hypothetical cyclohexatriene by 36 kcal/mol. That energy is called the *resonance energy* of benzene. Benzene is so stable that it is not oxidized by boiling $KMnO_4$ solution, and a temperature of 100°C or higher is required to reduce it with hydrogen to cyclohexane.

All the carbon-carbon bond lengths in benzene are the same, 1.39×10^{-10} m. That is in between the lengths of carbon-carbon single and double bonds. The hybridization of the atomic orbitals of carbon in benzene is *sp*^{2}, and the bond angles are 120°. The *p* orbitals, in which each carbon atom has a remaining electron, are overlapped sideways all around the ring to form a *circular pi orbital* with one phase on one side of the plane of the ring and the other phase on the other side of the ring.

The isomeric *disubstituted derivatives* of benzene are commonly designated by the prefixes *ortho*- (for 1,2-), *meta*(for 1,3-), and *para*- (for 1,4-), which are abbreviated *o*-, *m*-, and *p*- when written. If there are three or more substituents, locator numbers are used; they begin with the group of highest nomenclature priority as number 1. The benzene group, C_6H_5—, is called *phenyl,* and the group $C_6H_5CH_2$— is still frequently referred to as *benzyl*.

Coal tar is the chief source of aromatic hydrocarbons, but large quantities of benzene, toluene, and the xylenes are made by passing alkanes of appropriate numbers of carbon atoms over a platinum catalyst at 400°C. The process is called *reforming* or *platforming*. The *Friedel-Crafts reaction* is an important laboratory and industrial method of preparing *arenes* (alkylbenzenes).

The mechanism for *aromatic electrophilic substitution* is a two-step reaction involving an intermediate resonance-stabilized carbonium ion. The chief aromatic electrophilic substitution reactions are (a) *nitration,* (b) *sulfonation,* (c) *halogenation,* (d) *Friedel-Crafts alkylation,* and (e) *Friedel-Crafts acylation*. The acylbenzene produced by Friedel-Crafts acylation may be reduced to an alkylbenzene by the *Clemmensen reduction*.

The rate of reaction and the orientation of a second substituent on the benzene ring depend on the electron-attracting ability of the first substituent. *Electron-releasing substituents* further delocalize the positive charge of the intermediate carbonium ion; they thereby increase the ease of formation of the ion and *activate* the ring for further substitution. *Electron-withdrawing groups* tend to increase the magnitude of the positive charge on the intermediate carbonium ion; they thereby make formation of the ion more difficult and *deactivate* the ring for further substitution.

Electron release or withdrawal may involve *induction, resonance,* or both. *Electronegative atoms* have a *negative* (electron-withdrawing) *inductive effect* (−I effect). *Alkyl groups* have a *positive* (electron-releasing) *inductive effect* (+I effect). Substituents with an unshared pair of electrons are able to shift the electrons to a pi orbital in conjugation with the ring (*pi overlap* with the ring), so such groups exhibit a strong *positive* (electron-releasing) *resonance effect* (+R effect), which activates the ring. Substituents composed of electronegative atoms bonded together with some pi bonding (multiple bonds) will exhibit a strong *negative* (electron-withdrawing) *resonance effect* (−R effect). Most deactivating groups withdraw electrons by both induction and resonance (−I and −R effects) and are meta-directing. An exception is halogen, which is deactivating with a large −I effect but ortho,para-directing with a small +R effect. Any ortho,para-directing group is stronger in orienting effect than any competing meta-directing group.

Although the benzene ring itself is resistant to oxidation by boiling $KMnO_4$, organic substituents are subject to extensive oxidation to a carboxyl group.

Naphthalene is an aromatic hydrocarbon consisting of two *condensed* benzene rings. It is a resonance hybrid. The *Fries rule* is that the *most stable electronic form,* and the greatest contributor to the resonance hybrid of a *polycyclic hydrocarbon,* is that which contains the *greatest number of complete benzenoid rings*. The resonance energy of naphthalene is less per ring than that of benzene, which indicates that naphthalene is less stable and less aromatic than benzene is. Substitution in naphthalene takes place most readily in the 1 position.

Polycyclic aromatic compounds are formed and found in the smoke of any organic compound that is burned at about 600°C. Some of the polycyclic hydrocarbons formed in that way have been found to cause cancer, so inhalation of all smoke should be avoided.

To be aromatic, a ring must conform to the *Hückel rule,* which is that it must contain $(4n + 2)$ pi electrons, where $n = 0, 1, 2, 3, \ldots$. The most common type of aromatic ring is that in which $n = 1$, so that the ring contains six pi electrons. *Heterocyclic rings* are those which contain some other atoms in addition to the carbon atoms in the ring. The chief heteroatoms are nitrogen, oxygen, and sulfur. Some of the more important heterocyclic aromatic ring systems are pyrrole, furan, thiophene, imidazole, thiazole, pyridine, pyrimidine, indole, quinoline, and purine.

Key terms

Activating groups: substituents on the benzene ring which are electron-releasing and which thereby activate the ring for further substitution.

Arene: a derivative of benzene in which an alkyl group is substituted for hydrogen; an alkylbenzene.

Aromatic compound: an unsaturated, cyclic hydrocarbon in which the ring is essentially planar with strong pi overlap around the ring and with the p orbitals in phase; it contains $4n + 2$ pi electrons.

Aromatic electrophilic substitution: a chief reaction of benzene and other aromatic compounds; an electrophilic reagent is substituted for one or more hydrogens on the ring.

Aromatic polycyclic hydrocarbon: a compound consisting of two or more benzene rings condensed together.

Aromaticity: the special type of unsaturation possessed by aromatic compounds; it is defined by Hückel's rule.

Benzene: the simplest and best known aromatic compound, C_6H_6.

Circular pi orbital: the pi-bonding orbital formed by the lateral overlap of the $2p$ orbitals of adjacent carbon atoms in the benzene ring; one phase is above the plane of the ring, and the other phase is below the plane of the ring.

Clemmensen reduction: reduction of an acylbenzene with Zn(Hg)/HCl to yield a straight-chain alkylbenzene.

Condensed benzene ring system: a compound consisting of two or more benzene rings condensed together.

Deactivating groups: substituents on the benzene ring which are electron-withdrawing and which, for that reason, deactivate the ring for further substitution.

Electron-releasing substituents: atoms or groups which shift electron density toward the benzene ring and delocalize the positive charge of the intermediate carbonium ion; they activate the ring for further substitution.

Electron-withdrawing substituents: atoms or groups which shift electron density away from the benzene ring and concentrate the positive charge of the intermediate carbonium ion; they lessen the reactivity of the ring for further substitution.

Friedel-Crafts reaction: an aromatic electrophilic substitution reaction in which an alkyl group is introduced to the benzene ring by using an alkyl halide, alkene, or alcohol; or an acyl group is introduced by using an acyl halide or carboxylic acid anhydride as the acylating agent.

Fries rule: the most stable electronic form, and the greatest contributor to the resonance hybrid, of a condensed benzene ring system is that which contains the greatest number of complete benzenoid rings.

Heteroatom: an atom other than carbon which occurs in the rings of heterocyclic compounds; the most common heteroatoms are oxygen, sulfur, and nitrogen.

Heterocyclic compound: a cyclic compound which contains one or more atoms of elements other than carbon in the ring.

Hückel's rule: for a planar ring system to be aromatic, the pi electron system of the ring must consist of $(4n + 2)$ pi electrons, where $n = 0, 1, 2, 3, \ldots$.

Kekulé structural formula: a formula for benzene which shows a planar, cyclic structure of six carbon atoms with alternating double and single bonds and one hydrogen atom attached to each carbon atom; the actual structure is a resonance hybrid.

Meta: the prefix commonly used to denote 1,3 disubstitution on the benzene ring.

Meta-directing group: a group directing any further substituents on the ring to the meta (3) position; a deactivating group.

Negative inductive effect: the effect of an electron-withdrawing substituent, which pulls electron density away from the benzene ring. It is symbolized as $-\mathrm{I}$.

Negative resonance effect: the effect of a substituent composed of electronegative atoms with multiple bonds: it is electron-withdrawing and deactivates the ring. It is symbolized as $-\mathrm{R}$.

Ortho: the prefix commonly used to denote 1,2 disubstitution on the benzene ring.

Ortho,para-directing group: a group directing any further substituents on the ring to the ortho (2) and para (4) positions; an activating group, except halogen.

Para: the prefix commonly used to denote 1,4 disubstitution on the benzene ring.

Phenyl group: a benzene molecule from which a hydrogen atom has been abstracted; the C_6H_5— group.

Positive inductive effect: the effect of an electron-releasing substituent, which shifts electron density toward the benzene ring. Symbolized as +I.

Positive resonance effect: the effect of a substituent with an unshared pair of electrons which shifts to a pi orbital in conjugation with the ring. It activates the ring, and it is symbolized as +R.

Resonance energy: the amount by which the potential energy of a compound is lowered as a result of delocalization of its pi electrons through hybridization of two or more possible electronic forms for the compound.

Additional problems

5-10 Write structural formulas for the following compounds:

(a) 2,4-dinitrophenol
(b) 3,5-dinitrobenzoic acid or 3,5-dinitrobenzenecarboxylic acid
(c) *p*-aminophenol
(d) *p*-bromoaniline
(e) *p*-aminobenzenesulfonic acid or 4-aminobenzenesulfonic acid
(f) *p*-nitrotoluene or 1-methyl-4-nitrobenzene
(g) (1,1-dimethylethyl)benzene
(h) 2,4,6-trinitrotoluene (TNT) or 1-methyl-2,4,6-trinitrobenzene
(i) triphenylmethane
(j) 2,4-dichlorophenoxyacetic acid (2,4-D; a herbicide)

5-11 Name the following compounds.

(a) benzene ring with CH_3 and OH (ortho)

(b) $CH{=}CH_2$ on benzene ring, para NO_2

(c) CH_2Cl on benzene ring, para NO_2

(d) benzene ring with H—C=O and OH (ortho)

(Name as a hydroxybenzenecarbaldehyde.)

(e) three benzene rings on C—H

(f) benzene ring—$CH_2CH_2CH_2OH$

(g) benzene ring with NH_2 and CH_3 (meta)

5-12 Write an equation for the reaction of each of the following compounds with bromine. Use iron to generate the catalyst.

(a) aniline ($C_6H_5NH_2$)
(b) anisole ($C_6H_5OCH_3$)
(c) *p*-methylanisole
(d) salicylic acid,

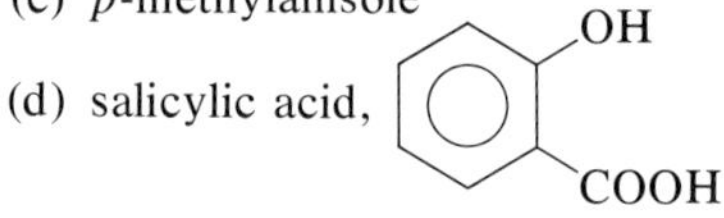

(e) *p*-nitroaniline
(f) *m*-nitroanisole

5-13 By using appropriate reference books such as the *Merck Index,* find the toxicity and hazards of the following compounds:

(a) benzene
(b) toluene
(c) aniline
(d) naphthalene

5-14 Write equations to show how the following conversions could be effected. (More than one reaction may be required.)

(a) toluene to *m*-nitrobenzoic acid
(b) benzene to *p*-nitroethylbenzene
(c) resorcinol,

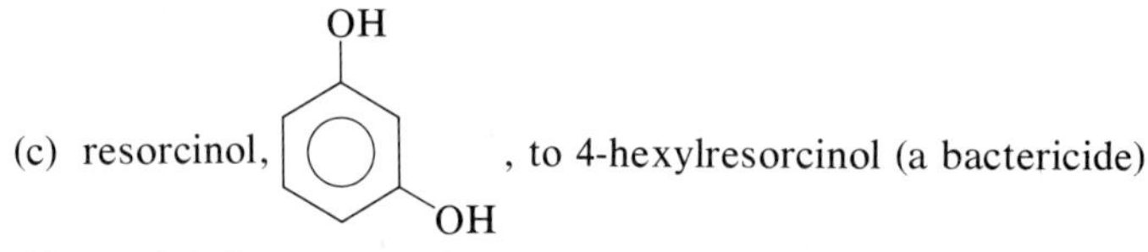

, to 4-hexylresorcinol (a bactericide)

(d) naphthalene to 4-aminonaphthalenesulfonic acid (*Hint:* An $—NO_2$ group can be reduced to an amino group with Sn and HCl.)
(e) toluene to 2,4,6-trinitrotoluene (TNT)
(f) toluene to *p*-aminobenzoic acid (a *biotic,* or growth-promoter)
(g) benzene to phenylcyclohexane
(h) 4-methylpyridine to 4-pyridinecarboxylic acid (isonicotinic acid)
(i) naphthalene to a mixture of 1,5- and 1,8-dinitronaphthalenes
(j) nicotine, N, N, CH_3, to 3-pyridinecarboxylic acid (nicotinic acid, part of vitamin B complex)

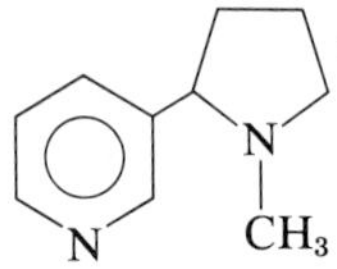

5-15 Write equations for the following reactions:

(a) 1-aminonaphthalene and HNO_3
(b) 1-nitronaphthalene and HNO_3
(c) furan and $Cl—\overset{\overset{\displaystyle O}{\|}}{C}—CH_3$, in the presence of $AlCl_3$ (goes to 2 position with numbering beginning with the heteroatom)
(d) benzene and chloromethylbenzene (benzyl chloride), in the presence of $AlCl_3$
(e) 1-chloro-4-chloromethylbenzene (*p*-chlorobenzyl chloride) and methylbenzene (toluene), in the presence of $AlCl_3$
(f) 1-chloromethyl-4-nitrobenzene (*p*-nitrobenzyl chloride) and benzene, in the presence of $AlCl_3$

(g) 4-methylaniline (*p*-toluidine) and aqueous bromine

(h) 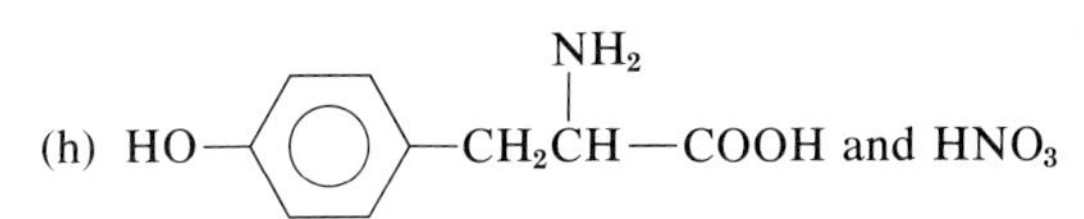

(i) octane heated at 400°C over a Pt catalyst
(j) hydrogenation of toluene at 200°C, 136 atm, over a nickel catalyst

5-16 (a) How many trimethyl benzenes are there?
(b) Which one of these would yield only one mononitro product?
(c) Which one could yield two mononitro products?
(d) Which one could yield three?

5-17 (a) Write all the resonance structures (five) for phenanthrene.
(b) According to the Fries rule, which should contribute most to the structure of the resonance hybrid?
(c) What can you speculate about the double bond character of the C-9–C-10 bond?

5-18 Draw structural formulas for the following compounds:

(a) *o*-chloroacetophenone
(b) 2-bromonaphthalene
(c) *o*-bromoanisole
(d) 2-phenylcyclohexanol
(e) 4-nitropyridine (N is number 1)
(f) 3-nitrophenanthrene

6

ORGANIC HALIDES

The chief use of organic halides is as reaction intermediates. In general, the halides are readily formed by the use of halogen, hydrohalogen, or other halogen-containing reagents. The halogen in the halides is then substituted by a great variety of nucleophilic reagents to produce alcohols, ethers, nitriles, alkynes, amines, and so on. However, some of the organic halides have an importance of their own. Fluorochlorocarbons, called Freons, are useful as refrigeration gases (Fig. 6-1). Alkyl chlorides

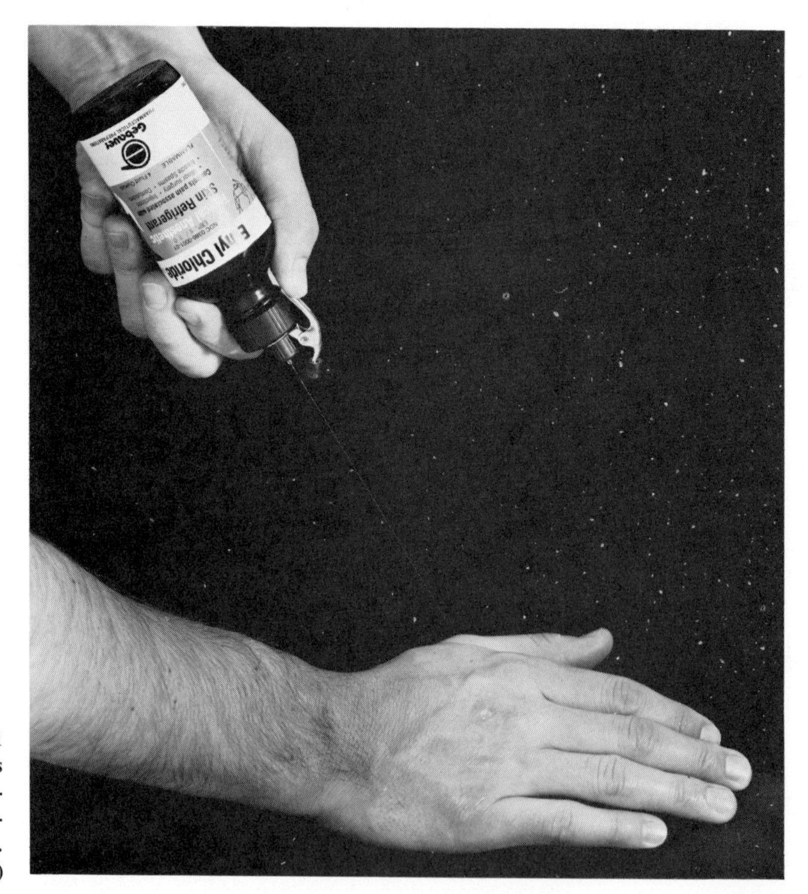

FIGURE 6-1
Ethyl chloride (chloroethane) is an organic halide that is frequently used as a topical anesthetic for injured athletes.
(Gebauer Chemical Company)

TABLE 6-1
Some useful organic halides

Formula	IUPAC name	Common name	Physical state	Use
CF_2Cl_2	Dichlorodifluoromethane	Freon-12	Gas	Refrigeration gas
CH_3CH_2Cl	Chloroethane	Ethyl chloride	Gas	External local anesthetic
CH_3Br	Bromomethane	Methyl bromide	Gas	Vermicide
Cl—⌬—Cl	1,4-Dichlorobenzene	*p*-Dichlorobenzene	Solid	Larvicide; moth repellant
$Cl_2C{=}CCl_2$	Tetrachloroethene	Tetrachloroethylene	Liquid	Dry-cleaning solvent
$(Cl{-}C_6H_4)_2CH{-}CCl_3$	1,1,1-Trichloro-2,2-bis(*p*-chlorophenyl)ethane*	DDT (di-*p*-chlorodiphenyl-trichloroethane)	Solid	Long-lasting insecticide

* In the IUPAC system, the prefix *bis*-, rather than the prefix *di*-, is used before a complex group. That is to make clear that the entire group is present in two locations.

have long been used as industrial and laboratory solvents and as dry-cleaning agents. A number of the organic halides serve as important insecticides, vermicides, and moth repellants; some examples are listed in Table 6-1. Both *alkyl halides* (RX) and *aryl halides* (aromatic halides, ArX) are included.

In general, the organic halides are not found in nature, but exceptions exist. Scientists at the Scripp Institution of Oceanography, La Jolla, California, have isolated small amounts of a number of halogenated terpenes from the sea hare *Aplysia californica*. Such compounds have also been found in red algae. The following formula is an example of one of them.

$$Br{-}CH_2{-}C(CH_3)(Cl){-}CH{=}CH{-}CH(Cl){-}C(CH_3)(Cl){-}CH{=}CH{-}Br$$

The significance of these compounds is as yet unknown.

6-1 Organic halide nomenclature

You have already been introduced to some organic halides as products of reactions of alkanes and alkenes, so the names are not strangers to you. Some principles of naming will be useful, however.

Naming of alkyl halides

Alkyl halides are commonly named by naming the alkyl group attached to halogen and adding the name of the halide; for example, $CH_3CH_2CH_2I$ is *n*-propyl iodide.

In the IUPAC system the halogen is named in its combining form, i.e., fluoro-, chloro-, bromo-, or iodo-, followed by the name of the longest continuous carbon chain holding the halogen atom. A locator number indicates the position of the halogen on the chain. For example, $CH_3CH_2CH_2I$ is 1-iodopropane and CH_3CHICH_3 is 2-iodopropane.

Common names in use for dihalides may be illustrated by the following examples: CH_2BrCH_2Br is ethylene bromide; $CH_3CHBrCH_2Br$ is propylene bromide; and $BrCH_2CH_2CH_2Br$ is trimethylene dibromide. The respective IUPAC names are 1,2-dibromoethane, 1,2-dibromopropane, and 1,3-dibromopropane.

Naming of halogenated cycloalkanes

If halogen is the only substituent, it is assumed to be at position number 1, but the 1 is omitted in the name. If there is more than one substituent, the substituent of highest priority (see Table 2-5) is numbered 1. Examples of halogenated cycloalkanes are

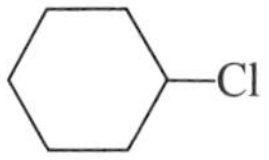

Cl
Chlorocyclohexane

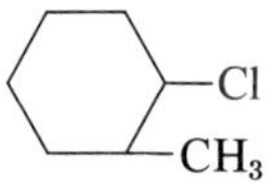

Cl
CH_3
2-Chloromethylcyclohexane

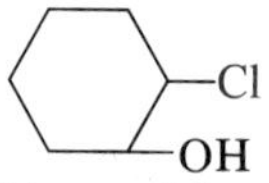

Cl
OH
2-Chlorocyclohexanol

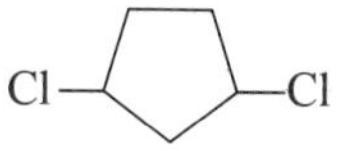

Cl Cl
1,3-Dichlorocyclopentane

The common and IUPAC names of these compounds are the same.

Naming of aryl halides

Aryl halides are compounds having halogen as a substituent on an aromatic ring. They are named as halogen-substituted benzene, toluene, phenol, aniline, and so on. Examples are as follows:

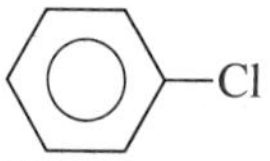

Cl
Chlorobenzene

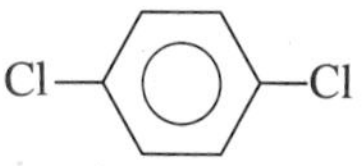

Cl Cl
1,4-Dichlorobenzene
(*p*-Dichlorobenzene)

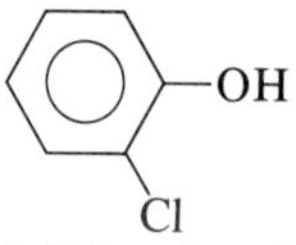

OH
Cl
2-Chlorophenol
(*o*-Chlorophenol)

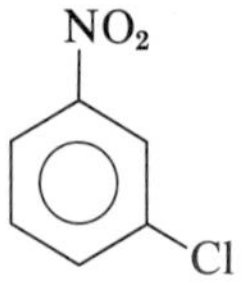

NO_2
Cl
3-Chloronitrobenzene
(*m*-Chloronitrobenzene)

Here the common name is given in parentheses below the IUPAC name.

Other names As you may recall from earlier chapters, hydrogen atoms in particular locations sometimes have special names. You have encountered *vinylic, allylic,* and *benzylic* hydrogens; they are shown in color in the following formulas:

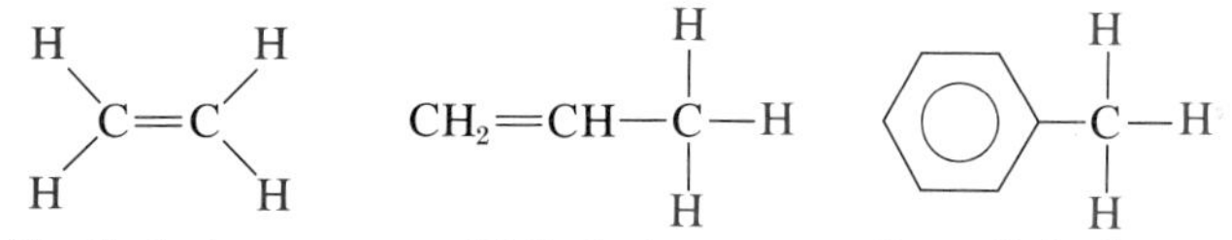

$H_2C{=}CH_2$ — Vinylic hydrogen

$CH_2{=}CH{-}CH_3$ — Allylic hydrogen

$C_6H_5{-}CH_3$ — Benzylic hydrogen

Substitution of halogen for these particular hydrogens results in halides with the common names vinyl chloride, allyl chloride, and benzyl chloride.

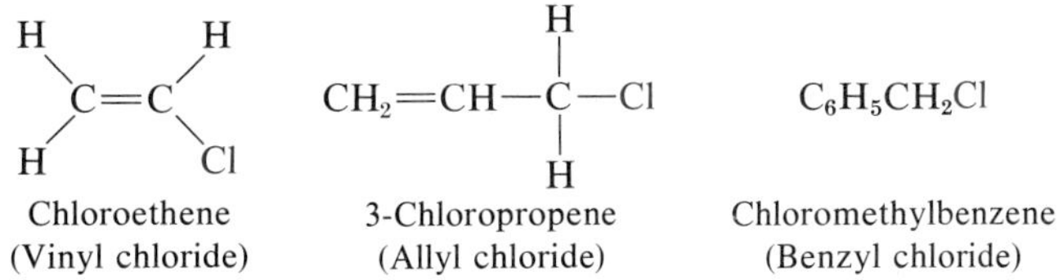

Chloroethene (Vinyl chloride) — 3-Chloropropene (Allyl chloride) — Chloromethylbenzene (Benzyl chloride)

PROBLEM 6-1 Give the IUPAC name for each of the following compounds.

(a) 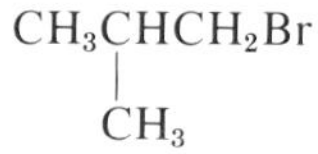 $CH_3CH(CH_3)CH_2Br$

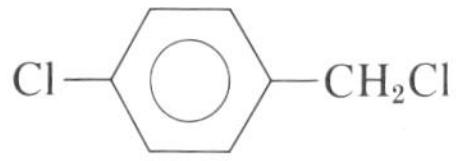

(b) 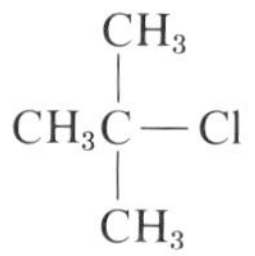 $CH_3C(CH_3)_2{-}Cl$

(c) $Cl{-}C_6H_4{-}CH_2Cl$ (para)

(d) 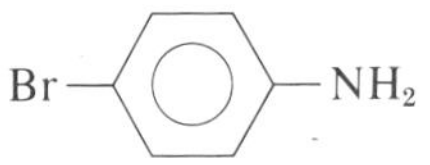 $Br{-}C_6H_4{-}NH_2$ (para)

Solution to (a) The IUPAC name for this compound is 1-bromo-2-methylpropane. It is commonly called isobutyl bromide.

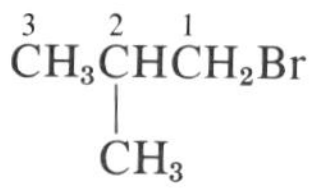

$$\overset{3}{C}H_3\overset{2}{C}H(CH_3)\overset{1}{C}H_2Br$$

PROBLEM 6-2 Give the structural formula for each of the following compounds. (a) 1,4-dichlorobutane, (b) *p*-bromobenzyl bromide, (c) 1,1,2-trichloroethane, (d) 3-chloroaniline.

Solution to (a) The structural formula for 1,4-dichlorobutane is

$$Cl\overset{1}{C}H_2\overset{2}{C}H_2\overset{3}{C}H_2\overset{4}{C}H_2Cl$$

6-2 Preparation of alkyl halides

The alkyl halides are prepared by (a) direct substitution of halogen for hydrogen, (b) addition of HX to alkenes, (c) addition of X_2 to alkenes to produce dihalides, and (d) from alcohols by use of HX, PX_3, or $SOCl_2$ (X can be Cl, Br, or I).

Direct halogenation Direct substitution of halogen for hydrogen is of limited application in most cases because of the formation of mixtures of polyhalogenated compounds. As you learned in Sec. 3-4, however, chloromethane can be prepared directly if methane is in excess.

$$\underset{\substack{\text{Methane}\\\text{in excess}}}{CH_4} + Cl_2 \xrightarrow{\text{light}} \underset{\substack{\text{Chloromethane}\\\text{(Methyl chloride)}}}{CH_3Cl} + HCl \tag{6-1}$$

Toluene and propylbenzene also can be halogenated directly:

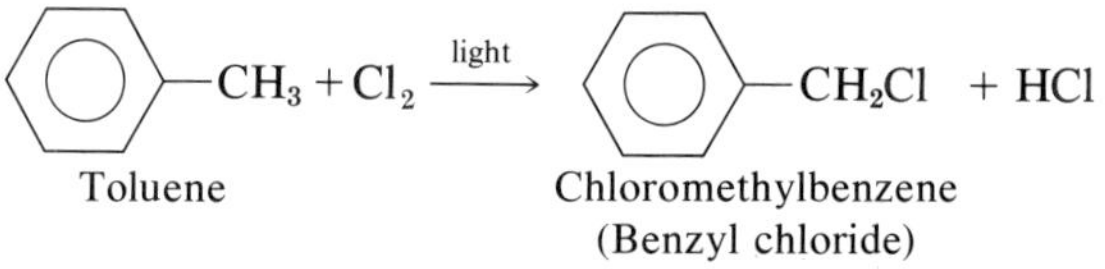

$$\underset{\text{Toluene}}{C_6H_5\text{—}CH_3} + Cl_2 \xrightarrow{\text{light}} \underset{\substack{\text{Chloromethylbenzene}\\\text{(Benzyl chloride)}}}{C_6H_5\text{—}CH_2Cl} + HCl \tag{6-2}$$

$$\underset{n\text{-Propylbenzene}}{C_6H_5\text{—}\overset{\alpha}{C}H_2CH_2CH_3} + Br_2 \xrightarrow{\text{light}} \underset{\text{1-Bromo-1-phenylpropane}}{C_6H_5\text{—}CH(Br)CH_2CH_3} + HBr \tag{6-3}$$

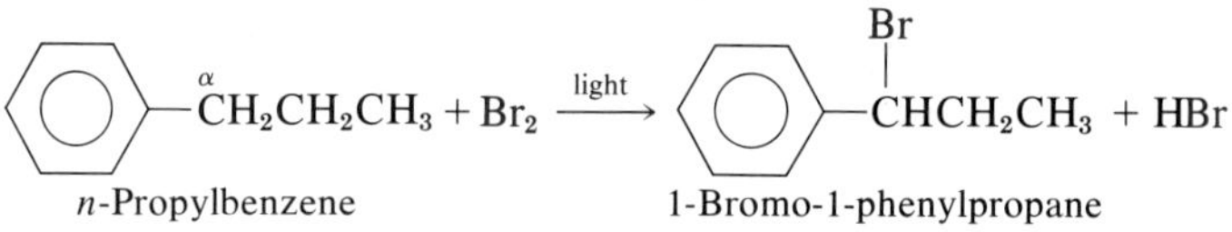

Note that, in the presence of light, substitution occurs in the side chain rather than in the ring as it does when an iron catalyst is used. Furthermore, it is the *alpha hydrogen* (i.e., the hydrogen atom on the carbon atom bonded to the ring) which is substituted. The intermediate free radical is more than a simple secondary free radical. It is a *benzylic-type free radical* in which the unpaired electron is delocalized. It is particularly stable and, in comparison with simple alkyl radicals, is readily formed. The mechanism for *benzylic substitution* is as follows:

$$Br_2 \xrightarrow{\text{light}} 2Br\cdot \tag{6-4}$$

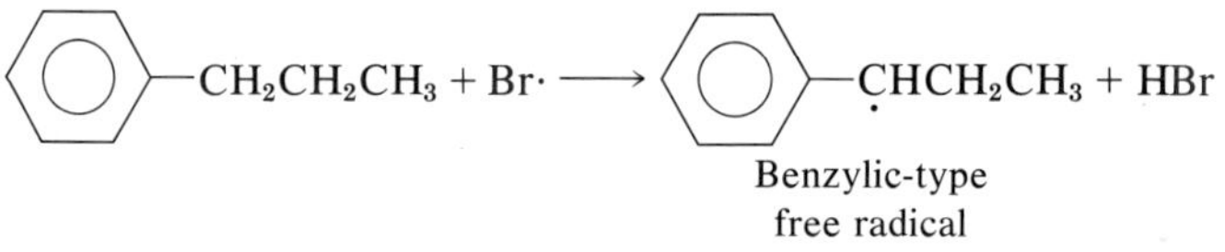

$$C_6H_5\text{—}CH_2CH_2CH_3 + Br\cdot \longrightarrow \underset{\substack{\text{Benzylic-type}\\\text{free radical}}}{C_6H_5\text{—}\dot{C}HCH_2CH_3} + HBr \tag{6-5}$$

$$C_6H_5\text{—}\dot{C}HCH_2CH_3 + Br_2 \longrightarrow C_6H_5\text{—}CH(Br)CH_2CH_3 + Br\cdot \tag{6-6}$$

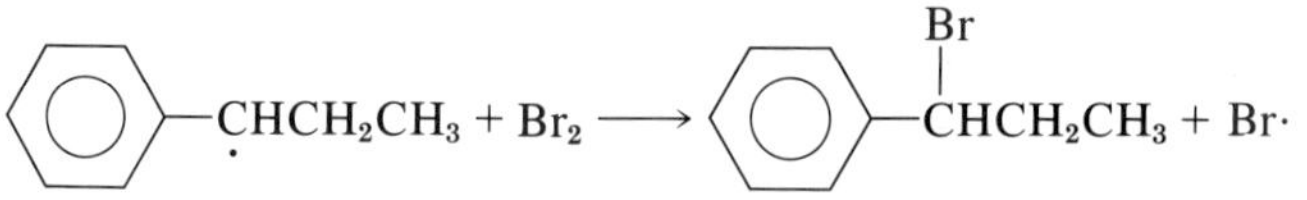

The benzylic-type free radical is a resonance hybrid of the following forms:

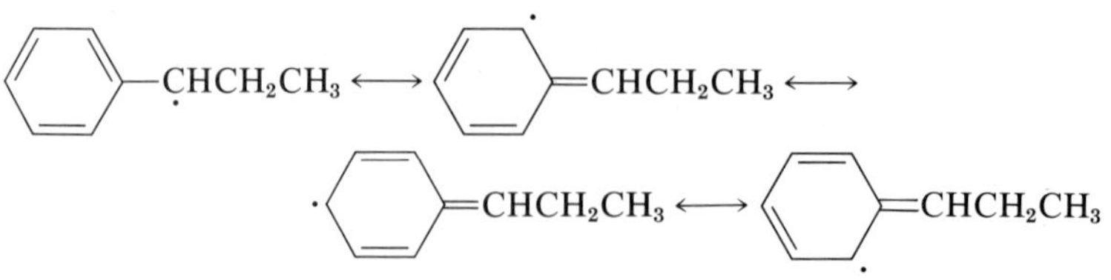

$$C_6H_5\text{—}\dot{C}HCH_2CH_3 \longleftrightarrow (o\text{-radical})C_6H_5\text{=}CHCH_2CH_3 \longleftrightarrow (p\text{-radical})C_6H_5\text{=}CHCH_2CH_3 \longleftrightarrow (o'\text{-radical})C_6H_5\text{=}CHCH_2CH_3 \tag{6-7}$$

The resonance hybrid might be represented by the single formula,

$$C_6H_5\text{---}CHCH_2CH_3$$

Benzylic substitution is analogous to the allylic substitution previously discussed [Eqs. (4-29) to (4-33)].

Addition of hydrohalogen to alkenes

This reaction between hydrohalogens and alkenes is generally limited by the direction of addition. Simple cases are governed by Markovnikov's rule. An example (see Fig. 6-2 also) is

$$\underset{\substack{\text{Propene}\\ \text{(Propylene)}}}{CH_3CH{=}CH_2} + HCl \longrightarrow \underset{\substack{\text{2-Chloropropane}\\ \text{(Isopropyl chloride)}}}{CH_3\underset{\displaystyle Cl}{\underset{|}{C}}HCH_3} \tag{6-8}$$

The addition of HBr to 1-phenyl-1-propene is interesting because the product is 1-bromo-1-phenylpropane. Little or no 2-bromo-1-phenylpropane is produced, although both products would involve secondary intermediate carbonium ions.

$$\underset{\text{1-Phenyl-1-propene}}{C_6H_5CH{=}CHCH_3} + HBr \longrightarrow \underset{\text{1-Bromo-1-phenylpropane}}{C_6H_5\overset{\displaystyle Br}{\overset{|}{C}}HCH_2CH_3} + HBr \tag{6-9}$$

The intermediate carbonium ion is much more than a simple secondary carbonium ion. It is a *benzylic-type carbonium ion* in which the positive charge is delocalized over the benzene ring. It is an especially stable and readily formed intermediate carbonium ion.

$$C_6H_5\text{—}CH{=}CHCH_3 + HBr \longrightarrow \underset{\text{A benzylic carbonium ion}}{C_6H_5\text{—}\overset{+}{C}H\text{—}CH_2CH_3} + Br^- \longrightarrow$$

$$C_6H_5\text{—}\underset{\displaystyle Br}{\underset{|}{C}}HCH_2CH_3 \tag{6-10}$$

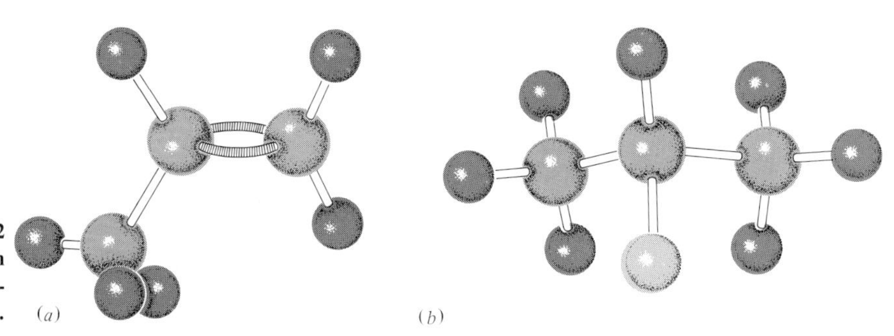

FIGURE 6-2
(*a*) Propene adds hydrogen chloride to yield (*b*) 2-chloropropane.

The benzylic-type carbonium ion is a resonance hybrid:

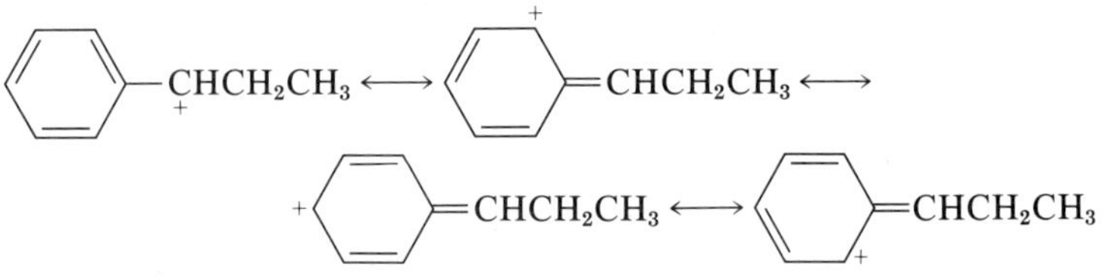

$$\text{C}_6\text{H}_5-\overset{}{\underset{+}{\text{C}}}\text{HCH}_2\text{CH}_3 \longleftrightarrow \text{(+)C}_6\text{H}_5{=}\text{CHCH}_2\text{CH}_3 \longleftrightarrow$$

$$\text{(+)C}_6\text{H}_5{=}\text{CHCH}_2\text{CH}_3 \longleftrightarrow \text{(+)C}_6\text{H}_5{=}\text{CHCH}_2\text{CH}_3 \qquad (6\text{-}11)$$

The resonance hybrid might be represented by the single formula,

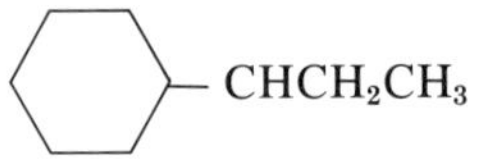

$$\text{C}_6\text{H}_5-\text{CHCH}_2\text{CH}_3$$

The benzylic *carbonium ion* delocalizes its *positive charge,* whereas the benzylic *radical* delocalizes an *unpaired electron* but has no charge.

Addition of halogen to alkenes

The addition of halogen to alkenes to produce dihalides was introduced as one of the important reactions of alkenes in Chap. 4; see Eqs. (4-10) to (4-12). The reaction produces *vicinal dihalides,* which means that the halogen atoms are on adjacent carbon atoms. An example is

$$\underset{\substack{\text{Propene}\\ \text{(Propylene)}}}{CH_3CH{=}CH_2} + Br_2 \xrightarrow{CCl_4} \underset{\substack{\text{1,2-Dibromopropane}\\ \text{(Propylene bromide)}}}{CH_3\underset{Br}{\underset{|}{C}}H-\underset{Br}{\underset{|}{C}}H_2} \qquad (6\text{-}12)$$

Vicinal dihalides are important intermediates for the introduction of a carbon-carbon triple bond into a molecule (production of an alkyne).

1,2-Dibromoethane (ethylene bromide) must be added to gasoline whenever tetraethyl lead is used to raise the octane rating. The ethylene bromide is necessary for the formation of lead bromide, which, although normally a solid, is volatile at engine temperature. Were it not formed, metallic lead deposits would accumulate in the cylinders of the engine and impede piston action.

Alkyl halides from alcohols

Alkyl halides may be produced from alcohols, which are generally readily available. Reagents which can be used for this process are PX_3, HX, and $SOCl_2$. The general equation for the reaction of an alcohol, ROH (remember that R represents an alkyl group), with PX_3 is

$$3ROH + PX_3 \longrightarrow 3RX + H_3PO_3 \qquad (6\text{-}13)$$

The general equation represents the reaction with PBr_3 and with PI_3; PCl_3 does produce some alkyl halide, but mostly it produces HCl and $P(OR)_3$.

An example of the reaction with PBr_3 is the production of 2-bromopropane from 2-propanol:

$$3CH_3CHOHCH_3 + PBr_3 \longrightarrow 3CH_3\underset{\displaystyle Br}{\underset{|}{C}}HCH_3 + H_3PO_3 \qquad (6\text{-}14)$$

Hydrohalogen may be used to produce an alkyl halide from an alcohol. In the general equation, X represents Cl, Br, or I. It should be noted that the use of HCl also requires the presence of $ZnCl_2$.

$$ROH + HX \longrightarrow RX + H_2O \qquad (6\text{-}15)$$

The order of reactivity of the hydrohalogens is HI > HBr > HCl. An example of the reaction with hydrogen bromide is

$$\underset{\text{1-Butanol}}{CH_3CH_2CH_2CH_2OH} + HBr \longrightarrow \underset{\text{1-Bromobutane}}{CH_3CH_2CH_2CH_2Br} + H_2O \qquad (6\text{-}16)$$

The HBr may be generated in the reaction itself by adding H_2SO_4 and NaBr, which react with each other to produce HBr and $NaHSO_4$.

Thionyl chloride, $SOCl_2$, may also be used to produce alkyl halides from alcohols. It is a liquid, and the resulting halide also is a liquid. The general equation is

$$ROH + \underset{\text{Thionyl chloride}}{SOCl_2} \longrightarrow RCl + SO_2(g) + HCl(g) \qquad (6\text{-}17)$$

An example of the reaction is as follows:

$$\underset{\substack{\text{Phenylmethanol} \\ \text{(Benzyl alcohol)}}}{C_6H_5CH_2OH} + SOCl_2 \longrightarrow \underset{\substack{\text{Chloromethylbenzene} \\ \text{(Benzyl chloride)}}}{C_6H_5CH_2Cl} + SO_2(g) + HCl(g) \qquad (6\text{-}18)$$

Note that the by-products of the reaction, SO_2 and HCl, are gases, and so they readily separate from the liquid halide product.

6-3 Preparation of aryl halides

Aryl halides are halides in which the halogen is attached *directly* to a benzene ring. They are prepared by (a) direct electrophilic aromatic substitution or by (b) substitution of the diazonium group.

Direct substitution

Direct electrophilic aromatic substitution may be represented by the chlorination of benzene.

$$C_6H_6 + Cl_2 \xrightarrow{Fe} C_6H_5Cl + HCl \qquad (6\text{-}19)$$

The iron reacts with chlorine to form iron(III) chloride, which is the actual catalyst. The following equations illustrate the formation of the catalyst.

$$2Fe + 3Cl_2 \longrightarrow 2FeCl_3 \quad \text{(a Lewis acid)} \qquad (6\text{-}20)$$

$$Cl_2 + FeCl_3 \longrightarrow Cl^+FeCl_4^- \qquad (6\text{-}21)$$

The electrophile is $Cl^+FeCl_4^-$.

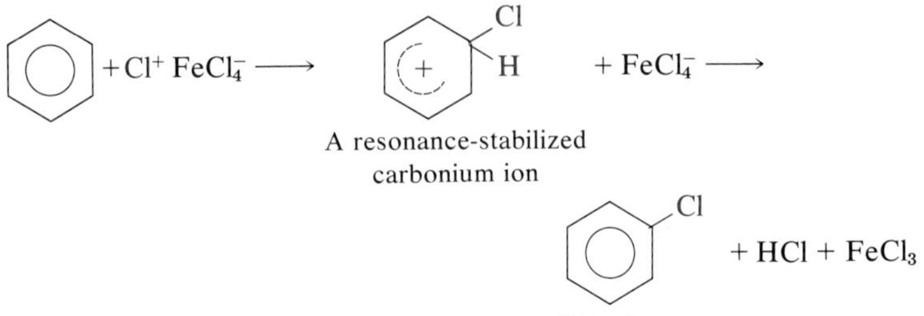

+ Cl⁺ FeCl₄⁻
Cl
H
+ FeCl₄⁻
Cl
+ HCl + FeCl₃

A resonance-stabilized carbonium ion

Chlorobenzene

(6-22)

Bromine reacts similarly. The reaction with fluorine is complex and is difficult to control, and the reaction with iodine is generally very poor.

Substitution for the diazonium group

Substitution of halogen for the diazonium group ($—N_2^+$) is often the best way to introduce halogen into the benzene ring, especially if the ring has other substituents. By proper choice of reagents and conditions, any of the four halogens may be introduced to the benzene ring.

An example of this preparation is

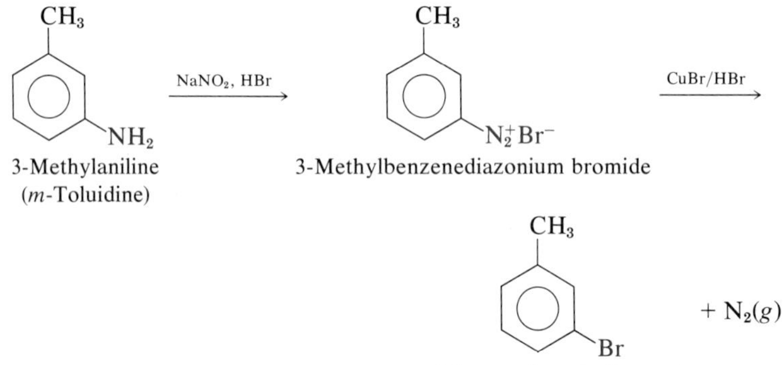

CH₃
NH₂
NaNO₂, HBr
CH₃
N₂⁺Br⁻
CuBr/HBr
CH₃
Br
+ N₂(g)

3-Methylaniline (*m*-Toluidine)

3-Methylbenzenediazonium bromide

1-Bromo-3-methylbenzene (*m*-Bromotoluene)

(6-23)

This general reaction will be discussed in more detail in Chap. 13.

PROBLEM 6-3 Write equations showing the preparation of the following halides from the starting materials indicated. Show any necessary conditions and the reagents required for the reactions.

(a) $C_6H_5CHBrCH_3$ from $C_6H_5CH_2CH_3$
(b) $C_6H_5CHBrCH_3$ from $C_6H_5CHOHCH_3$
(c) 2-bromo-2-methylpropane from 2-methylpropene

(d) 1-bromobutane from 1-butene
(e) 1,2-dibromoethane from ethene
(f) 2-chloropropane from 2-propanol
(g) 1-bromo-3-nitrobenzene from benzene (two reactions required)

Solution to (a) The overall reaction for this preparation is

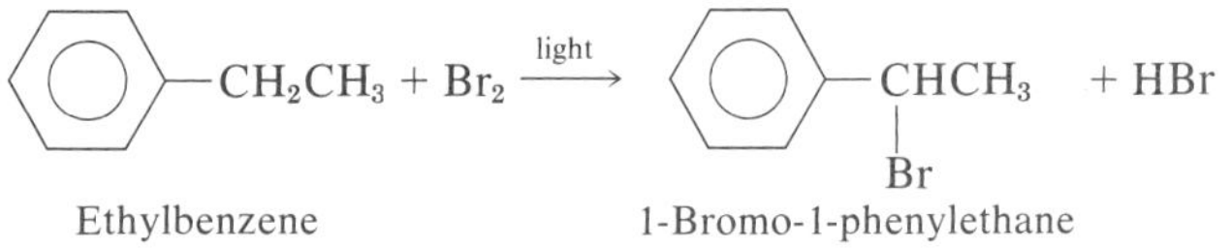

$$C_6H_5-CH_2CH_3 + Br_2 \xrightarrow{\text{light}} C_6H_5-CH(Br)CH_3 + HBr$$

Ethylbenzene → 1-Bromo-1-phenylethane

The mechanism is as follows:

$$Br_2 \xrightarrow{\text{light}} 2Br\cdot$$

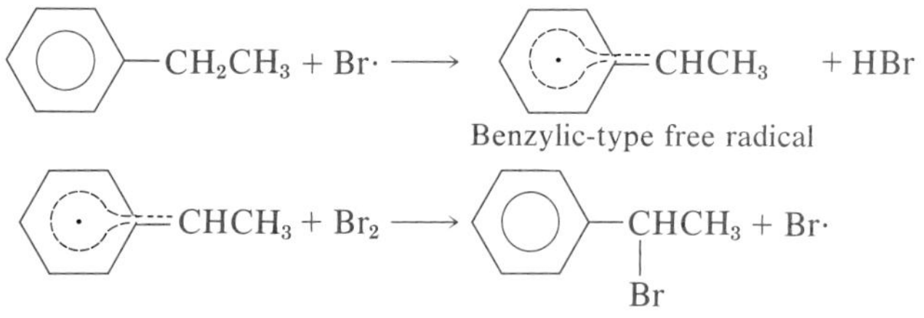

$$C_6H_5-CH_2CH_3 + Br\cdot \longrightarrow C_6H_5\text{=}\dot{C}HCH_3 + HBr$$

Benzylic-type free radical

$$C_6H_5\text{=}\dot{C}HCH_3 + Br_2 \longrightarrow C_6H_5-CH(Br)CH_3 + Br\cdot$$

6-4 Physical properties of organic halides

None of the organic halides are soluble in water, but all of them are soluble in organic solvents. The monochlorides are lighter than water, but the monobromides, the monoiodides, and all of the polyhalides are heavier than water. The presence of two or more chlorine atoms raises the density to more than 1. Chloromethane, bromomethane, and chloroethane are gases, but iodomethane and bromoethane are liquids.

6-5 Reactions of organic halides

The chief reaction of the alkyl halides is *nucleophilic substitution*, in which halide ion is displaced by a *nucleophilic reagent* or *nucleophile*. You have already encountered the term *electrophilic reagent* or electrophile and have learned that it represents an electron-deficient atom or group (such as H^+, X^+, $-\overset{|}{\underset{|}{C}}+$, $^+NO_2$, which therefore seeks additional electrons. Alkenes add electrophiles by using the pi electrons of their double bond, and aromatic hydrocarbons are substituted by electrophiles attracted to their pi molecular orbital.

A *nucleophilic reagent* or *nucleophile* is an *electron-rich atom* or *group* such as a negative ion or Lewis base (^-OH, ^-OR, ^-CN, ^-SH, NH_3, ROH, H_2O), which seeks a relatively positive nucleus with which to share its electrons (Fig. 6-3). If $Z:^-$ represents a nucleophile, then,

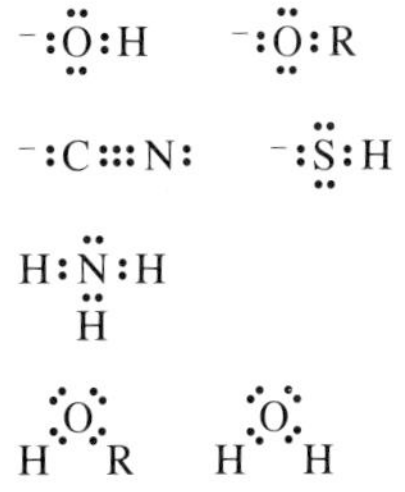

$$\underset{\text{Nucleophile, stronger base}}{Z:^-} + R-X \longrightarrow RZ + \underset{\text{Halide ion, weaker base}}{X^-} \qquad (6\text{-}24)$$

FIGURE 6-3
Nucleophilic reagents are electron-rich atoms or groups.

$^-:\ddot{\underset{..}{O}}:H$ $\quad$ $^-:\ddot{\underset{..}{O}}:R$

$^-:C:::N:$ $\quad$ $^-:\ddot{\underset{..}{S}}:H$

$H:\ddot{\underset{..}{N}}:H$ with H below N

$\ddot{O}$ bonded to H and R $\quad$ $\ddot{O}$ bonded to H and H

If we regard the reaction as a type of Lewis acid-base reaction, then we can understand that it tends to occur because of the formation of halide ion, which, as the conjugate base of a strong acid (HX), would be a weak base. Accordingly, a weak base like halide ion is said to be a good *leaving group*. The order of reactivity of the alkyl halides by type of halogen is RI > RBr > RCl > RF. Iodide ion, being the weakest base as the conjugate base of the strongest acid, HI, is the best leaving group. Any given halide is more reactive than the same halide of the higher homologs; that is, $CH_3X > C_2H_5X > C_3H_7X$.

There are other reactions of organic halides besides nucleophilic substitution. A summary of typical reactions is given in Table 6-2 at the end of Sec. 6-5. We will now examine the typical reactions in detail.

Nucleophilic substitution reactions, first order

The area of chemistry concerned with the *rates* of reactions is called *kinetics*. The rate of the nucleophilic substitution reaction of a *tertiary halide* is found to be proportional to the concentration of the organic halide and to be independent of the concentration of the nucleophile. The rate of such a reaction is said to be *first order;* the reaction is designated as S_N1 (S = substitution, N = nucleophilic, 1 = first order).

Since the nucleophile is obviously a participant in the overall reaction, the first-order rate indicates that the reaction consists of more than one step. The slow, rate-determining step does *not* involve the nucleophile, but it is unimolecular. An example of such a reaction is the hydrolysis of 2-chloro-2-methylpropane with water. The mechanism for that S_N1 reaction is considered to consist essentially of the following steps:

$$\underset{\substack{\text{2-Chloro-2-methylpropane}\\(\textit{tert}\text{-Butyl chloride})}}{CH_3C(CH_3)_2-Cl} \underset{}{\overset{\text{(slow)}}{\rightleftharpoons}} \underset{\text{Tertiary carbonium ion}}{(CH_3)_3C^+} + Cl^- \qquad (6\text{-}25)$$

$$(CH_3)_3C^+ + \underset{\text{Nucleophile}}{HOH} \xrightarrow{\text{(fast)}} (CH_3)_3C-\overset{+}{O}H_2 \xrightarrow[-H^+]{\text{(fast)}} \underset{\substack{\text{2-Methyl-2-propanol}\\(\textit{tert}\text{-Butyl alcohol})}}{(CH_3)_3C-OH} \qquad (6\text{-}26)$$

The rate of the first step, which is the rate-determining reaction, is given by

$$\text{Rate} = k_1[(CH_3)_3CCl] \qquad (6\text{-}27)$$

In this equation, k_1 represents a first-order rate constant and the brackets represent the concentration, in moles per liter, of the compound whose formula they enclose.

The order of reactivity of halides in the S_N1 reaction has been found to be 3° > 2° > 1° > CH_3X. That order is reasonable, since it is the order of stability of the intermediate carbonium ions formed in the slow, rate-determining step.

The reaction has been found to be subject to rearrangements when the intermediate carbonium ion can rearrange to a more stable carbonium ion. The following is an example of such a rearrangement:

$$CH_3-\underset{CH_3}{\overset{CH_3}{\underset{|}{\overset{|}{C}}}}-CH_2Br \xrightleftharpoons{\text{(slow)}} CH_3-\underset{CH_3}{\overset{CH_3}{\underset{|}{\overset{|}{C}}}}-CH_2^+ + Br^- \qquad (6\text{-}28)$$

1-Bromo-2,2-dimethyl propane (Neopentyl bromide) — 2,2-Dimethyl cation (Primary carbonium ion)

$$CH_3-\underset{CH_3}{\overset{CH_3}{\underset{|}{\overset{|}{C}}}}-CH_2^+ \xrightleftharpoons{\text{(fast)}} CH_3-\overset{CH_3}{\overset{|}{\underset{+}{C}}}-CH_2CH_3 \qquad (6\text{-}29)$$

$$CH_3-\overset{CH_3}{\overset{|}{\underset{+}{C}}}-CH_2CH_3 + HOH \xrightarrow{\text{(fast)}} CH_3-\underset{\overset{+}{O}H\,H}{\overset{CH_3}{\underset{|}{\overset{|}{C}}}}-CH_2CH_3 \xrightarrow{-H^+} CH_3-\underset{OH}{\overset{CH_3}{\underset{|}{\overset{|}{C}}}}-CH_2CH_3 \qquad (6\text{-}30)$$

2-Methyl-2-butanol (*tert* Amyl alcohol)

You will note in Eq. (6-29) that the primary carbonium ion rearranges, through the shift of a $—CH_3$ group, to produce a tertiary carbonium ion.

Nucleophilic substitution reactions, second order

The nucleophilic substitution reaction of methyl halides and of primary halides with sodium hydroxide solution is found to be *second order* (S_N2). That means the rate of the reaction is proportional to both the concentration of the halide and the concentration of the nucleophile.

$$CH_3Br + OH^- \longrightarrow CH_3OH + Br^- \qquad (6\text{-}31)$$

$$\text{Rate} = k_2[CH_3Br][OH^-] \qquad (6\text{-}32)$$

In Eq. (6-32), k_2 represents a second-order rate constant. The mechanism for the reaction is as follows:

$$HO^- + H_3C-Br \xrightarrow{S_N2} \left[\overset{\delta^-}{HO}\cdots \underset{H}{\overset{H\quad H}{C}} \cdots \overset{\delta^-}{Br}\right] \longrightarrow HO-CH_3 + Br^- \qquad (6\text{-}33)$$

Bromomethane (Methyl bromide) — Transition state — Methanol (Methyl alcohol)

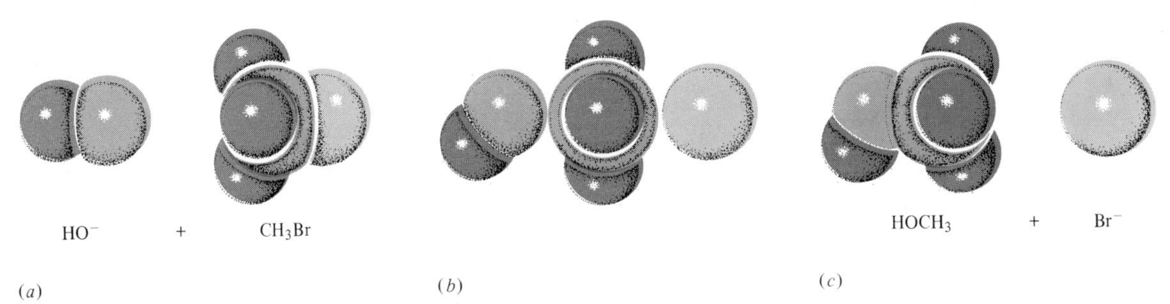

FIGURE 6-4 The mechanism for the reaction of bromomethane with hydroxide ion: (*a*) the hydroxide ion approaches from the rear; (*b*) the transition state; (*c*) the bromide ion leaves.

(See Fig. 6-4 also.) Note how the hydroxide ion attacks from the rear, away from the electronegative field of the bromine atom. As the hydroxide ion begins to bond to carbon from the rear, the bromine begins to leave as a bromide ion from the front. Groups larger than hydrogen tend to block the approach of the nucleophile, so that methyl halides are more reactive than primary halides.

It has been found that secondary halide reactions can be either S_N1 or S_N2. Stronger nucleophiles favor the S_N2 reaction; more polar solvents, such as water, favor the S_N1 reaction. Examples of reactions that take place by second-order nucleophilic substitution are these:

$$NaCN + \underset{\substack{\text{Bromoethane}\\\text{(Ethyl bromide)}}}{CH_3CH_2Br} \xrightarrow{S_N2} \underset{\substack{\text{Propanonitrile}\\\text{(Propionitrile or}\\\text{ethyl cyanide)}}}{CH_3CH_2CN} + NaBr \tag{6-34}$$

$$\underset{\text{Sodium propynylide}}{CH_3C{\equiv}C{:}^-Na^+} + CH_3CH_2Br \xrightarrow{S_N2} \underset{\text{2-Pentyne}}{CH_3C{\equiv}C{-}CH_2CH_3} + NaBr \tag{6-35}$$

$$\underset{\text{Sodium propoxide}}{CH_3CH_2CH_2O^-Na} + CH_3I \xrightarrow{S_N2} \underset{\substack{\text{1-Methoxypropane}\\\text{(Methyl }n\text{-propyl ether)}}}{CH_3CH_2CH_2OCH_3} + NaI \tag{6-36}$$

The order of reactivity of halides in the S_N2 reaction is $CH_3X > 1° > 2° > 3°$.

Elimination of hydrohalogen

Elimination of hydrohalogen from alkyl halides can proceed by either a first- or a second-order reaction. First-order elimination reaction is symbolized as E1 and the second-order elimination reaction as E2. You were actually introduced to the E2 mechanism in Chap. 4, Eq. (4-2), when elimination of hydrohalogen from an alkyl halide was discussed as a method of preparing alkenes. It is a concerted reaction. Since the rate of the reaction depends upon the concentration of both alkyl halide and base, it is a second-order reaction.

$$-\overset{\overset{X}{|}}{\underset{|}{C}}-\overset{|}{\underset{\underset{H\ \leftarrow\ {}^{-}OH}{|}}{C}}- \xrightarrow{\Delta,\ C_2H_5OH} -\underset{|}{C}=\overset{|}{C}- + HOH + X^- \qquad (6\text{-}37)$$

An additional requirement is that the halogen atom and the hydrogen atom eliminated be in an *anti coplanar conformation* (Fig. 6-5). Since the C—H and the C—X bonds in the above example eventually become the pi bond in the resulting alkene, it is understandable that they must be properly aligned in the same plane for the pi bond to form readily. The order of reactivity of types of halides is 3° > 2° > 1°, which is the order of stability of the resulting alkenes according to the Saytzeff rule.

Tertiary halides, and to some extent secondary halides, dehydrohalogenate via the E1 mechanism. The mechanism is two-step, and the first step is identical with the first step of the S_N1 reaction:

$$CH_3-\overset{\overset{CH_3}{|}}{\underset{\underset{CH_3}{|}}{C}}-X \xrightarrow[\text{(slow)}]{\text{E1}} CH_3-\overset{\overset{CH_3}{|}}{\underset{\underset{\underset{\underset{H}{|}}{HC-H}}{|}}{C^+}} \quad {}^{-}X \xrightarrow[{}^{-}OH]{\text{fast}} CH_3-\overset{\overset{CH_3}{|}}{\underset{\underset{CH_2}{\|}}{C}} + X^- + HOH \qquad (6\text{-}38)$$

The E1 reaction is a strong competitor of the S_N1 reaction, since its order of reactivity of halide types is the same, that is, 3° > 2° > 1°. The E1 reaction is favored by strong bases, higher temperatures, and tertiary halides. Because of the competition, the nucleophilic substitution reaction of an alkyl halide with sodium cyanide, a sodium acetylide, or a sodium alkoxide is generally successful only with primary halides. Tertiary halides eliminate hydrohalogen to form alkenes, and secondary ha-

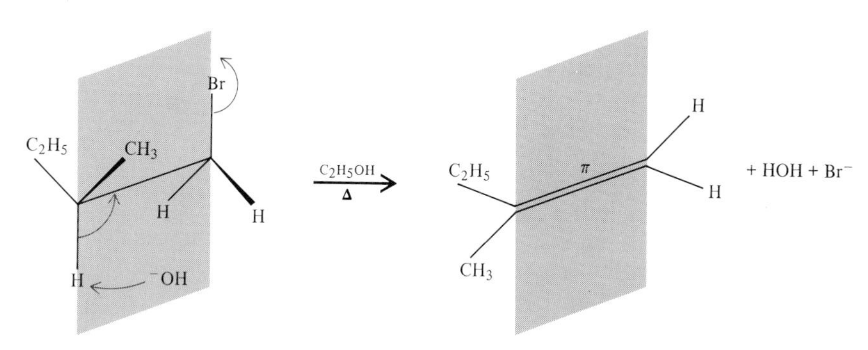

FIGURE 6-5
The E2 elimination reaction of 1-bromo-2-methylbutane from an anti coplanar conformation. The hydrogen atom being eliminated and the bromine atom are in an anti conformation with respect to each other and are in the same plane. The pi bond is then formed in this same plane.

lides exhibit both the S_N1 and the E1 reactions. Hydrolysis of 2-chloro-2-methylpropane (*tert*-butyl chloride) to produce 2-methylpropanol (*tert*-butyl alcohol) also produces much 2-methylpropene (isobutylene).

Elimination of hydrogen chloride from the insecticide DDT is catalyzed by the enzyme DDT dehydrochlorinase, which is developed by DDT-resistant insects. DDT is converted into a nonlethal derivative of an alkene.

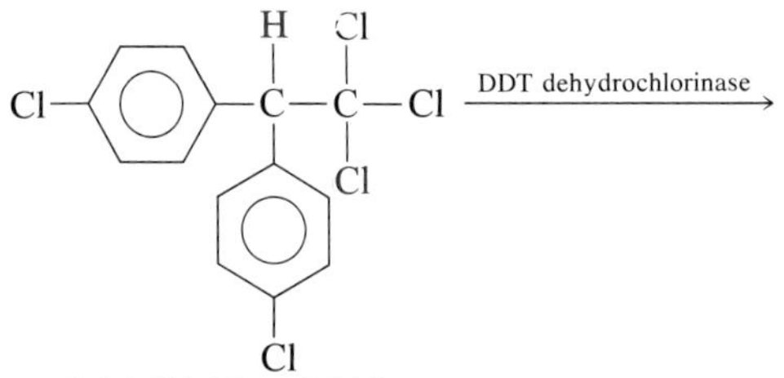

$$(p\text{-}ClC_6H_4)_2CH{-}CCl_3 \xrightarrow{\text{DDT dehydrochlorinase}} (p\text{-}ClC_6H_4)_2C{=}CCl_2 + HCl \qquad (6\text{-}39)$$

1,1,1-Trichloro-2,2-bis (*p*-chlorophenyl)ethane (DDT)

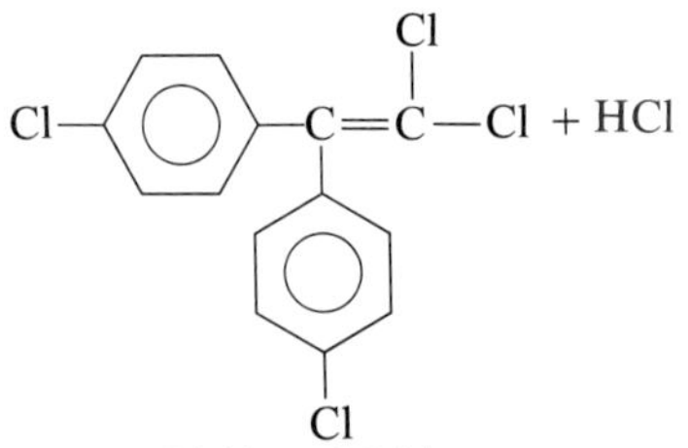

1,1-Dichloro-2,2-bis (*p*-chlorophenyl)ethene (DDE)

PROBLEM 6-4 Write an equation for the chief reaction of each of the following alkyl halides with the reagent indicated, and tell by what mechanism the reaction takes place.

(a) $CH_3CH_2Cl + NaCN$
(b) $CH_3CH_2Br + NaOCH_3$
(c) $C_6H_5CH_2Cl + NH_3$
(d) $CH_3CH_2CH_2Br + Na^{+-}C{\equiv}C{-}CH_3$
(e) $CH_3C(CH_3)_2Cl + NaCN$
(f) $(CH_3)_2C(Cl){-}CH_2CH_3 + NaOCH_3$

Solution to (a)

$$\underset{\text{Chloroethane}}{CH_3CH_2Cl} + NaCN \longrightarrow \underset{\text{Propanonitrile}}{CH_3CH_2CN} + NaCl$$

The reaction proceeds by a second-order nucleophilic substitution (S_N2) mechanism.

Reactions of aryl halides

The aryl halides are relatively unreactive; they do *not* react under ordinary conditions with NaOH, $NaOC_2H_5$, NaCN, NaSH, H_2O, NH_3, or $AgNO_3$. They may be *forced to react* under strenuous conditions of heat and pressure. They *will* react with magnesium to form Grignard reagents.

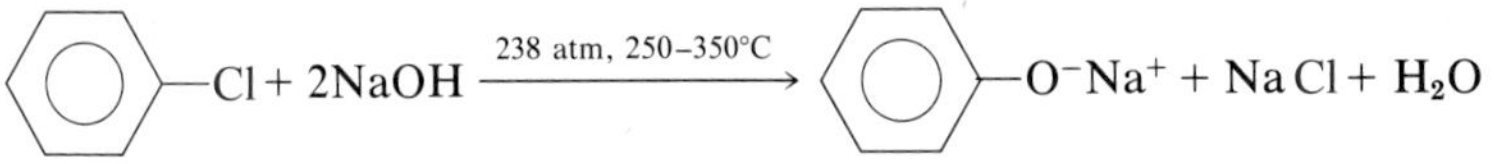

$$C_6H_5{-}Cl + 2NaOH \xrightarrow{238\text{ atm, }250\text{–}350^{\circ}C} C_6H_5{-}O^-Na^+ + NaCl + H_2O \qquad (6\text{-}40)$$

$$C_6H_5Cl + NH_3 \xrightarrow{51-58\ \text{atm},\ 175-225°C} C_6H_5NH_3^+Cl^- \qquad (6\text{-}41)$$

$$C_6H_5Br + Mg \xrightarrow{\text{ether}} \underset{\text{A Grignard reagent}}{C_6H_5MgBr} \qquad (6\text{-}42)$$

Having electron-withdrawing groups on the ring ortho and/or para to the halogen facilitates nucleophilic substitution of the halogen.

$$2,4\text{-}(NO_2)_2C_6H_3Cl + C_2H_5ONa \xrightarrow{100°C} 2,4\text{-}(NO_2)_2C_6H_3OC_2H_5 + NaCl \qquad (6\text{-}43)$$

The rate of this reaction has been found to be second-order; but unlike the nucleophilic substitution of alkyl halides, the particular halogen seems to make little difference. From that the conclusion is drawn that the breaking of the carbon-halogen bond is *not* involved in the rate-determining step. Therefore, the mechanism is considered to involve electron withdrawal by the nitro groups, which delocalizes the negative charge on the intermediate carbanion:

(a) $$\underset{\text{1-Chloro-2,4-dinitrobenzene}}{O_2N\text{-}C_6H_3(NO_2)\text{-}Cl} + {}^-OC_2H_5 \xrightleftharpoons{\text{(slow)}} \underset{\text{A carbanion}}{O_2N\text{-}C_6H_3^-(NO_2)(Cl)(OC_2H_5)} \qquad (6\text{-}44)$$

(b) $$O_2N\text{-}C_6H_3^-(NO_2)(Cl)(OC_2H_5) \xrightarrow{\text{(fast)}} \underset{\text{1-Ethoxy-2,4-dinitrobenzene}}{O_2N\text{-}C_6H_3(NO_2)\text{-}OC_2H_5} + Cl^- \qquad (6\text{-}45)$$

The carbanion is actually a resonance hybrid of the formula shown and other forms.

The mechanism for the reaction of other nucleophilic reagents in substituting for the halogen of aryl halides having electron-withdrawing groups is the same.

Table 6-2 summarizes the types of reactions undergone by organic halides.

6-6 Analysis of organic halides

The presence of halogen in an organic compound is readily detected by the *Beilstein test*. Preparation for the test consists of heating a small loop of copper wire in a flame until any impurities are consumed and the flame has become colorless. A drop or piece of the unknown is then placed on

TABLE 6-2 Reactions of organic halides

Nucleophilic substitution reactions of alkyl halides, first order

$$(CH_3)_3CCl + H_2O \longrightarrow (CH_3)_3COH + HCl$$
$$CH_2{=}CHCH_2Cl + H_2O \longrightarrow CH_2{=}CHCH_2OH + HCl$$
$$C_6H_5CH_2Cl + H_2O \longrightarrow C_6H_5CH_2OH + HCl$$

Nucleophilic substitution reactions of alkyl halides, second order

$$CH_3CH_2Br + NaCN \longrightarrow CH_3CH_2CN + NaBr$$
$$CH_3CH_2CH_2Cl + NaOH(aq) \longrightarrow CH_3CH_2CH_2OH + NaCl$$
$$CH_3CH_2CH_2CH_2Br + NaSH \longrightarrow CH_3CH_2CH_2CH_2SH + NaBr$$
$$CH_3CH_2CH_2CH_2Br + 2NH_3 \longrightarrow CH_3CH_2CH_2CH_2NH_2 + NH_4Br$$
$$CH_3CH_2CH_2Br + NaOCH_2CH_3 \longrightarrow CH_3CH_2CH_2OCH_2CH_3 + NaBr$$
$$CH_3CH_2Br + NaC{\equiv}C{-}CH_3 \longrightarrow CH_3CH_2C{\equiv}C{-}CH_3 + NaBr$$

Elimination of HX from alkyl halides

$$CH_3\overset{\displaystyle Br}{\underset{\displaystyle H}{\overset{|}{\underset{|}{C}}}}{-}\overset{\displaystyle H}{\underset{\displaystyle H}{\overset{|}{\underset{|}{C}}}}{-}H + KOH \xrightarrow{\text{alcohol}} CH_3CH{=}CH_2 + KBr + HOH$$

Nucleophilic substitution reactions of aryl halides

$$2,4\text{-}(NO_2)_2C_6H_3Cl + 2NaOH \longrightarrow 2,4\text{-}(NO_2)_2C_6H_4ONa + NaCl + H_2O$$

Formation of Grignard reagent

$$CH_3I + Mg \xrightarrow{\text{dry ether}} CH_3MgI$$
$$C_6H_5Br + Mg \xrightarrow{\text{dry ether}} C_6H_5MgBr$$

Friedel-Crafts reaction of alkyl halides

$$CH_3CH_2Cl + C_6H_6 \xrightarrow{AlCl_3} CH_3CH_2C_6H_5 + HCl$$

the loop, and the loop is again heated in the flame. A green flame is evidence of the presence of halogen. To ascertain which halogen is present, a small sample of the unknown is fused with sodium. That converts the halogen to a soluble sodium halide, which is then identified by the usual methods of inorganic qualitative analysis. A quantitative determination of halogen may be made by combustion of a weighed sample of the halide in a peroxide bomb. The combustion products are then subjected to the usual quantitative analysis procedures for the halides present.

Alkyl halides, benzyl halides, and allyl halides will react with alcoholic silver nitrate to give a precipitate of silver halide. However, aryl halides and vinyl halides will not react with alcoholic silver nitrate.

$$RX + AgNO_3 \xrightarrow{\text{alcohol}} AgX(s) + R^+ \longrightarrow \text{other products} \qquad (6\text{-}46)$$

The reaction will help in distinguishing alkyl halides from aryl halides and from vinyl halides. It will also help in distinguishing alkyl halides from hydrocarbons and from other types of organic compounds.

(*a*)

(*b*)

FIGURE 6-6
(*a*) Cornfield eaten by grasshoppers. (*Agricultural photography by Grant Heilman*) (*b*) Grasshoppers infesting Western rangeland. (*USDA photo*)

6-7 Chlorinated hydrocarbons as pesticides

About a third of the world's agricultural products are consumed or destroyed by insects. Both the increasing demand for food as the world's population expands and the present extent of malnutrition and starvation have intensified the need to reduce insect devastation of food supplies (Fig. 6-6). Insects damage not merely food crops but also livestock, timber, and fiber crops such as cotton. They transmit potentially deadly diseases such as malaria, yellow fever, bubonic palgue, and encephalitis (Fig. 6-7).

New insecticides are being developed, and so are new approaches to insect control that do not involve conventional chemical insecticides. The latter is important because most of the organochlorine compounds are being phased out for reasons that include the following:

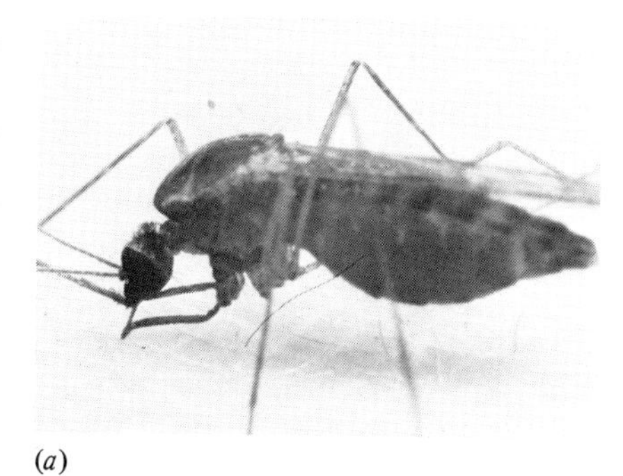

(*a*)

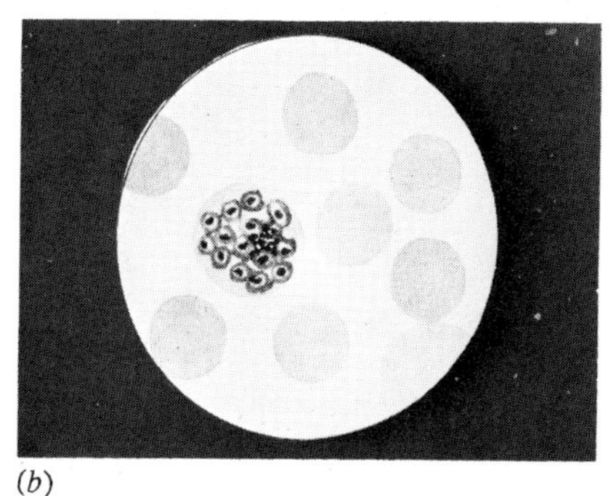

(*b*)

FIGURE 6-7
(*a*) *Anopheles* mosquito biting; note its swollen belly. (*b*) Blood cell containing *Plasmodium vivax;* the cell is about to burst. The *Anopheles* mosquito is the vector organism for malaria; it carries the actual disease-causing organism, *Plasmodium vivax,* from one person to another. (*Center for Disease Control*)

1 Some insects have increasing resistance to compounds that were once toxic to them.
2 The Environmental Protection Agency (EPA) has required that insect-control agents not merely control target insects but also be noninjurious to other organisms—humans, livestock, fish, crops, wildlife, and beneficial insects. Moreover, dangerous residues of the chemicals must not be present in food or livestock feed.
3 The EPA has restricted use of several effective and previously widely used insecticides that are believed to pose a serious threat to human life.

In December 1972 EPA placed a near-total ban on domestic use of DDT, which for many years was the most widely used insecticide in the United States. DDT has exceptional persistence—its half-life is 10 years. Because it degrades very slowly and is stored in the fat of living organisms, it tends to build up in the natural food chain and accumulate in the tissues of fish, wildlife, and humans. Opposition to its use intensified

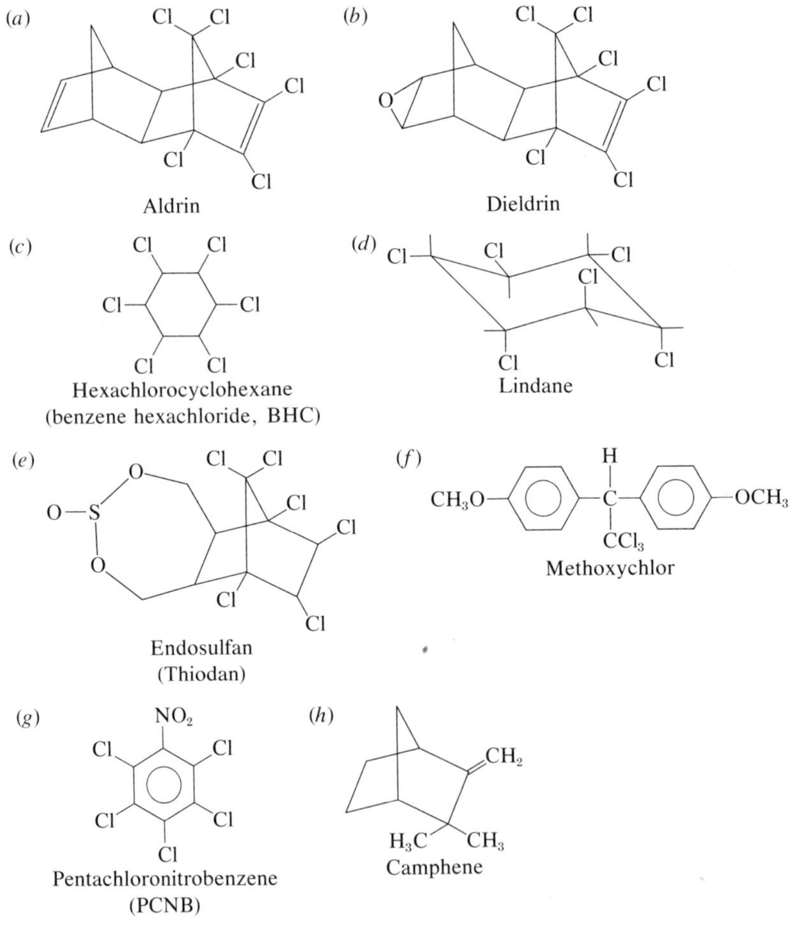

(*a*) Aldrin

(*b*) Dieldrin

(*c*) Hexachlorocyclohexane (benzene hexachloride, BHC)

(*d*) Lindane

(*e*) Endosulfan (Thiodan)

(*f*) Methoxychlor

(*g*) Pentachloronitrobenzene (PCNB)

(*h*) Camphene

FIGURE 6-8
Some chlorinated hydrocarbons are used as pesticides. Dieldrin (*b*) is one of the isomers of endrin. Hexachlorocyclohexane (*c*) has nine isomers, one of which is lindane (*d*). PCNB (*g*) is a fungicide used to treat seeds to prevent attack by soil organisms. Camphene (*h*) when chlorinated is the chief component of toxaphene.

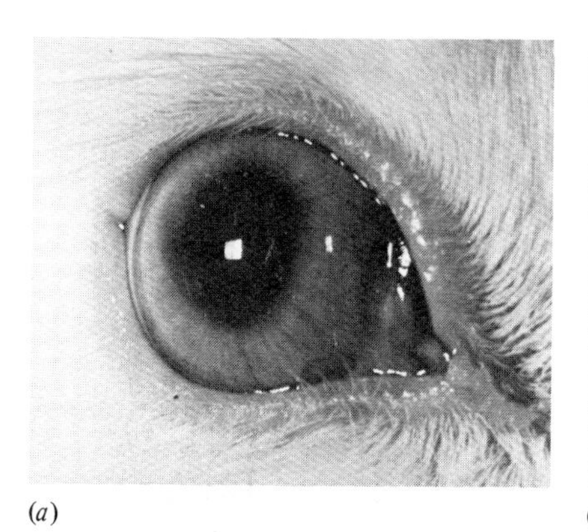
(*a*)

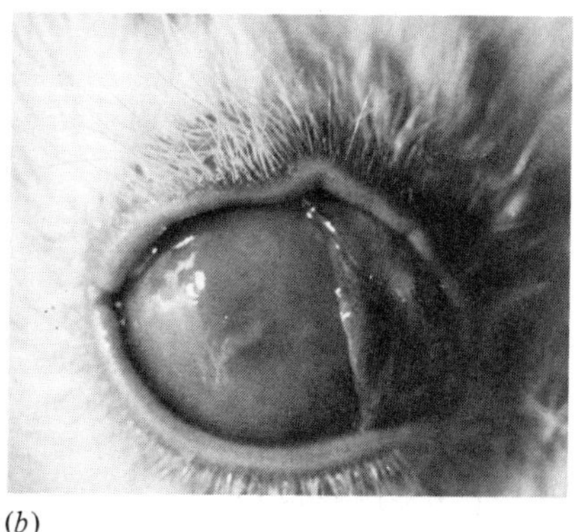
(*b*)

FIGURE 6-9
(*a*) Normal rabbit's eye. (*b*) Eye of the rabbit 7 days after laboratory exposure to a pesticide, *N*-alkyldimethylbenzylammonium chloride. Note that the membrane (conjunctiva) is swollen and bloodshot and that the cornea is so opaque that the pupil is barely visible. (U.S. Environment Protection Agency.)

when there were reports that it causes some birds to lay eggs with abnormally thin shells which break readily and cause the embryos to die. DDT was also found to cause tumors indicative of cancer in laboratory rats and mice. Despite those hazards, several foreign countries continue to use DDT for combatting mosquitoes that spread malaria. In the United States the chlorinated hydrocarbons aldrin and dieldrin also have been banned as insecticides. (See Figs. 6-8 and 6-9.)

Toxaphene has become the most widely used insecticide in the United States. Its primary application is the control of insects on cotton, but is also used on soybeans, corn, and other crops and on livestock. It is a mixture containing polychloro bicyclic terpenes in which chlorinated camphene predominates. Its approximate empirical formula is $C_{10}H_{10}Cl_8$. It has been under EPA review because of some evidence that it may cause cancer in test animals.

Three other organochlorine insecticides—benzene hexachloride (BHC), lindane (an isomer of BHC), and endrin—have been under EPA review for possible restriction. A few other organochlorine insecticides are being presently retained; they include methoxychlor and endosulfan (Thiodan). In 1977 an agricultural soil fumigant DBCP (1,2-dibromo-3-chloropropane) was found to be causing widespread reduction in sperm count of male industrial workers exposed to it. Manufacturers have suspended its production.

Some scientists at the U.S.D.A. Agricultural Research Service consider the restrictions on the use of organochlorine insecticides to be hasty and unwarranted. Production of food crops may suffer. Although the organochlorine insecticides can be replaced by less persistent insecticides (like the organophosphorus compounds), the substitutes are more expensive, require more frequent application, and are more toxic to humans and other mammals upon contact.

Obviously, the use of pesticides presents us with a dilemma. On the one hand, there is a very great need for effective insecticides to raise the quantity and quality of food needed by the world and to control epidemic insect-borne diseases. On the other hand, workers, farmers, consumers, and also wildlife need to be protected from the physiological effects of insecticides. Sterility, cancer, and liver damage are not to be readily dismissed. Trade-offs have to be made, each with its price, until such time as effective alternate methods of insect control are devised and implemented.

Summary

Although the chief use of *organic halides* is as *reaction intermediates,* some halides do have specific uses as solvents, anesthetics, refrigerants, vermicides, insecticides, and larvicides.

The organic halides are not generally found in nature. They are prepared by *direct substitution of halogen for hydrogen* (limited use), *addition of* HX *to alkenes, addition of* X_2 *to alkenes* to produce dihalides, and, by use of HX, PX_3, or $SOCl_2$, *from alcohols.*

Halogenation of the alkyl side chain of an alkylbenzene is catalyzed by light; it results in substitution of a *benzylic hydrogen* through a resonance-stabilized benzylic free-radical intermediate. The direction of *addition of* HX to a conjugated double bond of an *alkene side chain* on the benzene ring, $C_6H_5CH{=}CHR$, results in an α-halogen-substituted *arene,* $C_6H_5CHXCH_2R$, through a resonance-stabilized benzylic type of carbonium-ion intermediate.

Aryl halides are halides in which the halogen is attached directly to the benzene ring. They are prepared by *direct halogenation* or by *substitution of the diazonium group.*

All of the organic halides are insoluble in water, but they are soluble in organic solvents. The monochlorides are lighter than water, but the monobromides, monoiodides, and all of the polyhalides are heavier than water.

The chief reaction of the alkyl halides is *nucleophilic substitution* in which halide ion is displaced by a *nuclear-seeking reagent* (negative ion or a Lewis base). By type of halogen, the order of reactivity is RI > RBr > RCl > RF. Also, any given halide is more reactive than the same halide of the higher homologs. Tertiary halides, and to some extent, secondary halides, undergo nucleophilic substitution by a two-step *first-order mechanism* (S_N1) in which the rate of the reaction is proportional only to the concentration of the alkyl halide. The order of reactivity by type of halide is $3° > 2° > 1° > CH_3X$. The S_N1 reaction is subject to *rearrangements of the intermediate carbonium ion.*

Methyl halides, primary halides, and, to some extent, secondary halides undergo nucleophilic substitution reactions by a *second-order reaction mechanism* (S_N2) in which the rate is proportional to the concentrations of both halide and nucleophilic reagent. The latter approaches the carbon holding the halogen of the alkyl halide from the side opposite the halogen and forms a transition state by beginning to bond to carbon as the carbon-halogen bond begins to break. The order of reactivity by type of halide is $CH_3X > 1° > 2° > 3°$.

Stronger nucleophiles and higher concentrations of nucleophile favor the S_N2 reaction; more polar solvents favor the S_N1 reaction.

Elimination of HX from an alkyl halide to form an alkene can proceed by either a *first-order reaction* (E1) or a *second-order reaction* (E2). The order of reactivity for both is 3° > 2° > 1°. The E2 reaction is a one-step *concerted reaction*. Generally, an *anti coplanar conformation* of the halogen atom and the vicinal hydrogen atom is required.

Tertiary halides and, to some extent, secondary halides dehydrohalogenate via the E1 mechanism. The mechanism is two-step and involves formation of an intermediate carbonium ion. The E1 reaction is a strong competitor of the S_N1 reaction. It is favored by strong bases, higher temperatures (around 200°C), and tertiary halides. The strong bases, sodium alkoxide, sodium acetylide, and sodium cyanide, react with tertiary halides to give mainly a product of the elimination reaction, i.e., an alkene.

The *aryl halides* are *relatively unreactive;* they do not react under ordinary conditions with NaOH, $NaOC_2H_5$, NaCN, NaSH, H_2O, NH_3, or $AgNO_3$. They may be forced to react under strenuous conditions of heat and pressure. Like the alkyl halides, they will react with magnesium to form *Grignard reagents*. If an aryl halide has electron-withdrawing groups on the ring ortho and/or para to the halogen, nucleophilic substitution of the halogen is facilitated. The rate is *second-order*. The slow, rate-determining first step involves the formation of a *resonance-stabilized carbanion intermediate*. The electron-withdrawing groups help to delocalize the negative charge on the carbanion.

The presence of halogen in an organic compound is readily detected by the *Beilstein test,* which consists of heating a small sample on a loop of copper wire previously heated in a flame until the flame is without added color. If the flame is turned green by the sample, halogen is present. To determine which halogen is present, a sample is fused with sodium to convert the halogen to inorganic halides, which are then detected by appropriate methods.

Alkyl halides, benzyl halides, and allyl halides react with alcoholic $AgNO_3$ to give a precipitate of AgX, but aryl halides and vinyl halides will not react in that way.

Chlorinated hydrocarbons have played an important role as *insecticides* and *fungicides,* but there are problems connected with their use. Some of them have been found to be toxic to animals and humans; insects have developed a resistance to some of them; and all of them degrade so slowly that they have exceptional persistence and thus have great potential for polluting the environment.

Key terms

Alkyl halide: an alkane in which one or more halogen atoms are substituted for hydrogen atoms; the general formula is RX.

Alkylbenzene: a derivative of benzene in which one or more alkyl groups are substituted for hydrogen atoms. Also known as an **arene.**

α-Carbon atom: the number 1 carbon atom of an alkyl chain bonded to a benzene ring; also, the carbon atom which is bonded to the ring.

Anti coplanar conformation: the conformation in which two substituents on adjacent carbon atoms are anti to each other (on opposite sides of the molecule) and in the same plane.

Arene: a derivative of benzene in which alkyl groups are substituted for hydrogen; an alkylbenzene.

Aromatic nucleophilic substitution: nucleophilic substitution on an aromatic ring. It requires extreme conditions of heat and pressure or the presence of electron-withdrawing groups on the ring ortho and/or para to the halogen for the more common nucleophiles.

Aryl halide: an aromatic compound in which one or more ring hydrogen atoms are replaced by halogen atoms.

Beilstein test: a test for the presence of halogen in an organic compound; a small sample of the unknown carried in a loop of copper wire is heated in a flame, and a green flame is evidence of halogen.

Benzylic carbonium ion: a positively charged alkylbenzene ion in which the charge is delocalized from the α-carbon atom through the ring; it is a resonance hybrid.

Benzylic free radical: a free radical formed by removal of an α-hydrogen atom from an alkylbenzene; it is a resonance hybrid.

Benzylic hydrogen: a hydrogen atom bonded to the carbon atom in the α position on an alkyl chain attached to a benzene ring.

Benzylic substitution: substitution of another atom or group for a benzylic hydrogen atom.

Diazonium group: an ionic group bearing a positive charge consisting of two covalently bonded nitrogen atoms, generally attached to an aromatic ring.

E1 reaction: a first-order elimination reaction; the rate is proportional to the concentration of organic halide and is independent of the concentration of base.

E2 reaction: a second-order elimination reaction; the rate is proportional to both the concentration of organic halide and the concentration of base.

Elimination reaction: a reaction in which two groups on adjacent carbon atoms of a saturated chain are removed; the result is formation of a double bond between the two carbon atoms.

First-order reaction: a reaction in which the rate is proportional to the concentration of only one reactant and is independent of the concentrations of any other reactants; it is a multiple-step reaction.

Grignard reagent: an alkyl magnesium halide, RMgX.

Kinetics: the branch of chemistry concerned with the rates of reactions.

Leaving group: an atom or group which is displaced in a nucleophilic substitution; weak bases such as halide ions are good leaving groups.

Nucleophile: an electron-rich atom or group such as a negative ion or a Lewis base, which seeks a relatively positive nucleus with which to share its electrons. Also called a **nucleophilic reagent.**

Nucleophilic substitution: the chief reaction of the alkyl halides; halide ion is displaced by a nucleophile.

Resonance-stabilized carbanion: a carbanion (a negatively charged organic ion) which is a resonance hybrid. The delocalization of negative charge confers added stability.

S_N1 reaction: a first-order nucleophilic substitution reaction. The rate is proportional to the concentration of organic halide and is independent of the concentration of nucleophile.

S_N2 reaction: a second-order nucleophilic substitution reaction. The rate is proportional to both the concentration of organic halide and the concentration of nucleophile.

Second-order reaction: a reaction in which the rate is proportional to the concentration of two reactants.

Additional problems

6-5 Give the IUPAC name for each of the following compounds:

(a) $CH_3CH_2CHClCH_3$

(b) $C_6H_5CH_2CH_2CH_2Cl$

(c) 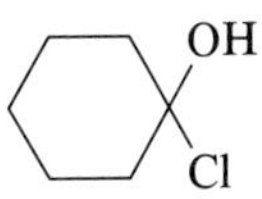 cyclohexane ring bearing OH and Cl on the same carbon

(d) 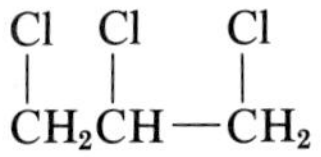 $\underset{\text{Cl}}{CH_2}\underset{\text{Cl}}{CH}-\underset{\text{Cl}}{CH_2}$ (Cl on each carbon)

6-6 Draw the structural formula for each of the following compounds.

(a) 1,2-dichloroethane (ethylene chloride)
(b) 3-bromopropene (allyl bromide)
(c) bromo-3-methylcyclohexane
(d) 2-bromooctane
(e) 1,2-dibromo-3-chloropropane (DBCP)

6-7 Write an equation or equations showing the preparation of each of the following halides from the starting materials indicated. Show any necessary conditions and reagents required for the reactions.

(a) $C_6H_5CHBrCH_3$ from $C_6H_5CH{=}CH_2$
(b) $C_6H_5CHBrCH_2Br$ from $C_6H_5CH_2CH_3$ (three reactions required)
(c) $C_6H_5CH_2CH_2Br$ from $C_6H_5CH_2CH_3$ (three reactions required)
(d) chlorocyclohexane from cyclohexane
(e) chlorocyclohexane from cyclohexene
(f) (4-chlorophenyl)cyclohexane from chlorocyclohexane and benzene (two reactions required)
(g) *m*-chlorobenzoic acid from toluene (two reactions required)
(h) *p*-chlorobenzyl chloride from toluene (two reactions required)

6-8 Write equations for the following reactions:

(a) $CH_3CH_2CH_2CH_2Br + NaOH(aq)$
(b) $C_6H_5CH_2Cl + H_2O$
(c) $CH_3CH_2CH_2Br + NaSH$
(d) $CH_3CH_2Br + Na^+ \ ^-C{\equiv}C{-}CH_3$
(e) $CH_3CH_2CH_2Br + Na^+ \ ^-OCH_2CH_3$

(f) $CH_3CH_2Br + NH_3$
(g) $C_6H_5CH_2Cl + AgNO_3 + H_2O$
(h) $C_6H_5CH_2Cl + NaCN$

(i) 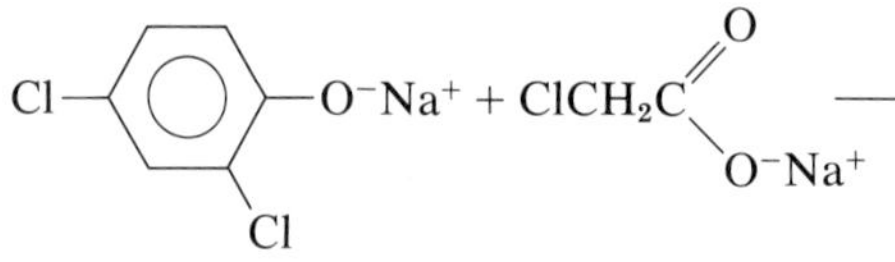 Cl—⟨2,4-dichlorophenyl⟩—$O^-Na^+ + ClCH_2C(=O)O^-Na^+$ ⟶ 2,4-dichlorophenoxyacetic acid (2,4-D, a herbicide)

(j) 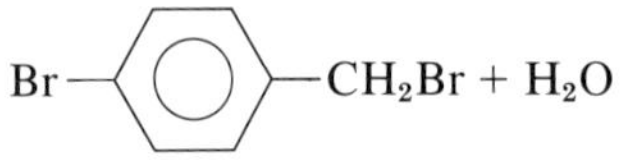 Br—C_6H_4—$CH_2Br + H_2O$

6-9 Write the equation for each of the following reactions. What mechanism accounts for all of these reactions?

(a) 2,4-dinitrochlorobenzene and sodium hydroxide
(b) 2,4-dinitrochlorobenzene and sodium phenoxide ($C_6H_5O^-Na^+$)
(c) 2,4-dinitrochlorobenzene and ammonia

6-10 Complete the equation for each of the following reactions and tell what mechanism accounts for the reaction.

(a) $CH_3CH_2CH(Br)CH_3 + KOH \xrightarrow{\Delta,\ \text{alcohol}}$

(b) $(CH_3)_3C—Br + Na^{+-}C{\equiv}C—CH_3 \xrightarrow{\Delta}$

(c) $(CH_3)_3C—Br + Na^+\ {}^-OC_2H_5 \xrightarrow{\Delta}$

6-11 Write the equation for each of the following reactions:

(a) benzene and benzyl chloride with $AlCl_3$ catalyst
(b) trichloromethane ($HCCl_3$, chloroform) and benzene (three equivalents), $AlCl_3$ catalyst
(c) benzene and chlorocyclohexane, $AlCl_3$ catalyst

6-12 Write the equations for the addition of 2 mol of chlorine to ethyne (acetylene) followed by a single dehydrohalogenation of that product.

6-13 Write equations for the conversion of toluene ($C_6H_5CH_3$) into 1-bromo-4-bromomethylbenzene (*p*-bromobenzyl bromide, Br—C_6H_4—CH_2Br).

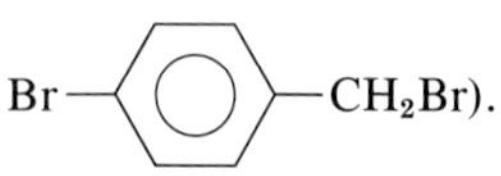

6-14 Write equations showing the *mechanism* of the following reaction:

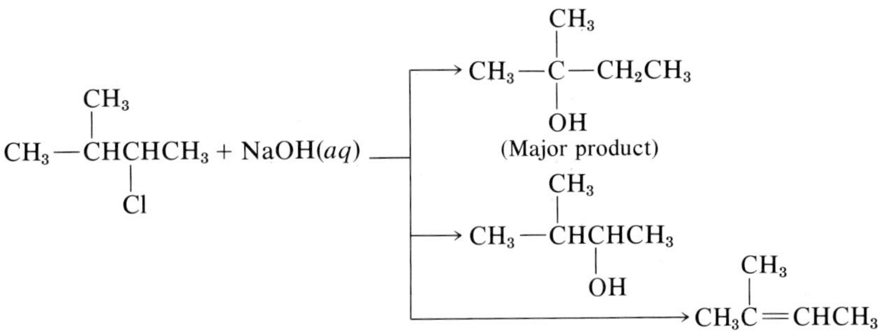

$$CH_3-\underset{Cl}{\underset{|}{\overset{CH_3}{\overset{|}{C}}}}HCHCH_3 + NaOH(aq) \longrightarrow \begin{cases} CH_3-\underset{OH}{\underset{|}{\overset{CH_3}{\overset{|}{C}}}}-CH_2CH_3 \quad \text{(Major product)} \\ CH_3-\overset{CH_3}{\overset{|}{C}}H\underset{OH}{\underset{|}{C}}HCH_3 \\ CH_3\overset{CH_3}{\overset{|}{C}}=CHCH_3 \end{cases}$$

6-15 Name a simple chemical test or reagent which will readily distinguish between each of the following pairs of compounds. Indicate which member of a pair gives the positive test or greater reaction.

(a) $CH_3CH_2CH_2CH_2CH_2CH_3$ and CH_3CH_2Br

(b) 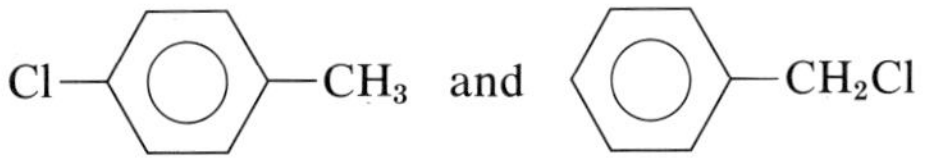 Cl–C_6H_4–CH_3 (p-chlorotoluene) and C_6H_5–CH_2Cl

(c) $ClCH_2CH{=}CH_2$ and $ClCH_2CH_2CH_3$

(d) $ClCH_2CH{=}CH_2$ and $CH_3CH{=}CHCl$

(e) 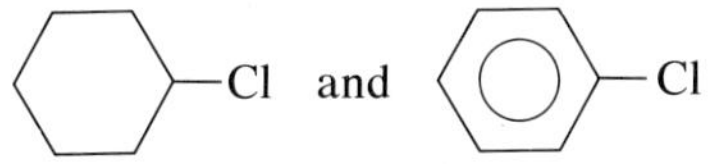 C_6H_{11}–Cl (cyclohexyl chloride) and C_6H_5–Cl

6-16 By using reference books, find the toxicity and health hazards of the following chlorinated hydrocarbons: (a) CCl_4, (b) CH_2Cl_2, (c) DDT, (d) TEL (tetraethyl lead), (e) $CHCl_3$, (f) CH_2BrCH_2Br, (g) Kepone.

ALCOHOLS, PHENOLS, AND ETHERS

The alcohols constitute one of the most important groups of compounds in organic chemistry. Their occurrence is widespread in nature, and they are also readily synthesized by industry and in the laboratory. They serve as solvents as well as important reaction intermediates. Hundreds of compounds can be produced beginning with ethyl alcohol. A great many people enjoy drinking beverages containing ethyl alcohol. Ethers are related compounds which serve as anesthetics and solvents. Phenols are weak acids, and they are important intermediates for the drug, dye, and plastics industries.

The *alcohols* can be considered the monoalkyl derivatives of water, the *phenols* the monoaryl derivatives of water, and the *ethers* the derivatives of water in which both hydrogen atoms of the water molecule have been substituted with alkyl groups, aryl groups, or both.

H—O—H	R—O—H	Ar—O—H	Ar—O—R	R—O—R	Ar—O—Ar
Water	Alcohol	Phenol		Ethers	

The alcohol or hydroxy group is —OH (Fig. 7-1).

FIGURE 7-1
Ball-and-stick models of (*a*) water, (*b*) methanol (methyl alcohol), and (*c*) methoxymethane (methyl ether).

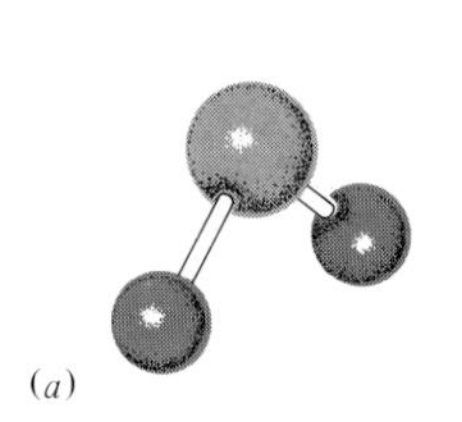

(*a*)

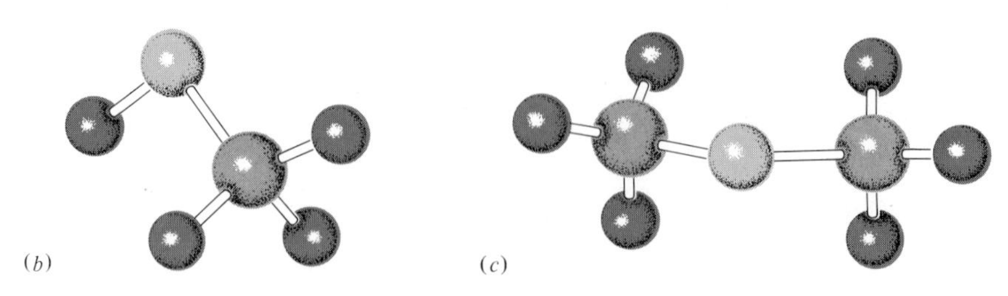

(*b*) (*c*)

----:Ö—H----:Ö—H----:Ö—H----
R R R

(*a*)

----:Ö—H----:Ö—H----:Ö—H----
Ar Ar Ar

(*b*)

FIGURE 7-2 Hydrogen bonding (*a*) in alcohols and (*b*) in phenols.

7-1 Physical properties

Hydrogen bonding is possible for phenols and alcohols just as it is for water (Fig. 7-2). It is not possible for ethers, which have no hydrogen at all attached to oxygen. The association of the molecules of alcohols and phenols through hydrogen bonding explains why the boiling points of alcohols and phenols are very much higher than those of hydrocarbons of corresponding molecular weight. On the other hand, the boiling points of ethers, with no hydrogen bonding, are not much higher than those of hydrocarbons of corresponding molecular weight. That is true even though ethers have a slight dipole attraction as a result of the nonlinear geometry of the ether molecule. Even the lowest homolog of the alcohols, CH_3OH, is a liquid, whereas the lowest homolog of the ethers, CH_3OCH_3, is a gas. The simplest phenol, C_6H_5OH, is a low-melting solid.

Alcohols, phenols, and ethers of about the same molecular weight are equally soluble in water. Since hydrogen of water can hydrogen bond to the oxygen atom of alcohols, phenols, or ethers, the lower-molecular-weight homologs of all three classes are soluble in water. The compounds containing three or fewer carbons are soluble in all proportions, and those with four carbons are partly soluble in water.

Tables 7-1, 7-2, and 7-3 provide information about common alcohols, phenols, and ethers. Figure 7-3 illustrates two complex and important alcohols.

7-2 Nomenclature

Many common names of alcohols, phenols, and ethers are still in use, so they, as well as the IUPAC names, must be learned.

Naming of alcohols

The common name of an alcohol is formed by naming the alkyl group attached to the —OH group and then adding the word *alcohol*. For example, CH_3OH is methyl alcohol, CH_3CH_2OH is ethyl alcohol, $CH_3CH_2CH_2CH_2OH$ is *n*-butyl alcohol, and $(CH_3)_2CHCH_2OH$ is isobutyl alcohol.

The IUPAC name of an alcohol is formed by selecting and naming the parent alkane, or longest continuous carbon chain containing the —OH

(*a*) HO

(*b*) OH HO

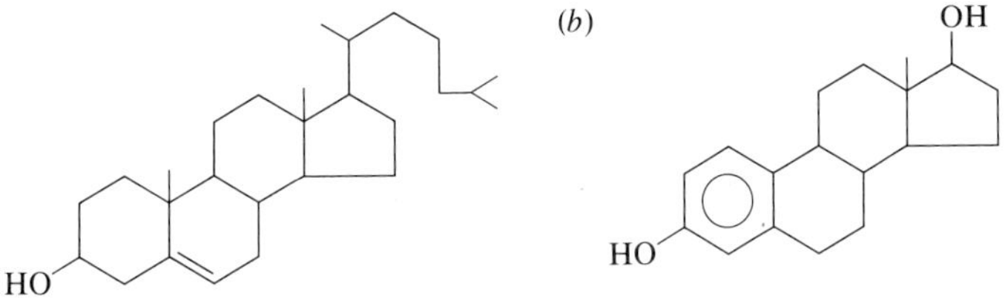

FIGURE 7-3
(*a*) Cholesterol, mp 148.5°C, and (*b*) estradiol, mp 220°C, a female hormone, are alcohols.

TABLE 7-1
Some selected alcohols

Structural formula	IUPAC name	Common name	Melting point, °C	Boiling point, °C
CH_3OH	Methanol	Methyl alcohol	−97.8	64.9
CH_3CH_2OH	Ethanol	Ethyl alcohol	−114.7	78.4
$CH_3CH_2CH_2OH$	1-Propanol	*n*-Propyl alcohol	−126.5	97.4
$CH_3CHOHCH_3$	2-Propanol	Isopropyl alcohol	−89.5	82.4
$CH_3CH_2CH_2CH_2OH$	1-Butanol	*n*-Butyl alcohol	−89.5	117.2
$CH_3CH_2CHOHCH_3$	2-Butanol	*sec*-Butyl alcohol	−114.7	99.5
$(CH_3)_2CHCH_2OH$	2-Methyl-1-propanol	Isobutyl alcohol	−108	108.4
$(CH_3)_3COH$	2-Methyl-2-propanol	*tert*-Butyl alcohol	25.5	82.2
$CH_3(CH_2)_4OH$	1-Pentanol	*n*-Amyl alcohol	−79	137.3
$C_6H_5CH_2OH$	Phenylmethanol	Benzyl alcohol	−15.3	205.3
CH_2OHCH_2OH	1,2-Ethanediol	Ethylene glycol	−17.4	198
$CH_2OHCHOHCH_2OH$	1,2,3-Propanetriol	Glycerol (glycerin)	20	290

TABLE 7-2
Some selected phenols

Structural formula	IUPAC name	Common name	Melting point, °C	Boiling point, °C
C_6H_5OH	Phenol	Phenol (carbolic acid)	43	181.7
o-$CH_3—C_6H_4OH$	2-Methylphenol	*o*-Cresol	30.9	190.9
m-$C_6H_4(OH)_2$	1,3-Dihydroxybenzene	Resorcinol	111	276
p-$C_6H_4(OH)_2$	1,4-Dihydroxybenzene	Hydroquinone	173	285

TABLE 7-3
Some selected ethers

Structural formula	IUPAC name	Common name	Melting point, °C	Boiling point, °C
CH_3OCH_3	Methoxymethane	Methyl ether	−138.5	−23
$CH_3CH_2OCH_3$	Methoxyethane	Ethyl methyl ether	. . .	10.8
$CH_3CH_2OCH_2CH_3$	Ethoxyethane	Ethyl ether	−116.6	34.5
$(CH_3CH_2CH_2CH_2)_2O$	Butoxybutane	Butyl ether	−95	142
$C_6H_5OCH_3$	Methoxybenzene	Anisole	−37.3	155
$C_6H_5OC_6H_5$	Phenoxybenzene	Phenyl ether	26.8	257.9
$H_2C—CH_2$ \ / O	1,2-Epoxyethane	Ethylene oxide	−111	10.7
$H_2C—CH_2$ | | H_2C CH_2 \ / O	1,4-Epoxybutane	Tetrahydrofuran	. . .	67
H—C—C—H ‖ ‖ H—C C—H \ / O	1,4-Epoxy-1,3-butadiene	Furan	. . .	31.3

group, and then changing the suffix *-e* to *-ol*. The suffix *-ol* designates the —OH group, and the group's position in the chain is indicated by locator numbers. For example, CH_3OH is methanol, CH_3CH_2OH is ethanol, $CH_3CH_2CH_2CH_2OH$ is 1-butanol, $CH_3CH_2CHOHCH_3$ is 2-butanol, and $(CH_3)_2CHCH_2OH$ is 2-methyl-1-propanol. Table 7-1 gives other examples.

Compounds with two —OH groups are commonly called *glycols;* in the IUPAC system they are *diols*. For example, $CH_3CHOHCH_2OH$ is commonly called propylene glycol; its IUPAC name is 1,2-propanediol.

Naming of phenols

The *phenols* are aromatic compounds having an —OH group as a substituent on the ring. The simplest one is C_6H_5OH. It is called phenol, so phenol is the name of a *specific compound* as well as the *name of a class*.

In the IUPAC system *monohydric* (one —OH group) *phenols* are named as derivatives of phenol and locator numbers are used to designate positions. The —OH group is assumed to be in the 1 position unless there is a substituent of higher priority, such as a —COOH group, on the ring. (See Table 2-5 for priority of groups.) *Dihydric* (two —OH groups) *phenols* are named by naming the —OH groups as hydroxy groups and indicating the positions of the hydroxy groups on the ring with locator numbers. The following are some examples of phenols and their IUPAC and common names.

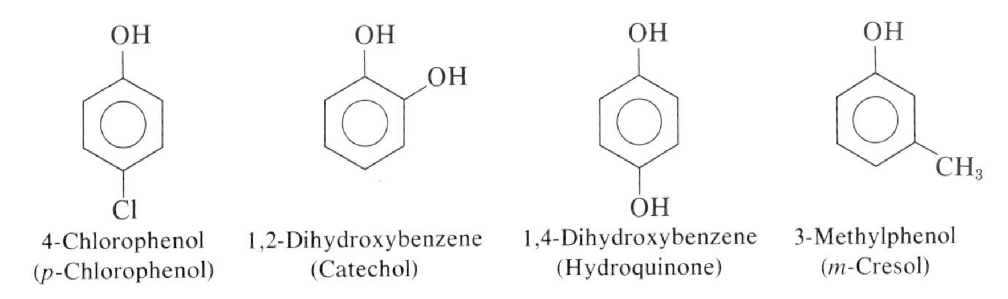

4-Chlorophenol (*p*-Chlorophenol) 1,2-Dihydroxybenzene (Catechol) 1,4-Dihydroxybenzene (Hydroquinone) 3-Methylphenol (*m*-Cresol)

Naming of ethers

An ether is commonly named by naming the two alkyl or aryl groups attached to the oxygen atom and adding the word *ether*. If both groups are the same, the name of the group is mentioned only once; the groups are assumed to be the same *without* the use of the prefix *di-*. For example, CH_3OCH_3 is methyl ether and $CH_3CH_2OCH_2CH_3$ is ethyl ether, but $CH_3OCH_2CH_3$ is ethyl methyl ether. (See Table 7-3 for other examples.)

In the IUPAC system the smaller of the two groups is combined with the oxygen and is named as an *alkoxy* or *aroxy group,* and it is then located as a substituent group on a parent alkane. For example, $CH_3OCH_2CH_3$ is methoxyethane, $C_6H_5OC_6H_5$ is phenoxybenzene, and $CH_3CH_2CH(OCH_3)CH_2CH_3$ is 3-methoxypentane.

Cyclic ethers are named as oxides or *epoxides*, and a few also have trivial names. The compound $H_2C{-}CH_2$ (bridged by O) is commonly called ethylene oxide; its IUPAC name is 1,2-epoxyethane. The cyclic ether (five-membered ring containing O) is commonly called tetrahydrofuran (THF), but its IUPAC name is 1,4-epoxybutane.

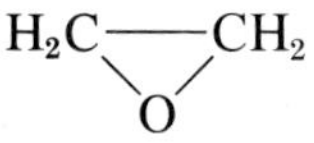

PROBLEM 7-1 Give the IUPAC name for each of the following compounds.

(a) $(CH_3)_3CCH_2OH$
(b) $CH_3CHOHCH_2CH_2CH_3$
(c) $C_6H_5CH_2CHOHCH_3$
(d) *o*-$CH_3C_6H_4OH$
(e) *p*-$NO_2C_6H_4OH$
(f) 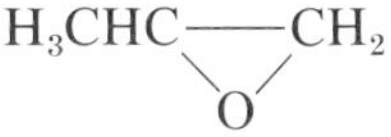 $H_3CHC{-}CH_2$ (bridged by O)
(g) $CH_3CH_2CH_2OCH_2CH_3$

Solution to (a) The correct IUPAC name for the compound is 2,2-dimethyl-1-propanol, as can be seen in the structural formula:

$$\overset{3}{C}H_3{-}\overset{2}{C}(CH_3)_2{-}\overset{1}{C}H_2OH$$

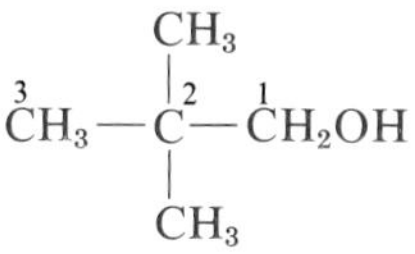

PROBLEM 7-2 Write a structural formula for each of the following compounds. If the common name is given here, write the IUPAC name and the structural formula.

(a) *p*-cresol
(b) ethyl phenyl ether
(c) *cis*-2,3-epoxybutane
(d) 2-ethyl-1-hexanol
(e) 2-methyl-1-pentanol
(f) *p*-chlorobenzyl alcohol
(g) *p*-aminophenol

Solution to (a) The structural formula for *p*-cresol is as shown below. The IUPAC name is 4-methylphenol.

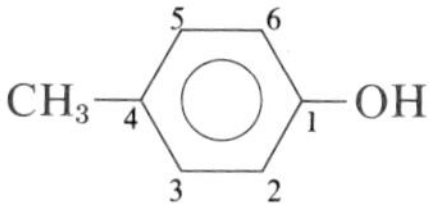

$$CH_3{-}C_6H_4{-}OH \quad (\text{ring positions } 1\text{–}6;\ OH \text{ at } 1,\ CH_3 \text{ at } 4)$$

7-3 Ethyl alcohol

The most commonly known of the alcohols is ethyl alcohol or *ethanol*. When the simple term "alcohol" is used without qualification, ethyl alcohol is intended. Ethyl alcohol is the beverage alcohol.

Toxicity

All the alcohols are toxic, but ethyl alcohol is least toxic. Fully half of the people of the world consume ethyl alcohol in beverages. Of those consumers as many as one in ten may become addicted to alcohol and drink alcoholic beverages excessively. Such an addict is called an alcoholic.

Ethyl alcohol is first oxidized in the body to acetaldehyde. Recent research has indicated that some of the acetaldehyde may be incorporated into the synthesis of amines with physiological and addictive reactions similar to those of the drugs from opium. The excess acetaldehyde produced in the blood from oxidation of ingested ethyl alcohol diminishes the elasticity of connective tissue. The result is premature aging of connective tissue, and that is a factor in the vascular disease arteriosclerosis and the lung disease emphysema. Ethyl alcohol itself constricts the small blood vessels going into the brain, and the result is oxygen starvation and death of some brain cells. Because ethyl alcohol depresses the higher nerve centers, people under the influence of alcohol suffer from diminished control of their judgment and actions, and so they may endanger themselves or others. Excessive use of alcohol permanently damages the liver and the brain.

Control

Because of the physiological and social effects of ethyl alcohol, governments have at times in the past prohibited the sale of alcoholic beverages. In general, ethyl alcohol is taxed heavily for both control and revenue. A gallon of it is worth about a dollar in the United States if it is purchased in large quantities, but the tax on the gallon is about 20 dollars. Only hospitals and nonprofit research or educational institutions may purchase ethyl alcohol tax-free. An industry that wishes to use ethyl alcohol as a solvent may avoid the tax by buying *denatured alcohol,* i.e., alcohol to which some denaturant has been added. The *denaturant* is a contaminant that makes the alcohol unsafe for drinking and is difficult to remove. For example, methyl alcohol, which can cause blindness and even death, has been used as a denaturant for ethyl alcohol. Rubbing alcohol is either denatured ethyl alcohol or isopropyl alcohol. The latter is the more likely.

Absolute alcohol

The usual ethyl alcohol is 95% alcohol and 5% water. That is a constant-boiling mixture (*azeotrope*), so the remaining water cannot be removed by fractional distillation. However, if benzene is added to the 95% alcohol, a *ternary azeotrope* is formed. When the mixture is fractionally distilled, the water is removed as part of the ternary azeotrope, which distills over as the lowest-boiling fraction. The 100% ethyl alcohol thereby produced is called *absolute alcohol.* It is required for some chemical reactions.

Ethyl alcohol from fermentation

For thousands of years the human species has utilized the natural fermentation of plant products containing carbohydrates to produce alcoholic beverages. The reaction is usually accomplished by the enzyme zymase in yeast.

$$\underset{\text{Glucose}}{C_6H_{12}O_6} \xrightarrow{\text{zymase}} 2CO_2 + \underset{\substack{\text{Ethanol} \\ \text{(Ethyl alcohol)}}}{2CH_3CH_2OH} \qquad (7\text{-}1)$$

Fermentation of *grains* produces alcoholic beverages called *beer* or *ale*. The beer may be distilled to produce whiskey, gin, or vodka. Fermentation of *fruits* and *berries* produces *wines*. The wine may be distilled to increase the concentration of alcohol and yield a brandy or cognac. The Japanese ferment rice to produce sake; the Mexicans ferment a species of cactus to produce mescal and, from it, tequila. Black strap molasses, a by-product of sugar refining, may be fermented and distilled to produce rum or, by further refinement, industrial alcohol.

The strength of beverage alcohol is expressed as *proof,* which is twice the volume percent of alcohol. *Proof alcohol,* or 100 proof beverages, are 50% alcohol by volume. The proof system is said to derive from the practice of proving the alcoholic concentration of a beverage by pouring a sample on a little pile of gunpowder and then attempting to ignite it. Successful ignition indicated proof alcohol.

A by-product of fermentation of carbohydrates to produce ethyl alcohol is a material called fusel oil. This is chiefly a mixture of 3-methyl-1-butanol, 2-methyl-1-butanol, and 2-methyl-1-propanol, along with some esters and aldehydes. About 60% of it boils at 122 to 138°C. Fusel oil is derived from the protein part of the grains being fermented and is considerably more toxic than ethyl alcohol. A proper fractional distillation is required to remove most of the fusel oil from a fermented mash to produce beverage alcohol, so would-be moonshiners beware!

7-4 Sources and preparation of alcohols

The *hydroxy group* which is characteristic of alcohols and phenols is of very common occurrence in nature. You may recall that the diterpene vitamin A, discussed in Chap. 4, is an alcohol. Note also that the natural products cholesterol and estradiol (Fig. 7-3) are alcohols.

n-Butyl alcohol from fermentation

The fermentation of starch with the bacterium *Clostridium acetobutylicum* yields a product consisting of 10% ethyl alcohol, 60% *n*-butyl alcohol, and 30% acetone. The process was developed in England during World War I to supply acetone for the manufacture of the gunpowder used by the British army. It was invented by Chaim Weizmann, a Russian-born Jewish biochemist, who became the first president of Israel in 1948. Weizmann was politically rewarded for his contribution by the Balfour Declaration, in 1917, that gave British support to the establishment of a national homeland for the Jews in Palestine. The process is now valued as a method for producing *n*-butyl alcohol for the plastics industry.

Alcohols from hydration of alkenes

With an acid catalyst, an alkene will add water to produce an alcohol. The direction of addition is governed by Markovnikov's rule. The general equation is

$$\gt C{=}C\lt \; + \; HOH \xrightarrow{H^+} -\overset{\overset{\displaystyle H}{|}}{\underset{|}{C}}-\overset{|}{\underset{\underset{\displaystyle OH}{|}}{C}}- \tag{7-2}$$

The mechanism for this reaction was discussed in Sec. 4-1. It is an example of electrophilic addition of water to an alkene. The method is important for the production of several alcohols:

$$CH_2{=}CH_2 + HOH \xrightarrow{H^+} \underset{\substack{\text{Ethanol} \\ \text{(Ethyl alcohol)}}}{CH_3CH_2OH} \tag{7-3}$$

$$CH_3CH{=}CH_2 + HOH \xrightarrow{H^+} \underset{\substack{\text{2-Propanol} \\ \text{(Isopropyl alcohol)}}}{CH_3CHOHCH_3} \tag{7-4}$$

$$CH_3-\overset{\overset{\displaystyle CH_3}{|}}{C}{=}CH_2 + HOH \xrightarrow{H^+} \underset{\substack{\text{2-Methyl-2-propanol} \\ (\textit{tert}\text{-Butyl alcohol})}}{CH_3-\overset{\overset{\displaystyle CH_3}{|}}{\underset{\underset{\displaystyle OH}{|}}{C}}-CH_3} \tag{7-5}$$

If sulfuric acid is used, the reaction may also be carried out by production and subsequent hydrolysis of the alkyl hydrogen sulfate.

$$CH_2{=}CH_2 + HOSO_2OH \longrightarrow \underset{\text{Ethyl hydrogen sulfate}}{CH_3CH_2OSO_2OH} \tag{7-6}$$

$$CH_3CH_2OSO_2OH \xrightarrow{H_2O} \underset{\text{Ethanol}}{CH_3CH_2OH} + H_2SO_4 \tag{7-7}$$

These reactions may be used in the chemical laboratory as well as by industry. Most alcohols are available by purchase from industry.

Alcohols from hydrolysis of alkyl halides

The hydrolysis of alkyl halides to produce alcohols is of use when the necessary halides are available. The general reaction is

$$RX + OH^- \longrightarrow ROH + X^- \tag{7-8}$$

Some examples of the reaction are the following:

$$\underset{\substack{\text{(Chloromethyl)benzene} \\ \text{(Benzyl chloride)}}}{C_6H_5CH_2Cl + H_2O} \xrightarrow{(S_N1)} \underset{\substack{\text{Phenylmethanol} \\ \text{(Benzyl alcohol)}}}{C_6H_5CH_2OH} + HCl \tag{7-9}$$

$$\underset{\substack{\text{2-Chloro-1-ethanol}\\ \text{(Ethylene chlorohydrin)}}}{\underset{\text{OH}}{CH_2}-\underset{\text{Cl}}{CH_2}} + H_2O \xrightarrow{\text{NaOH } (S_N2)} \underset{\substack{\text{1,2-Ethanediol}\\ \text{(Ethylene glycol)}}}{\underset{\text{OH}}{CH_2}-\underset{\text{OH}}{CH_2}} + NaCl \tag{7-10}$$

$$\underset{\substack{\text{1-Chloro-2-propene}\\ \text{(Allyl chloride)}}}{CH_2{=}CH-CH_2Cl} + H_2O \xrightarrow{(S_N1)} \underset{\substack{\text{2-Propen-1-ol}\\ \text{(Allyl alcohol)}}}{CH_2{=}CH-CH_2OH} + HCl \tag{7-11}$$

These reactions are examples of the *nucleophilic substitution* of OH^- or water in an alkyl halide, as discussed in Chap. 6. Benzyl and allyl chlorides; both of which form resonance-stabilized carbonium ions, will react with water, the weaker nucleophile, in an S_N1 reaction.

Alcohols from Grignard reactions

Alcohols can be prepared by the nucleophilic addition of a Grignard reagent to an aldehyde or a ketone. The Grignard reagent itself is prepared by the reaction of an alkyl or aryl halide with magnesium. Dry ether is used as a solvent.

$$\underset{\substack{\text{Alkyl}\\ \text{halide}}}{RBr} + Mg \xrightarrow{\text{ether}} \underset{\substack{\text{Grignard}\\ \text{reagent}}}{RMgBr} \tag{7-12}$$

$$\underset{\substack{\text{Aryl}\\ \text{halide}}}{C_6H_5Br} + Mg \xrightarrow{\text{ether}} \underset{\substack{\text{Grignard}\\ \text{reagent}}}{C_6H_5MgBr} \tag{7-13}$$

The mechanism of the reaction of a Grignard reagent with the *carbonyl group* of an aldehyde or a ketone involves nucleophilic addition followed by a hydrolysis step.

$$R^-Mg^{2+}X^- \; + \; \overset{\delta+}{\underset{|}{\overset{|}{C}}}{=}\overset{\delta-}{O} \longrightarrow R-\underset{|}{\overset{|}{C}}-O^-Mg^{2+}X^- \tag{7-14}$$

$$R-\underset{|}{\overset{|}{C}}-O^-Mg^{2+}X^- + H-O-H \longrightarrow R-\underset{|}{\overset{|}{C}}-OH + Mg(OH)X \tag{7-15}$$

The Grignard reagent may be reacted with (a) methanal (formaldehyde) to produce a primary alcohol, (b) any other aldehyde to produce a secondary alcohol, or (c) a ketone to produce a tertiary alcohol.

$$\text{(a)}\quad RMgBr + \underset{\substack{\text{Methanal}\\ \text{(Formaldehyde)}}}{H_2C{=}O} \longrightarrow RCH_2OMgBr \xrightarrow{H_2O} \underset{\text{Primary alcohol}}{RCH_2OH} \tag{7-16}$$

$$\text{(b)}\quad RMgBr + \underset{\text{An aldehyde}}{H-\underset{\substack{|\\ R'}}{C}{=}O} \longrightarrow R-\underset{\substack{|\\ R'}}{CH}OMgBr \xrightarrow{H_2O} \underset{\text{Secondary alcohol}}{R-\underset{\substack{|\\ R'}}{CH}OH} \tag{7-17}$$

$$\text{(c)}\quad RMgBr + R'\!-\!\underset{\underset{R''}{|}}{C}\!=\!O \longrightarrow R\!-\!\overset{\overset{R'}{|}}{\underset{\underset{R''}{|}}{C}}\!-\!OMgBr \xrightarrow{H_2O} R\!-\!\overset{\overset{R'}{|}}{\underset{\underset{R''}{|}}{C}}\!-\!OH \qquad (7\text{-}18)$$

A ketone — Tertiary alcohol

(See also Fig. 7-4.) Care must be taken in the initial addition step to exclude water and carbon dioxide, which also can react with the Grignard reagent at this point. The alkyl group R has some carbanion character; it is a strong base which will remove a proton from water to form an alkane, RH, which is a very much weaker acid than water.

$$\underset{\text{Stronger base}}{RMgX} + \underset{\text{Stronger acid}}{HOH} \longrightarrow \underset{\text{Weaker acid}}{RH} + \underset{\text{Weaker base}}{Mg(OH)X} \qquad (7\text{-}19)$$

Grignard reagents react similarly with other Brønsted-Lowry acids such as alcohols and ammonia.

Some specific examples of reactions of Grignard reagents with aldehydes and ketones are the following:

$$\underset{\text{Isobutyl magnesium bromide}}{CH_3\overset{\overset{CH_3}{|}}{C}HCH_2MgBr} + \underset{\text{Methanal (Formaldehyde)}}{H\overset{\overset{H}{|}}{C}\!=\!O} \xrightarrow{\text{ether}} CH_3\overset{\overset{CH_3}{|}}{C}HCH_2CH_2OMgBr \xrightarrow{H_2O} \underset{\text{3-Methyl-1-butanol}}{CH_3\overset{\overset{CH_3}{|}}{C}HCH_2CH_2OH} \qquad (7\text{-}20)$$

$$\underset{\text{Benzyl magnesium bromide}}{C_6H_5CH_2MgBr} + \underset{\text{Ethanal (Acetaldehyde)}}{CH_3\overset{\overset{H}{|}}{C}\!=\!O} \xrightarrow{\text{ether}} C_6H_5CH_2\overset{\overset{OMgBr}{|}}{C}HCH_3 \xrightarrow{H_2O} \underset{\text{1-Phenyl-2-propanol}}{C_6H_5CH_2\overset{\overset{OH}{|}}{C}HCH_3} \qquad (7\text{-}21)$$

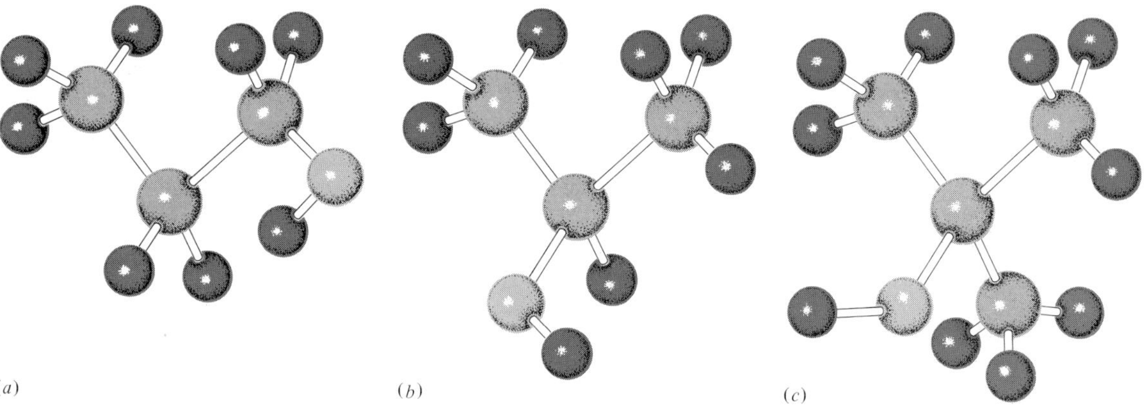

FIGURE 7-4
Models of (*a*) a primary alcohol, 1-propanol, (*b*) a secondary alcohol, 2-propanol, and (*c*) a tertiary alcohol, 2-methyl-2-propanol.

$$C_6H_5MgBr + CH_3\underset{\underset{CH_3}{|}}{C}{=}O \xrightarrow{\text{ether}} C_6H_5\overset{\overset{CH_3}{|}}{\underset{\underset{CH_3}{|}}{C}}{-}OMgBr \xrightarrow{H_2O} C_6H_5\overset{\overset{CH_3}{|}}{\underset{\underset{CH_3}{|}}{C}}{-}OH \qquad (7\text{-}22)$$

Phenyl magnesium bromide; Propanone (Acetone); 2-Phenyl-2-propanol

The reaction of Grignard reagents with carbonyl compounds is not generally used in industry because the reagents are too expensive. However, the preparation and use of Grignard reagents is one of the most generally useful methods available in the laboratory for synthesizing more complicated alcohols which are not available from commercial sources.

Alcohols by hydroboration-oxidation

Addition of water to a double bond can also be achieved in the laboratory by the use of boron hydride, $(BH_3)_2$, called *diborane*. Two reactions are involved. The first is the addition of boron hydride to the double bond; the second is the oxidation of the organoboron derivative to an alcohol and boric acid.

$$3\ R_2C{=}CR_2 + \tfrac{1}{2}(BH_3)_2 \longrightarrow \left(H{-}\underset{\underset{R}{|}}{\overset{\overset{R}{|}}{C}}{-}\underset{\underset{R}{|}}{\overset{\overset{R}{|}}{C}}{-}\right)_3 B \qquad (7\text{-}23)$$

$$\left(H{-}\underset{\underset{R}{|}}{\overset{\overset{R}{|}}{C}}{-}\underset{\underset{R}{|}}{\overset{\overset{R}{|}}{C}}{-}\right)_3 B \xrightarrow{H_2O_2,\ OH^-} 3H{-}\underset{\underset{R}{|}}{\overset{\overset{R}{|}}{C}}{-}\underset{\underset{R}{|}}{\overset{\overset{R}{|}}{C}}{-}OH + H_3BO_3 \qquad (7\text{-}24)$$

An example is the preparation of 1-propanol from propene:

$$3CH_3CH{=}CH_2 + \tfrac{1}{2}(BH_3)_2 \longrightarrow (CH_3CH_2CH_2)_3B \xrightarrow{H_2O_2,\ OH^-} 3CH_3CH_2CH_2OH + H_3BO_3 \qquad (7\text{-}25)$$

Propene (Propylene); 1-Propanol (*n*-Propyl alcohol)

The apparent addition of water in the product is anti-Markovnikov, which makes this a useful way to obtain some primary alcohols.

Alcohols from primary amines

Alcohols can also be prepared by the diazotization of primary amines. *Diazotization* is a process in which an amino group, $—NH_2$, reacts with nitrous acid, HNO_2, to form a very reactive intermediate ion called a *dia-*

zonium ion because it contains two nitrogen atoms (*azo* refers to nitrogen). The general reaction is

$$\underset{\substack{\text{Primary}\\\text{amine}}}{RNH_2} \xrightarrow{NaNO_2,\ H^+} \underset{\substack{\text{Alkyl diazonium}\\\text{ion}}}{RN_2^+} \xrightarrow{H_2O} \underset{\substack{\text{Primary}\\\text{alcohol}\\\text{and other products}}}{ROH} + N_2 \qquad (7\text{-}26)$$

The intermediate diazonium salts cannot be isolated, and they are spontaneously hydrolyzed to alcohol. This reaction will be discussed in more detail in Chap. 13.

Industrial synthesis of methanol

Methanol is synthesized in high purity by the high-temperature, high-pressure, catalytic reaction of carbon monoxide with hydrogen.

$$CO + 2H_2 \xrightarrow{ZnCr_2O_4,\ 370\text{–}375°C,\ 272\text{–}340\ atm} \underset{\substack{\text{Methanol}\\\text{(Methyl alcohol)}}}{CH_3OH} \qquad (7\text{-}27)$$

PROBLEM 7-3 Write equations showing the alcohols produced by the acid-catalyzed hydration of the following alkenes.

(a) propene
(b) 2-methylpropene
(c) 1-butene
(d) 2-butene
(e) cyclohexene
(f) 1-methylcyclohexene
(g) ethenylbenzene (vinyl benzene)
(h) 2-methyl-2-butene

Solution to (a)

$$\underset{\text{Propene}}{CH_3{-}CH{=}CH_2} + H_2O \xrightarrow{H^+} \underset{\text{2-Propanol}}{CH_3{-}\underset{\substack{|\\OH}}{CH}{-}CH_3}$$

PROBLEM 7-4 Write equations showing the alcohols produced by the hydrolysis of the following halides. (a) *p*-chlorobenzyl chloride, (b) 2-chloropentane, (c) chlorocyclohexane, (d) 2-chloro-1-ethanol.

Solution to (a)

$$p\text{-}ClC_6H_4CH_2Cl + H_2O \xrightarrow{S_N1} p\text{-}ClC_6H_4CH_2OH + HCl$$

1-Chloro-4-chloromethylbenzene (*p*-Chlorobenzyl chloride) → 4-Chlorophenylmethanol (*p*-Chlorobenzyl alcohol)

Benzyl halides are very reactive because formation of the intermediate carbonium ion is aided by delocalization of charge by resonance. Aryl halides (halogen attached directly to ring) are very unreactive.

PROBLEM 7-5 Write the equations for the syntheses of the following alcohols by using the Grignard reaction. Begin with benzene or toluene and any necessary aliphatic or inorganic reagents. (a) 1-phenyl-2-propanol, (b) 1-phenyl-1-propanol, (c) 2-phenyl-2-propanol, (d) 2-phenyl-1-propanol.

Solution to (a)

$$C_6H_5-CH_3 + Br_2 \xrightarrow{\text{light}} C_6H_5-CH_2Br + HBr$$

Toluene

$$C_6H_5-CH_2Br + Mg \xrightarrow{\text{ether}} C_6H_5-CH_2MgBr$$

$$C_6H_5-CH_2MgBr + CH_3\overset{\displaystyle H}{\overset{|}{C}}{=}O \xrightarrow{\text{ether}} C_6H_5-CH_2-\overset{\displaystyle OMgBr}{\overset{|}{C}}HCH_3$$

$$C_6H_5-CH_2-\overset{\displaystyle OMgBr}{\overset{|}{C}}HCH_3 \xrightarrow{H_2O} C_6H_5-CH_2-\overset{\displaystyle OH}{\overset{|}{C}}HCH_3 + Mg(OH)Br$$

1-Phenyl-2-propanol

7-5 Sources and preparation of phenols

Coal tar is a source of both phenol and the methylphenols (cresols). Several methods of synthesis also are available.

Fusion of a sodium arenesulfonate with NaOH

The general reaction of a sodium arenesulfonate with sodium hydroxide is

$$ArSO_2ONa \xrightarrow[-Na_2SO_3]{NaOH} ArONa \xrightarrow{HCl} \underset{\text{A phenol}}{ArOH} + NaCl \qquad (7\text{-}28)$$

Although no longer used to make phenol itself, this reaction is still used to make some phenols such as 1,3-dihydroxybenzene (resorcinol). Since a sulfonic acid group is meta-directing, a second introduced sulfonic acid group is meta to the first group to produce 1,3-benzenedisulfonic acid. The subsequent fusion of the salt of that acid with sodium hydroxide is an example of *nucleophilic aromatic substitution*.

$$C_6H_6 + \underset{\text{Fuming sulfuric acid}}{2H_2SO_4\cdot SO_3} \xrightarrow{\Delta} \underset{\text{1,3-Benzenedisulfonic acid}}{1,3\text{-}C_6H_4(SO_2OH)_2} \xrightarrow{2NaOH} 1,3\text{-}C_6H_4(SO_2O^-Na^+)_2 \qquad (7\text{-}29)$$

$$C_6H_4(SO_2O^-Na^+)_2 + 2NaOH \xrightarrow[-2Na_2SO_3]{\Delta} C_6H_4(O^-Na^+)_2 \xrightarrow{HCl} C_6H_4(OH)_2 + 2NaCl \quad (7\text{-}30)$$

1,3-Dihydroxybenzene
(Resorcinol)

One of the uses of resorcinol is to make hexylresorcinol, a germicide, by a Friedel-Crafts acylation reaction. The presence of the —OH groups as activating groups makes possible the use of a carboxylic acid as the acylating agent instead of an acyl halide.

$$C_6H_4(OH)_2 + CH_3(CH_2)_4COOH \xrightarrow{ZnCl_2} C_6H_3(OH)_2C(=O)(CH_2)_4CH_3 \quad (7\text{-}31)$$

1,3-Dihydroxybenzene (Resorcinol) — Hexanoic acid (Caproic acid)

$$C_6H_3(OH)_2C(=O)(CH_2)_4CH_3 \xrightarrow{Zn(Hg),\ HCl} C_6H_3(OH)_2(CH_2)_5CH_3 \quad (7\text{-}32)$$

1-Hexyl-2,4-dihydroxybenzene
(4-Hexylresorcinol)

The reaction shown in Eq. (7-32) is an example of the Clemmensen reduction.

Hydrolysis of aryl halides Phenol is made industrially by the high-temperature, high-pressure hydrolysis of chlorobenzene.

$$C_6H_5Cl + 2NaOH \xrightarrow{238\text{ atm},\ 250-350°C} C_6H_5O^-Na^+ \xrightarrow{H^+} C_6H_5OH \quad (7\text{-}33)$$

Chlorobenzene — Phenol

The process is known as the Dow method.

Hydrolysis of diazonium salts

Phenols are also available from the *diazotization* of aromatic primary amines.

$$\text{Aniline} \xrightarrow{NaNO_2,\ H^+} \text{Benzene diazonium salt } (C_6H_5N_2^+) \xrightarrow{H_2O} \text{Phenol } (C_6H_5OH) + N_2 \qquad (7\text{-}34)$$

This is actually the most general laboratory method of making phenols. It will be discussed further in Chap. 13.

Production of phenol from cumene hydroperoxide

In recent years some phenol has been produced industrially from cumene [(1-methylethyl)benzene, also known as isopropylbenzene]. The cumene is produced by a Friedel-Crafts alkylation of benzene with propene. Cumene is then oxidized to cumene hydroperoxide, which is subjected to an acid-catalyzed rearrangement and hydrolysis to phenol and propanone.

$$\underset{\text{(1-Methyl)benzene (Cumene)}}{C_6H_5CH(CH_3)_2} \xrightarrow{O_2} \underset{\text{(1-Methyl)benzene hydroperoxide (Cumene hydroperoxide)}}{C_6H_5C(CH_3)_2\text{—O—O—H}} \xrightarrow{H_2O,\ H^+} \underset{\text{Phenol}}{C_6H_5OH} + \underset{\text{Propanone (Acetone)}}{CH_3C(CH_3)\text{=O}} \qquad (7\text{-}35)$$

The mechanism for the acid-catalyzed rearrangement and hydrolysis of cumene hydroperoxide may be of interest. It involves protonation of the hydroperoxide followed by a rearrangement to form a resonance-stabilized cation. The cation then adds water to form a protonated *hemiacetal* (to be discussed in Chap. 8), which cleaves to give acetone and phenol.

$$\underset{\text{Cumene hydroperoxide}}{C_6H_5C(CH_3)_2\text{—O—O—H}} \xrightarrow{H^+} \underset{\text{Protonated cumene hydroperoxide}}{C_6H_5C(CH_3)_2\text{—O—}\overset{+}{O}H_2} \xrightarrow{(-H_2O)} \underset{\text{Oxonium ion}}{C_6H_5C(CH_3)_2\text{—}O^+} \longrightarrow \underset{\text{Cation + } H_2O}{(CH_3)_2C\text{=}\overset{+}{O}\text{—}C_6H_5 + H_2O} \longrightarrow$$

$$\underset{\text{A protonated hemiacetal}}{(CH_3)_2C(\overset{+}{O}H_2)\text{—O—}C_6H_5 \longrightarrow (CH_3)_2C(OH)\text{—}\overset{+}{O}H\text{—}C_6H_5} \rightleftharpoons \underset{\text{Phenol}}{C_6H_5OH} + \underset{\text{Propanone (Acetone)}}{CH_3\text{—}C(CH_3)\text{=O}} + H^+ \qquad (7\text{-}36)$$

This method of synthesis of phenol is economical because a profit can also be realized from the sale of the acetone which is produced.

7-6 Reactions of alcohols and phenols

Reactions of alcohols and phenols may involve breaking the O—H bond or the C—OH bond.

Reaction with active metals

The reaction of an alcohol or phenol with sodium is most common, although other active metals do react. The general equations are

$$\underset{\text{Alcohol}}{2ROH} + 2Na \longrightarrow \underset{\text{Sodium alkoxide}}{2RO^-Na^+} + H_2(g) \quad (7\text{-}37)$$

$$\underset{\text{Phenol}}{2ArOH} + 2Na \longrightarrow \underset{\text{Sodium phenoxide}}{2ArO^-Na^+} + H_2(g) \quad (7\text{-}38)$$

Examples of the reaction are the following:

$$\underset{\substack{\text{Ethanol} \\ \text{(Ethyl alcohol)}}}{2CH_3CH_2OH} + 2Na \longrightarrow \underset{\text{Sodium ethoxide}}{2CH_3CH_2O^-Na^+} + H_2(g) \quad (7\text{-}39)$$

$$\underset{\substack{\text{2-Methyl-2-propanol} \\ (\textit{tert}\text{-Butyl alcohol})}}{2CH_3{-}\overset{\overset{\large CH_3}{|}}{\underset{\underset{\large CH_3}{|}}{C}}{-}OH} + 2K \longrightarrow \underset{\substack{\text{Potassium} \\ \text{2-methyl-2-propoxide} \\ (\text{Potassium } \textit{tert}\text{-butoxide})}}{2CH_3{-}\overset{\overset{\large CH_3}{|}}{\underset{\underset{\large CH_3}{|}}{C}}{-}O^-K^+} + H_2(g) \quad (7\text{-}40)$$

This reaction has some use as a test for an alcohol, since the evolution of hydrogen is readily discerned. Also, the sodium alkoxides which are produced find use as strong bases; they are even stronger than sodium hydroxide. The reaction is vigorous and may be accompanied by a fire or explosion unless small amounts of reactants are used. The order of reactivity for types of alcohols in this reaction is $CH_3OH > 1° > 2° > 3°$.

Reaction of phenols with bases

Phenols are slightly acidic and, unlike alcohols, will turn litmus red, and neutralize and dissolve in aqueous sodium hydroxide.

$$C_6H_5OH + NaOH \longrightarrow C_6H_5O^-Na^+ + H_2O \quad (7\text{-}41)$$

The acidity of phenols is due to resonance stabilization with delocalization of the negative charge on the phenoxide anion.

(7-42)

However, the phenols are very weak acids; ordinarily, they are even weaker than carbonic acid. As a result, they do not generally dissolve in aqueous sodium bicarbonate with evolution of carbon dioxide as do the relatively stronger carboxylic acids. Instead, carbonic acid will displace a phenol from its salt.

$$\underset{\text{Stronger base}}{C_6H_5O^-Na^+} + \underset{\text{Stronger acid}}{H_2CO_3} \xrightarrow{H_2O} \underset{\text{Weaker base}}{NaHCO_3} + \underset{\text{Weaker acid}}{C_6H_5OH} \tag{7-43}$$

Thus although both phenols and carboxylic acids turn litmus red and dissolve in aqueous sodium hydroxide, only the carboxylic acids will dissolve and liberate carbon dioxide from aqueous sodium bicarbonate.

Difference in acidity can be used to separate phenol from other compounds. For example, if the acidic ingredients have been extracted from a mixture by use of aqueous sodium hydroxide, any benzoic acid would be in the aqueous solution as sodium benzoate, $C_6H_5COO^-Na^+$, and any phenol would be in the aqueous solution as sodium phenoxide, $C_6H_5O^-Na^+$. Saturating the aqueous solution with carbon dioxide would then form carbonic acid, which would liberate the free phenol (weaker acid), but not the benzoic acid (stronger acid).

When phenol is substituted with two or more meta-directing groups located ortho and para to the OH group, electron withdrawal is facilitated, so that the negative charge on the phenoxide anion is still further delocalized. That makes such substituted phenols as 2,4-dinitrophenol even stronger acids than carbonic acid. The substituted phenol will then dissolve in aqueous sodium bicarbonate with liberation of carbon dioxide. The effect is cumulative, so 2,4,6-trinitrophenol is still stronger and is referred to by its common name of picric acid.

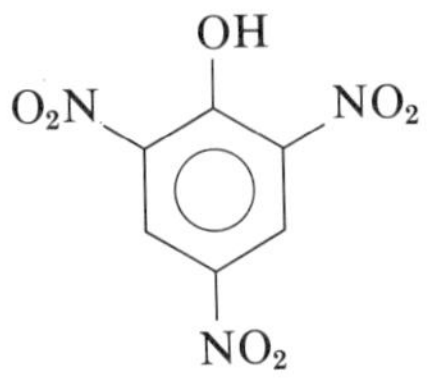

OH
O_2N NO_2
NO_2

Picric acid can be used as a yellow dye, and, like trinitrotoluene (TNT), it can be detonated as a high explosive.

Dehydration of alcohols to alkenes and to ethers

Dehydration of alcohols to alkenes was discussed in Chap. 4 and should be reviewed. Dehydration to ethers is discussed in Sec. 7-7.

Nucleophilic substitution reactions of alcohols

Production of alkyl halides. The reactions of alcohols with HX, PX_3, and $SOCl_2$ to prepare alkyl halides have already been discussed in Chap. 6. The order of reactivity of hydrohalogen is HI > HBr > HCl. Reaction of primary alcohols with HCl requires the addition of $ZnCl_2$ and use of heat. The order of reactivity of alcohols with HX is $3° > 2° > 1°$. That is the basis for the *Lucas test*. The Lucas reagent is concentrated hydrochloric acid and zinc chloride. Tertiary alcohols react immediately upon shaking with Lucas reagent to produce an immiscible upper layer of alkyl chloride. Secondary alcohols react in 2 or 3 min, and primary alcohols do not react unless the mixture is heated.

The mechanism for the reaction of HX with methanol or with primary alcohols is as follows:

$$\underset{\text{Primary alcohol}}{ROH} + HX \longrightarrow R-\overset{+}{\underset{}{O}}H_2 + X^- \xrightarrow{(S_N2)} \underset{\text{Alkyl halide}}{R-X} + H_2O \qquad (7\text{-}44)$$

The mechanism for the reaction of secondary or tertiary alcohols with HX is an S_N1 type.

$$ROH + HX \longrightarrow R\overset{+}{O}H_2 + X^- \xrightarrow{(S_N1)} H_2O + R^+ + X^- \longrightarrow H_2O + RX \qquad (7\text{-}45)$$

Note that, in both cases, water (a weak base) is the leaving group, *not* OH^- ion (a strong base).

Esterification. *Esterification* is another reaction in which the *carbon-oxygen bond* in the alcohol is broken *when an inorganic acid is used*. The compound formed by reaction of either an inorganic or an organic acid with an alcohol is called an *ester*. The mechanism is generally an S_N2 type similar to that of the reaction of an alcohol with hydrohalogen. If HOZ is used to represent an inorganic acid, the reaction is

$$R-OH + \underset{\text{Inorganic acid}}{HOZ} \longrightarrow R-\overset{+}{O}H_2 + OZ^- \xrightarrow{(S_N2)} \underset{\text{Ester of an inorganic acid}}{R-OZ} + H_2O \qquad (7\text{-}46)$$

The following two esterification reactions are S_N2 type.

$$\underset{\text{Methanol}}{CH_3OH} + \underset{\text{Sulfuric acid}}{HOSO_2OH} \xrightarrow[(S_N2)]{-H_2O} \underset{\text{Methyl hydrogen sulfate}}{CH_3OSO_2OH} \qquad (7\text{-}47)$$

$$\underset{\substack{\text{1-Dodecanol}\\\text{(Lauryl alcohol)}}}{CH_3(CH_2)_{10}CH_2OH} + HOSO_2OH \longrightarrow \underset{\substack{\text{1-Dodecanyl hydrogen sulfate}\\\text{(Lauryl hydrogen sulfate)}}}{CH_3(CH_2)_{10}CH_2OSO_2OH} + H_2O \qquad (7\text{-}48a)$$

$$\underset{\substack{\text{1-Dodecanyl hydrogen sulfate}\\\text{(Lauryl hydrogen sulfate)}}}{CH_3(CH_2)_{10}CH_2OSO_2OH} + NaOH \longrightarrow \underset{\substack{\text{Sodium 1-dodecanyl sulfate}\\\text{(Sodium lauryl sulfate)}}}{CH_3(CH_2)_{10}CH_2OSO_2O^-Na^+} \qquad (7\text{-}48b)$$

Sodium lauryl sulfate is a synthetic detergent that can be used in hard water because its calcium and magnesium salts are soluble.

Another important esterification reaction is the following:

$$\underset{\substack{\text{1,2,3-Propanetriol} \\ \text{(Glycerol)}}}{\underset{\substack{| \\ OH}}{CH_2}-\underset{\substack{| \\ OH}}{CH}-\underset{\substack{| \\ OH}}{CH_2}} + 3HONO_2 \xrightarrow{H_2SO_4} \underset{\substack{\text{Propane-1,2,3-triyl trinitrate} \\ \text{(Glycerol trinitrate,} \\ \text{or nitroglycerin)}}}{\underset{\substack{| \\ O \\ | \\ NO_2}}{CH_2}-\underset{\substack{| \\ O \\ | \\ NO_2}}{CH}-\underset{\substack{| \\ O \\ | \\ NO_2}}{CH_2}} + 3H_2O \qquad (7\text{-}49)$$

Glycerol trinitrate is an explosive used to make dynamite. It is also used as a vasodilator (dilates blood vessels) in treating spasms of the coronary artery (angina pectoris).

Amyl acetate is an example of an ester formed from an organic acid, acetic acid:

$$\underset{\substack{\text{1-Pentanol} \\ \text{(Amyl alcohol)}}}{CH_3CH_2CH_2CH_2CH_2OH} + \underset{\substack{\text{Ethanoic acid} \\ \text{(Acetic acid)}}}{CH_3C\begin{matrix} \diagup\!\!\!\diagup O \\ \diagdown OH \end{matrix}} \xrightarrow{H^+}$$

$$\underset{\substack{\text{1-Pentyl ethanoate} \\ \text{(Amyl acetate)}}}{CH_3CH_2CH_2CH_2CH_2O-\overset{\substack{O \\ \|}}{C}-CH_3} + H_2O \qquad (7\text{-}50)$$

See Fig. 7-5 for a model of methyl acetate.

Many of the esters of alcohols and organic acids have pleasant odors, and some of them are used in perfumes and artificial flavors. Amyl acetate has an odor like bananas. The mechanism for this type of esterification will be considered in Chap. 9.

Oxidation of alcohols

If conditions are not too drastic, primary alcohols can be oxidized to aldehydes. The aldehyde can then be oxidized further to a carboxylic acid (Fig. 7-6). Secondary alcohols are oxidized to ketones, and tertiary alco-

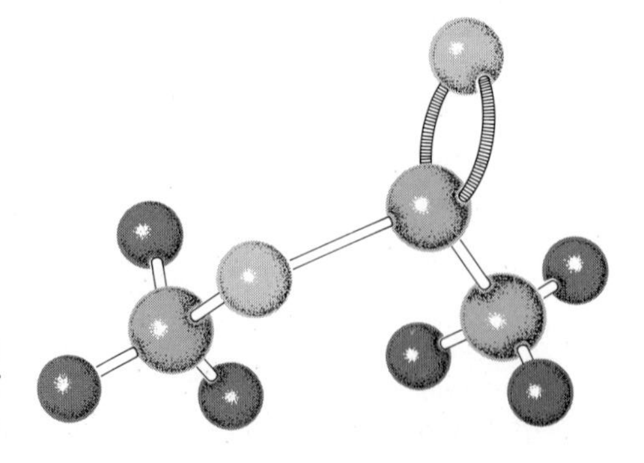

FIGURE 7-5
Ball-and-spring model of methyl acetate, an ester.

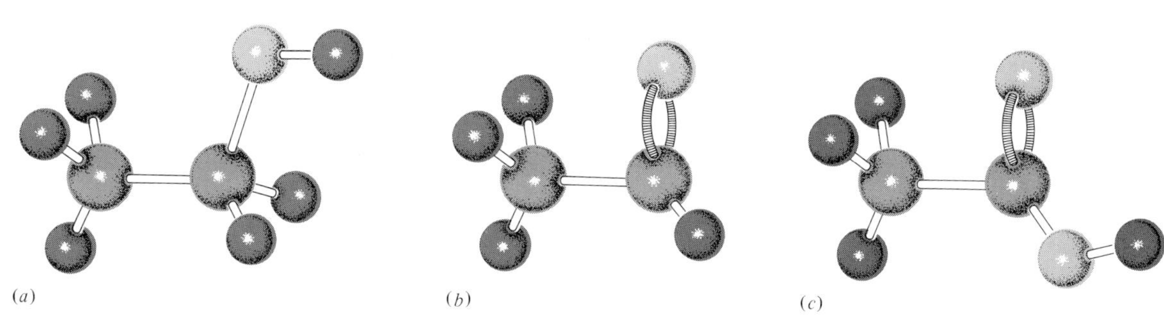

FIGURE 7-6
Oxidation of (*a*) a primary alcohol, ethanol, yields (*b*) an aldehyde, ethanal. Further oxidation yields (*c*), a carboxylic acid, ethanoic acid.

hols are *not* readily oxidized. That is because the carbon bearing the oxygen has no carbon-hydrogen bonds left to be broken; it has only carbon-carbon bonds. (Remember that a tertiary alcohol is R_3COH.) The general reactions for primary and secondary alcohols are

$$\underset{\text{Primary alcohol}}{RCH_2OH} \xrightarrow{\text{CuO, 200°C, or } CrO_3\text{/pyridine}} \underset{\text{Aldehyde}}{RC(H){=}O} \xrightarrow{KMnO_4} \underset{\text{Carboxylic acid}}{RC({=}O)OH} \qquad (7\text{-}51)$$

$$\underset{\text{Primary alcohol}}{RCH_2OH} \xrightarrow{KMnO_4} \underset{\text{Carboxylic acid}}{RC({=}O)OH} \qquad (7\text{-}52)$$

$$\underset{\text{Secondary alcohol}}{R{-}C(R)(H){-}OH} \xrightarrow{KMnO_4 \text{ or } Na_2Cr_2O_7 \text{ or CuO at 200°C}} \underset{\text{Ketone}}{R{-}C(R){=}O} \qquad (7\text{-}53)$$

Specific examples of the oxidation of alcohols (Fig. 7-7) are the following:

$$\underset{\substack{\text{Ethanol}\\ \text{(Ethyl alcohol)}}}{CH_3CH_2OH} \xrightarrow{\text{CuO, 200°C}} \underset{\substack{\text{Ethanal}\\ \text{(Acetaldehyde)}}}{CH_3C(H){=}O} \qquad (7\text{-}54)$$

$$\underset{\substack{\text{1-Butanol}\\ (n\text{-Butyl alcohol})}}{CH_3CH_2CH_2CH_2OH} \xrightarrow{KMnO_4} \underset{\substack{\text{Butanoic acid}\\ (n\text{-Butyric acid})}}{CH_3CH_2CH_2COOH} \qquad (7\text{-}55)$$

$$\underset{\substack{\text{2-Butanol}\\ (sec\text{-Butyl alcohol})}}{CH_3CH_2CHOHCH_3} \xrightarrow{Na_2Cr_2O_7} \underset{\substack{\text{Butanone}\\ \text{(Ethyl methyl ketone)}}}{CH_3CH_2C({=}O){-}CH_3} \qquad (7\text{-}56)$$

Difference in susceptibility to oxidation may be used to distinguish primary and secondary alcohols, which are readily oxidized, from tertiary al-

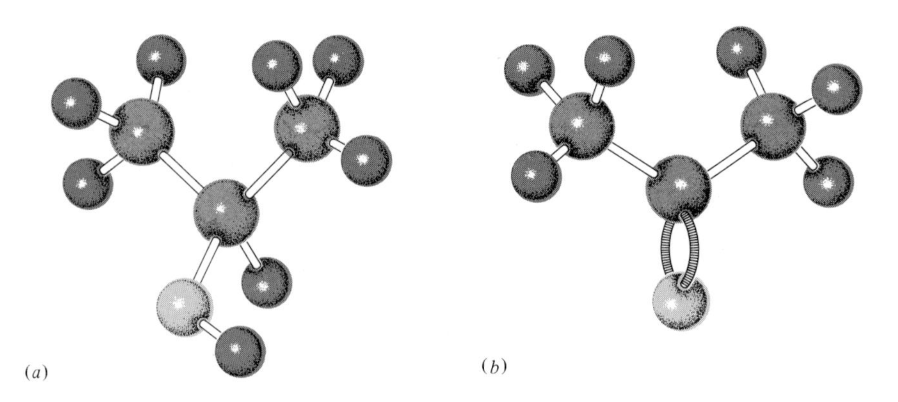

FIGURE 7-7
Oxidation of (*a*) a secondary alcohol, 2-propanol, yields (*b*) a ketone, propanone (acetone).

cohols, which are not. For the purpose, chromic acid (H_2CrO_4, made by mixing CrO_3 and H_2SO_4) is added to the alcohol. If the alcohol is a primary or a secondary one, it is oxidized and the chromium is reduced and changed in color from yellow to a greenish blue.

$$\text{Primary or secondary alcohol} + \underset{\text{(Yellow)}}{H_2CrO_4} \longrightarrow$$
$$\text{aldehyde or ketone} + \underset{\text{(Green)}}{Cr^{3+}} + Cr^{6+} \qquad (7\text{-}57)$$

Tertiary alcohols are not readily oxidized, so they give no such color change.

Oxidation of vicinal glycols with periodic acid

Vicinal glycols are oxidized by periodic acid (HIO_4) to fragments consisting of aldehydes, ketones, or acids. The general equation is

$$\underset{\text{A vicinal glycol}}{-\underset{OH}{\overset{|}{\underset{|}{C}}}-\underset{OH}{\overset{|}{\underset{|}{C}}}-} + \underset{\text{Periodic acid}}{HIO_4} \longrightarrow \underset{\text{Aldehydes or ketones}}{-\underset{O}{\overset{|}{\underset{\|}{C}}} \quad \underset{O}{\overset{|}{\underset{\|}{C}}}-} + \underset{\text{Iodic acid}}{HIO_3} + H_2O \qquad (7\text{-}58)$$

An example of the reaction is

$$\underset{\substack{\text{1,2-Propanediol} \\ \text{(Propylene glycol)}}}{CH_3\underset{OH}{\overset{H}{\underset{|}{\overset{|}{C}}}}-\underset{OH}{\overset{H}{\underset{|}{\overset{|}{C}}}}H} + HIO_4 \longrightarrow HIO_3 + \underset{\substack{\text{Ethanal} \\ \text{(Acetaldehyde)}}}{CH_3\overset{H}{\overset{|}{C}}=O} + \underset{\substack{\text{Methanal} \\ \text{(Formaldehyde)}}}{O=\overset{H}{\overset{|}{C}}H} + H_2O \qquad (7\text{-}59)$$

The isomeric 1,3-propanediol does not react, since it is not a vicinal glycol. Evidence for reaction taking place can be obtained by adding silver nitrate, which reacts with HIO_3 to form a precipitate of silver iodate ($AgIO_3$).

The reaction of glycerol with periodic acid is another example.

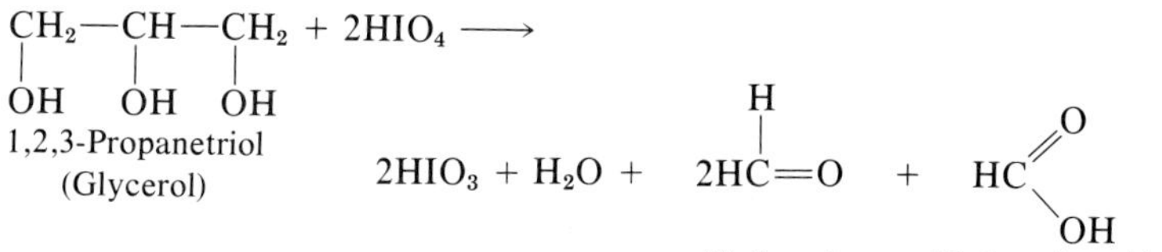

$$\underset{\substack{\text{1,2,3-Propanetriol} \\ \text{(Glycerol)}}}{\underset{|\quad\ \ |\quad\ \ |}{CH_2-CH-CH_2}\atop{OH\quad OH\quad OH}} + 2HIO_4 \longrightarrow 2HIO_3 + H_2O + \underset{\substack{\text{Methanal} \\ \text{(Formaldehyde)}}}{2H\overset{H}{\overset{|}{C}}=O} + \underset{\substack{\text{Methanoic acid} \\ \text{(Formic acid)}}}{HC(=O)OH} \qquad (7\text{-}60)$$

The 2 mol of methanal comes from the end carbons, and the mole of methanoic acid comes from the center carbon. (Cleave all carbon-carbon bonds of carbon atoms bearing vicinal —OH groups; add —OH groups to bonds which are cleaved; and eliminate a molecule of water to obtain cleavage fragments.)

PROBLEM 7-6 Write equations for the following reactions of alcohols or phenols.

(a) 2-methyl-2-propanol (*tert*-butyl alcohol) and sodium
(b) phenol and potassium hydroxide
(c) 2-methylphenol (*o*-cresol) and sodium carbonate
(d) 2-butanol, H_2SO_4, 200°C
(e) 2,4,6-trinitrophenol and $NaHCO_3$
(f) 1-pentanol and nitric acid, H_2SO_4 catalyst
(g) 2,3-butanediol and HIO_4
(h) 1-propanol, CrO_3/pyridine

Solution to (a)

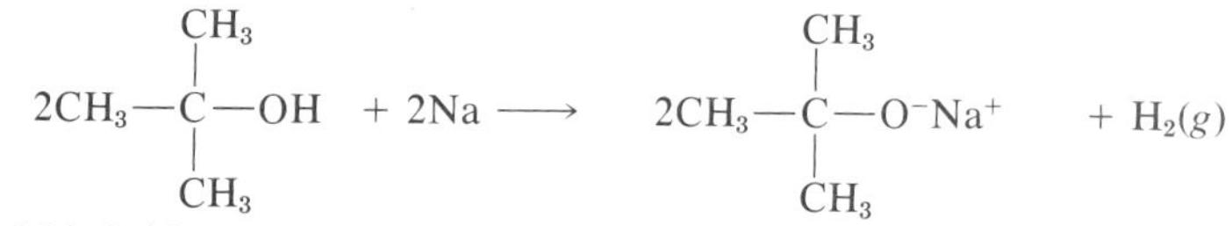

$$\underset{\substack{\text{2-Methyl-2-propanol} \\ (\textit{tert}\text{-Butyl alcohol})}}{2CH_3-\overset{CH_3}{\overset{|}{\underset{CH_3}{\underset{|}{C}}}}-OH} + 2Na \longrightarrow \underset{\substack{\text{Sodium 2-methyl-2-propoxide} \\ (\text{Sodium } \textit{tert}\text{-butoxide})}}{2CH_3-\overset{CH_3}{\overset{|}{\underset{CH_3}{\underset{|}{C}}}}-O^-Na^+} + H_2(g)$$

7-7 Preparation of ethers

Two methods for preparation of ethers will be considered.

Intermolecular dehydration of alcohols

The intermolecular dehydration method is used for making the simple lower-molecular-weight ethers. The *simple ethers* are those in which both alkyl groups attached to the oxygen atom are the same. The reaction amounts to an *intermolecular* dehydration reaction (Fig. 7-8).

$$2ROH \longrightarrow R-O-R + H_2O \qquad (7\text{-}61)$$

An example is

$$\underset{\substack{\text{Ethanol} \\ \text{(Ethyl alcohol)}}}{2CH_3CH_2OH} \xrightarrow{H_2SO_4,\ 120-140°C} \underset{\substack{\text{Ethoxyethane} \\ \text{(Ethyl ether)}}}{CH_3CH_2-O-CH_2CH_3} + H_2O \qquad (7\text{-}62)$$

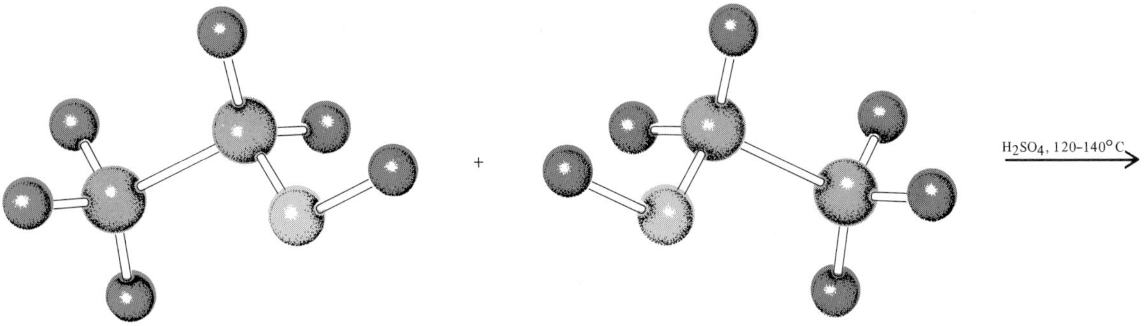

H_2SO_4, 120–140° C

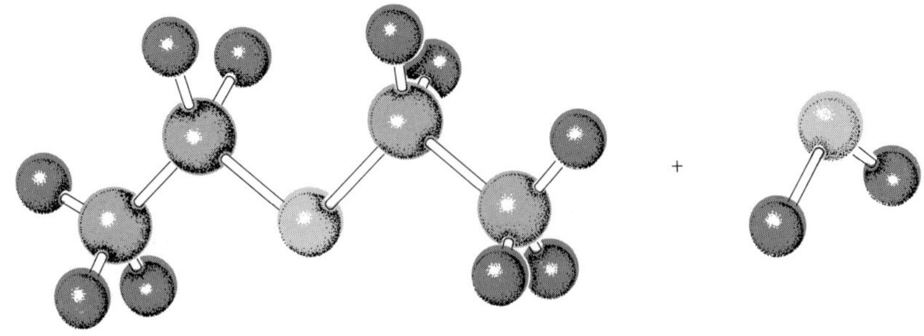

FIGURE 7-8
The intermolecular dehydration of ethanol yields ethoxyethane (ethyl ether) and water.

An elimination reaction to produce an alkene, a substitution reaction to produce a sulfate ester, and an oxidation reaction to produce an aldehyde or ketone are competing reactions. The mechanism is S_N1 or S_N2 depending on the type of alcohol. Methanol and primary alcohols tend to react by the S_N2 mechanism, and secondary and tertiary alcohols tend to react by the S_N1 mechanism. Sulfate ester formation is favored by high concentrations of H_2SO_4, and elimination reactions are favored by higher temperatures. Figure 7-9 shows the competing reactions.

Ethyl ether (ethoxyethane) is prepared by the action of sulfuric acid on ethanol. It has been used as an anesthetic since the 1840s. Vinyl ether (ethenyloxyethene, $CH_2{=}CH{-}O{-}CH{=}CH_2$) also is used as an anesthetic. It is prepared by the action of sulfuric acid on ethylene chlorohydrin (2-chloroethanol, $\underset{\text{Cl}}{\underset{|}{CH_2}}{-}\underset{\text{OH}}{\underset{|}{CH_2}}$ followed by dehydrohalogenation with alcoholic KOH.

The Williamson synthesis

The Williamson synthesis method of preparation of ethers is the most general method in use, since with it ethers in which the groups attached to the oxygen atom differ from each other can be prepared. It involves the S_N2 type of reaction of an alkoxide or of a phenoxide with an alkyl halide. It does not work with aryl halides or vinyl halides because those halides are too unreactive. It also does not work with tertiary halides, which too

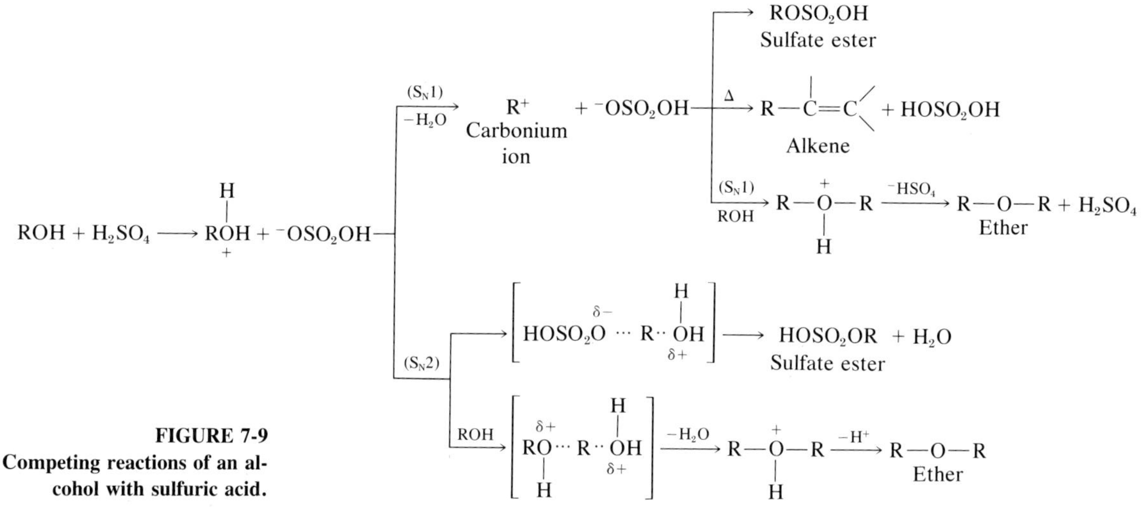

$$ROH + H_2SO_4 \longrightarrow R\overset{+}{O}H_2 + {}^-OSO_2OH$$

(S_N1), $-H_2O$: R^+ (Carbonium ion) $+ {}^-OSO_2OH$ →
- $ROSO_2OH$ (Sulfate ester)
- Δ: $R-C=C< + HOSO_2OH$ (Alkene)
- (S_N1), ROH: $R-\overset{+}{O}(H)-R \xrightarrow{-HSO_4^-} R-O-R + H_2SO_4$ (Ether)

(S_N2):
- $[HOSO_2\overset{\delta-}{O} \cdots R \cdots \overset{\delta+}{O}H_2] \longrightarrow HOSO_2OR + H_2O$ (Sulfate ester)
- ROH: $[R\overset{\delta+}{O}(H) \cdots R \cdots \overset{\delta+}{O}H_2] \xrightarrow{-H_2O} R-\overset{+}{O}(H)-R \xrightarrow{-H^+} R-O-R$ (Ether)

FIGURE 7-9
Competing reactions of an alcohol with sulfuric acid.

readily undergo dehydrohalogenation to form an alkene. The general reaction is

$$\underset{\text{An alkoxide}}{R-O^-} + R'-X \xrightarrow{(S_N2)} R-O-R' + X^- \qquad (7\text{-}63)$$

$$\underset{\text{A phenoxide}}{Ar-O^-} + R-X \xrightarrow{(S_N2)} Ar-O-R + X^- \qquad (7\text{-}64)$$

The following reactions are examples of the Williamson synthesis:

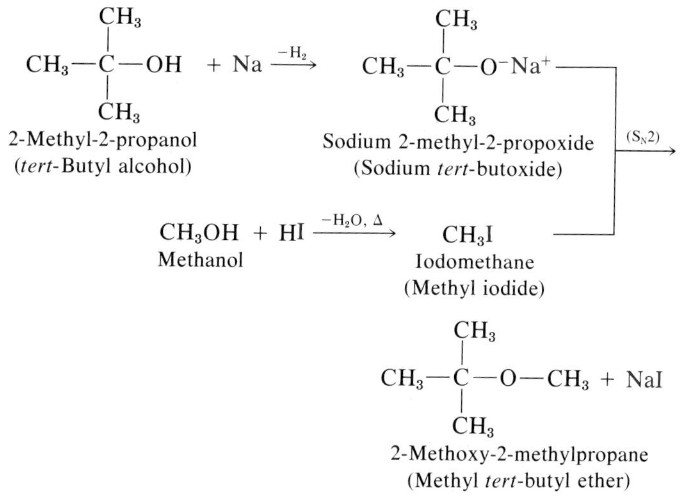

$$\underset{\substack{\text{2-Methyl-2-propanol}\\ (\textit{tert}\text{-Butyl alcohol})}}{(CH_3)_3C-OH} + Na \xrightarrow{-H_2} \underset{\substack{\text{Sodium 2-methyl-2-propoxide}\\ (\text{Sodium } \textit{tert}\text{-butoxide})}}{(CH_3)_3C-O^-Na^+}$$

$$\underset{\text{Methanol}}{CH_3OH} + HI \xrightarrow{-H_2O,\ \Delta} \underset{\substack{\text{Iodomethane}\\ (\text{Methyl iodide})}}{CH_3I}$$

$$\xrightarrow{(S_N2)} \underset{\substack{\text{2-Methoxy-2-methylpropane}\\ (\text{Methyl } \textit{tert}\text{-butyl ether})}}{(CH_3)_3C-O-CH_3} + NaI \qquad (7\text{-}65)$$

$$C_6H_5OH + NaOH \xrightarrow{-H_2O} C_6H_5O^-Na^+$$
Phenol — Sodium phenoxide

$$CH_3CH_2OH + HBr \xrightarrow{-H_2O,\ \Delta} CH_3CH_2Br$$
Ethanol — Bromoethane

$$\xrightarrow{(S_N2)} C_6H_5—O—CH_2CH_3 + NaBr \qquad (7\text{-}66)$$
Ethoxybenzene
(Ethyl phenyl ether)

Dimethyl sulfate, $CH_3O—SO_2—OCH_3$, instead of CH_3I, is sometimes used with NaOH for making methyl ethers.

$$2\ —\overset{|}{\underset{|}{C}}—OH + 2NaOH + CH_3OSO_2OCH_3 \longrightarrow 2\ —\overset{|}{\underset{|}{C}}—OCH_3 + Na_2SO_4 + 2H_2O \qquad (7\text{-}67)$$

PROBLEM 7-7 Write equations for the preparation of the following ethers beginning with appropriate alcohols or phenols. After the first three, more than one reaction will be required.

(a) 1-butoxybutane (*n*-butyl ether)
(b) isopropyl ether
(c) benzyl ether
(d) Methoxybenzene (methyl phenyl ether)
(e) 2-methoxybutane (*s*-butyl methyl ether)
(f) 1-methoxypropane (methyl propyl ether)
(g) 2-methyl-2-methoxybutane
(h) (4-ethoxytoluene) (*p*-cresyl ethyl ether)

Solution to (a)

$$2CH_3CH_2CH_2CH_2OH \xrightarrow{H_2SO_4,\ 120-140°C} CH_3CH_2CH_2CH_2—O—CH_2CH_2CH_2CH_3 + H_2O$$
1-Butanol (*n*-Butyl alcohol) — 1-Butoxybutane (*n*-Butyl ether)

7-8 Reactions of ethers

Because ethers are in general very unreactive and also because many organic compounds are soluble in them, they are often used as solvents for organic reactions. They are immune to the action of strong oxidizing agents like $KMnO_4$ and also to the action of strong bases. They do, however, undergo a few characteristic reactions.

Formation of oxonium salts with strong acids

Ethers will react with and dissolve in strong acids at room temperature because of the formation of oxonium salts:

$$\underset{\text{An ether}}{R{-}O{-}R} + H_2SO_4 \rightleftharpoons \underset{\text{An oxonium salt}}{R{-}\overset{+}{\underset{\underset{H}{|}}{O}}{-}R} + HOSO_2O^- \qquad (7\text{-}68)$$

This reaction will take place with alcohols also. An alcohol that is only slightly soluble in water will dissolve in a strong acid because of the formation of an oxonium salt.

$$R{-}OH + H_2SO_4 \rightleftharpoons R{-}\overset{+}{\underset{\underset{H}{|}}{O}}H + HOSO_2O^- \qquad (7\text{-}69)$$

Addition of water readily hydrolyzes an oxonium salt to regenerate the ether or the alcohol.

$$R{-}\overset{+}{\underset{\underset{H}{|}}{O}}{-}R + HOH \longrightarrow R{-}O{-}R + H_3O^+ \qquad (7\text{-}70)$$

These reactions may be used to distinguish alcohols or ethers from alkanes and alkyl halides because the latter do not dissolve in concentrated sulfuric acid.

Cleavage of ethers by strong acids

If an ether is heated with a *strong* acid, and particularly with a hydrohalogen acid, it is cleaved. The order of reactivity of the hydrohalogen acids is the same as the order of strengths of the acids, HI > HBr > HCl, so HI is the most reactive in the cleavage reaction.

$$2ROR' + 2HI \xrightarrow{\Delta} \left[\begin{array}{c} ROH + R'I \\ R'OH + RI \end{array}\right] \xrightarrow{2HI} 2RI + 2R'I + 2H_2O \qquad (7\text{-}71)$$

An excess of hydriodic acid is used to simplify the products so that only iodides appear as products. The overall reaction is then

$$ROR' + 2HI \xrightarrow{\Delta} RI + R'I + H_2O \qquad (7\text{-}72)$$

The mechanism is an S_N2 type of substitution of iodide ion for alcohol.

$$R{-}\ddot{\underset{\cdot\cdot}{O}}{-}R' + H{-}I \longrightarrow R{-}\overset{\overset{H}{|}}{\underset{+}{\ddot{O}}}{-}R' \;\; I^- \xrightarrow{(S_N2)} R{-}\ddot{\underset{\cdot\cdot}{O}}{-}H + R'I \qquad (7\text{-}73a)$$

$$R—\ddot{O}—H + H—I \longrightarrow R—\overset{+}{\ddot{O}}(H)—H \quad I^- \xrightarrow{(S_N2)} RI + H_2O \qquad (7\text{-}73b)$$

Aryl alkyl ethers react with only one equivalent of HI, since phenols do not react with HI.

$$Ar—O—R + HI \xrightarrow{\Delta} Ar—OH + RI \qquad (7\text{-}74)$$

The following are some specific examples of cleavage of ethers with HI.

$$CH_3O—CH_2CH_2CH_3 + 2HI \xrightarrow{\Delta} CH_3I + CH_3CH_2CH_2I + H_2O \qquad (7\text{-}75)$$

1-Methoxypropane (Methyl *n*-propyl ether) — Iodomethane (Methyl iodide) — Iodopropane (*n*-Propyl iodide)

$$C_6H_5—O—CH_3 + HI \xrightarrow{\Delta} C_6H_5—OH + CH_3I \qquad (7\text{-}76)$$

Methoxybenzene (Methyl phenyl ether, anisole) — Phenol — Iodomethane (Methyl iodide)

Such cleavage of ethers has been of use in the degradation of natural products containing methoxy groups.

Formation of peroxides

An undesirable and potentially hazardous reaction with ethers is peroxide formation. When an ether is exposed to oxygen, a small amount of a very explosive hydroperoxide is formed (Fig. 7-10).

$$CH_3CH_2—O—CH_2CH_3 + O_2 \longrightarrow CH_3\underset{\displaystyle O—O—H}{CH}—O—CH_2CH_3 \qquad (7\text{-}77)$$

Ethoxyethane (Ethyl ether) — Ethoxyethane hydroperoxide (Ethyl ether hydroperoxide)

For that reason it is hazardous to attempt to recover the last traces of solvent ether by distillation to dryness. After a bottle of ether has been opened for a short time, it should be suspected of containing a trace of peroxide. Laboratory manuals should be consulted for methods of identifying and removing the peroxide before such ether is used. *To store for a long time a container of ether which has been opened is dangerous*. The cap on such a container should not be disturbed.

Flammability of ethers

The ethers are in general very flammable. They vaporize readily, and the vapors have a low kindling temperature and ignite at temperatures not much over 100°C. Carelessness in the use of ethers, especially ethyl ether, has been a common cause of fire in laboratories. Open flames and smoking should be prohibited in any areas where ethyl ether is being used.

FIGURE 7-10 Laboratory damage after ether explosion. The ether was stored in the refrigerator on the right. ***(Courtesy of Norman Steere)***

Ethers as anesthetics

Ethyl ether has been used for many years as a general anesthetic. It produces unconsciousness by depressing the activity of the central nervous system. It does have an irritating effect on the respiratory passages, and it produces postanesthetic nausea. Methyl propyl ether (1-methoxypropane, Neothyl) is a more potent anesthetic than ethyl ether and is less irritating to the respiratory passages. Vinyl ether also has been used.

In the first stage of anesthesia there is a high degree of analgesia (incapacity to feel pain) without loss of consciousness. The second stage involves loss of consciousness and is accompanied by irregular deep respiration and sometimes by purposeless muscular activity. In the third stage the necessary muscular relaxation is achieved (sometimes by supplementing the general anesthetic with a muscle relaxant such as a curare drug) and surgery can be performed. The fourth stage is characterized by paralysis of the respiratory muscles, and death follows if corrective action is not taken. The anesthesiologist carefully monitors alterations in vital signs to determine and control the level of anesthesia.

PROBLEM 7-8

Write the equation for the reaction of excess HI with each of the following ethers. (a) ethoxyethane (ethyl ether), (b) 1-methoxypropane (methyl *n*-propyl ether), (c) methoxybenzene (anisole, $C_6H_5OCH_3$), (d) 1,4-epoxybutane.

Solution to (a)

$$\underset{\substack{\text{Ethoxyethane} \\ \text{(Ethyl ether)}}}{CH_3CH_2{-}O{-}CH_2CH_3} + 2HI \longrightarrow \underset{\substack{\text{Iodoethane} \\ \text{(Ethyl iodide)}}}{2CH_3CH_2I} + H_2O$$

7-9 Epoxides

An epoxide is a special type of cyclic ether that usually has a three-atom ring consisting of two carbon atoms and an oxygen atom.

$$-\overset{|}{C}\underset{O}{\diagdown\!\diagup}\overset{|}{C}- \qquad H_2C\underset{O}{\diagdown\!\diagup}CH_2 \qquad CH_3\overset{H}{\overset{|}{C}}\underset{O}{\diagdown\!\diagup}CH_2$$

An epoxide — 1,2-Epoxyethane (Ethylene oxide) — 1,2-Epoxypropane (Propylene oxide)

Probably because of the strain involved in a three-atom ring (recall the reactivity, of cyclopropane), epoxides are more reactive than other ethers (Fig. 7-11). They are cleaved not only by strong acids but also by dilute acids and bases.

Preparation

Epoxides may be prepared by an *intramolecular* type of Williamson synthesis. An S_N2 type of reaction is involved.

$$>C=C< + Cl_2/H_2O \longrightarrow -\underset{O-H}{\overset{|}{\underset{|}{C}}}-\overset{Cl}{\overset{|}{\underset{|}{C}}}- \xrightarrow{^-OH} -\overset{|}{C}\underset{O^-}{\diagdown\!\diagup}\overset{Cl}{\overset{|}{\underset{|}{C}}}- + H_2O \xrightarrow{(S_N2)} -\overset{|}{C}\underset{O}{\diagdown\!\diagup}\overset{|}{C}- + Cl^- \qquad (7\text{-}78)$$

An example of this preparation is

$$CH_3CH=CH_2 + Cl_2/H_2O \longrightarrow CH_3\overset{H}{\overset{|}{\underset{OH}{\underset{|}{C}}}}-\overset{Cl}{\overset{|}{\underset{H}{\underset{|}{C}}}}-H \xrightarrow{Ca(OH)_2} \underset{CH_3}{}\overset{H}{\overset{|}{C}}\underset{O}{\diagdown\!\diagup}\overset{H}{\overset{|}{C}}\underset{H}{} + CaCl_2 + H_2O \qquad (7\text{-}79)$$

Propene (Propylene) — 1-Chloro-2-propanol (Propylene chlorohydrin) — 1,2-Epoxypropane (Propylene oxide)

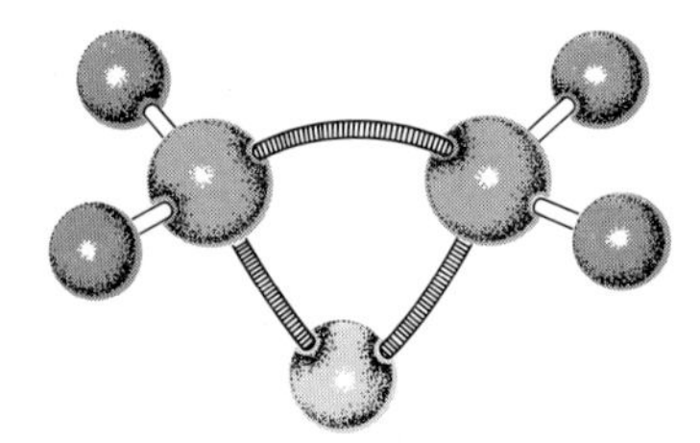

FIGURE 7-11
A ball-and-stick model of 1,2-epoxyethane (ethylene oxide).

Epoxides may be prepared by air-oxidation of the simple alkenes over a silver oxide catalyst.

$$2\ \rangle C{=}C\langle\ +\ O_2\ (\text{Air}) \xrightarrow{Ag_2O} 2\ -\overset{|}{C}\underset{O}{\diagdown\!\!\diagup}\overset{|}{C}- \qquad (7\text{-}80)$$

An example is the oxidation of ethene:

$$\underset{\substack{\text{Ethene}\\(\text{Ethylene})}}{2CH_2{=}CH_2} + \underset{(\text{Air})}{O_2} \xrightarrow{Ag_2O} \underset{\substack{\text{1,2-Epoxyethane}\\(\text{Ethylene oxide})}}{2H_2C\underset{O}{\diagdown\!\!\diagup}CH_2} \qquad (7\text{-}81)$$

Epoxides may also be prepared by the reaction of alkenes with a peroxyacid.

$$\rangle C{=}C\langle\ +\ R\overset{O}{\overset{\|}{C}}OOH \longrightarrow -\overset{|}{C}\underset{O}{\diagdown\!\!\diagup}\overset{|}{C}-\ +\ R\overset{O}{\overset{\|}{C}}OH \qquad (7\text{-}82)$$

An example is

$$\underset{\textit{cis}\text{-2-Butene}}{CH_3\overset{H}{\overset{|}{C}}{=}\overset{H}{\overset{|}{C}}{-}CH_3} + \underset{\substack{\text{Peroxybenzene}\\\text{carboxylic acid}\\(\text{Peroxybenzoic acid})}}{C_6H_5\overset{O}{\overset{\|}{C}}OOH} \longrightarrow$$

$$\underset{\textit{cis}\text{-2,3-Epoxybutane}}{CH_3\overset{H}{\overset{|}{C}}\underset{O}{\diagdown\!\!\diagup}\overset{H}{\overset{|}{C}}{-}CH_3} + \underset{\substack{\text{Benzenecarboxylic acid}\\(\text{Benzoic acid})}}{C_6H_5\overset{O}{\overset{\|}{C}}{-}OH} \qquad (7\text{-}83)$$

This latter reaction is a popular laboratory method for synthesizing epoxides.

Reactions

Most reactions of epoxides involve nucleophilic ring opening.

$$-\overset{|}{C}\underset{O}{\diagdown\!\!\diagup}\overset{|}{C}-\ +\ {}^{-}OH \xrightarrow{(S_N2)} -\underset{O^-}{\underset{|}{\overset{|}{C}}}-\underset{|}{\overset{OH}{\overset{|}{C}}}- \xrightarrow{H_2O} -\underset{OH}{\underset{|}{\overset{|}{C}}}-\underset{|}{\overset{OH}{\overset{|}{C}}}-\ +\ {}^{-}OH \qquad (7\text{-}84)$$

If the reaction is acid-catalyzed, weaker nucleophiles—even water—may be used.

$$-\overset{|}{C}\underset{O}{\diagdown\!\!\diagup}\overset{|}{C}-\ +\ H^+ \longrightarrow -\overset{|}{C}\underset{\underset{H}{|}}{\underset{{}^{+}O}{\diagdown\!\!\diagup}}\overset{|}{C}- \xrightarrow[(S_N2)]{:Z^-} -\underset{OH}{\underset{|}{\overset{|}{C}}}-\underset{|}{\overset{Z}{\overset{|}{C}}}- \qquad (7\text{-}85)$$

In Eq. (7-85), $:Z^-$ represents a nucleophile. An example of nucleophilic ring opening is the following reaction:

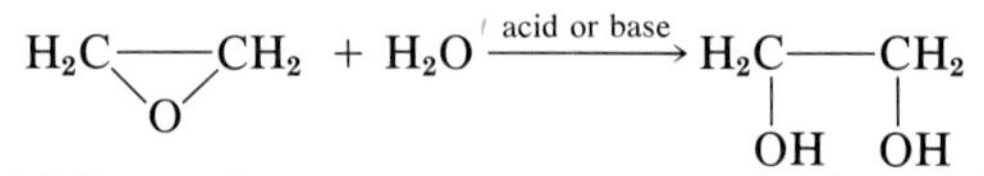

$$\underset{\text{(Ethylene oxide)}}{\underset{\text{1,2-Epoxyethane}}{H_2C\overset{O}{\frown}CH_2}} + H_2O \xrightarrow{\text{acid or base}} \underset{\text{(Ethylene glycol)}}{\underset{\text{1,2-Ethanediol}}{\underset{OH\quad OH}{H_2C{-}CH_2}}} \tag{7-86}$$

Ethylene glycol is used as an antifreeze and coolant liquid for automobile radiators because of its high boiling point, low freezing point, and great solubility in water.

Another example of ring opening is

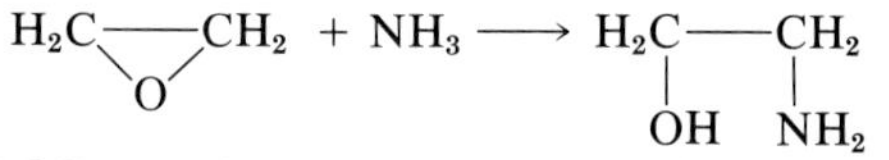

$$\underset{\text{(Ethylene oxide)}}{\underset{\text{1,2-Epoxyethane}}{H_2C\overset{O}{\frown}CH_2}} + NH_3 \longrightarrow \underset{\text{(Ethanolamine)}}{\underset{\text{1-Aminoethanol}}{\underset{OH\quad NH_2}{H_2C{-}CH_2}}} \tag{7-87}$$

Epoxides react with Grignard reagents to form alcohols.

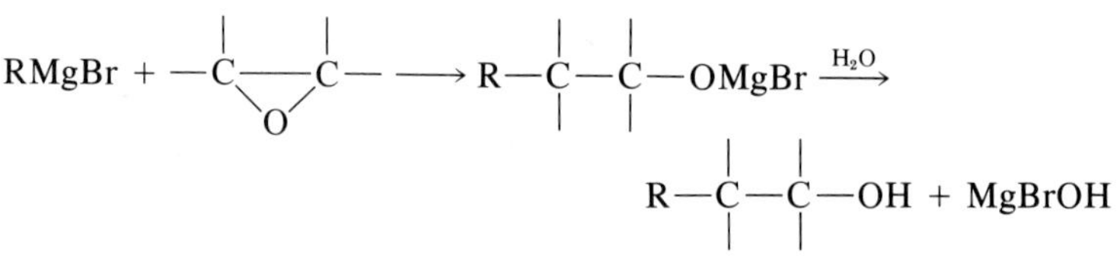

$$RMgBr + {-}\overset{|}{C}\overset{O}{\frown}\overset{|}{C}{-} \longrightarrow R{-}\overset{|}{\underset{|}{C}}{-}\overset{|}{\underset{|}{C}}{-}OMgBr \xrightarrow{H_2O} R{-}\overset{|}{\underset{|}{C}}{-}\overset{|}{\underset{|}{C}}{-}OH + MgBrOH \tag{7-88}$$

An example of the reaction is the following:

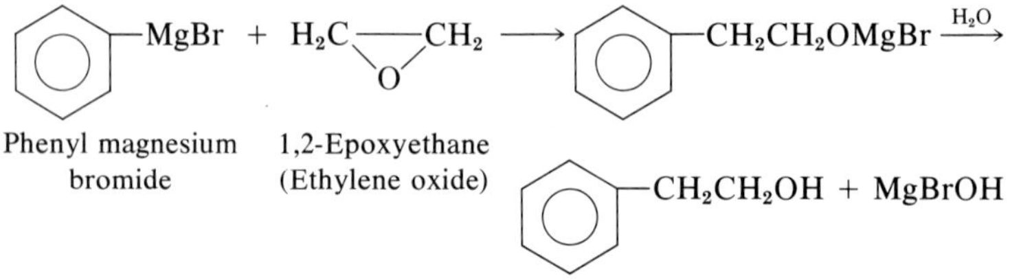

$$\underset{\text{bromide}}{\underset{\text{Phenyl magnesium}}{C_6H_5{-}MgBr}} + \underset{\text{(Ethylene oxide)}}{\underset{\text{1,2-Epoxyethane}}{H_2C\overset{O}{\frown}CH_2}} \longrightarrow C_6H_5{-}CH_2CH_2OMgBr \xrightarrow{H_2O} \underset{\text{2-Phenylethanol}}{C_6H_5{-}CH_2CH_2OH} + MgBrOH \tag{7-89}$$

An interesting application of ethylene oxide is hospital sterilization of respiratory therapy equipment after use.

Summary

The *alcohols* can be considered the *monoalkyl derivatives of water,* the *phenols* the *monoaryl* derivatives of water, and the *ethers* the derivatives of water in which both hydrogen atoms of the water molecule have been substituted by alkyl or aryl groups.

Hydrogen bonding is possible for *phenols* and *alcohols* as it is for water, but to a lesser extent. It is *not* possible for *ethers,* since the ethers have no hydrogen attached to the oxygen atom. Hydrogen bonding accounts for the *abnormally high boiling points* of alcohols and phenols. Since hydrogen of water can hydrogen bond to the oxygen atom of an alcohol, phenol, or ether, the lower-molecular-weight homologs of all three classes are *soluble in water.*

The common names of alcohols, phenols, and ethers are still in wide use. In the IUPAC system, an *alcohol* is named by replacing the terminal *-e* of the corresponding alkane with the suffix *-ol*, and a locator number indicates the position of the —OH group. A *phenol* containing one —OH group is named as a *derivative of phenol* by using locator numbers. A phenol containing two or more —OH groups is named as a *hydroxybenzene,* again by using locator numbers. In the IUPAC system, *ethers* are named as *alkoxy alkanes. Cyclic ethers* are named as *epoxides* of the corresponding alkanes.

All alcohols are toxic; ethyl alcohol is simply least toxic. It is heavily taxed to control its use and consumption. *Denatured alcohol* contains a contaminant which is difficult to remove and which makes the alcohol unsafe to drink. Such alcohol may be purchased tax-free.

Absolute alcohol is 100% alcohol. The usual solvent alcohol is an *azeotrope* (a constant-boiling mixture) consisting of 95% alcohol and 5% water.

Fermentation of plant products containing *carbohydrates,* catalyzed by the enzyme *zymase* from yeast, produces beverage alcohols. The concentration of alcohol in distilled beverages is expressed as *proof,* which is twice the volume percent of alcohol.

Alcohols can be prepared by *fermentation* (ethanol and 1-butanol), by *hydration of alkenes,* by *hydrolysis of alkyl halides,* by reaction of *Grignard reagents* with aldehydes or ketones, and by *diazotization of primary amines.* Methanol is prepared industrially from CO and H_2.

Phenols can be prepared by *fusion* of a *sodium arenesulfonate with* NaOH and by *hydrolysis* of *aryl halides* and of *diazonium salts.* Ethers are prepared by several methods. *Simple ethers* are prepared by the *intermolecular dehydration of alcohols* with H_2SO_4. *Mixed ethers* are made by the *Williamson synthesis. Epoxides* are prepared by *air-oxidation of alkenes,* from *halohydrins,* or from *alkenes and peroxyacids.*

Alcohols undergo many reactions. *Cleavage of the* O—H *bond* can produce *alkoxides* (reaction with active metals), *ethers,* and *esters* of organic acids. *Cleavage of the* C—O *bond* produces *alkyl halides* and *esters* of inorganic acids. *Dehydration* of alcohols produces *alkenes.* Primary alcohols can be *oxidized* to *aldehydes* and *carboxylic acids,* whereas secondary alcohols can be oxidized to *ketones. Vicinal glycols* undergo oxidative cleavage with *periodic acid.*

Phenols react with strong bases to form salts. They also undergo ring substitution ortho and para to the —OH group.

Ethers can react with strong acids to form *oxonium salts,* and they can also be *cleaved* by HX. They tend to form very explosive *peroxides* upon exposure to oxygen, and they also readily undergo combustion. Because of their extreme flammability and potential explosiveness, they must be handled with care. *Cyclic ethers* (epoxides) can be *hydrolyzed* with acid or base.

Because of their effect on the central nervous system, ethyl ether, methyl propyl ether, and vinyl ether are useful *anesthetics*.

Key terms

Absolute alcohol: 100% ethanol; it is required for some chemical reactions.

Alcohol: a monoalkyl derivative of water. Alcohols have the general formula R—OH.

Alkoxide: a strong base resulting from the reaction of an active metal with an alcohol. An example is $NaOCH_3$, sodium methoxide; the strong base resulting from the loss of hydrogen from the —OH group of an alcohol.

Alkoxy alkane: an ether. In the IUPAC system, an ether is named as an alkoxy alkane.

Alkoxy group: an alkyl group bearing an oxygen atom able to form an additional bond; RO—.

Aroxy group: a benzene ring bearing an oxygen atom on the ring; the oxygen is able to form an additional bond, ArO—.

Azeotrope: a constant-boiling mixture; it cannot be separated by fractional distillation.

Denatured alcohol: alcohol to which a contaminant has been added to render it unsafe for drinking. The contaminant used is one difficult to remove.

Diazonium salt: a salt consisting of the positive diazonium ion, RN_2^+ or ArN_2^+, and an anion from an acid, e.g., $ArN_2^+Cl^-$. *Diazo* signifies two nitrogen atoms.

Diazotization: a reaction of an amino group ($—NH_2$) with nitrous acid (HNO_2) to form a very reactive intermediate called a diazonium ion, RN_2^+ or ArN_2^+.

Diol: the IUPAC name for a compound that contains two —OH groups. Commonly called a **glycol.**

Dow method: the industrial preparation of phenol by the high-temperature, high-pressure hydrolysis of chlorobenzene.

Epoxide: a cyclic ether, usually one that has a three-atom ring consisting of two carbon atoms and an oxygen atom.

Ester: a compound formed by the reaction of an alcohol with an inorganic or an organic acid. See Chap. 9.

Esterification: the reaction between an alcohol and an inorganic or organic acid. When an inorganic acid is used, the C—O bond in the alcohol is broken.

Ether: a compound, of the general formula R—O—R, Ar—O—R, or Ar—O—Ar, which may be considered as a derivative of water in which both hydrogen atoms of the water have been substituted with alkyl groups, aryl groups, or both.

Fermentation: a biochemical reaction of an enzyme on carbohydrate material to produce ethanol, carbon dioxide, and other products. The enzyme is produced by a microorganism, generally a yeast.

Glycol: the common name for a compound with two —OH groups. Called a **diol** in the IUPAC system.

Grignard reactions: in this chapter, reaction of a Grignard reagent, RMgBr or ArMgBr, with an aldehyde or a ketone to produce an alcohol.

Hydroxybenzene: a phenol. In the IUPAC system, phenols containing two or more —OH groups are named as hydroxybenzenes.

Lucas test: a test used to determine whether an alcohol is primary, secondary, or tertiary. The alcohol is mixed with concentrated HCl and $ZnCl_2$. A tertiary alcohol reacts immediately to produce an immiscible layer of alkyl chloride; a secondary alcohol reacts in 2 to 3 min; a primary alcohol reacts only upon heating.

Mixed ether: an ether in which two different alkyl groups are attached to the oxygen atom.

Oxonium salt: a salt consisting of an oxygen-containing compound, such as an alcohol or ether, in which the oxygen atom has been protonated by a proton acid to produce a cation with the positive charge on the oxygen atom, e.g., $R\overset{+}{O}H_2\ HSO_4^-$.

Phenol: an aromatic compound with an —OH group as a substituent on the ring. "Phenol" is the name of a specific compound, C_6H_5OH, as well as the name of a class of compounds.

Phenoxide: the salt, e.g., $C_6H_5O^-Na^+$, resulting from neutralizing a phenol with a strong base. Phenoxide ions themselves act as weak bases: they are readily hydrolyzed by water to produce phenols and hydroxide ions.

Primary alcohol: an alcohol in which the carbon atom bearing the —OH group is bonded to one other carbon atom. The general formula is RCH_2OH.

Proof alcohol: beverage alcohol which is 50% alcohol by volume. In general, the proof is twice the volume percent of alcohol.

Secondary alcohol: an alcohol in which the carbon atom bearing the —OH group is bonded to two other carbon atoms. The general formula is R_2CHOH.

Simple ether: an ether in which both alkyl groups attached to the oxygen atom are the same.

Ternary azeotrope: a constant-boiling mixture of three compounds.

Tertiary alcohol: an alcohol in which the carbon atom bearing the —OH group is bonded to three other carbon atoms. The general formula is R_3COH.

Williamson synthesis: the most general preparation of ethers: an S_N2 type of reaction of an alkoxide or a phenoxide with a primary or secondary alkyl halide.

Additional problems

7-9 Give the IUPAC name for each of the following compounds.

(a) $CH_2OHCH_2CH_2CH_2OH$
(b) $CH_3CHOHCHOHCH_3$
(c) $CH_3CHOHCH_2Cl$
(d) $o\text{-}C_6H_4(OH)_2$
(e) $p\text{-}CH_3C_6H_4OCH_3$

7-10 Write a structural formula for each of the following compounds. If the common name is given here, write the IUPAC name with the structural formula.

(a) *p*-methoxybenzyl alcohol
(b) isobutyl ether
(c) 2-ethoxy-1-ethanol
(d) 2,3-butanediol
(e) cyclohexanol
(f) 2-methoxycyclohexanol
(g) 2-hydroxypyridine

7-11 Write equations for the following reactions of alcohols.

(a) ethanol and potassium
(b) 2-methyl-2-propanol (*tert*-butyl alcohol) and aluminum
(c) 1-butanol and thionyl chloride
(d) 2-methylcyclohexanol, H_2SO_4, 200°C
(e) 2-propanol, CuO, 200°C
(f) 1-propanol, $KMnO_4$, heat
(g) 2-pentanol, PBr_3
(h) 2-pentanol, $Na_2Cr_2O_7$

7-12 Indicate a *simple chemical* test or reagent which may be used to readily distinguish between the members of each of the following pairs. Indicate which member of the pair gives the positive test or greater reaction.

(a) CH_3CH_2OH and $CH_3CHOHCH_3$
(b) CH_3OH and $(CH_3)_2C{=}O$
(c) 1,2-cyclopentanediol and 1,3-cyclopentanediol
(d) 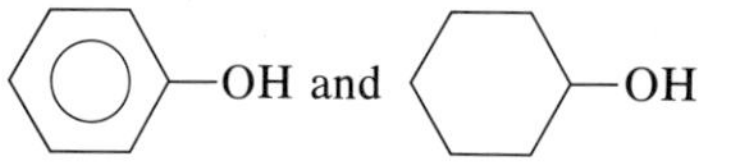 C_6H_5—OH and C_6H_{11}—OH (benzene ring—OH and cyclohexane ring—OH)
(e) 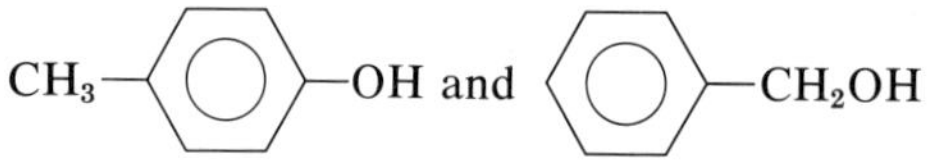 CH_3—C_6H_4—OH and C_6H_5—CH_2OH
(f) 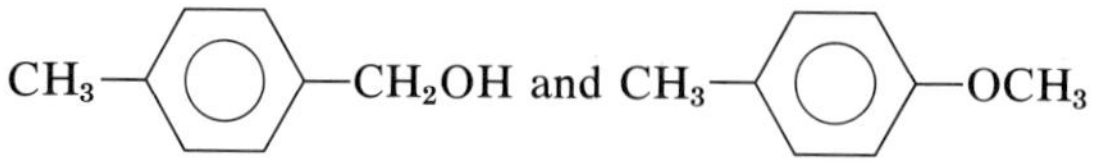 CH_3—C_6H_4—CH_2OH and CH_3—C_6H_4—OCH_3
(g) $CH_3CH_2{-}O{-}CH_2CH_3$ and $CH_3CH_2CH_2CH_2CH_3$
(h) 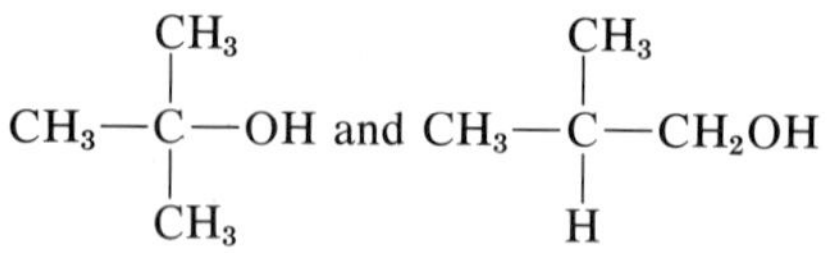 $CH_3{-}C(CH_3)_2{-}OH$ and $CH_3{-}CH(CH_3){-}CH_2OH$
(i) 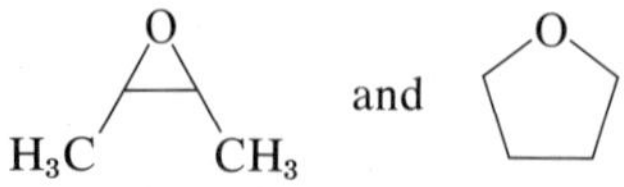 2,3-dimethyloxirane (epoxide ring O with H_3C and CH_3) and tetrahydrofuran (five-membered ring with O)
(j) $CH_3CH_2{-}O{-}CH_2CH_3$ and $CH_3CH_2CH{=}CHCH_3$

(k) 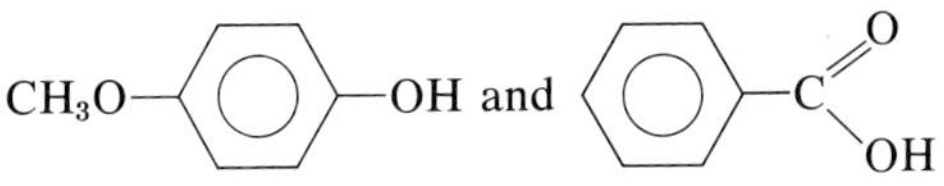 $CH_3O-C_6H_4-OH$ and $C_6H_5-C(=O)OH$

7-13 Beginning with benzene or toluene and alcohols of four or fewer carbons, write equations for the synthesis of each of the following *alcohols* by utilizing the Grignard reaction: (a) 2-methyl-1-butanol, (b) 2,3,3-trimethyl-2-butanol, (c) 1-phenyl-2-propanol, (d) 2-methyl-1-phenyl-1-propanol.

7-14 Give the structural formulas and names of the chief products expected from the reaction (if any) of cyclopentanol with

(a) cold, dilute $KMnO_4$
(b) cold, conc. H_2SO_4
(c) conc. H_2SO_4, heat
(d) CrO_3, H_2SO_4
(e) Br_2/CCl_4
(f) conc. $HBr(aq)$
(g) Na
(h) $P + I_2$
(i) H_2, Ni
(j) CH_3MgI
(k) $NaOH(aq)$

7-15 Cholesterol is a high-molecular-weight alcohol which is a normal constituent of the human body. Its structure is given in Fig. 7-3. Write the structural formulas for the reaction of cholesterol with the reagents indicated. (Note that cholesterol also has a carbon-carbon double bond.)

(a) Br_2/CCl_4
(b) CrO_3/acetic acid, heat
(c) H_2/Ni
(d) Na
(e) product of (d) and CH_3I
(f) cold, dil. $KMnO_4(aq)$
(g) conc. $HBr(aq)$
(h) conc. H_2SO_4, heat
(i) cold, conc. H_2SO_4

7-16 Write equations for the following reactions:

(a) 1,2-ethanediol with conc. HNO_3/H_2SO_4

(b)  tetrahydrofuran (five-membered ring with O) + HI (excess)

(c) 1-propanol heated at 120 to 140°C with conc. H_2SO_4
(d) 2-propanol and PCl_3
(e) cyclohexanol with H_2SO_4 at 200°C

(f) 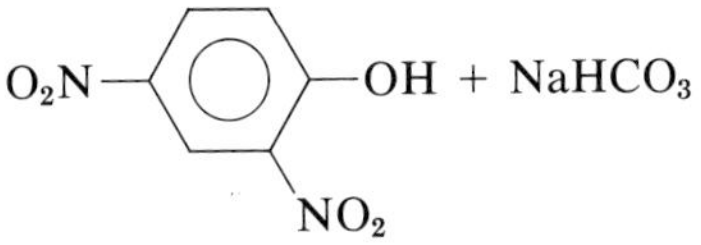 $O_2N-C_6H_3(NO_2)-OH + NaHCO_3$ (2,4-dinitrophenol, NO_2 ortho to OH)

(g) 1-butanol, conc. H_2SO_4, 40°C

7-17 Beginning with benzene or toluene and alcohols of four or fewer carbons, write equations for the synthesis of each of the following *ethers:*

(a) 1-ethoxybututane (*n*-butyl ethyl ether)
(b) 2-methoxy-2-methylpropane (*tert*-butyl methyl ether)
(c) propoxybenzene (phenyl *n*-propyl ether)
(d) phenoxycyclohexane (cyclohexyl phenyl ether)
(e) (phenylmethoxy)benzene (benzyl phenyl ether)

7-18 Write equations for the following reactions for the production of ethenyloxyethene (vinyl ether).

(a) ethene and Cl_2/H_2O to give 2-chloroethanol (ethylene chlorohydrin)
(b) 2-chloroethanol heated with H_2SO_4 at 140°C to produce a dichlorodiethyl ether
(c) dichlorodiethyl ether heated with alcoholic KOH to produce the ethenyloxyethene (vinyl ether)

7-19 A liquid compound, A, reacts with sodium to give off hydrogen, does *not* react with the Lucas reagent at room temperature, and is neutral to litmus. It is colorless and somewhat soluble in water but completely soluble in concentrated H_2SO_4. When heated to 180°C with concentrated H_2SO_4, it gives off a flammable gas which has a density of 2.5 g/liter at standard conditions. Write the structural formula for compound A.

7-20 By using reference books, find the health and occupational hazards and toxicity of the following compounds: (a) methanol, (b) isopropyl ether, (c) phenol, (d) 2,4,6-trinitrophenol (picric acid).

7-21 By using library resources, determine the commercial uses of the following:

(a) 1-hexyl-2,4-dihydroxybenzene (hexyl resorcinol)
(b) 1,2-ethanediol (ethylene glycol)
(c) ethoxyethane (ethyl ether)
(d) 1,2-epoxyethane (ethylene oxide)
(e) picric acid salts (picrates)

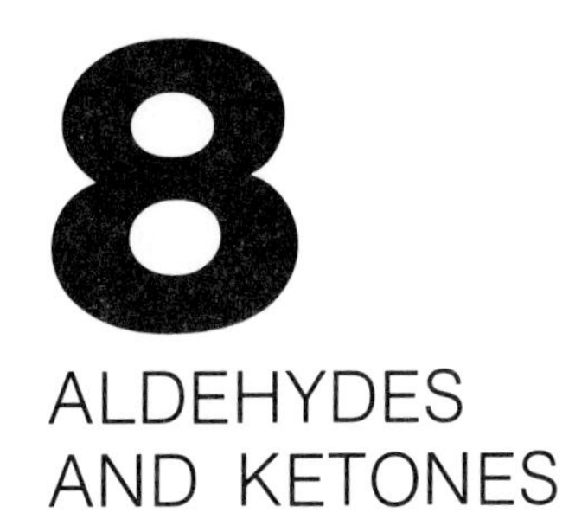

8 ALDEHYDES AND KETONES

An *aldehyde* is characterized by having *one* alkyl or aryl group and one hydrogen atom attached to the *carbonyl group,* $\rangle C{=}O$ (Fig. 8-1*a*).

R(H)C=O	Ar(H)C=O
Aliphatic aldehyde	Aromatic aldehyde

A ketone is characterized by having *two* alkyl or aryl groups attached to the carbonyl group (Fig. 8-1*b*). The groups need not be the same.

R(R′)C=O	Ar(R)C=O	Ar(Ar′)C=O
Aliphatic ketone	Aromatic aliphatic ketone	Aromatic ketone

A study of the chemistry of aldehydes, especially acetal formation, is important in understanding the structure and chemistry of carbohydrates, DNA, and RNA. The ketone carbonyl group is a functional group found in some of the steroid hormones that control sexual development and

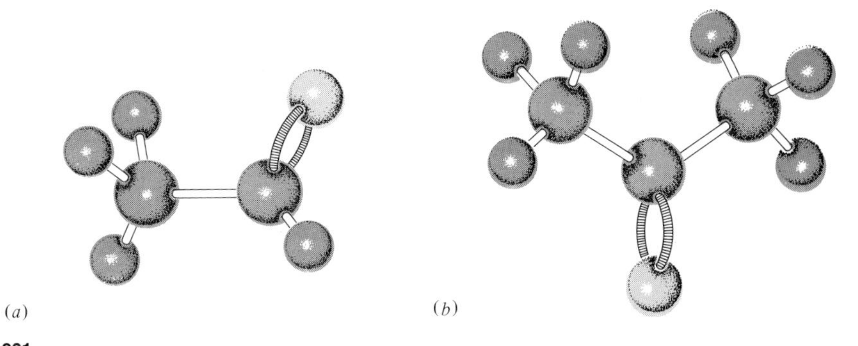

FIGURE 8-1
Models of (*a*) a representative aldehyde, ethanal (acetaldehyde), and (*b*) a representative ketone, propanone (acetone).

function. Some fragrant compounds used in perfumes also contain the carbonyl group. The simple ketone acetone (propanone) is an important solvent and reaction intermediate. Aldehydes and ketones in general enter into a great number of interesting reactions.

8-1 Structure of the carbonyl group

The structure of the carbonyl group of an aldehyde or ketone resembles that of an alkene in that the carbonyl carbon atom is sp^2-hybridized and forms sigma bonds to three other atoms. Thus it has a planar, trigonal geometry with bond angles of about 120° and a pi bond formed by lateral overlap of a p orbital on the carbonyl carbon atom with a p orbital on the oxygen atom.

A detailed orbital description of the carbonyl group is shown in Fig. 8-2. Note that the oxygen atom of the carbonyl group has its two unshared pairs of electrons in sp^2 hybrid orbitals. The carbon-oxygen double bond consists of a sigma bond and a pi bond. However, the carbon-oxygen double bond of the carbonyl group differs significantly from the carbon-carbon double bond of an alkene in that it is *highly polarized*. That is because the oxygen atom is much more electronegative than the carbon atom. The carbonyl group may be considered a resonance hybrid of the two main contributing forms. In the following diagram, R represents an alkyl group and R′ represents a hydrogen atom in the case of an aldehyde and an alkyl group in the case of a ketone.

$$\underset{R\quad R'}{\overset{O}{\overset{\|}{C}}} \longleftrightarrow \underset{R\quad R'}{\overset{O^-}{\overset{|}{C^+}}} \qquad \text{equivalent to} \qquad \underset{R\quad R'}{\overset{O^{\delta-}}{\overset{\|}{C^{\delta+}}}} \qquad (8\text{-}1)$$

This polarization leaves the carbon atom of the carbonyl group partially positive and therefore subject to attack by a nucleophilic reagent. The planar geometry facilitates the approach of the nucleophile to react with the carbonyl carbon atom. The approach can occur from above or below the plane of the group.

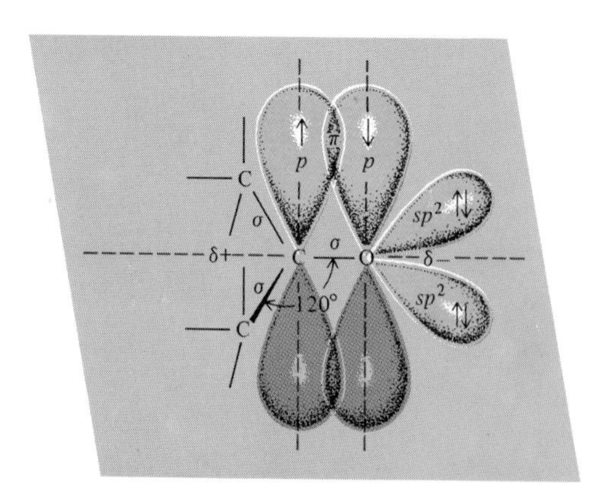

FIGURE 8-2
The structure of the carbonyl group.

8-2 Properties of aldehydes and ketones

The carbonyl group, as the functional group, is largely responsible for both the chemical and the physical properties of aldehydes and ketones. In general, the compounds are quite reactive. The high reactivity of the carbonyl group enables aldehydes and ketones to function more as intermediates than as end products in metabolism or synthesis.

Relative reactivity

For both *steric* (spatial) and *electronic reasons,* aldehydes are more vulnerable than ketones to attack by nucleophilic reagents. They are more reactive for steric reasons because a nucleophilic reagent is better able to approach the carbonyl group of an aldehyde (only one carbon group attached) than it is to approach the carbonyl group of a ketone (two carbon groups attached). The space around the carbonyl group of a ketone is more fully occupied. Aldehydes are more reactive for electronic reasons because they have only one attached alkyl group with its positive inductive effect (electron release). The positive inductive effect in part delocalizes the partial positive charge on the carbonyl carbon and thereby lessens its reactivity. In a ketone, two alkyl groups contribute to the delocalization of charge, and hence the reactivity of a ketone is less than that of an aldehyde.

$$\underset{\text{Aldehyde (more reactive)}}{\text{R(H)}\overset{\delta+}{\text{C}}=\overset{\delta-}{\text{O}}} \qquad \underset{\text{Ketone (less reactive)}}{\text{R}_2\overset{\delta+}{\text{C}}=\overset{\delta-}{\text{O}}}$$

Aromatic aldehydes are less reactive than aliphatic aldehydes because the carbonyl group of an aromatic aldehyde is conjugated with the benzene ring. That results in electron release by a positive resonance effect which lessens the positive nature and hence the reactivity of the carbonyl carbon of an aromatic aldehyde. Equation (8-2) illustrates the resonance of an aromatic aldehyde:

$$C_6H_5-\overset{H}{\underset{}{C}}=O \longleftrightarrow {}^{+}C_6H_5=\overset{H}{C}-O^- \longleftrightarrow {}^{+}C_6H_5=\overset{H}{C}-O^- \longleftrightarrow {}^{+}C_6H_5=\overset{H}{C}-O^- \qquad (8\text{-}2)$$

Intermolecular attractions and boiling points

Since the aldehydes and ketones do not have hydrogen atoms bonded to fluorine, oxygen, or nitrogen, they do not exhibit intermolecular hydrogen bonding between their own molecules. However, because of the very polar carbonyl group, they do exhibit strong dipole-dipole intermolecular attractions. (For example, acetone has a dipole moment of 2.88 Debye units, Table 1-3.) As a result, aldehydes and ketones have boiling points which are higher than those of hydrocarbons, alkyl halides, and ethers of

FIGURE 8-3
Aldehydes and ketones used as flavoring agents or as fragrances.

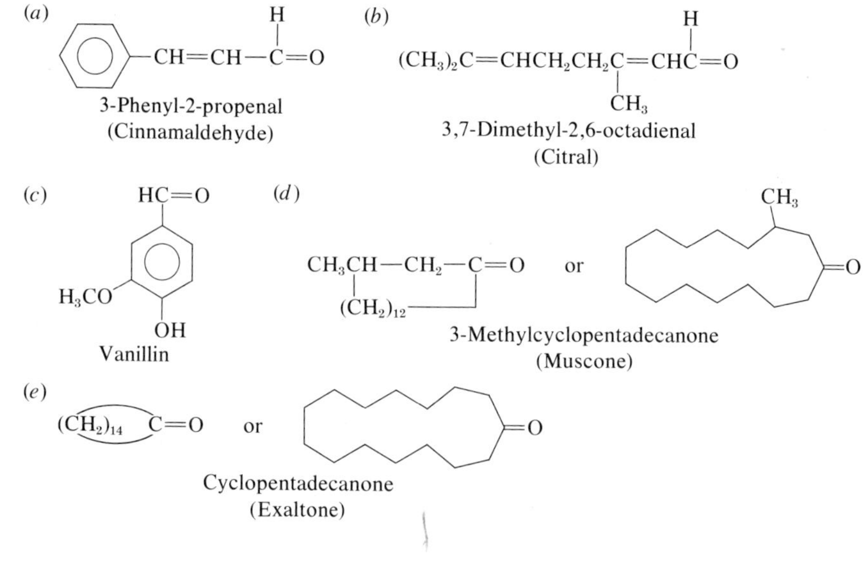

comparable molecular weight but lower than those of alcohols and carboxylic acids, which are capable of hydrogen bonding.

Since the hydrogen atoms of water can hydrogen bond to the oxygen atom of the carbonyl group, aldehydes and ketones have about the same solubility in water as alcohols and ethers of the same molecular weight. Those with three or fewer carbon atoms are miscible (soluble in all proportions) with water. The solvation of carbonyl compounds through hydrogen bonding of proton solvents can be represented with the proton solvent water:

$$\overset{\delta+}{C}=\overset{\delta-}{O}\cdots H-O-H$$

Alcohols also are proton solvents and will hydrogen bond to carbonyl compounds in a similar way.

Fragrance of aldehydes and ketones

Lower-molecular-weight aldehydes generally have disagreeable odors, but most ketones and some aldehydes have fragrant, pleasant odors. Cinnamaldehyde is responsible for the odor and flavor of cinnamon, and citral has a lemonlike odor. Vanillin has the odor and flavor of vanilla. Muscone (3-methylcyclopentadecanone), from the musk deer of the Himalayas, has a musk odor and is highly valued as an odor carrier and fixative in perfume formulation. Cyclopentadecanone is a synthetic compound called exaltone; it also has a musk odor and is used in making perfumes (Fig. 8-3).

PROBLEM 8-1 All the following compounds have about the same molecular weight. Taking into account polarity and hydrogen bonding, number them in order of decreasing boiling point; that is, the compound with the highest boiling point will be number 1. ___$CH_3OCH_2CH_3$, ___$CH_3CH_2CH_2CH_3$, ___$CH_3CHOHCH_3$, ___$CH_3\overset{\overset{\displaystyle O}{\|}}{C}{-}CH_3$, ___$CH_3CH_2\overset{\overset{\displaystyle O}{\|}}{C}{-}H$

8-3 Nomenclature

An aldehyde is commonly named by using the name of the corresponding carboxylic acid in which the suffix *-ic* is replaced by *-aldehyde*. For example, HCHO is formaldehyde (related to formic acid), CH_3CHO is acetaldehyde (related to acetic acid), and CH_3CH_2CHO is propionaldehyde (related to propionic acid).

In the IUPAC system, the stem or root for the name of an aldehyde comes from the name of the parent alkane, and to the stem the suffix *-al* (denoting an aldehyde group) is added. For example, HCHO is methanal, CH_3CHO is ethanal, and CH_3CH_2CHO is propanal.

The simplest aromatic aldehyde is C_6H_5CHO. It is commonly called benzaldehyde, but its more systematic IUPAC name is benzenecarbaldehyde. Likewise, $CH_3OC_6H_5CHO$ is *p*-methoxybenzaldehyde or 4-methoxybenzenecarbaldehyde.

The common name for a ketone is obtained by naming the groups attached to the carbonyl group and then, similarly to the common way of naming an ether, adding the word *ketone*. For example, $CH_3CH_2\overset{\overset{\displaystyle O}{\|}}{C}{-}CH_3$ is ethyl methyl ketone, $CH_3\overset{\overset{\displaystyle O}{\|}}{C}{-}CH_3$ is dimethyl ketone (even more commonly known as acetone), and $C_6H_5\overset{\overset{\displaystyle O}{\|}}{C}{-}CH_3$ is methyl phenyl ketone (also known as acetophenone).

A ketone is named in the IUPAC system by using the name of the parent alkane as the stem or root of the name and then adding the suffix *-one*. If the ketone has more than four carbon atoms, the position of the carbonyl group is indicated by a locator number. For example, $CH_3CH_2\overset{\overset{\displaystyle O}{\|}}{C}{-}CH_2CH_3$ is 3-pentanone, $C_6H_5\overset{\overset{\displaystyle O}{\|}}{C}{-}CH_2CH_2CH_3$ is 1-phenyl-1-butanone, and $CH_3{-}\langle\text{cyclohexane ring}\rangle{=}O$ is 4-methylcyclohexanone. Study Table 8-1 for other examples.

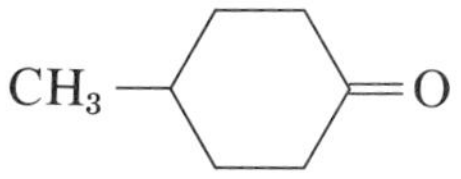

PROBLEM 8-2 Give the IUPAC name and another acceptable name for each of the following compounds.

(a) $CH_3\overset{\overset{\displaystyle O}{\|}}{C}{-}\overset{\overset{\displaystyle CH_3}{|}}{C}HCH_3$

(b) $CH_3\overset{\overset{\displaystyle CH_3}{|}}{C}HCH_2\overset{\overset{\displaystyle H}{|}}{C}{=}O$

(c) $C_6H_5C(CH_2CH_3){=}O$

(d) 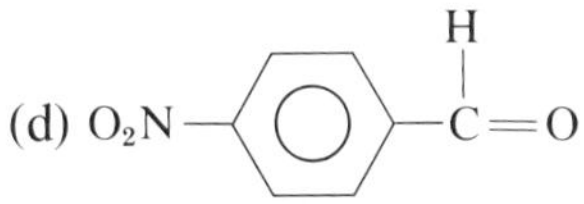 $O_2N{-}C_6H_4{-}C(H){=}O$

Solution to (a) The IUPAC name for this compound is 3-methyl-2-butanone. The common name is isopropyl methyl ketone.

$$\underset{1}{CH_3}\underset{2}{C}(=O)-\underset{3}{CH}(CH_3)-\underset{4}{CH_3}$$

TABLE 8-1
Some aldehydes and ketones

Structural formula	IUPAC name	Common name	Boiling point, °C
$HC(H){=}O$	Methanal	Formaldehyde	−21
$CH_3C(H){=}O$	Ethanal	Acetaldehyde	20
$CH_3CH_2C(H){=}O$	Propanal	Propionaldehyde	49
$CH_3CH_2CH_2C(H){=}O$	Butanal	*n*-Butyraldehyde	75
$CH_3CH(CH_3){-}C(H){=}O$	2-Methylpropanal	Isobutyraldehyde	63
$C_6H_5{-}C(H){=}O$	Benzenecarbaldehyde	Benzaldehyde	178
o-$HO{-}C_6H_4{-}C(H){=}O$	2-Hydroxybenzene-carbaldehyde	Salicylaldehyde	197
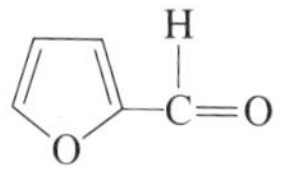 (furan ring, O)$-C(H){=}O$		Furfural	161
$(CH_3)_2C{=}O$	Propanone	Acetone, dimethyl ketone	56
$CH_3CH_2C(CH_3){=}O$	Butanone	Ethyl methyl ketone	79
$(CH_3CH_2)_2C{=}O$	3-Pentanone	Diethyl ketone	101
$CH_3CH_2CH_2C(CH_3){=}O$	2-Pentanone	Methyl *n*-propyl ketone	102
$C_6H_5{-}C(CH_3){=}O$	Phenylethanone	Acetophenone, methyl phenyl ketone	202
$(C_6H_5)_2C{=}O$	Diphenylmethanone	Benzophenone, diphenyl ketone	306

PROBLEM 8-3 Write the structural formula for each of the following compounds. (a) 2-methylbutanal, (b) 2-methyl-2-pentenal, (c) hexanal, (d) 3-hexanone.

Solution to (a) The structural formula for 2-methylbutanal is

$$\overset{4}{CH_3}-\overset{3}{CH_2}-\underset{\displaystyle H_3C}{\overset{2}{C}}-\underset{\displaystyle H}{\overset{1}{C}}=O$$

8-4 Preparation of aldehydes and ketones

Both aldehydes and ketones can be prepared by the oxidation of appropriate alcohols. Aromatic aldehydes can be prepared by partial oxidation of arenes (alkylbenzenes). Ketones are prepared by Friedel-Crafts acylation or by pyrolysis of metal salts of carboxylic acids.

Oxidation of alcohols

Oxidation of *primary* alcohols yields *aldehydes* (Fig. 8-4); oxidation of *secondary* alcohols yields *ketones* (Fig. 8-5). Those reactions were discussed in Chap. 7 as reactions of alcohols. In the reactions given there, water was a product. The general equations are

$$\underset{\text{Primary alcohol}}{R-CH_2OH} \xrightarrow{(O)} \underset{\text{Aldehyde}}{R-\underset{\displaystyle H}{C}=O} + H_2(g) \quad \text{or} \quad H_2O \tag{8-3}$$

$$\underset{\text{Secondary alcohol}}{R_2-CHOH} \xrightarrow{(O)} \underset{\text{Ketone}}{R-\underset{\displaystyle R}{C}=O} + H_2(g) \quad \text{or} \quad H_2O \tag{8-4}$$

If the alcohol is passed over copper powder or over copper chromite ($CuCr_2O_4 \cdot CuO$) at 250°C without air, it is *dehydrogenated* to an aldehyde or ketone with evolution of hydrogen. Indeed, the name "aldehyde" is given to this family to describe the members as products of *al*cohol *de-*

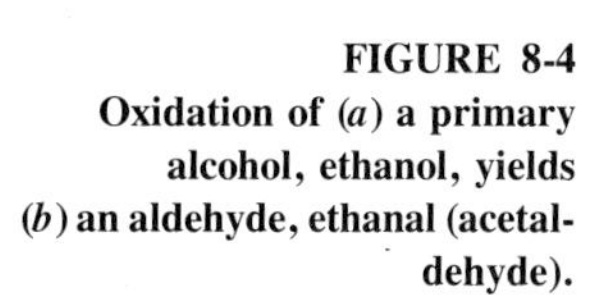

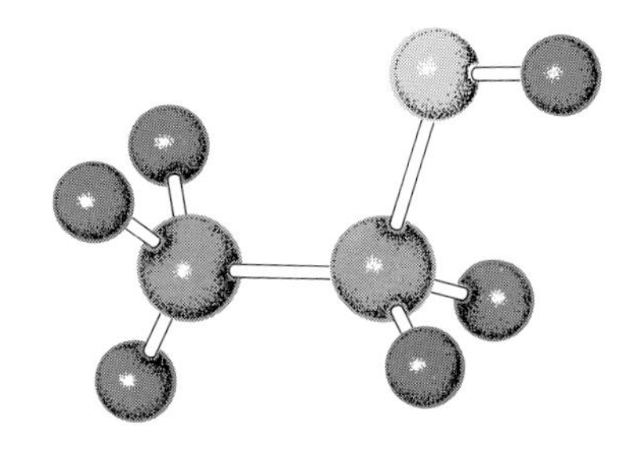

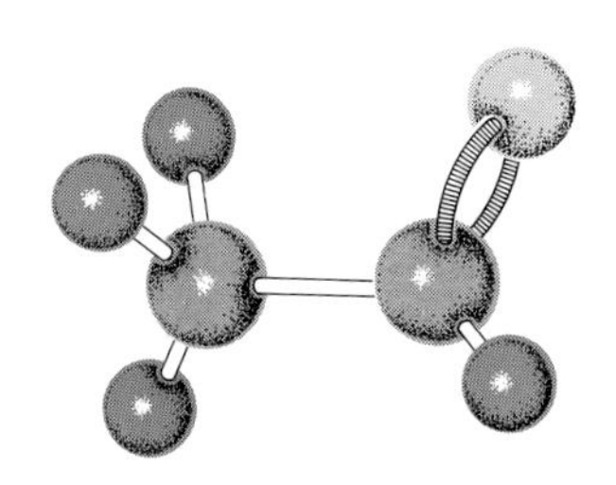

FIGURE 8-4 **Oxidation of (*a*) a primary alcohol, ethanol, yields (*b*) an aldehyde, ethanal (acetaldehyde).**

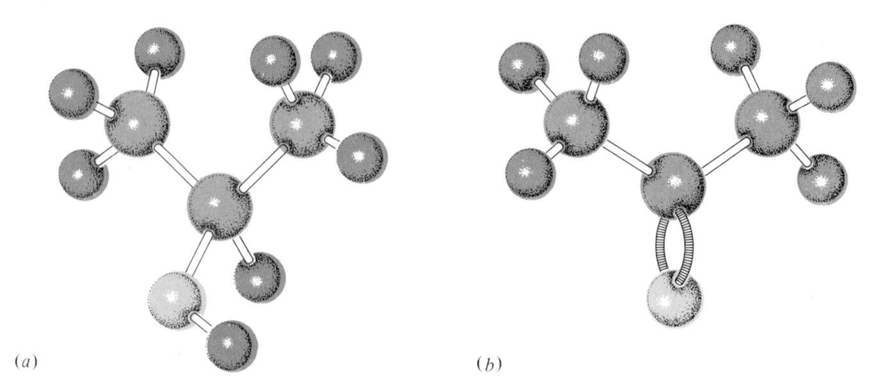

FIGURE 8-5 **Oxidation of (*a*) a secondary alcohol, 2-propanol, yields (*b*) a ketone, propanone (acetone).**

*hyd*rogenation. Of course, dehydrogenation can be considered as a type of oxidation. An example of the reaction is

$$(CH_3)_2CHCH_2CH_2OH \xrightarrow{CuCr_2O_4\cdot CuO,\ 250^\circ C} (CH_3)_2CHCH_2\overset{\displaystyle H}{\overset{|}{C}}{=}O + H_2(g) \qquad (8\text{-}5)$$

3-Methyl-1-butanol (Isoamyl alcohol) → 3-Methylbutanal (Isovaleraldehyde)

This particular reaction gives a 95% yield of the aldehyde. In general, the method is a good one.

Partial oxidation of methylbenzenes

Aromatic aldehydes can be prepared by the partial oxidation of methylbenzenes. The general equation may be given as

$$Ar{-}CH_3 \xrightarrow{(O)} Ar{-}\overset{\displaystyle H}{\overset{|}{C}}{=}O + H_2O \qquad (8\text{-}6)$$

An example is the partial oxidation of toluene to benzenecarbaldehyde:

$$C_6H_5{-}CH_3 \xrightarrow{CrO_3/CH_3COOH,\ \Delta} C_6H_5{-}\overset{\displaystyle H}{\overset{|}{C}}{=}O + H_2O \qquad (8\text{-}7)$$

Toluene → Benzenecarbaldehyde (Benzaldehyde)

A variation of the procedure is to halogenate the arene in the side chain and then hydrolyze the resulting halide to yield aldehyde. Halogenation can be considered as a type of oxidation. An example of the reaction is

$$C_6H_5{-}CH_3 + 2Cl_2 \xrightarrow{light} 2HCl + C_6H_5{-}\overset{\displaystyle H}{\overset{|}{\underset{\underset{\displaystyle Cl}{|}}{C}}}{-}Cl \xrightarrow{H_2O} C_6H_5{-}\overset{\displaystyle H}{\overset{|}{C}}{=}O \qquad (8\text{-}8)$$

Toluene; (Dichloromethyl)benzene (Benzal chloride); Benzenecarbaldehyde (Benzaldehyde)

The reaction has had some industrial application. One may think of the dichloride as being hydrolyzed to a diol which then eliminates a molecule of water to form the aldehyde.

Preparation of aliphatic aromatic ketones

Aliphatic aromatic ketones can be synthesized by using the Friedel-Crafts acylation reaction. The general equation is

$$\mathrm{ArH} + \mathrm{Cl{-}\overset{\overset{\displaystyle O}{\|}}{C}{-}R} \xrightarrow{AlCl_3} \underset{\text{Aliphatic aromatic ketone}}{\mathrm{Ar{-}\overset{\overset{\displaystyle O}{\|}}{C}{-}R}} + \mathrm{HCl} \tag{8-9}$$

Acylation preferably takes place para to an electron-releasing substituent (an ortho,para-directing group). It does not occur if the ring has a meta-directing substituent. An example of the reaction is

$$\underset{\text{Toluene}}{\mathrm{CH_3{-}C_6H_4{-}H}} + \underset{\substack{\text{Propanoyl chloride}\\ \text{(Propionyl chloride)}}}{\mathrm{Cl{-}\overset{\overset{\displaystyle O}{\|}}{C}CH_2CH_3}} \xrightarrow{AlCl_3}$$

$$\underset{\substack{\text{1-(4-Methylphenyl)-1-propanone}\\ \text{(Ethyl } p\text{-toluyl ketone)}}}{\mathrm{CH_3{-}C_6H_4{-}\overset{\overset{\displaystyle O}{\|}}{C}CH_2CH_3}} + \mathrm{HCl} \tag{8-10}$$

Another, more complicated, example is the reaction of toluene with succinic anhydride:

$$\underset{\text{Toluene}}{\mathrm{CH_3{-}C_6H_5}} + \underset{\substack{\text{Butanedioic anhydride}\\ \text{(Succinic anhydride)}}}{\begin{matrix} \mathrm{H_2C{-}C{=}O} \\ | \quad\quad \backslash \\ | \quad\quad\; \mathrm{O} \\ | \quad\quad / \\ \mathrm{H_2C{-}C{=}O} \end{matrix}} \xrightarrow{AlCl_3} \underset{\text{3-(}p\text{-Methylbenzoyl)-propanoic acid}}{\mathrm{CH_3{-}C_6H_4{-}\overset{\overset{\displaystyle O}{\|}}{C}CH_2CH_2C({=}O)OH}} \tag{8-11}$$

Such Friedel-Crafts reactions as (8-10) and (8-11) have been extremely useful in the laboratory for preparing a great number and variety of ketones for research purposes and for reaction intermediates. When benzene or toluene is the reactant, it can serve as its own solvent. If a solid aromatic hydrocarbon is a reactant, nitrobenzene can be used as the solvent because it does not react.

Ketones by pyrolysis of metal salts of acids

The *pyrolysis* of metal salts of acids may be used to produce ketones. Pyrolysis means a cleavage reaction achieved by heating. The general equation is

$$(RCOO^-)_2M^{2+} \xrightarrow{300°C} R-\overset{\overset{O}{\|}}{C}-R + MCO_3 \qquad (8\text{-}12)$$

Acetic acid vapor, when passed through a heated tube packed with an oxide of calcium, barium, manganese, or thorium, produces acetone. The metallic salt of acetic acid which is first produced immediately decomposes into a metallic carbonate and acetone. The metallic carbonate of the heavier metal then decomposes back into the oxide and liberates carbon dioxide.

$$2CH_3C(=O)OH + CaO \longrightarrow \left[(CH_3COO^-)_2\,Ca^{2+}\right] + H_2O \xrightarrow{300°C} CH_3\overset{\overset{O}{\|}}{C}CH_3 + CaCO_3 \qquad (8\text{-}13)$$

Ethanoic acid (Acetic acid) — Calcium acetate — Propanone (Acetone)

$$2CH_3C(=O)OH \xrightarrow{MnO,\ 300°C} CH_3\overset{\overset{O}{\|}}{C}CH_3 + H_2O + CO_2 \qquad (8\text{-}14)$$

Ethanoic acid (Acetic acid) — Propanone (Acetone)

The reaction is a general one, and it may be extended to dicarboxylic acids to yield cyclic ketones.

$$(CH_2)_4(COO^-)_2\,Ba^{2+} \xrightarrow{300°C} (CH_2)_4C{=}O + BaO + CO_2(g) \qquad (8\text{-}15)$$

Barium hexanedioate (Barium adipate) — Cyclopentanone

PROBLEM 8-4 Write equations for the preparation of the following aldehydes and ketones beginning with an alcohol, benzene, or toluene, if appropriate, and any needed aliphatic and inorganic reagents.

Hint for (*d*)*:* Obtain the precursor alcohol by using a Grignard reaction.

(a) $C_6H_5C(=O)-CH_2CH_2CH_3$

(b) $C_6H_5C(=O)-CH_2CH_2COOH$

(c) $CH_3CH_2CH_2C(H)=O$

(d) $C_6H_5CH_2C(=O)-CH_3$

(e) *p*-chlorobenzaldehyde

(f) acetaldehyde

(g) acetone

(h) 2-methylbutanal

(i) diethyl ketone

(j) methanal

Solution to (a)

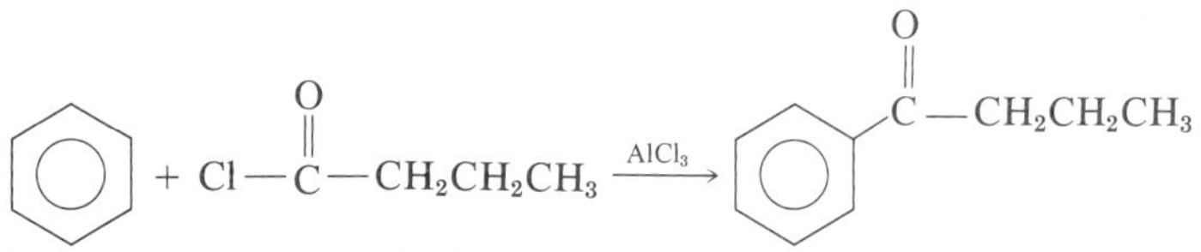

$$C_6H_6 + Cl-C(=O)-CH_2CH_2CH_3 \xrightarrow{AlCl_3} C_6H_5-C(=O)-CH_2CH_2CH_3$$

Benzene Butanoyl chloride 1-Phenyl-1-butanone

8-5 Reactions of aldehydes and ketones

The reactions of aldehydes and ketones include *oxidation, reduction,* and a large number of *nucleophilic addition reactions*. Table 8-2 gives the general equation for each of those types of reaction. In addition, it gives examples of a number of specific reactions. All of the reaction types, with other examples of aldehydes and ketones and some reagents that can be used, are discussed in the following sections. Table 8-2 is *not* a complete summary of the reactions of aldehydes and ketones. It will, however, give you an overview of the reactions and be a useful reference after you have studied the reactions in detail.

Oxidation

Aldehydes are readily oxidized by all ordinary oxidizing agents—even air. If a bottle of an aldehyde has been opened to the air for a day or so, some of it will have been oxidized to the corresponding carboxylic acid (Fig. 8-6).

$$R-C(=O)H \xrightarrow{O_2 \text{ from air}} R-C(=O)OH \quad (8\text{-}16)$$

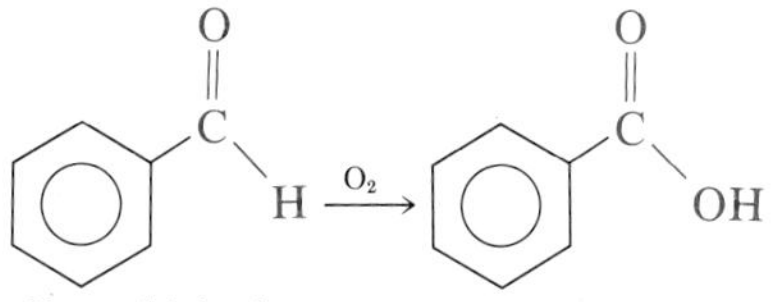

$$C_6H_5-C(=O)H \xrightarrow{O_2} C_6H_5-C(=O)OH \quad (8\text{-}17)$$

Benzaldehyde Benzoic acid

The carbon-carbon bond of a ketone is more difficult to oxidize than the carbonyl carbon-hydrogen bond of an aldehyde; so ketones are not readily oxidized. The difference can be used to distinguish an aldehyde from

TABLE 8-2
Some reactions of aldehydes and ketones

Oxidation

$$R{-}C(=O){-}H \xrightarrow{(O)} R{-}C(=O){-}OH$$

Examples
Catalytic

$$\underset{\text{Ethanal}}{CH_3C(=O){-}H} \xrightarrow{O_2,\ MnO,\ \Delta} \underset{\text{Ethanoic acid}}{CH_3C(=O){-}OH}$$

Tollens' reagent

$$\underset{\text{Propanal}}{CH_3CH_2C(=O){-}H} + 2Ag(NH_3)_2^+ + 2OH^- \longrightarrow \underset{\text{Ammonium propanoate}}{CH_3CH_2C(=O){-}O^-NH_4^+} + 2Ag(s) + H_2O + 3NH_3$$

Haloform reaction

$$\underset{\text{Butanone}}{CH_3CH_2C(=O){-}CH_3} + 3I_2 + 4NaOH \longrightarrow \underset{\text{Sodium propanoate}}{CH_3CH_2C(=O)ONa} + \underset{\text{Iodoform}}{CHI_3} + 3H_2O + 3NaI$$

Reduction

$$R{-}C(=O){-}H \xrightarrow{(H)} R{-}CH_2{-}OH$$

$$R{-}C(=O){-}R \xrightarrow{(H)} R{-}CH(R){-}OH$$

Examples
Catalytic addition of hydrogen

$$\underset{\text{2-Ethyl-3-hydroxyhexanal}}{CH_3CH_2CH_2CHOHCH(CH_2CH_3)CH{=}O} + H_2 \xrightarrow{Pd} \underset{\text{2-Ethyl-1,3-hexanediol}}{CH_3CH_2CH_2CHOHCH(CH_2CH_3)CH_2OH}$$

With lithium aluminum hydride

$$\underset{\text{Phenylethanone}}{4C_6H_5C(=O){-}CH_3} + LiAlH_4 \longrightarrow (C_6H_5CH(CH_3){-}O)_4AlLi \xrightarrow{H_2O} \underset{\text{1-Phenyl-1-ethanol}}{4C_6H_5CH(CH_3)OH} + LiOH + Al(OH)_3$$

TABLE 8-2 (continued)

Nucleophilic addition

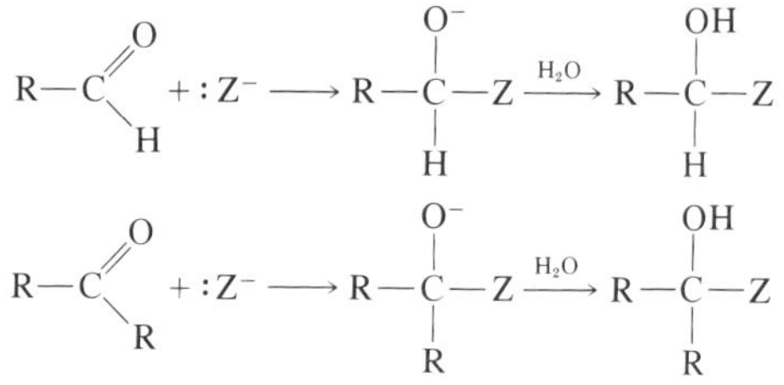

$$R{-}C(=O){-}H + :Z^- \longrightarrow R{-}C(O^-)(H){-}Z \xrightarrow{H_2O} R{-}C(OH)(H){-}Z$$

$$R{-}C(=O){-}R + :Z^- \longrightarrow R{-}C(O^-)(R){-}Z \xrightarrow{H_2O} R{-}C(OH)(R){-}Z$$

Examples

Formation of cyanohydrins

$$C_6H_5C(H){=}O + HCN \longrightarrow C_6H_5CHOHCN$$

Benzenecarbaldehyde (Benzaldehyde) → Benzaldehyde cyanohydrin

Addition of water

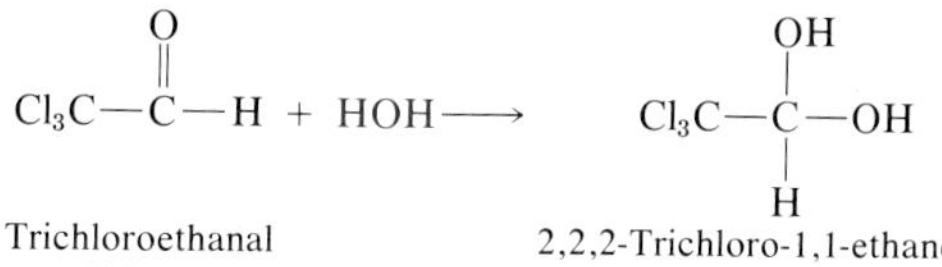

$$Cl_3C{-}C(=O){-}H + HOH \longrightarrow Cl_3C{-}C(OH)(H){-}OH$$

Trichloroethanal (Chloral) → 2,2,2-Trichloro-1,1-ethanediol (Chloral hydrate)

Addition of alcohol

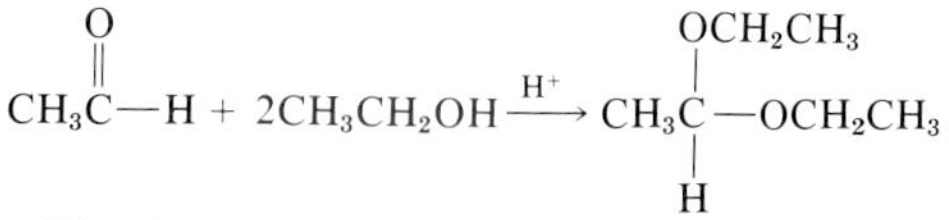

$$CH_3C(=O){-}H + 2CH_3CH_2OH \xrightarrow{H^+} CH_3C(OCH_2CH_3)(H){-}OCH_2CH_3$$

Ethanal; Ethanol → Ethyl acetal of ethanal

Addition of sodium hydrogen sulfite

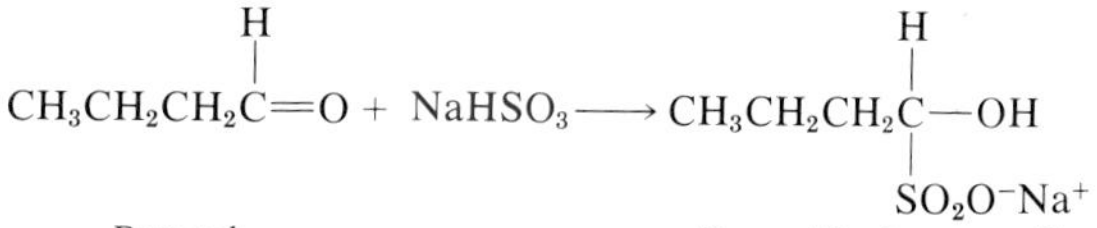

$$CH_3CH_2CH_2C(H){=}O + NaHSO_3 \longrightarrow CH_3CH_2CH_2C(H)(SO_2O^-Na^+){-}OH$$

Butanal → Butanal hydrogen sulfite addition compound

Addition of hydroxylamine

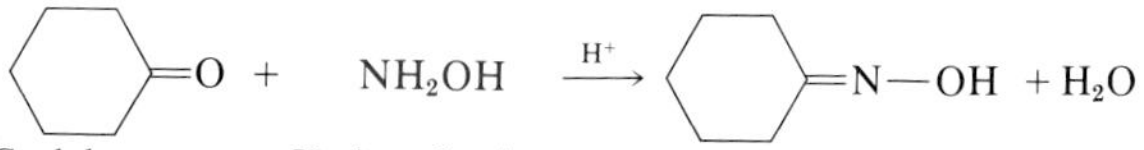

$$C_6H_{10}{=}O + NH_2OH \xrightarrow{H^+} C_6H_{10}{=}N{-}OH + H_2O$$

Cyclohexanone; Hydroxylamine → Cyclohexanone oxime

TABLE 8-2 (continued)

Addition of phenylhydrazine

$$CH_3CH_2\overset{\overset{H}{|}}{C}{=}O + NH_2NHC_6H_5 \xrightarrow{H^+} CH_3CH_2CH{=}N{-}NHC_6H_5 + H_2O$$

Propanal — Phenylhydrazine — Propanal phenylhydrazone

Addition of semicarbazide

$$(CH_3CH_2)_2C{=}O + NH_2NH\overset{\overset{O}{\|}}{C}{-}NH_2 \xrightarrow{H^+} (CH_3CH_2)_2C{=}N{-}NH{-}\overset{\overset{O}{\|}}{C}{-}NH_2$$

3-Pentanone — 3-Pentanone semicarbazone

Addition of Grignard reagent

$$C_6H_5\overset{\overset{H}{|}}{C}{=}O + C_6H_5CH_2MgBr \xrightarrow{(1)\text{ ether, }(2)\ H_2O} C_6H_5\overset{\overset{OH}{|}}{\underset{\underset{H}{|}}{C}}{-}CH_2C_6H_5$$

Benzaldehyde — 1,2-Diphenyl-1-ethanol

Aldol condensation

$$2CH_3CH_2CH_2\overset{\overset{H}{|}}{C}{=}O \xrightarrow{NaOH} CH_3CH_2CH_2\underset{\underset{OH}{|}}{C}H{-}\underset{\underset{CH_2CH_3}{|}}{C}H\overset{\overset{H}{|}}{C}{=}O$$

Butanal — 2-Ethyl-3-hydroxyhexanal

Cannizzaro reaction

$$2C_6H_5\overset{\overset{H}{|}}{C}{=}O + NaOH \longrightarrow C_6H_5CH_2OH + C_6H_5COONa$$

Benzaldehyde — Phenylmethanol (Benzyl alcohol) — Sodium benzoate

its isomeric ketone. One reaction employed for the purpose is carried out with *Tollens' reagent*. The reagent consists of ammoniacal silver nitrate, and it is made by dissolving silver nitrate in an excess of ammonium hydroxide:

$$AgNO_3 + 3NH_4OH \longrightarrow Ag(NH_3)_2OH + NH_4NO_3 + 2H_2O \qquad (8\text{-}18)$$

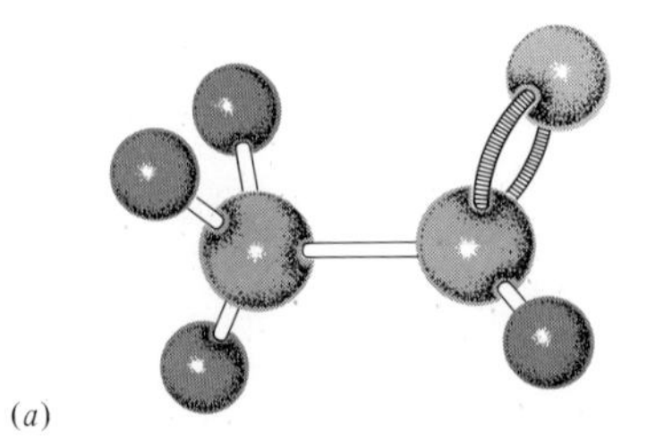

(*a*)

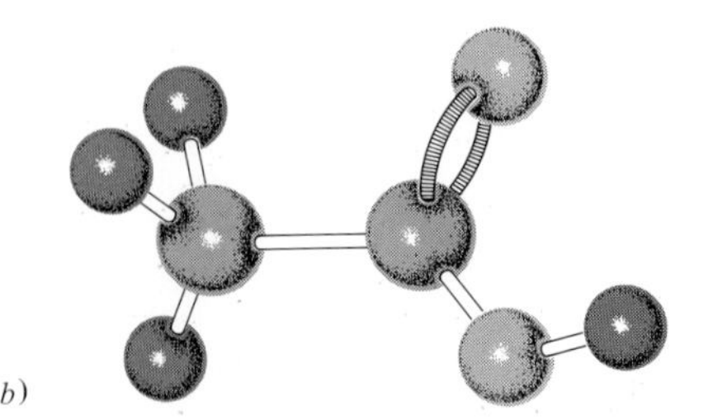

(*b*)

FIGURE 8-6
An aldehyde such as (*a*) ethanal (acetaldehyde) is readily oxidized to a carboxylic acid such as (*b*) ethanoic acid (acetic acid).

The oxidation-reduction reaction with an aldehyde produces free silver which adheres to the glass walls of the container (if clean) to form a silver mirror. The commercial manufacture of mirrors uses methanal (formaldehyde) for the purpose. The general equation is

$$R-C(=O)H + 2Ag(NH_3)_2^+ + 2OH^- \longrightarrow R-C(=O)O^-NH_4^+ + 2Ag(s) + 3NH_3 + H_2O \qquad (8\text{-}19)$$

In other test reagents copper(II) ion is the oxidizing agent. In *Fehling's solution* the tartrate ion forms a complex copper(II) ion soluble in basic solution. In *Benedict's solution* citrate ion forms a complex with the copper(II) ion. In either case, the copper(II) ion is reduced by the aldehyde to copper(I) ion and on further reduction to free copper.

$$R-C(=O)H + Cu^{2+}\ (\text{complex}) \xrightarrow{OH^-} R-C(=O)O^- + Cu^+ \text{ or } Cu \qquad (8\text{-}20)$$

The result seen is a change from the dark blue color of the reagent through shades of olive green to the orange-red color produced by the fine precipitate of copper(I) ion and/or free copper.

Ketones and aldehydes which contain the $CH_3C{=}O$ group are oxidized by the oxidizing agent NaOX, where X = Cl, Br, or I. The reaction is called the *haloform reaction* because a haloform (CHX_3) is one of the products of the reaction. The NaOX, called *hypohalite,* is produced in the reaction vessel by adding sodium hydroxide and halogen to the methyl ketone. The hypohalite is produced by the reaction

$$2NaOH + X_2 \longrightarrow NaOX + NaX + H_2O \qquad (8\text{-}21)$$

When the reaction is used as a test for a methyl ketone, iodine is used for the halogen because the iodoform (triiodomethane, HCI_3) produced is a yellow solid with a characteristic odor and melting point which makes it easy to detect. The general equations for the reaction are

$$\underset{\text{A methyl ketone}}{R-C(=O)-CH_3} + 3I_2 + 3NaOH \longrightarrow R-C(=O)-CI_3 + 3H_2O + 3NaI \qquad (8\text{-}22)$$

$$R-C(=O)-CI_3 + NaOH \longrightarrow R-C(=O)-O^-Na^+ + \underset{\text{Iodoform}}{CHI_3(s)} \qquad (8\text{-}23)$$

Secondary alcohols which can be oxidized to give the $CH_3\overset{|}{C}{=}O$ group, and also the *primary alcohol* ethanol (which when oxidized gives $CH_3\overset{H}{\overset{|}{C}}{=}O$), also will react with NaOX to produce haloform.

$$\underset{\text{Secondary alcohol}}{R{-}\overset{OH}{\overset{|}{C}}HCH_3} \xrightarrow{NaOI} \underset{\text{A methyl ketone}}{R{-}\overset{O}{\overset{\|}{C}}{-}CH_3} \xrightarrow{3I_2,\ 3NaOH} R{-}\overset{O}{\overset{\|}{C}}{-}CI_3 + 3H_2O + 3NaI \quad (8\text{-}24)$$

$$R{-}\overset{H}{\overset{\|}{C}}{-}CI_3 + NaOH \longrightarrow R{-}\overset{O}{\overset{\|}{C}}{-}O^-Na^+ + \underset{\text{Iodoform}}{CHI_3(s)} \quad (8\text{-}23)$$

When the reaction is utilized for synthesis of a carboxylic acid from a methyl ketone, chlorine is used because it is cheaper than iodine. An example is the synthesis of 2-naphthoic acid.

$$\underset{\text{2-Acetylnaphthalene}}{C_{10}H_7{-}\overset{O}{\overset{\|}{C}}{-}CH_3} \xrightarrow{NaOH/Cl_2} \underset{\text{Sodium 2-naphthoate}}{C_{10}H_7{-}\overset{O}{\overset{\|}{C}}{-}O^-Na^+} + \underset{\substack{\text{Trichloromethane}\\ \text{(Chloroform)}}}{CHCl_3} \quad (8\text{-}25)$$

The haloform reaction can be used to distinguish readily between such isomeric ketones as 2-pentanone, which gives iodoform, and 3-pentanone, which does not (Fig. 8-7). It can also be used to distinguish between 1-propanol and 2-propanol and between 2-pentanol and 3-pentanol. Acetaldehyde is the only aldehyde which will give the haloform reaction.

Reduction Reduction of *aldehydes* produces *primary alcohols*, and reduction of *ketones* produces *secondary alcohols*. The reduction might be a *catalytic*

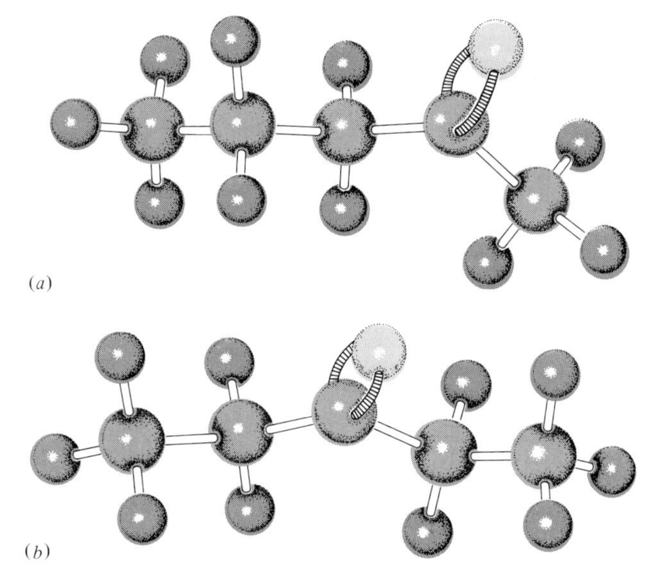

FIGURE 8-7
(*a*) 2-Pentanone, a methyl ketone, gives a positive reaction in the iodoform test. (*b*) Its isomer, 3-pentanone, does not give a positive reaction.

addition of hydrogen to the carbonyl group. The general reactions are

$$\underset{\text{Aldehyde}}{R-\overset{\displaystyle H}{\overset{|}{C}}=O} + H_2 \xrightarrow{\text{Pt, Pd, Ni, or } CuCr_2O_4} \underset{\text{Primary alcohol}}{R-\underset{\underset{\displaystyle H}{|}}{\overset{\overset{\displaystyle H}{|}}{C}}-OH} \tag{8-26}$$

$$\underset{\text{Ketone}}{R-\overset{\displaystyle O}{\overset{\|}{C}}-R} + H_2 \xrightarrow{\text{Pt, Pd, Ni, or } CuCr_2O_4} \underset{\text{Secondary alcohol}}{R-\underset{\underset{\displaystyle H}{|}}{\overset{\overset{\displaystyle OH}{|}}{C}}-R} \tag{8-27}$$

An example of this reaction is the reduction of cyclopentanone to cyclopentanol:

$$\underset{\text{Cyclopentanone}}{\text{C}_5\text{H}_8{=}O} + H_2 \xrightarrow{\text{Ni, 50°C, 68 atm}} \underset{\text{Cyclopentanol}}{\text{C}_5\text{H}_8\!<^{OH}_{H}} \tag{8-28}$$

The reduction may also be accomplished by using *sodium borohydride* or *lithium aluminum hydride*. The mechanism is nucleophilic addition to the carbonyl group.

$$R-\overset{\displaystyle H}{\overset{|}{C}}=O + \underset{\text{Sodium borohydride}}{H-\underset{\underset{\displaystyle H}{|}}{\overset{\overset{\displaystyle H}{|}}{B^-}}-H\ \ Na^+} \longrightarrow BH_3 + R-\underset{\underset{\displaystyle H}{|}}{\overset{\overset{\displaystyle H}{|}}{C}}-O^-Na^+ \xrightarrow{H_2O}$$

$$R-CH_2OH + NaOH \tag{8-29}$$

An example of this reaction is

$$4CH_3-\overset{O}{\overset{\|}{C}}-O-C_6H_4-\overset{O}{\overset{\|}{C}}-CH_3 + NaBH_4 + 3H_2O \longrightarrow$$

4-Ethanoylphenyl ethanoate
(*p*-Acetylphenyl acetate)

$$4CH_3\overset{O}{\overset{\|}{C}}-O-C_6H_4-\overset{OH}{\overset{|}{C}}HCH_3 + NaOB(OH)_2 \qquad (8\text{-}30)$$

4-(1-Hydroxyethyl)phenyl ethanoate

Sodium borohydride is a milder reducing agent than lithium aluminum hydride. It will reduce aldehydes and ketones, but not acids or esters. The reaction can even be carried out in aqueous solution.

The reduction using lithium aluminum hydride requires ether as a solvent for the first step. The general equation for reduction of an aldehyde with lithium aluminum hydride is

$$R-\overset{H}{\overset{|}{C}}=O + H-\underset{H}{\underset{|}{\overset{H}{\overset{|}{Al^-}}}}-H\ Li^+ \xrightarrow{\text{ether}} R-\underset{H}{\underset{|}{\overset{H}{\overset{|}{C}}}}-O^-AlH_3Li^+ \xrightarrow{H_2O}$$

Aldehyde — Lithium aluminum hydride

$$R-\underset{H}{\underset{|}{\overset{H}{\overset{|}{C}}}}-OH + Al^{3+} + Li^+ \qquad (8\text{-}31)$$

Primary alcohol

An example of reduction of a ketone with lithium aluminum hydride is the reduction of acetophenone to 1-phenyl-1-ethanol:

$$4C_6H_5\overset{O}{\overset{\|}{C}}-CH_3 + LiAlH_4 \xrightarrow{\text{ether}} (C_6H_5\underset{CH_3}{\underset{|}{\overset{H}{\overset{|}{C}}}}-O)_4AlLi \xrightarrow{H_2O}$$

Phenylethanone (Acetophenone) — Lithium aluminum hydride

$$4C_6H_5\underset{CH_3}{\underset{|}{C}}HOH + LiOH + Al(OH)_3 \qquad (8\text{-}32)$$

1-Phenyl-1-ethanol

The *Clemmensen reduction,* which employs amalgamated zinc and concentrated hydrochloric acid to reduce the carbonyl group to a methylene group, was mentioned in Chap. 5. The general reaction is

$$R-\overset{O}{\overset{\|}{C}}-R \xrightarrow{Zn(Hg),\ HCl} R-\underset{H}{\underset{|}{\overset{H}{\overset{|}{C}}}}-R \qquad (8\text{-}33)$$

An alternative to the Clemmensen reduction for an acid-sensitive ketone is the *Wolff-Kishner reduction,* which employs hydrazine (N_2H_4) and potassium hydroxide. The solvent is 1,2-ethanediol (ethylene glycol).

$$R-\overset{\overset{\displaystyle O}{\|}}{C}-R \xrightarrow{H-\overset{H}{\overset{|}{N}}-\overset{H}{\overset{|}{N}}-H,\ KOH,\ 150^\circ C} R-\overset{H}{\overset{|}{\underset{\underset{H}{|}}{C}}}-R \qquad (8\text{-}34)$$

An example of the Wolff-Kishner reduction is the following:

$$CH_3O-C_6H_4-\overset{\overset{\displaystyle O}{\|}}{C}-CH_3 \xrightarrow{NH_2NH_2,\ KOH,\ 150^\circ C} CH_3O-C_6H_4-CH_2CH_3 \qquad (8\text{-}35)$$

(4-Methoxyphenyl)ethanone (*p*-Methoxyacetophenone) — 1-Ethyl-4-methoxybenzene

Remember that catalytic hydrogenation or reaction with a hydride reduces an aldehyde or ketone to an alcohol. The Clemmensen reduction or the Wolff-Kishner reduction reduces the carbonyl group to a methylene ($-CH_2-$) group.

Addition reactions

The chief reaction of aldehydes and ketones is *nucleophilic addition* to the *partially positive carbon* of the carbonyl group. The general reactions are as follows:

$$\underset{\text{Aldehyde}}{R(H)C{=}O} + \underset{\text{Nucleophile}}{:Z^-} \rightleftharpoons R-\overset{O^-}{\overset{|}{\underset{\underset{H}{|}}{C}}}-Z \overset{H_2O}{\rightleftharpoons} R-\overset{OH}{\overset{|}{\underset{\underset{R}{|}}{C}}}-Z \qquad (8\text{-}36)$$

$$\underset{\text{Ketone}}{R_2C{=}O} + \underset{\text{Nucleophile}}{:Z^-} \rightleftharpoons R-\overset{O^-}{\overset{|}{\underset{\underset{R}{|}}{C}}}-Z \overset{H_2O}{\rightleftharpoons} R-\overset{OH}{\overset{|}{\underset{\underset{R}{|}}{C}}}-Z \qquad (8\text{-}37)$$

The nucleophile, $:Z^-$, can be ^-OH, ^-OR, ^-CN, NH_3, H_2O, ROH, NH_2OH, $C_6H_5NHNH_2$, RMgX, $HO\overset{\overset{O}{\|}}{S}ONa$, or $R\overset{-}{C}{=}\overset{\overset{O}{\|}}{C}-$.

Acid catalysis facilitates the reaction of the weaker nucleophiles, such as water, alcohol, and ammonia, by protonating and increasing the positive nature of the carbonyl group.

$$R_2C{=}\ddot{O}: + H^+ \rightleftharpoons R_2C{=}\overset{+}{O}H \longleftrightarrow R_2\overset{+}{C}{-}OH \xrightarrow{:ZH} R{-}\underset{R}{\overset{OH}{C}}{-}\overset{+}{Z}H \longrightarrow R{-}\underset{R}{\overset{OH}{C}}{-}Z + H^+ \qquad (8\text{-}38)$$

The general reaction shown in Eq. (8-38) is that for a ketone. For an aldehyde, the reaction is the same except that one of the R groups is simply a hydrogen atom.

Specific nucleophilic substitution reactions are discussed in the following sections.

Formation of cyanohydrins. Hydrogen cyanide adds to an aldehyde or a ketone to form a *cyanohydrin*. The value of a cyanohydrin is that it can be hydrolyzed to an α-hydroxycarboxylic acid by heating it with aqueous acid.

$$\underset{\text{Aldehyde or ketone}}{R_2C{=}O} + H{-}CN \longrightarrow \underset{\text{A cyanohydrin}}{R{-}\underset{R}{\overset{OH}{C}}{-}CN} \xrightarrow{H_2SO_4(aq),\ \Delta} \underset{\text{An }\alpha\text{-hydroxycarboxylic acid}}{R{-}\underset{R}{\overset{OH}{C}}{-}C(=O)OH} \qquad (8\text{-}39)$$

The reactions beginning with ethanal (acetaldehyde) produce 2-hydroxypropanoic acid, commonly known as lactic acid.

$$\underset{\substack{\text{Ethanal}\\\text{(Acetaldehyde)}}}{H_3C(H)C{=}O} + HCN \longrightarrow \underset{\substack{\text{2-Hydroxypropanenitrile}\\\text{(Acetaldehyde cyanohydrin)}}}{CH_3{-}\overset{OH}{CH}{-}CN} \xrightarrow{H_2SO_4(aq)} \underset{\substack{\alpha\text{-Hydroxypropanoic acid}\\\text{(Lactic acid)}}}{CH_3\overset{OH}{CH}{-}C(=O)OH} \qquad (8\text{-}40)$$

Lactic acid occurs in sour milk.

Addition of water. The possibility exists for Lewis base–type molecules with unshared pairs of electrons, such as water, alcohol, and ammonia, to serve as nucleophiles in addition reactions to carbonyl compounds. Although water may add to a slight extent, the reaction is reversible and the equilibrium is unfavorable except in a few cases. It may be recalled that the *gem-diols* which would be produced are generally unstable.

$$\underset{R'}{\overset{R}{\diagdown}}C{=}O + HOH \rightleftharpoons R-\underset{R}{\overset{O^-}{\overset{|}{C}}}-\overset{H}{\overset{|}{\underset{+}{O}}}H \rightleftharpoons R-\underset{R}{\overset{OH}{\overset{|}{C}}}-OH \tag{8-41}$$

A gem diol

The reaction is favorable for the following case.

$$\underset{H}{\overset{Cl_3C}{\diagdown}}C{=}O + HOH \rightleftharpoons Cl_3C-\underset{H}{\overset{OH}{\overset{|}{C}}}-OH \tag{8-42}$$

Trichloroethanal (Chloral) — 2,2,2-Trichloro-1,1-ethanediol (Chloral hydrate)

Chloral hydrate is a *soporific* commonly known as "knock-out drops." First prepared by Liebig in 1832 and introduced into medicine in 1869, it still is widely used as a sleep producer.

Another case in which the equilibrium is favorable for the reaction is

$$\text{Cyclopropanone} + HOH \rightleftharpoons \text{1,1-Cyclopropanediol} \tag{8-43}$$

Cyclopropanone — 1,1-Cyclopropanediol

The electron withdrawal (negative inductive effect) by the chlorine stabilizes the product in the case of chloral. In the case of cyclopropanone, addition of water changes the hybridization of the carbonyl carbon atom from sp^2 (normal angle 120°) to sp^3 (normal angle 109.5°) and thereby relieves some of the bond angle strain in the ring and stabilizes the diol relative to the ketone.

Addition of alcohols. In the presence of dry hydrogen chloride as a catalyst, alcohols will react with aldehydes and to some extent with the less reactive ketones.

$$R-\overset{H}{\overset{|}{C}}{=}O + ROH \xrightleftharpoons{HCl(g)} R-\underset{OH}{\overset{H}{\overset{|}{C}}}-OR \xrightleftharpoons{HCl(g),\ ROH} R-\underset{OR}{\overset{H}{\overset{|}{C}}}-OR + H_2O \tag{8-44}$$

Aldehyde — Alcohol — Hemiacetal — Acetal

The similar final product obtained from a ketone is called a *ketal*.

$$CH_3-\underset{CH_3}{\overset{}{C}}{=}O + \begin{matrix} HO-CH \\ HO-CH \end{matrix} \xrightarrow{HCl(g)} \underset{H_3C}{\overset{H_3C}{\diagdown}}C\begin{matrix} O-CH \\ O-CH \end{matrix} \tag{8-45}$$

Acetone, a ketone — A diol — A ketal of acetone

The mechanism involves protonation of the carbonyl oxygen atom followed by nucleophilic addition of an alcohol to the carbonyl carbon atom. A loss of a proton at this point produces a *hemiacetal*. A shift of a proton, rather than its loss, enables a molecule of water to be lost. That is followed by nucleophilic addition of a second mole of alcohol and loss of a proton to produce the *acetal* (or a *ketal* when a ketone is used).

$$\mathrm{R(R)C{=}O + HCl}(g) \rightleftharpoons \mathrm{R(H)C{=}\overset{+}{O}HCl^-} \xrightleftharpoons{\mathrm{ROH}} \mathrm{R{-}C(OH)(H){-}\overset{+}{O}(H)R} \xrightleftharpoons{-\mathrm{H}^+} \underset{\text{A hemiacetal}}{\mathrm{R{-}C(OH)(H){-}OR}}$$

$$\mathrm{R{-}C(H)(\overset{+}{O}H_2){-}OR} \xrightleftharpoons{-\mathrm{H_2O}} \mathrm{R{-}C(H){=}\overset{+}{O}R} \xrightleftharpoons{\mathrm{ROH}} \mathrm{R{-}C(H)(\overset{+}{O}HR){-}OR} \xrightleftharpoons{-\mathrm{H}^+} \underset{\text{An acetal}}{\mathrm{R{-}C(H)(OR){-}OR}} \qquad (8\text{-}46)$$

The intermediate *hemiacetal* resulting from addition of one mole of the alcohol is not stable for simple aldehydes, but the reaction goes on to eliminate water and add another molecule of the alcohol. The result is a special type of ether, a *gem-dialkoxyalkane* called an *acetal*. An example is the reaction of ethanal and ethanol:

$$\underset{\substack{\text{Ethanal}\\ \text{(Acetaldehyde)}}}{\mathrm{CH_3C(H){=}O}} + \underset{\substack{\text{Ethanol}\\ \text{(Ethyl alcohol)}}}{\mathrm{2CH_3CH_2OH}} \xrightarrow{\mathrm{HCl}(g)} \underset{\substack{\text{1,1-Diethoxyethane}\\ \text{(Diethyl acetal of acetaldehyde)}}}{\mathrm{CH_3C(H)(OCH_2CH_3){-}OCH_2CH_3}} \qquad (8\text{-}47)$$

Like other ethers, acetals are good solvents. They are stable to bases and oxidizing agents, but they are cleaved by acids to regenerate the aldehyde and alcohol. Acetals are more easily cleaved than regular ethers; they are cleaved even by dilute acids. The mechanism is just the reverse of that for the formation of the acetal.

$$\underset{\text{1,1-Diethoxyethane}}{\mathrm{CH_3C(H)(O{-}CH_2CH_3){-}OCH_2CH_3}} + \mathrm{H_2O} \xrightarrow{\mathrm{HCl}} \underset{\text{Ethanal}}{\mathrm{CH_3C(H){=}O}} + \underset{\text{Ethanol}}{\mathrm{2CH_3CH_2OH}} \qquad (8\text{-}48)$$

Acetal formation is sometimes used to protect a carbonyl group from reacting with a strong reagent; the acetal is then hydrolyzed back to the aldehyde. The hemiacetal and acetal groups are of much importance in carbohydrate chemistry. A plastic laminate for safety glass is an acetal of butyraldehyde and polyvinyl alcohol.

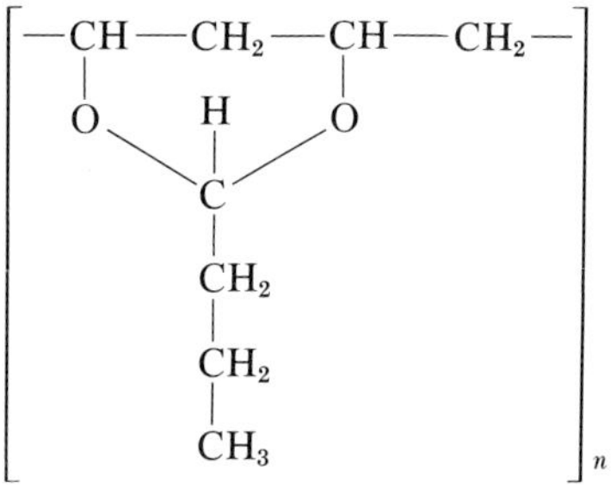

$$\left[\begin{array}{c} -CH-CH_2-CH-CH_2- \\ \quad | \qquad\qquad | \\ \quad O \quad H \quad O \\ \qquad \diagdown \; | \; \diagup \\ C \\ | \\ CH_2 \\ | \\ CH_2 \\ | \\ CH_3 \end{array} \right]_n$$

This polymer is called polyvinylbutyral.

Addition of sodium hydrogen sulfite. Aldehydes and methyl ketones will add sodium hydrogen sulfite to form a water-soluble, crystalline hydrogen sulfite addition compound. Ketones with *both* alkyl groups larger than methyl groups will not react because the space about the carbonyl group becomes too crowded to permit the bulky bisulfite ion in to react. That is another example of an effect due to spatial crowding.

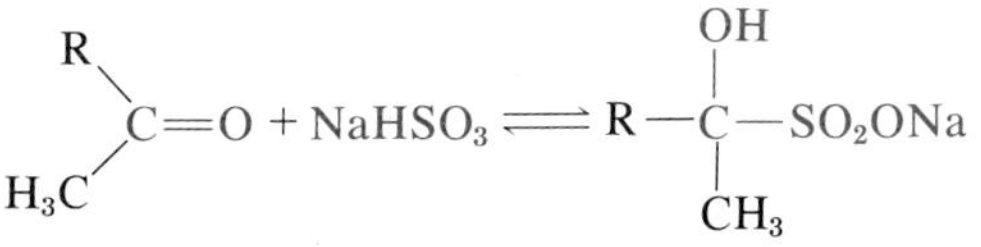

$$\begin{array}{c} R \\ \diagdown \\ C{=}O \\ \diagup \\ H_3C \end{array} + NaHSO_3 \rightleftharpoons R-\overset{\displaystyle OH}{\underset{\displaystyle CH_3}{\overset{|}{\underset{|}{C}}}}-SO_2ONa \tag{8-49}$$

The mechanism involves nucleophilic addition of the sulfur atom to the carbonyl carbon atom.

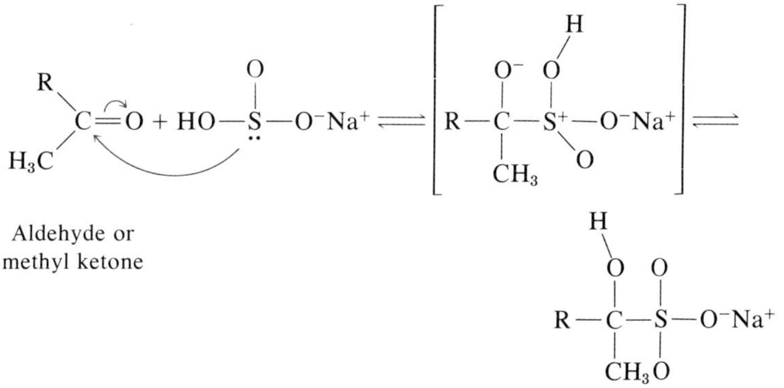

$$\begin{array}{c} R \\ \diagdown \\ C{=}O \\ \diagup \\ H_3C \end{array} + HO-\overset{\displaystyle O}{\overset{|}{\underset{\cdot\cdot}{S}}}-O^-Na^+ \rightleftharpoons \left[R-\overset{\displaystyle O^-}{\underset{\displaystyle CH_3}{\overset{|}{\underset{|}{C}}}}-\overset{\displaystyle \overset{H}{/} \atop O}{\overset{|}{S^+}}\!\!\underset{\diagdown O}{}-O^-Na^+ \right] \rightleftharpoons$$

Aldehyde or methyl ketone

$$R-\overset{\displaystyle \overset{H}{\diagdown} \atop O}{\underset{\displaystyle CH_3}{\overset{|}{\underset{|}{C}}}}-\overset{\displaystyle O}{\underset{\displaystyle O}{\overset{|}{\underset{|}{S}}}}-O^-Na^+ \tag{8-50}$$

Sodium hydrogen sulfite addition compound

The reaction may be reversed by the addition of either acid or base to destroy the sodium hydrogen sulfite and liberate the aldehyde.

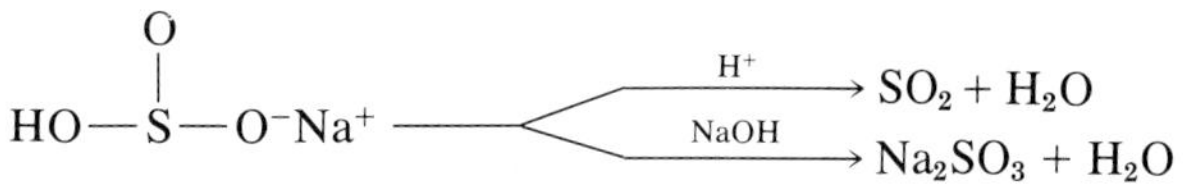

$$HO-\overset{\displaystyle O}{\overset{|}{S}}-O^-Na^+ \;\text{—}\!\!\!< \begin{array}{l} \xrightarrow{H^+} SO_2 + H_2O \\ \xrightarrow{NaOH} Na_2SO_3 + H_2O \end{array} \tag{8-51}$$

Addition of sodium hydrogen sulfite is employed in the purification of an aldehyde. Converted to the water-soluble hydrogen sulfite addition compound, the aldehyde can be extracted from ether-soluble organic contaminants. Then it can be regenerated by addition of either acid or base.

Addition of ammonia and derivatives of ammonia. Like the addition of water to aldehydes and ketones, the addition of ammonia itself is reversible with an unfavorable equilibrium and is not generally of much significance. The product, $R_2C{=}NH$, is called an *imine*.

$$R_2C{=}O + NH_3 \rightleftharpoons \left[R{-}\underset{R}{\overset{O^-}{C}}{-}\overset{+}{N}H_3 \right] \rightleftharpoons R{-}\underset{R}{\overset{OH}{C}}{-}NH_2 \rightleftharpoons \underset{\text{An imine}}{R{-}\underset{R}{C}{=}NH} + H_2O \qquad (8\text{-}52)$$

However, the addition of certain derivatives of ammonia is successful and is of interest. The solid derivatives produced may aid in the identification of an unknown aldehyde or ketone. An example is the reaction of propanone with hydroxylamine (NH_2OH) and an acid catalyst:

$$\underset{\substack{\text{Propanone}\\\text{(Acetone)}}}{(H_3C)_2C{=}O} + \underset{\substack{\text{Hydroxylamine,}\\\text{acid catalyst}}}{H^+{:}\underset{H}{\overset{H}{N}}{-}OH} \rightleftharpoons \left[CH_3{-}\underset{CH_3}{\overset{HO}{C}}{-}\underset{H}{\overset{H}{\overset{+}{N}}}{-}OH \right] \xrightarrow{-H_3O^+} \underset{\substack{\text{2-Propanone oxime, mp, 59°C}\\\text{(Acetone oxime)}}}{CH_3{-}\underset{CH_3}{C}{=}\ddot{N}{-}OH} \qquad (8\text{-}53)$$

Other ammonia derivatives which can be used are phenylhydrazine and semicarbazide. Note that these reactions are usually acid-catalyzed.

$$\underset{\substack{\text{Ethanal}\\\text{(Acetaldehyde)}}}{H_3C(H)C{=}O} + \underset{\text{Phenylhydrazine}}{H{-}\overset{H}{N}{-}\overset{H}{N}{-}C_6H_5} \xrightarrow{H^+} \left[CH_3{-}\underset{H}{\overset{OH}{C}}{-}\underset{H}{\overset{H}{N}}{-}\overset{H}{N}{-}C_6H_5 \right]^+ \xrightarrow{-H_3O^+} \underset{\substack{\text{Acetaldehyde phenyl-}\\\text{hydrazone, mp, 63°C}}}{CH_3{-}\underset{H}{C}{=}N{-}\overset{H}{N}{-}C_6H_5} \qquad (8\text{-}54)$$

$$(CH_3CH_2)_2C{=}O + H_2N{-}NH{-}\overset{O}{\overset{\|}{C}}{-}NH_2 \xrightarrow{H^+}$$

2-Pentanone (Diethyl ketone) — Semicarbazide

$$\left[CH_3CH_2\overset{OH}{\underset{CH_2CH_3}{C}}{-}NH_2^{+}{-}NH{-}\overset{O}{\overset{\|}{C}}NH_2\right]^+ \xrightarrow{-H_3O^+} CH_3CH_2\underset{CH_2CH_3}{C}{=}N{-}NH{-}\overset{O}{\overset{\|}{C}}{-}NH_2 \qquad (8\text{-}55)$$

Diethyl ketone semicarbazone, mp, 139°C

The reagent 2,4-dinitrophenylhydrazine is often used in place of phenylhydrazine because it gives higher-melting derivatives.

Addition of Grignard reagents. The general reaction for the addition of Grignard reagents to aldehydes and ketones is

$$RMgX + \gt C{=}O \longrightarrow R{-}\overset{|}{\underset{|}{C}}{-}OMgX \xrightarrow{H_2O} R{-}\overset{|}{\underset{|}{C}}{-}OH \qquad (8\text{-}56)$$

This reaction was discussed in Chap. 7 as a preparation of alcohols.

Addition of carbanions. The very polar carbonyl group increases the acidity of the C—H bond on an adjacent or *α-carbon* atom. Removal of the so-called α-hydrogen atom as a proton by a base results in a *resonance-stabilized anion.* The anion may then serve as a nucleophilic reagent in attack on the carbonyl group of another molecule of a carbonyl compound.

$$R{-}\overset{H}{\underset{|}{C}}{-}\overset{O}{\overset{\|}{C}}{-}R' + :B^- \longrightarrow$$

Aldehyde or ketone with one or more α-hydrogen atoms — Base

$$\left\{ HB + R{-}\overset{\cdot\cdot -}{\underset{|}{C}}{-}\overset{O}{\overset{\|}{C}}{-}R' \longleftrightarrow R{-}\underset{|}{C}{=}\overset{O^-}{\overset{|}{C}}{-}R' \right\} \; R{-}\overset{-}{\underset{|}{C}}{-}\overset{O}{\overset{\|}{C}}{-}R' \xrightarrow{\gt C=O} {-}\overset{O^-}{\overset{|}{\underset{|}{C}}}{-}\overset{R}{\overset{|}{\underset{|}{C}}}{-}\overset{O}{\overset{\|}{C}}{-}R' \qquad (8\text{-}57)$$

Resonance-stabilized carbanion

Applied to ethanal (acetaldehyde), the equations would be

$$CH_3-\overset{\overset{\displaystyle O}{\|}}{C}-H + {}^-OH \longrightarrow H-\underset{\underset{\displaystyle H}{|}}{\overset{-}{C}}=\overset{\overset{\displaystyle O}{\|}}{C}-H + HOH \tag{8-58}$$

$$CH_3-\overset{\overset{\displaystyle O}{\|}}{C}-H + H-\underset{\underset{\displaystyle H}{|}}{\overset{-}{C}}=\overset{\overset{\displaystyle O}{\|}}{C}-H \longrightarrow CH_3-\underset{\underset{\displaystyle H}{|}}{\overset{\overset{\displaystyle O^-}{|}}{C}}-CH_2-\overset{\overset{\displaystyle O}{\|}}{C}-H \xrightarrow{HOH}$$

Ethanal
(Acetaldehyde)

$$CH_3-\underset{\underset{\displaystyle H}{|}}{\overset{\overset{\displaystyle OH}{|}}{C}}-CH_2-\overset{\overset{\displaystyle O}{\|}}{C}-H + {}^-OH \tag{8-59}$$

3-Hydroxybutanal
(Aldol)

A model of 3-hydroxybutanal is shown in Fig. 8-8. Because the product resulting from the reaction of Eq. (8-59) has both the alcohol and the aldehyde functional groups, the name *aldol* was given to both the particular compound and compounds of the same class. The reaction is referred to as the *aldol condensation*. Other examples of the reaction are

$$CH_3CH_2\overset{\overset{\displaystyle O}{\|}}{C}-H + CH_3CH_2\overset{\overset{\displaystyle O}{\|}}{C}-H \xrightarrow{NaOH} CH_3CH_2\underset{\underset{\displaystyle H}{|}}{\overset{\overset{\displaystyle OH}{|}}{C}}-\underset{\underset{\displaystyle H}{|}}{\overset{\overset{\displaystyle CH_3}{|}}{C}}-C\begin{matrix}\nearrow O\\ \searrow H\end{matrix} \tag{8-60}$$

Propanal
(Propionaldehyde)

3-Hydroxy-2-methyl-propanal

$$CH_3-\underset{\underset{\displaystyle CH_3}{|}}{\overset{\overset{\displaystyle O}{\|}}{C}} + CH_3-\overset{\overset{\displaystyle O}{\|}}{C}-CH_3 \xrightarrow{Ba(OH)_2} CH_3-\underset{\underset{\displaystyle CH_3}{|}}{\overset{\overset{\displaystyle OH}{|}}{\overset{\beta}{C}}}-\overset{\alpha}{C}H_2\overset{\overset{\displaystyle O}{\|}}{C}-CH_3 \tag{8-61}$$

Propanone
(Acetone)

4-Hydroxy-4-methyl-2-pentanone
(Diacetone alcohol)

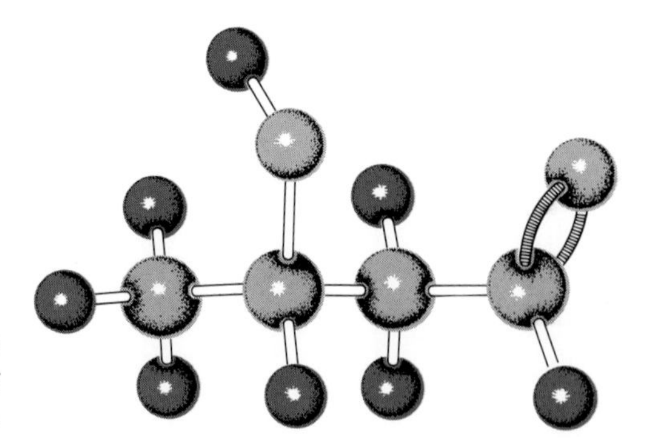

FIGURE 8-8
Ball-and-spring model of 3-hydroxybutanal (aldol).

Observe that, in these products of the aldol condensation, the hydroxy group is always on the *second* or *beta* (β)-carbon atom from the carbonyl group. (Beta is the second letter in the Greek alphabet.) The product of the aldol reaction is a *β-hydroxyaldehyde or ketone.*

$$C_6H_5\text{-}\overset{H}{\overset{|}{C}}{=}O + CH_3\overset{O}{\overset{\|}{C}}{-}H \xrightarrow{NaOH} \left[C_6H_5\text{-}\overset{OH}{\overset{|}{\underset{H}{\underset{|}{C}}}}{-}CH_2\overset{H}{\overset{|}{C}}{=}O \right] \xrightarrow{-H_2O} C_6H_5\text{-}CH{=}CH{-}\overset{H}{\overset{|}{C}}{=}O \quad (8\text{-}62)$$

Benzenecarbaldehyde (Benzaldehyde) — Ethanal (Acetaldehyde) — 3-Phenylpropenal (Cinnamaldehyde)

The third example is called a *crossed aldol condensation.* Note that the aldehyde or ketone furnishing the carbanion must have an alpha hydrogen, but the aldehyde or ketone it reacts with need not have one. Some aldol could also be expected from the reaction of the carbanion with acetaldehyde instead of benzaldehyde. Note also that if dehydration of the β-hydroxyaldehyde or ketone results in a carbon-carbon double bond conjugated with the ring, it occurs spontaneously.

The products of other aldol reactions may be readily dehydrated by heating with a little acid or iodine.

$$CH_3{-}\overset{OH}{\overset{|}{\underset{H}{\underset{|}{C}}}}{-}CH_2\overset{H}{\overset{|}{C}}{=}O \xrightarrow{H^+ \text{ or } I_2,\ \Delta} CH_3CH{=}CH{-}\overset{H}{\overset{|}{C}}{=}O + H_2O \quad (8\text{-}63)$$

3-Hydroxybutanal (Aldol) — 2-Butenal (Crotonaldehyde)

These are important reactions for synthesis of a wide variety of compounds. The active ingredient of some insect repellants, 2-ethyl-1,3-hexanediol, is prepared by such reactions.

$$2CH_3CH_2CH_2\overset{H}{\overset{|}{C}}{=}O \xrightarrow{NaOH} CH_3CH_2CH_2\overset{OH}{\overset{|}{C}H}{-}\underset{\underset{CH_3}{\underset{|}{CH_2}}}{\underset{|}{CH}}\overset{H}{\overset{|}{C}}{=}O \xrightarrow{H_2/Ni} CH_3CH_2CH_2\overset{OH}{\overset{|}{C}H}\underset{\underset{CH_3}{\underset{|}{CH_2}}}{\underset{|}{CH}}CH_2OH \quad (8\text{-}64)$$

Butanal (*n*-Butyraldehyde) — 2-Ethyl-1,3-hexanediol

Cannizzaro reaction. In the presence of concentrated alkali, an aldehyde without an α hydrogen (which cannot undergo the aldol condensation) will undergo an oxidation-reduction reaction, called the *Cannizzaro reaction*. In that reaction, one mole of aldehyde is reduced to the corresponding alcohol while another mole of aldehyde is oxidized to the corresponding carboxylic acid. Since the first step involves nucleophilic addition of the hydroxide ion to the carbonyl group, the reaction has been grouped with the addition reactions. The following mechanism has been suggested:

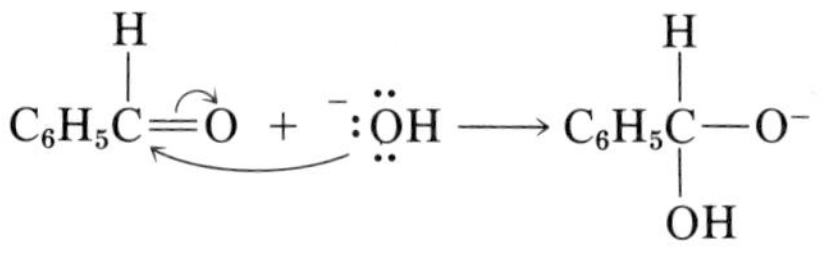

$$C_6H_5CH{=}O + {}^{-}{:}\ddot{O}H \longrightarrow C_6H_5CH(OH){-}O^- \tag{8-65}$$

$$C_6H_5CH(OH){-}O^- + C_6H_5CH{=}O \longrightarrow C_6H_5C(=O){-}O^- + C_6H_5CH_2{-}O^- \xrightarrow{H_2O} C_6H_5CH_2OH + OH^- \tag{8-66}$$

(HO:⁻ removes H from the OH group)

Hydride shift

The overall reaction for benzaldehyde is written as follows:

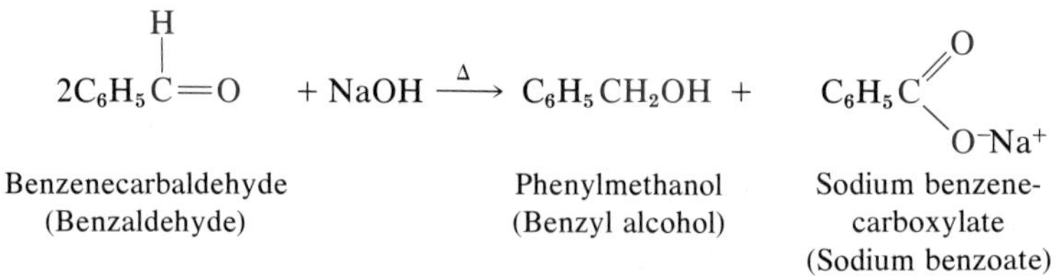

$$2C_6H_5CH{=}O + NaOH \xrightarrow{\Delta} C_6H_5CH_2OH + C_6H_5C(=O)O^-Na^+ \tag{8-67}$$

Benzenecarbaldehyde (Benzaldehyde) — Phenylmethanol (Benzyl alcohol) — Sodium benzenecarboxylate (Sodium benzoate)

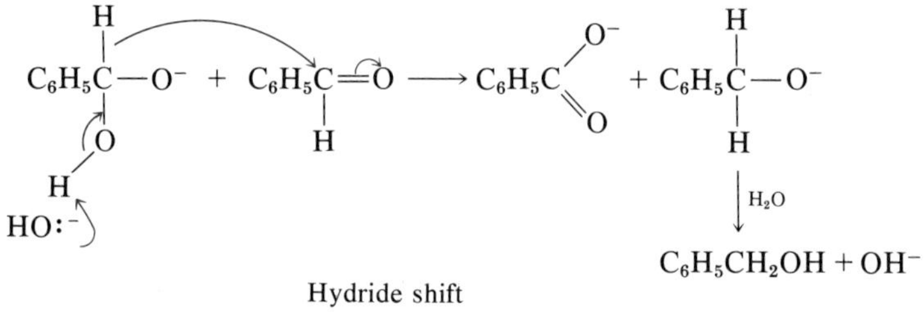

A *crossed Cannizzaro reaction* also is possible; in it the 2 moles of aldehyde which react are different. Methanal (formaldehyde) is the most easily oxidized.

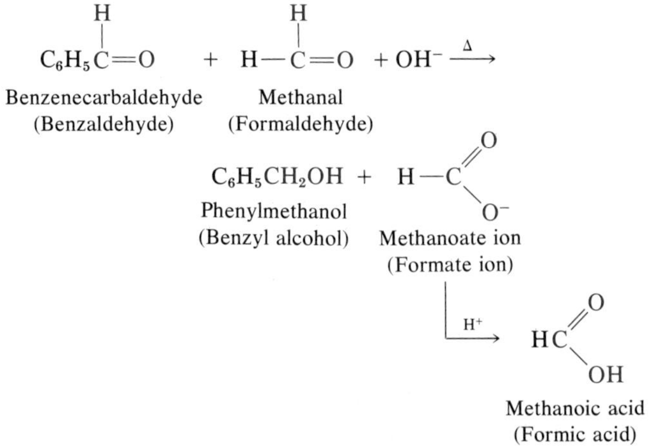

$$C_6H_5CH{=}O + H{-}CH{=}O + OH^- \xrightarrow{\Delta}$$

Benzenecarbaldehyde (Benzaldehyde) — Methanal (Formaldehyde)

$$C_6H_5CH_2OH + H{-}C(=O){-}O^-$$

Phenylmethanol (Benzyl alcohol) — Methanoate ion (Formate ion)

$$\xrightarrow{H^+} HC(=O)OH \tag{8-68}$$

Methanoic acid (Formic acid)

PROBLEM 8-5 Write equations for the following reactions.

(a) 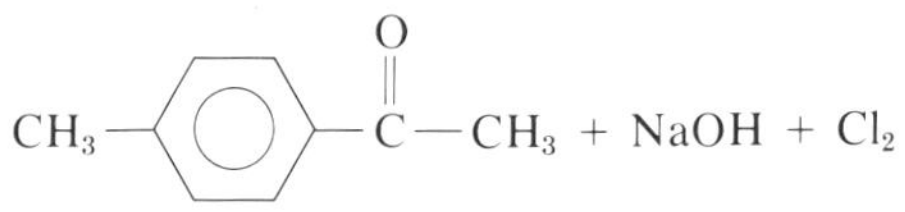 $CH_3-C_6H_4-\overset{O}{\overset{\|}{C}}-CH_3 + NaOH + Cl_2$

(b) $CH_2CH{=}CH-\overset{H}{\overset{|}{C}}{=}O + H_2/Ni$

(c) $CH_3CH{=}CH-\overset{H}{\overset{|}{C}}{=}O + LiAlH_4$, then H_2O

(d) $CH_3\overset{O}{\overset{\|}{C}}-CH_3 + HCN$

(e) $CH_3\overset{H}{\overset{|}{C}}{=}O + 2CH_3CH_2OH \xrightarrow{H^+}$

(f) $CH_3\overset{O}{\overset{\|}{C}}-CH_3 + NaHSO_3$

(g) $C_6H_5\overset{O}{\overset{\|}{C}}-CH_2CH_2COOH + Zn(Hg), HCl$

Solution to (a)

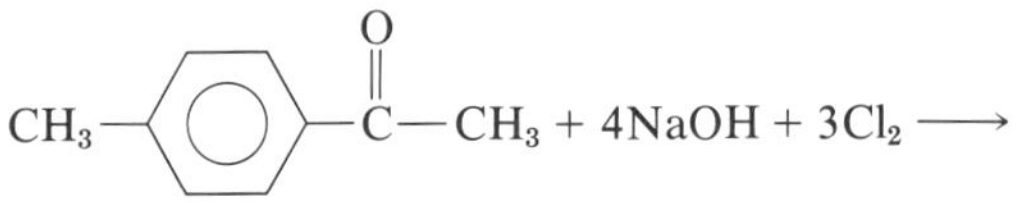

$$CH_3-C_6H_4-\overset{O}{\overset{\|}{C}}-CH_3 + 4NaOH + 3Cl_2 \longrightarrow$$

(4-Methylphenyl)ethanone
(*p*-Methylacetophenone)

$$CH_3-C_6H_4-\overset{O}{\overset{\|}{C}}-ONa \quad + \quad CHCl_3 \quad + 3H_2O + 3NaCl$$

Sodium 4-methylbenzenecarboxylate (Sodium *p*-methylbenzoate) — Trichloromethane (Chloroform)

This is an example of the haloform reaction.

8-6 Analysis of aldehydes and ketones

The presence of a carbonyl group is indicated by a characteristic and strong carbon-oxygen stretching frequency at about 1700 cm^{-1} in the infrared absorption spectrum (Fig. 8-9). The presence of the carbonyl group of an aldehyde or ketone also is indicated by reaction with 2,4-dinitrophenylhydrazine to form an orange-red crystalline derivative. Aldehydes may then be distinguished from ketones by the Tollens' reaction, since aldehydes reduce the silver ion to form a silver mirror and ketones do not. The iodoform reaction may be used to test for acetaldehyde or a methyl ketone. Suitable derivatives may then be prepared for final identification. See Fig. 8-10.

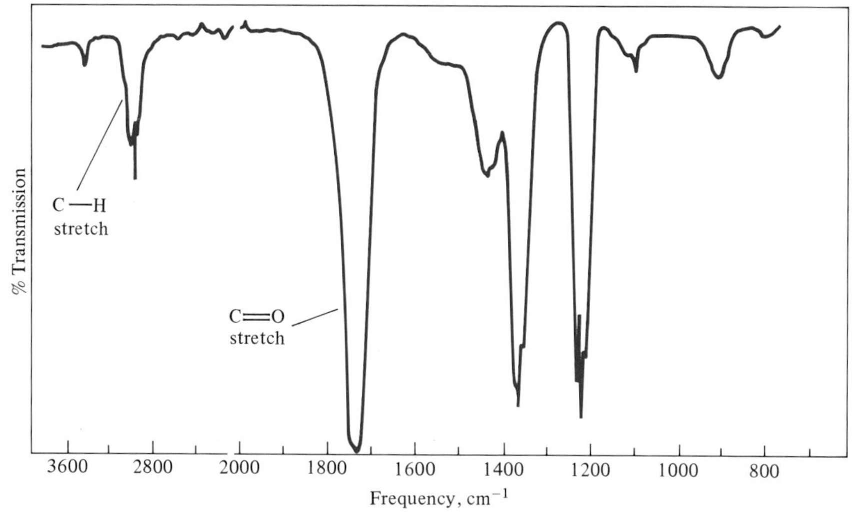

C—H stretch

C═O stretch

% Transmission

3600 2800 2000 1800 1600 1400 1200 1000 800

Frequency, cm^{-1}

FIGURE 8-9 The infrared spectrum of propanone (acetone), $CH_3—\overset{\overset{\displaystyle O}{\|}}{C}—CH_3$.

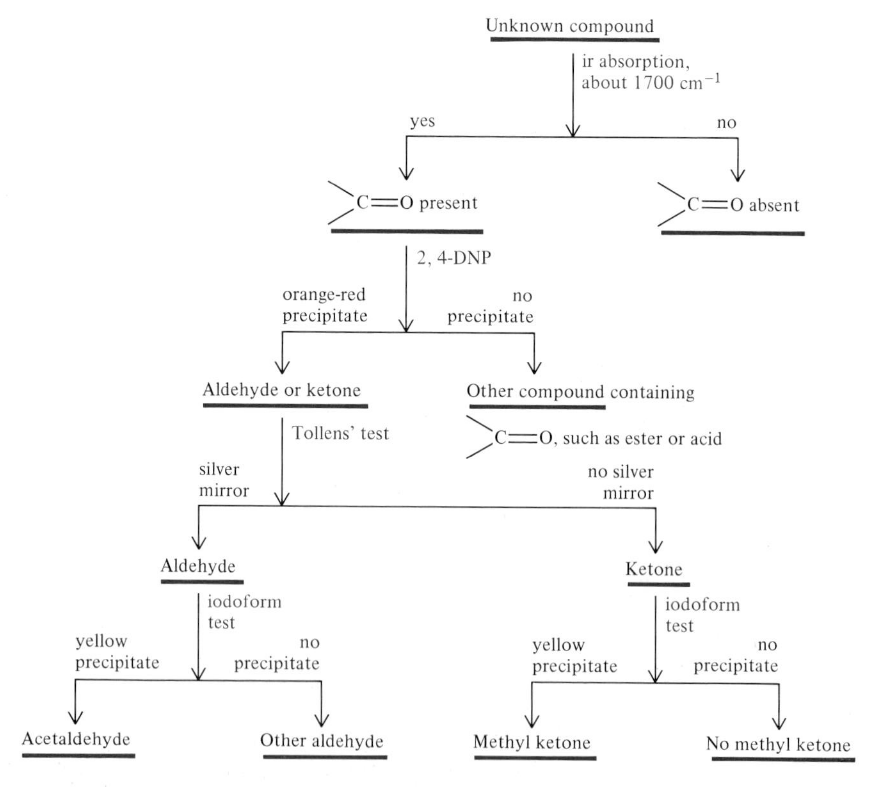

FIGURE 8-10 Analysis scheme for aldehydes and ketones.

PROBLEM 8-6 Indicate a *simple chemical test* or reagent which will *readily* distinguish between the members of each of the following pairs of compounds. Indicate which member of the pair gives the positive test or greater reaction.

(a) $C_6H_5C(=O)—CH_2CH_3$ and $C_6H_5CH_2C(=O)—CH_3$

(b) $CH_3CH_2CH_2C(H)=O$ and $CH_3CH=CHC(H)=O$

(c) 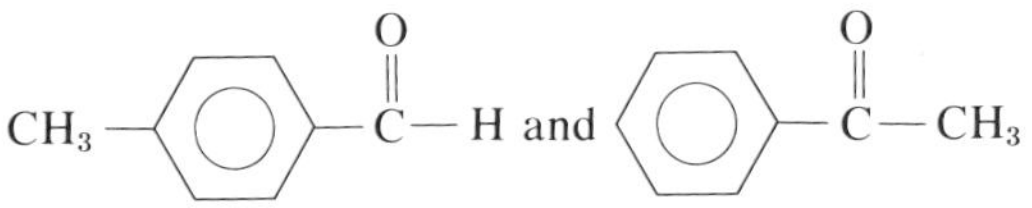 $CH_3—C_6H_4—C(=O)—H$ and $C_6H_5—C(=O)—CH_3$

(d) $CH_3CH_2OCH_2CH_3$ and $CH_3CH_2CH_2C(H)=O$

(e) $CH_3O—CH_2—O—CH_2CH_3$ and $CH_3O—CH_2CH_2—O—CH_3$

Solution to (a) The iodoform test can be used to distinguish 1-phenyl-1-propanone (ethyl phenyl ketone) from 1-phenyl-2-propanone (benzyl methyl ketone). Only *methyl* ketones and ethanal (acetaldehyde) will give the yellow precipitate of iodoform. The reaction of 1-phenyl-2-propanone is

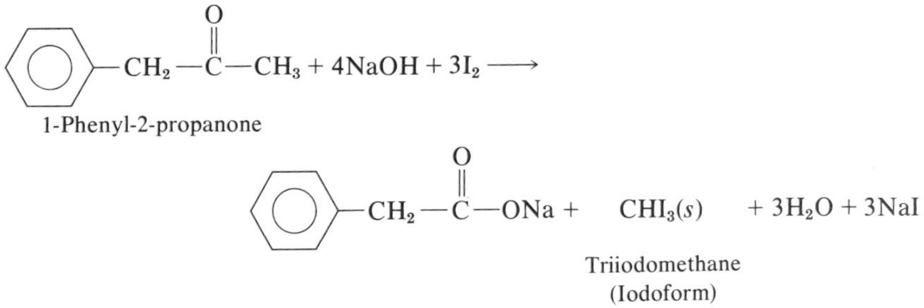

$$C_6H_5—CH_2—C(=O)—CH_3 + 4NaOH + 3I_2 \longrightarrow$$

1-Phenyl-2-propanone

$$C_6H_5—CH_2—C(=O)—ONa + CHI_3(s) + 3H_2O + 3NaI$$

Triiodomethane (Iodoform)

Summary

An *aldehyde* has *one alkyl* or *aryl group* and *one hydrogen atom* attached to a *carbonyl group*, $R(R')C=O$ or $Ar(R)C=O$. A *ketone* has *two alkyl* or *aryl groups* attached to a carbonyl group.

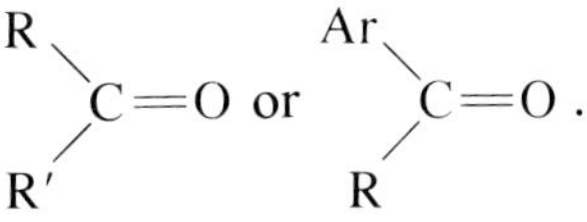

The *carbonyl group* of an aldehyde or ketone has on its carbon atom sp^2 *orbitals* forming *sigma bonds* with three other atoms in a *planar trigonal geometry* with bond angles of approximately 120°. A *pi bond* is then formed by lateral overlap of a *p* orbital on the carbonyl carbon atom with a *p* orbital on the oxygen atom. The *carbon-oxygen double bond* differs from the carbon-carbon double bond in that it is *highly polarized* because of the electronegativity of the oxygen atom. The polarization of the carbonyl group leaves the carbon atom with a partial positive charge and therefore subject to attack by a nucleophilic reagent.

The order of *decreasing reactivity* of aldehydes and ketones is aliphatic aldehydes > aromatic aldehydes > methyl ketones > ketones. The reasons are both *steric* and *electronic*.

Aldehydes and ketones are unable to hydrogen bond to their own molecules, but *hydrogen donors* like water and alcohols can *hydrogen bond* to the oxygen of the carbonyl group. That results in aldehydes and ketones being more *soluble in water* than hydrocarbons of similar molecular weight. Because of the relatively large *dipole moment* of the *carbonyl group,* aldehydes and ketones have a relatively strong dipole-dipole attraction, which gives them *higher boiling points* than hydrocarbons and ethers of similar molecular weight.

Most aldehydes and some ketones have fragrant, pleasant odors and are used in perfumery. However, the lower-molecular-weight aldehydes have disagreeable, unpleasant odors.

The trivial or common name for an aldehyde is derived from the common name of the corresponding carboxylic acid. The trivial or common name for a ketone is formed by naming the groups attached to the carbonyl group and adding the word *ketone*. The *IUPAC name* takes the stem for the name from the *parent alkane* and adds the suffix *-al* for an *aldehyde* or *-one* for a *ketone*. The aldehyde group will generally be assumed to be in the number 1 position, but the location of the ketone group is indicated by a locator number.

Oxidation or *dehydrogenation* of *primary alcohols* yields *aldehydes;* further oxidation produces *carboxylic acids*. Oxidation or dehydrogenation of *secondary alcohols* yields *ketones*. *Aromatic aldehydes* may be prepared by the partial *oxidation* of *arenes*. *Aliphatic aromatic ketones* can be synthesized by using the *Friedel-Crafts acylation* reaction. The *pyrolysis* of *metal salts* of acids may be used to produce *ketones*.

Aldehydes are *readily oxidized* by all ordinary oxidizing agents. *Ketones* are *not* readily oxidized, so that difference can be used to distinguish an aldehyde from its isomeric ketone. An oxidizing agent such as Ag^+ (*Tollens'* reagent) or Cu^{2+} (*Fehling's* or *Benedict's* solutions) is used.

The *haloform reaction* uses halogen and sodium hydroxide to produce haloform, HCX_3, and the sodium salt of a carboxylic acid from a *methyl ketone* or ethanal (acetaldehyde) or from alcohols which produce those compounds upon oxidation.

Reduction of *aldehydes* gives *primary alcohols,* and reduction of *ketones* gives *secondary alcohols*. The aldehydes may be reduced by a hydride or by using molecular hydrogen and a catalyst. Reaction of a carbonyl group with amalgamated zinc and concentrated hydrochloric acid, called the *Clemmensen reduction,* reduces the *carbonyl group* to a *methylene group,* $—CH_2—$.

The chief reaction of aldehydes and ketones is *nucleophilic addition* to the *carbonyl group*. The nucleophile can be —CN, H_2O, ROH, NH_3,

NH_2OH, $C_6H_5NHNH_2$, $NaHSO_3$, RMgX, or a resonance-stabilized carbanion. The respective products are *cyanohydrins, gem-diols, hemiacetals* and *acetals, imines, oximes, phenylhydrazones, hydrogen sulfite addition compounds, alcohols,* and *β-hydroxyaldehydes* or *ketones*. In forming oximes and phenylhydrazones, the product of nucleophilic addition of ammonia or its derivatives to the carbonyl group spontaneously dehydrates to form a *carbon-nitrogen double bond.*

In the presence of concentrated alkali, an aldehyde without α hydrogen undergoes an *oxidation-reduction reaction* in which one mole of aldehyde is oxidized to a carboxylate anion and another mole of aldehyde is reduced to a primary alcohol. That is called the *Cannizzaro reaction.*

The presence of a carbonyl group is indicated by a strong absorption band at about 1700 cm^{-1} in the infrared spectrum. The presence of a carbonyl group in an aldehyde or ketone is indicated by reaction with 2,4-dinitrophenylhydrazine to form an orange-red crystalline derivative. Aldehydes may be distinguished from ketones by reaction with Tollens' reagent; aldehydes reduce the *reagent* to free silver and ketones do not. The *iodoform test* may be used to test for ethanal (acetaldehyde) or a methyl ketone.

Key terms

Acetal: a special type of ether, a *gem*-dialkoxyalkane, of the general formula $RCH(OR)_2$. It is produced by the addition of 2 mol of alcohol to 1 mol of aldehyde.

Aldehyde: a compound in which one alkyl or one aryl group and one hydrogen atom are attached to the carbonyl group; the general formulas are

$$\underset{H}{\overset{R}{>}}C{=}O \quad \text{and} \quad \underset{H}{\overset{Ar}{>}}C{=}O.$$

Aldol: a β-hydroxy aldehyde or ketone; also, the compound 3-hydroxybutanal.

Aldol condensation: the condensation of 2 mol of aldehyde or ketone, in the presence of a base, to produce a β-hydroxy aldehyde or ketone. The reacting aldehyde or ketone must have an α-hydrogen atom.

α-Carbon atom: in this chapter, a carbon atom adjacent to the carbonyl carbon.

Benedict's solution: a solution of a complex ion of copper(II) with citrate. The Cu^{2+} ion is reduced to Cu^{+} and free copper by an aldehyde, and the reaction is used as a test for an aldehyde.

β-Hydroxyaldehyde: an aldehyde with a hydroxy group on the number 3 carbon (counting the carbonyl carbon as number 1).

Cannizzaro reaction: an oxidation-reduction reaction, in the presence of concentrated alkali, of an aldehyde without an α-hydrogen atom. In it one mole of aldehyde is reduced to the corresponding alcohol and a second mole of aldehyde is oxidized to the corresponding carboxylic acid.

Carbonyl group: a carbon atom double-bonded to an oxygen atom and single-bonded to two other groups or atoms: $>C{=}O$.

Clemmensen reduction: the reduction of a carbonyl group, $>C{=}O$, to a methylene group, $-CH_2-$. It uses a zinc-mercury amalgam and concentrated hydrochloric acid.

Crossed aldol condensation: an aldol condensation in which an aldehyde or ketone is condensed with a different aldehyde or ketone. Only one reactant need have an α-hydrogen atom.

Crossed Cannizzaro reaction: a Cannizzaro reaction in which the 2 mol of aldehyde which react are different.

Cyanohydrin: a compound of the general formula $R_2C(OH)-CN$ (OH on the carbon bearing CN).

Fehling's solution: a solution of a complex ion of copper(II) with tartrate. The Cu^{2+} ion is reduced to Cu^{+} and free copper by an aldehyde, and the reaction is used as a test for an aldehyde.

***Gem*-diol:** a compound bearing two hydroxy groups on the same carbon atom; the general formula is $R_2C(OH)_2$.

Haloform reaction: the reaction of NaOX, hypohalite, with ethanal, methyl ketone, or an alcohol which can be oxidized to ethanal or a methyl ketone. It produces haloform, CHX_3.

Hemiacetal: a compound of the general formula $R-CH(OH)-OR$. It is produced by the addition of 1 mol of alcohol to 1 mol of aldehyde. Unstable for the simple aldehydes.

Hypohalite: NaOX produced from the reaction of sodium hydroxide with halogen; it is used in the haloform reaction.

Imine: a compound of the general formula $R_2C{=}NH$; it is produced by the reaction of ammonia with an aldehyde or ketone.

Iodoform test: the haloform reaction using NaOI. Reaction of the reagent with ethanal, a methyl ketone, or an alcohol which can be oxidized to ethanal or a methyl ketone produces a yellow precipitate of iodoform (HCI_3, triiodomethane), and it is used as a test for methyl ketones.

Ketal: a compound produced by the reaction of a ketone with a diol, or with 2 mol of an alcohol. It is analogous to an acetal from an aldehyde.

Ketone: a compound in which two alkyl or aryl groups are attached to the carbonyl group; the general formulas are $R_2C{=}O$, $Ar(R)C{=}O$, and $Ar_2C{=}O$.

Oxime: a compound of the general formula $R_2C{=}N{-}OH$. It can be produced by the reaction of hydroxylamine, NH_2OH, with an aldehyde or a ketone in the presence of an acid catalyst, and it is useful in the identification of aldehydes and ketones.

Phenylhydrazone: a compound of the general formula

$$R_2C{=}N{-}\overset{\displaystyle H}{\overset{|}{N}}{-}C_6H_5$$

It can be produced by the reaction of an aldehyde or a ketone with phenylhydrazine, $NH_2NH{-}C_6H_5$, in the presence of an acid catalyst, and it is useful in identification of aldehydes and ketones.

Pyrolysis: a cleavage reaction achieved by heating.

Semicarbazone: a compound of the general formula

$$R_2C{=}N{-}\overset{\displaystyle H}{\overset{|}{N}}{-}\overset{\displaystyle O}{\overset{\|}{C}}{-}NH_2$$

It can be prepared by the reaction of semicarbazide,

$$NH_2NH{-}\underset{\displaystyle O}{\underset{\|}{C}}{-}NH_2$$

with an aldehyde or ketone, in the presence of an acid catalyst, and it is useful in identification of aldehydes and ketones.

Steric effect: an effect caused by spatial factors such as crowding around a group or atom.

Tollens' reagent: ammoniacal silver nitrate, $Ag(NH_3)_2OH$, prepared by dissolving silver nitrate in excess ammonium hydroxide. Its oxidation-reduction reaction with an aldehyde produces free silver, and it is used as a test for aldehydes and in the manufacture of mirrors.

Wolff-Kishner reduction: a reaction used to reduce the carbonyl group of an acid-sensitive ketone to a methylene group. It uses hydrazine, N_2H_4, and potassium hydroxide in 1,2-ethanediol as a solvent.

Additional problems

8-7 Give an acceptable name for each of the following compounds.

(a) $CH_3\overset{\displaystyle O}{\overset{\|}{C}}{-}CH_2Br$

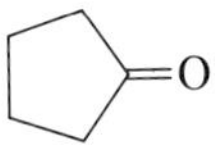

(b) 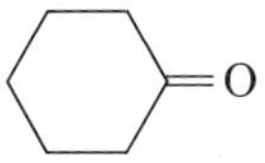 (cyclohexane ring)=O

(c) (cyclopentane ring)=O

(d) 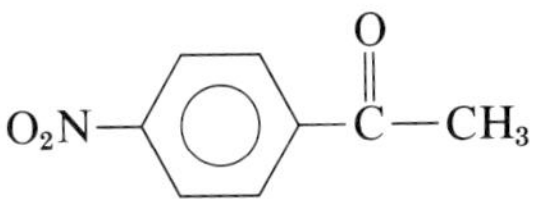 $O_2N{-}C_6H_4{-}\overset{\displaystyle O}{\overset{\|}{C}}{-}CH_3$

8-8 Write the structural formula for each of the following compounds: (a) 4-methoxybenzenecarbaldehyde (*p*-methoxybenzaldehyde), (b) cyclodecanone, (c) 2-ethyl-3-hydroxyhexanal, (d) 1,3-diphenyl-2-propenone.

8-9 Beginning with benzene or toluene and an alcohol of four or fewer carbons, write equations for the synthesis of each of the following:

(a) butanone
(b) cyclohexanone
(c) phenylethanone (acetophenone)

(*Hint:* $CH_3CH_2OH \xrightarrow{(O)} CH_3C(=O)OH \xrightarrow{SOCl_2} CH_3C(=O)Cl$)

(d) 3-phenylpropenal (cinnamaldehyde)
(e) 1,1-dimethoxypropane

8-10 Write equations for the following reactions:

(a) cyclohexanone and hydroxylamine
(b) 2-pentanone and phenylhydrazine
(c) propanal and 2,4-dinitrophenylhydrazine
(d) 3-pentanone and methyl magnesium iodide, then water
(e) 2-methylpropanal (isobutyraldehyde), aqueous sodium hydroxide
(f) benzenecarbaldehyde (benzaldehyde), 2-propanone (acetone), hydroxide ion

8-11 Indicate a *simple chemical test* or reagent which will readily distinguish between the members of each of the following pairs of compounds. Indicate which member of the pair gives the positive test or greater reaction.

(a) butanal and 1-butanol
(b) butanal and butanone
(c) 2-pentanone and 3-pentanone
(d) ethoxyethane (ethyl ether) and 2-butanone (ethyl methyl ketone)
(e) ethanol and methanol
(f) 1,1-diethyoxyethane and 1,2-diethoxyethane
(g) 2-pentene and butanone
(h) $CH_3—C(=O)—CH_2CH_3$ and $CH_3—C(=O)—O—CH_3$

8-12 Write equations for the reaction of each of the following sets of reactants:

(a) butanal and hydroxylamine
(b) benzenecarbaldehyde (benzaldehyde) and semicarbazide
(c) butanone and 2,4-dinitrophenylhydrazine
(d) phenylethanone (acetophenone) and HCN
(e) 2-furancarbonal (furfural) heated with concentrated NaOH
(f) ethanal (acetaldehyde) heated with hydrogen and a hydrogenation catalyst
(g) phenylethanal (phenylacetaldehyde) and $NaBH_4$

8-13 Complete the following table:

Name	Structural formula	Uses	Hazards and toxicity
Methanal (Formaldehyde)			
	C_6H_5CHO		
Ethanal (Acetaldehyde)			
	$CH_2{=}CH{-}CHO$		
	$C_6H_5\overset{\overset{\Large O}{\|}}{C}{-}CH_2Cl$		

8-14 Write equations for each of the following reactions:

(a) cyclohexanone, H_2, Pd catalyst
(b) cyclopropanone and water
(c) 2-propanone (acetone) and sodium bisulfite
(d) cyclohexanone and phenyl magnesium bromide, then water
(e) propanal (propionaldehyde), excess methanol, HCl(*g*)
(f) butanal warmed with dilute, aqueous NaOH
(g) furfural, ethanal (acetaldehyde), dilute, aqueous NaOH

8-15 Complete the equations for the reactions of the following:

(a) 2-pentanone (diethyl ketone) warmed with $Ba(OH)_2$
(b) cyclopentanone warmed with $Ba(OH)_2$
(c) $CH_3CH_2C(=O)OH$ reacted with $Ba(OH)_2$ and the salt then heated at 300°C
(d) benzenecarbaldehyde (benzaldehyde) and phenylethanone (acetophenone) warmed together with base

8-16 You have now studied reactions which involve the following types of reaction mechanisms:

1 free-radical substitution of alkanes
2 electrophilic addition to alkenes
3 free-radical addition to alkenes
4 1,4 electrophilic addition to dienes
5 allylic substitution
6 aromatic electrophilic substitution
7 aromatic nucleophilic substitution
8 S_N2-type nucleophilic substitution
9 S_N1-type nucleophilic substitution
10 E1-type elimination reaction
11 E2-type elimination reaction
12 benzylic substitution
13 nucleophilic addition to the carbonyl group

For each of the following reactions, identify the applicable reaction mechanism from the above list. Then write the mechanism for the reaction. Underline your structure for the reaction intermediate (if any), i.e., free radical, cation, or anion.

(a) $C_6H_5C(CH_3)(H)—Cl + NaOH(aq) \longrightarrow C_6H_5C(CH_3)(H)—OH + NaCl$

(b) $CH_2{=}CH—CH{=}CH—CH_3 + Br_2 \longrightarrow BrCH_2CH{=}CH—CH(Br)—CH_3$

(c) cyclohexene $+ Cl_2 \xrightarrow{500°C}$ 3-chlorocyclohexene (ring—Cl) $+ HCl$

(d) $CH_3C(CH_3){=}CH_2 + HCl \longrightarrow CH_3C(Cl)(CH_3)—CH_3$

(e) $CH_3(CH_2)_{10}CH_2OH + H_2SO_4 \longrightarrow CH_3(CH_2)_{10}CH_2OSO_2OH + H_2O$

(f) $CH_3CHOHCH_3 \xrightarrow{H_2SO_4,\ 200°C} CH_3CH{=}CH_2 + H_2O$

(g) $CH_3C(=O)—CH_3 + HCN \longrightarrow CH_3—C(OH)(CN)—CH_3$

(h) cyclohexane $+ Cl_2 \xrightarrow{light}$ cyclohexane—Cl $+ HCl$

(i) $C_6H_5—CH{=}CH—CH_3 + HBr \xrightarrow{peroxide} C_6H_5—CH_2CHBrCH_3$

(j) 1-Cl-2,4-$(NO_2)_2C_6H_3$ $+ 2NaOH \xrightarrow{\Delta}$ 1-ONa-2,4-$(NO_2)_2C_6H_3$ $+ NaCl + H_2O$

(k) $C_6H_5Cl + 2HNO_3 \xrightarrow{H_2SO_4,\ \Delta}$ 1-Cl-2,4-$(NO_2)_2C_6H_3$

(l) $CH_3C(CH_3)_2—Br + H_2O \longrightarrow CH_3—C(CH_3)_2—OH + HBr$

(m) $CH_3C(CH_3)_2—Br + NaOCH_3 \longrightarrow CH_3—C(CH_3){=}CH_2 + CH_3OH + NaBr$

(n) $CH_3CH_2CH_2Br + NaOCH_3 \longrightarrow CH_3CH_2CH_2—O—CH_3 + NaBr$

9

CARBOXYLIC ACIDS AND DERIVATIVES

The functional group of the *carboxylic acids* is the *carboxyl group*, written $-\overset{\displaystyle O}{\overset{\|}{C}}-OH$, —COOH, or —$CO_2H$. The group was so named because it can be considered a combination of a *carb*onyl group, $>C=O$, and a hydr*oxy* group, —OH (Figs. 9-1 and 9-2).

$$R-\overset{\displaystyle O}{\overset{\|}{C}}-OH \qquad Ar-\overset{\displaystyle O}{\overset{\|}{C}}-OH$$

Aliphatic carboxylic acid — Aromatic carboxylic acid

The carboxylic acids are widespread in nature and are very interesting and important compounds. Organic acids are responsible for the pleasant acidic taste of many of the foods we eat, and they are involved in our metabolic processes. The organic chemist uses them and their derivatives synthetically, and they are involved in the manufacture of soaps, plastics, fibers, herbicides, and pharmaceuticals.

9-1 Structure of carboxylic acids

Like the carbonyl group of aldehydes and ketones, the carboxyl group is essentially *planar trigonal* with sp^2 hybrid orbitals for the central carbon

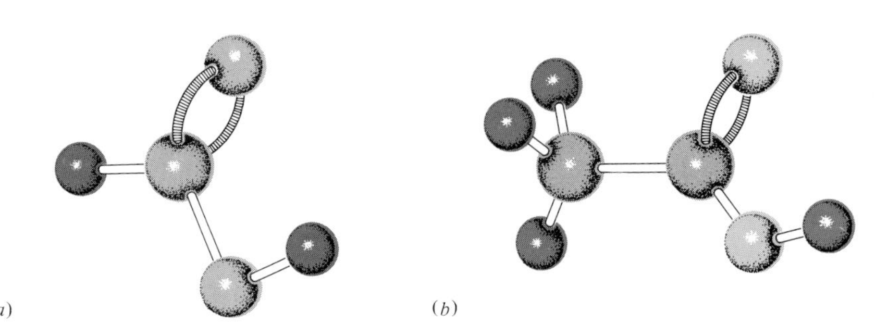

FIGURE 9-1
(*a*) Methanoic acid (formic acid) and (*b*) ethanoic acid (acetic acid) are the simplest aliphatic carboxylic acids.

FIGURE 9-2 A space-filling model of benzoic acid, the simplest aromatic carboxylic acid.

atom that result in bond angles of about 120°. The orbital structure of the carboxyl group is shown in Fig. 9-3.

The carboxyl group is actually a *resonance hybrid* to which an electronic form with separation of charge makes a small contribution.

$$-\mathrm{C}\begin{matrix}\nearrow\!\!\!\!/\,\mathrm{O}\\ \searrow\,\ddot{\mathrm{O}}\mathrm{H}\end{matrix} \longleftrightarrow -\mathrm{C}\begin{matrix}/\,\mathrm{O}^{-}\\ \searrow\!\!\!\!\backslash\,\overset{+}{\ddot{\mathrm{O}}}\mathrm{H}\end{matrix} \quad \text{or} \quad -\mathrm{C}\begin{matrix}\mathrm{O}^{\delta+}\\ \mathrm{OH}^{\delta-}\end{matrix} \qquad (9\text{-}1)$$

The result of this resonance effect is that, because the carbon atom is less positive, the carbonyl portion of a carboxyl group is much less reactive than the carbonyl group of an aldehyde or ketone toward nucleophilic addition.

Because a hydrogen atom is attached to oxygen in the carboxyl group, hydrogen bonding is possible. In the liquid state the molecules of a carboxylic acid occur as *dimers,* in which the two units are held together by a pair of hydrogen bonds.

$$\mathrm{R{-}C}\begin{matrix}\mathrm{O{-}{-}{-}{-}H{-}O}\\ \mathrm{O{-}H{-}{-}{-}{-}O}\end{matrix}\mathrm{C{-}R}$$

Dimer of a carboxylic acid

9-2 Properties of carboxylic acids

Because of dimerization through hydrogen bonding, the *boiling points* of carboxylic acids are even *higher* than those of the alcohols of corresponding molecular weight; e.g., compare the boiling points of the following, all of which have about the same molecular weight (Fig. 9-4).

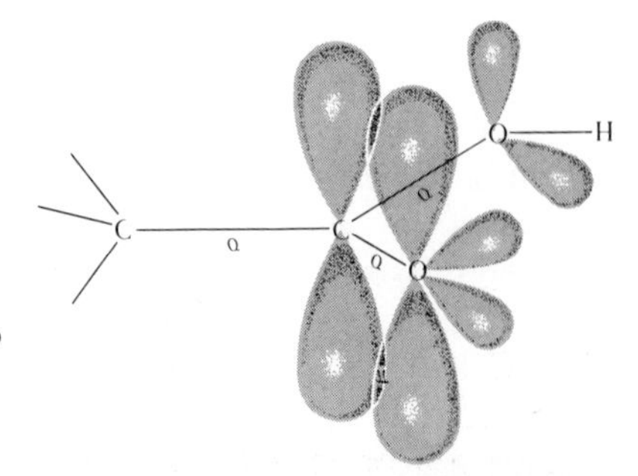

FIGURE 9-3 The orbital structure of the carboxyl group. Because the carbon atom of the carbonyl group is sp^2-hybridized, the two carbon atoms and the two oxygen atoms are in the same plane. Each oxygen atom has two pairs of unshared electrons.

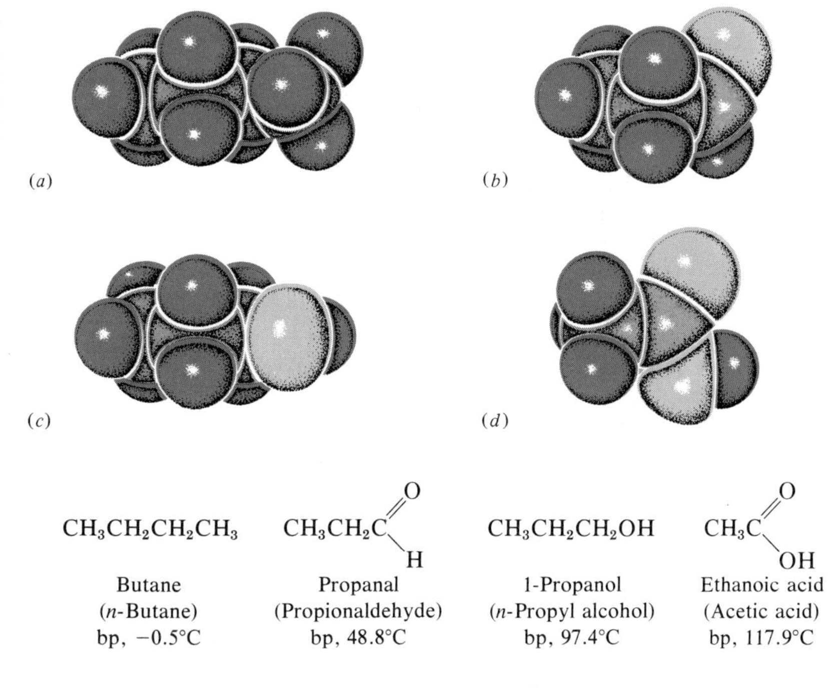

FIGURE 9-4
(*a*) Butane, (*b*) propanal, (*c*) 1-propanol, and (*d*) ethanoic acid have about the same molecular weight. Their boiling points differ significantly, however.

Even methanoic acid (formic acid), $H{-}C(=O)OH$, the acid with the lowest molecular weight, is a liquid. Because of the possibility of hydrogen bonding with water, all carboxylic acids with four or fewer carbon atoms are soluble in water.

9-3 Nomenclature

A large number of carboxylic acids have widely used common names which need to be learned. Those with an even number of carbon atoms ranging from 4 to 22 may be obtained by hydrolysis of animal and vegetable fats and oils. They are referred to as *fatty acids,* and they have common names derived from various sources. Formic acid derives its name from the Latin word for ants, because it is one of the toxic ingredients of the secretion injected by a stinging ant. Butanoic acid (butyric acid) derives its name from butter, in which it is found when the butter becomes rancid. Caproic, caprylic, and capric acids are involved in the odor of a goat, and their names derive from the Latin word, *caper,* for goat.

Table 9-1 lists the formulas, IUPAC and common names, and melting and boiling points of some of the most important carboxylic acids. The acids are important biologically, commercially, or both. Note that the dicarboxylic acids have relatively high melting points. That is because of the greatly increased possibility of hydrogen bonding with a second carboxyl group. The compounds with two or more polar groups, such as —OH, NH_2, or —COOH, are much more soluble in water and less soluble in organic solvents.

The IUPAC names are formed by adding the suffix *-oic* to the stem from

TABLE 9-1
Some carboxylic acids

Structural formula	IUPAC name	Common name	mp, °C	bp, °C
HCOOH	Methanoic acid	Formic acid	8	100.5
CH_3COOH	Ethanoic acid	Acetic acid	16.6	118
CH_3CH_2COOH	Propanoic acid	Propionic acid	−22	141
$CH_3CH_2CH_2COOH$	Butanoic acid	*n*-Butyric acid	−6	164
$(CH_3)_2CHCOOH$	2-Methylpropanoic acid	Isobutyric acid	−46.1	153.2
$CH_3(CH_2)_3COOH$	Pentanoic acid	*n*-Valeric acid	−34	187
$CH_3(CH_2)_4COOH$	Hexanoic acid	Caproic acid	−3	205
$CH_3(CH_2)_6COOH$	Octanoic acid	Caprylic acid	16	239
$CH_3(CH_2)_8COOH$	Decanoic acid	Capric acid	31	269
$CH_3(CH_2)_{10}COOH$	Dodecanoic acid	Lauric acid	44	
$CH_3(CH_2)_{12}COOH$	Tetradecanoic acid	Myristic acid	54	
$CH_3(CH_2)_{14}COOH$	Hexadecanoic acid	Palmitic acid	63	
$CH_3(CH_2)_{16}COOH$	Octadecanoic acid	Stearic acid	70	
$CH_3(CH_2)_7\overset{\overset{\displaystyle H}{\mid}}{C}=\overset{\overset{\displaystyle H}{\mid}}{C}(CH_2)_7COOH$	*cis*-9-Octadecenoic acid	Oleic acid	16	
$CH_3(CH_2)_4\overset{\overset{\displaystyle H}{\mid}}{C}=\overset{\overset{\displaystyle H}{\mid}}{C}CH_2\overset{\overset{\displaystyle H}{\mid}}{C}=\overset{\overset{\displaystyle H}{\mid}}{C}(CH_2)_7COOH$	*cis,cis*-9,12-Octadecadienoic acid	Linoleic acid	−5	
$CH_3CH_2\overset{\overset{\displaystyle H}{\mid}}{C}=\overset{\overset{\displaystyle H}{\mid}}{C}CH_2\overset{\overset{\displaystyle H}{\mid}}{C}=\overset{\overset{\displaystyle H}{\mid}}{C}CH_2\overset{\overset{\displaystyle H}{\mid}}{C}=\overset{\overset{\displaystyle H}{\mid}}{C}(CH_2)_7COOH$	*cis,cis,cis*-9,12,15-Octadecatrienoic acid	Linolenic acid	−11	
C_6H_5COOH	Benzenecarboxylic acid	Benzoic acid	122	250
o-$C_6H_4(COOH)_2$	1,2-Benzenedicarboxylic acid	Phthalic acid	231	
m-$C_6H_4(COOH)_2$	1,3-Benzenedicarboxylic acid	Isophthalic acid	348	
p-$C_6H_4(COOH)_2$	1,4-Benzenedicarboxylic acid	Terephthalic acid	300*	

the name of the parent alkane and then adding the word *acid*. For example, HCOOH is methanoic acid, CH_3COOH is ethanoic acid, $CH_3CH_2CH_2COOH$ is butanoic acid, and $CH_3CH_2CH_2\underset{\underset{\displaystyle CH_3}{\mid}}{C}HCOOH$ is 2-methylpentanoic acid. In the last example, in which a locator number is used, you will note that numbering begins with the carboxyl carbon atom as number 1.

In the common system of naming substituted carboxylic acids, the positions on the skeletal chain are indicated by Greek letters beginning with alpha on the carbon atom *adjacent to* the carboxyl carbon atom and continuing from there.

$$-\underset{6}{\overset{\epsilon}{C}}-\underset{5}{\overset{\delta}{C}}-\underset{4}{\overset{\gamma}{C}}-\underset{3}{\overset{\beta}{C}}-\underset{2}{\overset{\alpha}{C}}-COOH$$

Thus, $CH_3\underset{\underset{\displaystyle Cl}{\mid}}{C}HCOOH$ is named 2-chloropropanoic acid, or α-chloro-

TABLE 9-1
(continued)

Structural formula	IUPAC name	Common name	mp, °C	bp, °C
o-HO—C_6H_4COOH	2-Hydroxybenzenecarboxylic acid	Salicylic acid	159	
COOH N	3-Pyridinecarboxylic acid	Nicotinic acid	237	
HOOC—COOH	Ethanedioic acid	Oxalic acid	189	
$HOOCCH_2COOH$	1,3-Propanedioic acid	Malonic acid	136	
$HOOCCH_2CH_2COOH$	1,4-Butanedioic acid	Succinic acid	185	
$HOOCCH_2CH_2CH_2COOH$	1,5-Pentanedioic acid	Glutaric acid	98	
$HOOC(CH_2)_4COOH$	1,6-Hexanedioic acid	Adipic acid	151	
HOOC—C(H)=C(H)—COOH	*cis*-2-Butene-1,4-dioic acid	Maleic acid	130.5	
HOOC—C(H)=C(H)—COOH	*trans*-2-Butene-1,4-dioic acid	Fumaric acid	302	
CH_3—C(OH)(H)—COOH	(*R*)-2-Hydroxypropanoic acid	D(−)-Lactic acid	53	103
$HOC(CH_2COOH)_2COOH$	2-Hydroxy-1,2,3-propanetricarboxylic acid	Citric acid	153	
$CH_3C(=O)COOH$	2-Oxopropanoic acid	Pyruvic acid	13.6	165
$BrCH_2CH_2CH_2COOH$	4-Bromobutanoic acid	γ-Bromobutyric acid	33	

* sublimes

propionic acid, but the parts of the two names should never be mixed. The IUPAC names have not achieved wide acceptance except for the carboxylic acids having odd numbers of carbon atoms, which are not as commonly found in nature. Note the many examples of names given in Table 9-1. The aromatic carboxylic acid, C_6H_5COOH, is commonly named benzoic acid; the IUPAC name benzenecarboxylic acid has not been popular.

Observe in Table 9-1 how the dicarboxylic acids are named; observe also that they have common names. For example, HOOC—COOH has the common name oxalic acid and the IUPAC name ethanedioic acid; HOOC—CH_2—COOH is commonly called malonic acid and it is 1,3-propanedioic acid in the IUPAC system.

PROBLEM 9-1 Give an IUPAC name and another acceptable name for each of the following compounds:

(a) $CH_3CH(CH_3)CH_2COOH$

(b) $CH_3CH{=}CHCOOH$

(c) $CH_3CHOHCOOH$

(d) $HOOC{-}(CH_2)_4COOH$

(e) $HOCH_2CH_2CH_2COOH$

(f) $CH_3CH_2CH(Br)COOH$

(g) $CH_3CH(NH_2)COOH$

(h) 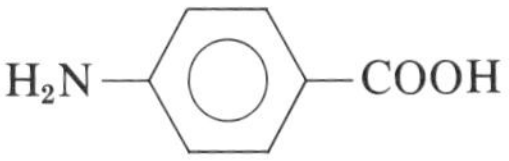 $H_2N{-}C_6H_4{-}COOH$ (para)

(i) 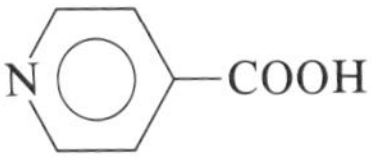 pyridine ring (N)—COOH (see Table 9-1)

(j) 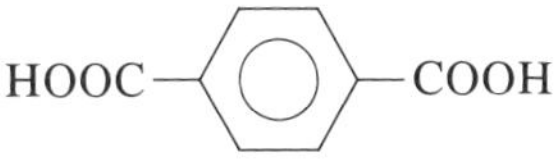 $HOOC{-}C_6H_4{-}COOH$ (para)

Solution to (a) The IUPAC name for this compound is 3-methylbutanoic acid.

$$\overset{4}{C}H_3\overset{3}{C}H(CH_3)\overset{2}{C}H_2\overset{1}{C}OOH$$

Its trivial name is isovaleric acid. It is also known as isopropylacetic acid.

PROBLEM 9-2 Write the structural formula and give the IUPAC name for each of the following compounds:

(a) malonic acid
(b) ethylmalonic acid
(c) 6-aminohexanoic acid
(d) α-chloroacetic acid
(e) trichloroacetic acid
(f) cyanoacetic acid
(g) oleic acid
(h) caproic acid
(i) linoleic acid
(j) α-hydroxypropionic acid
(k) 2-hydroxypropanoic acid
(l) *p-tert*-butylbenzoic acid

Solution to (a) The structural formula for malonic acid is

$$HOO\overset{1}{C}{-}\overset{2}{C}H_2{-}\overset{3}{C}OOH$$

and the IUPAC name is 1,3-propanedioic acid.

9-4 Preparation of carboxylic acids

The most general methods for the preparation of carboxylic acids, in both industry and the laboratory, involve the oxidation of primary alcohols or aldehydes, the oxidation of arenes, the hydrolysis of nitriles, and the hydrolysis of natural esters such as fats and oils. The reaction of a Grignard reagent with carbon dioxide is sometimes used as a laboratory method.

Oxidation of primary alcohols or aldehydes

Oxidation of primary alcohols or of aldehydes is a general way to prepare carboxylic acids. The general equations are

$$RCH_2OH \xrightarrow{KMnO_4,\ \Delta} RC(=O)OH \qquad (9\text{-}2)$$

$$\mathrm{R\overset{H}{\overset{|}{C}}{=}O \xrightarrow{KMnO_4,\ \Delta} RC(=O)OH} \qquad (9\text{-}3)$$

Two examples of the preparation are the following:

$$\underset{\substack{\text{2-Methyl-1-propanol}\\ \text{(Isobutyl alcohol)}}}{\mathrm{CH_3\overset{CH_3}{\overset{|}{C}}HCH_2OH}} \xrightarrow{KMnO_4,\ \Delta} \underset{\substack{\text{2-Methylpropanoic acid}\\ \text{(Isobutyric acid)}}}{\mathrm{CH_3\overset{CH_3}{\overset{|}{C}}HCOOH}} \qquad (9\text{-}4)$$

$$\underset{\substack{\text{Ethanal}\\ \text{(Acetaldehyde)}}}{\mathrm{CH_3C(H)=O}} \xrightarrow{O_2,\ Mn^{2+},\ \Delta} \underset{\substack{\text{Ethanoic acid}\\ \text{(Acetic acid)}}}{\mathrm{CH_3COOH}} \qquad (9\text{-}5)$$

Since ethanal is readily prepared by hydration of ethyne (acetylene) (Sec. 4-5), the oxidation of ethanal to ethanoic acid (acetic acid) opens the door to the preparation of acetic acid and its derivatives from ethyne.

Oxidation of arenes Oxidation of arenes is a good method for preparing aromatic acids. Two examples are

$$\underset{\text{Toluene}}{\mathrm{C_6H_5CH_3}} \xrightarrow{KMnO_4,\ \Delta} \underset{\substack{\text{Benzenecarboxylic acid}\\ \text{(Benzoic acid)}}}{\mathrm{C_6H_5C(=O)OH}} \qquad (9\text{-}6)$$

$$\underset{\substack{\text{1,4-Dimethylbenzene}\\ (p\text{-Xylene})}}{\mathrm{CH_3{-}C_6H_4{-}CH_3}} \xrightarrow{KMnO_4,\ \Delta\ (\text{or } O_2,\ V_2O_5)} \underset{\substack{\text{1,4-Benzenedicarboxylic acid}\\ \text{(Terephthalic acid)}}}{\mathrm{HO(O{=})C{-}C_6H_4{-}C(=O)OH}} \qquad (9\text{-}7)$$

This is the chief industrial method of preparing terephthalic acid, which is an intermediate in the production of the polyester Dacron.

Hydrolysis of nitriles Hydrolysis of nitriles with either acid or base is a good method whenever the nitrile can be readily prepared. Primary alkyl cyanides (*nitriles*) can be prepared by an S_N2 reaction of sodium cyanide with a primary alkyl halide. The general equation is

$$\mathrm{RX + NaCN \xrightarrow{(S_N2)} RCN + NaX} \qquad (9\text{-}8)$$

Secondary and tertiary halides are not used because they tend to dehydrohalogenate when heated with sodium cyanide.

The general reactions for hydrolysis of a nitrile with acid or base are as follows:

$$RCN + H_2O \xrightarrow{H^+, \Delta} RCOOH + NH_4^+ \tag{9-9}$$

$$RCN + {}^-OH(aq) \longrightarrow RCOO^- + NH_3 \tag{9-10}$$

Some specific examples of the reaction are

$$\underset{\substack{\text{Phenylethanenitrile} \\ \text{(Phenylacetonitrile,} \\ \text{benzyl cyanide)}}}{C_6H_5CH_2CN} \xrightarrow{H_2SO_4, \Delta} \underset{\substack{\text{Phenylethanoic acid} \\ \text{(Phenylacetic acid)}}}{C_6H_5CH_2COOH} + (NH_4)_2SO_4 \tag{9-11}$$

$$\underset{\substack{\text{Ethanal} \\ \text{(Acetaldehyde)}}}{CH_3\overset{\overset{\displaystyle H}{|}}{C}{=}O} + HCN \longrightarrow \underset{\substack{\text{2-Hydroxypropanenitrile} \\ \text{(Acetaldehyde cyanohydrin)}}}{CH_3CHOHCN} \xrightarrow{(1)\ NaOH,\ \Delta,\ (2)\ H^+} \underset{\substack{\text{2-Hydroxypropanoic acid} \\ \text{(Lactic acid)}}}{CH_3CHOHCOOH} \tag{9-12}$$

$$\underset{\substack{\text{Hexanedinitrile} \\ \text{(Adiponitrile)}}}{NC{-}(CH_2)_4CN} \xrightarrow{H_2SO_4, \Delta} \underset{\substack{\text{Hexanedioic acid} \\ \text{(Adipic acid)}}}{HOOC(CH_2)_4COOH} + (NH_4)_2SO_4 \tag{9-13}$$

The phenylethanoic acid produced in the first example has an odor like a perspiring horse. It has some plant hormone activity, and a dilute aqueous solution will initiate root growth on plant cuttings. The hydrolysis of hexanedinitrile to hexanedioic acid, Eq. (9-13), is an important step in one of the industrial preparations of nylon.

Grignard reagent reaction with CO_2

The reaction of a Grignard reagent with carbon dioxide may be used to prepare a carboxylic acid with one more carbon atom than the alkyl halide used to make the Grignard reagent.

$$RMgX + CO_2 \longrightarrow RC\genfrac{}{}{0pt}{}{\diagup\!\!\diagup O}{\diagdown OMgX} \xrightarrow{H_2O} RC\genfrac{}{}{0pt}{}{\diagup\!\!\diagup O}{\diagdown OH} \tag{9-14}$$

The reaction involves nucleophilic addition similar to that of a Grignard reagent to the carbonyl group of an aldehyde or ketone.

$$R^-Mg^{2+}X^- + O{=}C{=}O \longrightarrow R{-}\overset{\overset{\displaystyle O}{\|}}{C}{-}O^-Mg^{2+}X^- \xrightarrow{H_2O} R{-}\overset{\overset{\displaystyle O}{\|}}{C}{-}OH + MgX(OH) \tag{9-15}$$

The method has been used to produce 1-naphthoic acid.

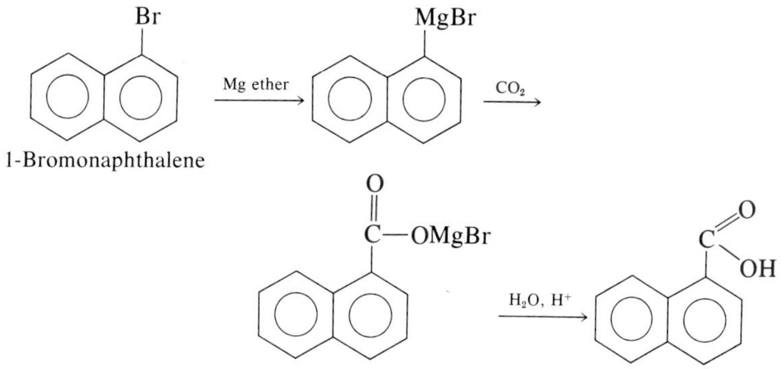

1-Bromonaphthalene $\xrightarrow{\text{Mg ether}}$ 1-naphthylmagnesium bromide (MgBr) $\xrightarrow{CO_2}$ $C_{10}H_7$—C(=O)—OMgBr $\xrightarrow{H_2O,\ H^+}$ $C_{10}H_7$—C(=O)—OH (9-16)

1-Naphthoic acid

This particular example, which is described in "Organic Syntheses,"[1] is an excellent way to prepare 1-naphthoic acid for research purposes. The carbon dioxide may be conveniently added to the ether solution in its solid state (dry ice).

Hydrolysis of naturally occurring esters

Hydrolysis of naturally occurring esters such as fats and oils is a good source of some of the higher-molecular-weight carboxylic acids.

$$\begin{array}{l} CH_2—OOC—R \\ | \\ CH—OOC—R \\ | \\ CH_2—OOC—R \end{array} + 3NaOH \longrightarrow \begin{array}{l} CH_2OH \\ | \\ CHOH \\ | \\ CH_2OH \end{array} + 3Na^+\ {}^-OOC—R \xrightarrow{H^+} 3HOOC—R \qquad (9\text{-}17)$$

Oil or fat; Glycerol; A soap; A fatty acid

An example of the reaction is

$$\begin{array}{l} CH_2—OOC(CH_2)_{10}CH_3 \\ | \\ CH—OOC(CH_2)_{10}CH_3 \\ | \\ CH_2—OOC(CH_2)_{10}CH_3 \end{array} + 3NaOH \longrightarrow$$

1,2,3-Propanetriol tridodecanoate (Glycerol trilaurate)

$$\begin{array}{l} CH_2OH \\ | \\ CHOH \\ | \\ CH_2OH \end{array} + 3Na^+\ {}^-OOC(CH_2)_{10}CH_3 \xrightarrow{H^+} 3HOOC(CH_2)_{10}CH_3 \qquad (9\text{-}18)$$

1,2,3-Propanetriol (Glycerol); Sodium dodecanoate (Sodium laurate); Dodecanoic acid (Lauric acid)

[1] Blatt, "Organic Syntheses, Collective Volume II," Wiley, New York, 1943, p. 425.

PROBLEM 9-3 Write the equations for a synthesis of each of the following acids from the starting materials indicated:

(a) 2-methylpropanoic acid (isobutyric acid) beginning with 2-propanol and utilizing a Grignard reaction
(b) phenylacetic acid, $C_6H_5CH_2COOH$, beginning with toluene and proceeding through nitrile formation
(c) 4-pyridine carboxylic acid (isonicotinic acid) from 4-methylpyridine
(d) 2-phenylpropanoic acid beginning with ethylbenzene and utilizing a Grignard reaction
(e) stearic acid (octadecanoic acid) from hydrolysis of glycerol tristearate

Solution to (a) Proceeding backwards from desired product to designated starting material is a useful technique.

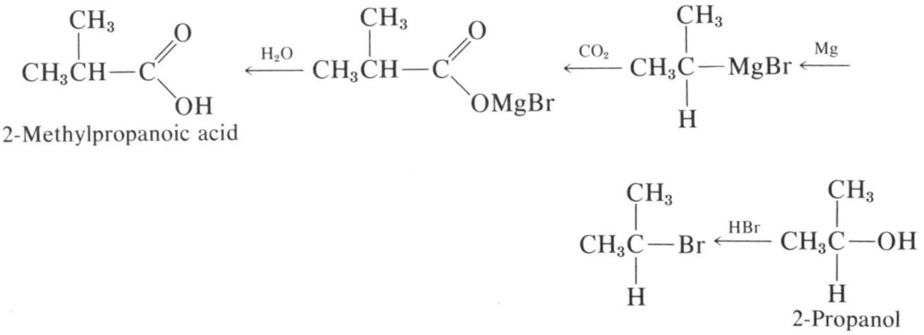

$$\underset{\text{2-Methylpropanoic acid}}{CH_3\overset{CH_3}{\overset{|}{C}}H-C(=O)OH} \xleftarrow{H_2O} CH_3\overset{CH_3}{\overset{|}{C}}H-C(=O)OMgBr \xleftarrow{CO_2} CH_3\underset{H}{\underset{|}{\overset{CH_3}{\overset{|}{C}}}}-MgBr \xleftarrow{Mg}$$

$$CH_3\underset{H}{\underset{|}{\overset{CH_3}{\overset{|}{C}}}}-Br \xleftarrow{HBr} \underset{\text{2-Propanol}}{CH_3\underset{H}{\underset{|}{\overset{CH_3}{\overset{|}{C}}}}-OH}$$

9-5 Reactions of carboxylic acids

The carboxylic acids may react *as acids,* in which case the *hydrogen-oxygen bond* of the carboxyl group is broken, $R-\overset{O}{\overset{\|}{C}}-O\nmid H$. The reaction of a carboxylic acid with an alcohol to produce an ester and the conversion of a carboxylic acid into its acid halide, $R-\overset{O}{\overset{\|}{C}}-X$, involve breaking the *carbon-oxygen bond* of the carboxyl group, $R-\overset{O}{\overset{\|}{C}}\nmid O-H$. The carboxyl group may also be reduced.

Reaction with bases

The characteristic feature of the carboxyl group is that the hydrogen is *acidic*. The hydrogen-oxygen bond is weakened by electron withdrawal of the adjacent carbonyl part. Removal of a proton by a base results in an anion which is stabilized by resonance with delocalization of the negative charge.

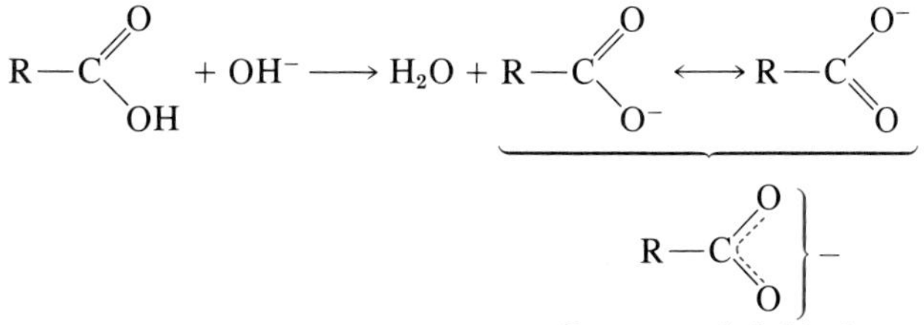

$$\text{R—C}\begin{matrix}\nearrow\!\!\!\!/\text{O} \\ \searrow\text{OH}\end{matrix} + OH^- \longrightarrow H_2O + \text{R—C}\begin{matrix}\nearrow\!\!\!\!/\text{O} \\ \searrow\text{O}^-\end{matrix} \longleftrightarrow \text{R—C}\begin{matrix}\nearrow\text{O}^- \\ \searrow\!\!\!\!\backslash\text{O}\end{matrix} \qquad (9\text{-}19)$$

$$\left[\text{R—C}\begin{matrix}\text{O} \\ \text{O}\end{matrix}\right]^-$$

Resonance hybrid of the carboxylate anion

The reaction with water as a base is represented by Eq. (9-20) for the equilibrium reaction involved. In most cases the position of equilibrium lies to the left as the equation is written, and only a very low concentration of hydronium ion is present at equilibrium. In other words, carboxylic acids are generally weak acids.

$$RCOOH + HOH \rightleftharpoons RCOO^- + H_3O^+ \qquad (9\text{-}20)$$

The expression for the equilibrium constant is

$$\frac{[RCOO^-][H_3O^+]}{[RCOOH][H_2O]} = K \qquad (9\text{-}21)$$

where K is the equilibrium constant. The brackets denote concentration, in moles per liter.

In dilute solutions, the concentration of water remains almost constant and is about

$$\frac{1000 \text{ g}}{(1 \text{ liter})(18 \text{ g/mol})} = 55.5 \text{ mol/liter}$$

If we then substitute K_a for $55.5K$, the equation becomes

$$\frac{[RCOO^-][H_3O^+]}{[RCOOH]} = K_a \qquad (9\text{-}22)$$

where K_a represents the ionization constant for a weak acid in water. The anion $RCOO^-$ is really a resonance hybrid which may be represented by

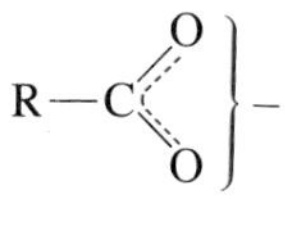

$$\left[\text{R—C}\begin{matrix}\text{O} \\ \text{O}\end{matrix}\right]^-$$

Any electron-withdrawing group attached to the alkyl group R further delocalizes the negative charge on the anion. That enhances the anion formation and the strength of the acid. On the other hand, electron-releasing groups make the acid weaker. The effect is greatest when the electron-withdrawing group is closest to the carboxyl group where the charge is mostly located. Table 9-2 illustrates how the strength of acids varies with

TABLE 9-2
Acid strengths of some carboxylic acids

Structural formula	IUPAC name	Common name	K_a
HCOOH	Methanoic acid	Formic acid	17.7×10^{-5}
CH_3COOH	Ethanoic acid	Acetic acid	1.75×10^{-5}
$CH_3ClCOOH$	Chloroethanoic acid	Chloroacetic acid	136×10^{-5}
$CHCl_2COOH$	Dichloroethanoic acid	Dichloroacetic acid	$5{,}530 \times 10^{-5}$
CCl_3COOH	Trichloroethanoic acid	Trichloroacetic acid (TCA)	$23{,}200 \times 10^{-5}$
$CH_2BrCOOH$	Bromoethanoic acid	Bromoacetic acid	125×10^{-5}
CH_2ICOOH	Iodoethanoic acid	Iodoacetic acid	67×10^{-5}
$CH_3CH_2CH_2COOH$	Butanoic acid	Butyric acid	1.52×10^{-5}
$CH_3CH_2CHClCOOH$	2-Chlorobutanoic acid	α-Chlorobutyric acid	139×10^{-5}
$CH_3CHClCH_2COOH$	3-Chlorobutanoic acid	β-Chlorobutyric acid	89×10^{-5}
$CH_2ClCH_2CH_2COOH$	4-Chlorobutanoic	γ-Chlorobutyric	2.96×10^{-5}
C_6H_5COOH	Benzenecarboxylic acid	Benzoic acid	6.3×10^{-5}
p-$O_2NC_6H_4COOH$	4-Nitrobenzene-carboxylic acid	*p*-Nitrobenzoic acid	36×10^{-5}
o-$O_2NC_6H_4COOH$	2-Nitrobenzene-carboxylic acid	*o*-Nitrobenzoic acid	670×10^{-5}
o-ClC_6H_4COOH	2-Chlorobenzene-carboxylic acid	*o*-Chlorobenzoic acid	120.0×10^{-5}
$C_6H_5CH_2COOH$	Phenylethanoic acid	Phenylacetic acid	4.9×10^{-5}

the substituents. Note that the strength of the halogen-substituted acetic acids is exactly in accord with the electronegativity of the halogens. That is, fluoroacetic acid is stronger than chloroacetic acid, which is stronger than bromoacetic acid, which is stronger than iodoacetic acid.

Electron-withdrawal effects of ring substituents on the acidity of aromatic carboxylic acids involve resonance effects as well as inductive effects. They are more complex and will not be discussed here.

The reaction of a carboxylic acid with a base *stronger* than water produces a salt of the carboxylic acid. For example,

$$\underset{\text{Ethanoic acid (Acetic acid)}}{CH_3COOH} + KOH \longrightarrow \underset{\text{Potassium ethanoate (Potassium acetate)}}{CH_3COO^-K^+} + H_2O \quad (9\text{-}23)$$

$$\underset{\text{Butanoic acid } (n\text{-Butyric acid})}{CH_3CH_2CH_2COOH} + NH_3 \longrightarrow \underset{\text{Ammonium butanoate (Ammonium } n\text{-butyrate)}}{CH_3CH_2CH_2COO^-NH_4^+} \quad (9\text{-}24)$$

$$\underset{\text{Butanoic acid } (n\text{-Butyric acid})}{CH_3CH_2CH_2COOH} + NaHCO_3 \longrightarrow \underset{\text{Sodium butanoate (Sodium } n\text{-butyrate)}}{CH_3CH_2CH_2COO^-Na^+} + CO_2 + H_2O \quad (9\text{-}25)$$

Butyric acid (butanoic acid) has the unpleasant odor of rancid butter; valeric acid (pentanoic acid) is said to smell like stinking feet. The last two reactions, shown with butyric acid as an example, may be used to help clean up and get rid of the odor of the unpleasant acids in vomit in the event of an unexpected result from nausea.

The sodium salts of the longer-chain fatty acids are soaps.

Reaction with alcohols

A carboxylic acid reacts with an alcohol in the presence of an acid catalyst to produce a derivative of a carboxylic acid called an *ester* (Fig. 9-5).

$$R-C(=O)-OH + HOR' \underset{}{\overset{H^+}{\rightleftharpoons}} \underset{\text{An ester}}{R-C(=O)-OR'} + HOH \qquad (9\text{-}26)$$

The esterification reaction is reversible and reaches an equilibrium. In the case of ethanoic acid (acetic acid) and ethanol, the equilibrium is reached when conversion to ester is about two-thirds complete. The reaction can be driven toward completion by using an excess of one of the reactants and by removing one of the products (generally water) from the reaction to slow down or prevent the reverse reaction.

Use of isotopic oxygen (designated by an asterisk in the equations which follow) has revealed that the oxygen which forms water comes from the acid rather than from the alcohol. The following mechanism has

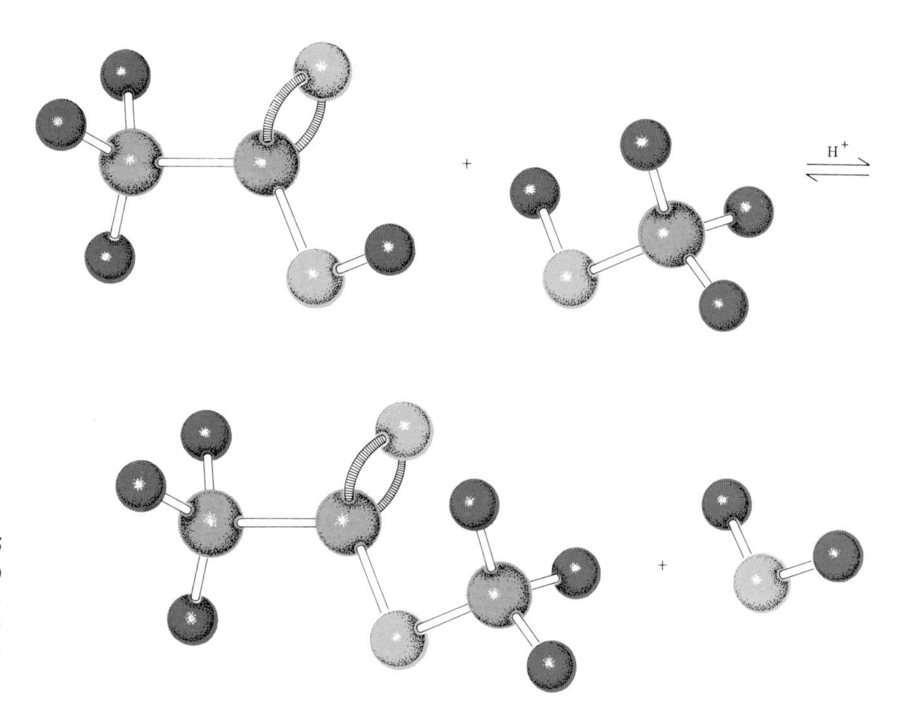

FIGURE 9-5
Ethanoic acid (acetic acid) and methanol react in the presence of an acid catalyst to produce methyl ethanoate (methyl acetate) and water.

been deduced for the esterification of a carboxylic acid with an alcohol. The reaction of ethanoic acid (acetic acid) with ethanol containing isotopic oxygen is used as an example.

$$CH_3C(=O)OH + H^+ \rightleftharpoons \left[CH_3C(\text{---}OH)(\text{---}OH)\right]^+ \overset{C_2H_5\overset{*}{O}H}{\rightleftharpoons} CH_3C(OH)(OH)\text{—}\overset{+}{\overset{*}{O}}H(C_2H_5) \rightleftharpoons$$

Esterification of an acid ⇄ Hydrolysis of an ester

$$CH_3C(HO)(\overset{*}{O}C_2H_5)\text{—}\overset{+}{O}H_2 \rightleftharpoons HOH + \left[CH_3C(\text{---}OH)(\text{---}\overset{*}{O}C_2H_5)\right]^+ \rightleftharpoons CH_3C(=O)\overset{*}{O}C_2H_5 + H_3O^+ \quad (9\text{-}27)$$

Ethyl ethanoate
(Ethyl acetate)

The *reverse reactions* represent the mechanism for the *acid-catalyzed hydrolysis of an ester.*

Conversion into acid halides

In general, a carboxylic acid may be converted into an *acid halide* by substitution of halogen for the hydroxy part of the carboxyl group by using any one of the following reagents: PCl_3, PBr_3, PI_3, PCl_5, or $SOCl_2$ (Fig. 9-6).

$$R\text{—}C(=O)OH \xrightarrow{PX_3,\ PCl_5,\ \text{or}\ SOCl_2} R\text{—}C(=O)X \quad (9\text{-}28)$$

An acid halide

The reagents indicated are those used to convert an alcohol to an alkyl halide, except that hydrohalogen acid is not used here.

An acid halide is one of the *acyl derivatives* of carboxylic acids which are discussed in Sec. 9-6. The acid halides are of value as chemical intermediates for syntheses. Some specific examples of their preparation from carboxylic acids follow.

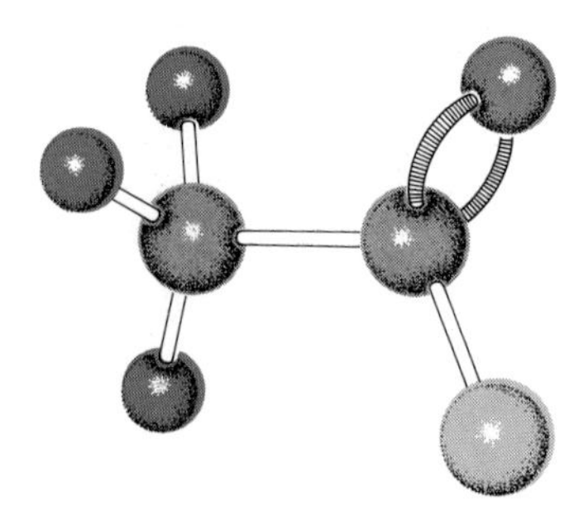

FIGURE 9-6
Ball-and-spring model of ethanoyl chloride (acetyl chloride), an acid halide.

$$C_6H_5C(=O)OH + SOCl_2 \xrightarrow{\Delta} C_6H_5C(=O)Cl + SO_2 + HCl \quad (9\text{-}29)$$

Benzenecarboxylic acid (Benzoic acid); Thionyl chloride; Benzenecarbonyl chloride (Benzoyl chloride)

$$3,5\text{-}(O_2N)_2C_6H_3COOH + PCl_5 \xrightarrow{\Delta}$$

3,5-Dinitrobenzenecarboxylic acid (3,5-Dinitrobenzoic acid) — Phosphorus pentachloride

$$3,5\text{-}(O_2N)_2C_6H_3COCl + POCl_3 + HCl \qquad (9\text{-}30)$$

3,5-Dinitrobenzenecarbonyl chloride (3,5-Dinitrobenzoyl chloride) — Phosphoryl chloride

$$3CH_3CH_2COOH + PCl_3 \longrightarrow 3CH_3CH_2COCl + H_3PO_3 \qquad (9\text{-}31)$$

Propanoic acid — Phosphorus trichloride — Propanoyl chloride — Phosphorus acid

Thionyl chloride ($SOCl_2$) is especially convenient because the by-products sulfur dioxide and hydrogen chloride are gases. They are easily separated from the product acid chloride. Any excess of the low-boiling (79°C) thionyl chloride is easily removed by distillation.

Conversion into acid anhydrides

An *acid anhydride* is the product resulting from loss of water by an acid. Addition of water to the acid anhydride re-forms the acid. For example, strongly heating sulfuric acid drives off water and produces sulfur trioxide. Sulfur trioxide is the anhydride of sulfuric acid.

$$H_2SO_4 \overset{\Delta}{\rightleftharpoons} SO_3 + H_2O \qquad (9\text{-}32)$$

Likewise, carbon dioxide is the anhydride of carbonic acid. However, carboxylic acids generally may *not* be converted into their anhydrides simply by heating them. Also, the anhydride of a carboxylic acid derives from the *intermolecular dehydration* of *two molecules of acid* rather than loss of water from just one molecule. An anhydride of a carboxylic acid may be derived from two molecules of the *same acid* to form a *simple* or *symmetrical anhydride*, $R{-}\overset{O}{\overset{\|}{C}}{-}O{-}\overset{O}{\overset{\|}{C}}{-}R$, or from two molecules of *different acids* to form a *mixed* or *unsymmetrical* anhydride with the general formula $R{-}\overset{O}{\overset{\|}{C}}{-}O{-}\overset{O}{\overset{\|}{C}}{-}R'$.

A general method applicable to the formation of both types involves the addition of the hydroxy part of the carboxyl group of one molecule of acid to the carbonyl group of an acid halide (nucleophilic addition) in the presence of the organic base pyridine (C_5H_5N) as a catalyst.

$$R{-}C(=O){-}\ddot{O}H + O{=}C(Cl){-}R' \longrightarrow R{-}C(=O){-}O^{+}(H){-}C(O^{-})(Cl){-}R' \qquad (9\text{-}33)$$

Acid — Acid anhydride

$$R{-}C(=O){-}O^{+}(H){-}C(O^{-})(Cl){-}R' \;\;(C_5H_5N:) \longrightarrow R{-}\overset{O}{\overset{\|}{C}}{-}O{-}\overset{O}{\overset{\|}{C}}{-}R' + C_5H_5NH^{+}Cl^{-} \qquad (9\text{-}34)$$

Acid halide — Pyridinium chloride

An example is

$$CH_3CH_2C(=O)OH + O{=}C(Cl){-}CH_3 \xrightarrow{C_5H_5N} CH_3CH_2\overset{O}{\overset{\|}{C}}{-}O{-}\overset{O}{\overset{\|}{C}}{-}CH_3 \qquad (9\text{-}35)$$

Propanoic acid (Propionic acid) — Ethanoyl chloride (Acetyl chloride) — Ethanoic propanoic anhydride (Acetic propionic anhydride)

Actually, the only anhydride of a monocarboxylic acid that is widely used, in quantity, is acetic anhydride. It is used for synthesizing acetate esters and amides. It is prepared industrially by the addition of acetic acid to the very reactive, unsaturated, gaseous aldehyde ketene (ethenal). Ketene itself is prepared by passing acetic acid over an aluminum phosphate catalyst at 500°C or by pyrolysis of acetone over a hot tungsten wire.

$$CH_3COOH \xrightarrow{AlPO_4,\ 500°C} H_2C{=}C{=}O + H_2O \qquad (9\text{-}36)$$

Ethanoic acid (Acetic acid) — Ethenal (Ketene)

$$CH_3\overset{O}{\overset{\|}{C}}{-}CH_3 \xrightarrow{600°C} CH_4 + H_2C{=}C{=}O \qquad (9\text{-}37)$$

Propanone (Acetone) — Methane — Ethenal (Ketene)

$$H_2C{=}C{=}O + CH_3C(=O)OH \longrightarrow CH_3\overset{O}{\overset{\|}{C}}{-}O{-}\overset{O}{\overset{\|}{C}}{-}CH_3 \qquad (9\text{-}38)$$

Ethenal (Ketene) — Ethanoic acid (Acetic acid) — Ethanoic anhydride (Acetic anhydride)

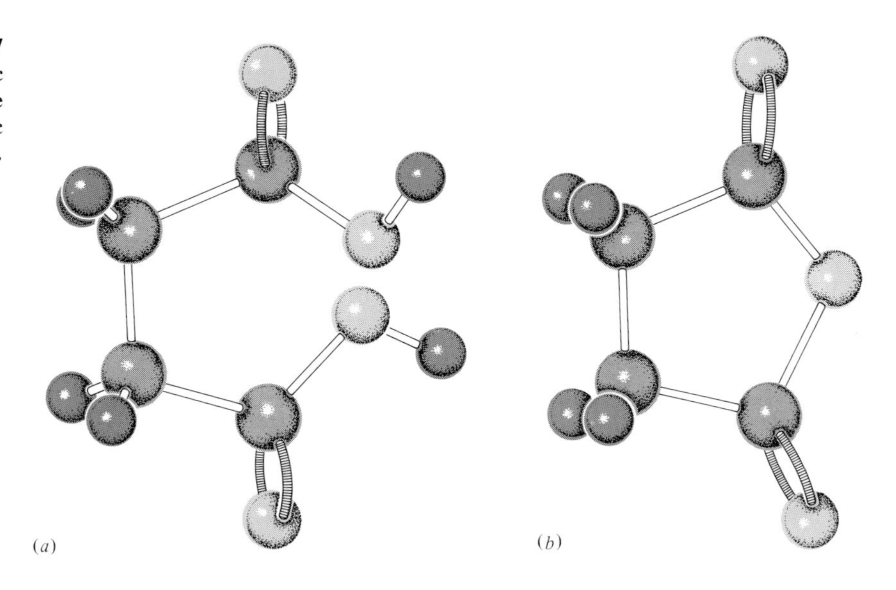

FIGURE 9-7
Upon heating, (*a*) succinic acid splits out a molecule of water to form (*b*) succinic anhydride.

Some *dicarboxylic* acids *do* readily lose water by simple heating to form *cyclic anhydrides*. The reaction is most likely if five- or six-atom rings are formed, since smaller rings evidently have too much bond angle strain. Formation of larger rings is in competition with formation of *linear polyanhydrides* or with the formation of *cyclic ketones* if base is present (Sec. 8-4). Some examples are given below (see Fig. 9-7 for another such reaction):

$$\begin{array}{l} H_2C-C(=O)OH \\ \;| \\ H_2C-C(=O)OH \end{array} \xrightarrow{\Delta} H_2O + \begin{array}{l} H_2C-C(=O) \\ \;|\qquad\quad \backslash O \\ H_2C-C(=O) \end{array} \tag{9-39}$$

Butanedioic acid (Succinic acid) → Butanedioic anhydride) (Succinic anhydride)

$$C_6H_4(COOH)_2 \xrightarrow{\Delta} C_6H_4(CO)_2O + H_2O \tag{9-40}$$

1,2-Benzenedicarboxylic acid (Phthalic acid) → 1,2-Benzenedicarboxylic anhydride (Phthalic anhydride)

$$\begin{matrix} \text{H} & & \text{O} \\ & \text{C}-\text{C} & \\ & \| & \text{OH} \\ & \| & \text{OH} \\ & \text{C}-\text{C} & \\ \text{H} & & \text{O} \end{matrix} \xrightarrow{\Delta} \begin{matrix} \text{H} & & \text{O} \\ & \text{C}-\text{C} & \\ & \| & \text{O} \\ & \text{C}-\text{C} & \\ \text{H} & & \text{O} \end{matrix} + H_2O \qquad (9\text{-}41)$$

cis-2-Butene-1,4-dioic acid (Maleic acid) → *cis*-2-Butene-1,4-dioic anhydride (Maleic anhydride)

The trans isomer of maleic acid, known as fumaric acid, will not form a cyclic anhydride upon heating because of geometry that is unfavorable for ring closure. For the same reason, isophthalic acid (1,3-benzenedicarboxylic acid) and terephthalic acid (1,4-benzenedicarboxylic acid) do not form cyclic anhydrides upon heating.

Reduction

Reduction of the carboxyl group can be readily accomplished with lithium aluminum hydride.

$$4RCOOH \xrightarrow{LiAlH_4} (RCH_2O)_4AlLi \xrightarrow{H_2O} 4RCH_2OH + LiOH + Al(OH)_3 \qquad (9\text{-}42)$$

The reaction involves nucleophilic substitution (Sec. 9-6) of a hydride ion for the hydroxide ion (a weaker base than hydride ion). The product is an aldehyde, which is then further reduced to a primary alcohol [Sec. 8-5, Eq. (8-31)]. The mechanism for the reaction is as shown in Eqs. (9-43) and (9-44):

$$R-C(=O)-OH + H-AlH_2(H^-)-H\ Li^+ \xrightarrow{ether} R-C(O^-)(H)-OH\ AlH_3Li^+ \qquad (9\text{-}43)$$

$$R-C(O^-)(H)-OH\ AlH_3Li^+ \longrightarrow R-C(=O)-H + AlH_3(OH)Li \longrightarrow \longrightarrow R-CH_2-OH \qquad (9\text{-}44)$$

An aldehyde; A primary alcohol

An example of this reaction is the reduction of oleic acid to oleyl alcohol.

$$4CH_3(CH_2)_7CH{=}CH-(CH_2)_7COOH \xrightarrow{LiAlH_4,\ ether}$$

Oleic acid

$$[CH_3(CH_2)_7CH{=}CH-(CH_2)_7CH_2O]_4AlLi \xrightarrow{H_2O}$$

$$4CH_3(CH_2)_7CH{=}CH-(CH_2)_7CH_2OH \qquad (9\text{-}45)$$

Oleyl alcohol

PROBLEM 9-4 Write the structural formula for the product of the reaction of benzoic acid with each of the following reagents.

(a) ethanol, H^+
(b) concentrated nitric acid
(c) thionyl chloride ($SOCl_2$)
(d) ammonium hydroxide
(e) sodium hydroxide and then fuse that product with sodium hydroxide
(f) bromine, iron catalyst
(g) lithium aluminum hydride and then water

Solution to (a) The product of the reaction of benzoic acid with ethanol, in the presence of an acid catalyst, is ethyl benzenecarboxylate (ethyl benzoate), an ester:

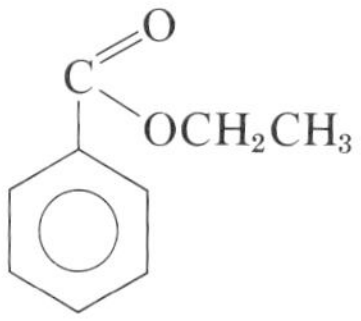

C_6H_5–C(=O)OCH_2CH_3

The reaction is

$$C_6H_5COOH + CH_3CH_2OH \xrightarrow{H^+} H_2O + C_6H_5COOCH_2CH_3$$

PROBLEM 9-5 Write equations for the reaction of each of the following compounds with the reagents indicated.

(a) phthalic acid and $LiAlH_4$
(b) butanoic acid and thionyl chloride
(c) ethanol, butanoic acid, H^+, heat
(d) pentanoic acid and ammonium hydroxide
(e) sodium benzoate and hydrochloric acid
(f) lactic acid (2-hydroxypropanoic acid) and sodium bicarbonate
(g) glutaric acid (pentanedioic acid), heat
(h) 4-hydroxybutanoic acid, H^+, heat (it forms a cyclic ester with itself)
(i) butanoic acid and PCl_3
(j) *p*-bromobenzoic acid and PCl_5

Solution to (a)

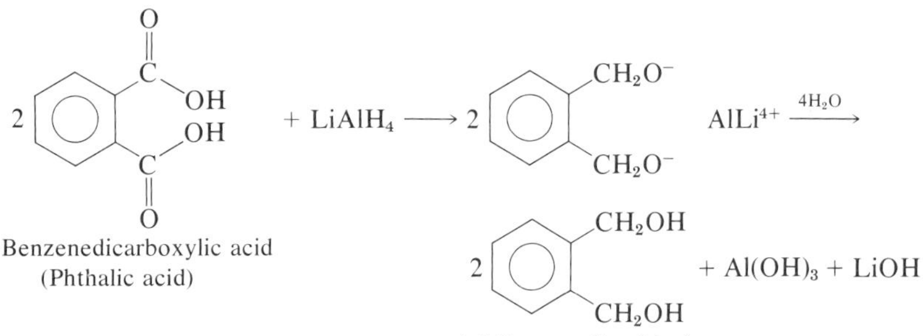

$$2\,C_6H_4(COOH)_2 + LiAlH_4 \longrightarrow 2\,C_6H_4(CH_2O^-)_2\ AlLi^{4+} \xrightarrow{4H_2O}$$

1,2-Benzenedicarboxylic acid
(Phthalic acid)

$$2\,C_6H_4(CH_2OH)_2 + Al(OH)_3 + LiOH$$

1,2-Benzenedicarbinol
(Phthalyl alcohol)

9-6 Acyl derivatives of carboxylic acids

The *acyl group* is $R{-}C{\overset{\displaystyle O}{\diagup\!\!\diagup}}\diagdown$, so an acyl derivative of a carboxylic acid is a derivative in which the —OH part of the carboxyl group has been substituted by another group. The general formula is

$$R{-}C(=O){-}W$$

where W stands for the substituent (Fig. 9-8). The important acyl derivatives are shown in Table 9-3.

FIGURE 9-8 Representative acyl derivatives of (*a*) acetic acid are (*b*) acetyl chloride, (*c*) acetic anhydride, (*d*) methyl acetate, and (*e*) acetamide.

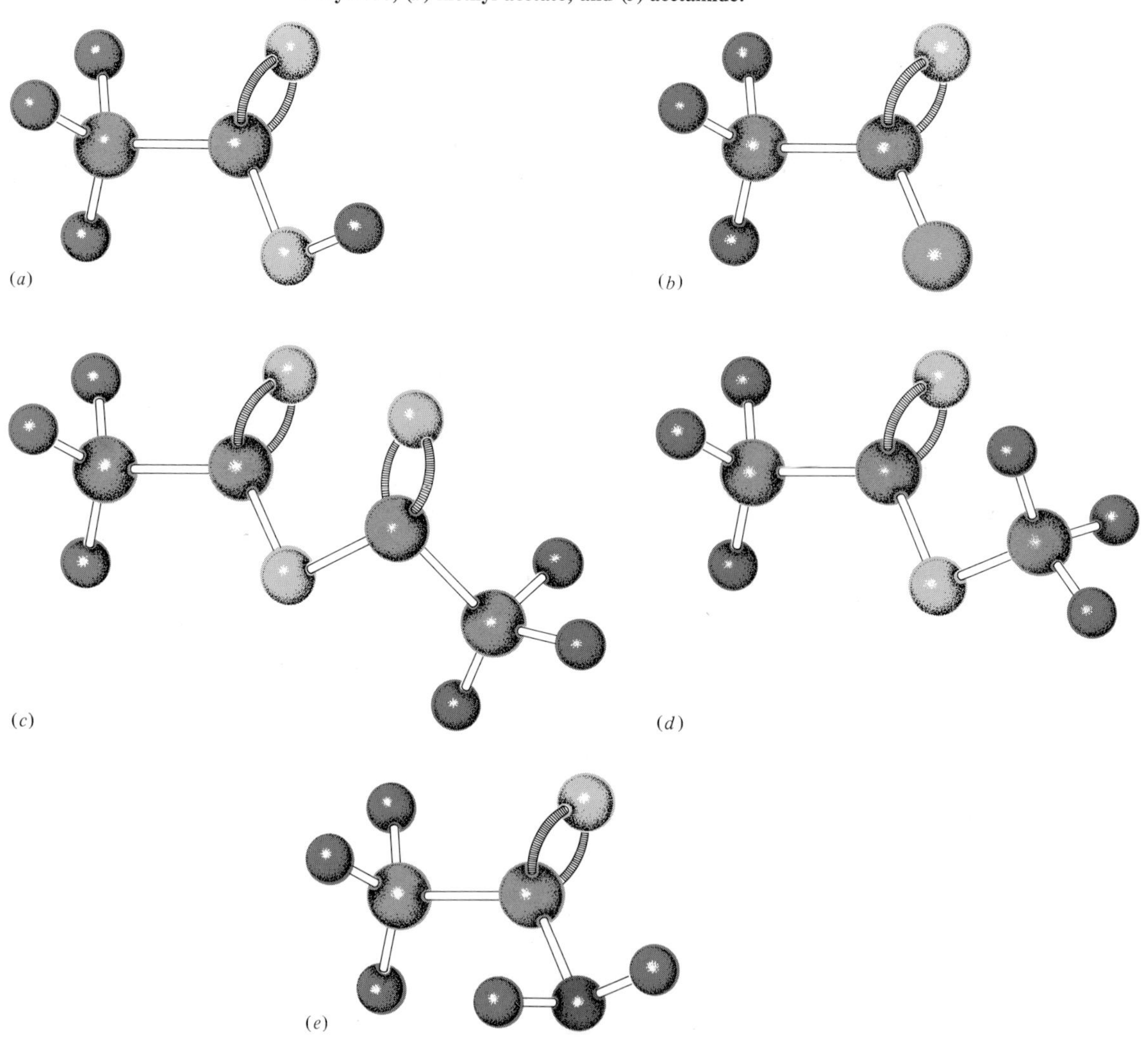

TABLE 9-3
Important acyl derivatives,

Substituent, W	Name	General formula
—OH	Carboxylic acid (itself)	$R{-}C(=O){-}OH$
—X	Acid halide or acyl halide	$R{-}C(=O){-}X$
$-O-\overset{O}{\overset{\|}{C}}-R$	Acid anhydride	$R{-}C(=O){-}O{-}\overset{O}{\overset{\|}{C}}{-}R$
—OR′	Ester	$R{-}C(=O){-}OR'$
$-NH_2$	Amide	$R{-}C(=O){-}NH_2$

Nomenclature

Acid halides. The common name for an acid halide is formed from the name for the *corresponding acid* to which it may be hydrolyzed. The suffix *-ic* is replaced by the suffix *-yl*, and the name of the halide is added. For example, $CH_3C(=O)Cl$ is acetyl chloride, $CH_3CH_2CH_2C(=O)Br$ is *n*-butyryl bromide, and $C_6H_5C(=O)Cl$ is benzoyl chloride. The IUPAC name is formed from the name of the *parent alkane*. The final *-e* is replaced by the suffix *-oyl,* and the name of the halide is added. For example, $CH_3C(=O)Cl$ is ethanoyl chloride and $CH_3CH_2CH_2C(=O)Br$ is butanoyl bromide.

Acid anhydrides. The common name for an acid anhydride is formed from the name for the corresponding acid, and the word *anhydride* is added. For example, $CH_3\overset{O}{\overset{\|}{C}}-O-\overset{O}{\overset{\|}{C}}-CH_3$ is acetic anhydride and $CH_3CH_2\overset{O}{\overset{\|}{C}}-O-\overset{O}{\overset{\|}{C}}-CH_2CH_3$ is propionic anhydride. The IUPAC name is formed by adding the word "anhydride" to the IUPAC name for the acid. Thus, $CH_3\overset{O}{\overset{\|}{C}}-O-\overset{O}{\overset{\|}{C}}-CH_3$ is ethanoic anhydride and

$CH_3CH_2C(=O)—O—C(=O)—CH_2CH_3$ is propanoic anhydride.

Esters. The common name of an ester is formed by naming the alkyl group related to the alcohol from which the ester could be formed, and that is followed by the common name for the acid part with the *-ic* suffix changed to *-ate* as it is for salts. Thus $CH_3—O—C(=O)—CH_3$ is methyl acetate and $CH_3CH_2—O—C(=O)—H$ is ethyl formate. The IUPAC name is similar, but it utilizes the IUPAC name for the acid part. Thus $CH_3O—C(=O)—CH_3$ is methyl ethanoate and $CH_3CH_2—O—C(=O)—H$ is ethyl methanoate.

Amides. The common name of an amide is related to the common name for the acid from which the amide is derived, and the *-ic* suffix is replaced by *-amide*. Thus, $CH_3C(=O)NH_2$ is acetamide, $CH_3CH_2C(=O)NH_2$ is propionamide, and $HC(=O)NH_2$ is formamide. The IUPAC name is similar, but it is based on the IUPAC name for the related acid part of the name. $HC(=O)NH_2$ is methanamide; $CH_3C(=O)NH_2$ is ethanamide; and $CH_3CH_2C(=O)NH_2$ is propanamide.

Salts. A salt is named by naming the cation and then naming the acid anion by using the stem of the name of the acid (either common or IUPAC) and adding the suffix *-ate*. For example, $NH_4^+ \, ^-O—CH(=O)$ is ammonium formate (common) or ammonium methanoate (IUPAC).

PROBLEM 9-6 Give an IUPAC name and a common name for each of the following compounds.

(a) CH_3COONH_4 (b) $(HCOO)_2Ca$

(c) $NH_4OOC(CH_2)_4COONH_4$

(d) $NH_2C(=O)—CH_2C(=O)—NH_2$

(e) $CH_3CH_2O—C(=O)—CH_2CH_2C(=O)—O—CH_2CH_3$

(f) COO^- | COO Ca^{2+}

(g) C_6H_5COONa

(h) $CH_3(CH_2)_{14}CH_2O—C(=O)—(CH_2)_{14}CH_3$

(i) $NH_2C(=O)—(CH_2)_4C(=O)—NH_2$

Solution to (a) The IUPAC name for CH_3COONH_4 is ammonium ethanoate. The common name is ammonium acetate.

PROBLEM 9-7 Write the structural formula for and give the IUPAC name for each of the following compounds.

(a) sodium propionate
(b) zinc stearate
(c) potassium oleate
(d) sodium trifluoracetate
(e) nicotinamide
(f) dimethyl phthalate
(g) ethyl benzoate
(h) magnesium palmitate
(i) disodium glutarate
(j) diethyl carbonate

Solution to (a) The IUPAC name is sodium propanoate.

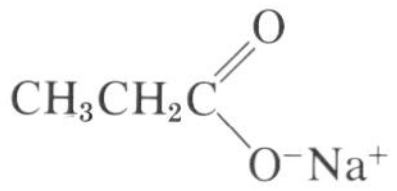

$$CH_3CH_2C(=O)O^-Na^+$$

Nucleophilic substitution The chief reaction of the acyl derivatives is *nucleophilic substitution* on the acyl carbon. The most important of these reactions are *hydrolysis*, *alcoholysis*, and *ammonolysis*. Those reactions also constitute important methods of preparation of the acyl derivatives other than the derivative being reacted. The first step in the nucleophilic substitution reaction is identical with that for nucleophilic addition to an aldehyde or ketone. (Z represents H_2O, ROH, or NH_3.)

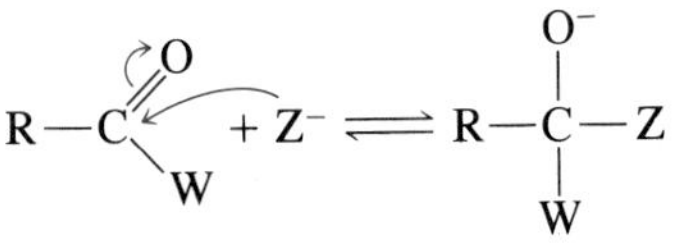

$$R—C(=O)—W + Z^- \rightleftharpoons R—C(O^-)(W)—Z \qquad (9\text{-}46)$$

The second step then involves the leaving of the group W.

$$R-\overset{\overset{\displaystyle O^-}{|}}{\underset{\underset{\displaystyle W}{|}}{C}}-Z \longrightarrow R-C\begin{matrix}\nearrow O \\ \searrow Z\end{matrix} + W^- \qquad (9\text{-}47)$$

The order of ease of displacement of W^- is

$$X^- > {}^-O-\overset{\overset{\displaystyle O}{\|}}{C}-R > {}^-OH > {}^-OR > {}^-NH_2 \gg {}^-{:}H \text{ or } {}^-{:}R$$

We may think of these groups as somewhat like bases and their reactions as like acid-base reactions. We see from the above listing that the acid halide is the most reactive for ease of displacement. That is what would be expected if the reaction is regarded as an acid-base reaction, since halide ion is the weakest base of the groups listed. Indeed, the decreasing order of ease of displacement is the same as the order of increasing strength as bases. (Remember that acid-base reactions tend to take place in the direction that favors the formation of the weaker acid and the weaker base.)

If W is H or R, then we have an aldehyde or ketone, which will undergo the first step of nucleophilic addition. It generally does *not* undergo the second step, since the leaving group would have to be a strong base—hydride ion ($:H^-$) or carbanion ($:R^-$).

The following are some examples of nucleophilic substitution of acid halides and acid anhydrides. The first is hydrolysis of acetyl chloride:

$$CH_3-C\begin{matrix}\nearrow O \\ \searrow Cl\end{matrix} + HOH \longrightarrow CH_3-\overset{\overset{\displaystyle O^-}{|}}{\underset{\underset{\displaystyle Cl}{|}}{C}}-O^+\begin{matrix}\nearrow H \\ \searrow H\end{matrix} \qquad (9\text{-}48)$$

$$CH_3-\overset{\overset{\displaystyle O^-}{|}}{\underset{\underset{\displaystyle Cl}{|}}{C}}-O^+\begin{matrix}\nearrow H \\ \searrow H\end{matrix} \longrightarrow CH_3-C\begin{matrix}\nearrow O \\ \searrow OH\end{matrix} + HCl \qquad (9\text{-}49)$$

Another example is the hydrolysis of acetic anhydride:

$$\begin{matrix} CH_3-C\begin{matrix}\nearrow O \\ \searrow \end{matrix} \\ O + HOH \\ CH_3-C\begin{matrix}\nearrow \\ \searrow O\end{matrix} \end{matrix} \longrightarrow \begin{matrix} CH_3-\overset{\overset{\displaystyle O^-}{|}}{C}-O^+\begin{matrix}\nearrow H \\ \searrow H\end{matrix} \\ O \\ CH_3-C\begin{matrix}\nearrow \\ \searrow O\end{matrix} \end{matrix} \qquad (9\text{-}50)$$

$$CH_3-\underset{\underset{\underset{\overset{\|}{O}}{CH_3-C}}{O}}{\overset{O^-}{C}}-\overset{+}{O}H_2 \longrightarrow CH_3-C(=O)OH + CH_3-C(OH)=O \qquad (9\text{-}51)$$

The ammonolysis of acetic anhydride is as follows:

$$CH_3-C(=O)-O-C(=O)-CH_3 + :NH_3 \longrightarrow CH_3-\overset{O^-}{C}(-\overset{+}{N}H_3)-O-C(=O)-CH_3 \qquad (9\text{-}52)$$

$$CH_3-\overset{O^-}{C}(-\overset{+}{N}H_3)-O-C(=O)-CH_3 \longrightarrow \underset{\text{Acetamide}}{CH_3-\overset{O}{\overset{\|}{C}}-NH_2} + CH_3-C(OH)=O \xrightarrow{NH_3} \underset{\text{Ammonium acetate}}{CH_3-\underset{\underset{O}{\|}}{C}-O^-NH_4^+} \qquad (9\text{-}53)$$

If W is —OR or —NH_2 (making the derivative an ester or an amide), nucleophilic substitution does not take place readily without the aid of an acid or the use of a strong base. Acid catalysis helps in the nucleophilic substitution reactions of esters and amides because it makes it possible for the leaving groups to be the weaker bases, alcohol and ammonia, rather than the strong bases, ^-OR and $^-NH_2$. The mechanism for the acid-catalyzed hydrolysis of an ester is just the reverse of that for esterification. It was given in Eq. (9-27). The acid-catalyzed hydrolysis of acetamide is an example for an amide.

$$\underset{\text{Acetamide}}{CH_3-C(=O)NH_2} + H^+ \longrightarrow CH_3-\overset{+}{C}(OH)NH_2 \xrightarrow{HOH} CH_3-\underset{\underset{NH_2}{|}}{\overset{\overset{OH}{|}}{C}}-\overset{+}{O}H_2 \qquad (9\text{-}54)$$

$$CH_3-\underset{\underset{NH_2}{|}}{\overset{\overset{OH}{|}}{C}}-\overset{+}{O}H_2 \longrightarrow CH_3-\overset{+}{C}(O-H)OH + :NH_3 \longrightarrow \underset{\text{Acetic acid}}{CH_3-C(=O)OH} + NH_4^+ \qquad (9\text{-}55)$$

The hydrolysis of esters and amides using base actually substitutes the stronger nucleophile hydroxide ion for water, so that the reaction occurs more readily. The hydrolysis of an ester with a strong base like NaOH is as follows:

$$R-C(=O)-OR' + {}^{-}OH \longrightarrow R-C(O^-)(OH)-OR' \tag{9-56}$$

$$R-C(O^-)(OH)-OR' \longrightarrow R-C(=O)-OH + {}^{-}OR' \longrightarrow R-C(=O)-O^- + HOR' \tag{9-57}$$

Alkaline hydrolysis is sometimes referred to as *saponification,* since the alkali metal salts resulting from the reaction are *soaps* if higher-molecular-weight fatty acids are involved.

$$\underset{\text{Methyl palmitate}}{CH_3(CH_2)_{14}COOCH_3} + NaOH \longrightarrow \underset{\substack{\text{Sodium palmitate}\\ \text{(a soap)}}}{CH_3(CH_2)_{14}COO^-Na^+} + \underset{\text{Methanol}}{CH_3OH} \tag{9-58}$$

Hydrolysis of the amide, acetamide, with aqueous NaOH is as follows:

$$CH_3-C(=O)-NH_2 + {}^{-}OH \longrightarrow CH_3-C(O^-)(OH)-NH_2 \tag{9-59}$$

$$CH_3-C(O^-)(NH_2)-O-H \longrightarrow CH_3-C(=O)-O^- + NH_3 \tag{9-60}$$

The interrelations among the acid derivatives are summarized in Table 9-4. Specific examples are used, but the reactions are general.

The alcoholysis of esters is also known as *transesterification.* The general reaction is

$$R-C(=O)-OR + H^+ \rightleftharpoons R-C^+(OH)-OR \overset{R'OH}{\rightleftharpoons} R-C(OH)(RO)-\overset{+}{O}(H)-R' \rightleftharpoons$$

$$R-C(O-H)(\overset{+}{R}OH)-OR' \rightleftharpoons R-C(=O)-OR' + ROH + H^+ \tag{9-61}$$

The example of transesterification given in Table 9-4 is

TABLE 9-4 Reactions of the acyl derivatives

Acyl derivative	Product of reaction with H_2O Hydrolysis	C_2H_5OH Alcoholysis	NH_3 Ammonolysis
$CH_3C(=O)Cl$ Ethanoyl chloride (Acetyl chloride)	$CH_3C(=O)OH + HCl$ Ethanoic acid (Acetic acid)	$CH_3C(=O)OC_2H_5 + HCl$ Ethyl ethanoate (Ethyl acetate)	$CH_3C(=O)NH_2 + NH_4Cl$ Ethanamide (Acetamide)
$CH_3C(=O)OC(=O)CH_3$ Ethanoic anhydride (Acetic anhydride)	$2CH_3C(=O)OH$ Ethanoic acid (Acetic acid)	$CH_3C(=O)OC_2H_5 + CH_3C(=O)OH$ Ethyl ethanoate, Ethanoic acid	$CH_3C(=O)NH_2 + CH_3C(=O)O^-NH_4^+$ Ethanamide, Ammonium ethanoate
$CH_3C(=O)OCH_3$ Methyl ethanoate (Methyl acetate)	$CH_3C(=O)OH + HOCH_3$ Ethanoic acid, Methanol	$CH_3C(=O)OC_2H_5 + HOCH_3$ Ethyl ethanoate, Methanol	$CH_3C(=O)NH_2 + HOCH_3$ Ethanamide, Methanol
$CH_3C(=O)NH_2$ Ethanamide (Acetamide)	$CH_3C(=O)OH + NH_3$ Ethanoic acid, Ammonia	$CH_3C(=O)OC_2H_5 + NH_3$ Ethyl ethanoate, Ammonia	

$$CH_3C(=O)OCH_3 + CH_3CH_2OH \overset{H^+}{\rightleftharpoons} CH_3C(=O)OCH_2CH_3 + CH_3OH \quad (9\text{-}62)$$

Methyl ethanoate (Methyl acetate) + Ethanol (Ethyl alcohol) ⇌ Ethyl ethanoate (Ethyl acetate) + Methanol (Methyl alcohol)

An excess of alcohol is used to drive the equilibrium toward the right.

A practical example of transesterification is the production of the polyester Dacron. The reaction is that of methyl terephthalate with ethylene glycol.

$$nCH_3O{-}C(=O){-}C_6H_4{-}C(=O){-}OCH_3 + n\,HOCH_2CH_2OH \xrightarrow{H^+,\ \Delta}$$

$$\left[{-}C(=O){-}C_6H_4{-}C(=O){-}OCH_2CH_2O{-}\right]_n + 2nCH_3OH \quad (9\text{-}63)$$

This polyester is called Dacron when it is formed into fibers and Mylar when it is formed into a sheet or film. Mylar is a very tough, strong plastic sheeting. One of its uses is as a base for movie film.

Use in Friedel-Crafts acylation

Acylation is a very useful reaction. Some examples were given in Chap. 8 as a method of preparing alkyl-aryl ketones. Such ketones may then be reduced to alkyl benzenes by using the *Clemmensen reduction*. Not only acid halides but also acid anhydrides and even carboxylic acids may sometimes be used as acylating reagents in that reaction. Some further examples are

$$C_6H_6 + R{-}C(=O)Cl \xrightarrow{AlCl_3} C_6H_5{-}C(=O){-}R \xrightarrow{ZnHg/HCl} C_6H_5CH_2R \qquad (9\text{-}64)$$

$$C_6H_6 + CH_3C(=O){-}O{-}C(=O){-}CH_3 \xrightarrow{AlCl_3} C_6H_5C(=O){-}CH_3 + HO{-}C(=O){-}CH_3 \qquad (9\text{-}65)$$

Benzene; Ethanoic anhydride (Acetic anhydride); Phenylethanone (Acetophenone); Ethanoic acid (Acetic acid)

$$C_6H_6 + \text{(succinic anhydride)} \xrightarrow{AlCl_3} C_6H_5C(=O)CH_2CH_2C(=O)OH \xrightarrow{Zn(Hg)/HCl} C_6H_5CH_2CH_2CH_2C(=O)OH \qquad (9\text{-}66)$$

Benzene; Butanedioic anhydride (Succinic anhydride); 4-Oxo-4-phenylbutanoic acid (3-Benzoylpropanoic acid); 4-Phenylbutanoic acid

$$C_6H_5CH_2CH_2CH_2C(=O)OH \xrightarrow{\text{polyphosphoric acid, } \Delta} \text{1-tetralone} + H_2O \qquad (9\text{-}67)$$

4-Phenylbutanoic acid; 1-Oxo-1,2,3,4-tetrahydronaphthalene (1-Tetralone)

The last two equations illustrate the utility of Friedel-Crafts acylation for the synthesis of polycyclic ring systems. The acylation is a key reaction in the synthesis of substituted naphthalene and phenanthrene compounds.

Hydrogenolysis of esters

Under pressures of 136 to 204 atm and temperatures of 250 to 350°C, molecular hydrogen with a nickel or copper chromite catalyst will reduce esters to alcohols. The general reaction is

$$RCOOR' + 2H_2 \xrightarrow{\text{catalyst}} RCH_2OH + HOR' \tag{9-68}$$

The reaction has practical application in the preparation of higher-molecular-weight alcohols for making sodium alkyl sulfates for detergents.

$$\underset{\text{Methyl laurate}}{CH_3(CH_2)_{10}COOCH_3} \xrightarrow{CuCr_2O_4 \cdot CuO,\ 250°C,\ 170\ atm} CH_3OH + \underset{\text{Lauryl alcohol}}{CH_3(CH_2)_{10}CH_2OH} \tag{9-69}$$

$$\underset{\text{Lauryl alcohol}}{CH_3(CH_2)_{10}CH_2OH} + HOSO_2OH \longrightarrow \underset{\text{Lauryl hydrogen sulfate}}{CH_3(CH_2)_{10}CH_2OSO_2OH} \xrightarrow{NaOH} \underset{\text{Sodium lauryl sulfate (a synthetic detergent)}}{CH_3(CH_2)_{11}OSO_2O^-Na^+} \tag{9-70}$$

Esters may also be reduced to alcohols by using lithium aluminum hydride.

$$R-C(=O)-OR' \xrightarrow{LiAlH_4} \cdots \xrightarrow{H_2O,\ H^+} R-CH_2-OH + H-O-R' \tag{9-71}$$

$$\underset{\text{Methyl oleate}}{CH_3(CH_2)_7CH{=}CH(CH_2)_4COOCH_3} \xrightarrow{LiAlH_4} \cdots \xrightarrow{H_2O,\ H^+} \underset{\text{Oleyl alcohol}}{CH_3(CH_2)_7CH{=}CH(CH_2)_7CH_2OH} + \underset{\text{Methanol}}{CH_3OH} \tag{9-72}$$

The mechanism and intermediate products are not given here because they are similar to those for the reduction of carboxylic acids with lithium aluminum hydride, Eqs. (9-43) and (9-44). Observe that the carbon-carbon double bond is retained with this type of reduction.

Claisen condensation of esters

The α-hydrogen atoms of an ester such as ethyl acetate are weakly acidic. In the presence of a strong base, such as sodium ethoxide, they react to produce a resonance-stabilized carbanion. The carbanion is a strong nucleophile which may then attack the carbonyl carbon of a second ester molecule. The final product is a β-keto ester, ethyl acetoacetate in this case. The reaction is called the *Claisen condensation* (Ludwig Claisen, 1851–1930, University of Kiel). It bears a resemblance to the aldol condensation except that the initial addition is followed by an elimination, so that the result is nucleophilic substitution.

An example of the Claisen condensation is

$$C_2H_5O^- + H-\underset{H}{\overset{H}{C}}-\overset{O}{\overset{\|}{C}}-OC_2H_5 \rightleftharpoons C_2H_5OH + H-\underset{H}{C}=\overset{O^-}{C}-OC_2H_5 \qquad (9\text{-}73)$$

Ethoxide ion — Ethyl ethanoate (Ethyl acetate) — Ethanol — Resonance-stabilized carbanion

$$CH_3\overset{O}{\overset{\|}{C}}-OC_2H_5 + H-\underset{H}{C}=\overset{O^-}{C}-OC_2H_5 \rightleftharpoons CH_3\underset{OC_2H_5}{\overset{O^-}{C}}-CH_2\overset{O}{\overset{\|}{C}}-OC_2H_5 \rightleftharpoons$$

Ethyl ethanoate — Carbanion

$$CH_3\overset{O}{\overset{\|}{C}}CH_2\overset{O}{\overset{\|}{C}}OC_2H_5 + C_2H_5O^- \qquad (9\text{-}74)$$

Ethyl 3-oxobutanoate (Ethyl acetoacetate)

The β-keto esters are useful in a number of organic syntheses, especially of ketones.

Ethyl acetoacetate is an unusual compound in that it has not only the properties of an ester and of a ketone but also the definite properties of an enol, $-\overset{|}{C}=\overset{|}{C}-OH$. That is due to a type of isomerism called *tautomerism*. *Tautomers* are equilibrium isomers; in this case they are the *keto* and *enol* forms of ethyl acetoacetate. Such tautomers actually exist in equilibrium in a definite percentage. The tautomers of ethyl acetoacetate (Fig. 9-9) are

$$H_3C-\overset{O}{\overset{\|}{C}}-\overset{\alpha}{CH_2}-\overset{O}{\overset{\|}{C}}-OC_2H_5 \rightleftharpoons CH_3-\overset{O-H\cdots}{C}=\underset{H}{C}-\overset{\cdots O}{\overset{\|}{C}}-OC_2H_5 \qquad (9\text{-}75)$$

Keto form, 92.5% — Enol form, 7.5%

The keto form of ethyl acetoacetate has a large dipole moment, reacts with phenylhydrazine, and has the higher boiling point. The enol form reacts with Br_2 and gives a red color with $FeCl_3$.

The enol form of ethyl acetoacetate exhibits *intramolecular hydrogen bonding*. The reason for the term is that the hydrogen of the hydroxy group hydrogen bonds to the carbonyl part of the ester group, and all the atoms are part of the same molecule. The *intra*molecular bonding limits *inter*molecular hydrogen bonding. Thus the vapor pressure is higher and the boiling point is lower than would otherwise be true of a compound with an —OH group.

Any carbonyl compound with an α-hydrogen atom can exist as keto and enol tautomers. (The simple aldehydes and ketones, however, have very

FIGURE 9-9
Ball-and-spring models of (*a*) the keto form and (*b*) the enol form of ethylacetoacetate.

little enol form at equilibrium.) The enolization of the keto form may be catalyzed by acid or base. The mechanism for acid-catalyzed enolization is

$$\text{H}-\overset{|}{\underset{|}{\text{C}}}-\overset{\text{O}}{\overset{\|}{\text{C}}}- + \text{H}^+ \rightleftharpoons -\overset{\text{H}}{\overset{|}{\underset{|}{\text{C}}}}-\overset{\overset{+}{\text{O}}\text{H}}{\overset{\|}{\text{C}}}- \rightleftharpoons -\text{C}=\overset{\text{OH}}{\overset{|}{\text{C}}}- + \text{H}^+ \tag{9-76}$$

With base, enolization is promoted as follows:

$$-\overset{\text{H}}{\overset{|}{\underset{|}{\text{C}}}}-\overset{\text{O}}{\overset{\|}{\text{C}}}- + \text{OH}^- \rightleftharpoons -\overset{-}{\underset{|}{\text{C}}}-\overset{\text{O}}{\overset{\|}{\text{C}}}- \longleftrightarrow -\underset{|}{\text{C}}=\overset{\text{O}^-}{\overset{|}{\text{C}}}- + \text{H}_2\text{O} \rightleftharpoons -\underset{|}{\text{C}}=\overset{\text{OH}}{\overset{|}{\text{C}}}- + \text{OH}^- \tag{9-77}$$

When ethyl acetoacetate is hydrolyzed and heated, it *decarboxylates* (loses a carboxyl group as CO_2) to yield acetone and carbon dioxide.

$$\underset{\text{Ethyl acetoacetate}}{\text{CH}_3\overset{\text{O}}{\overset{\|}{\text{C}}}\text{CH}_2\overset{\text{O}}{\overset{\|}{\text{C}}}\text{OC}_2\text{H}_5} + \text{H}_2\text{O} \xrightarrow[(-\text{C}_2\text{H}_5\text{OH})]{\text{H}^+,\ \Delta} \underset{\text{Acetoacetic acid}}{\text{CH}_3\overset{\text{O}}{\overset{\|}{\text{C}}}\text{CH}_2\overset{\text{O}}{\overset{\|}{\text{C}}}-\text{OH}} \xrightarrow{\Delta} \text{CO}_2 + \underset{\text{Acetone}}{\text{CH}_3\overset{\text{O}}{\overset{\|}{\text{C}}}\text{CH}_3} \tag{9-78}$$

Acetoacetic acid in very small amounts is a normal constituent of the body. In the disease diabetes, a pathological condition in which abnormally large amounts of acetone and acetoacetic acid (ketone bodies) are found in the blood and urine may develop. The condition is called *ketosis*.

Amides, polyamides, and cyclic amides

With two hydrogen atoms attached to nitrogen, the simple amides may undergo extensive hydrogen bonding. They have relatively high boiling points and melting points, and they are generally solids. As shown in Table 9-4, amides may be prepared by the ammonolysis of either acid halides or acid anhydrides.

$$R-C(=O)Cl + 2NH_3 \longrightarrow R-C(=O)NH_2 + NH_4Cl \qquad (9\text{-}79)$$

$$R-\overset{O}{\overset{\|}{C}}-O-\overset{O}{\overset{\|}{C}}-R + 2NH_3 \longrightarrow R-C(=O)NH_2 + R-C(=O)O^- \;\; NH_4^+ \qquad (9\text{-}80)$$

These ammonolysis reactions are rapid, convenient methods of preparing amides as derivatives for identification of acids or amines (ammonia derivatives). An amino group may be deactivated or protected by amide formation and then later restored by hydrolysis of the amide. Proteins are essentially polyamides, so amides are very important.

Amides from an ammonium salt of a carboxylic acid. Strongly heating an ammonium salt of a carboxylic acid eliminates water and produces an amide.

$$R-C(=O)O^-NH_4^+ \xrightarrow{\Delta} R-C(=O)NH_2 + H_2O \qquad (9\text{-}81)$$

For example, acetamide may be prepared by heating ammonium acetate.

$$\underset{\substack{\text{Ammonium ethanoate}\\ \text{(Ammonium acetate)}}}{CH_3-C(=O)O^-NH_4^+} \xrightarrow{\Delta} \underset{\substack{\text{Ethanamide}\\ \text{(Acetamide)}}}{CH_3-C(=O)NH_2} + H_2O \qquad (9\text{-}82)$$

The reaction is also used commercially to synthesize urea, $H_2N-\overset{O}{\overset{\|}{C}}-NH_2$, which may be considered the amide of carbonic acid, $HO-\overset{O}{\overset{\|}{C}}-OH$ (Fig. 9-10).

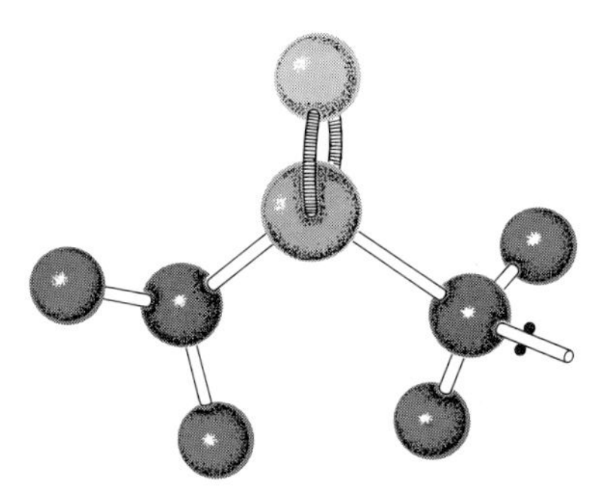

FIGURE 9-10
Ball-and-spring model of urea (carbamide).

$$CO_2 + NH_3 \longrightarrow \underset{\text{Ammonium carbamate}}{H_2N{-}\overset{\overset{\displaystyle O}{\|}}{C}{-}O^-NH_4^+} \xrightarrow{\Delta} \underset{\text{Urea (carbamide)}}{H_2N{-}\overset{\overset{\displaystyle O}{\|}}{C}{-}NH_2} + H_2O \qquad (9\text{-}83)$$

Urea is the chief nitrogenous waste product in urine. It is useful as a fertilizer, as a livestock protein supplement, and as a chemical intermediate.

Polyamides. The synthesis of nylon 66, a polyamide, is another example of preparing an amide by heating an ammonium salt. Hexanedioic acid (adipic acid) is mixed with 1,6-hexanediamine (1,6-diaminohexane) to produce the substituted ammonium salt, 1,6-hexanediammonium hexanedioate. The salt is then isolated and heated to produce the corresponding polyamide.

$$\underset{\text{1,6-Hexanediamine}}{H_2N(CH_2)_6NH_2} + \underset{\substack{\text{Hexanedioic acid}\\ \text{(Adipic acid)}}}{HO\overset{\overset{\displaystyle O}{\|}}{C}(CH_2)_4\overset{\overset{\displaystyle O}{\|}}{C}OH} \longrightarrow H_3\overset{+}{N}(CH_2)_6\overset{+}{N}H_3\ {}^{-}O\overset{\overset{\displaystyle O}{\|}}{C}(CH_2)_4\overset{\overset{\displaystyle O}{\|}}{C}O^- \qquad (9\text{-}84)$$

$$\underset{\text{1,6-Hexanediammonium hexanedioate}}{n\,H_3\overset{+}{N}(CH_2)_6\overset{+}{N}H_3\ {}^{-}O\overset{\overset{\displaystyle O}{\|}}{C}(CH_2)_4\overset{\overset{\displaystyle O}{\|}}{C}O^-} \xrightarrow{\Delta} \underset{\text{Nylon}}{\left[{-}\overset{\overset{\displaystyle H}{|}}{N}(CH_2)_6\overset{\overset{\displaystyle H}{|}}{N}{-}\overset{\overset{\displaystyle O}{\|}}{C}(CH_2)_4\overset{\overset{\displaystyle O}{\|}}{C}{-}\right]_n} + nH_2O \qquad (9\text{-}85)$$

The structure of nylon is somewhat like that of silk. Silk is a protein, and proteins are naturally occurring polyamides.

Amides may be dehydrated by a strong dehydrating agent like phosphorus pentoxide, P_2O_5, to give a nitrile.

$$R{-}C\begin{matrix}{\nearrow}^{\!\!\!\!/} O\\ \searrow NH_2\end{matrix} \xrightarrow{P_2O_5,\ \Delta} R{-}C{\equiv}N + H_2O \qquad (9\text{-}86)$$

The reaction is one of the methods for production of the intermediates for making nylon. Air-oxidation of cyclohexane over a cobalt salt produces adipic acid. Part of the adipic acid may then be converted into adipamide, dehydrated to adiponitrile, and then reduced to the diamine.

$$\underset{\text{Cyclohexane}}{C_6H_{12}} \xrightarrow{O_2\text{ (air), Co salt}} \underset{\substack{\text{Hexanedioic acid}\\ \text{(Adipic acid)}}}{HOOC(CH_2)_4COOH} \xrightarrow{NH_3}$$

$$\underset{\substack{\text{Diammonium hexanedioate}\\ \text{(Diammonium adipate)}}}{NH_4^+\ {}^-OOC(CH_2)_4COO^-NH_4^+} \xrightarrow{-2H_2O} \underset{\substack{\text{Hexanediamide}\\ \text{(Adipamide)}}}{H_2N{-}\overset{\overset{\displaystyle O}{\|}}{C}(CH_2)_4\overset{\overset{\displaystyle O}{\|}}{C}{-}NH_2} \xrightarrow{P_2O_5,\ \Delta,\ -2H_2O}$$

$$\underset{\substack{\text{Hexanedinitrile}\\ \text{(Adiponitrile)}}}{N{\equiv}C(CH_2)_4C{\equiv}N} \xrightarrow{H_2/Ni} \underset{\text{1,6-Hexanediamine}}{H_2N(CH_2)_6NH_2} \qquad (9\text{-}87)$$

The required hexanedinitrile may also be prepared by an S_N2 reaction of sodium cyanide with 1,4-dichlorobutane. Research and commercial development have made nylon one of our cheapest and most versatile synthetic fibers.

Cyclic amides and barbiturates. A cyclic amide results if one of the hydrogen atoms of the $—NH_2$ group of an amide is substituted by the skeletal chain that contains the amide group. Such a cyclic amide is called a *lactam*. An example is caprolactam. Upon hydrolysis with aqueous sodium hydroxide it yields sodium 6-aminohexanoate (sodium ϵ-aminocaproate).

$$\underset{\text{Caprolactam}}{\text{(cyclic } CH_2{-}C({=}O){-}NH{-}CH_2{-}CH_2{-}CH_2{-}H_2C{-})} + NaOH(aq) \longrightarrow \underset{\text{Sodium 6-aminohexanoate}}{H_2N(CH_2)_5COONa} \qquad (9\text{-}88)$$

Hydrolysis of this lactam with acid produces 6-aminohexanoic acid, which upon heating eliminates water to form a polyamide known as nylon 6.

$$n\ \text{Caprolactam} \xrightarrow{H^+} \underset{\text{6-Aminohexanoic acid}}{n\,H_2N(CH_2)_5COOH} \longrightarrow \underset{\text{Nylon 6}}{\left[{-}\overset{\overset{\displaystyle H}{|}}{N}{-}(CH_2)_5\overset{\overset{\displaystyle O}{\|}}{C}{-}\right]_n} \qquad (9\text{-}89)$$

Important cyclic-type amides called *barbiturates* are produced by condensing urea with various substituted malonic esters. (Barbiturates were synthesized by Johann F. W. A. von Baeyer, who allegedly named them after a friend, Barbara.) Phenobarbital, a sleep-inducing agent, will serve as an example.

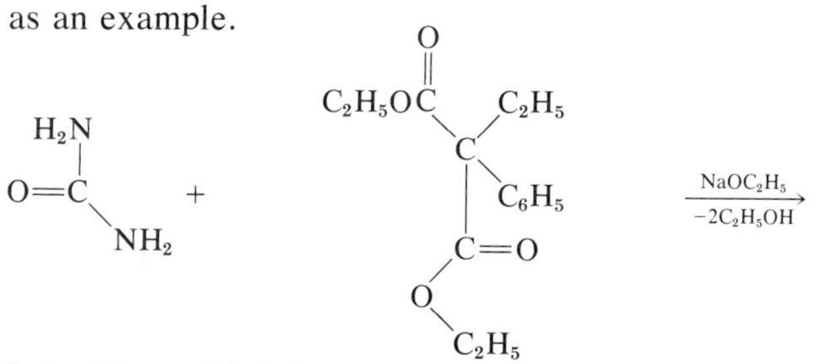

$$H_2N-C(=O)-NH_2 \quad + \quad (C_2H_5OC(=O))(C_2H_5)C(C_6H_5)-C(=O)-O-C_2H_5 \xrightarrow[-2C_2H_5OH]{NaOC_2H_5}$$

Carbamide (Urea)

Diethyl 2-ethyl-2-phenylpropanedioate (Diethyl ethylphenylmalonate)

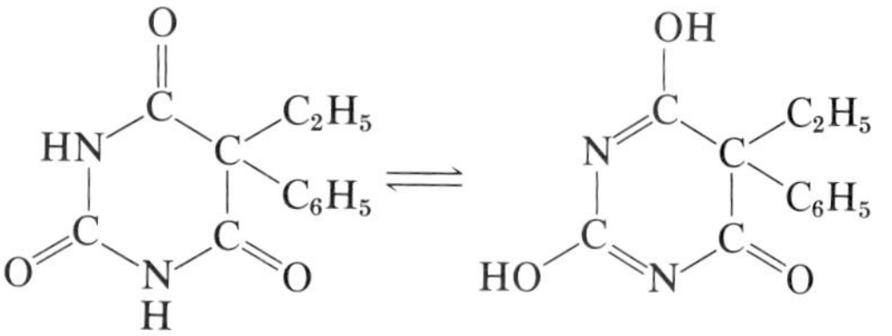

$$\text{Phenobarbital (keto form)} \rightleftharpoons \text{Phenobarbital (enol form)} \qquad (9\text{-}90)$$

Phenobarbital (keto form)

Phenobarbital (enol form)

5-Ethyl-5-phenyl-2,4,6-trioxypyrimidine

PROBLEM 9-8 Write the equations showing how the following conversions can be effected. Use any necessary reagents.

(a) ethyl acetate into acetamide
(b) acetic anhydride into ethyl acetate
(c) ethyl acetate into methyl acetate
(d) acetic acid into acetamide
(e) lauric acid into lauryl alcohol (1-dodecanol)
(f) acetic anhydride and benzene into acetophenone ($C_6H_5C(=O)-CH_3$)
(g) methyl palmitate into 1-hexadecanol
(h) ammonium adipate into the polyamide, nylon 66
(i) glycerol trilinoleate into glycerol tristearate
(j) methyl laurate into sodium laurate
(k) 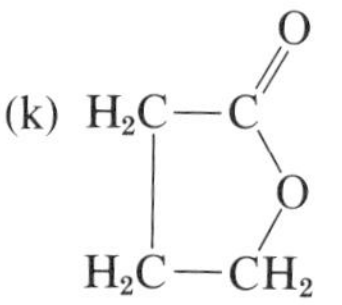 $H_2C-C(=O)-O-CH_2-H_2C$ (ring) into 4-hydroxybutanoic acid
(l) 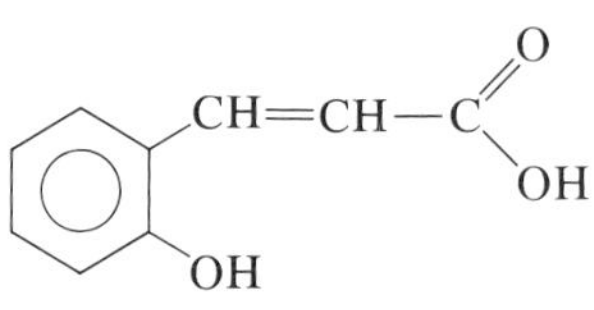 o-$HOC_6H_4-CH=CH-C(=O)OH$ into C_6H_4 fused ring $-CH=CH-C(=O)-O-$ (ring)

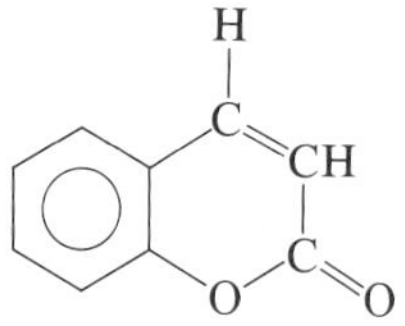

Solution to (a) Ammonolysis is the process used to convert ethyl acetate into acetamide. The equation is

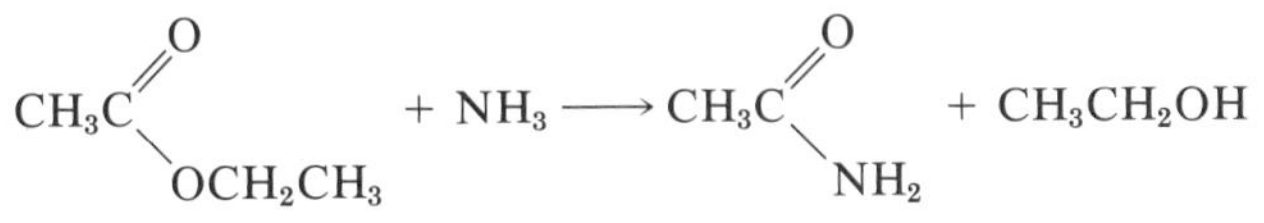

$$CH_3C(=O)OCH_2CH_3 + NH_3 \longrightarrow CH_3C(=O)NH_2 + CH_3CH_2OH$$

Ethyl ethanoate (Ethyl acetate) — Ethanamide (Acetamide) — Ethanol

Summary

The functional group of a *carboxylic acid* is the *carboxyl group*, —COOH. The general formulas for aliphatic and aromatic carboxylic acids are respectively

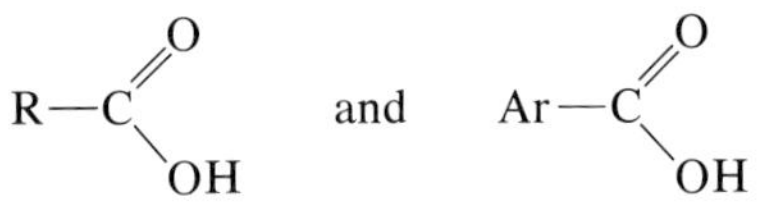

$$R{-}C(=O)OH \quad \text{and} \quad Ar{-}C(=O)OH$$

The central carbon atom of the carboxyl group has *sp^2 hybrid orbitals*. The geometry of the group is essentially *planar trigonal* with bond angles of 120° about the central carbon atom. The group is actually a *resonance hybrid,* which makes the carbonyl portion of the group *less vulnerable* to nucleophilic addition than the carbonyl group of an aldehyde or a ketone.

Hydrogen bonding between molecules of a carboxylic acid is possible. In the liquid state, carboxylic acids occur as *dimers,* in which two units of acid are held together by a pair of hydrogen bonds. For that reason the *boiling points* of carboxylic acids are *higher* than those of alcohols of corresponding molecular weight. All the homologs with four or fewer carbon atoms are *soluble* in water.

Many carboxylic acids have widely used common names. The IUPAC names of the carboxylic acids are formed by adding the suffix *-oic* to the stem of the name of the parent alkane and then adding the word *acid.* Many *dicarboxylic acids* also have common names. Their IUPAC names are formed by attaching the suffix *-dioic* to the stem of the parent alkane and adding the word *acid.* In the common naming system the position of a substituent is indicated by using a Greek letter beginning with α for the carbon atom immediately attached to the carboxyl group. In the IUPAC system, a locator number is used to indicate the position of a substituent. Numbering of the chain begins with the *carboxyl carbon atom as number 1.*

Carboxylic acids can be prepared by *oxidation* of an *aldehyde* (RCHO) or a *primary alcohol* (RCH_2OH), *oxidation* of an *arene* (Ar-R), *hydrolysis* of a *nitrile* (RCN), reaction of a *Grignard reagent* with CO_2, and *hydrolysis* of *natural esters.*

The carboxyl group may react either by *rupture of the oxygen-hydrogen bond* (reacting as an *acid*) or by *rupture of the carbon-oxygen bond* as in *nucleophilic substitution reactions.*

The *carboxylate anion,* $RCOO^-$, is a *resonance hybrid. Electron-*

withdrawing substituents on the carbon chain *increase the stability* of the resonance hybrid of the anion by increased electron delocalization. The effect makes the acid *stronger*. The strength of acids is in accord with the electronegativity of the substituent. *Electron-releasing substituents,* such as alkyl groups, *decrease the acidity*.

The carboxylic acids react *with alcohols* in the presence of an acid catalyst to produce *esters*. The reaction is reversible. Carboxylic acids may be converted into *acid halides*, $R{-}C\begin{smallmatrix}{\nearrow O}\\{\searrow X}\end{smallmatrix}$, by using PX_3, PCl_5, or $SOCl_2$. *Acid anhydrides* of monocarboxylic acids result from the *intermolecular dehydration* of two molecules of acid (not achieved directly). *Simple* or *symmetrical anhydrides* are derived from two molecules of the *same acid; mixed* or *unsymmetrical anhydrides* are derived from two molecules of *different acids*. Only acetic anhydride is used widely; it is prepared by the reaction of acetic acid with ketene. *Dicarboxylic acids* form *cyclic anhydrides* upon simple heating *if* five- or six-atom rings are possible. The carboxyl group can be *reduced* to a *primary alcohol* by using lithium aluminum hydride.

The *acyl group* is $R{-}C\begin{smallmatrix}{\nearrow O}\\{\searrow}\end{smallmatrix}$. *Acyl derivatives* of carboxylic acids are those in which the —OH part of the carboxyl group has been substituted by some other group. The most important acyl derivatives are *acid halides* (also called *acyl halides*), *acid anhydrides, esters,* and *amides*.

The *common name* for an *acid halide* is formed from the name of the *acid* by replacing the suffix *-ic* with the suffix *-yl* and adding the name of the halide. The IUPAC name is formed from the stem of the name of the parent *alkane* by adding the suffix *-oyl* and then the name of the halide. The common name for an *acid anhydride* is formed from the name of the corresponding *acid* followed by the word *anhydride*. The IUPAC name is formed by adding the word *anhydride* to the IUPAC name for the acid. The common name of an *ester* is formed by naming the alkyl group related to the alcohol followed by the stem of the acid part with the *-ic* suffix changed to *-ate*. The IUPAC name is similar, but uses the IUPAC name for the acid part. The common name of an *amide* is related to the common name for the acid from which the amide is derived; the *-ic* suffix is replaced by *-amide*. The IUPAC name is similar, but the name of the parent *alkane* is used and *-amide* is added as a suffix. A *salt* is named by naming the cation and then naming the acid anion by using the stem of the name of the acid and adding the suffix *-ate*.

The chief reaction of the *acyl derivatives* of the carboxylic acids is *nucleophilic substitution;* the *order of reactivity* is acid halide > acid anhydride > acid > ester > amide > aldehyde or ketone. The chief nucleophilic substitution reactions of the acyl derivatives are *hydrolysis, alcoholysis,* and *ammonolysis*.

Acid halides, acid anhydrides, and sometimes carboxylic acids are used in *Friedel-Crafts acylation reactions* to produce *aryl alkyl ketones.* The *α-hydrogen atoms* of *esters* are *weakly acidic,* and in the presence of a strong base they react to produce a *resonance-stabilized carbanion.* That ion is then involved in nucleophilic attack on the carbonyl carbon atom of another molecule of the ester, and the final product is a *β-keto ester*. The reaction is known as the *Claisen condensation*. An important product of it is ethyl acetoacetate, which exists as an equilibrium mixture of *tautomers*. Because of *intramolecular hydrogen bonding,* the *enol* form boils at a lower, temperature than the *keto* form.

The *amides,* with two hydrogen atoms attached to nitrogen, may undergo extensive *hydrogen bonding*. Thus they have relatively high boiling points and melting points, and they are generally solids. An amide may be prepared by the *ammonolysis* of either an acid halide or an acid anhydride or by heating the ammonium salt of carboxylic acid. *Urea* is produced industrially by the reaction of CO_2 and NH_3 to form ammonium carbamate, which is then heated to produce urea. A *polyamide fiber* called nylon 66 is produced by forming and heating the substituted ammonium salt, 1,6-hexanediammonium adipate. Important *cyclic-type amides* called *barbiturates* are produced by condensing urea with various substituted malonic esters.

Key terms

Acid anhydride: the product resulting from loss of water by an acid. It is produced by the intermolecular dehydration of two molecules of a monocarboxylic acid or the intramolecular dehydration of a dicarboxylic acid, and it has the general formula $R{-}C({=}O){-}O{-}C({=}O){-}R$

Acid halide: a compound of the general formula $R{-}C({=}O){-}X$. It is produced by the reaction of a carboxylic acid with PCl_3, PBr_3, PI_3, PCl_5, or $SOCl_2$. Also called an **acyl halide.**

Acyl derivative: a derivative of a carboxylic acid in which the —OH part of the carboxyl group has been substituted by another group. The most important acyl derivatives are acid halides, acid anhydrides, esters, and amides.

Acyl group: the group $R{-}C({=}O){-}$.

Alcoholysis: reaction with alcohol. The alcoholysis of an acyl derivative produces an ester.

Amide: an acyl derivative in which the —OH group of the carboxylic acid is replaced by $-NH_2$; the general formula is $R{-}C({=}O){-}NH_2$.

Ammonolysis: reaction with ammonia. The ammonolysis of an acyl derivative produces an amide.

Barbiturate: a cyclic-type amide produced by condensing urea with one of several substituted malonic esters.

β-Keto ester: an ester containing a carbonyl group (keto) in the β position; it has the general formula $R{-}\overset{\overset{\displaystyle O}{\|}}{C}{-}CH_2COOR'$.

Carboxyl group: the $-C\begin{matrix}{\diagup\!\!\!\!\!\diagup O}\\{\diagdown OH}\end{matrix}$ group, which can be considered a combination of a carbonyl group, $\rangle C{=}O$, and a hydroxy group, $-OH$.

Carboxylate anion: the ion $RCOO^-$; it is a resonance-stabilized hybrid ion.

Carboxylic acid: an organic compound containing the carboxyl group as the functional group. The general formula is $R{-}C\begin{matrix}{\diagup\!\!\!\!\!\diagup O}\\{\diagdown OH}\end{matrix}$ for an aliphatic carboxylic acid and $Ar{-}C\begin{matrix}{\diagup\!\!\!\!\!\diagup O}\\{\diagdown OH}\end{matrix}$ for an aromatic carboxylic acid.

Claisen condensation: the condensation of two molecules of an ester, in the presence of a strong base; it produces a β-keto ester

$$R{-}\overset{\overset{\displaystyle O}{\|}}{C}{-}CH_2{-}COOR'$$

Cyclic acid anhydride: an acid anhydride produced by the dehydration of a dicarboxylic acid; is most likely to form if the product contains a five- or six-atom ring.

Decarboxylate: to lose a carboxyl group as CO_2.

Dicarboxylic acid: an organic acid containing two carboxyl groups.

Dimer: a unit of two molecules. In the liquid state, a carboxylic acid exists as a dimer in which two units are held together by a pair of hydrogen bonds.

Enol: a compound that contains a double bond bearing a hydroxy group, $-\overset{|}{C}{=}\overset{|}{C}{-}OH$, and is capable of rearranging into a keto form, $-\overset{|}{C}H{-}\overset{|}{C}{=}O$; a tautomer in equilibrium with a keto form.

Ester: a compound of the general formula $RCOOR'$. It is produced by the reaction of a carboxylic acid with an alcohol in the presence of an acid catalyst.

Fatty acid: a carboxylic acid with an even number of carbon atoms, ranging from 4 to 22, that is obtained by hydrolysis of an animal or vegetable fat or oil.

Hell-Volhard-Zelinsky reaction: the reaction of an alkyl carboxylic acid containing an α-hydrogen atom with chlorine in the presence of red phosphorus to produce an α-chloroacid. (See Prob. 9-13.)

Hydrolysis: cleavage reaction with water; the hydrolysis of an acyl derivative produces a carboxylic acid.

Keto form: a ketone capable of rearranging into, and being in equilibrium with, an enol, $-\overset{|}{C}=\overset{|}{C}-OH$; see **tautomers.**

Ketosis: a pathological condition in which abnormally large amounts of acetone and acetoacetic acid (ketone bodies) are found in the blood and urine. It may develop into diabetes.

Lactam: a cyclic amide in which one of the hydrogen atoms of the $-NH_2$ group of the amide is substituted by the same skeletal chain that contains the amide group.

Lactone: an intramolecular cyclic ester resulting from heating a γ- or δ-hydroxy acid (see Prob. 9-17).

Mixed anhydride: an acid anhydride derived from two molecules of different carboxylic acids; also called an **unsymmetrical anhydride.**

Nitrile: an organic cyanide; the general formula is $R-C\equiv N$.

Polyamide: a polymer with linkages composed of many amide groups; the polymer linkages are $-R-\underset{H}{\underset{|}{N}}-\underset{O}{\underset{\|}{C}}-R'-$. Nylon is a polyamide.

Polyester: a polymer formed by the transesterification of the ester of a dicarboxylic acid by a dihydroxy alcohol. An example is Dacron (fiber) or Mylar (the same polymer in sheet form).

Simple anhydride: an acid anhydride derived from two molecules of the same carboxylic acid. Also called a **symmetrical anhydride.**

Tautomers: isomers which exist in equilibrium in a definite percentage; examples are the keto and enol forms of ethyl acetoacetate.

Transesterification: the alcoholysis of esters. The alcohol on the original ester is displaced by the alcohol used in the reaction, and an excess of reactant alcohol is used to drive the equilibrium to the right.

Additional problems

9-9 Draw the structural formulas and give the IUPAC name of each of the following compounds:

(a) p-aminobenzoic acid
(b) α-chlorobutyric acid
(c) 2-chloropropanoic acid
(d) 4-pyridinecarboxylic acid (isonicotinic acid)
(e) calcium oxalate
(f) zinc stearate

9-10 Write the equations for a synthesis of each of the following acids from the starting materials indicated:

(a) hexanoic acid (caproic acid) from 1-hexanol

(b) 1,6-hexanedioic acid (adipic acid) from 1,4-epoxybutane (tetrahydrofuran) via cleavage with HI and a nitrile synthesis
(c) 2-butenoic acid beginning with ethanal (acetaldehyde) via aldol condensation, dehydration, and so on
(d) 2-hydroxy-2-methylpropanoic acid beginning with addition of HCN to propanone (acetone)

9-11 Mandelic acid, α-phenyl-α-hydroxyacetic acid, may be prepared by addition of HCN to benzaldehyde followed by hydrolysis of the resulting cyanohydrin. Write equations for the reactions.

9-12 Beginning with benzene or toluene and alcohols of four or fewer carbon atoms, write the equations for the synthesis of each of the following:

(a) *m*-bromobenzoic acid
(b) *p*-$CH_3C_6H_4COOH$
(c) crotonic acid ($CH_3CH{=}CHCOOH$)
(d) methyl methacrylate, $CH_2{=}C\begin{smallmatrix}\diagup CH_3\\ \diagdown COOCH_3\end{smallmatrix}$ *Hint:* Begin by adding HCN to propanone (acetone) from oxidation of 2-propanol (isopropyl alcohol).

(e) pyruvic acid ($CH_3\overset{\overset{O}{\|}}{C}COOH$)

9-13 α-Chloroacids are made by reacting an alkyl carboxylic acid with α-hydrogen atoms with chlorine in the presence of red phosphorus as a catalyst. The reaction is known as the Hell-Volhard-Zelinsky reaction. Write the equation for the preparation of α-chlorobutyric acid (2-chlorobutanoic acid).

9-14 Salicylic acid (*o*-hydroxybenzoic acid) is produced by the reaction of CO_2 with sodium phenoxide under heat and pressure.

(a) Write the equation for the production of salicylic acid.
(b) When salicylic acid is acetylated with acetic anhydride, acetylsalicylic acid, or aspirin, is produced. Write the equation for the reaction.
(c) Write the equation for the reaction of aspirin with sodium bicarbonate. (This is the chief reaction that takes place when Alka-Seltzer is added to water. The effervescence is due to escaping carbon dioxide.)
(d) Salicylic acid may be esterified with methanol in the presence of an acid catalyst to give methyl salicylate (oil of wintergreen). Write the equation for the reaction.

9-15 Draw the structural formulas for each of the following compounds:

(a) diethyl malonate
(b) diammonium succinate
(c) glutaric anhydride
(d) dimethyl terephthalate
(e) 3,5-dinitrobenzoyl chloride
(f) propionic anhydride
(g) maleic anhydride
(h) sodium palmityl sulfate
(i) *p*-bromobenzoyl bromide

9-16 Number the following compounds in order of decreasing acidity; that is, the most acidic compound will be number 1.

____ fluoroacetic acid	____ 3-chlorobutanoic acid
____ trichloroacetic acid	____ butanoic acid
____ 4-chlorobutanoic acid	____ chloroacetic acid
____ dichloroacetic acid	____ acetic acid

9-17 When heated, γ- and δ-hydroxy acids react with themselves to eliminate a molecule of water and form intramolecular cyclic esters called *lactones*. Write the equations illustrating that reaction for the following:

(a) $HOCH_2CH_2CH_2COOH$, γ-hydroxybutyric acid (4-hydroxybutanoic acid)
(b) $HOCH_2CH_2CH_2CH_2COOH$, δ-hydroxyvaleric acid (5-hydroxypentanoic acid)

9-18 Coumarin is a naturally occurring lactone with a pleasant grassy odor. Write the equation for the reaction of coumarin when it is heated with aqueous NaOH.

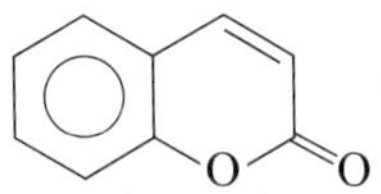

[Coumarin structure] + NaOH ⟶

Coumarin

9-19 Indicate a *simple chemical test* or *reagent* which could be used to distinguish between the members of each of the following pairs. Indicate which member of the pair gives the positive test or greater reaction.

(a) C_6H_5COOH and p-$CH_3C_6H_4OH$

(b) $CH_3C(=O)Cl$ and $CH_3C(=O)OC(=O)CH_3$

(c) $CH_3C(=O)NH_2$ and $CH_3C(=O)O^-NH_4^+$

(d) $CH_3C(=O)OC(=O)CH_3$ and $CH_3C(=O)OCH_2CH_2CH_3$

(e) $CH_3CH(OH)COOH$ and $CH_3CH(Cl)COOH$

10

FATS, OILS, WAXES; SOAP AND DETERGENTS

Natural products which are soluble in ether, chloroform, and other water-immiscible organic solvents, but insoluble in water, are known as *lipids*. The lipids are important constituents of all plant and animal tissue. Fats and oils are the most abundant of the lipids, and they are one of our three important classes of food (the other two are carbohydrates and proteins). The edible fats and oils make up approximately 40% of the American diet. Certain vegetable oils are important ingredients of paints and varnishes. They also serve as raw materials for the preparation of soap and other important commodities.

10-1 Fats and oils

Chemically, the animal and vegetable fats and oils are esters of carboxylic acids of 4 to 22 carbon atoms and the trihydric alcohol glycerol (1,2,3-propanetriol). They are sometimes referred to as *glycerides*. The chief physical difference between a fat and an oil is that the fat is solid or semisolid and oil is liquid. Chemically that difference is related to the degree of unsaturation. The oils contain more carbon-carbon double bonds in the acid part of their structures than do the fats.

As a consequence of the biosynthetic pathway by which fats and oils are produced from stepwise addition of acetyl groups, only acids with an even number of carbon atoms are generally found in the natural fats and oils. The acids commonly encountered in the natural fats and oils are included in Table 9-1. Since a great number of triglyceride combinations are possible with them, the fats and oils are never found as pure compounds. They occur as complex mixtures of many different triglycerides each of which is derived from two or three different acids. Such a molecule is re-

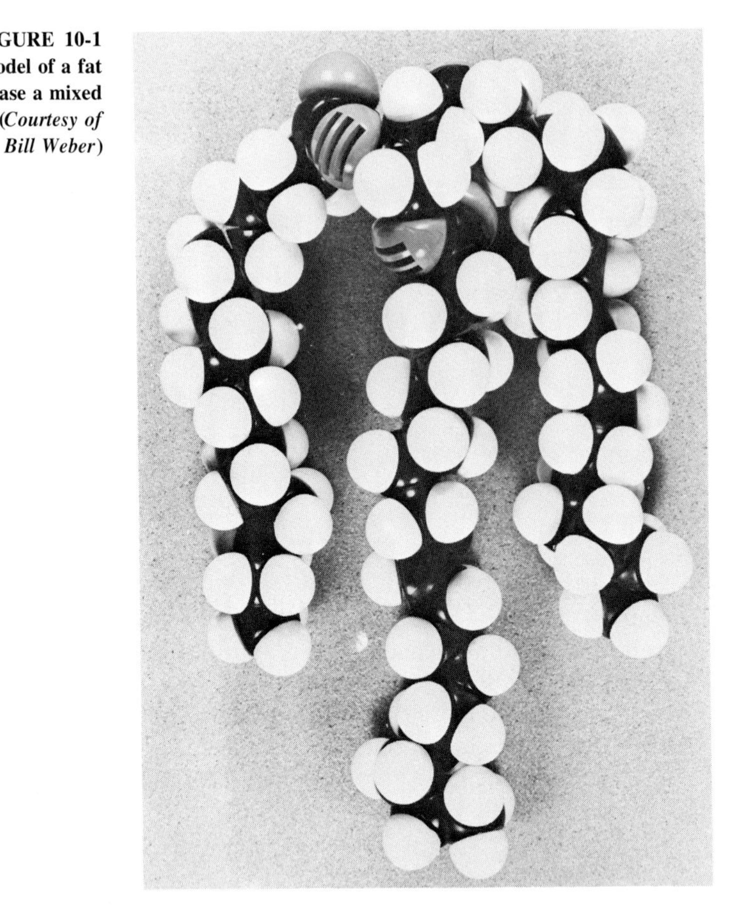

FIGURE 10-1
Space-filling model of a fat molecule; in this case a mixed triglyceride. (*Courtesy of Bill Weber*)

ferred to as a *mixed triglyceride* (Fig. 10-1). The general formula for a mixed triglyceride representing a fat or an oil is

$$
\begin{array}{l}
\mathrm{H}-\overset{\displaystyle \mathrm{H}}{\overset{|}{\mathrm{C}}}-\mathrm{O}-\overset{\displaystyle \mathrm{O}}{\overset{\|}{\mathrm{C}}}-\mathrm{R} \\
\mathrm{H}-\overset{|}{\mathrm{C}}-\mathrm{O}-\overset{\displaystyle \mathrm{O}}{\overset{\|}{\mathrm{C}}}-\mathrm{R'} \\
\mathrm{H}-\underset{\displaystyle \mathrm{H}}{\underset{|}{\overset{|}{\mathrm{C}}}}-\mathrm{O}-\overset{\displaystyle \mathrm{O}}{\overset{\|}{\mathrm{C}}}-\mathrm{R''}
\end{array}
$$

PROBLEM 10-1 Write the structural formula for each of the following glycerides.

(a) glycerol 1-palmitate 2,3-distearate
(b) glycerol 1-laurate 2-palmitate 3-linoleate
(c) glycerol 1-oleate 2-myristate 3-linoleate

Solution to (a)

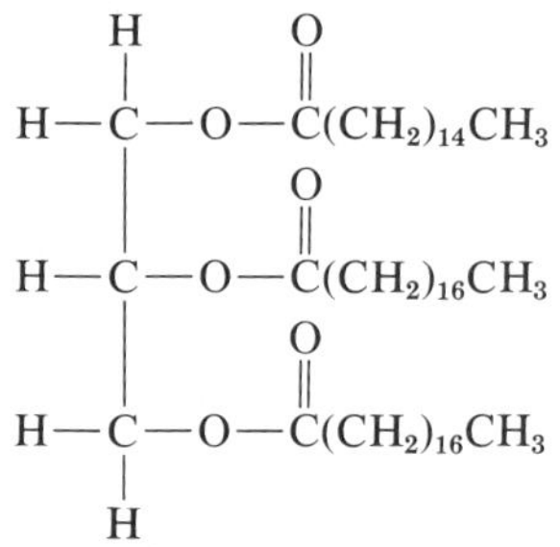

$$\begin{array}{l} \quad\;\; H \qquad\quad O \\ \quad\;\; | \qquad\quad \| \\ H-C-O-C(CH_2)_{14}CH_3 \\ \quad\;\; | \qquad\quad O \\ \quad\;\; | \qquad\quad \| \\ H-C-O-C(CH_2)_{16}CH_3 \\ \quad\;\; | \qquad\quad O \\ \quad\;\; | \qquad\quad \| \\ H-C-O-C(CH_2)_{16}CH_3 \\ \quad\;\; | \\ \quad\;\; H \end{array}$$

1,2,3-Propanetriol 1-hexadecanoate 2,3-dioctadecanoate
(Glycerol 1-palmitate 2,3-distearate)

Simple triglycerides in which all of the fatty acid moieties (parts or fragments) are the same can be *synthesized* and used for model compounds:

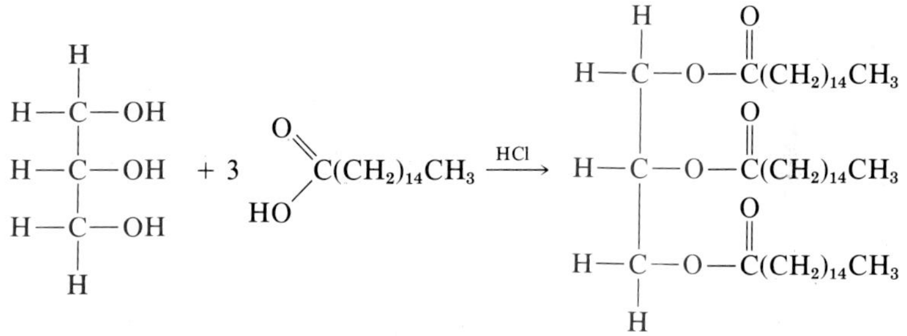

$$\begin{array}{l} \quad\;\; H \\ \quad\;\; | \\ H-C-OH \\ \quad\;\; | \\ H-C-OH \\ \quad\;\; | \\ H-C-OH \\ \quad\;\; | \\ \quad\;\; H \end{array} + 3\; HO-\overset{O}{\overset{\|}{C}}(CH_2)_{14}CH_3 \xrightarrow{HCl} \begin{array}{l} \quad\;\; H \qquad\quad O \\ \quad\;\; | \qquad\quad \| \\ H-C-O-C(CH_2)_{14}CH_3 \\ \quad\;\; | \qquad\quad O \\ \quad\;\; | \qquad\quad \| \\ H-C-O-C(CH_2)_{14}CH_3 \\ \quad\;\; | \qquad\quad O \\ \quad\;\; | \qquad\quad \| \\ H-C-O-C(CH_2)_{14}CH_3 \\ \quad\;\; | \\ \quad\;\; H \end{array} \tag{10-1}$$

1,2,3-Propanetriol (Glycerol) — Hexadecanoic acid (Palmitic acid) — 1,2,3-Propanetriol hexadecanoate (Glycerol tripalmitate)

For convenience, the formula of a simple glyceride may be used to represent a fat or an oil in a reaction. An example is *saponification* (*alkaline hydrolysis*) of a fat or an oil:

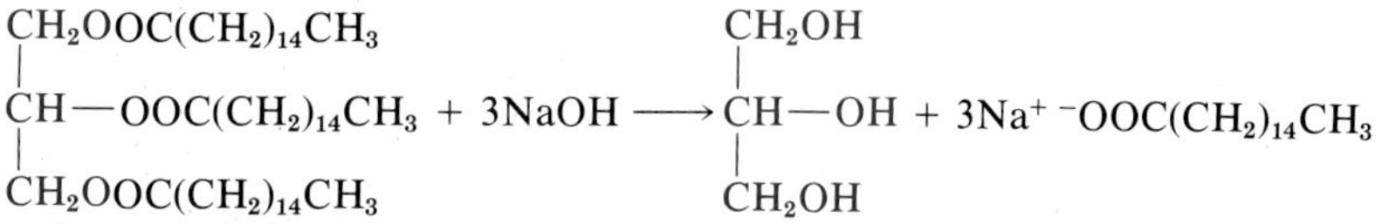

$$\begin{array}{l} CH_2OOC(CH_2)_{14}CH_3 \\ | \\ CH-OOC(CH_2)_{14}CH_3 \\ | \\ CH_2OOC(CH_2)_{14}CH_3 \end{array} + 3NaOH \longrightarrow \begin{array}{l} CH_2OH \\ | \\ CH-OH \\ | \\ CH_2OH \end{array} + 3Na^{+}\;{}^{-}OOC(CH_2)_{14}CH_3 \tag{10-2}$$

Glycerol tripalmitate (Tripalmitin) — Glycerol (Glycerin) — Sodium palmitate (A soap)

The average length of the chain of the fatty acids incorporated into a fat or an oil is expressed by the *saponification number:* the number of milligrams of potassium hydroxide required to *saponify* one gram of a fat or oil. The number varies inversely with the average length of the fatty acid chain. Consequently, butter, which has a rather large amount of the butanoic (butyric) acid moiety in it, has the highest saponification number of all of the fats and oils. Tallow, with a goodly amount of stearic acid (C_{18}), has a relatively low saponification number, and the saponification numbers of the unsaturated vegetable oils with relatively large amounts of the C_{18} unsaturated acids also are relatively low.

PROBLEM 10-2 Write the equation for the saponification of and calculate the saponification number for each of the following fats: (a) glycerol tristearate (stearin), (b) glycerol trioleate, (c) glycerol tributyrate.

Solution to (a) The equation for the saponification of glycerol tristearate is

$$\begin{array}{l} H_2C{-}OOC(CH_2)_{16}CH_3 \\ \quad | \\ H{-}C{-}OOC(CH_2)_{16}CH_3 \\ \quad | \\ H_2C{-}OOC(CH_2)_{16}CH_3 \\ \text{Glycerol tristearate} \end{array} + 3NaOH \longrightarrow \begin{array}{l} H_2C{-}OH \\ \quad | \\ H{-}C{-}OH \\ \quad | \\ H_2C{-}OH \\ \text{Glycerol} \end{array} + \begin{array}{c} 3NaOOC(CH_2)_{16}CH_3 \\ \text{Sodium stearate} \end{array}$$

The molecular formula is $C_{57}H_{110}O_6$, so the molecular weight is

$$\frac{57(12\text{ g C})}{\text{mol}} + \frac{110(1\text{ g H})}{\text{mol}} + \frac{6(16\text{ g O})}{\text{mol}} = \frac{684 + 110 + 96}{\text{mol}} = 890\text{ g per mol stearin}$$

The formula weight of KOH is 39.1 g K per mol + 16.0 g O per mol + 1.0 g H per mol = 56.1 g per mol KOH. The saponification number is the number of milligrams of KOH required to saponify 1 g of fat, so the saponification number of glycerol tristearate is given by

$$\text{Sap. no.} = \frac{(1\text{ mol stearin})(3\text{ mol KOH})(56.1\text{ g KOH})(1000\text{ mg})}{(890\text{ g stearin})(1\text{ mol stearin})(1\text{ mol KOH})(1\text{ g})} = 189\text{ mg KOH per g stearin}$$

10-2 Soap and other detergents

For hundreds of years soap has been produced and used as a cleansing agent or detergent. The pioneers made their own soap by saving animal fats and cooking them up with potash (KOH) or lye (NaOH) in a big iron kettle. The crude soap made by that method had a lot of residual hydroxide in it, so it was rather harsh on the skin of the user. The making of soap is now a highly refined industrial process. Dyes and perfumes are incorporated to make toilet soaps more attractive. Abrasives may be added to produce scouring soaps or disinfectants to produce germicidal soaps. Potassium soaps are liquid; they are used for shaving soaps.

The cleansing action of a soap is a consequence of the dual nature of soap molecules, which makes soap a *surface-active agent*—one that can operate at the interface between the phases of a heterogeneous mixture. The soap molecule has a water-soluble ionic end and a long oil-soluble hydrocarbon chain. The soap molecules become oriented in a layer or film one molecule thick about an oil droplet with the oil-soluble part extending down into the oil and the water-soluble part in the adjoining water. That

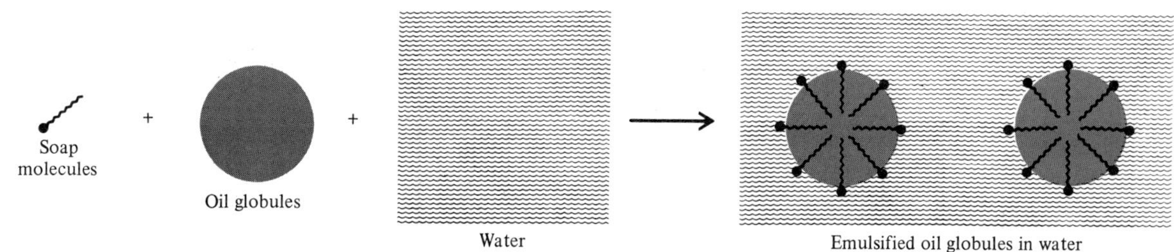

FIGURE 10-2 In the emulsifying action of soap, the polar end of the soap molecule faces the water and the nonpolar end reaches into the nonpolar oil. *(Adapted by permission from H. L. Helmprecht and L. T. Friedman,* **Basic Chemistry for the Life Sciences,** *McGraw-Hill Book Company, New York,* **1977, p. 360.)**

protective film serves to keep the oil droplets from sticking to each other or to other objects. In other words, the soap *emulsifies* the oil. It facilitates formation of a stable dispersion of the oil and dirt particles so the particles can be more readily dislodged and carried away by the wash water. See Fig. 10-2.

Hard water interferes with the action of soap because it contains calcium, magnesium, and iron(III) ions which displace the sodium ions. The result is calcium, magnesium, or iron(III) soaps, which are not soluble in water and therefore precipitate:

$$2NaOOC(CH_2)_{14}CH_3 + Ca^{2+} \longrightarrow Ca[OOC(CH_2)_{14}CH_3]_2(s) + 2Na^+ \quad (10\text{-}3)$$

Soap is thereby removed from solution, and more soap is required for cleansing to occur. Water softeners function by removing the undesirable calcium, magnesium, and iron(III) ions from the water prior to the use of soap.

Synthetic cleansing agents, whose calcium, magnesium, and iron(III) salts are soluble, give no problem with hard water. The sodium alkyl sulfates, $ROSO_2O^-Na^+$, which have already been mentioned, are one type of synthetic detergent, or "syndet." Cheaper and more common syndets are the sodium alkylbenzenesulfonates. They were made for several years by alkylating benzene with a tetramer of propene and then sulfonating and making the sodium salt.

$$4CH_3CH{=}CH_2 \xrightarrow{H^+} (H_3C)(H)C{=}CH(-\underset{|}{\overset{CH_3}{C}}H-CH_2-)_3H \quad (10\text{-}4)$$

$$C_6H_6 + (H_3C)(H)C{=}CH(-\overset{CH_3}{C}H-CH_2-)_3H \xrightarrow{HF}$$

$$C_6H_5-\overset{CH_3}{C}H-CH_2(-\overset{CH_3}{C}H-CH_2-)_3H \xrightarrow{(1)\ H_2SO_4,\ (2)\ NaOH} Na^+\,{}^-OO_2S-C_6H_4-\overset{CH_3}{C}HCH_2(-\overset{CH_3}{C}HCH_2-)_3H \quad (10\text{-}5)$$

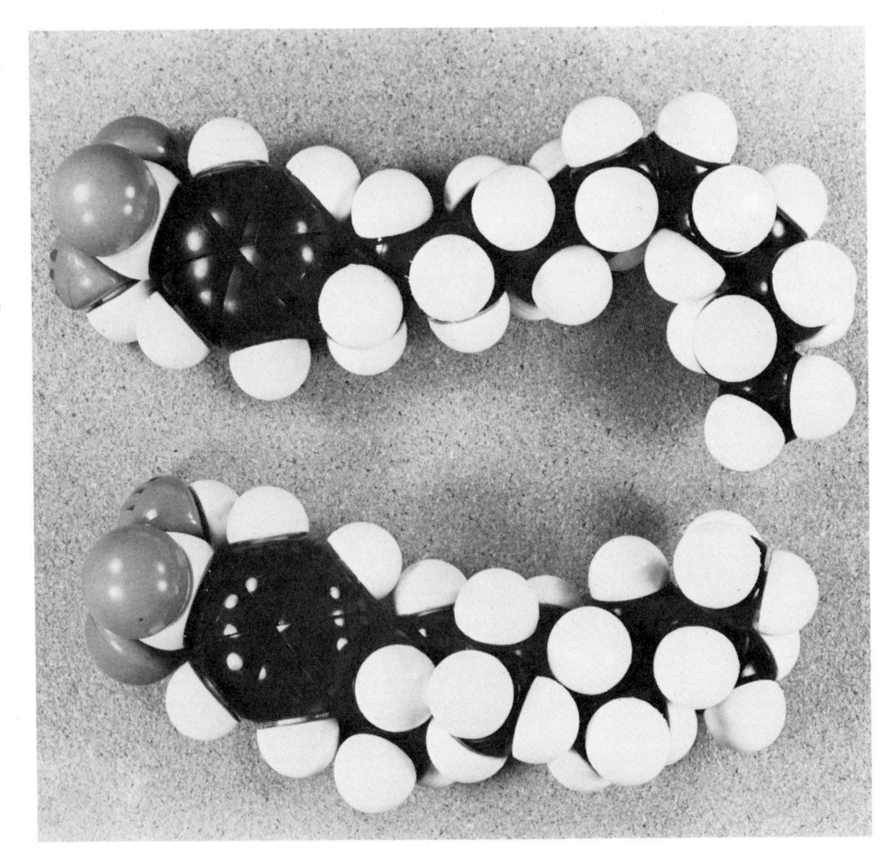

FIGURE 10-3 Space-filling models of detergent molecules. The top model, with the continuous, unbranched chain, is an example of a biodegradable detergent, while the bottom model, with the branched side chain, shows a nonbiodegradable detergent. (*Courtesy of Bill Weber*)

Although this alkylbenzenesulfonate was an effective detergent, it was found to be *nonbiodegradable*. The microbiological reactions that normally take place in sewage would not break it down. The results were excessive foaming and destruction of the microflora of sewage treatment plants. Chemists found that the difficulty was caused by the *branched side chain* on the benzene ring. When benzene was alkylated with a long *continuous-chain* alkene (produced by polymerizing ethene with metal alkyl catalysts), the resulting detergent became acceptably biodegradable (Fig. 10-3).

10-3 Hydrogenation of vegetable oils

Hydrogenation of vegetable oils was exploited to utilize cottonseed oil and provide cheaper substitutes for lard and butter. The process consists in partially hydrogenating a vegetable oil until the desired hardness is achieved. (Complete hydrogenation would result in a very hard fat containing much stearic acid, like tallow.) The reaction is illustrated by using glycerol trilinoleate as a model.

$$\begin{array}{l} \text{HCOOC(CH}_2)_7\text{CH}{=}\text{CHCH}_2\text{CH}{=}\text{CH(CH}_2)_4\text{CH}_3 \\ \text{HCOOC(CH}_2)_7\text{CH}{=}\text{CHCH}_2\text{CH}{=}\text{CH(CH}_2)_4\text{CH}_3 + 4\text{H}_2 \xrightarrow{\text{Ni, 175°C, 3 atm}} \\ \text{HCOOC(CH}_2)_7\text{CH}{=}\text{CHCH}_2\text{CH}{=}\text{CH(CH}_2)_4\text{CH}_3 \end{array}$$

(each row shown with H atoms above the double-bonded carbons; glycerol backbone H at top and bottom)

Glycerol trilinoleate

$$\begin{array}{l} \text{HCOOC(CH}_2)_7\text{CH}{=}\text{CH(CH}_2)_7\text{CH}_3 \\ \text{HCOOC(CH}_2)_{10}\text{CH}{=}\text{CH(CH}_2)_4\text{CH}_3 \\ \text{HCOOC(CH}_2)_{16}\text{CH}_3 \end{array} \qquad (10\text{-}6)$$

(And variations)

If the product is to be margarine, then salt, yellow dye, cultured milk, vitamins A and D, and the flavoring agents 2,3-butanedione and 3-hydroxy-2-butanone are added. The "soft margarines" have not been hydrogenated as much. They are more unsaturated and must be kept refrigerated to keep them solid. They are usually made from corn, safflower, or soybean oil.

10-4 Unsaturated oils

The degree of unsaturation of a fat or an oil is described by the *iodine number,* which is defined as the number of grams of iodine that will combine with 100 g of a fat or oil. Table 10-1 gives iodine numbers for some of the fats and oils.

The chief reason for the production of relatively more unsaturated soft margarines is that a dietary factor has been experimentally demonstrated in the vascular disease *atherosclerosis* by using test animals such as chickens and primates and, to some extent, humans. A high intake of saturated fats (which come chiefly from animal sources) is associated with

TABLE 10-1
Iodine numbers of some fats and oils

Fat or oil	Iodine number	Fat or oil	Iodine number
Butter	25–40	Corn	115–130
Beef tallow	30–45	Soybean	125–140
Lard	45–70	Safflower	130–150
Olive	75–95	Linseed	175–205
Peanut	85–100	Tung	160–180
Cottonseed	100–117	Fish	120–180

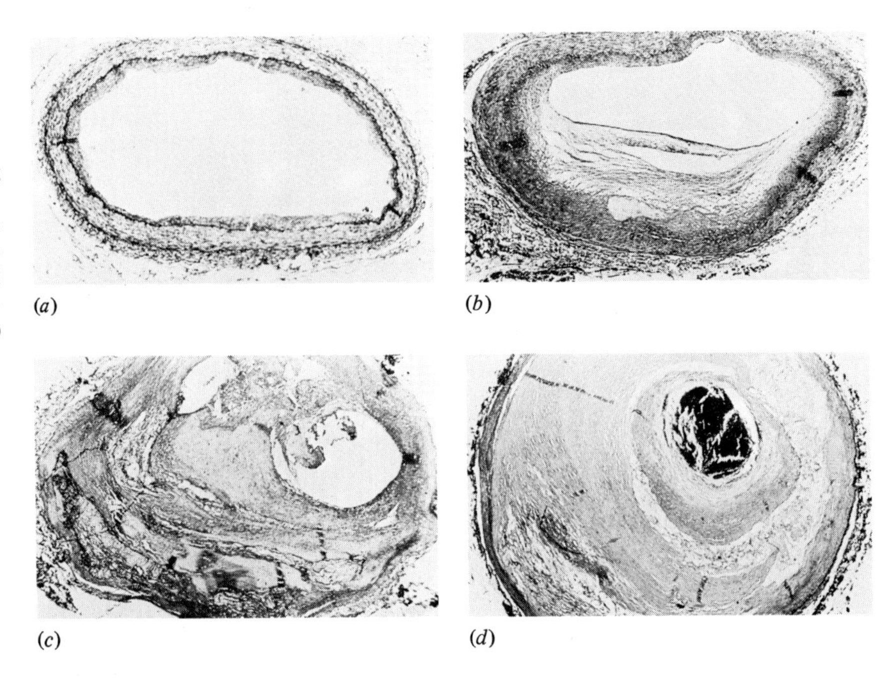

FIGURE 10-4 **Photomicrographs of cross sections of arterial walls in progressive stages of atherosclerosis. (*a*) Healthy coronary artery; (*b*) deposits starting to form in inner lining; (*c*) hardened deposits; note how narrow the channel has become; (*d*) narrowed channel is easily blocked by a blood clot. (*American Heart Association*)**

higher levels of blood serum cholesterol and glycerides and attendant higher incidence of atherosclerosis. Unsaturated fats do not have that effect (Fig. 10-4). Atherosclerosis is characterized by a thickening and stiffening of the arterial walls to such an extent that the flow of blood is impeded or may even be stopped by clot formation at the site of the restriction.

The result of atherosclerosis is the pain of angina pectoris from restricted flow, heart damage or failure, or brain damage (stroke) from an actual blockage of critical arteries. The condition is also aggravated by such other factors as lack of muscle tone, age, obesity, stress, heredity, and smoking. Since fish and fowl have relatively less saturated fat and more unsaturated fat than beef, pork, or mutton, they may be recommended for people who have atherosclerosis or who are more likely to get it. (However, shellfish, unlike fish, are rich in cholesterol.) The ancient Israelites were forbidden to eat the fat of animals; it would seem for good reason.

Unsaturated fats can develop *oxidative rancidity,* which involves the reaction of oxygen at or near the double bonds in the fatty acid chains of an unsaturated fat. The reaction produces oxygenated short-chain aldehydes, ketones, and acids that often have unpleasant odors and tastes. Although *hydrolytic rancidity* (involving hydrolysis of the ester groups) may occur in any fat, oxidative rancidity can develop only in unsaturated fats. Thus vegetable oils, richer in unsaturated acids, are more susceptible to oxidative rancidity than animal fats are. To prevent the damage, both natural and synthetic antioxidants are added in the preparation of foods,

TABLE 10-2
Constituents of oil-base paints and varnishes

Paint	Varnish
Vehicle: turpentine or a petroleum fraction	Vehicle: turpentine or a petroleum fraction
Drying oil	Drying oil
White pigment: titanium oxide (for covering power)	Resin—natural or synthetic
Colored pigments (for colored paints)	

oils, and cosmetics. The antioxidant stops the oxidation chain reaction of the unsaturated parts of the fat or oil. Some herbs and spices, such as clove, sage, allspice, and oregano, have antioxidant properties in food. However, their flavor may not be desired and food processing often destroys their antioxidant properties, so synthetic antioxidants have been developed. They are primarily complex phenols which are readily oxidized. Two common examples of synthetic antioxidants are BHA and BHT, whose structural formulas and common names are as follows.

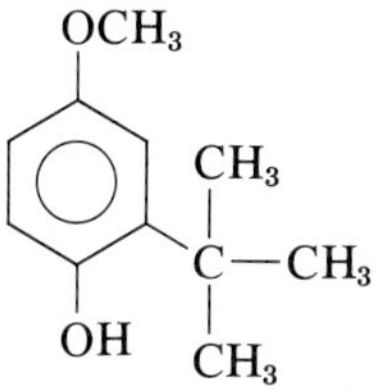

Butylated hydroxyanisole
(BHA)

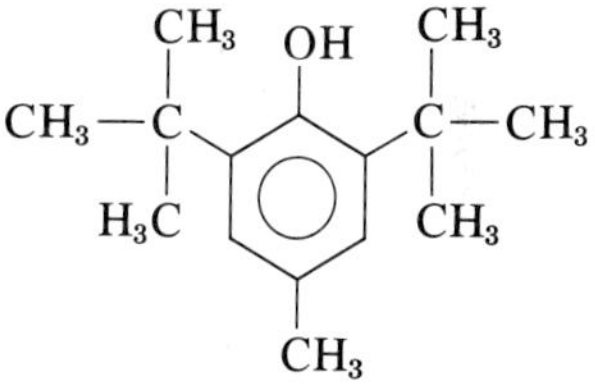

Butylated hydroxytoluene
(BHT)

Certain highly unsaturated oils, called *drying oils,* are important ingredients of paints and varnishes. Actually, they do not dry; instead, they harden by air-catalyzed polymerization. Tung oil, from the seeds of the tung tree of China, contains 75 to 91% eleosteric acid (9,11,13-octadecatrienoic acid). That acid differs from the other C_{18} unsaturated acids listed in Table 10-1 in that its double bonds are conjugated, which makes it polymerize, or ''dry,'' much faster. The chief unsaturated acid in linseed oil (which comes from flax seed) is linolenic acid. The double bonds in linolenic acid are not conjugated, so raw linseed oil hardens very slowly. That is corrected by boiling the raw linseed oil with lead salt catalysts which isomerize the double bonds to a conjugated form. The resulting boiled linseed oil is then a good drying oil. Of the oils listed in Table 10-1, linseed, tung, and fish oils are the best drying oils. Cottonseed, corn, soybean, and safflower oils are semidrying oils.

The constituents of paints and varnishes are given in Table 10-2.

PROBLEM 10-3 Write the equation for the reaction of each of the following glycerides with iodine and then calculate the iodine number. (Write the equation as though the reagent were pure iodine—which it actually is not.) (a) glycerol trioleate (olein), (b) glycerol trilinoleate (linolein), (c) glycerol trilinolenate (linolenin).

Solution for (a) The equation for the reaction of glycerol trioleate with iodine is

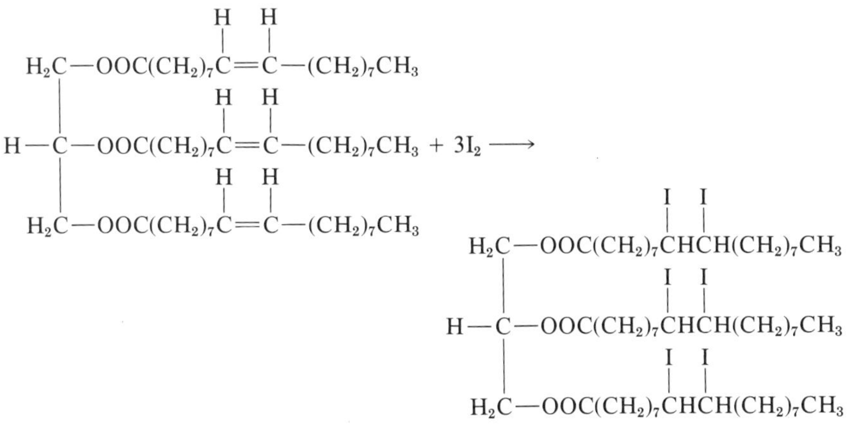

$$\begin{array}{l} H_2C{-}OOC(CH_2)_7CH{=}CH{-}(CH_2)_7CH_3 \\ H{-}C{-}OOC(CH_2)_7CH{=}CH{-}(CH_2)_7CH_3 \\ H_2C{-}OOC(CH_2)_7CH{=}CH{-}(CH_2)_7CH_3 \end{array} + 3I_2 \longrightarrow \begin{array}{l} H_2C{-}OOC(CH_2)_7CHICHI(CH_2)_7CH_3 \\ H{-}C{-}OOC(CH_2)_7CHICHI(CH_2)_7CH_3 \\ H_2C{-}OOC(CH_2)_7CHICHI(CH_2)_7CH_3 \end{array}$$

The molecular formula of glycerol trioleate is $C_{57}H_{104}O_6$, so the molecular weight is 884. The molecular weight of iodine is 2(127) = 254. The iodine number is the number of grams of iodine required for 100 g of fat, so the iodine number of glycerol trioleate is given by

$$\frac{(1 \text{ mol olein})(3 \text{ mol } I_2)(254 \text{ g } I_2)(100)}{(884 \text{ g olein})(1 \text{ mol olein})(1 \text{ mol } I_2)} = 86.2 \text{ g } I_2 \text{ per } 100 \text{ g olein}$$

10-5 Waxes

The waxes are simple esters of the longer-chain fatty acids with higher-molecular-weight alcohols. Examples are beeswax, carnauba wax, and spermacetti wax. The waxes are mixtures of esters and may even contain some ketones and hydrocarbons. Spermacetti wax, from the sperm whale, consists mainly of cetyl palmitate (hexadecanyl hexadecanoate), $CH_3(CH_2)_{15}OOC(CH_2)_{14}CH_3$. Waxes are found associated with many insects and plants. Their function appears to be prevention of loss of fluid by evaporation; many desert plants have waxy leaves. Carnauba wax, a hard wax from a Brazilian palm, finds much use in floor and automobile polishes. It consists mainly of C_{24} and C_{28} fatty acids and C_{32} and C_{34} normal alcohols.

10-6 Essential fatty acids and prostaglandins

Polyunsaturated acids, such as linoleic and linolenic acids, have been shown in nutritional experiments to be essential to the rat. Although deficiency disease has not yet been demonstrated for humans, it has been established that humans are unable to synthesize such polyunsaturated acids (sometimes referred to as vitamin F).

Humans do convert some *cis,cis*-9,12-octadecadienoic acid (linoleic acid) to *cis,cis,cis*-9,12,15-eicosanoic acid (bishomo-γ-linolenic acid),[1]

[1] *Homo* means next higher homolog, i.e., one more $—CH_2—$; *bishomo* means two more $—CH_2—$ groups in the molecule.

FIGURE 10-5
The conversion of linoleic acid to prostaglandin E_1.

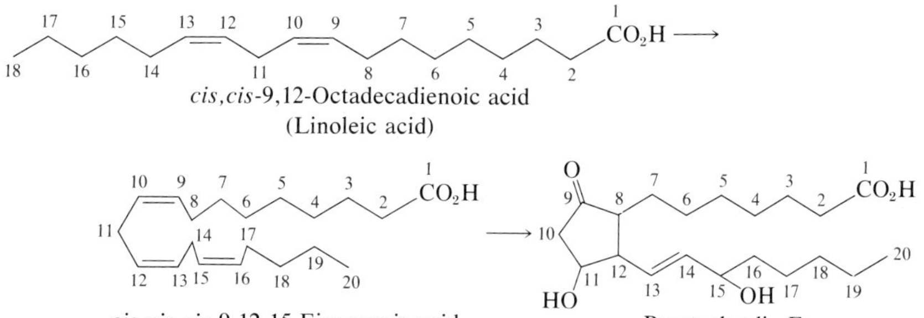

cis,cis-9,12-Octadecadienoic acid
(Linoleic acid)

cis,cis,cis-9,12,15-Eicosanoic acid
(Bishomo-γ-linolenic acid, C_{20})

Prostaglandin E_1

which is then converted to a C_{20} lipid acid called prostaglandin E_1 (Fig. 10-5).

Prostaglandin E_1 is only one of a number of closely related C_{20} unsaturated acids that are hormone-like substances originally isolated from prostate gland secretions and referred to as *prostaglandins*. At least sixteen are now known. They have been found in extremely minute amounts in a variety of body tissues and biological fluids including menstrual fluid. The prostaglandins are among the most potent biological substances known. They have many potential uses, including regulating menstruation and fertility, preventing conception, inducing labor, and relieving asthma and nasal congestion.

Summary

Lipids are natural products which are soluble in nonpolar solvents but insoluble in water. The *fats* and *oils* are the most abundant of the lipids. Fats and oils are *esters* of *glycerol* and *higher-molecular-weight fatty acids* that are referred to as *glycerides*. Oils are liquids and contain more carbon-carbon double bonds than do fats, which are solids. The *saponification number* of a fat or an oil is the number of milligrams of KOH required to *saponify* one gram of the fat or oil. Its magnitude varies inversely with the average length of fatty acid chains in the molecule.

Alkaline hydrolysis of a fat or an oil, called *saponification*, produces glycerol and sodium salts of the fatty acids (soap). *Soap* is a *detergent* or *surface-active agent*—one that can operate at the interface between the phases of a heterogeneous mixture to *emulsify* the mixture. The Ca^{2+}, Mg^{2+}, and Fe^{3+} ions in hard water displace the sodium ions to form insoluble Ca, Mg, and Fe soaps which precipitate and remove the soap from action. *Sodium alkyl sulfates*, $ROSO_2O^-Na^+$, are *synthetic detergents* whose Ca and Mg salts are soluble, so hard water is not a problem in their use. *Alkyl benzene sulfonates* are another class of synthetic detergents which work in hard water. The *straight-chain* synthetic detergents are *biodegradable* in sewage systems, whereas the *branched-chain* ones are not.

Making a vegetable oil *semisolid* by *partial hydrogenation* of its carbon-carbon double bonds is a commercial reaction for producing margarine and the solid vegetable shortenings. The *iodine number* of a fat or oil is the number of grams of iodine that will combine with 100 g of the fat or oil. The higher the iodine number the more unsaturated is the fat or oil. A high intake of saturated fats is associated with higher levels of blood serum cholesterol and glycerides and attendant higher incidence of atherosclerosis in humans.

Drying oils are certain *highly unsaturated* oils which are important ingredients of paints and varnishes. The drying oils harden by *air-catalyzed polmerization.*

Waxes are chiefly *mixtures of esters* of the longer-chain fatty acids with higher-molecular-weight alcohols.

Linoleic and linolenic acids are said to be *essential fatty acids* in our diet. Humans convert some linoleic acid into a C_{20} lipid acid called *prostaglandin* E_1. This is one of a number of closely related unsaturated acids. They have important *regulatory functions* in the body.

Key terms

Air-catalyzed polymerization: the process by which drying oils in paints and varnishes harden, or "dry."

Biodegradable: capable of being broken down by microbiological reactions.

Detergent: a washing or cleansing agent, generally with a surface-active or emulsifying action. Soap is one type; syndets are synthetic detergents of any kind.

Drying oil: a highly unsaturated oil used in paint or varnish. It hardens by air-catalyzed polymerization.

Emulsify: to convert into a suspension of small droplets of one liquid, such as oil, in another liquid, such as water, with which the first liquid is not miscible.

Essential fatty acid: a polyunsaturated fatty acid, such as linoleic acid or linolenic acid, that is essential to human metabolism but which humans cannot synthesize.

Fat: an animal or vegetable compound which is an ester of carboxylic acids of 4 to 22 carbon atoms and glycerol. It may be solid or semisolid, and it is more saturated than an oil.

Glyceride: an ester of glycerol and carboxylic acids of 4 to 22 carbon atoms. The glycerides comprise the fats and oils. Sometimes called a **triglyceride.**

Iodine number: the number of grams of iodine that will combine with 100 g of a fat or an oil. It is used to describe the degree of unsaturation of the fat or oil.

Lipid: a natural product which is soluble in ether, chloroform, and other water-immiscible organic solvents but is insoluble in water. The most abundant lipids are fats and oils.

Margarine: a partially hydrogenated vegetable oil to which salt, yellow dye, cultured milk, vitamins A and D, and flavoring agents are added to produce a substitute for butter.

Mixed triglyceride: a triglyceride derived from two or three different fatty acids.

Oil: an animal or vegetable compound which is an ester of carboxylic acids of 4 to 22 carbon atoms and glycerol. An oil, a liquid, is more unsaturated than a fat, a solid or semisolid.

Paint, oil-base: a mixture containing turpentine or a petroleum fraction as a vehicle, a drying oil, a white pigment, and possibly also a colored pigment.

Prostaglandin: a hormone-like unsaturated acid found in extremely minute amounts in a variety of body tissues and fluids. The prostaglandins have many potential uses in regulating reproductive functions and in relieving asthma and nasal congestion.

Rancidity: a condition of rank odor or taste of a fat or oil. Rancidity can involve hydrolysis of ester groups (hydrolytic rancidity) or the reaction of oxygen at or near any double bond to produce short-chain carbonyl compounds of unpleasant odor and taste (oxidative rancidity).

Saponification: the alkaline hydrolysis of esters of high-molecular-weight fatty acids to produce soap and alcohol. In general, alkaline hydrolysis of any ester or amide.

Saponification number: the number of milligrams of KOH required to saponify one gram of a fat or an oil; its magnitude varies inversely with the average length of the fatty acid chains.

Soap: a metal salt of a high-molecular-weight fatty acid. It is a type of detergent, a surface-active agent.

Surface-active agent: a compound which can operate at the interface between the phases of a heterogeneous mixture. The molecules have a dual nature, such as that of soap, which has a water-soluble ionic end and a long, oil-soluble hydrocarbon chain.

Triglyceride: an ester of glycerol and carboxylic acids of 4 to 22 carbon atoms. The triglycerides are fats and oils. Also called **glycerides.**

Varnish: a mixture of turpentine or a petroleum fraction as a vehicle, a drying oil, and a natural or synthetic resin.

Wax: a simple ester of a longer-chain fatty acid with a higher-molecular-weight alcohol. The waxes are mixtures of such esters.

Additional problems

10-4 Define and give an example of each of the following:

(a) fat
(b) oil
(c) fatty acid
(d) wax
(e) soap
(f) detergent
(g) lipid
(h) drying oil

10-5 Write the equation for the methanolysis of glycerol trilaurate. Use HCl(*g*) as a catalyst.

10-6 If a sample of safflower oil has a saponification number of 190 and an iodine number of 140, (a) what is the average molecular weight of the oil? (b) What is the average number of double bonds per molecule of the oil?

10-7 Indicate which member of each of the following pairs would have the *higher* saponification number: (a) butter or margarine, (b) tallow (a fat) or beeswax.

10-8 Indicate which member of each of the following pairs would have the *higher* iodine number: (a) olive oil or corn oil, (b) butter or margarine, (c) Crisco or cottonseed oil, (d) spermacetti wax or tallow.

10-9 Explain why the synthetic detergents function better in hard water than soap does.

11

STEREOISOMERISM AND OPTICAL ACTIVITY

In Chap. 2 we had an introduction to isomerism. We considered chain or skeletal isomerism, position isomerism, functional group isomerism, and stereoisomerism. You learned that *stereoisomers* are isomers which differ from each other only in the arrangements of their atoms in space. Stereoisomers which *do not* interconvert rapidly under normal conditions, and are therefore stable enough to be separated, are called *configurational isomers*. Stereoisomers that *do* interconvert rapidly by rotation about single covalent bonds are called *conformational isomers*. They were discussed in Chap. 2.

Geometric isomers are one of two types of configurational isomers. They exist because rotation about a carbon-carbon covalent bond is prevented either by a pi bond, as in an alkene, or by a ring structure, as in a cycloalkane. Geometric isomers were introduced in Chap. 2, and the use of cis and trans in designating their structures was discussed. A more exact system of designating their structures will be given in this chapter.

Chiral isomers are the other type of configurational isomers. In this chapter, we shall consider their nature, their properties, the ways of depicting their structures, and the ways of separating and identifying them. Many molecules of biological importance are chiral. Knowledge of the particular configuration of a molecule is often necessary to understand a biological reaction fully.

The diagram in Fig. 11-1 is a classification of the various types of stereoisomers. The terms and symbols used in the diagram will be discussed in subsequent sections of this chapter.

11-1 Chiral isomers

Chiral isomers are configurational isomers; they have *chiral* or *disymmetric molecules*. If two molecules are *identical,* they are *superimposable,* one upon the other, so that they coincide at all points. But a chiral molecule has a *mirror-image isomer,* called an *enantiomer,* which is *not superimposable* on it. Hence, the two are not identical, but are a special type of stereoisomer.

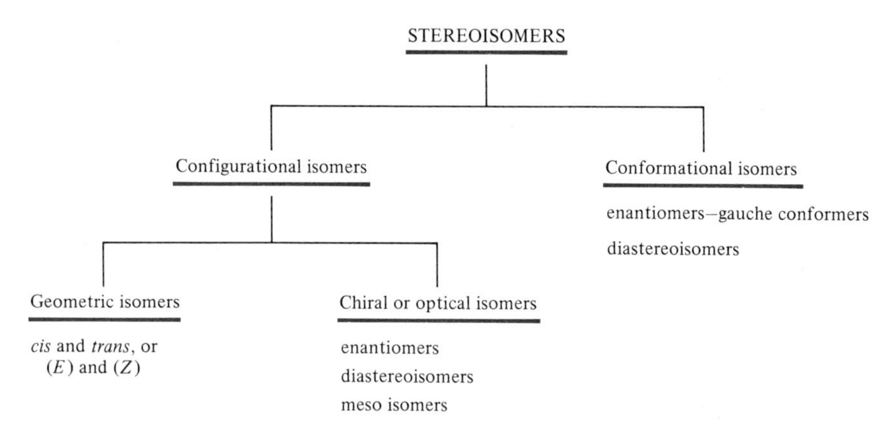

FIGURE 11-1 Classification of stereoisomers. Note that conformational isomers are not generally isolable.

Molecules that lack *reflection symmetry*—meaning that they are not identical with their mirror images—are chiral. ''Chiral'' is derived from the Greek word *cheir,* which means ''hand.'' Chirality, or ''handedness,'' is a property of dissymmetric (not symmetric) molecules. Molecules that do possess reflection symmetry—meaning they are identical with their mirror images—are said to be *achiral.* Enantiomers are not possible for them.

Enantiomeric pairs

An *enantiomeric pair* consists of two isomeric compounds whose molecules are *nonidentical mirror images*. By way of analogy, we may observe that there are objects much larger than molecules which occur as *enantiomorphic pairs* (nonidentical mirror-image objects): a person's hands, feet, ears, eyes, shoes, or gloves (see Fig. 11-2).

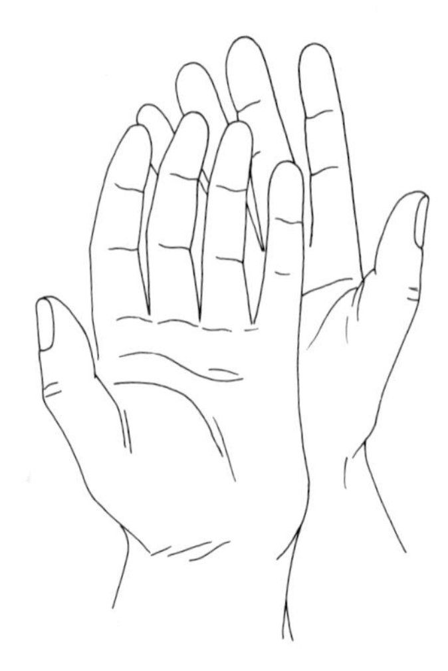

FIGURE 11-2 Right and left hands are not superimposable; they are examples of an enantiomorphic pair.

PROBLEM 11-1 Consider carefully each of the following familiar objects. Indicate whether it is chiral or achiral. (You may find it helpful to sketch the object.)

(a) wood screw
(b) nail
(c) hammer
(d) drill (bit)
(e) stocking
(f) one-arm student chair with writing desk
(g) boots
(h) automobile controls

Solution to (a) A wood screw is chiral because of the direction of its threads. Most screws have right-handed threads; the enantiomer is a left-handed screw.

Models of the enantiomers of lactic acid are shown in Figs. 11-3 and 11-4 as examples of chiral molecules—nonidentical mirror images. Their nonsuperimposability is depicted in Fig. 11-5.

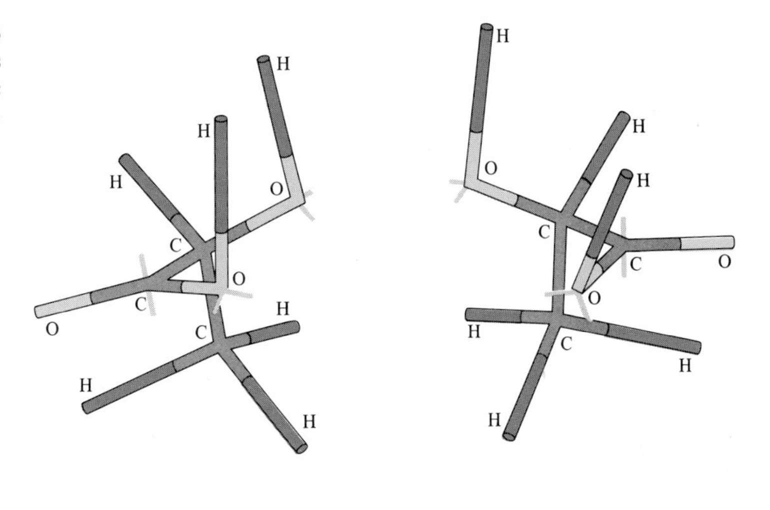

FIGURE 11-3
Framework Molecular Models of the enantiomers of lactic acid.

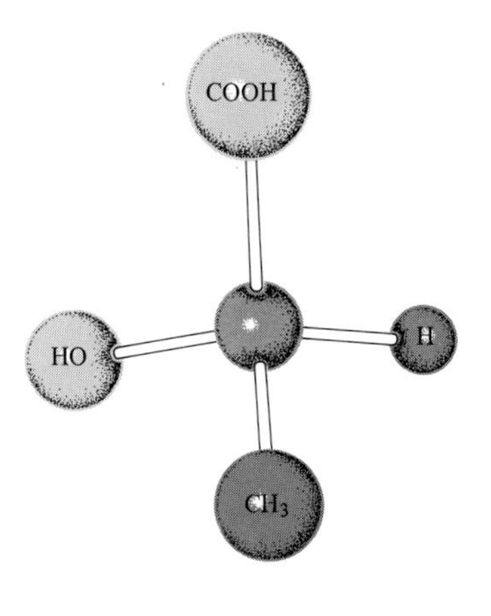

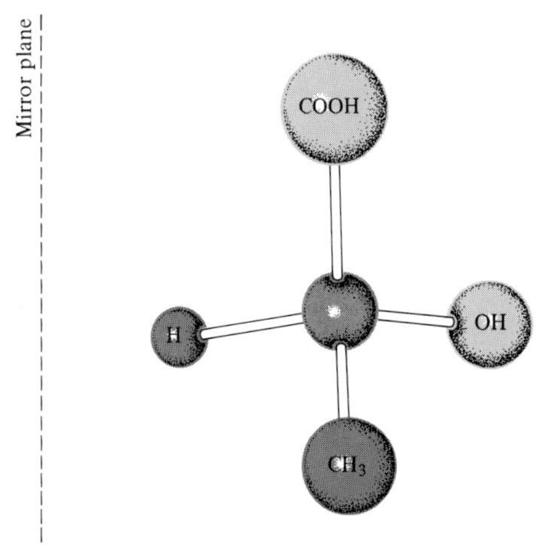

FIGURE 11-4
Simplified ball-and-stick models of the enantiomers of lactic acid.

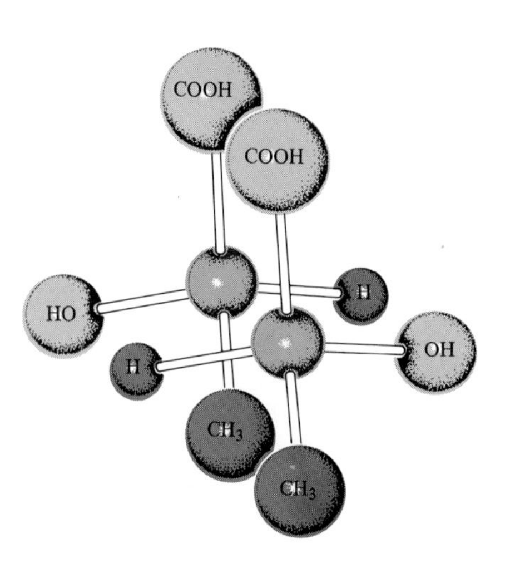

FIGURE 11-5
The enantiomers of lactic acid cannot be superimposed upon each other.

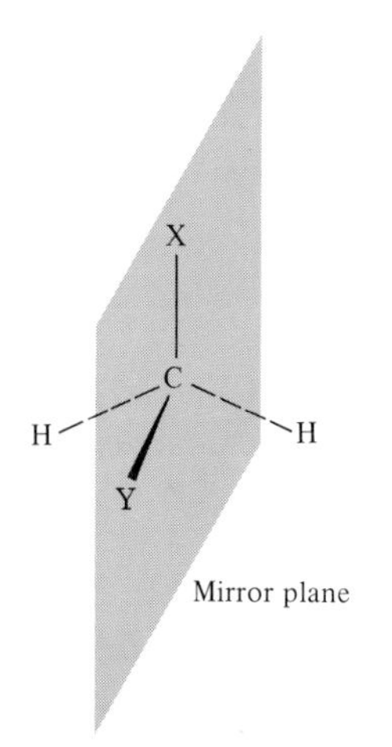

FIGURE 11-6
When two of the groups attached to a carbon atom are the same, the molecule has a plane of symmetry and is achiral.

Since Scheele discovered it in sour milk in 1780, lactic acid has been extensively investigated. It has been found to be produced by the bacterial fermentation of milk sugar (lactose) and other naturally occurring sugars.

$$\underset{\text{Lactose}}{C_{12}H_{22}O_{11}} + H_2O \xrightarrow[\text{enzymes}]{\text{bacterial}} \underset{\text{Lactic acid}}{4CH_3CHOHCOOH} \qquad (11\text{-}1)$$

In 1807 Berzelius discovered that a lactic acid could be extracted from muscle.

The chiral center

The most common origin of chirality in molecules is the presence of one or more atoms (usually carbon atoms), each of which forms noncoplanar bonds to *four different atoms* or *groups*. Thus any tetrahedral atom with four different groups attached to it is a *chiral center*. A carbon atom that carries four different substituents is described as an *asymmetric carbon atom;* it is one type of chiral center. A molecule with a chiral center has no plane of symmetry. That is the case with lactic acid. The second carbon atom along the lactic acid chain is bonded to four different groups: carboxyl, hydrogen, methyl, and hydroxy. Therefore, the lactic acid molecule is chiral and a pair of enantiomers is possible.

If a carbon atom has only three different groups, that is, if two groups are the same, then the molecule has a plane of symmetry and is achiral as shown in Fig. 11-6.

In evaluating a chemical structure for chirality, you should look for carbon atoms carrying *four different* attached groups. Also, there may be more than one asymmetric carbon atom. You should be alert to the fact that structural differences in the attached groups do not necessarily show up at the first, or even the second, position along a chain.

PROBLEM 11-2 Write the structural formula for each of the following compounds and identify any asymmetric carbon atoms in it.

(a) 2-butanol
(b) 2-methyl-1-butanol
(c) 2-hydroxypropanenitrile (acetaldehyde cyanohydrin)
(d) 2,3-butanediol
(e) 1,2-propanediol
(f) 2-chlorocyclohexanol

Solution to (a)

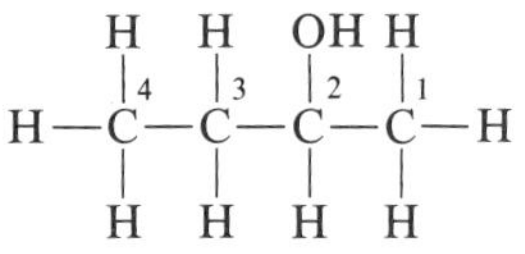

```
     H   H   OH  H
     |4  |3  |2  |1
H —  C — C — C — C — H
     |   |   |   |
     H   H   H   H
```

Carbon atom 2 is an asymmetric carbon atom, or a chiral center. It has attached to it a hydrogen atom, a methyl group, an ethyl group, and a hydroxy group.

Configurational formulas

To represent the isomerism of chiral molecules, ordinary structural formulas are inadequate. One must use *configurational formulas* that depict the molecules' three-dimensional configurations. *Perspective formulas* may be used; they are shown for the enantiomers of lactic acid in Fig. 11-7.

Configurational formulas may be more simply represented with the usual structural formulas, called *Fischer projection formulas,* if a convention is followed in their interpretation. According to convention, the asymmetric carbon atom is to be considered in the plane of the paper; the groups attached to it on the left and right are to be considered as projecting out of the plane of the paper *toward the viewer;* and the groups attached to it above and below are to be considered as projecting back of the plane of the paper. It is permissible to rotate the Fischer formula of a molecule 180° (but not 90°) in the plane of the paper to ascertain if it is superimposable on another formula. It is *not* permissible to remove it from the plane of the paper to check superimposability. If it were, the convention could be violated and the formulas could appear to coincide when they really do not. The following are the Fischer projection for-

COOH, C, HO, H, CH_3 — (*R*)-(−)-Lactic acid | Mirror plane | COOH, C, H, OH, CH_3 — (*S*)-(+)-Lactic acid

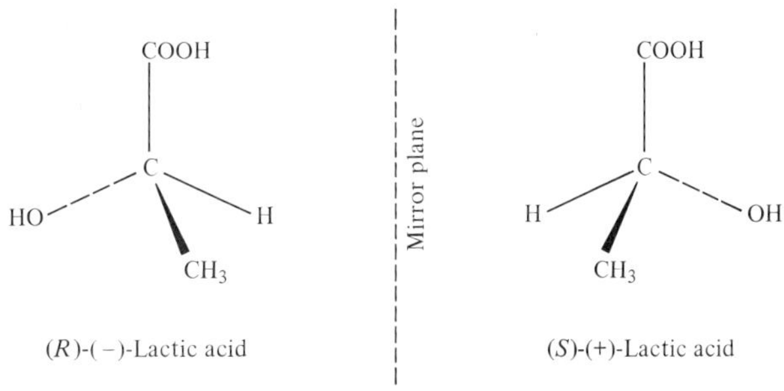

FIGURE 11-7 Perspective formulas for the enantiomers of lactic acid. Solid lines represent bonds in the plane of the paper; dashed lines represent bonds behind the plane of the paper; wedge lines represent bonds in front of the plane of the paper.

mulas for the enantiomers of lactic acid and of 2-butanol. (The *R* and *S* designations will be explained in the next section.)

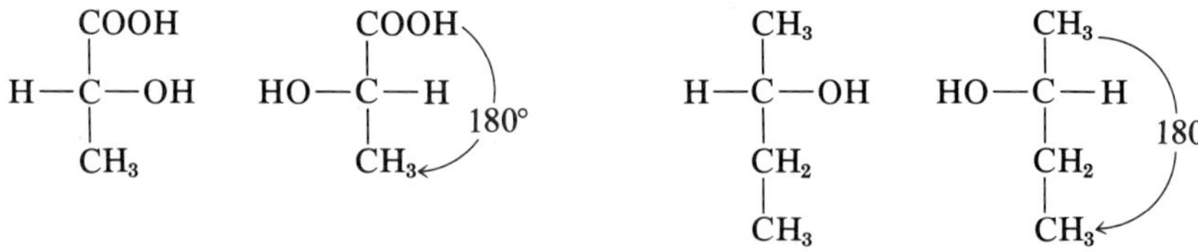

$$\begin{array}{c} COOH \\ | \\ H-C-OH \\ | \\ CH_3 \end{array} \qquad \begin{array}{c} COOH \\ | \\ HO-C-H \\ | \\ CH_3 \end{array} \curvearrowright 180^\circ \qquad \begin{array}{c} CH_3 \\ | \\ H-C-OH \\ | \\ CH_2 \\ | \\ CH_3 \end{array} \qquad \begin{array}{c} CH_3 \\ | \\ HO-C-H \\ | \\ CH_2 \\ | \\ CH_3 \end{array} \curvearrowright 180^\circ$$

(*R*)-Lactic acid (*S*)-Lactic acid (*S*)-2-Butanol (*R*)-2-Butanol

Now let us rotate the formula of one of each pair 180°, as indicated by the arrows, and test its superimposability on the other member of the pair.

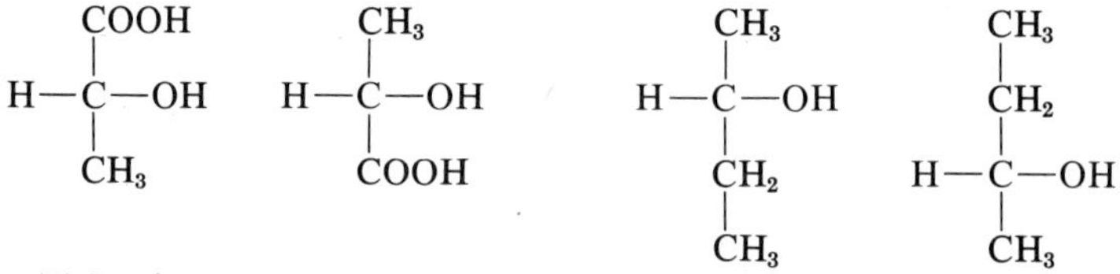

$$\begin{array}{c} COOH \\ | \\ H-C-OH \\ | \\ CH_3 \end{array} \qquad \begin{array}{c} CH_3 \\ | \\ H-C-OH \\ | \\ COOH \end{array} \qquad \begin{array}{c} CH_3 \\ | \\ H-C-OH \\ | \\ CH_2 \\ | \\ CH_3 \end{array} \qquad \begin{array}{c} CH_3 \\ | \\ CH_2 \\ | \\ H-C-OH \\ | \\ CH_3 \end{array}$$

(*R*)-Lactic acid (*S*)-Lactic acid (*S*)-2-Butanol (*R*)-2-Butanol

We observe that the formulas are *still nonsuperimposable* and represent true enantiomers. Now let us try the same thing with two mirror-image representations for 2-propanol.

$$\begin{array}{c} CH_3 \\ | \\ H-C-OH \\ | \\ CH_3 \end{array} \qquad \begin{array}{c} CH_3 \\ | \\ HO-C-H \\ | \\ CH_3 \end{array} \curvearrowright 180^\circ \qquad\qquad \begin{array}{c} CH_3 \\ | \\ H-C-OH \\ | \\ CH_3 \end{array} \qquad \begin{array}{c} CH_3 \\ | \\ H-C-OH \\ | \\ CH_3 \end{array}$$

Before rotation After rotation

After rotation of the second formula, the two representations are obviously superimposable. Thus they are identical.

Designating configuration

You have now seen how configurational isomers can be represented by formulas. It is also useful to designate isomeric configurations by *names*. That is the function of the *R* and the *S* used in the names of the examples which have been given. *R* stands for *rectus* (right), and *S* stands for *sinister* (left). The terms relate to the *Cahn-Ingold-Prelog system* of designating the *absolute configuration* about a chiral center such as an asymmetric carbon atom. The system requires the following of a definite procedure:

First, *assign an order of priority* to the four atoms or groups attached to the chiral center. The order is based on atomic number. The atom with the highest atomic number has the highest priority, if hydrogen is one of the atoms attached, it has the lowest priority. If two of the atoms attached are

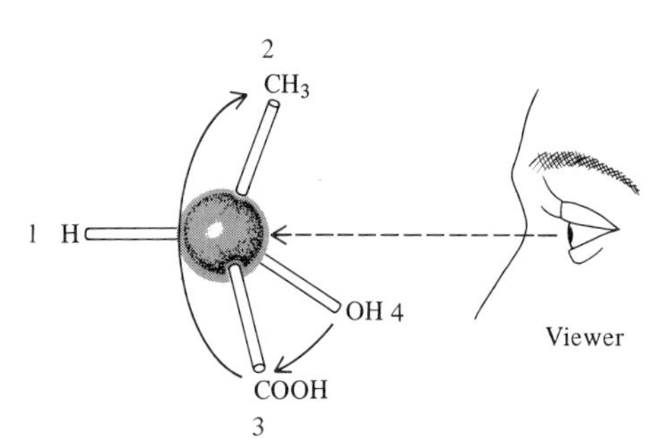

FIGURE 11-8
Determining the Cahn-Ingold-Prelog designation for an enantiomer of lactic acid. The arrows are clockwise for the *R* enantiomer.

of the same element, carbon for example, the higher priority is assigned the atom with the highest-priority elements attached to it. Multiple bonds count for multiple attachments. The priority of two like-situated isotopic atoms is determined by the mass number; the atom with the higher mass number has the higher priority.

For lactic acid the priorities of the four attached groups are

$$\underset{4}{-\text{OH}} > \underset{3}{-\text{C}(=\text{O})\text{OH}} > \underset{2}{-\text{CH}_3} > \underset{1}{-\text{H}}$$

Highest ... Lowest

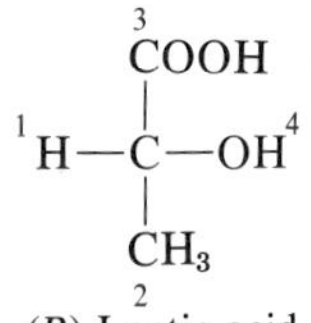

$$\begin{array}{c} \overset{3}{\text{COOH}} \\ | \\ {}^{1}\text{H}-\text{C}-\text{OH}^{4} \\ | \\ \underset{2}{\text{CH}_3} \end{array}$$

(*R*)-Lactic acid

Next, with the asymmetric carbon atom *in the plane of the paper,* mentally rotate the molecule until the group of *lowest* priority is directly behind the chiral center. In the case of lactic acid, that is the hydrogen atom. Rotation leaves the other three groups projecting out toward you. Now locate the group of *highest* priority; in the case of lactic acid, it is the —OH group. Determine in which direction the group of *next highest* priority lies—to the right (clockwise) or to the left (counterclockwise). If it is to the right, the configuration is rectus (*R*); if to the left, the configuration is sinister (*S*). Figure 11-8 illustrates the method for determining the *R* designation for the enantiomer of lactic acid shown.

If rotation to the side is necessary to have the group with lowest priority directly back of the asymmetric atom, the group on the right will be transposed to the left side in the rotated projection. If rotation of the formula to the other side is necessary, the group on the left will be transposed to the right side (Fig. 11-9*a*). However, if rotation of the formula either up or down is necessary, no transposition occurs and all three remaining groups retain their same relative positions in the new projection (Fig. 11-9*b*).

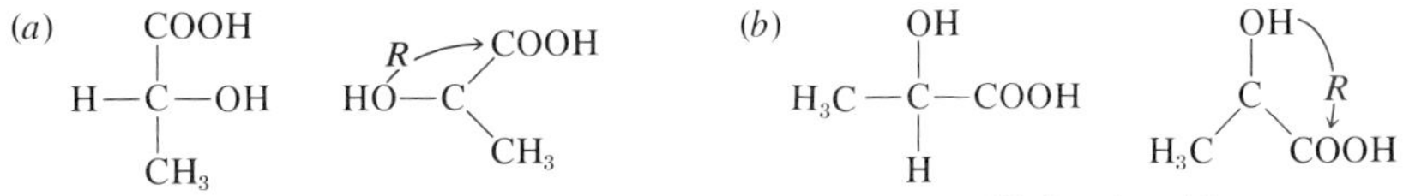

(*a*)

$$\begin{array}{c} \text{COOH} \\ | \\ \text{H}-\text{C}-\text{OH} \\ | \\ \text{CH}_3 \end{array} \qquad \text{HO}-\text{C}(\text{COOH})(\text{CH}_3) \quad (R: \text{HO} \rightarrow \text{COOH})$$

(*R*)-Lactic acid

(*b*)

$$\begin{array}{c} \text{OH} \\ | \\ \text{H}_3\text{C}-\text{C}-\text{COOH} \\ | \\ \text{H} \end{array} \qquad \text{C}(\text{OH})(\text{H}_3\text{C})(\text{COOH}) \quad (R: \text{OH} \rightarrow \text{COOH})$$

(*R*)-Lactic acid

FIGURE 11-9
(*a*) Rotation to the side is necessary so that the group of lowest priority (H) is behind the asymmetric carbon atom. As a result, the OH group is transposed from the right side to the left side. (*b*) When rotation is up or down, no transposition occurs.

That, of course, is a consequence of the geometry implied in the Fischer projection formula.

The objective is to enable you to determine the *R-S* designation simply by looking at a Fischer projection formula on paper, because physical models, although extremely helpful, are not always at hand when needed. Visualizing the whole process may be difficult. We have found that if we can rotate the formula of a stereoisomer 180° in the plane of the paper and then superimpose it on another formula, the two formulas must be identical. However, if the two cannot be superimposed after one of them is rotated 180°, it is still possible that they are identical. One way to determine identicalness is to build a model for each one and see if the two are superimposable by rotating them in various ways. That is a very instructive process, so you are urged to acquire a model set and work with it. A more convenient way is to take each model through the Cahn-Ingold-Prelog system to see whether it is an *R* or an *S* configuration. If two stereoisomers of a compound have the same *R-S* designation, they are identical; if they are of opposite *R-S* designation, they are enantiomers.

An example of the application of the Cahn-Ingold-Prelog system to the designation of configuration in the naming of chiral molecules is 3-chloro-1-butene:

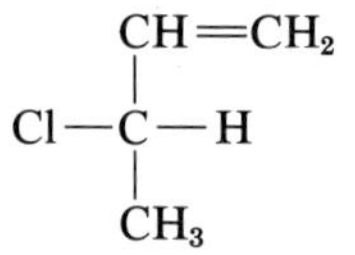

$$\begin{array}{c} CH{=}CH_2 \\ | \\ Cl{-}C{-}H \\ | \\ CH_3 \end{array}$$

Priority

4 —Cl
3 —$CH{=}CH_2$
2 —CH_3
1 —H

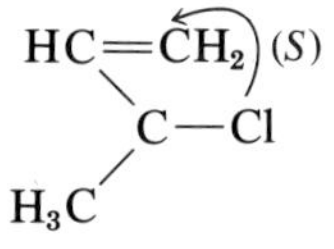

$HC{=}CH_2$ (*S*)
C—Cl
H_3C

The name is, therefore, (*S*)-3-chloro-1-butene. The formula for the same isomer could have been written

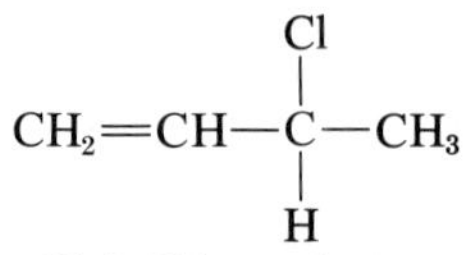

$$\begin{array}{c} Cl \\ | \\ CH_2{=}CH{-}C{-}CH_3 \\ | \\ H \end{array}$$

(*S*)-3-Chloro-1-butene

(*S*) Cl
C
$CH_2{=}HC$ CH_3

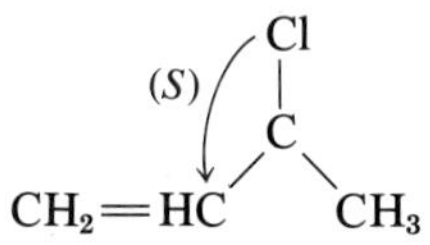

Other examples of the application of the Cahn-Ingold-Prelog system are the following:

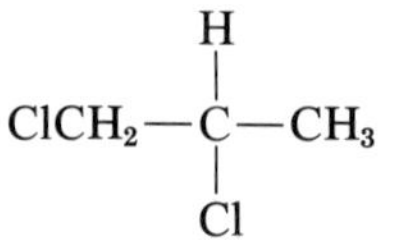

$$\begin{array}{c} H \\ | \\ ClCH_2{-}C{-}CH_3 \\ | \\ Cl \end{array}$$

(*R*)-1,2-Dichloropropane

Priority

4 —Cl
3 —CH_2Cl
2 —CH_3
1 —H

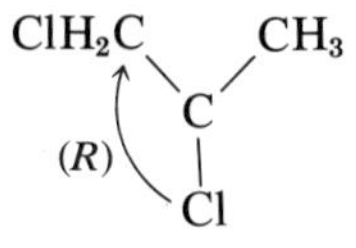

ClH_2C CH_3
C
(*R*)
Cl

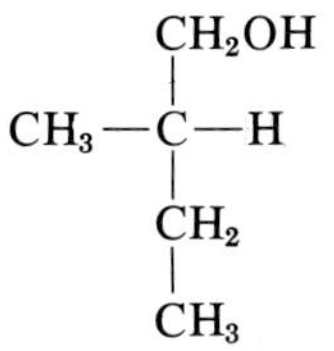

$$\begin{array}{c} CH_2OH \\ | \\ CH_3-C-H \\ | \\ CH_2 \\ | \\ CH_3 \end{array}$$

(*S*)-2-Methyl-1-butanol

Priority

4 $-CH_2OH$
3 $-CH_2CH_3$
2 $-CH_3$
1 $-H$

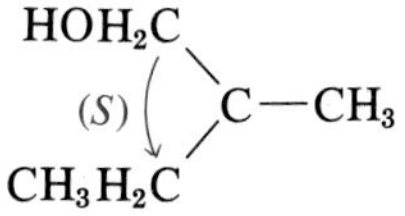

HOH_2C, CH_3H_2C $C-CH_3$ (*S*)

PROBLEM 11-3 Give the *R* or *S* designation and the IUPAC name for each of the following structures:

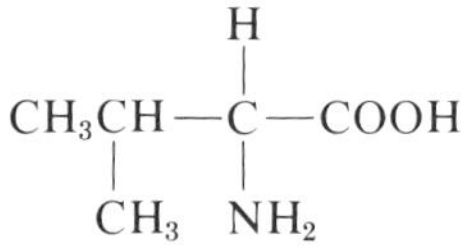

(a) $CH_3CH_2C(Br)(H)-CH_3$

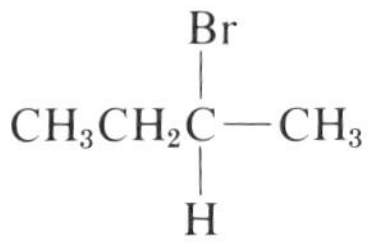

(b) $CH_3CH(CH_3)-C(H)(NH_2)-COOH$

(c) $HO-C(HC{=}O)(CH_2OH)-H$

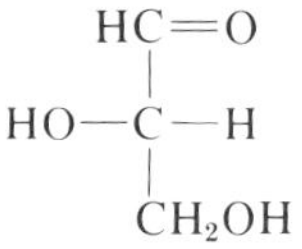

(d) $H-C(COOH)(C_6H_5)-OH$

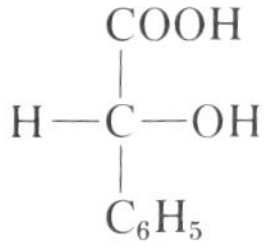

Solution to (a)

$CH_3CH_2-C(Br)(H)-CH_3$

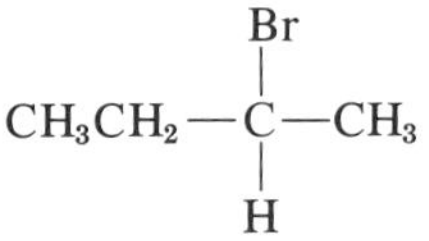

Priority

4 $-Br$
3 $-CH_2CH_3$
2 $-CH_3$
1 $-H$

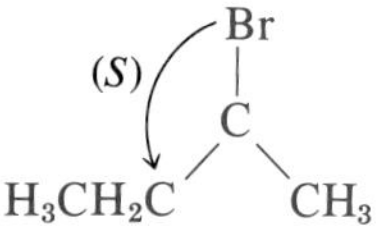

(*S*) Br, C, H_3CH_2C, CH_3

The correct name for this compound is (*S*)-2-bromobutane.

PROBLEM 11-4 Write configurational formulas for the enantiomers of the following compounds. Then determine the Cahn-Ingold-Prelog *R* or *S* designation of the configuration of each isomer.

(a) 1,2-propanediol
(b) 1-chloro-2-propanol
(c) 2-aminopropanoic acid
(d) 3-methyl-1-pentene
(e) 3-methyl-1-pentanol
(f) 2,3-dihydroxypropanal

Solution to (a) Configurational formulas of the two enantiomers of 1,2-propanediol are

1 $HOH_2C-C(OH)(H)-CH_3$ 2 $CH_3-C(OH)(H)-CH_2OH$

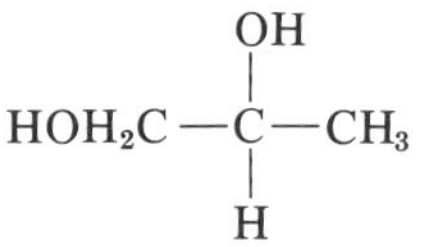

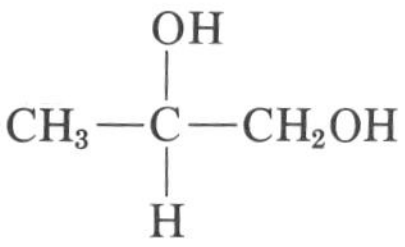

The order of priority is (4) $-OH$, (3) $-CH_2OH$, (2) $-CH_3$, (1) $-H$.

Rotation of the formulas so the group of lowest priority (—H) is directly behind the asymmetric carbon gives

1: HOH_2C, OH, CH_3 around C, arrow *S*   2: H_3C, OH, CH_2OH around C, arrow *R*

Thus the two compounds are

1: $HOH_2C—C(OH)(H)—CH_3$ (*S*)-1,2-Propanediol

2: $CH_3—C(OH)(H)—CH_2OH$ (*R*)-1,2-Propanediol

The Cahn-Ingold-Prelog system has been applied to the naming of geometric isomers, and it has been officially adopted by *Chemical Abstracts* to replace the use of cis and trans. It is especially useful in the naming of geometric isomers which are so substituted that the use of cis or trans is ambiguous. Consider, for example, the two isomeric 1-bromo-2-methyl-1-butenes.

H_3CH_2C, H_3C C=C Br, H — (*Z*)-1-Bromo-2-methyl-1-butene

H_3CH_2C, H_3C C—C H, Br — (*E*)-1-Bromo-2-methyl-1-butene

There may be uncertainty whether the hydrogen atom or the bromine atom should be considered as either cis or trans to the ethyl group.

The *E* and *Z* designations, which are used in place of cis and trans, are determined by first assigning Cahn-Ingold-Prelog priorities to the two groups attached to each end of the double bond. (In this example, *bromine* is of higher priority than hydrogen on the one end and the *ethyl group* is of higher priority than the methyl group on the other end.) If the groups of *highest priority* are on the *same side* of the double bond, the isomer is designated *Z* (from the German word, *zusammen*, meaning together). If the groups of highest priority are on opposite sides of the double bond, the isomer is designated *E* (from the German word, *entgegen*, meaning opposite).

The *E* or *Z* designation does not necessarily correspond to cis or trans usage. Consider the *E* and *Z* designations for the compounds *cis*-2-butene and *cis*-2-chloro-2-butene.

H_3C, H C=C CH_3, H — *cis*-2-Butene (*Z*)-2-Butene

H_3C, Cl C=C CH_3, H — *cis*-2-Chloro-2-butene (*E*)-2-Chloro-2-butene

The groups of higher priority are shown in color.

PROBLEM 11-5 Give the E or Z designation and the IUPAC name for each of the following alkenes.

(a) H_3CH_2C, H_3C / C=C \ Cl, Br

(c) H_3CH_2C, Cl / C=C \ CH_2OH, CH_3

(b) ClH_2CH_2C, Cl / C=C \ CH_3, CH_2CH_3

(d) H_3C, ClH_2C / C=C \ CH_3, Cl

Solution to (a) In the following formula the groups of higher priority in the E–Z system are shown in black:

H_3CH_2C, H_3C / C=C \ **Cl**, **Br**

The correct designation of this compound is (E)-1-bromo-1-chloro-2-methyl-1-butene.

11-2 Optical activity

Enantiomers are said to be *optically active* because they are able to rotate the *plane* of *plane-polarized light*. In the name of an enantiomer a (+) sign indicates that the enantiomer rotates the plane of polarized light to the right (*dextrorotatory*) and a (−) sign indicates rotation of the plane of polarized light to the left (*levorotatory*). A pair of enantiomers will rotate the plane of polarized light exactly the same amount but in opposite directions. Optical rotation is useful in identifying a particular enantiomer, because it is the only physical property in which enantiomers differ; boiling points, melting points, indexes of refraction, densities, and colors are the same. However, enantiomers may differ in odor and taste. Since chiral molecules are optically active, their type of stereoisomerism has long been known as *optical isomerism*.

Plane polarization of light

A beam of light can be considered, in one of its aspects, as a transverse electromagnetic wave vibrating in all planes perpendicular to its axis of propagation (see Fig. 11-10). When light is reflected, it becomes partially plane-polarized; that is, the components of all the wave motions vibrating in a particular plane are preferentially reflected. Thus moonlight, sky light, and light reflected from a pavement are examples of partially plane-polarized light.

Certain transparent substances have the ability to almost completely plane polarize the light they transmit. Calcite, a crystalline form of calcium carbonate, exhibits the phenomenon of *double refraction*—you see

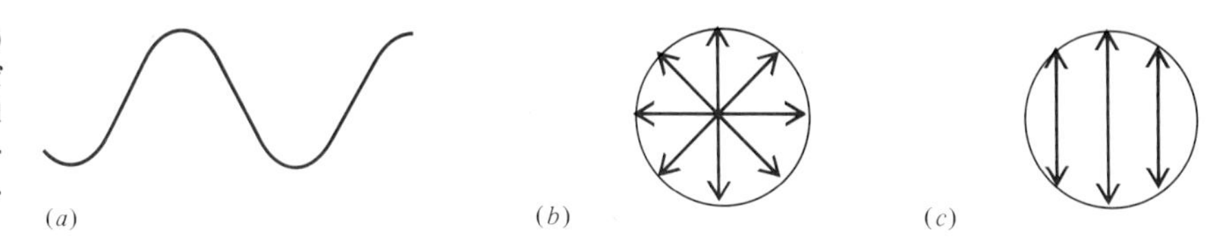

FIGURE 11-10
Representations of a beam of light: (*a*) side view; (*b*) end view; (*c*) end view of a plane-polarized beam of light.

a double image through a piece of it. It also has the ability to plane polarize light when two prisms are properly cut from it and then cemented together to form what is known as a *Nicol prism.* If two Nicol prisms are given the same orientation, light which is plane-polarized by one of them will pass essentially undiminished through the second one. However, if the second Nicol prism is rotated 90° with respect to the first one, extinction occurs: the plane-polarized light is *not* transmitted through the second prism. See Fig. 11-11 for a representation of the phenomenon.

Rotation of the plane of plane-polarized light

Experiments with plane-polarized light led to the discovery that certain transparent compounds have the ability to *rotate the plane* of plane-polarized light when placed in the path of the light. Such rotation thus requires that the second prism be oriented at some angle with respect to the first prism other than 0° (or 180°) for maximum transmission or at some angle other than 90° (or 270°) for maximum extinction. The property of rotating the plane of plane-polarized light is called *optical activity,* and compounds which rotate the plane of polarized light are said to be *optically active.*

Among the first substances found to be optically active was tourmaline, a type of quartz which crystallizes in *enantiomorphic crystals* (chiral or mirror-image crystals, Fig. 11-12). One of the enantiomorphs would rotate the plane of polarized light a certain amount in one direction, and its

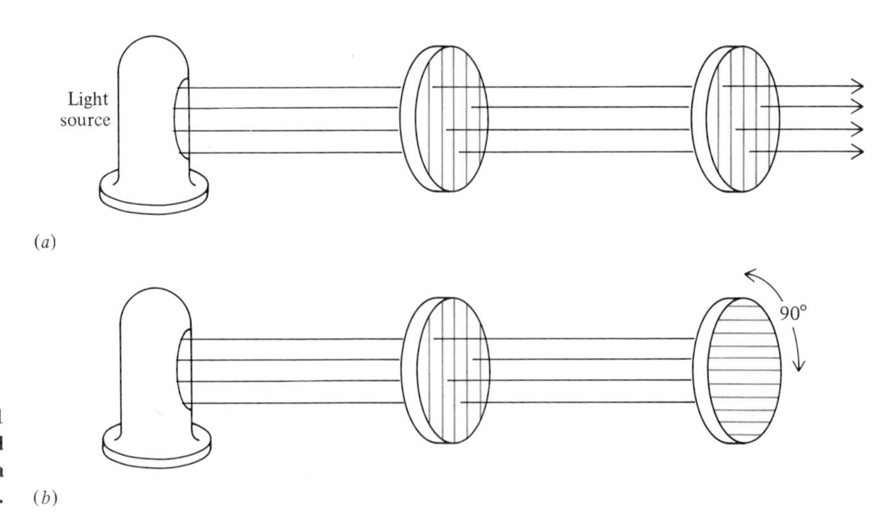

FIGURE 11-11
(*a*) Maximum transmission and (*b*) maximum extinction of a beam of plane-polarized light.

FIGURE 11-12
Enantiomorphic quartz crystals.

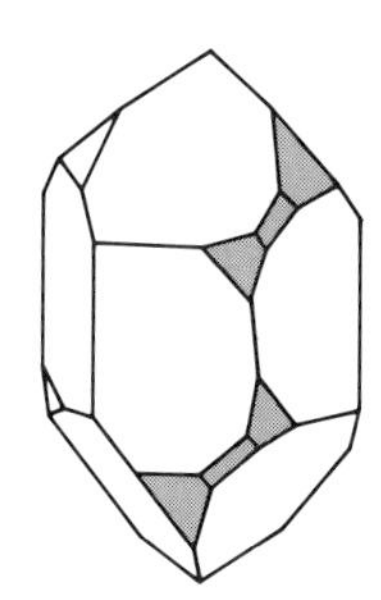
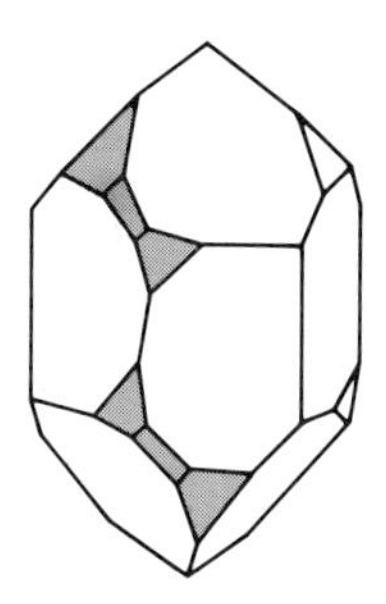

mirror-image crystal would rotate the plane by the same amount but in the opposite direction. A French physicist named Biot discovered that some organic compounds also would rotate the plane of polarized light when they were in liquid form or even when in solution. The effect suggested that the phenomenon was occurring at the molecular level.

Pasteur and tartaric acid

Louis Pasteur, a student of Biot's, was blessed with serendipity when, in 1848, he obtained enantiomorphic crystals of a salt of tartaric acid. With the aid of a magnifying glass and a pair of tweezers he was able to separate the ''right-handed'' crystals from the ''left-handed'' crystals. He found that separate solutions of the two enantiomorphs rotated the plane of polarized light the same amount but in opposite directions. If the two forms were then mixed in equal amounts in solution, the rotation was zero. Since the physical properties of enantiomers are the same except for the direction of rotation of the plane of plane-polarized light, separation of such enantiomers is very difficult. Pasteur's mechanical separation of the crystals of enantiomers was the first such separation, which is called *resolution*.

Unfortunately, very few enantiomers crystallize in enantiomorphic crystals so that resolution by mechanical separation is possible. Resolution has also been effected by a biochemical method in which an enzyme attacks one of the enantiomers and not the other. The method permits the recovery of one of the enantiomers, but the other enantiomer is changed or destroyed. Enantiomers also are alike in all of their *chemical* properties *except* in their reactions with other optically active compounds. Enzymes are always optically active compounds, so they do not react the same way with the enantiomers in a *racemic mixture* (a 50:50 mixture of enantiomers). An enzyme in *Penicillium glaucum* will react with (+)-tartaric acid and leave the (−)-tartaric acid in a mixture of the two.

The most generally applicable method for resolution of a racemic mixture is discussed in Sec. 11-5.

Explanation of optical activity

Molecules of *non–optically active* compounds are *achiral;* that is, they have an element of symmetry so that identical molecules can serve as their own mirror images. A usual random distribution of such molecules results in equal numbers of molecules oriented in such a way that they are opposed as mirror images of each other. Polarized light penetrating into such an arrangement of molecules will then be rotated as much in one direction as it is in the opposite direction and the net result will be zero rotation. For example, molecules of bromochloromethane can serve as their own mirror images, but they are still identical, superimposable mirror images (Fig. 11-13).

FIGURE 11-13 **These molecules of bromochloromethane are mirror images. However, when (*b*) is rotated 180°, it is superimposable on (*a*), and hence the molecules are identical.**

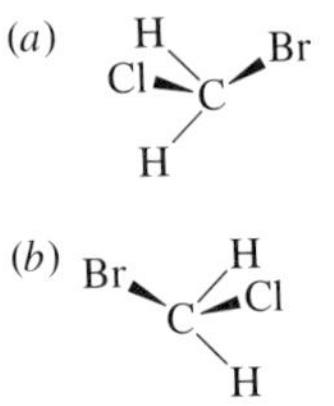

(*a*) H, Br, Cl, C, H

(*b*) Br, H, Cl, C, H

However, molecules of *optically active compounds* are *chiral*. One molecule can *never* serve as the mirror image of another identical molecule, *no matter how it is oriented*. Polarized light is, therefore, rotated in the same direction by a particular stereoisomer. The mirror-image molecules, which rotate the light the same amount but in the opposite direction, must be molecules of the enantiomer, a different compound.

Measurement of optical rotation

The direction and amount of rotation caused by an optically active compound are determined by an instrument called a *polarimeter*. The instrument consists of a long tubular optical system containing a Nicol prism at the light-source end (the *polarizer*), an opening in the tube in which a *polarimeter tube* containing the compound or its solution can be inserted, and another Nicol prism (the *analyzer*) attached to an eyepiece which can be rotated (Fig. 11-14). Rotation of the eyepiece moves an indicator across a circular scale calibrated in degrees. The observer places a sample in the polarimeter tube and rotates the analyzer to the right or left as necessary to again balance light intensity in the field of vision. The amount of rotation is read in degrees on the circular scale.

The polarization of light is essentially a *refraction phenomenon*. Refraction is the bending of a ray or wave that is passed obliquely from one

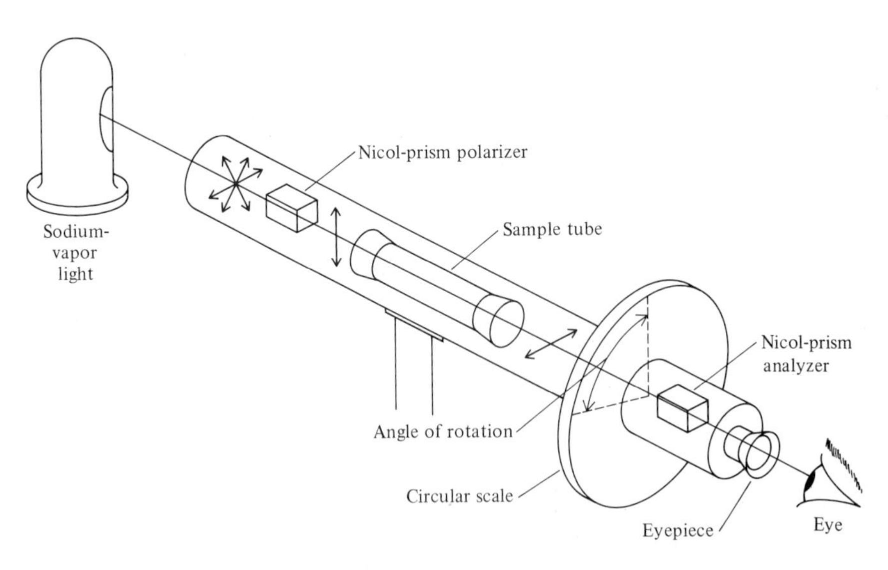

FIGURE 11-14 **A polarimeter.**

medium to another medium of different density. The change of density alters the speed of the wave front and results in a change of direction, i.e., a bending, of the ray or wave. Because refraction is a function of density, the amount of rotation depends on the density of the liquid (or concentration of the solution), the amount of material the light passes through (as represented by the length of the containing tube), and the temperature. In order to standardize conditions so that comparisons of optical rotation are meaningful, the observed rotation is converted to a specific rotation by the formula

$$[\alpha]_{\mathrm{D}}^{t} = \frac{\alpha}{lc} \tag{11-2}$$

where α = observed rotation
$[\alpha]$ = specific rotation
l = length of tube, dm (decimeters)
c = density of liquid or concentration of solution, g/ml
D = D lines of sodium vapor light, 589.3 nanometers (nm)
t = temperature, usually 25°C, formerly 20°C

Since refraction also results in dispersion of light into a spectrum of colors, the observation is made more accurate by using a *monochromatic* source of light (light of single, or almost single, wavelength). The most popular and convenient light source is the sodium vapor lamp, which emits an intense yellow light.

PROBLEM 11-6 Calculate the specific rotations of each of the following solutions measured in a 2.0-dm tube with the sodium vapor light at 25°C.

(a) A liquid of density 1.25 g/ml, with an observed rotation of +50.0°.
(b) A solution of 10.0 g per 50.0 ml of solution, with an observed rotation of −4.0°.
(c) A solution of 10 g per 100 ml of solution, with an observed rotation of +8.4°.

Solution to (a)

$$\text{Specific rotation} = [\alpha]_{589.3\,\text{nm}}^{25^\circ\text{C}} = \frac{\alpha}{lc}$$

$$= \frac{+50.0^\circ}{(2.0\ \text{dm})(1.25\ \text{g/ml})} = +20.0^\circ$$

11-3 Molecules with more than one asymmetric carbon atom

The tartaric acids occur in grape juice and wine, and they were well known to and studied by Pasteur and other French chemists in the early 1800s. They are stereoisomers of 2,3-dihydroxysuccinic acid, and study of them eventually led to an understanding of the principles of stereochemistry. Observe that the tartaric acid molecule contains two asymmetric carbon atoms; the general structural formula is

HOOC—CHOH—CHOH—COOH

van't Hoff rule for number of stereoisomers

When a molecule has more than one asymmetric carbon atom, the number of possible stereoisomers is increased accordingly. If all the asymmetric carbon atoms in the molecule are differently substituted, then the van't Hoff rule gives the number of possible stereoisomers as 2^n, where n is the number of unlike or differently substituted asymmetric carbon atoms (J. H. van't Hoff, 1852–1911, Dutch physical chemist). If any of the asymmetric carbon atoms are alike, however, the total number of stereoisomers is reduced.

The meso stereoisomer and diastereoisomers

Since tartaric acid has two asymmetric carbon atoms, application of the van't Hoff rule indicates $2^2 = 4$ stereoisomers. The configurations would be as follows:

```
     COOH               COOH             COOH               COOH
      |                  |                |                  |
  H—C—OH            HO—C—H          H—C—OH            HO—C—H
      |                  |                |                  |
 HO—C—H              H—C—OH         H—C—OH            HO—C—H
      |                  |                |                  |
     COOH               COOH             COOH               COOH
(R,R)-(+)-Tartaric  (S,S)-(−)-Tartaric  (R,S)-Tartaric  or  (S,R)-Tartaric
      acid               acid             acid                acid
      \_________________________/        \_________________________/
             Enantiomers                      Meso tartaric acid
                                              (optically inactive)
```

Since the two asymmetric carbon atoms are substituted by exactly the same groups, there are *fewer* than 2^n isomers. Actually, there is only one pair of enantiomers (the *R*,*R*- and *S*,*S*-tartaric acids).

The other pair shown (the *R*,*S* and the *S*,*R* configurations) consists of one and the same compound. If the *S*,*R* isomer is rotated 180° in the plane of the paper, it is superimposable on the *R*,*S* isomer, so the two are identical. Furthermore, in the conformations shown, a plane of symmetry exists.

```
      COOH            COOH
       |               |
   H—C—OH        HO—C—H
 ------+---------------+------------ plane of symmetry
   H—C—OH        HO—C—H
       |               |
      COOH            COOH
   R,S isomer      S,R isomer
```

Thus the molecule is *achiral*, and hence it is *optically inactive* despite the fact that it contains two asymmetric carbon atoms.

A stereoisomer that has multiple asymmetric centers but is an achiral molecule and so is optically inactive is called *meso*. Meso tartaric acid is a diastereoisomer of the other two isomeric tartaric acids. Stereoisomers of a given molecular formula which are not enantiomers (mirror-image isomers) are called *diastereoisomers*. They have different physical and chemical properties, and so they can be separated by such usual techniques as fractional distillation, fractional crystallization, and chromatographic adsorption.

The Cahn-Ingold-Prelog *R* or *S* designation is given to the configuration about each asymmetric carbon atom in a stereoisomer with multiple asymmetric carbon atoms. If there is any ambiguity about which asymmetric carbon atoms are being designated, locator numbers should also be used with the *R* or *S* designation. For example, note how the following compounds are named as to their *R*-*S* designation.

```
        CH2Cl                        H—C=O
          |                            |
     H—C—Cl                     HO—C—H
          |                            |
     Cl—C—H                       H—C—OH
          |                            |
        CH3                          H—C—OH
                                       |
                                     CH2OH
```

(2*S*,3*S*)-1,2,3-Trichlorobutane

(2*S*,3*R*,4*R*)-2,3,4,5-Tetrahydroxypentanal (D-Arabinose)

If one of the hydroxy groups in the tartaric acid molecule is changed to an —NH_2 group, then the two asymmetric carbon atoms are no longer alike. There are then $2^2 = 4$ stereoisomers consisting of two pairs of enantiomers.

```
      Enantiomers          Diastereoisomers          Enantiomers
    COOH             COOH              COOH                COOH
     |                |                 |                   |
 H—C—OH         HO—C—H             H—C—OH              HO—C—H
     |                |                 |                   |
 H2N—C—H          H—C—NH2            H—C—NH2             H2N—C—H
     |                |                 |                   |
    COOH             COOH              COOH                COOH
```

(2*R*,3*R*)-2-Hydroxy-3-aminosuccinic acid; (2*S*,3*S*)-2-Hydroxy-3-aminosuccinic acid; (2*R*,3*S*)-2-Hydroxy-3-aminosuccinic acid; (2*S*,3*R*)-2-Hydroxy-3-aminosuccinic acid

These examples serve to illustrate the increasing numbers of stereoisomers that can arise with multiple chiral centers. The sugar D-glucose (dextrose), which will be studied in Chap. 12, has five asymmetric carbon atoms in its cyclic form, so there are $2^5 = 32$ stereoisomers. Although not all the members of such a group of stereoisomers are apt to be of equal importance, the properties of the more important members depend on unique structure, so structures must be understood.

11-4 Absolute configuration

In 1949 J. M. Bijvoet, of the University of Utrecht, determined by using x-ray diffraction the actual arrangement in space of the atoms of a salt of (+)-tartaric acid. He thus made the first determination of the *absolute configuration* of an optically active substance, and he found that the substance actually did have the configuration that had previously been assigned it. Prior to that time all configurations that had been assigned were only relative to each other.

If one stereoisomer can be converted into another stereoisomer without breaking any bonds to the asymmetric carbon atom, it can be assumed to have the same configuration. It may or may not have the same *R-S* designation, since the priority order in the new compound might be different than that in the old compound. For example, (*S*)-3-hexanol is obtained by hydrogenation of (*R*)-1-hexene-3-ol.

$$\begin{array}{c} CH_2 \\ \| \\ H—C \\ | \\ H—C—OH \\ | \\ CH_2CH_2CH_3 \end{array} + H_2 \xrightarrow{Pd} \begin{array}{c} CH_3 \\ | \\ CH_2 \\ | \\ H—C—OH \\ | \\ CH_2CH_2CH_3 \end{array} \qquad (11\text{-}3)$$

(*R*)-1-Hexene-3-ol — (*S*)-3-Hexanol

The two compounds have the same configuration but different *R-S* designations. The group $—CH{=}CH_2$ is of higher priority than $—CH_2CH_2CH_3$, but $—CH_2CH_2CH_3$ is of higher priority than $—CH_2CH_3$.

The direction and amount of rotation depend on the density of the compound and the nature of the substituents on the chiral center, as well as on the configuration. The direction of rotation may change in a conversion even though the configuration does not. The following conversion is an example:

$$\begin{array}{c} CH_3 \\ | \\ H—C—CH_2OH \\ | \\ C_2H_5 \end{array} + HCl \xrightarrow{ZnCl_2} \begin{array}{c} CH_3 \\ | \\ H—C—CH_2Cl \\ | \\ C_2H_5 \end{array} \qquad (11\text{-}4)$$

(*S*)-(−)-2-Methyl-1-butanol — (*S*)-(+)-1-Chloro-2-methylbutane

The configuration is retained in this reaction, but the direction of optical rotation changes.

PROBLEM 11-7 Write the configurational formulas for all the isomers resulting from the chlorination of (*R*)-2-chlorobutane to produce isomeric dichlorobutanes. Name each isomer and label each diastereoisomer with its appropriate *R-S* designation. State how many fractions could be obtained in a fractional distillation and tell whether each fraction as obtained would be optically active or inactive.

$$\begin{array}{c} CH_3 \\ | \\ Cl—C—H \\ | \\ CH_2 \\ | \\ CH_3 \end{array} + Cl_2 \xrightarrow{light} C_4H_8Cl_2 + HCl$$

(*R*)-2-Chlorobutane

PROBLEM 11-8 Supply the information requested in Prob. 11-7 for the following reactions.

(a) 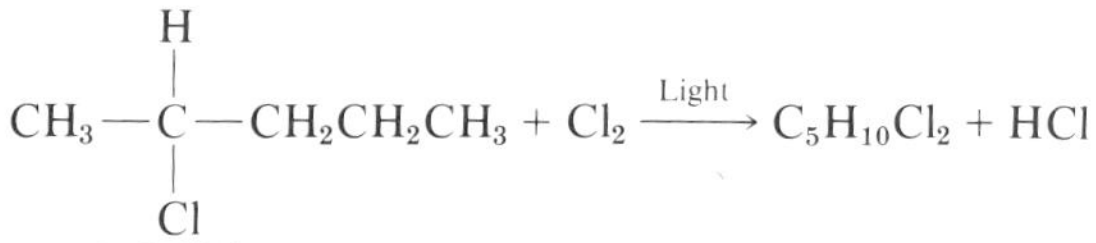
$$CH_3-\underset{\underset{Cl}{|}}{\overset{\overset{H}{|}}{C}}-CH_2CH_2CH_3 + Cl_2 \xrightarrow{\text{Light}} C_5H_{10}Cl_2 + HCl$$

(*S*)-2-Chloropentane

(b) 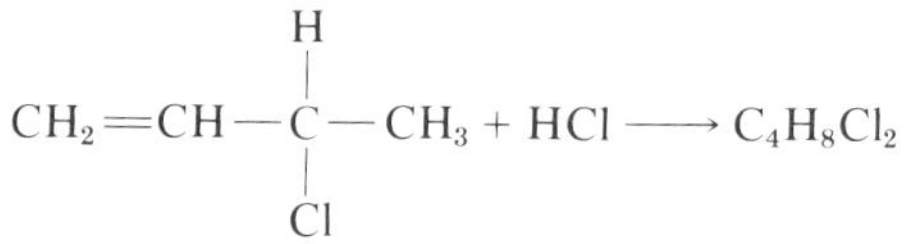
$$CH_2{=}CH-\underset{\underset{Cl}{|}}{\overset{\overset{H}{|}}{C}}-CH_3 + HCl \longrightarrow C_4H_8Cl_2$$

(*R*)-3-Chloro-1-butene

PROBLEM 11-9 Cyclopentene reacts with peroxyformic acid to give a racemic mixture of *trans*-1,2-cyclopentanediol. Write the structural formulas for the two enantiomers. After resolution, they exhibit optical activity.

Cyclopentene reacts with cold dilute $KMnO_4$ to produce *cis*-1,2-cyclopentanediol. Write its structural formula. Can this cis isomer be resolved into optically active enantiomers? Why or why not?

11-5 Resolution of a racemic mixture

We are now in a position to consider the general method for resolution of a racemic mixture of a pair of enantiomers. The methods and limitations of mechanical and biochemical resolution were mentioned earlier.

Reaction of a racemic mixture with an optically active reagent results in the formation of diastereoisomers, which may be separated by fractional distillation, fractional crystallization, or chromatographic techniques. After separation has been accomplished, the diastereoisomers may than be converted back into the original enantiomers. For example, racemic lactic acid, $CH_3CHOHCOOH$, might be reacted with the optically active base (−)-brucine to produce a mixture of diastereoisomeric brucine lactate salts.

$$\left.\begin{array}{r}(+)\text{-Lactic acid}\\(-)\text{-Lactic acid}\end{array}\right\} + (-)\text{-brucine} \longrightarrow \left\{\begin{array}{l}[(-)\text{-brucine-}(+)\text{-lactic acid] salt}\\{[(-)\text{-brucine-}(-)\text{-lactic acid] salt}}\end{array}\right. \quad (11\text{-}5)$$

The diastereoisomeric salts could then be separated by fractional crystallization. Lactic acid is regenerated by reaction with hydrochloric acid.

$$[(-)\text{-Brucine-}(+)\text{-lactic acid] salt} + HCl \longrightarrow (+)\text{-lactic acid} + (-)\text{-brucine}\cdot HCl \quad (11\text{-}6)$$

$$[(-)\text{-Brucine-}(-)\text{-lactic acid] salt} + HCl \longrightarrow (-)\text{-lactic acid} + (-)\text{-brucine}\cdot HCl \quad (11\text{-}7)$$

The diastereoisomeric relation can be seen more clearly with the structural formulas:

$$
\underbrace{\underset{(R)\text{-Lactic acid}}{\begin{array}{c} CH_3 \\ | \\ HO{-}C{-}H \\ | \\ C{=}O \\ | \\ HO \end{array}} + \underset{(S)\text{-Lactic acid}}{\begin{array}{c} CH_3 \\ | \\ H{-}C{-}OH \\ | \\ C{=}O \\ | \\ HO \end{array}}}_{\text{Racemic mixture}} + 2\underset{(S)\text{-base}}{\begin{array}{c} NH_2 \\ | \\ CH_3{-}C{-}H \\ | \\ R \end{array}} \longrightarrow
$$

$$
\underbrace{\underset{(R,S)\text{-Salt}}{\begin{array}{c} CH_3 \\ | \\ HO{-}C{-}H \\ | \\ C{=}O \\ | \\ O^- \\ NH_3^+ \\ | \\ CH_3{-}C{-}H \\ | \\ R \end{array}} + \underset{(S,S)\text{-Salt}}{\begin{array}{c} CH_3 \\ | \\ H{-}C{-}OH \\ | \\ C{=}O \\ | \\ O^- \\ NH_3^+ \\ | \\ CH_3{-}C{-}H \\ | \\ R \end{array}}}_{\text{Diastereoisomers (separable)}} \qquad (11\text{-}8)
$$

This general method of resolution of a racemic mixture is very important. It is used by suppliers of biochemicals to prepare pure enantiomers for biochemical and medical research. Resolution is an expensive process. Racemic 2-aminopropanoic acid (alanine) was listed by one supplier in 1978 at $4.10 per 100 g, and the *R* and *S* enantiomers from resolution were listed respectively at $94.50 per 100 g and $56.00 per 100 g.

Summary

Stereoisomers differ from each other only in the *different spatial arrangements* of the atoms in their molecules. *Configurational isomers* are stereoisomers that do *not* interconvert rapidly under normal conditions. They are stable enough to be separated. *Conformational isomers* are stereoisomers that *do* interconvert by rotation about single covalent bonds.

Geometric isomers are one type of configurational isomer; *chiral isomers* are the other type. Chiral isomers are configurational isomers which have *chiral molecules*. Chiral molecules *lack reflection symmetry*—they are *not* identical with their mirror images. A chiral molecule has a *mirror-image isomer* called an *enantiomer*, which is not superimposable on it. Molecules that do *possess reflection symmetry*—meaning they are identical with their mirror images—are said to be *achiral*. Enantiomers are not possible for them.

An *enantiomeric pair* is a pair of *compounds* whose molecules are nonidentical mirror images. An *enantiomorphic pair* is a pair of *objects* that are nonidentical mirror images.

Any tetrahedral atom with *four different groups* attached to it is a *chiral center*. A carbon atom that carries four different substituents is described as an *asymmetric carbon atom*. It is one type of chiral center.

Configurational formulas depict a molecule's three-dimensional configuration; *Fischer projection formulas* are the most common ones. In using a Fischer projection formula, the asymmetric carbon atom is considered to be in the plane of the paper. The groups attached to it on the left and right are considered as projecting out of the plane of the paper toward the viewer. The groups attached to the asymmetric carbon atom above and below are considered as projecting back of the plane of the paper. In checking superimposability of two configurational formulas, it is permissible to rotate a formula 180° in the plane of the paper, but it is not permissible to remove the formula from the plane of the paper.

The *Cahn-Ingold-Prelog system* is a method of designating the absolute configuration of a molecule with *one or more chiral centers*. A priority based on atomic numbers is assigned to the four atoms or groups attached to the chiral center. The formula is then rotated so that the group of lowest priority is directly behind the chiral center and the other three groups project toward the viewer. The group of highest priority is then located. The direction in which the group of next highest priority lies is then determined. If it is *clockwise*, the designation is *R* (rectus); if it is *counterclockwise*, the designation is *S* (sinister).

The configuration of a *geometric isomer* is determined by assigning Cahn-Ingold-Prelog priorities to the two groups attached to each end of the double bond. If the groups of highest priority are on the *same side* of the double bond, the designation is *Z* (zusammen); if the groups of highest priority are on *opposite sides* of the double bond, the designation of the isomer is *E* (entgegen).

Enantiomers are said to be *optically active* because they are able to *rotate the plane of plane-polarized light*. If an enantiomer rotates the plane to the *right*, it is said to be *dextrorotatory*, and that is indicated by (+) in its name. If rotation is to the *left*, the enantiomer is said to be *levorotatory*, and that is indicated by (−) in its name.

A pair of enantiomers will rotate the plane of polarized light exactly the *same amount* but in *opposite directions*. That fact is useful in identifying a particular enantiomer of a pair, because the members of the pair are alike in almost all of their other physical properties.

Separation of the members of a pair of enantiomers from a mixture of the two is called *resolution*. Since very few enantiomers crystallize in *enantiomorphic crystals*, *mechanical resolution* is seldom possible. Another method of resolution utilizes a *biochemical reaction* of the mixture with an *enzyme*. An enzyme is itself an optically active compound, and so it will react with only one of the enantiomers and leave the other one intact. A *50:50 mixture* of enantiomers is referred to as a *racemic mixture*.

Molecules of *non–optically active* compounds are *achiral;* that is, they have an element of symmetry such that identical molecules can serve as their own mirror images. Thus a usual random distribution of such molecules results in equal amounts so oriented as to be essentially opposing mirror images of each other. Polarized light penetrating such molecules is rotated as much in one direction as the other, so the net rotation is zero. However, chiral molecules can never serve as their own mirror images, so that any polarized light passing through is rotated in the same direction.

A *polarimeter* is an instrument for measuring the *direction* and *amount of rotation* of plane-polarized light by an optically active compound. *Specific rotation* is equal to the observed rotation divided by the length of the sample tube in decimeters and by the density (or concentration, if a solution) in grams per milliliter.

Molecules may contain *multiple chiral centers* (usually asymmetric carbon atoms). Each chiral center is given an *R-S* designation. The *van't Hoff rule* states that *the number of stereoisomers is* 2^n, where n is the number of differently substituted asymmetric carbon atoms. If any of the asymmetric carbon atoms are alike, the total number of stereoisomers is reduced.

A *meso form* is a stereoisomer with *two or more chiral centers* but with such a configuration that the *molecule itself is achiral* in one of its conformations. That is, the molecule has a *plane of symmetry*. *Diastereoisomers* are stereoisomers of a given molecular formula which are *not enantiomers*. They differ in both physical and chemical properties, and so they can be separated by the usual techniques.

If one stereoisomer can be converted into another stereoisomer without breaking any bonds to the asymmetric carbon atom, it can be assumed that the configuration about that asymmetric carbon atom will be retained. Since J. M. Bijvoet determined the actual configuration of the atoms of a salt of tartaric acid by x-ray diffraction, any configuration known relative to tartaric acid is now an *absolute configuration*.

The *general method for resolution* of a racemic mixture is to react the mixture with an *optically active reagent* to convert the enantiomers into a mixture of diastereoisomers, which are then separated. After separation, each enantiomer is recovered by hydrolysis or other appropriate reaction of the separated diastereoisomers.

Key terms

Absolute configuration: the actual arrangement in space of the atoms of a molecule.

Achiral molecules: molecules which possess reflection symmetry; they are identical with their mirror images, and they do not have enantiomers.

Asymmetric carbon atom: a carbon atom that carries four different substituents; it is one type of chiral center.

Biochemical resolution: separation of enantiomers by reaction with an appropriate enzyme. One enantiomer is destroyed or changed, and the other is left behind.

Cahn-Ingold-Prelog system: a system used to designate the absolute configuration about a chiral center. It is also used instead of cis and trans to designate the configuration of geometric isomers.

Chiral center: any tetrahedral atom with four different atoms or groups attached to it.

Chiral isomers: configurational isomers which exist if a compound consists of chiral molecules; they are mirror images of each other.

Chiral molecules: molecules that lack reflection symmetry; they are not identical with their mirror images. Also called **dissymmetric molecules.**

Chirality: "handedness"; an absence of reflection symmetry and a property of dissymmetric molecules.

Configurational formula: a formula that depicts the three-dimensional configuration of a molecule.

Configurational isomers: stereoisomers that do not interconvert rapidly under normal conditions; they include geometric isomers and chiral isomers.

Conformational isomers: stereoisomers that interconvert rapidly by rotation about single covalent bonds.

Dextrorotatory: the designation of an enantiomer that rotates the plane of polarized light to the right. It is indicated in the name of the compound by (+).

Diastereoisomers: stereoisomers of a given molecular formula which are not enantiomers (mirror-image isomers). They have different physical and chemical properties.

Dissymmetric molecules: molecules which lack reflection symmetry; they are not identical with their mirror images. Also called **chiral molecules.**

Double refraction: the differential bending of a ray of light entering a crystal whose axes are not the same in every direction. The light is split into two separate components with different velocities and paths. An object seen through such a crystal is perceived as a double image.

Enantiomeric pair: a pair of compounds whose molecules are nonidentical mirror images.

Enantiomers: mirror-image isomers of a chiral molecule. They lack reflection symmetry and are not superimposable.

Enantiomorphic crystals: crystals which are nonidentical mirror images; "left-handed" and "right-handed" crystals.

Enantiomorphic pair: a pair of objects which are nonidentical mirror images.

Entgegen (*E*): used in the Cahn-Ingold-Prelog system to designate configuration about a double bond or in a ring structure. When the groups of highest priority are on opposite sides of the double bond (or of the ring), the isomer is designated as *E*.

Fischer projection formula: a simple type of configurational formula; definite conventions must be followed in its use.

Geometric isomers: isomers which exist because rotation about a carbon-carbon covalent bond is restricted either by a double bond or by a ring structure.

Levorotatory: the designation of an enantiomer that rotates the plane of polarized light to the left. It is indicated in the name of the compound by (−).

Mechanical resolution: the separation of enantiomers by literally picking out the "right-handed" crystals from the "left-handed" crystals. It is possible only for the few enantiomers that crystallize in enantiomorphic crystals.

Meso form: a stereoisomer that has multiple asymmetric centers but is an achiral molecule and is therefore optically inactive.

Monochromatic light: light of a single, or almost single, wavelength.

Nicol prism: a prism formed by cementing together two properly cut prisms of calcite. It has the ability to plane-polarize light, and it is used in the construction of the polarimeter.

Observed rotation: the actual rotation of the plane of polarized light observed in a polarimeter with a particular sample. It is converted to specific rotation to get standard values for comparison purposes.

Optical activity: the capacity of chiral molecules to rotate the plane of polarized light.

Optical isomerism: the form of stereoisomerism that results from chiral molecules.

Optical rotation: the rotation of the plane of polarized light by enantiomers. It is useful in identifying a particular enantiomer.

Optically active compound: a compound which can rotate the plane of polarized light.

Perspective formula: one type of configurational formula.

Plane-polarized light: light in which all of the wave motions are vibrating in a particular plane.

Polarimeter: an instrument used to determine the direction and amount of rotation caused by an optically active compound.

Racemic mixture: a 50:50 mixture of enantiomers.

Rectus (*R*): used in the Cahn-Ingold-Prelog system to designate absolute configuration about a chiral center. When, following the rules of the system, the direction from the group of highest priority to the group of next highest priority is to the right (clockwise), the configuration is designated as *R*.

Reflection symmetry: symmetry within an object or a molecule such that the mirror image of the object or molecule is identical with the object or molecule: the mirror image and the object are identical.

Refraction: bending of a ray or wave as it passes obliquely from one medium to another of different density in which its speed is different.

Resolution: separation of enantiomers. The most general method is to react a racemic mixture with an optically active reagent and thereby form diastereoisomers. The diastereoisomers can be separated and then reconverted to the original enantiomers.

Sinister (*S*): used in the Cahn-Ingold-Prelog system to designate absolute configuration about a chiral center. When, following the rules of the system, the direction from the group of highest priority to the group of next highest priority is to the left (counterclockwise), the configuration is designated as *S*.

Specific rotation: the optical rotation calculated by dividing the observed rotation caused by an optically active compound by the length of the polarimeter tube, in decimeters, and by the density of the liquid (or concentration of solution), in grams per milliliter. Standard condition is 25°C, and the sodium line at 589.3 nanometers (nm) is used as the light source.

Stereoisomers: isomers which differ from each other only in the arrangement in space of the atoms in their molecules; they include configurational and conformational isomers.

Superimposable molecules: molecules which are identical; when superimposed upon each other, they coincide at all points.

Tartaric acids: stereoisomers of 2,3-dihydroxysuccinic acids. The enantiomers of tartaric acid were the first to be separated, mechanically; the absolute configuration of (+)-tartaric acid was the first to be determined.

van't Hoff rule: if all of the asymmetric carbon atoms in a molecule are differently substituted, the number of possible stereoisomers is 2^n, where n is the number of unlike or differently substituted asymmetric carbon atoms.

Zusammen (Z): used in the Cahn-Ingold-Prelog system to designate configuration about a double bond or in a ring structure. When the groups of highest priority are on the same side of the double bond (or of the ring), the isomer is designated as *Z*.

Additional problems

11-10 Which of the following compounds contain a chiral center because of an asymmetric carbon atom?

(a) $CH_3CH(NH_2)COOH$
(b) $CH_3CH(CH_3)CH_2OH$
(c) $CH_3CH_2CHOHCH_3$
(d) $CH_3CH_2CH(CH_3)CH_2OH$
(e) $(CH_3)_2CHCH_2CH_2OH$

(f)

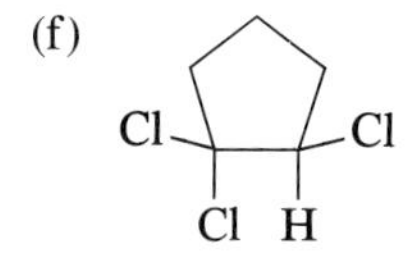

11-11 Remembering possible stereoisomers, write the structural formulas and names for all of the isomeric butanols.

11-12 Write configurational formulas and the names for all of the configurational isomers of 2,3,4-trihydroxybutanal.

11-13 Calculate the specific rotation of a sugar whose 5.0 g in 50.0 ml solution produces a rotation of −18.4° when used to fill a 2.0-dm polarimeter tube.

11-14 Which of the following could be resolved into enantiomers? Which would be a meso form or exist as nonresolvable conformational enantiomers?

(a) *cis*-1,4-cyclohexanediol
(b) *trans*-1,4-cyclohexanediol
(c) *cis*-1,2-cyclohexanediol
(d) *trans*-1,2-cyclohexanediol
(e) *cis*-1,3-cyclohexanediol
(f) *trans*-1,3-cyclohexanediol

11-15 How many asymmetric carbon atoms are there in the following structural formulas? How many stereoisomers could there be?

(a) 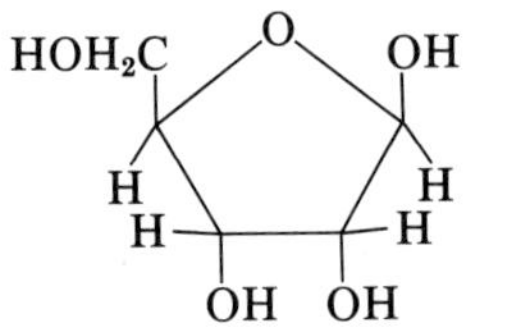 HOH_2C, O, OH, H, H, H, H, OH, OH

(b) 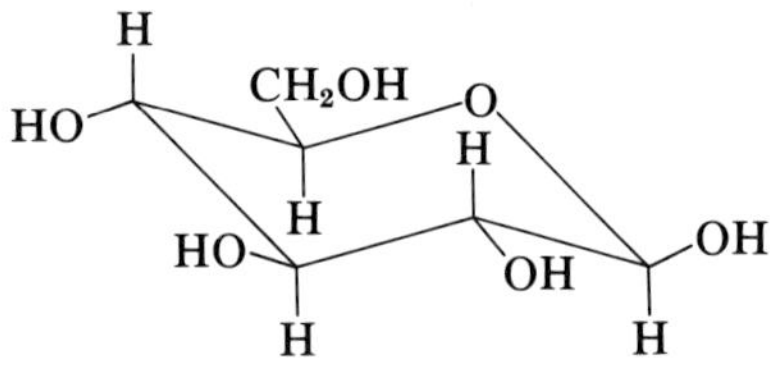 H, CH_2OH, O, HO, H, H, HO, OH, OH, H, H

11-16 The following formula represents galactaric (or mucic) acid. Is galactaric acid optically active? Why or why not? Draw structures for all of the stereoisomers of galactaric acid. Indicate pairs of enantiomers and any meso forms.

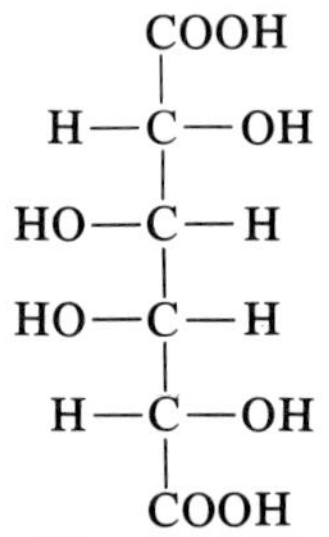

$$\begin{array}{rcl} & COOH & \\ & | & \\ H- & C & -OH \\ & | & \\ HO- & C & -H \\ & | & \\ HO- & C & -H \\ & | & \\ H- & C & -OH \\ & | & \\ & COOH & \end{array}$$

Galactaric acid

11-17 Write configurational formulas for the following compounds:

(a) (R)-2-pentanol
(b) (E)-1,2-diphenylethene
(c) (Z)-(4S)-4-chloro-2-pentene

11-18 Which of the following reactions would proceed with retention of configuration and optical activity?

(a) 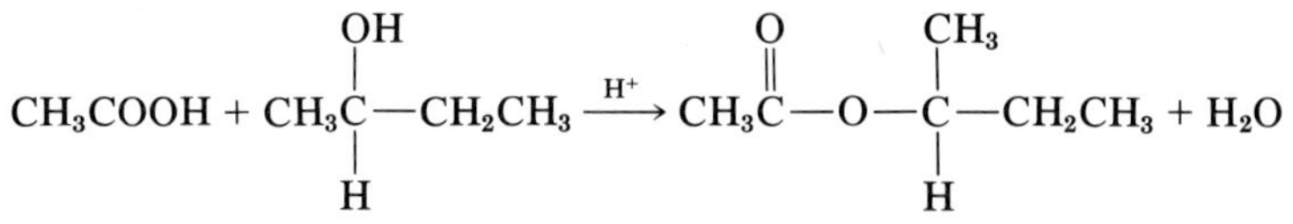 $$CH_3COOH + CH_3\overset{OH}{\underset{H}{C}}-CH_2CH_3 \xrightarrow{H^+} CH_3\overset{O}{\overset{\|}{C}}-O-\overset{CH_3}{\underset{H}{C}}-CH_2CH_3 + H_2O$$

(b) $$CH_3\overset{O}{\overset{\|}{C}}-Cl + CH_3\overset{OH}{\underset{H}{C}}-CH_2CH_3 \longrightarrow CH_3\overset{O}{\overset{\|}{C}}-O-\overset{CH_3}{CH}CH_2CH_3 + HCl$$

(c) $$CH_3\overset{OH}{\underset{H}{C}}-CN + H_2O \xrightarrow{H^+} CH_3\overset{OH}{\underset{H}{C}}-COOH + NH_4Cl$$

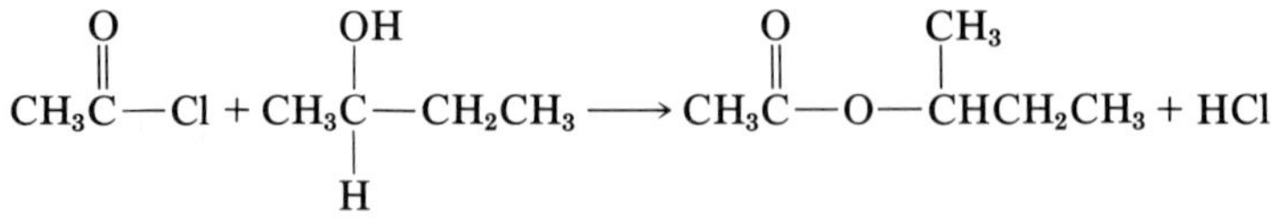

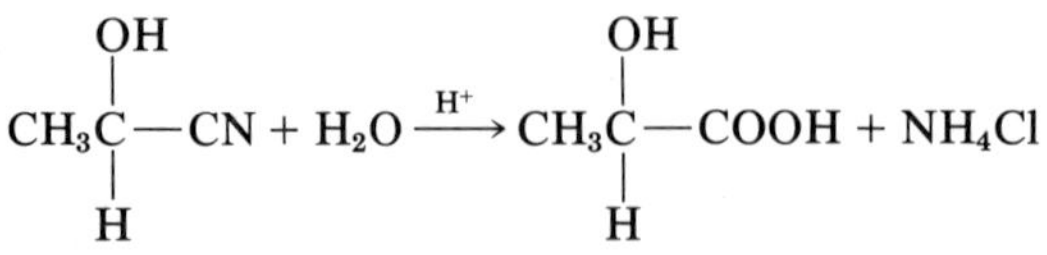

(d) $$CH_3CH_2\underset{\displaystyle H}{\overset{\displaystyle CH_3}{\overset{|}{\underset{|}{C}}}}-O^-Na^+ + CH_3I \longrightarrow CH_3CH_2\underset{\displaystyle H}{\overset{\displaystyle CH_3}{\overset{|}{\underset{|}{C}}}}-O-CH_3 + NaI$$

(e) $$CH_3CH_2\underset{\displaystyle H}{\overset{\displaystyle CH_3}{\overset{|}{\underset{|}{C}}}}-Br + CH_3O^-Na^+ \longrightarrow CH_3CH_2\overset{\displaystyle CH_3}{\overset{|}{C}}H-OCH_3 + NaBr$$

12 CARBOHYDRATES

The *carbohydrates,* one of the three main classes of foods, include sugars, starches, cellulose, gums, pectins, and their derivatives. They are important not only as food but also as clothing and shelter in the form of cotton, linen, and rayon for fabrics and wood for building houses.

The carbohydrates were so named because the formulas of many of them could be written as *hydrates of carbon.* For example, the formula for the six-carbon sugar $C_6H_{12}O_6$ could be written $C_6(H_2O)_6$. When concentrated sulfuric acid is added to sugar, a very exothermic reaction takes place. In it water is given off and a porous black carbon residue remains. However, that reaction, unlike the dehydration reaction of a true hydrate, is irreversible. The water given off is a product of oxidation—not water of hydration.

Chemical tests indicate that the functional groups of the carbohydrates are carbonyl groups and hydroxy groups. Thus *carbohydrates* may be defined as *polyhydroxy aldehydes* or *ketones* (sugars) and their derivatives or compounds which can be hydrolyzed to such. The latter part of the definition is pertinent because the high-molecular-weight carbohydrates cellulose and starch can be hydrolyzed to sugars.

12-1 Classification of carbohydrates

The carbohydrates may be classified as polysaccharides, disaccharides, and monosaccharides. The *polysaccharides* are the *polymeric carbohydrates.* They can be hydrolyzed to disaccharides and ultimately to many monosaccharide units. The chief examples are cellulose, starch, and glycogen, all of which are polymers of the sugar glucose.

The *disaccharides* are *dimers* of *monosaccharides,* to which they can be hydrolyzed. The chief examples are the sugars maltose, cellobiose, lactose, and sucrose.

The *monosaccharides* are *sugars*—referred to as *simple sugars*—and their derivatives. They are not further divisible by hydrolysis. The chief examples are glucose, fructose, mannose, galactose, xylose, arabinose, and ribose.

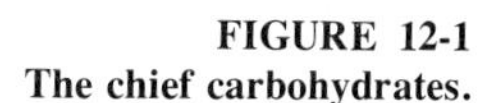

FIGURE 12-1
The chief carbohydrates.

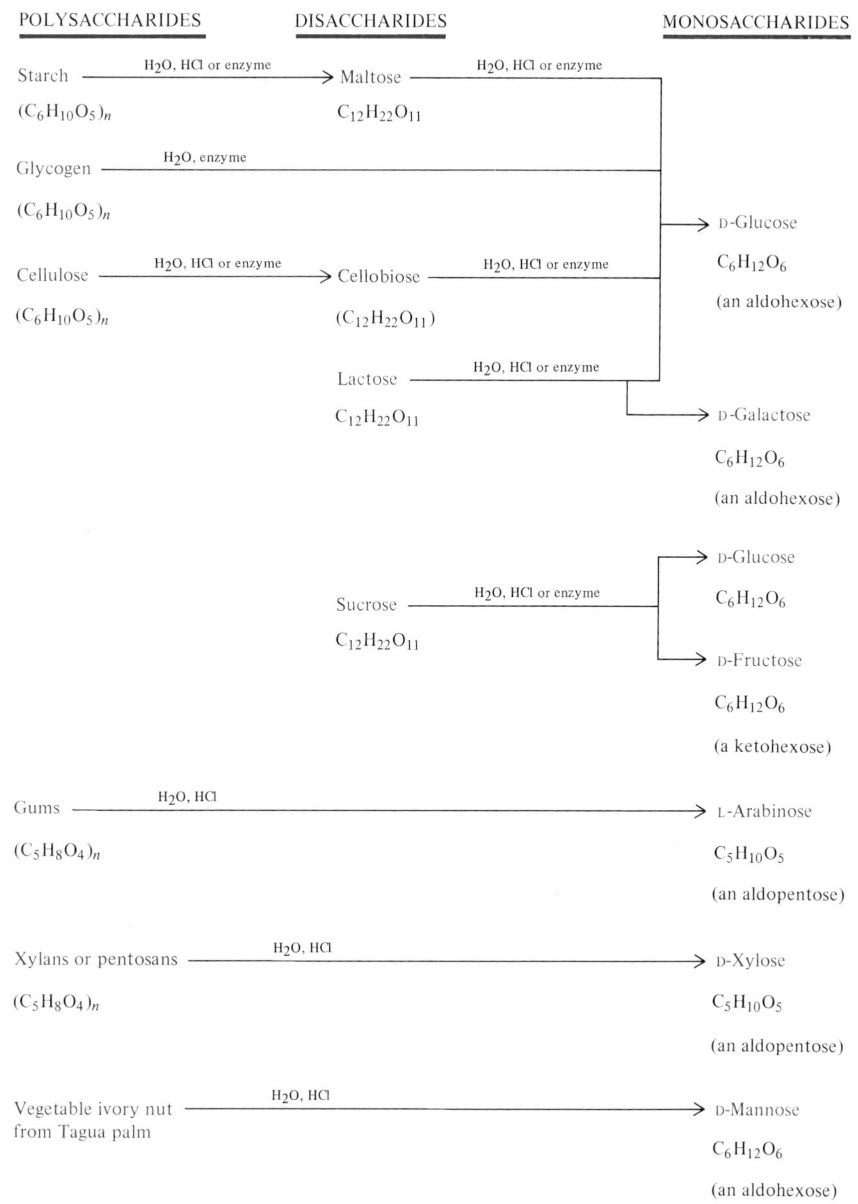

The polysaccharides are high-molecular-weight semicrystalline solids which are not very soluble in water. The sugars are colorless, crystalline solids which are very soluble in water, taste sweet, and may decompose at their melting points. The sugars may be monosaccharides, disaccharides, or even a trisaccharide (raffinose).

Figure 12-1 gives an overview of carbohydrate classification. It indicates the major groups of carbohydrates and their relations. You will note that hydrolysis of polysaccharides and disaccharides yields particular monosaccharides. The classes of carbohydrates and the individual compounds shown in Fig. 12-1 will be discussed in subsequent sections of the chapter. You may find it useful to refer to the illustration frequently.

12-2 Nomenclature

The *polysaccharides* are chiefly represented by their *common names,* such as starch, glycogen, and cellulose.

The *sugars* are given *common names* ending in the suffix *-ose*. Sugars which contain an aldehyde group are referred to (but not named) as *aldoses,* and those containing a ketone group are referred to as *ketoses*. A sugar may also be referred to by a name reflecting the number of carbon atoms in the molecule. For example, a *hexose* is a six-carbon sugar, and a *pentose* is a five-carbon sugar. A *ketohexose* is a six-carbon sugar with a ketone group, and an *aldohexose* is a six-carbon sugar with an aldehyde group.

Sugars are optically active stereoisomers, and the optical rotation is indicated by either (+) or (−) in front of the name. The common name of a sugar represents the configuration of all of its asymmetric carbon atoms except the last one. The configuration of the last asymmetric carbon atom is indicated by either D or L in front of the name; an example is D-(+)-glucose. The meaning of D and L will be explained when the structures are discussed.

12-3 D-Glucose

The most important of all of the simple sugars, or monosaccharides, is *glucose,* specifically, D-glucose. It is blood sugar; it is one of the monosaccharides found in honey; it is corn sugar; and it is the hydrolysis product of starch, cellulose, and glycogen, or animal starch. Oxidation of it is a

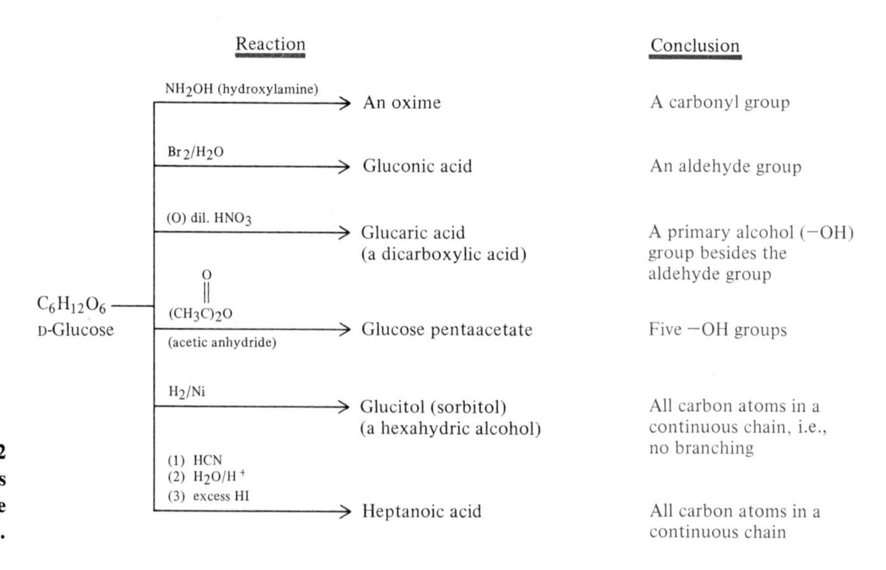

FIGURE 12-2
These chemical reactions give clues to the structure of glucose.

source of cellular energy in living things. It is optically active, and it rotates the plane of polarized light to the right. That makes its more complete name D-(+)-glucose. It is also known as *dextrose*.

Structure Chemical analysis and molecular-weight determination of glucose indicate the molecular formula to be $C_6H_{12}O_6$, and the chemical reactions shown in Fig. 12-2 give clues as to the structure. Study the illustration carefully. From those reactions it can be deduced that the structure of D-glucose is

```
 H   O
  \ //
   1C
    |
   2CHOH
    |
   3CHOH
    |
   4CHOH
    |
   5CHOH
    |
   6CH2OH
```

Note that we are following the convention of carbohydrate chemists and writing the formula vertically.

Configuration Carbon atoms 2 to 5 in the D-glucose structure are asymmetric, but at this point we do not know their configurations. The number of possible stereoisomers is given by the formula 2^n, where n represents the number of asymmetric carbon atoms. Since there are four asymmetric carbon atoms, we can have 2^4 or 16 stereoisomers. The major question, then, is which of the 16 possible configurations is D-glucose? That problem was solved in 1891 by Emil Fischer and his coworkers at the University of Würzburg. The configurations of the stereoisomers of D-glucose also were worked out. For his work, Fischer received the Nobel Prize in Chemistry in 1902.

The three most important aldohexoses of the 16 stereoisomers are D-(+)-glucose, D-(+)-mannose, and D-(+)-galactose. The most important ketohexose is D-(−)-fructose. You should learn their structures.

```
    H—C=O             H—C=O             H—C=O              CH2OH
      |                 |                 |                  |
    H—C—OH           HO—C—OH            H—C—OH              C=O
      |                 |                 |                  |
   HO—C—H            HO—C—H            HO—C—H            HO—C—H
      |                 |                 |                  |
    H—C—OH            H—C—OH           HO—C—OH            H—C—OH
      |                 |                 |                  |
    H—C—OH            H—C—OH            H—C—OH            H—C—OH
      |                 |                 |                  |
     CH2OH             CH2OH             CH2OH              CH2OH
```

D-(+)-Glucose	D-(+)-Mannose	D-(+)-Galactose	D-(−)Fructose
$[\alpha]_D$ + 53°	$[\alpha]_D$ + 14°	$[\alpha]_D$ + 80°	$[\alpha]_D$ − 92°

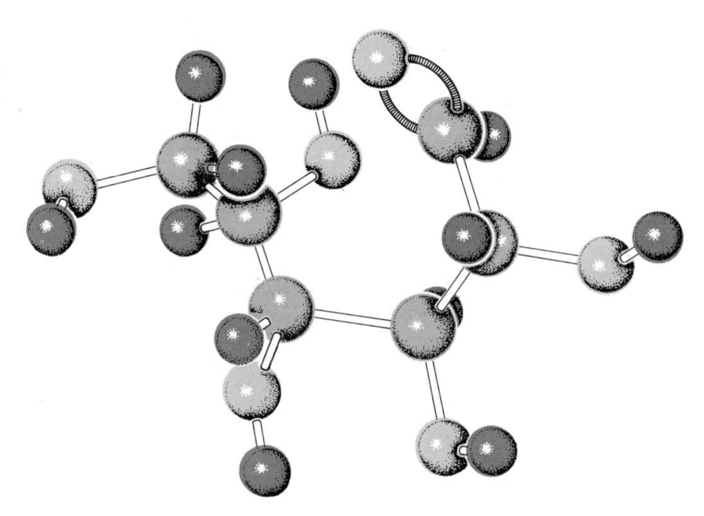

FIGURE 12-3
A ball-and-spring model of D-glucose.

These are *Fischer projection formulas*. The carbon atoms are to be *individually considered* in the plane of the paper; the groups to the right and to the left of each carbon atom are to be considered as projecting out of the plane of the paper toward the viewer; the groups attached above and below each carbon atom, as written, are to be considered as extending back of the plane of the paper (Fig. 12-3).

Each of the above compounds has a possible mirror-image isomer or *enantiomer*. [For example, the enantiomer of D-(+)-glucose is L-(−)-glucose.] The 16 possible stereoisomeric aldohexoses consist of 8 pairs of enantiomers. Carbon atom 5 is used to distinguish the members of each pair. It is the first asymmetric carbon atom from the bottom as the formula is written. The member of each enantiomeric pair which has *the —OH group* of that carbon atom *to the right* is arbitrarily said to belong to the D family. It is designated as a D-*sugar*. If the —OH group on carbon atom 5 is *to the left*, the sugar belongs to the L *family*. It is designated as an L-*sugar*.

Almost all of the sugars found in nature belong to the D family. Theoretically, every D-sugar has an enantiomer which is an L-sugar, but the L-sugars are not commonly found in nature.

The origin of the use of D and L relates to the fact that the D-sugars have the same configuration about their last asymmetric carbon atom—the one farthest from the top as written—as does the triose D-glyceraldehyde. D-Glyceraldehyde is dextrorotatory. The L-sugars have a configuration similarly related to that of the levorotatory L-glyceraldehyde. See Fig. 12-4.

Thus, D and L *refer to configurational families,* and not to optical rotation. Note that D-fructose is actually levorotatory. The direction of rotation of polarized light is determined experimentally in a polarimeter and is designated by (+) or (−).

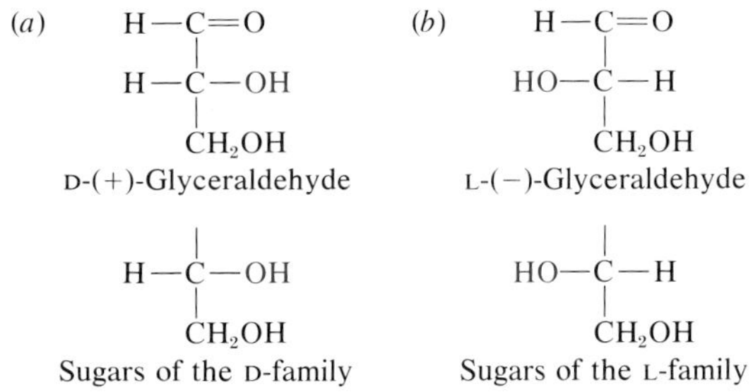

FIGURE 12-4
The structures of (*a*) D-(+)-glyceraldehyde and sugars of the D family and (*b*) L-(−)-glyceraldehyde and sugars of the L family.

(*a*)

$$\begin{array}{c} H-C=O \\ | \\ H-C-OH \\ | \\ CH_2OH \end{array}$$

D-(+)-Glyceraldehyde

$$\begin{array}{c} | \\ H-C-OH \\ | \\ CH_2OH \end{array}$$

Sugars of the D-family

(*b*)

$$\begin{array}{c} H-C=O \\ | \\ HO-C-H \\ | \\ CH_2OH \end{array}$$

L-(−)-Glyceraldehyde

$$\begin{array}{c} | \\ HO-C-H \\ | \\ CH_2OH \end{array}$$

Sugars of the L-family

Figure 12-5 shows the D-sugars related to D-glyceraldehyde. You will note that the formulas are *abbreviated structural formulas*. The vertical lines represent the carbon chains with the asymmetric carbon atoms at the intersections. The horizontal lines represent the —OH groups. The presence of the remaining hydrogen atoms is assumed.

In studying Fig. 12-5, you will note that D-mannose differs from D-glucose only in the configuration about C-2. *Non*–mirror-image stereoisomers of a compound are generally called *diastereoisomers*. This particular type of diastereoisomer in which the difference in configuration relates to *only one* of the asymmetric carbon atoms (in the case of mannose and glucose, C-2), is called an *epimer,* so D-glucose and D-mannose are epimers of each other. D-Mannose is a product of hydrolysis of certain polysaccharides, one of which is vegetable ivory nut from the Tagua palm.

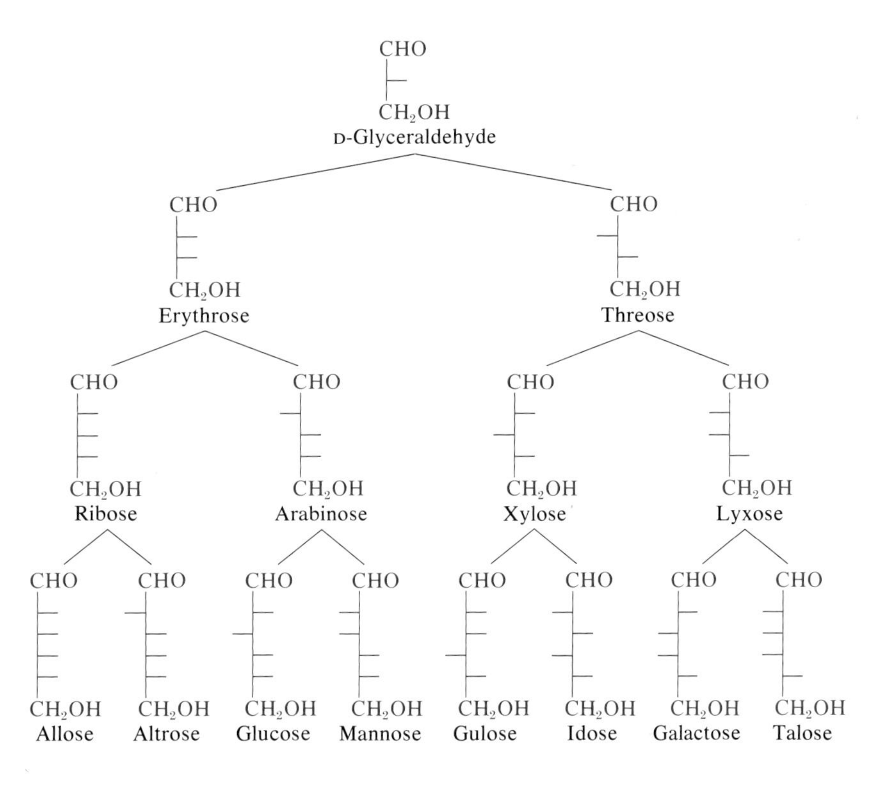

FIGURE 12-5
D-Sugars related to D-glyceraldehyde. An enantiomeric L-sugar is possible for each one of the sugars shown.

D-Galactose differs from D-glucose only in the configuration about C-4, and it is thus an epimer of D-glucose. It is obtained, along with D-glucose, when lactose (milk sugar) is hydrolyzed.

D-Fructose is a ketohexose with the same configuration about carbon atoms 3, 4, and 5 as in D-glucose. It is obtained, along with D-glucose, when common sugar (sucrose) is hydrolyzed. It can also be obtained from the Jerusalem artichoke. D-Fructose is the sweetest of all of the natural sugars. Because it is very levorotatory (−92°), it is also called levulose.

PROBLEM 12-1 Write Fischer projection formulas for each of the following compounds and its enantiomer.

(a) D-glucose
(b) D-mannose
(c) D-galactose
(d) D-fructose
(e) D-arabinose
(f) D-ribose
(g) D-xylose
(h) D-gulose

Solution to (a)

$$\begin{array}{rcl}
\mathrm{H}- & \mathrm{C} & =\mathrm{O} \\
 & | & \\
\mathrm{H}- & \mathrm{C} & -\mathrm{OH} \\
 & | & \\
\mathrm{HO}- & \mathrm{C} & -\mathrm{H} \\
 & | & \\
\mathrm{H}- & \mathrm{C} & -\mathrm{OH} \\
 & | & \\
\mathrm{H}- & \mathrm{C} & -\mathrm{OH} \\
 & | & \\
 & \mathrm{CH_2OH} & \\
\end{array}
\qquad
\begin{array}{rcl}
\mathrm{O}= & \mathrm{C} & -\mathrm{H} \\
 & | & \\
\mathrm{HO}- & \mathrm{C} & -\mathrm{H} \\
 & | & \\
\mathrm{H}- & \mathrm{C} & -\mathrm{OH} \\
 & | & \\
\mathrm{HO}- & \mathrm{C} & -\mathrm{H} \\
 & | & \\
\mathrm{HO}- & \mathrm{C} & -\mathrm{H} \\
 & | & \\
 & \mathrm{CH_2OH} & \\
\end{array}$$

D-Glucose L-Glucose

PROBLEM 12-2 Honey is essentially an equal mixture of D-glucose and D-fructose. Such a mixture is referred to as *invert sugar*. If the specific rotation of D-glucose is +53° and that of D-fructose is −92°, what is the rotation of invert sugar? (The specific rotation of the mixture will lie halfway along the difference between the two.)

PROBLEM 12-3 The structural formula for one of the key metabolic intermediates, fructose-6-phosphate, is

$$\begin{array}{rcl}
 & \mathrm{CH_2OH} & \\
 & | & \\
 & \mathrm{C}=\mathrm{O} & \\
 & | & \\
\mathrm{HO}- & \mathrm{C} & -\mathrm{H} \\
 & | & \\
\mathrm{H}- & \mathrm{C} & -\mathrm{OH} \\
 & | & \\
\mathrm{H}- & \mathrm{C} & -\mathrm{OH} \\
 & | & \\
 & \mathrm{CH_2O} & -\overset{\overset{\mathrm{O}}{\|}}{\underset{\underset{\mathrm{OH}}{|}}{\mathrm{P}}}-\mathrm{OH} \\
\end{array}$$

The compound is converted, with the aid of an enzyme, to fructose-1,6-diphosphate. Write the structural formula for fructose-1,6-diphosphate.

12-4 Reactions and derivatives of the simple sugars

The most reactive groups on a molecule of a simple sugar are the *carbonyl group* and the *primary alcohol group*. The carbonyl group may be reduced to a primary alcohol group, or it may undergo nucleophilic addition with a reagent such as HCN. The aldehyde group is readily oxidized to a carboxyl group. The primary alcohol group may also be oxidized to a carboxyl group.

Aldonic acids

An *aldonic acid* is produced by oxidizing the aldehyde group of an *aldose* with bromine water. The oxidizing agent is the OBr^- ion, which is formed in bromine water by the following reaction:

$$Br_2 + H_2O \rightleftharpoons HBr + HOBr \tag{12-1}$$

The hypobromite ion, OBr^-, is a mild oxidizing agent. It will oxidize the aldehyde group, but not the primary alcohol group.

The aldonic acids are named by replacing the suffix *-ose* in the name of the sugar with the suffix *-onic* and adding the word *acid*. Thus oxidation of D-glucose with bromine water produces D-gluconic acid.

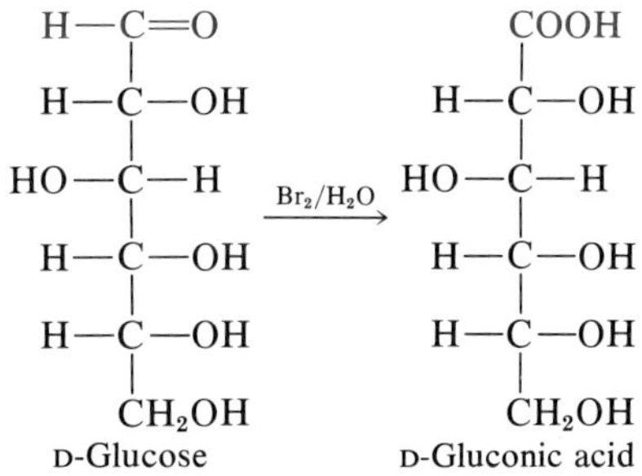

$$\begin{array}{c} H-C{=}O \\ | \\ H-C-OH \\ | \\ HO-C-H \\ | \\ H-C-OH \\ | \\ H-C-OH \\ | \\ CH_2OH \\ \text{D-Glucose} \end{array} \xrightarrow{Br_2/H_2O} \begin{array}{c} COOH \\ | \\ H-C-OH \\ | \\ HO-C-H \\ | \\ H-C-OH \\ | \\ H-C-OH \\ | \\ CH_2OH \\ \text{D-Gluconic acid} \end{array} \tag{12-2}$$

Aldonic acids have been of use in identifying sugars. The ammonium salt of gluconic acid is used as a catalyst in textile printing. The magnesium salt of gluconic acid is used as an antispasmodic.

Aldaric acids

An *aldaric acid* is produced by oxidation of an *aldose* with nitric acid. Both the aldehyde group and the primary alcohol group are oxidized, so the aldaric acid produced is a *dicarboxylic acid*.

The aldaric acids are named by replacing the suffix *-ose* in the name of the sugar with the suffix *-aric* and adding the word *acid*. For example, oxidation of D-galactose with nitric acid produces D-galactaric acid.

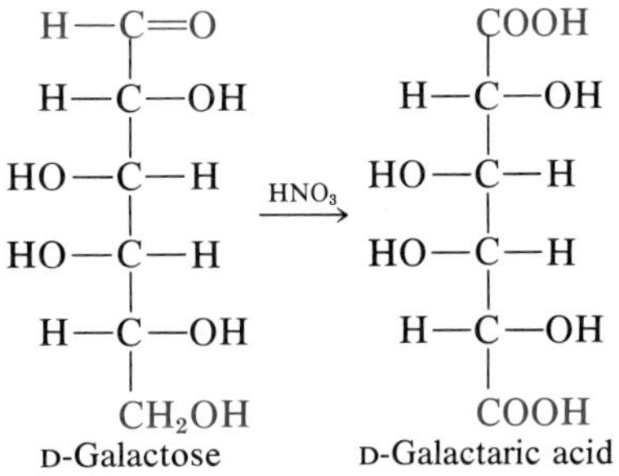

$$\begin{array}{c} H-C{=}O \\ | \\ H-C-OH \\ | \\ HO-C-H \\ | \\ HO-C-H \\ | \\ H-C-OH \\ | \\ CH_2OH \\ \text{D-Galactose} \end{array} \xrightarrow{HNO_3} \begin{array}{c} COOH \\ | \\ H-C-OH \\ | \\ HO-C-H \\ | \\ HO-C-H \\ | \\ H-C-OH \\ | \\ COOH \\ \text{D-Galactaric acid} \end{array} \tag{12-3}$$

D-Galactaric acid is also known as mucic acid. Mucic acid has been of use in identifying galactose. It has been proposed as a replacement for potassium tartrate in baking powder and as the acid component of granular effervescing salts.

Alduronic acids

An *alduronic acid* can be produced by microbiological oxidation of the primary alcohol group only, which leaves the aldehyde group intact.

The alduronic acids are named by replacing the suffix *-ose* in the name of the sugar by the suffix *-uronic* and adding the word *acid.* For example, such oxidation of D-galactose produces D-galacturonic acid.

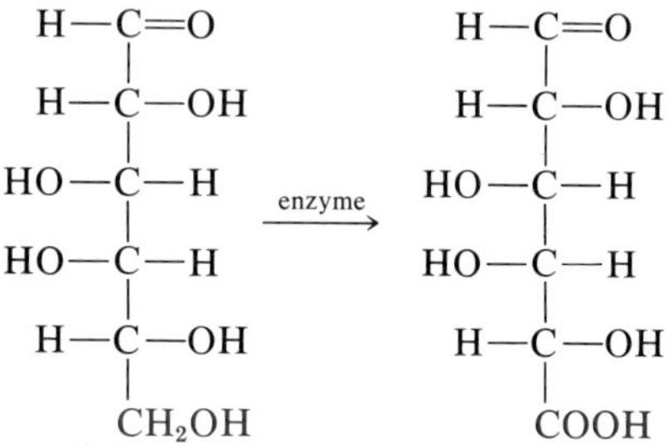

$$\begin{array}{c} H-C=O \\ | \\ H-C-OH \\ | \\ HO-C-H \\ | \\ HO-C-H \\ | \\ H-C-OH \\ | \\ CH_2OH \\ \text{D-Galactose} \end{array} \xrightarrow{\text{enzyme}} \begin{array}{c} H-C=O \\ | \\ H-C-OH \\ | \\ HO-C-H \\ | \\ HO-C-H \\ | \\ H-C-OH \\ | \\ COOH \\ \text{D-Galacturonic acid} \end{array} \qquad (12\text{-}4)$$

Alduronic acids are also prepared by irradiation of an aldose in dilute aqueous solution. The human body uses glucuronic acid to detoxify poisonous hydroxy-containing substances in the liver by forming glucuronides, a type of glycoside. D-Galacturonic acid is obtained by hydrolysis of pectin, in which it is present as polygalacturonic acid.

Alditols

An *alditol* is produced by reduction of an aldose with hydrogen in the presence of a nickel catalyst, with sodium borohydride, or with sodium and alcohol. The aldehyde group of the aldose is reduced to a primary alcohol group to produce a *polyhydroxy alcohol.*

Alditols are named by replacing the suffix *-ose* in the name of the sugar by the suffix *-itol.* For example, reduction of the sugar D-mannose produces D-mannitol.

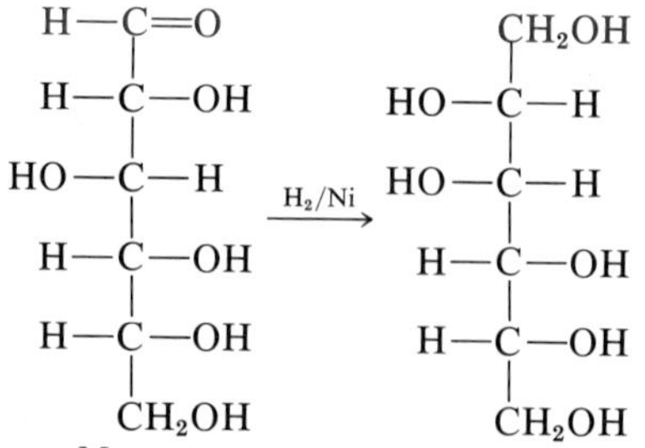

$$\begin{array}{c} H-C=O \\ | \\ H-C-OH \\ | \\ HO-C-H \\ | \\ H-C-OH \\ | \\ H-C-OH \\ | \\ CH_2OH \\ \text{D-Mannose} \end{array} \xrightarrow{H_2/Ni} \begin{array}{c} CH_2OH \\ | \\ HO-C-H \\ | \\ HO-C-H \\ | \\ H-C-OH \\ | \\ H-C-OH \\ | \\ CH_2OH \\ \text{D-Mannitol} \end{array} \qquad (12\text{-}5)$$

Mannitol can be obtained from seaweed. It is used in pharmacy as a dil-

TABLE 12-1 Derivatives of the simple sugars

Compound	General name	Specific examples
Simple sugar	Aldose	Glucose; galactose
Monocarboxylic acid	Aldonic acid	Gluconic acid; galactonic acid
Dicarboxylic acid	Aldaric acid	Glucaric acid; galactaric acid
Aldhydo acid	Alduronic acid	Glucuronic acid; galacturonic acid
Polyhydroxy alcohol	Alditol	Glucitol (sorbitol); galactitol

uent for solids and liquids. It can be nitrated to form mannitol hexanitrate, an explosive.

Glucitol, better known as sorbitol, occurs in many berries and in cherries, plums, pears, apples, seaweed, and algae. It is used as a moisture conditioner for printing rolls, leather, and tobacco. It is also used in the manufacture of candy.

The four derivatives of aldoses are summarized in Table 12-1.

Base-catalyzed isomerization and oxidation

Sugars are very sensitive to alkali, which promotes *enolization*. The *enol form* not only is in equilibrium with other isomers but is also more sensitive to oxidation of a carbon-carbon bond. For that reason, alkaline oxidation of aldoses is very extensive and goes much further than merely oxidizing an aldehyde group. Ketoses also are oxidized in alkaline solution.

Adding a little sodium hydroxide solution to D-glucose produces an equilibrium mixture of D-glucose, D-mannose, and D-fructose. The mixture is very difficult to crystallize and separate. The common enol intermediate produced can revert to any of the three sugars.

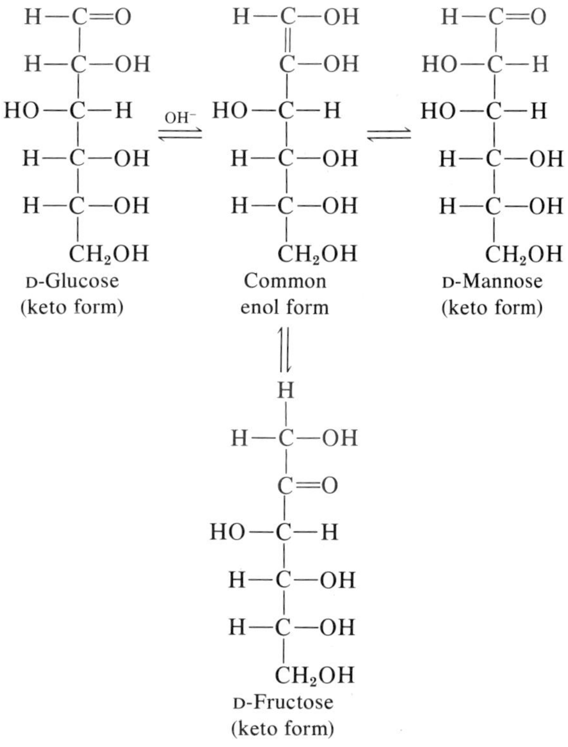

$$
\begin{array}{ccccc}
\begin{array}{c}
\mathrm{H{-}C{=}O} \\ | \\ \mathrm{H{-}C{-}OH} \\ | \\ \mathrm{HO{-}C{-}H} \\ | \\ \mathrm{H{-}C{-}OH} \\ | \\ \mathrm{H{-}C{-}OH} \\ | \\ \mathrm{CH_2OH} \\ \text{D-Glucose} \\ \text{(keto form)}
\end{array}
& \underset{}{\overset{\mathrm{OH^-}}{\rightleftharpoons}} &
\begin{array}{c}
\mathrm{H{-}C{-}OH} \\ \| \\ \mathrm{C{-}OH} \\ | \\ \mathrm{HO{-}C{-}H} \\ | \\ \mathrm{H{-}C{-}OH} \\ | \\ \mathrm{H{-}C{-}OH} \\ | \\ \mathrm{CH_2OH} \\ \text{Common} \\ \text{enol form}
\end{array}
& \rightleftharpoons &
\begin{array}{c}
\mathrm{H{-}C{=}O} \\ | \\ \mathrm{HO{-}C{-}H} \\ | \\ \mathrm{HO{-}C{-}H} \\ | \\ \mathrm{H{-}C{-}OH} \\ | \\ \mathrm{H{-}C{-}OH} \\ | \\ \mathrm{CH_2OH} \\ \text{D-Mannose} \\ \text{(keto form)}
\end{array} \\
& & \Updownarrow & & \\
& &
\begin{array}{c}
\mathrm{H} \\ | \\ \mathrm{H{-}C{-}OH} \\ | \\ \mathrm{C{=}O} \\ | \\ \mathrm{HO{-}C{-}H} \\ | \\ \mathrm{H{-}C{-}OH} \\ | \\ \mathrm{H{-}C{-}OH} \\ | \\ \mathrm{CH_2OH} \\ \text{D-Fructose} \\ \text{(keto form)}
\end{array}
& &
\end{array}
\qquad (12\text{-}6)
$$

Most of the sugars reduce Fehling's, Benedict's, or Tollens' solutions. Such sugars are said to be *reducing sugars*. Since the reagents are alkaline solutions which catalyze enolization, the oxidation of sugars is complex and is accompanied by the breaking of carbon-carbon bonds.

Tollens' reagent is a solution of silver nitrate in dilute aqueous ammonia. The silver ion and ammonia molecules form the complex ion $Ag(NH_3)_2^+$, which is stable in the presence of base. (Otherwise, silver oxide, Ag_2O, would form and precipitate.) The silver ion is the oxidizing agent and is reduced to free silver, being deposited as a silver mirror on the inside of the test tube.

$$\underset{\text{Sugar}}{\sim\overset{\overset{\displaystyle O}{\|}}{C}-H} \xrightarrow{AgNO_3,\ NH_3,\ H_2O} \underset{\substack{\text{Aldonic acid}\\ \text{and other products}}}{\sim\overset{\overset{\displaystyle O}{\|}}{C}-OH} + \underset{\text{Silver mirror}}{Ag(s)} \tag{12-7}$$

Both Benedict's and Fehling's solutions consist of copper(II) hydroxide in aqueous solution. In each case, a special complexing agent is necessary, since copper(II) hydroxide by itself is insoluble. The complexing agent in Benedict's solution is sodium citrate; in Fehling's it is sodium tartrate. Benedict's solution is used more frequently than Fehling's because it is more stable and can be stored without deterioration. Fehling's solution decomposes when it is stored, and so it must be prepared just prior to use. Both Benedict's and Fehling's solutions contain a Cu^{2+} complex ion and are a bright, deep blue color. A positive test consists in the reduction of the Cu^{2+} ion to Cu_2O, copper(I) oxide.

$$\underset{\text{Sugar}}{\sim\overset{\overset{\displaystyle O}{\|}}{C}-H} \xrightarrow{Cu^{2+}\text{ complex, }OH^-} \underset{\substack{\text{Aldonic acid}\\ \text{and other products}}}{\sim\overset{\overset{\displaystyle O}{\|}}{C}-OH} + Cu_2O(s) \tag{12-8}$$

Copper(I) oxide is brick red in color and is insoluble. Hence, in a positive test, the blue color will disappear and a reddish solid will appear. The change takes place in a few minutes if the test solution is warmed.

Fehling's and Benedict's solutions have been used for many years to test for glucose in urine (glucosuria). Normally, urine does not contain glucose, but the sugar does appear in certain conditions such as diabetes. A positive test for glucose in urine may vary from development of a bright green color (0.25% glucose), to yellow-orange (1% glucose), to brick red (over 2% glucose) with 5 ml of Benedict's reagent added to 0.5 ml of urine. Commercially prepared test papers have now supplanted the use of Benedict's solution for routine testing of urine.

The formation of osazones with phenylhydrazine

All of the simple sugars, both aldoses and ketoses, react with phenylhydrazine. The reaction is more extensive than formation of a phenylhydrazone as with a simple aldehyde or ketone, and 3 mol of phenylhydrazine is consumed. D-Glucose, D-mannose, and D-fructose, because of their similarity in structure, give the same product (glucosazone) with phenylhydrazine. The reaction with D-glucose is illustrated. The first step is the usual nucleophilic addition of phenylhydrazine to the carbonyl group followed by dehydration to form a phenylhydrazone. The second step involves oxidation of the neighboring —OH group to a ketone with reduction of a mole of phenylhydrazine to aniline and ammonia. The third step results in the final product called an *osazone*. (The intermediates are placed in brackets to indicate that they are not isolated.)

Step 1

$$\underset{\text{D-Glucose}}{\begin{array}{c} H-C=O \\ | \\ H-C-OH \\ | \\ HO-C-H \\ | \\ H-C-OH \\ | \\ H-C-OH \\ | \\ CH_2OH \end{array}} + NH_2NHC_6H_5 \longrightarrow \underset{\text{D-Glucose phenylhydrazone}}{\left[\begin{array}{c} H-C=NNHC_6H_5 \\ | \\ H-C-OH \\ | \\ HO-C-H \\ | \\ H-C-OH \\ | \\ H-C-OH \\ | \\ CH_2OH \end{array}\right]} + H_2O \qquad (12\text{-}9a)$$

Step 2

$$\underset{\text{D-Glucose phenylhydrazone}}{\left[\begin{array}{c} H-C=NNHC_6H_5 \\ | \\ H-C-OH \\ | \\ HO-C-H \\ | \\ H-C-OH \\ | \\ H-C-OH \\ | \\ CH_2OH \end{array}\right]} + NH_2NHC_6H_5 \xrightarrow{(O)} \left[\begin{array}{c} H-C=NNHC_6H_5 \\ | \\ C=O \\ | \\ HO-C-H \\ | \\ H-C-OH \\ | \\ H-C-OH \\ | \\ CH_2OH \end{array}\right] + \underset{\text{Aniline}}{C_6H_5NH_2} + NH_3 \qquad (12\text{-}9b)$$

Step 3

$$\left[\begin{array}{c} H-C=NNHC_6H_5 \\ | \\ C=O \\ | \\ HO-C-H \\ | \\ H-C-OH \\ | \\ H-C-OH \\ | \\ CH_2OH \end{array}\right] + NH_2NHC_6H_5 \longrightarrow \underset{\text{D-Glucosazone}}{\begin{array}{c} H-C=NNHC_6H_5 \\ | \\ C-NNHC_6H_5 \\ | \\ HO-C-H \\ | \\ H-C-OH \\ | \\ H-C-OH \\ | \\ CH_2OH \end{array}} + H_2O \qquad (12\text{-}9c)$$

The osazones are yellow, crystalline solids with well-defined melting points and crystalline structure. They are useful in the identification of sugars.

PROBLEM 12-4 The reactions of the simple sugars with hydrogen cyanide, HCN, and with hydroxylamine, NH_2OH, are the same as those of simple aldehydes and ketones. Write the structural formula for the product of reaction of each of the following compounds.

(a) D-glucose and NH_2OH
(b) D-arabinose and HCN (two diastereoisomers produced)
(c) D-galacturonic acid and NH_2OH
(d) D-fructose and excess phenylhydrazine
(e) D-mannose and bromine water

Solution to (a)

```
  H—C=O                                    H—C=N—OH
    |                                        |
  H—C—OH                                   H—C—OH
    |                                        |
 HO—C—H                    H+             HO—C—H
    |      +    NH2OH     ——→                |          + H2O
  H—C—OH                                   H—C—OH
    |                                        |
  H—C—OH                                   H—C—OH
    |                                        |
   CH2OH                                    CH2OH
 D-Glucose    Hydroxylamine              D-Glucose oxime
```

PROBLEM 12-5 What simple chemical test could be used to distinguish D-gluconic acid from D-glucuronic acid?

PROBLEM 12-6 Heating D-gluconic acid with a trace of acid results in loss of a mole of water and production of an inner ester, or 1,4-lactone. Write the structural formula for the lactone. The lactone can be reduced with sodium amalgam in a carbon dioxide atmosphere to D-glucose. Write the equation for the reaction. [Merely indicate the sodium amalgam, Na(Hg), and CO_2 over the arrow.]

PROBLEM 12-7 Write the equation for the reduction of D-galactose with hydrogen over a nickel catalyst. Is the product optically active? Why or why not? Would L-galactose give the same product?

12-5 Ring structures of sugars

Emil Fischer had not long elucidated the configuration of D-glucose before the outcome of certain reactions made it evident that all was still not right with the structural formula. For example, D-glucose would not give a bisulfite addition compound, as would ordinary aldehydes. Also, it would not give the *Schiff test,* which involves the formation of a red color by reaction of an aldehyde with a reduced, colorless form of a dye.

Furthermore, two different crystalline forms of glucose were isolated; they had different melting points and different optical rotations. The more common form of glucose, designated α, crystallized out of cold water, had a melting point of 146°C, and a specific rotation of +112°. However, over a period of time a *mutarotation* of its solution occurred until a value of $+53°$ was reached. (Mutarotation means a change in rotation.) The other form of glucose, designated β and obtained by crystallization from water at temperatures above 98°C, had a melting point of 150°C and a specific rotation of $+19°$. The solution of β-D-glucose also underwent mutarotation over a period of time. Its specific rotation gradually increased until it reached the value of $+53°$. It thus appeared that D-glucose was actually two interconvertible compounds which were in equilibrium with each other in water solution. The final value of $+53°$ for the specific rotation represented the specific rotation of the equilibrium mixture.

It was also found that D-glucose would react with only *one* mole of methanol, with dry hydrogen chloride as a catalyst, to produce two diastereoisomeric compounds designated as methyl α-D-glucoside and methyl β-D-glucoside. Although those derivatives corresponded to α-D-glucose and β-D-glucose, they did not exhibit mutarotation as did the glucoses. Also, they had the properties of *acetals* even though only one mole of methanol had reacted per mole of glucose in forming them.

You should remember from Chap. 8 that a mole of an aldehyde reacts with a mole of an alcohol in the presence of an acid catalyst to form a hemiacetal in a reversible reaction. The hemiacetal of a simple aldehyde is not very stable, but it reacts with another mole of alcohol to form an acetal.

$$R{-}\underset{\displaystyle |}{\overset{\displaystyle H}{\overset{|}{C}}}{=}O + HOR' \overset{H^+}{\rightleftharpoons} R{-}\underset{\underset{\displaystyle OR'}{|}}{\overset{\overset{\displaystyle H}{|}}{C}}{-}OH \tag{12-10}$$

A hemiacetal

$$R{-}\underset{\underset{\displaystyle OR'}{|}}{\overset{\overset{\displaystyle H}{|}}{C}}{-}OH + HOR' \overset{H^+}{\rightleftharpoons} R{-}\underset{\underset{\displaystyle OR'}{|}}{\overset{\overset{\displaystyle H}{|}}{C}}{-}OR' + H_2O^+ \tag{12-11}$$

An acetal

Ketones form ketals in the same way. Acetals and ketals are *gem*-dialkoxy compounds and are a special type of ether. Like ethers, they are stable to bases, but they are hydrolyzed by acids.

D-Glucose as a cyclic hemiacetal

The facts discussed above led to the discovery that glucose and the other simple sugars exist mostly as *cyclic intramolecular hemiacetals;* that is, the aldehyde group forms a hemiacetal with one of the molecule's own —OH groups. The preferred —OH group for the aldohexoses is the one on C-5. That creates a six-atom ring consisting of five carbon atoms and one oxygen atom. Such a ring is called a *pyran ring,* so that form of a sugar is called the *pyranose* form. D-Glucose to a slight extent, and some other sugars to a greater extent, involve the —OH group on C-4 to form a

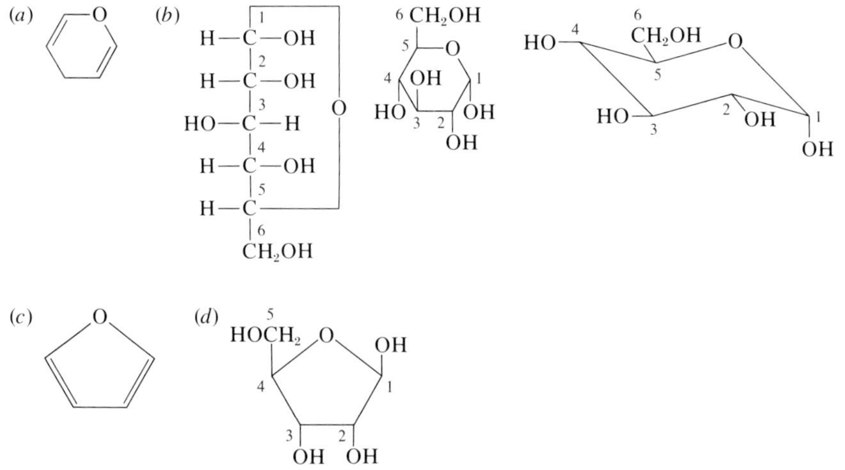

FIGURE 12-6 **(*a*) Pyran; (*b*) three representations of α-D-glucopyranose; (*c*) furan; (*d*) β-D-ribofuranose.**

five-atom ring. Such a ring is called a *furan ring,* so that form of sugar is referred to as the *furanose* form. See Fig. 12-6.

Whenever oxygen of the —OH group bonds to the carbon atom of the planar carbonyl group to form the cyclic hemiacetal, it may bond from either side of the plane to produce two diastereoisomers. When in solution, the diastereoisomers are in equilibrium through the free-aldehyde form (Fig. 12-7). Sugars which differ only in the configuration of the carbon atom of their hemiacetal group are called *anomers*.

Carbohydrate acetals in general are called *glycosides*. Acetals of glucose are glucosides; acetals of mannose are mannosides; ketals of fructose are fructosides; and so forth. Since the methyl D-glucosides have six-membered rings, they are properly named methyl α-D-glucopyranoside and methyl β-D-glucopyranoside. The bonding of the anomeric carbon atom through an oxygen atom to the carbon atom of another molecule to form an acetal is referred to as a *glycosidic linkage*. It may be either α or β.

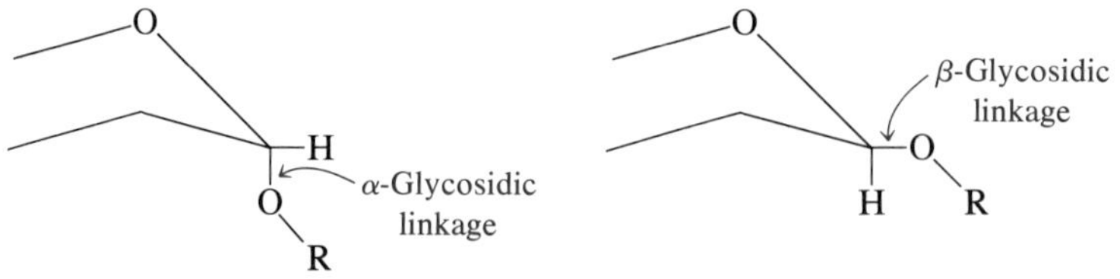

Since glycosides are acetals, they do not exist in equilibrium with an open-chain aldehyde or ketone form in neutral or basic aqueous solution.

In the D series, the —OH or other substituent on the anomeric carbon is designated α when it is "down" in the cyclic formula (or to the right in the Fischer projection formula) and β when it is "up" (or to the left).

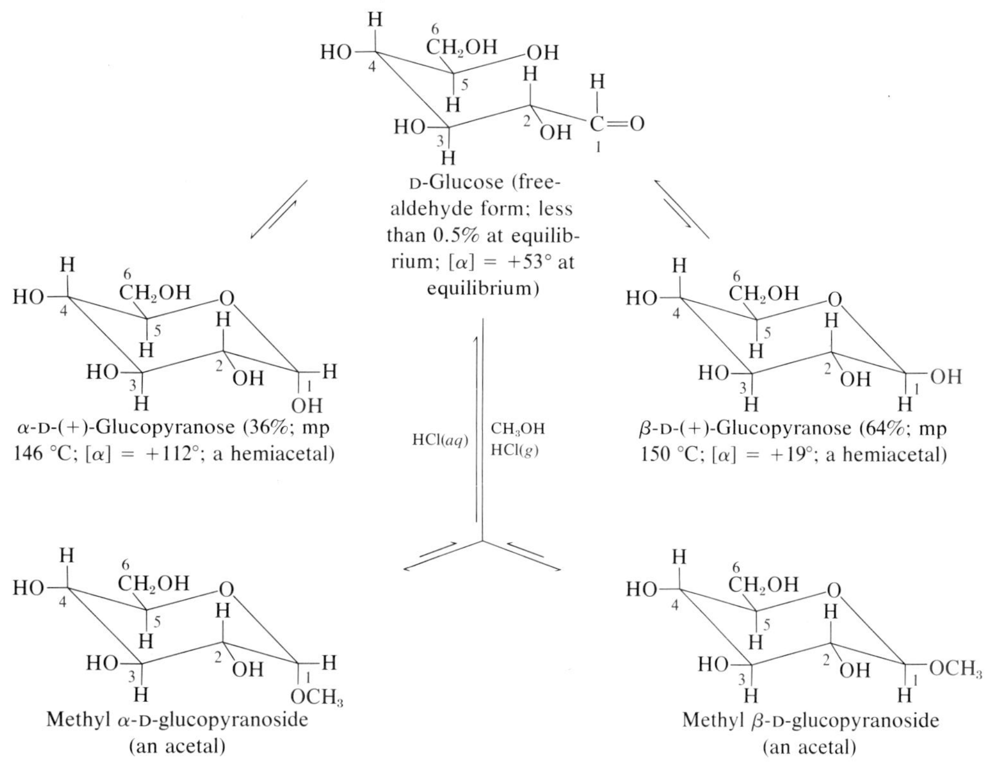

FIGURE 12-7 **α-D-(+)-Glucopyranose and β-D-(+)-glucopyranose are anomers of D-glucose which are in equilibrium through the free-aldehyde form. Note that, in the α form, the —OH group on C-1 is axial and "down" and that, in the β form, the —OH on C-1 is equatorial and "up." Methyl α-D-glucopyranoside and methyl β-D-glucopyranoside are acetals of glucose.**

The chair conformation is preferred for the pyranose ring just as it is for cyclohexane. Note that β-D-(+)-glucopyranose has a slightly higher melting point than α-D-(+)-glucopyranose and exists at equilibrium in greater amount. That suggests it may be more stable. We can observe (Fig. 12-7) that β-D-(+)-glucopyranose must be the only aldohexopyranose (except for its uncommon L enantiomer) which has all of its large attached groups (—OH's and —CH_2OH) connected by equatorial bonds. The groups have more space and hence less intramolecular repulsion to lessen the stability of the molecule. That leaves the more crowded axial positions occupied only by the much smaller hydrogen atoms. The special stability may account for the common occurrence and importance of β-D-glucopyranose and its α anomer, which with only its —OH group on the C-1 axial, is almost as stable. See Fig. 12-8 for models of β-D-glucopyranose.

In ring formation an additional asymmetric carbon atom is generated. Thus there are $2^5 = 32$ stereoisomeric pyranoses for the aldohexoses;

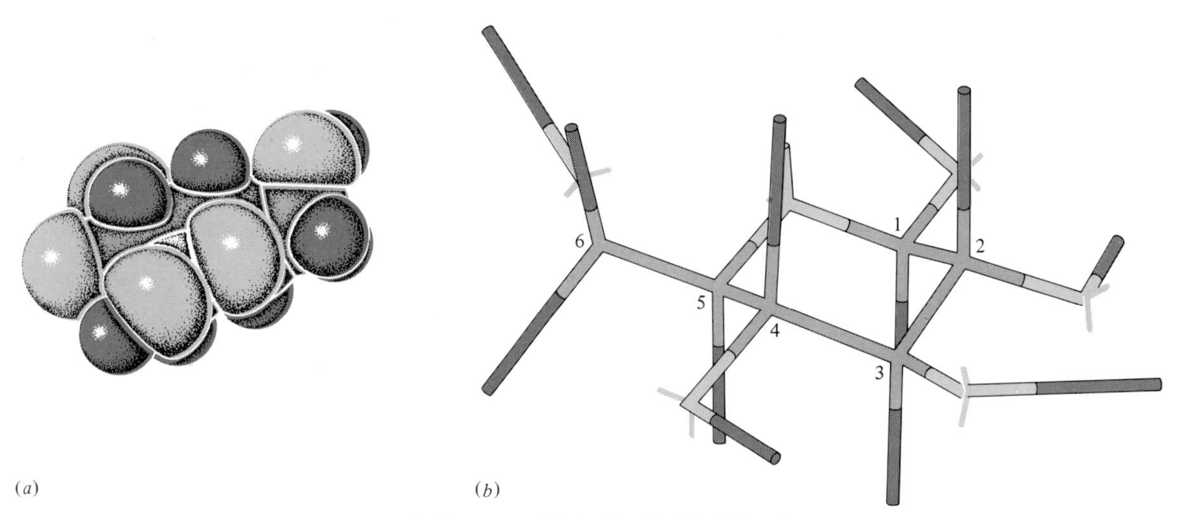

FIGURE 12-8 (*a*) **Space-filling model and** (*b*) **Framework Molecular Model of β-D-glucopyranose.**

they consist of 16 pairs of enantiomers. The enantiomer of α-D-(+)-glucopyranose is α-L-(−)-glucopyranose; the enantiomer of β-D-(+)-glucopyranose is β-L-(−)-glucopyranose.

The methyl D-glucopyranosides

The formation of the methyl glucosides also is shown in Fig. 12-7 for each anomer of D-glucopyranose. Not only do the methyl glucosides not show mutarotation; they also are not reducing sugars. They will not reduce either Fehling's or Benedict's solution because they are no longer in equilibrium with a free-aldehyde form. However, like other acetals they can be readily hydrolyzed back to glucose and methanol by heating with hydrochloric acid. Formation of methyl glucosides requires only one mole of methanol because the sugar is already existing as a hemiacetal. It is in hemiacetal form because of intramolecular reaction with one of its own —OH groups, the one on C-5 in the case of the aldohexopyranoses.

D-Galactopyranose and D-mannopyranose

Although they are not nearly as common or as important as D-glucose, D-galactose and D-mannose are two other aldohexoses which have some interest and importance. D-Galactose is found combined with D-glucose in the disaccharide lactose. Our bodies cannot metabolize galactose directly, but must first enzymatically isomerize it to D-glucose. An inherited disease of infants, galactosemia, is characterized by inability to metabolize galactose because of lack of the isomerizing enzyme. Galactose accumulates in various tissues and causes cell damage. Milk must be excluded from the diet of such infants.

D-Galactose and D-mannose exist mostly as the cyclic hemiacetal anomers of D-galactopyranose and D-mannopyranose, respectively. It is easy to write chair conformation structural formulas for them. You will now learn to write the chair conformation for β-D-glucopyranose by placing hydrogen atoms on all axial bonds and hydroxy groups on all the equatorial bonds except C-5 which holds the —CH_2OH group. Let us go through it stepwise.

1 Draw a flat V and connect oxygen on the right side.

2 Draw another V so that its right side is parallel with the right side of the first V.

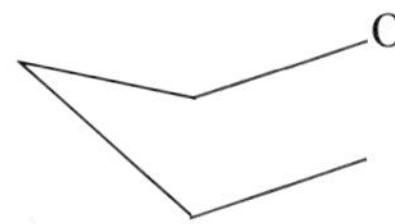

3 Now connect a third and final V so that its right side is parallel with the left side of the second V.

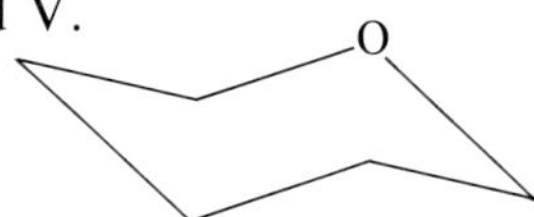

4 Attach axial bonds parallel to a hypothetical center axis. Attach hydrogen atoms to the axial bonds.

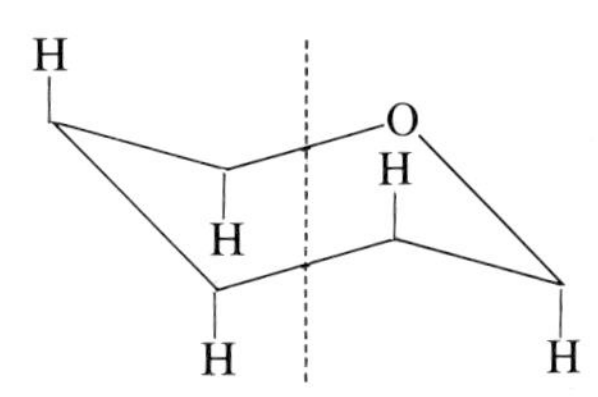

5 Attach equatorial bonds holding —OH groups and a —CH_2OH group. The equatorial bonds should be directed *toward* the ends of the structure (except for the bonds on the end carbons), and parallel with the opposite side of the structure. The result is the conformational structure of β-D-glucopyranose.

If the conformational structure of β-D-galactopyranose is desired, the positions of the —H and —OH on C-4 are interchanged, so that the —OH is on the equatorial bond. If the conformational structure of β-D-mannopyranose is desired, the positions of the —H and —OH on C-2 are interchanged.

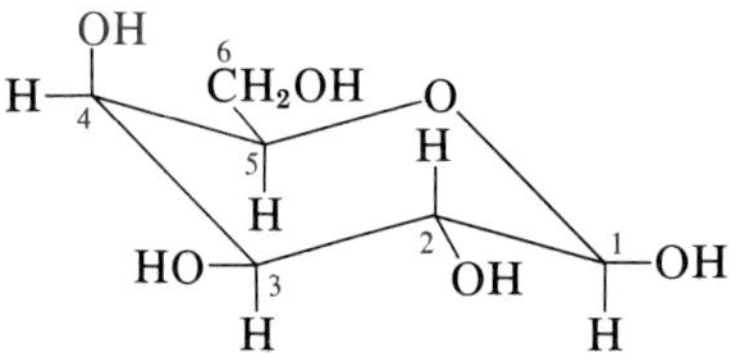

β-D-(+)-Galactopyranose 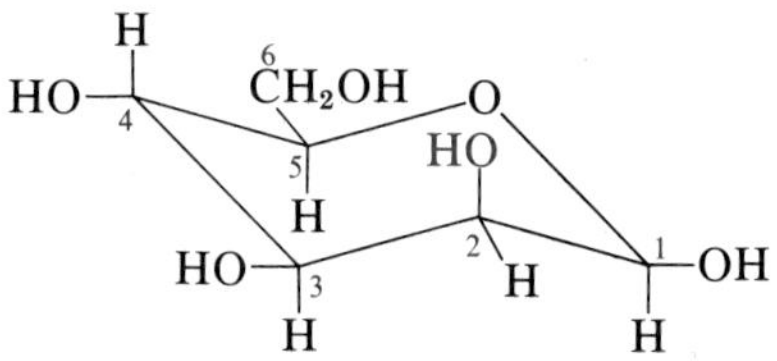 β-D-(+)-Mannopyranose

If the α form of any one of these three β-pyranose forms is desired, the formula of the corresponding β form is simply altered by interchanging the —H and the —OH on the anomeric carbon, C-1.

Haworth cyclic formulas The structural formulas first suggested by the English chemist Sir W. N. Haworth for representing the cyclic forms of the sugars are preferred by many people. The molecules are drawn as planar, hexagonal slabs with darkened edges toward the viewer. Ring carbon atoms and the hydrogen atoms directly attached to them are usually not shown. Any group of atoms written to the right in the Fischer projection formula appears below the plane of the ring, and any group written to the left appears above the plane. (That is true only if the usual convention is followed: the molecule is oriented with the ring oxygen to the back and C-1 to the right.) The anomeric —OH or other group is then down for the α anomer and up for the β anomer. The rules are applicable to furanose rings as well as pyranose rings.

Examples of Haworth cyclic formulas, and their relations to the Fischer projection formulas, are the following:

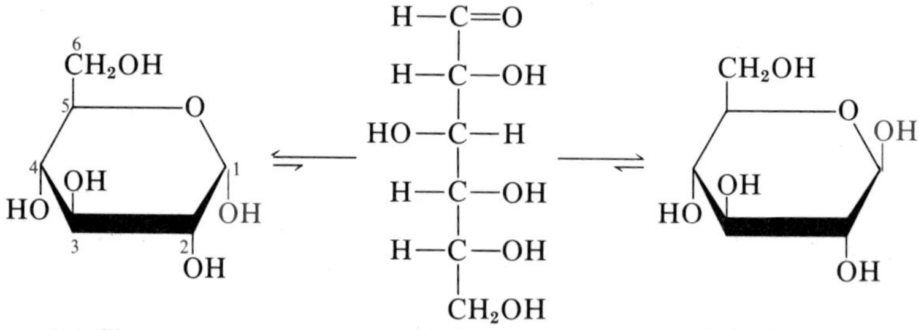

α-D-(+)-Glucopyranose ⇌ D-(+)-Glucose ⇌ β-D-(+)-Glucopyranose (12-12)

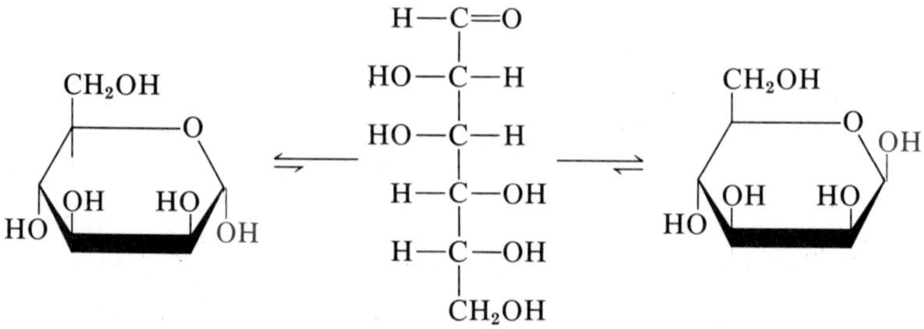

α-D-(+)-Mannopyranose ⇌ D-(+)-Mannose ⇌ β-D-(+)-Mannopyranose (12-13)

α-D-(+)-Galactopyranose ⇌ D-(+)-Galactose ⇌ β-D-(+)-Galactopyranose (12-14)

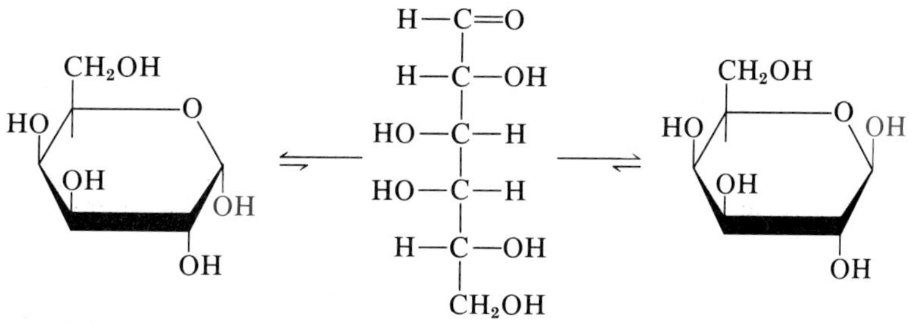

Fructose forms both fructopyranoses and fructofuranoses. The reactions may be depicted as

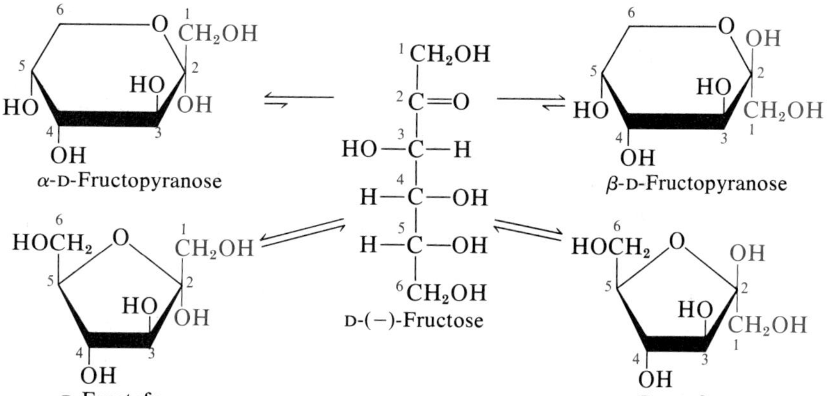

α-D-Fructopyranose ⇌ D-(−)-Fructose ⇌ β-D-Fructopyranose

α-D-Fructofuranose ⇌ D-(−)-Fructose ⇌ β-D-Fructofuranose

(12-15)

In Sec. 12-7, we shall see that, as a component of the disaccharide sucrose, D-fructose is in the *β-furanose* form. D-Fructose in water solution, however, exists mostly as an equilibrium mixture of the α- and β-*pyranose* forms.

PROBLEM 12-8 Write Haworth and chair conformational formulas for methyl α-D-mannopyranoside.

PROBLEM 12-9 What simple chemical test would distinguish between α-D-glucopyranose and methyl α-D-glucopyranoside?

12-6 Aldopentoses

The aldopentoses contain three asymmetric carbon atoms which give rise to $2^3 = 8$ isomeric aldopentoses consisting of four pairs of enantiomers. However, they also form cyclic hemiacetals, so each one of the 8 isomers has an α and a β form for a total of 16 stereoisomers. The aldopentoses occur in both six- and five-atom rings. The five-atom ring consists of four carbon atoms and one oxygen atom—the furan ring.

The most important aldopentoses are D-ribose, D-deoxyribose, D-2-

FIGURE 12-9
Four important aldopentoses.

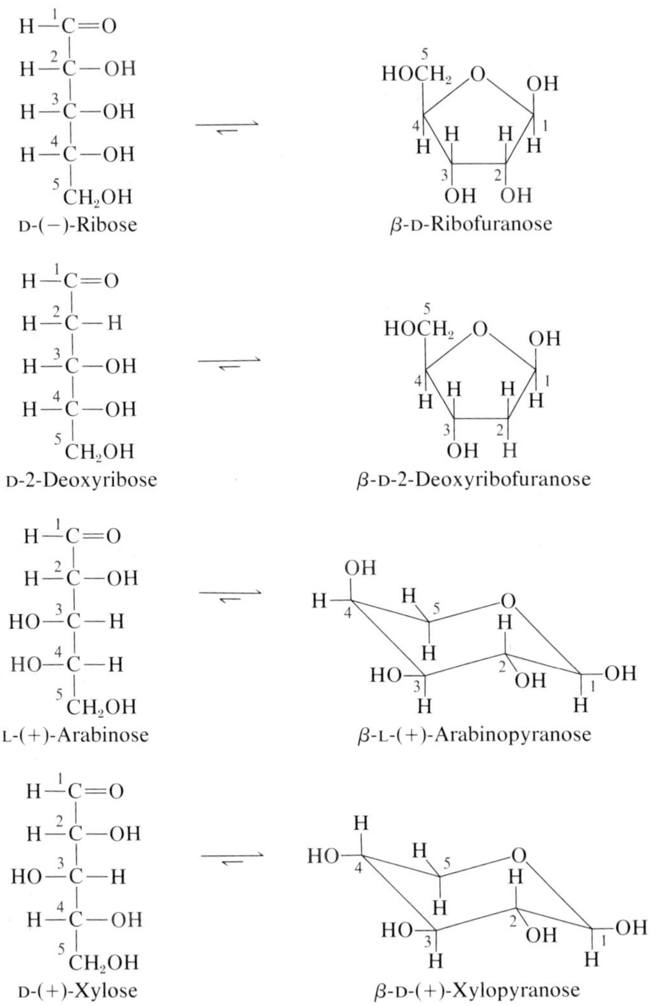

D-(−)-Ribose ⇌ β-D-Ribofuranose

D-2-Deoxyribose ⇌ β-D-2-Deoxyribofuranose

L-(+)-Arabinose ⇌ β-L-(+)-Arabinopyranose

D-(+)-Xylose ⇌ β-D-(+)-Xylopyranose

deoxyribose, L-arabinose, and D-xylose. They are shown in Fig. 12-9. D-(−)-Ribose occurs in the nucleic acid RNA (ribonucleic acid). D-2-Deoxyribose occurs in the nucleic acid DNA (deoxyribonucleic acid). L-(+)-Arabinose is obtained by hydrolysis of plant gums such as gum arabic. D-(+)-Xylose is obtained by hydrolysis of the polysaccharides, called *pentosans,* found in such plant products as corn cobs and oat hulls.

12-7 Disaccharides

The important sugars called disaccharides include maltose, cellobiose, lactose, and sucrose. Their hydrolysis products are the following:

$$\text{Maltose} + H_2O \xrightarrow{\text{HCl or maltase}} 2\text{D-glucose} \tag{12-16}$$

$$\text{Cellobiose} + H_2O \xrightarrow{\text{HCl or emulsin}} 2\text{D-glucose} \tag{12-17}$$

$$\text{Lactose} + H_2O \xrightarrow{\text{HCl or emulsin}} \text{D-galactose} + \text{D-glucose} \qquad (12\text{-}18)$$

$$\text{Sucrose} + H_2O \xrightarrow{\text{HCl or sucrase}} \text{D-glucose} + \text{D-fructose} \qquad (12\text{-}19)$$

The sugars maltose, cellobiose, and lactose are reducing sugars. That is, they reduce Fehling's solution, which indicates that they have a hemiacetal part of the molecule in equilibrium with a free-aldehyde group. They also exhibit mutarotation, and they can be oxidized by bromine water to aldonic acids. On the other hand, sucrose is a nonreducing sugar, so it must not have a hemiacetal part in its molecule. The two monosaccharides comprising a disaccharide are bonded together by a glycosidic linkage.

Maltose

The enzyme *maltase* effects hydrolysis of methyl α-D-glucopyranoside, but not of the β anomer. Since maltose also is hydrolyzed by maltase but not by the enzyme *emulsin,* it can be reasoned by analogy that maltose contains an α-glycosidic linkage like that of methyl α-D-glucopyranoside. Additional evidence indicates that the two glucose units are connected with the α linkage through an oxygen atom which joins the C-1 of one glucose unit to the C-4 of the other glucose unit. The structural formula and name of maltose would then be as shown below. (Ring H's have been omitted.)

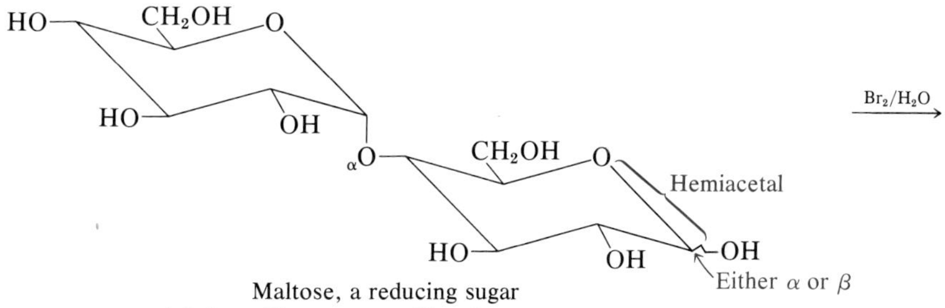

Maltose, a reducing sugar
4-O-[α-D-Glucopyranosyl]-D-glucopyranose

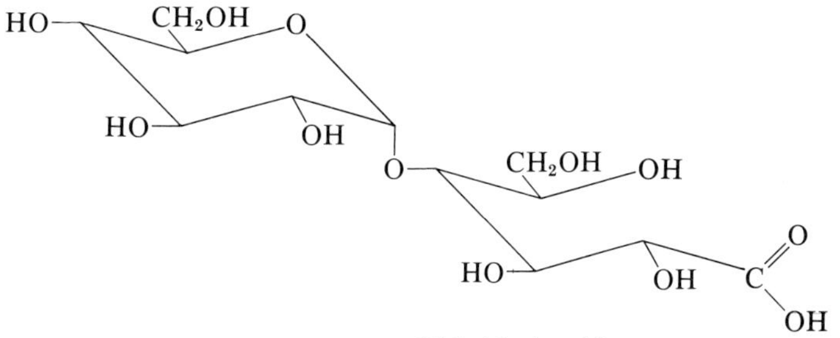

$\xrightarrow{Br_2/H_2O}$

Maltobionic acid
(Maltonic acid)

(12-20)

The purpose of the capital O in the chemical name is to indicate that the group in brackets is attached to C-4 through oxygen, and not directly. The hemiacetal nature of one ring of maltose is shown by this reaction, in which maltose is oxidized to maltobionic acid by bromine water. Hydrolysis of maltobionic acid yields D-glucose and D-gluconic acid.

Maltose is malt sugar. It is formed by the action of the enzyme diastase from malt (germinated grain, usually barley) on starch. Maltose has a characteristic malt flavor and is not as sweet as the other common sugars.

Cellobiose

Cellobiose, a sugar obtained from the partial hydrolysis of cellulose, may be hydrolyzed by the action of emulsin, but not by maltase. The enzyme specificity is shared by methyl β-D-glucopyranoside. The conclusion, therefore, is that cellobiose has its two glucose units connected by a β-glycosidic linkage. The structural formula and name are as follows:

Either α or β

Cellobiose
4-O-[β-D-Glucopyranosyl]-D-glucopyranose

Determination of the structure of cellobiose was an aid in determination of the structure of cellulose. Cellobiose itself is not of much importance.

Lactose

Lactose is milk sugar. It is found to the extent of about 5% by weight in the milk of most mammals. Since it is hydrolyzed by the action of emulsin, a β-glycosidic linkage is indicated. Both D-glucose and D-galactose are obtained by the hydrolysis, so the question that arises is whether lactose is a 4-O-[β-D-glucopyranosyl]-D-galactopyranose or a 4-O-[β-D-galactopyranosyl]-D-glucopyranose. Lactose reacts with phenylhydrazine to form a lactosazone which, upon hydrolysis, yields D-galactose and D-glucosazone. Also, oxidation with bromine water gives the aldonic acid, lactobionic acid, which upon hydrolysis yields D-galactose and D-gluconic acid. Therefore, the structure and name of lactose must be

Either α or β

Lactose
4-O-[β-D-Galactopyranosyl]-D-glucopyranose

The equations for the reactions of lactose are given in Fig. 12-10.

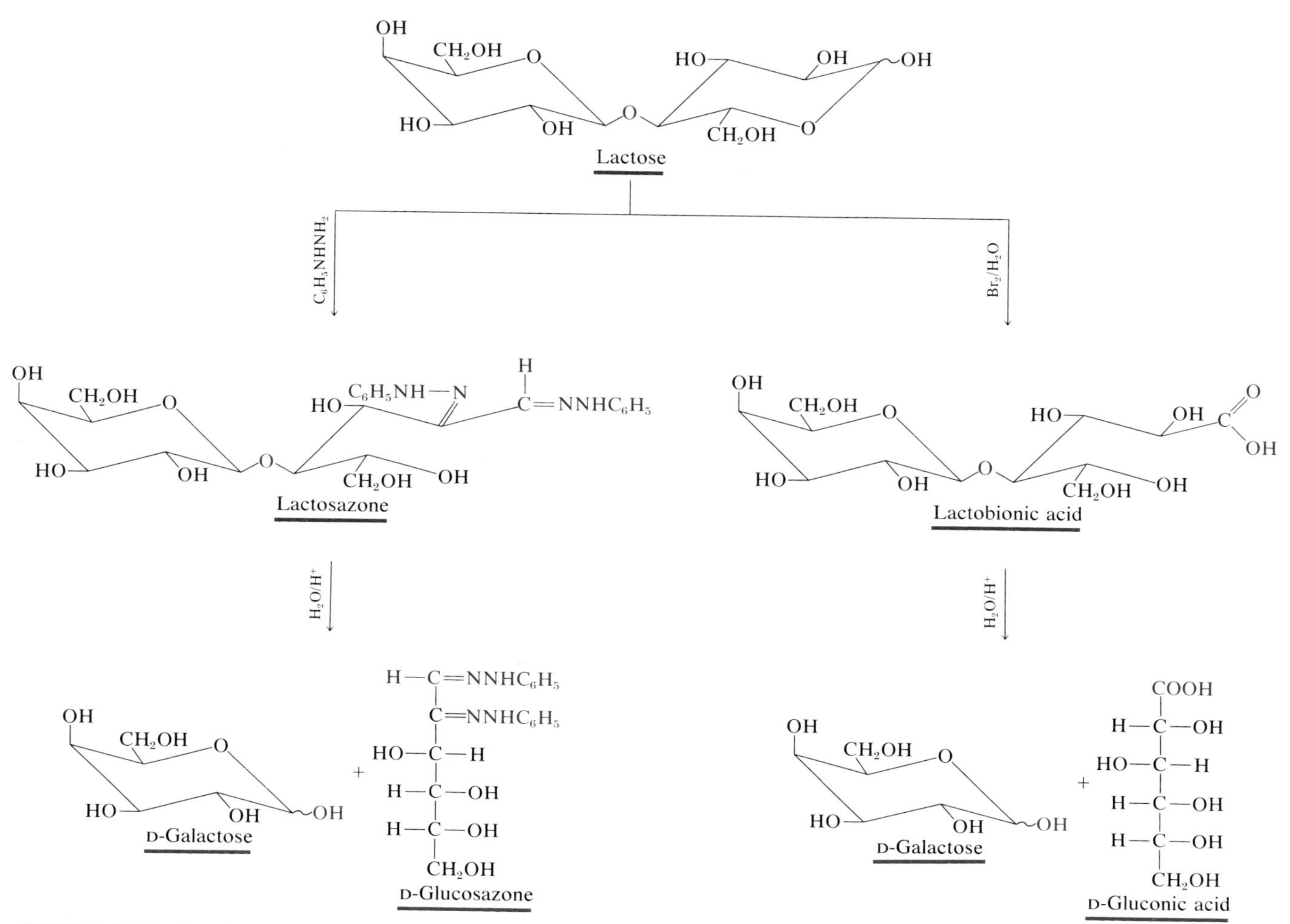

FIGURE 12-10 Reactions of lactose.

Sucrose Sucrose is our common sugar. It is obtained from sugar beets or sugar cane, and it is highly refined to a very pure compound. It is our most common sweetening agent, and as such it is an important ingredient of confections and candy. It is a nonreducing, nonmutarotating sugar which does not react with phenylhydrazine. That leads to the conclusion that the D-glucose and D-fructose units which comprise the molecule are connected in an acetal or glycosidic linkage involving the carbon atoms of the potential carbonyl groups of *both* monosaccharides. That is, the acetal or glycosidic linkage must be from C-1 of D-glucose to C-2 of the ketohexose, fructose. Such a linkage precludes the possibility of a hemiacetal linkage in any other part of the molecule, so there is no potential aldehyde group in equilibrium to react with phenylhydrazine or reduce Fehling's solution. Additional evidence indicates that sucrose is the α form of D-glucose linked with the β form of D-fructose and that the fructose is in the furanose (five-atom) ring form. The structural formula and chemical name for sucrose are

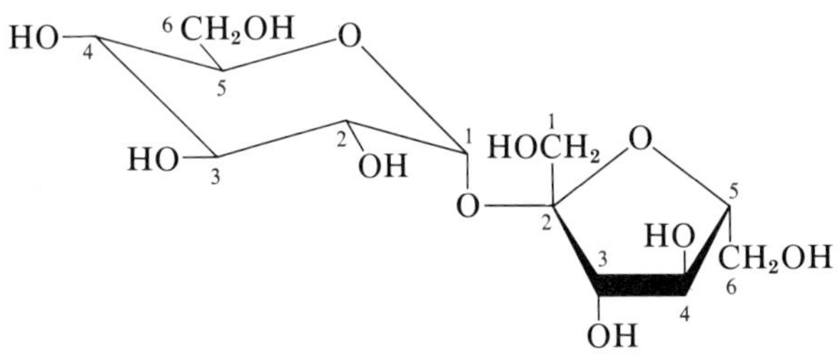

Sucrose
α-D-Glucopyranosyl-β-D-fructofurano*side*

As sucrose is hydrolyzed, its specific rotation of +66.5° begins to decrease. The rotation finally becomes negative at −19.9° as D-glucose and D-fructose are produced, since D-fructose is more levorotatory than D-glucose is dextrorotatory. Because of the inversion of specific rotation, the mixture of products is called *invert sugar*. Honey is essentially invert sugar.

Ancient peoples did not have refined sucrose; they had to rely on fruits, honey, and milk for their sweet foods. After tea and coffee were introduced into Europe about the middle of the seventeenth century, sucrose came to be used as a sweetening agent for those drinks. Then sucrose began to be used to make candy, confections, and jams and to sweeten many other foods. Today the average person probably consumes far more of it than is desirable for reasons of health. It is a big factor in the formation of dental caries (tooth decay); it contributes to obesity; and it has been suspected as a contributing factor in atherosclerosis and in maturity-onset diabetes.

PROBLEM 12-10 Draw Haworth and conformational structures for each of the following disaccharides or derivatives.

(a) 6-O-(β-D-glucopyranosyl)-β-D-glucopyranose
(b) 4-O-(α-D-glucopyranosyl)-D-gluconic acid
(c) 6-O-(α-D-galactopyranosyl)-β-D-fructofuranose

Solution to (a)

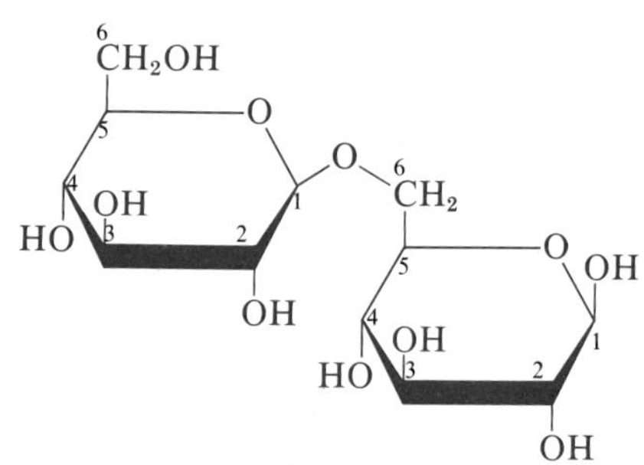

Haworth structure of
6-O-(β-D-glucopyranosyl)-β-D-glucopyranose

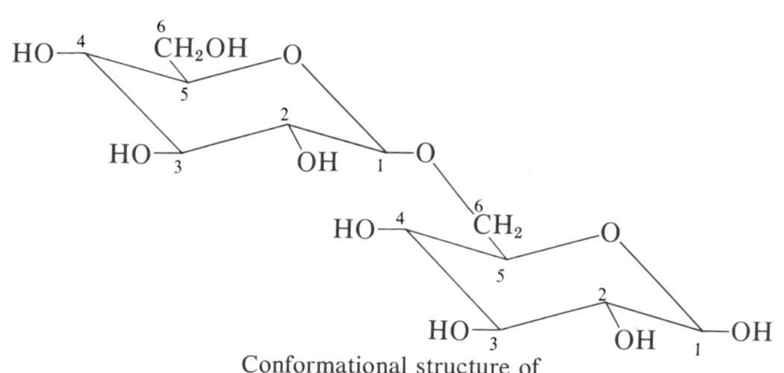

Conformational structure of
6-O-(β-D-glucopyranosyl)-β-D-glucopyranose

PROBLEM 12-11 What structural features are necessary if a disaccharide is to be a reducing sugar? Draw the structure of a hypothetical *nonreducing* disaccharide composed of two aldopentoses.

PROBLEM 12-12 Trehalose is a nonreducing disaccharide that yields glucose upon hydrolysis. Its systematic name is α-D-glucopyranosyl-α-D-glucopyranoside. Draw the structural formula of trehalose.

PROBLEM 12-13 Give the common name of each of the following systematically named sugars.

(a) 4-O-(β-D-galactopyranosyl)-D-glucopyranose
(b) 4-O-(α-D-glucopyranosyl)-D-glucopyranose
(c) 4-O-(β-D-glucopyranosyl)-D-glucopyranose
(d) α-D-glucopyranosyl-β-D-fructofuranoside

Solution to (a) The common name for 4-O-(β-D-galactopyranosyl)-D-glucopyranose is lactose.

PROBLEM 12-14 How could you distinguish between the members of each of the following pairs of sugars? (a) sucrose and invert sugar, (b) mannose and maltose, (c) sucrose and maltose.

Solution to (a) The simplest way to distinguish between sucrose and invert sugar would be to determine the specific rotation. The specific rotation of sucrose is $+66.5°$; that of invert sugar is $-19.9°$. Invert sugar, a mixture of D-glucose and D-fructose, is the hydrolysis product of sucrose. Also, Fehling's solution is reduced by invert sugar, but not by sucrose.

12-8 Polysaccharides

The polysaccharides include starches, glycogen, cellulose, gums, pectins, and pentosans. The first three will be discussed in more detail. The pentosans have already been mentioned briefly. The pectins are found in fruits; they are the factor which makes jelly gel. Chemically they are polymers of galacturonic acid.

Starch

Starch, a natural polymer of glucose, is the storage form of glucose in plants; it occurs in seeds, roots, and tubers. The plants that serve us as our chief sources of edible starch are corn, potatoes, rice, wheat, and other grains.

About 20% of starch is somewhat water-soluble and can be separated out into a fraction called *amylose*. The remaining 80% is called *amylopectin*. Although complete hydrolysis of starch gives only D-glucose, partial hydrolysis gives some maltose but never any cellobiose. Hence, starch must be an α-polyglycoside of D-glucose. The molecular weight of amylose indicates about 1000 units of D-glucose in the chain.

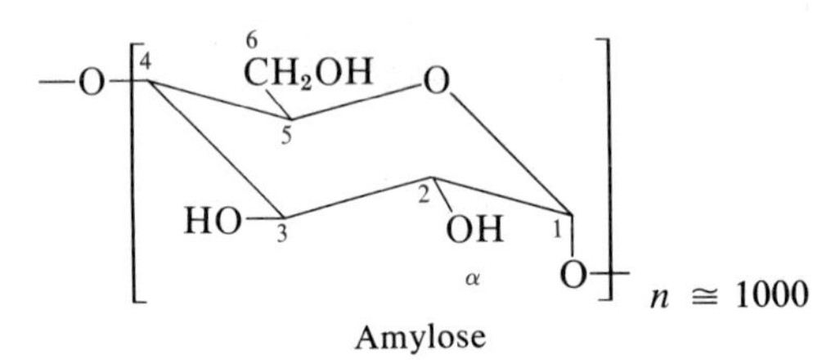

Amylose

Observe that the units of D-glucose are connected from position C-4 of one unit through oxygen to C-1 of the unit to its left as the formula is written. The polymer can be referred to as a C-4 to C-1 α-polyglucoside.

By a process of methylation and hydrolysis, it has been determined that amylopectin has a similar chain of C-4 to C-1 α-glucosidic linkages. In addition, it has chains of about 20 or more units branching off from C-6 of some units. The molecular weight of amylopectin is much higher than that of amylose and indicates 1000 to 6000 units of glucose. Figure 12-11 shows the basic unit of the amylopectin polymer. Figure 12-12 shows larger sections of both the amylose and amylopectin polymers.

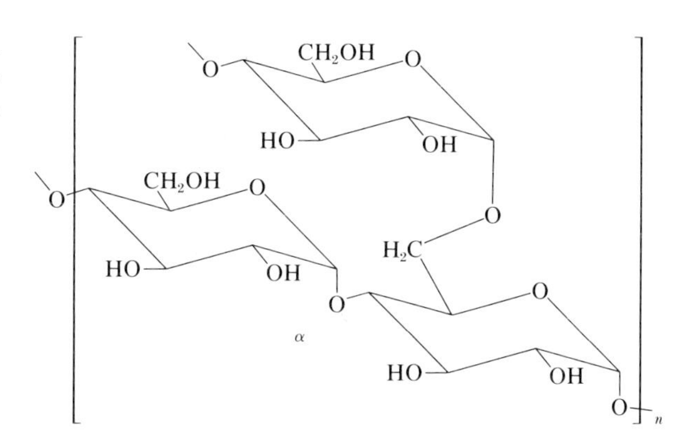

FIGURE 12-11 **The basic unit of the amylopectin polymer.**

The hydrolysis of starch (amylose and amylopectin) produces, in three successive stages, dextrins, maltose, and D-glucose. *Dextrins* are glucose polysaccharides of intermediate size. They become sticky when wet, and so they are used as adhesives on stamps, labels, and envelopes. Because they are more easily digested than starch, they are used in the commercial preparation of baby foods. Starch can be hydrolyzed by heating it with dilute acid, or it can be degraded in stages by enzymes called *amylases*.

$$\text{Starch} \xrightarrow[\text{or amylase}]{H^+,\ \Delta,} \text{dextrins} \xrightarrow[\text{or amylase}]{H^+,\ \Delta,} \text{maltose} \xrightarrow[\text{or maltase}]{H^+,\ \Delta,} \text{D-glucose} \quad (12\text{-}21)$$

When treated with iodine, starch produces a very deep, dark blue color. As starch is hydrolyzed to dextrins, the dark blue color with iodine changes to a reddish-blue and then to a pink. Finally, the products of hydrolysis produce no color at all with iodine. Experimental evidence indicates that the amylose chain is actually coiled like a spring. When coiled, the molecule has just enough room in its core to accommodate an iodine molecule. The characteristic color that starch gives with iodine is due to the formation of the amylose-iodine complex.

Glycogen

Glycogen is the polysaccharide form in which animals store glucose; it is sometimes referred to as animal starch. In structure it is similar to amylopectin, but it is of higher molecular weight (about 7000 glucose units). It has a greater number of branches, but the branches are of somewhat shorter chain length (12 to 18 glucose units). It is stored in the liver and in the muscles.

Cellulose

Since its partial hydrolysis produces some cellobiose but never any maltose, cellulose can be considered to be a C-4 to C-1 β-polyglucoside. There are no branching chains; rather there is a long, linear polymer of 2000 to 6000 units of D-glucose.

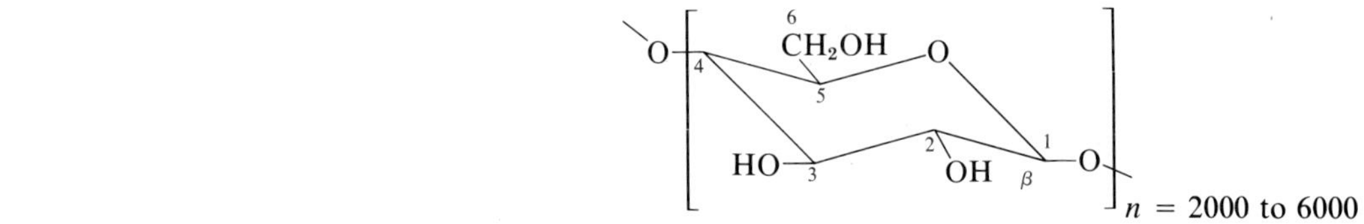

$n = 2000$ to 6000

Cellulose

Unlike starch, cellulose is very insoluble in water. It is the chief structural material of plants. The human digestive system is not capable of hydrolyzing it even though it is a polymer of D-glucose similar to starch. However, many animals, especially ruminants, have digestive tracts which, with the aid of enzymes from bacteria, can hydrolyze cellulose. Of course, human beings benefit from cellulose as a food indirectly, because human beings may eat the animals which utilize the cellulose.

Because its structure allows close packing of its long, linear chains, cellulose is relatively tough, strong, and insoluble. It comprises 40 to 50% of wood. Wood pulp, which has had the lignin extracted out with sodium hydroxide–sodium bisulfite solution, is 93% cellulose and is used to make paper. Cotton and linen also are essentially cellulose.

Examination of the structural formula for cellulose shows two secondary alcohol groups and one primary alcohol group for each glucose unit. Some or all of the alcohol groups can be caused to react to produce ester or ether groups. Partial acetylation using a mixture of acetic anhydride and acetic acid produces a *cellulose acetate*. The product can be fabricated into fibers for clothing or into sheets for a relatively nonflammable base for home movie film.

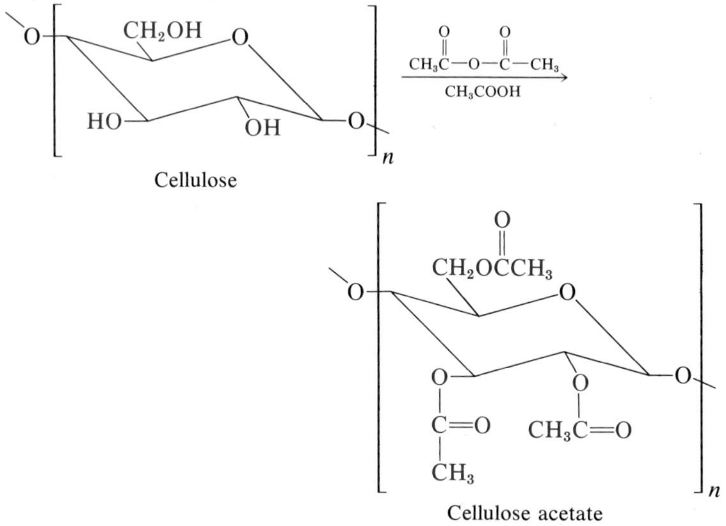

Cellulose $\xrightarrow[CH_3COOH]{CH_3C(=O)-O-C(=O)-CH_3}$ Cellulose acetate (12-22)

Partial nitration of cellulose using nitric and sulfuric acids produces *cellulose nitrate,* which is soluble in ether and acetone. The more highly ni-

(a)

(b)

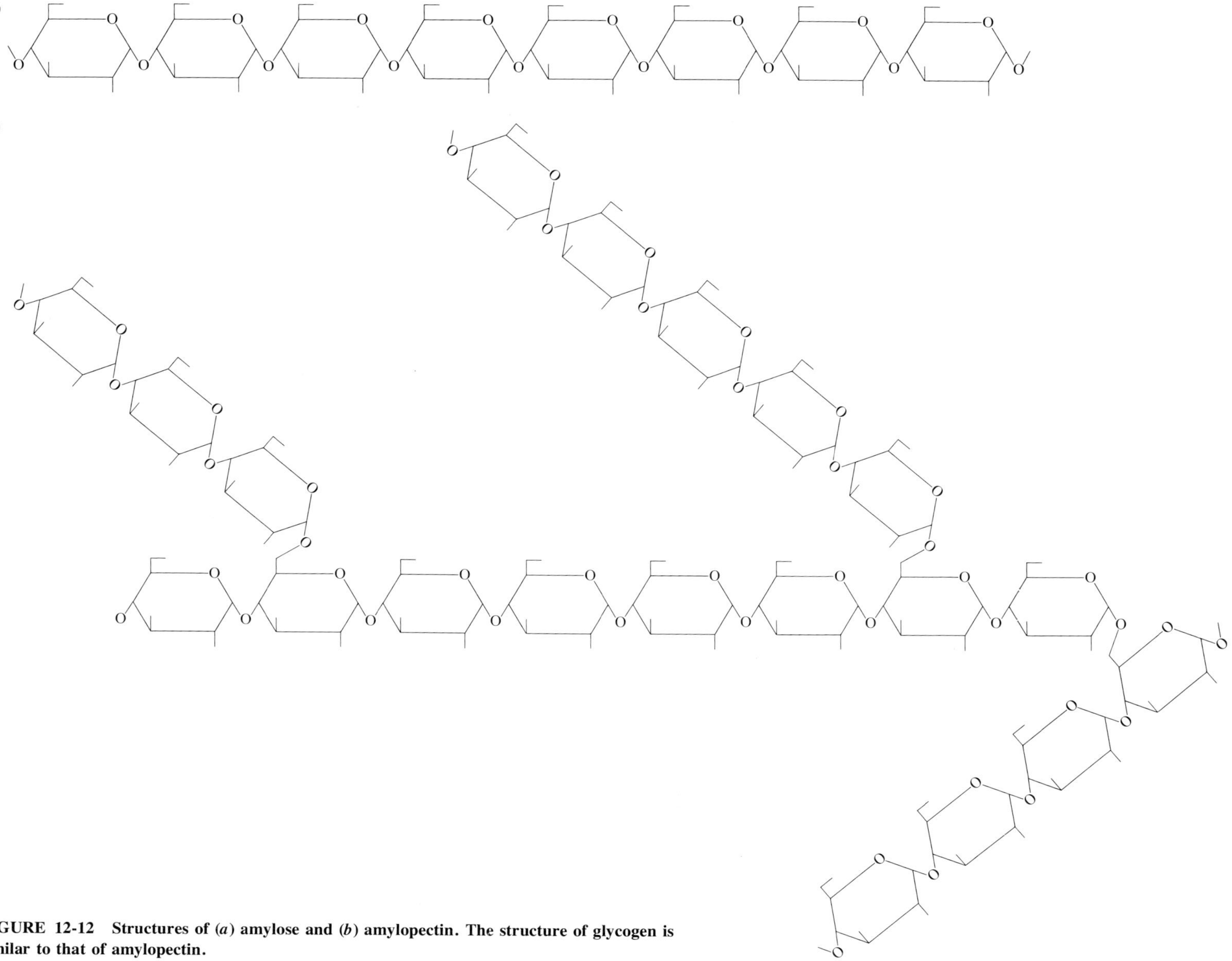

FIGURE 12-12 **Structures of (*a*) amylose and (*b*) amylopectin. The structure of glycogen is similar to that of amylopectin.**

trated product is used for making smokeless gunpowder. The less highly nitrated product is plasticized with camphor to make a type of plastic called Celluloid or Pyralin.

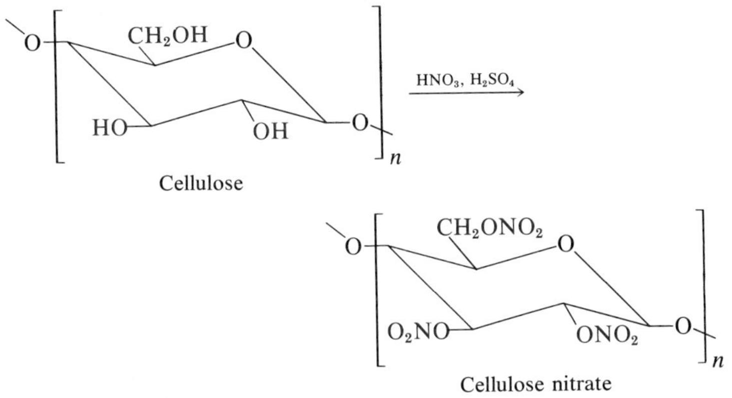

CH_2OH O HO OH O n

Cellulose

$\xrightarrow{HNO_3,\ H_2SO_4}$

CH_2ONO_2 O O_2NO ONO_2 O n

Cellulose nitrate

(12-23)

Viscose rayon is produced by reacting cellulose, in the form of wood pulp, with sodium hydroxide and carbon disulfide. The product is a very viscous solution of a compound called cellulose xanthate:

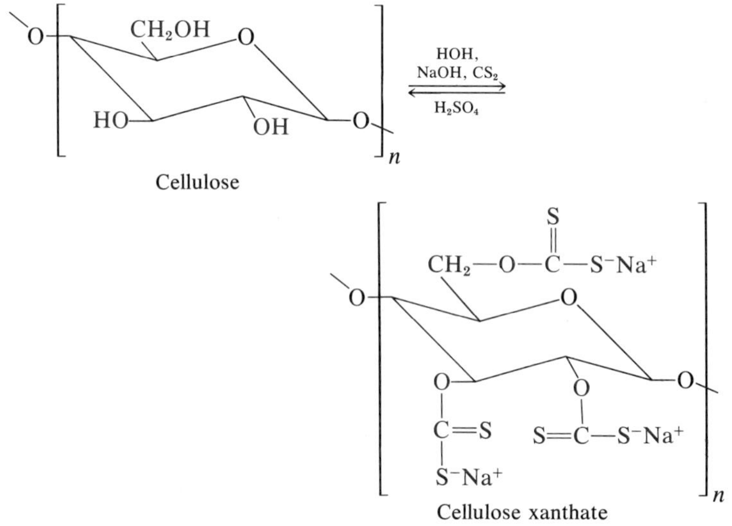

CH_2OH O HO OH O n

Cellulose

$\underset{H_2SO_4}{\overset{HOH,\ NaOH,\ CS_2}{\rightleftarrows}}$

$CH_2-O-\overset{S}{\overset{\|}{C}}-S^-Na^+$ O $O-C(=S)-S^-Na^+$ $O-\overset{}{C}(=S)-S^-Na^+$ O n

Cellulose xanthate

(12-24)

A *xanthate* may be considered a half ester half salt of thiocarbonic acid ($HO-\overset{S}{\overset{\|}{C}}-SH$). The viscous solution is forced through small holes in a device called a *spinneret* into a sulfuric acid solution. There the cellulose is immediately regenerated as a solid, insoluble fiber of cellulose, but now of somewhat shorter chain length owing to partial hydrolysis of the chain. The fibers are spun into thread and made into a fabric called viscose rayon. If the cellulose xanthate is extruded and used as a film instead of a fiber, it is called *cellophane*—a transparent wrapping material.

12-9 Photosynthesis

The carbohydrates are produced in plants by the biochemical reaction of *photosynthesis*. Water and carbon dioxide are combined in the presence of the green-colored pigment chlorophyll as a catalyst and with energy provided by sunlight. The products of photosynthesis include glucose, starch, cellulose, and other carbohydrates. The process is complex, and many steps and intermediates are involved. The work of Melvin Calvin (University of California, Berkeley, Nobel Prize in 1961), along with that of many other scientists, has now revealed many of the intermediate steps involved. The simplified overall reaction for photosynthesis is

$$6CO_2 + 6H_2O \xrightarrow{\text{Light, chlorophyll}} \underset{\substack{\text{D-Glucose and}\\ \text{polysaccharides}}}{C_6H_{12}O_6} + 6O_2 \qquad (12\text{-}25)$$

Summary

The *carbohydrates* include sugars, starches, cellulose, gums, and pectins. Carbohydrates are *polyhydroxy aldehydes* or *ketones* and their derivatives or compounds which can be hydrolyzed to such.

Polysaccharides are *polymeric carbohydrates* which may be hydrolyzed to disaccharides and ultimately to many monosaccharide units. *Disaccharides* are dimers of monosaccharides, to which they may be hydrolyzed. *Monosaccharides* are *simple sugars* and their derivatives. They are not further divisible by hydrolysis.

Sugars are given *common names* ending in *-ose*. A sugar that contains an aldehyde group is referred as an *aldose;* one that contains a ketone group is referred to as a *ketose*. A sugar may also be referred to by a name reflecting the number of carbon atoms in the molecule. For example, a *hexose* is a six-carbon-atom sugar and a *pentose* is a five-carbon-atom sugar. Sugars are *optically active stereoisomers,* and the direction of optical rotation is designated by either (+) or (−) in front of a name. The common name of a sugar represents the configuration of all of the sugar's asymmetric carbon atoms except the last one, whose configuration is indicated by either D or L in front of the name.

D-(+)-Glucose ($C_6H_{12}O_6$) is the most important of all the simple sugars. It is blood sugar, corn sugar, and one of the sugars in honey, and it is the hydrolysis product of starch, glycogen, and cellulose. Because it is dextrorotatory, it is also known as *dextrose*. Since the aldehyde form of D-glucose has four asymmetric carbon atoms, there are $2^4 = 16$ stereoisomers consisting of 8 pairs of enantiomers. Their configurations were elucidated by Emil Fischer and his coworkers. The enantiomer of each pair which has the —OH group of the first asymmetric carbon atom from the bottom (as the formula is normally written) to the right is said to belong to the D *family* and is a D-*sugar*. If the —OH group is to the left, the sugar belongs to the L *family*. Almost all of the naturally occurring sugars and their polymers and derivatives belong to the D family.

Epimers are a pair of *diastereoisomers* in which the difference in configuration relates to only *one* of the asymmetric carbon atoms. D-Galactose differs from D-glucose only in the configuration about carbon atom 4, and it is thus an epimer of D-glucose. D-Mannose and D-glucose differ only in the configuration about C-2, so they also are epimers.

D-Fructose (levulose) is a *ketose* with the same configuration about C-3, C-4, and C-5 as D-glucose. It is obtained along with D-glucose when common sugar (sucrose) is hydrolyzed. It is found in honey and many fruits, and it is the sweetest of all the sugars.

An *aldonic acid* is a *monocarboxylic acid* produced by oxidizing the *aldehyde* group of an *aldose* with bromine water. An *aldaric acid* is a *dicarboxylic acid* produced by oxidation of an *aldose* with HNO_3. An *alduronic acid* is produced by enzymatic oxidation of only the *primary alcohol* group to a carboxyl group to produce an *aldehydo acid*. An *alditol* is a *polyhydroxy alcohol* produced by reduction of an aldose with hydrogen and a metal catalyst, with sodium and alcohol, or with sodium borohydride.

Sugars are very sensitive to alkali because the alkali promotes *enolization,* which in turn facilitates extensive oxidation. Sugars which reduce such reagents as Tollens', Benedict's, or Fehling's solutions are called *reducing sugars*. All of the simple sugars, and also the disaccharides which are reducing sugars, react with 3 mol of phenylhydrazine to form *osazones*. The osazones are yellow, crystalline solids with well-defined melting points. D-Glucose, D-mannose, and D-fructose give the same osazone, which demonstrates they have like configurations except for C-1 and C-2.

There are α and β forms of D-glucose with different melting points and specific rotations. When placed in solution, each form *mutarotates* to the same equilibrium value of $[\alpha] = +53°$. D-Glucose reacts with only one mole of methanol to form two *diastereoisomeric acetals* called methyl α-D-glucoside and methyl β-D-glucoside. The *methyl glucosides* are *cyclic acetals*. In solution they are not in equilibrium with free aldehyde, so they do not exhibit mutarotation and they are nonreducing. D-Glucose forms two diastereoisomeric *six-atom cyclic hemiacetals* with the hydroxy group of its own C-5. They are called *α-D-glucopyranose* and *β-D-glucopyranose*. In solution they are in equilibrium with each other through the aldehyde form. Sugars which differ only in the configuration of their *hemiacetal group* are called *anomers*. The chair conformation is the preferred one for the *pyranose ring* forms.

The other aldohexoses and pentoses also form pyranose rings, but D-glucose is the most stable of the aldohexoses. It is the only one with all of its —OH groups and its —CH_2OH group in the more roomy equatorial positions. The sugars also may form *five-atom cyclic hemiacetal* structures that are referred to as *furanose* forms. They are less common than the pyranose forms for simple sugars.

Haworth formulas for sugars are drawn as planar, hexagonal slabs with darkened edges toward the viewer. Any group of atoms written to the right in the Fischer projection formula appears below the plane of the ring, and any group written to the left appears above the plane.

The *aldopentoses* occur as *cyclic hemiacetals* in both five- and six-atom rings. The most important ones are D-ribose (in RNA), D-2-deoxyribose (in DNA), L-arabinose (in gum arabic), and D-xylose (in pentosans).

The most important *disaccharides* are maltose, cellobiose, lactose, and sucrose. The systematic name of *maltose* is 4-O-(α-D-glucopyranosyl)-D-glucopyranose. It exhibits mutarotation; it is a reducing sugar; and it can be oxidized by bromine water to an aldonic acid. *Cellobiose,* a sugar obtained from partial hydrolysis of cellulose, is 4-O-(β-D-glucopyranosyl)-D-glucopyranose. It exhibits mutarotation; it is a reducing sugar; and it can be oxidized to an aldonic acid. *Lactose,* milk sugar, is hydrolyzed by the action of emulsin, which indicates a β-glycosidic linkage. Its name is 4-O-(β-D-galactopyranosyl)-D-glucopyranose. It exhibits mutarotation and is oxidized by bromine water to an aldonic acid.

Sucrose, our common sugar, is obtained from sugar beets and sugar cane. It is a nonreducing sugar and does not exhibit mutarotation. Its systematic name is α-D-glucopyranosyl-β-D-fructofuranoside. There is no free-aldehyde form in equilibrium with the cyclic acetal forms, which involve the anomeric carbon atoms of both monosaccharide parts. Hydrolysis of sucrose produces an equimolar mixture of D-glucose and D-fructose referred to as *invert sugar* because the positive rotation of sucrose changes to negative rotation as hydrolysis is accomplished.

Starch is a natural C-4 to C-1 *α-polyglucoside*. It is the storage form of glucose in plants. Starch is about 20% a semisoluble form called *amylose* consisting of a continuous-chain polyglucoside and about 80% an insoluble, highly branched α-polyglucoside called *amylopectin*.

Cellulose is an insoluble C-4 to C-1 β-polyglucoside of continuous chain. The polymer consists of about 2000 to 6000 units of glucose. It is the chief structural material of plants. It is more difficult to hydrolyze than starch. Cellulose may be nitrated with a mixture of nitric and sulfuric acids to produce *cellulose nitrate,* which is used in making smokeless gunpowder. It may be acetylated with a mixture of acetic anhydride and acetic acid to form *cellulose acetate,* which is used as a base for home movie film and as a fiber for clothing. A reconstituted cellulose can be produced as a fiber (viscose rayon) or as a film (cellophane).

The carbohydrates are produced in plants by the biochemical reactions of *photosynthesis,* whereby water and carbon dioxide are combined to form sugars and their polymers.

Key terms

Acetal: a special type of ether, a *gem*-dialkoxyalkane, of the general formula $RCH(OR)_2$, ordinarily produced by the addition of 2 mol of alcohol to 1 mol of aldehyde. Simple sugars, which normally exist as intramolecular cyclic hemiacetals, react with 1 mol of alcohol to produce glycosides, which are acetals.

Aldaric acid: a polyhydroxy dicarboxylic acid produced by oxidizing the aldehyde group and the primary alcohol group of an aldose with nitric acid.

Alditol: a polyhydroxy alcohol produced by reduction of the aldehyde group of an aldose.

Aldohexose: a six-carbon sugar with an aldehyde group. The most important aldohexoses are D-(+)-glucose, D-(+)-mannose, and D-(+)-galactose.

Aldonic acid: a polyhydroxy monocarboxylic acid produced by oxidizing the aldehyde group of an aldose with bromine water.

Aldopentose: a five-carbon sugar with an aldehyde group. The most important aldopentoses are D-ribose, D-2-deoxyribose, L-arabinose, and D-xylose.

Aldose: a sugar which contains an aldehyde group.

Alduronic acid: a polyhydroxy aldehydo carboxylic acid produced by the microbiological oxidation of the primary alcohol group of an aldose.

Amylopectin: the fraction comprising about 80% of starch. It is an insoluble, highly branched polymer of 1000 to 6000 glucose units.

Amylose: the fraction comprising about 20% of starch. It is a semisoluble, continuous-chain polymer of about 1000 glucose units.

Anomers: cyclic sugars which differ only in the configuration of the carbon atoms of their hemiacetal groups. Also, cyclic diastereoisomers in carbohydrate chemistry which differ only in the configuration about the glycosidic carbon atom. They are designated α or β.

Benedict's solution: a solution of a complex ion of copper(II) with citrate. The Cu^{2+} ion is reduced to Cu^{+} by an aldehyde, and that is used as a test for reducing sugars.

Carbohydrates: polyhydroxy aldehydes or ketones, their derivatives, or compounds which can be hydrolyzed to such. They include sugars, starches, cellulose, gums, pectins, and their derivatives.

Cellulose: a long, linear β-glycosidic polymer of 2000 to 6000 units of D-glucose that is the chief structural material of plants. It is very insoluble in water.

Dextrin: a glucose polysaccharide of intermediate size produced in the first stage of hydrolysis of starch.

Diastereoisomers: non–mirror-image stereoisomers of a compound.

Disaccharides: dimers of monosaccharides, to which they can be hydrolyzed.

Enantiomers: mirror-image isomers.

Enol: a tautomer formed from a corresponding carbonyl form in which one carbon atom of a carbon-carbon double bond bears a hydroxy group, $-\overset{|}{C}=\overset{\overset{\text{OH}}{|}}{C}-$. Formation of enols from sugars is especially catalyzed by bases.

Enolization: a reaction in which a carbonyl compound partially converts to its enol tautomer to form an equilibrium mixture of the two or more tautomeric forms possible. Also, a reaction of sugars readily catalyzed by bases.

Epimer: a particular type of diastereoisomer in which the difference in configuration relates to only one of the asymmetric carbon atoms.

Fehling's solution: a solution of a complex ion of copper(II) with tartrate. The Cu^{2+} ion is reduced to Cu^{+} by an aldehyde, and that is used as a test for reducing sugars.

Fischer projection formula: a type of configurational formula often used to represent carbohydrates. Definite conventions must be followed in its use.

Furanose ring: a five-atom ring produced by intramolecular hemiacetal formation by the aldehyde group and the —OH group on C-4, or by the ketone group and the —OH group on C-5, of a simple sugar.

D-Glucose: the most important of all the simple sugars; its oxidation is the source of cellular energy in living things. Also known as **dextrose.**

Glycogen: the polysaccharide form in which animals store glucose; it contains about 7000 glucose units.

Glycoside: a carbohydrate acetal. Acetals of glucose are glucosides; those of mannose are mannosides; and so on.

Glycosidic linkage: the bonding of the anomeric carbon atom of a cyclic sugar through an oxygen atom to a carbon atom of another molecule to form an acetal. The two monosaccharide units of a disaccharide are bonded by a glycosidic linkage.

Gum: any one of various viscid, amorphous exudations from plants. Chemically some gums are polysaccharides. An example is gum arabic, which yields L-arabinose upon hydrolysis.

Haworth formula: a formula for representing the cyclic form of a sugar. The molecule is drawn as a planar, hexagonal slab with darkened edges toward the viewer. Definite conventions apply to its use.

Hemiacetal: a compound of the general formula R—CH(OH)—OR produced by the addition of one mole of alcohol to one mole of aldehyde. The simple sugars exist mostly as intramolecular cyclic hemiacetals.

Hexose: a six-carbon sugar.

Invert sugar: an equimolar mixture of D-glucose and D-fructose. It is the hydrolysis product of sucrose.

Ketohexose: a six-carbon sugar with a ketone group. The most important ketohexose is D-(−)-fructose.

Ketose: a sugar which contains a ketone group.

Monosaccharides: simple sugars and their derivatives; they are not further divisible by hydrolysis.

Mutarotation: the spontaneous gradual change in the specific rotation of reducing sugars when they are placed in aqueous solution. The change is caused by partial conversion of a reducing sugar to its anomer.

Osazone: the product of the reaction of a simple sugar with phenylhydrazine. In an osazone, C-1 and C-2 are substituted with $=NNHC_6H_5$. The osazones are useful in the identification of sugars.

Pectins: polysaccharides found in fruits; they are the factor which makes jelly gel.

Pentosans: polysaccharides found in such plant products as corn cobs and oat hulls; they yield D-xylose on hydrolysis.

Pentose: a five-carbon sugar.

Photosynthesis: the biochemical process in plants whereby water and carbon dioxide are combined in the presence of chlorophyll and sunlight to produce carbohydrates.

Polysaccharides: polymeric carbohydrates; they can be hydrolyzed to disaccharides and monosaccharides.

Pyranose ring: a six-atom ring produced by intramolecular hemiacetal formation by the aldehyde group and the —OH group on C-5 of an aldohexose or of the keto group and the —OH on C-6 of a ketohexose. It is the form in which simple sugars exist most of the time.

Reducing sugar: a sugar which reduces Fehling's, Benedict's, or Tollens' solution to produce an aldonic acid and other products. A reducing sugar must have a free carbonyl group in equilibrium with the cyclic form.

Simple sugars: the monosaccharides, which are not further divisible by hydrolysis.

Starch: a natural polymer of glucose that is the storage form of glucose in plants. It is an α-polyglucoside.

D-Sugar: a sugar in which the —OH group is to the right on the first asymmetric carbon atom from the bottom as the formula is normally written. Almost all naturally occurring sugars are D-sugars.

L-Sugar: a sugar in which the —OH group is to the left on the first asymmetric carbon atom from the bottom as the formula is normally written.

Tollens' solution: ammoniacal silver nitrate, $Ag(NH_3)_2OH$, prepared by dissolving silver nitrate in excess ammonium hydroxide. Its oxidation-reduction reaction with an aldehyde produces free silver, and that is used as a test for reducing sugars.

Additional problems

12-15 What structural relation is indicated by the term D-sugar? Why are both (+)-glucose and (−)-fructose classified as D-sugars?

12-16 Write equations, using configurational formulas, for each of the following reactions:

(a) D-glucose oxidized with bromine water
(b) D-mannose reacted with excess phenylhydrazine
(c) D-fructose reacted with HCN
(d) D-galactose oxidized with HNO_3
(e) reduction of D-glucose with hydrogen-Ni catalyst

12-17 (a) Would you expect D-mannitol to be optically active?
(b) Write Fischer projection formulas for all of the D-aldohexoses that would yield optically *inactive* alditols.

12-18 The alditol, D-xylitol, is a natural sugar alcohol obtained from birch wood, but it also occurs in many fruits and vegetables. Studies suggest that xylitol is much less likely to cause dental caries than sucrose and may even prevent their formation. Some studies suggest that xylitol is carcinogenic; others that it is not. Further research on both effects of xylitol is being undertaken. D-Xylitol is reported to be about as sweet as sucrose and have a pleasant, cool taste. (a) Give the structural formula for D-xylitol. (b) Write an equation for the reaction whereby D-xylitol could be prepared from D-xylose.

12-19 Two naturally occurring ketohexoses are psicose and tagatose. D-Psicose yields the same phenylosazone as D-allose or D-altrose; D-tagatose yields the same osazone as D-galactose or D-talose. What are the structures of D-psicose and D-tagatose?

12-20 Ketones cannot be oxidized by mild oxidizing agents, yet fructose is oxidized by Benedict's solution. Explain.

12-21 Write equations, using configurational formulas, for each of the following reactions:

(a) heating of D-gluconic acid with HCl to form the γ-lactone
(b) oxidation of maltose with bromine water
(c) reaction of lactose with excess phenylhydrazine
(d) hydrolysis of lactosazone
(e) reaction of D-galactose with methanol, HCl(g) catalyst
(f) methylation of methyl β-D-glucopyranoside with dimethyl sulfate and NaOH

12-22 How could you distinguish between each of the following: (a) glucose and starch, (b) glucose and fructose, (c) maltose and cellobiose?

12-23 Indicate a test, in addition to melting-point determination, for distinguishing between each of the following pairs: (a) D-glucose and D-galactose, (b) D-glucose and sucrose, (c) sucrose and lactose.

12-24 Look up the structure of the antibiotic streptomycin. Determine whether the compound is a mono-, di-, or trisaccharide.

12-25 An enzymatic process for isomerizing D-glucose to D-fructose to produce high-fructose corn syrup has been commercially developed. What advantage would the product have over regular corn syrup?

12-26 Write equations, using configurational formulas, for each of the following reactions: (a) nitration of cellulose, (b) acetylation of cellulose with acetic anhydride, (c) complete hydrolysis of starch.

13
AMINES, DYES, AND ALKALOIDS

Just as the alcohols and ethers can be considered to be the mono- and dialkyl derivatives of water, so the *amines* can be considered to be the *alkyl* and *aryl derivatives of ammonia*. The chemical and physical properties of amines somewhat resemble those of ammonia. The amines are widespread in nature, and they are important compounds.

13-1 Classification and structure of amines

The structure of a molecule of an amine is like that of a molecule of ammonia, but with one, or two, or all three of the hydrogen atoms substituted by alkyl or aryl groups. The orbitals on nitrogen are sp^3 *hybrid orbitals* with an unshared pair of electrons in one of them. The bond angles are a little less than the tetrahedral bond angle, being about 107° (Fig. 13-1).

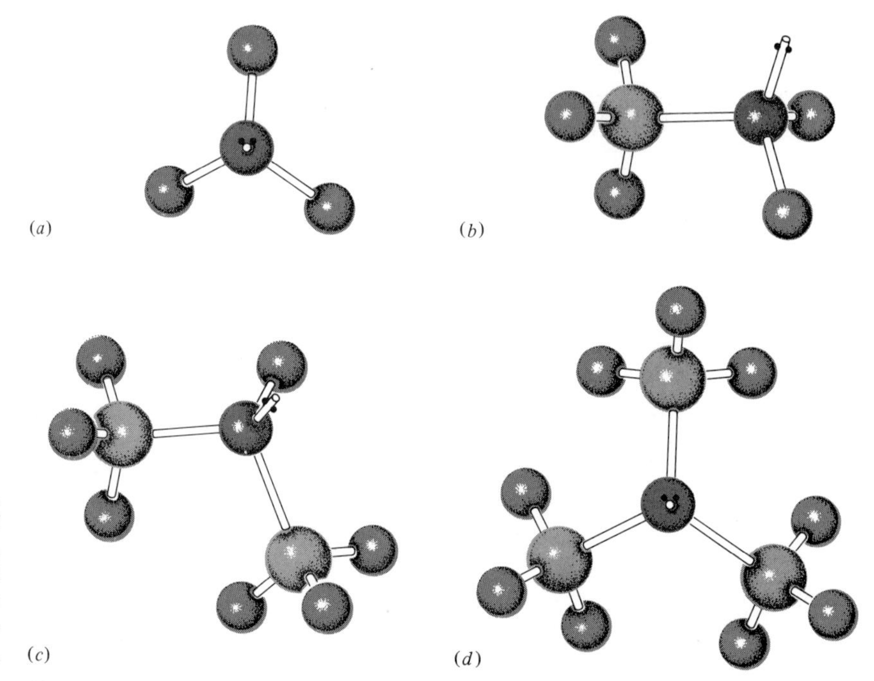

FIGURE 13-1
Ball-and-stick models of (*a*) ammonia, (*b*) methylamine, a primary amine, (*c*) dimethyl amine, a secondary amine, and (*d*) trimethylamine, a tertiary amine.

TABLE 13-1
Some selected amines

Structural formula	IUPAC name	Common name	bp, °C
CH_3NH_2	Aminomethane	Methylamine	−6.3
$(CH_3)_2NH$		Dimethylamine	7.4
$(CH_3)_3N$		Trimethylamine	2.8
$CH_3CH_2NH_2$	Aminoethane	Ethylamine	16.6
$CH_3CH_2CH_2NH_2$	1-Aminopropane	*n*-Propylamine	47.8
$CH_3CH_2CH_2CH_2NH_2$	1-Aminobutane	*n*-Butylamine	77.8
$HOCH_2CH_2NH_2$	2-Aminoethanol	Ethanolamine	170
$H_2N(CH_2)_4NH_2$	1,4-Diaminobutane	Putrescine (1,4-Butanediamine)	158
$H_2N(CH_2)_5NH_2$	1,5-Diaminopentane	Cadaverine (1,5-Pentanediamine)	178
$H_2N(CH_2)_6NH_2$	1,6-Diaminohexane	Hexamethylenediamine (1,6-Hexanediamine)	204
$C_6H_5CH_2NH_2$		Benzylamine	185
$C_6H_5NH_2$		Aniline	184
$C_6H_5N(CH_3)_2$		*N,N*-Dimethylaniline	194
N		Pyridine	115
N—H		Piperidine	106

Amines are classified as *primary, secondary,* or *tertiary* depending on whether they have one, two, or three organic groups substituted for the hydrogen atoms of ammonia:

$\ddot{N}$ (H, H, H; 107°)	$\ddot{N}$ (R, H, H)	$\ddot{N}$ (R, R, H)	$\ddot{N}$ (R, R, R)
Ammonia	Primary amine	Secondary amine	Tertiary amine

13-2 Physical properties

The *primary* and *secondary* amines have hydrogen attached to nitrogen, so they, like ammonia, can exhibit hydrogen bonding. As a consequence, their boiling points are somewhat higher than those of alkanes of similar molecular weight but still much lower than those of alcohols. Methylamine, dimethylamine, trimethylamine, ethylamine, and ethylmethylamine are gases. *n*-Propylamine is a liquid at room temperature. The limit of water solubility of the amines is reached with about six carbon atoms. The amines have odors somewhat like the odor of ammonia; but as the number of carbon atoms increases, they acquire a putrid odor. The formulas, names, and boiling points of some selected amines are given in Table 13-1.

13-3 Nomenclature of amines

The simpler *aliphatic amines* are commonly named by naming the alkyl groups attached to the nitrogen atom and adding the ending *-amine*. For

FIGURE 13-2
A space-filling model of aniline, $C_6H_5NH_2$.

example, CH_3NH_2 is methylamine (a primary amine), $(CH_3)_2CHNH_2$ is isopropylamine (also a primary amine), $(CH_3)_2NH$ is dimethylamine (a secondary amine), $(CH_3CH_2)_3N$ is triethylamine (a tertiary amine), and $(CH_3)_3CNH_2$ is *tert*-butylamine (a primary amine).

In the IUPAC system, the amino group (—NH_2) may be treated as a substituent and named as an amino group on the major structure. For example, $HOCH_2CH_2NH_2$ is 2-aminoethanol (its common name is ethanolamine), $H_2N(CH_2)_5NH_2$ is 1,5-diaminopentane (commonly, 1,5-pentanediamine or cadaverine), and $H_2NCH_2CH_2CH_2CO_2H$ is 4-aminobutanoic acid (γ-aminobutyric acid).

Aryl amines are named as derivatives of the parent member of the family, aniline. (See also Fig. 13-2.)

NH_2 NH_2 NH_2 NO_2 OCH_3

Aniline 3-Nitroaniline or *m*-nitroaniline 4-Methoxyaniline or *p*-methoxyaniline

The *acyl derivatives* of ammonia ($R—\overset{O}{\overset{\|}{C}}—NH_2$) are called *amides*. They were considered as derivatives of the carboxylic acids in Chap. 9. The *diacyl derivatives* of ammonia are called *imides*, and the *sulfonyl derivatives* of ammonia are *sulfonamides*. For example,

$CH_2—C$ O NH $CH_2—C$ O CH_3 SO_2NH_2

Succinimide *p*-Toluenesulfonamide

p-Aminobenzenesulfonamide is known as sulfanilamide. It and its derivatives constitute the sulfa drugs. Two other examples are sulfapyridine and sulfathiazole.

NH_2 NH_2 NH_2 SO_2NH_2 SO_2NH N SO_2NH N S

Sulfanilamide Sulfapyridine Sulfathiazole

Domagk, a German scientist, discovered that sulfanilamide and its derivatives were effective in curing many streptococcal and staphylococcal infections. In 1939 he was honored with the Nobel Prize in Medicine. Discovery and development of the sulfa drugs was of great significance in the development of modern chemotherapy. The drugs were especially important during the 1930s and during World War II before the development of penicillins. The action of the sulfa drugs will be explained in Chap. 16.

A *secondary* or *tertiary aryl amine,* which has substituents on the nitrogen atom, is named by specifying the substituents. A preceding *N* denotes that attachment is to nitrogen. Examples are

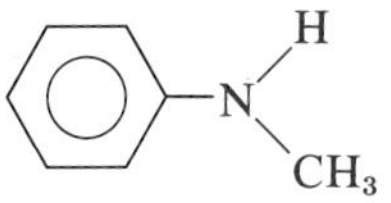

$C_6H_5-NH(CH_3)$ *N*-Methylaniline

$C_6H_5-N(CH_3)_2$ *N,N*-Dimethylaniline

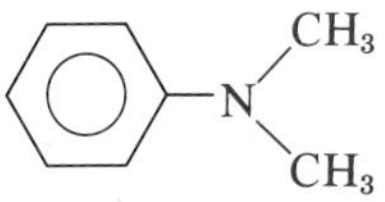

Alkyl substituents on the nitrogen atom of an amide also are indicated by using *N* before the names of the substituents.

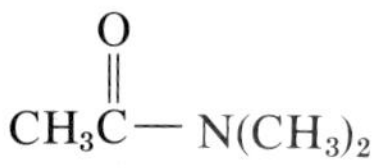

$CH_3C(=O)-N(CH_3)_2$ *N,N*-Dimethylacetamide

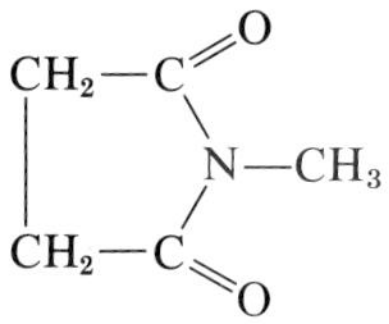

$CH_2-C(=O)-N(CH_3)-C(=O)-CH_2$ (ring) *N*-Methylsuccinimide

$(CH_3)_2N-C_6H_4-SO_2OH$ *N,N*-Dimethylaminobenzenesulfonamide

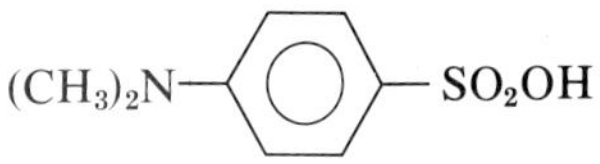

PROBLEM 13-1 Name each of the following compounds and classify the amino function as primary, secondary, or tertiary.

(a) 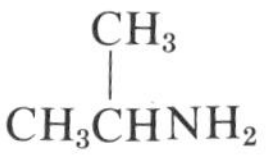 $CH_3CH(CH_3)NH_2$ (b) 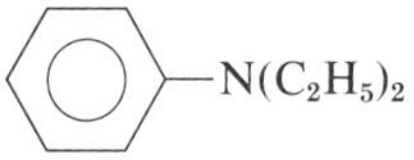 $C_6H_5-N(C_2H_5)_2$ (c) $CH_3N(H)-CH_2CH_2CH_3$

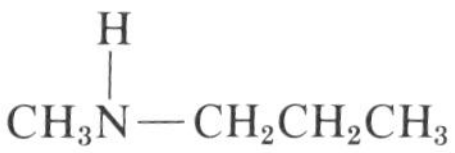

(d) 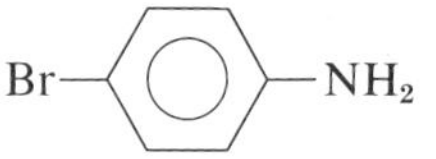 $Br-C_6H_4-NH_2$ (e) 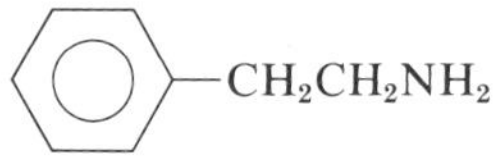 $C_6H_5-CH_2CH_2NH_2$

Solution to (a) The common name of $CH_3CH(CH_3)NH_2$ is isopropylamine; the IUPAC name is 2-aminopropane. The compound is a primary amine.

PROBLEM 13-2 Give the structural formula for each of the following compounds.

(a) 2-aminohexane
(b) diisopropylamine
(c) *p*-aminobenzoic acid
(d) *N*-*n*-butylphthalimide
(e) diethylamine
(f) aminocyclohexane

Solution to (a) The structural formula of 2-aminohexane is

$$\overset{1}{CH_3}-\overset{\overset{NH_2}{|}}{\underset{2}{C}H}-\overset{3}{CH_2}-\overset{4}{CH_2}-\overset{5}{CH_2}-\overset{6}{CH_3}$$

13-4 Preparation of amines

Several good methods are available for preparing amines, particularly primary amines. The method used depends upon the availability of the necessary starting materials and the purity of product desired.

Reduction Reduction of an organic nitrogen compound possessing nitrogen in a higher oxidation state, such as RNO_2, $R{-}N{=}O$, $RCH{=}NOH$, or $RC{\equiv}N$, is a good general method of preparing an amine. Examples are

$$\underset{\text{Nitrobenzene}}{C_6H_5NO_2} \xrightarrow{H_2/Ni \text{ or, } Sn \text{ or } Fe/HCl} \underset{\text{Aniline}}{C_6H_5NH_2} \qquad (13\text{-}1)$$

$$\underset{\text{Hexanedinitrile}}{N{\equiv}C(CH_2)_4C{\equiv}N} + 4H_2 \xrightarrow{Ni} \underset{\text{1,6-Diaminohexane}}{H_2N(CH_2)_6NH_2} \qquad (13\text{-}2)$$

Ammonium hydrogen sulfide and ammonia will reduce only one of the nitro groups in dinitrobenzene. The reaction is sometimes useful.

$$\underset{m\text{-Dinitrobenzene}}{C_6H_4(NO_2)_2} \xrightarrow{NH_4SH,\ NH_3} \underset{m\text{-Nitroaniline}}{C_6H_4(NH_2)(NO_2)} \qquad (13\text{-}3)$$

m-Nitroaniline is a dyestuff intermediate. It is very toxic. Acute exposure can cause methemoglobinemia and cyanosis; chronic exposure can cause liver damage.

Reductive amination Reductive amination of aldehydes or ketones is an excellent method of preparing amines. By using an excess of ammonia, good yields of primary amines may be obtained free of the contamination of secondary amines. The general equation is

$$\underset{\text{Aldehyde or ketone}}{\overset{R}{\underset{R}{>}}C{=}O} + :NH_3 \rightleftharpoons R-\overset{\overset{R}{|}}{\underset{\underset{H}{|}}{\underset{N-H}{\underset{|}{C}}}}-OH \xrightarrow{-H_2O} \underset{\text{An imine}}{R-\overset{\overset{R}{|}}{C}{=}NH} \xrightarrow{H_2,\ \Delta,\ \text{pressure, catalyst}} \underset{\text{A primary amine}}{R-\overset{\overset{R}{|}}{\underset{\underset{H}{|}}{C}}-NH_2} \qquad (13\text{-}4)$$

The reaction is completed without isolating the intermediates. Using an excess of ammonia serves to ensure that the product, primary amine, does not get to react with the aldehyde or ketone to produce a secondary amine. An example of the reaction is

$$\underset{\substack{\text{Propanone}\\\text{(Acetone)}}}{CH_3\overset{\overset{O}{\|}}{C}-CH_3} + \underset{\substack{\text{Ammonia}\\\text{(excess)}}}{NH_3} + H_2 \xrightarrow{CuCr_2O_4\cdot CuO,\ \Delta,\ \text{pressure}} \underset{\substack{\text{2-Aminopropane}\\\text{(Isopropylamine)}}}{CH_3\overset{\overset{NH_2}{|}}{C}H-CH_3} \qquad (13\text{-}5)$$

Reductive amination can be used by industry as well as by the research laboratory to produce many desired amines for product development or research. The required aldehydes and ketones are usually readily available.

Ammonolysis of alkyl halides

Ammonolysis of alkyl halides is of limited use as a means of preparing amines. The general reaction is

$$\underset{\substack{\text{Ammonia}\\\text{(nucleophile)}}}{H_3N:} + \underset{\text{Alkyl halide}}{R-X} \xrightarrow{(S_N2)} H_3N-R + X^- \xrightarrow{NH_3} \underset{\text{Amine}}{H_2NR} + NH_4^+X^- \qquad (13\text{-}6)$$

A specific example is the following:

$$\underset{\substack{\text{Bromoethane}\\\text{(Ethyl bromide)}}}{CH_3CH_2Br} + \underset{\text{(Excess)}}{NH_3} \longrightarrow \underset{\substack{\text{Aminoethane}\\\text{(Ethylamine)}}}{CH_3CH_2NH_2} + NH_4Br \qquad (13\text{-}7)$$

With an excess of ammonia, the ammonolysis reaction may be used to produce fairly good yields of a primary amine from a primary halide.

If an excess of ammonia is not used, the product primary amine tends to react further with the alkyl halide to form secondary amine, again to form tertiary amine, and finally again to form a *quaternary salt*. The resulting mixture of products makes the method an unsatisfactory one for preparing amines, but the quaternary salts are useful compounds.

$$H-\underset{H}{\overset{H}{N}}: + CH_3I \xrightarrow{(S_N2)} \left[H-\underset{H}{\overset{H}{N}}-CH_3\right]^+ I^- \xrightarrow{NH_3} \underset{\text{Methylamine}}{H\overset{H}{\ddot{N}}-CH_3} + NH_4^+I^- \qquad (13\text{-}8)$$

$$H-\underset{CH_3}{\overset{H}{N}}: + CH_3I \xrightarrow{(S_N2)} \left[H-\underset{CH_3}{\overset{H}{N}}-CH_3\right]^+ I^- \xrightarrow{NH_3} \underset{\text{Dimethylamine}}{H\underset{CH_3}{\ddot{N}}-CH_3} + NH_4^+I^- \qquad (13\text{-}9)$$

$$\underset{\substack{|\\ CH_3}}{\overset{\substack{H\\ |}}{CH_3N:}} + CH_3I \xrightarrow{(S_N2)} \left[\underset{\substack{|\\ CH_3}}{\overset{\substack{H\\ |}}{CH_3N}}{-}CH_3\right]^{+} I^- \xrightarrow{NH_3} \underset{\substack{|\\ CH_3\\ \text{Trimethylamine}}}{CH_3\ddot{N}}{-}CH_3 + NH_4^+I^- \quad (13\text{-}10)$$

$$\underset{\substack{|\\ CH_3}}{\overset{\substack{CH_3\\ |}}{CH_3N:}} + CH_3I \xrightarrow{(S_N2)} \underset{\substack{\text{Tetramethylammonium iodide}\\ \text{(a quaternary salt)}}}{\left[\underset{\substack{|\\ CH_3}}{\overset{\substack{CH_3\\ |}}{CH_3N}}{-}CH_3\right]^{+} I^-} \quad (13\text{-}11)$$

The quaternary salt (Fig. 13-3) can be reacted with silver hydroxide ($Ag_2O + H_2O$) to give a precipitate of silver halide and a solution of the *quaternary base*. Since such a base exists only in the ionic form, it is very strong and is comparable to NaOH and KOH.

$$\underset{\substack{\text{Tetramethyl-}\\ \text{ammonium iodide}}}{(CH_3)_4N^+I^-} + AgOH \longrightarrow AgI(s) + \underset{\substack{\text{Tetramethyl-}\\ \text{ammonium hydroxide}}}{(CH_3)_4N^+OH^-} \quad (13\text{-}12)$$

Higher homologs of such quaternary bases have a water-soluble anion and an oil-soluble cation (just the reverse of soap). They have a detergent action, and they are referred to as *invert soaps* because of the reversal of cation-anion function. Some of them have been found to have germicidal activity.

The quaternary ammonium ion *choline* and its acetate ester *acetylcholine* are of special importance in human physiology.

$$\underset{\text{Choline}}{CH_3{-}\underset{\substack{|\\ CH_3}}{\overset{\substack{CH_3\\ +|}}{N}}{-}CH_2CH_2OH} \qquad \underset{\text{Acetylcholine}}{CH_3{-}\underset{\substack{|\\ CH_3}}{\overset{\substack{CH_3\\ +|}}{N}}{-}CH_2CH_2O{-}\overset{\substack{O\\ \|}}{C}{-}CH_3}$$

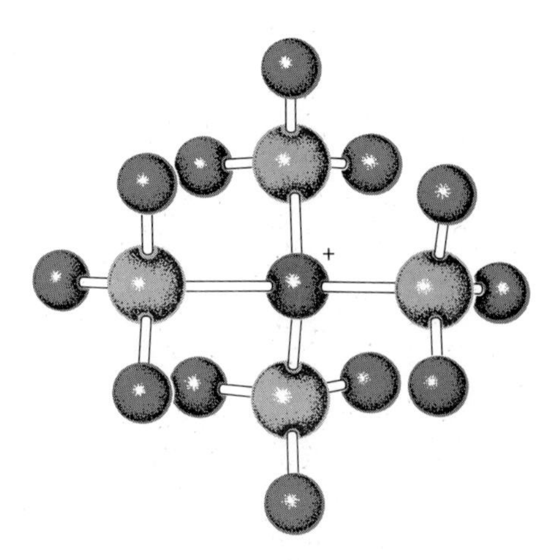

FIGURE 13-3
The tetramethylammonium ion.

Choline is a molecular part of a group of phospholipids known as *lecithins,* which are important in the construction of cell membranes. Acetylcholine is a transmitter substance in certain motor neurons responsible for initiating contraction of voluntary muscles. It is stored in synaptic vesicles and released in response to electrical activity of the neuron; it diffuses across the synapse and interacts with receptor sites on a neighboring neuron to cause transmission of a nerve impulse. After that interaction, it is deactivated through hydrolysis catalyzed by the enzyme cholinesterase.

$$CH_3-\overset{\overset{CH_3}{|}}{\underset{\underset{CH_3}{|}}{N}}-CH_2CH_2-O-\overset{\overset{O}{\|}}{C}-CH_3 + H_2O \xrightarrow{\text{cholinesterase}}$$

Acetylcholine

$$CH_3-\overset{\overset{CH_3}{+|}}{\underset{\underset{CH_3}{|}}{N}}-CH_2CH_2OH + CH_3\overset{\overset{O}{\|}}{C}-O^- + H^+ \qquad (13\text{-}13)$$

Choline Acetate

The inhibition of this enzymatic reaction will be discussed in Chap. 16.

Hofmann degradation

Primary amines may also be prepared by the *Hofmann degradation reaction,* whereby an amide is degraded to a primary amine with one less carbon atom.

$$R-C\begin{matrix}\nearrow O \\ \searrow NH_2\end{matrix} \xrightarrow{\text{NaOH, Br}_2,\ \Delta} R-NH_2 + Na_2CO_3 + NaBr \qquad (13\text{-}14)$$

The reaction is believed to proceed through the following steps.

$$2NaOH + Br_2 \longrightarrow Na^{+-}OBr + Na^+Br^- + H_2O \qquad (13\text{-}15)$$

$$R-\overset{\overset{O}{\|}}{C}-\overset{\overset{H}{|}}{\ddot{N}}-H + {}^-O-Br \longrightarrow R-\overset{\overset{O}{\|}}{C}-N\begin{matrix}\nearrow Br \\ \searrow H\end{matrix} + OH^- \longrightarrow$$

$$\left[R-\overset{\overset{O}{\|}}{C}-\ddot{N}:\right] + HOH + Br^- \qquad (13\text{-}16)$$

$$\left[R-\overset{\overset{O}{\|}}{C}-\ddot{N}:\right] \longrightarrow R-N=C=O \xrightarrow{\text{NaOH}} R-NH_2 + Na_2CO_3 \qquad (13\text{-}17)$$

An isocyanate Primary amine

An example of the Hofmann degradation reaction is

$$\text{2-Naphthalene carboxamide } (C_{10}H_7CONH_2) \xrightarrow{NaOH,\ Br_2} \text{2-Aminonaphthalene } (C_{10}H_7NH_2) + Na_2CO_3 + NaBr \qquad (13\text{-}18)$$

PROBLEM 13-3 Write equations for the preparation of each of the following compounds from the starting materials indicated.

(a) 3-aminopentane from 3-pentanone
(b) *m*-nitroaniline beginning with benzene
(c) choline beginning with 2-aminoethanol and iodomethane (methyl iodide)
(d) benzylamine beginning with toluene
(e) 1,4-diaminobutane beginning with 1,2-ethanediol (ethylene glycol)
(f) 1-aminopropane (*n*-propylamine) beginning with butanoic acid (*n*-butyric acid)
(g) 2-aminopropanoic acid beginning with 2-chloropropanoic acid

Solution to (a)

$$\underset{\text{3-Pentanone}}{CH_3CH_2\overset{O}{\overset{\|}{C}}CH_2CH_3} + NH_3 + H_2 \xrightarrow{\Delta,\ \text{pressure, catalyst}} \underset{\text{3-Aminopentane}}{CH_3CH_2\overset{NH_2}{\overset{|}{C}}HCH_2CH_3}$$

13-5 Reactions of amines

Like ammonia, the amines react as bases with water and acids, and they react with acid anhydrides and acid chlorides to form amides. Those are their important important reactions.

Reaction with water

An amine reacts as a base with water just as ammonia does.

$$\underset{\text{Ammonia}}{H{-}\overset{H}{\overset{|}{\underset{H}{\underset{|}{N}}}}:} + HOH \rightleftharpoons \underset{\text{Ammonium hydroxide}}{NH_4^+ + OH^-} \qquad (13\text{-}19)$$

$$\underset{\text{Methylamine}}{CH_3\overset{H}{\overset{|}{\underset{H}{\underset{|}{N}}}}:} + HOH \rightleftharpoons \underset{\text{Methylammonium hydroxide}}{CH_3\overset{H}{\overset{|}{\underset{H}{\underset{|}{N^+}}}}{-}H + OH^-} \qquad (13\text{-}20)$$

$$C_6H_5CH_2\overset{\displaystyle H}{\underset{\displaystyle H}{\overset{|}{\underset{|}{N}}}}: + HOH \rightleftharpoons C_6H_5CH_2NH_3^+ + OH^- \tag{13-21}$$

Benzylamine — Benzylammonium hydroxide

$$C_6H_5\overset{\displaystyle H}{\underset{\displaystyle H}{\overset{|}{\underset{|}{N}}}}: + HOH \rightleftharpoons C_6H_5NH_3^+ + OH^- \tag{13-22}$$

Aniline — Anilinium hydroxide

Such aqueous solutions of amines, just like aqueous solutions of ammonia, are weakly basic. They will turn moist litmus blue. Because of a positive inductive effect of alkyl groups, the positive charge on the alkyl-substituted ammonium ion is somewhat delocalized by alkyl groups. The result is an *increased basicity,* i.e., an increased tendency of the amine to accept a proton from water to form alkyl ammonium and hydroxide ions. Thus the alkyl amines are, in general, slightly stronger bases than ammonia.

$$R \rightrightarrows \overset{\displaystyle H}{\underset{\displaystyle H}{\overset{|}{\underset{|}{N^+}}}}—H$$

Positive inductive effect

The general equilibrium reaction for an amine and water is

$$R—\overset{\displaystyle H}{\underset{\displaystyle H}{\overset{|}{\underset{|}{N}}}}: + H—O—H \rightleftharpoons RNH_3^+ + OH^- \tag{13-23}$$

The *basicity constant,* K_b, is given by the expression

$$K_b = \frac{[RNH_3^+][OH^-]}{[RNH_2]} \tag{13-24}$$

The larger the value of K_b, the greater is the tendency of the amine to accept a proton from water and the greater is the concentration of RNH_3^+ and OH^- in the solution. Larger values of K_b, therefore, are associated with the amines that are stronger bases. The value of K_b is sometimes expressed as pK_b, which is defined as equal to $-\log K_b$. Values of K_b and pK_b for some selected amines are given in Table 13-2. Because pK_b is a negative logarithm, the smaller the value of pK_b the stronger the base and the larger the value of pK_b the weaker the base. Ammonia has been included for comparison.

Note that aniline is about a million times weaker as a base than ammonia. The unshared pair of electrons on the nitrogen atom of aniline tends

TABLE 13-2 Basicities of some selected amines

Formula	Common name	K_b	$pK_b = -\log K_b$
NH_3	Ammonia	1.8×10^{-5}	4.74
CH_3NH_2	Methylamine	4.4×10^{-4}	3.36
$CH_3CH_2NH_2$	Ethylamine	4.7×10^{-4}	3.33
$CH_3CH_2CH_2NH_2$	Propylamine	5.1×10^{-4}	3.29
$(CH_3)_2NH$	Dimethylamine	5.2×10^{-4}	3.28
$(CH_3CH_2)_2NH$	Diethylamine	9.6×10^{-4}	3.02
$(CH_3)_3N$	Trimethylamine	5.0×10^{-5}	4.30
$(CH_3CH_2)_3N$	Triethylamine	5.7×10^{-4}	3.24
$C_6H_5NH_2$	Aniline	4.2×10^{-10}	9.38
$C_6H_5NHCH_3$	*N*-Methylaniline	5.0×10^{-10}	9.30
$C_6H_5N(CH_3)_2$	*N,N*-Dimethylaniline	1.1×10^{-9}	8.94
p-$CH_3C_6H_4NH_2$	*p*-Toluidine	1.2×10^{-9}	8.92
p-$NO_2C_6H_4NH_2$	*p*-Nitroaniline	1.0×10^{-13}	13.00
2,4-$(NO_2)_2C_6H_3NH_2$	2,4-Dinitroaniline	3.0×10^{-19}	18.52

to participate in the resonance hybrid of the aromatic structure through pi overlap with the pi orbital of the ring. That, of course, makes the unshared pair of electrons much less available for reaction as a base.

:NH_2 ⟷ $+NH_2$ ⟷ $+NH_2$ ⟷ $+NH_2$ (13-25)

Aniline

Electron-releasing groups on the ring make aniline a stronger base (note pK_b for *p*-toluidine) by slightly increasing the electron density available on nitrogen. Electron-withdrawing groups make aniline a weaker base. 2,4-Dinitroaniline is such a weak base that it will not even dissolve in aqueous acids.

Reaction with acids

Amines react with acids to form salts. Even the amines which are very weak bases may still react with strong *nonaqueous* acids such as concentrated sulfuric acid or hydrogen chloride dissolved in dry ether. The general reaction is

$$R{-}\ddot{N}H_2 + H^+ \longrightarrow R{-}\overset{+}{N}H_3 \qquad (13\text{-}26)$$

An amine — Acid — Alkylammonium ion

Examples of the reaction are the following:

$$\underset{\text{Methylamine}}{CH_3NH_2} + HCl \longrightarrow \underset{\text{Methylammonium chloride}}{CH_3NH_3^+Cl^-} \qquad (13\text{-}27)$$

$$C_6H_5NH_2 + H_2SO_4 \longrightarrow C_6H_5NH_3^+HSO_4^- \qquad (13\text{-}28)$$

Aniline → Anilinium hydrogen sulfate

$$C_5H_5N + HCl \longrightarrow C_5H_5NH^+ \; Cl^- \qquad (13\text{-}29)$$

Pyridine → Pyridinium chloride

The reaction is of great utility in the chemical separation of amines from neutral and acidic compounds which are not soluble in acid solution.

Reaction with acid anhydrides or halides

Acylation of amines with acid anhydrides or acid halides results in the formation of amides and salts. The general reaction is

$$R{-}\ddot{N}H_2 + CH_3{-}\overset{O}{\overset{\|}{C}}{-}Cl \rightleftharpoons \left[CH_3{-}\overset{O^-}{\overset{|}{C}}{-}Cl \atop \overset{+}{H}NHR \right] \longrightarrow CH_3{-}\overset{O}{\overset{\|}{C}}{-}\ddot{N}H{-}R + HCl \qquad (13\text{-}30)$$

Amine, Acid halide → Amide

The reaction of amines with acid anhydrides or acid halides is like the reaction of ammonia with acid anhydrides or acid halides, Sec. 9-6. You may wish to review that discussion.

Tertiary amines do *not* react, since there is no hydrogen atom which can leave its attachment to nitrogen to complete the reaction following the original reversible addition of the amine to the carbonyl group.

$$R_3N: + CH_3\overset{O}{\overset{\|}{C}}{-}Cl \rightleftharpoons \left[CH_3{-}\overset{O^-}{\overset{|}{C}}{-}Cl \atop R\overset{+}{N}R{-}R \right] \qquad (13\text{-}31)$$

Tertiary amine, Acid halide

Primary amines react to form N-*substituted amides,* and secondary amines react to form N,N-*disubstituted amides*.

$$2CH_3NH_2 + CH_3\overset{O}{\overset{\|}{C}}{-}O{-}\overset{O}{\overset{\|}{C}}{-}CH_3 \longrightarrow$$

Methylamine, Acetic anhydride

$$CH_3\overset{O}{\overset{\|}{C}}{-}\overset{H}{\overset{|}{N}}CH_3 + CH_3NH_3^+ \, {}^-O{-}\overset{O}{\overset{\|}{C}}{-}CH_3 \qquad (13\text{-}32)$$

N-Methyl acetamide, Methylammonium acetate

$$2C_6H_5N(H)-CH_3 + Cl-C(=O)-CH_3 \longrightarrow$$

N-Methylaniline   Acetyl chloride

$$C_6H_5N(CH_3)-C(=O)CH_3 \quad + \quad C_6H_5-NH_2^+-CH_3\,Cl^- \qquad (13\text{-}33)$$

N-Methyl-*N*-phenylacetamide   *N*-Methylanilinium chloride

N,N-Dimethylformamide, $(CH_3)_2N-C(=O)-H$, is a liquid, and it is a very useful polar solvent for some reactions. *N,N*-Diethyl-3-methylbenzamide, $m\text{-}CH_3C_6H_4C(=O)-N(CH_2CH_3)_2$, is an excellent insect repellent.

Reaction with sulfonyl chlorides

Primary and secondary amines react with sulfonyl chlorides to yield N-*substituted sulfonamides*. Again, tertiary amines do *not* react. Sodium hydroxide catalyzes the reaction; furthermore, the sulfonamides of primary amines are soluble in aqueous sodium hydroxide, since the electron withdrawal by the sulfonic group delocalizes the negative charge of the resulting anion. In other words, the sulfonamides of primary amines are slightly acidic.

$$C_6H_5-\ddot{N}(H)-H + C_6H_5SO_2-Cl \longrightarrow C_6H_5SO_2-N(H)-C_6H_5 \quad + HCl \qquad (13\text{-}34)$$

Aniline   Benzenesulfonyl chloride   *N*-Phenylbenzenesulfonamide

$$C_6H_5SO_2-N(H)-C_6H_5 + Na^+OH^- \longrightarrow C_6H_5SO_2-\ddot{N}^-Na^+-C_6H_5 \;+ H_2O \qquad (13\text{-}35)$$

Sodium salt of *N*-phenylbenzenesulfonamide (soluble)

Treatment of the sodium salt with aqueous acid restores the sulfonamide.

$$C_6H_5SO_2-\ddot{N}^-Na^+-C_6H_5 + HCl(aq) \longrightarrow C_6H_5SO_2-\ddot{N}(H)-C_6H_5 + NaCl \qquad (13\text{-}36)$$

N-Phenylbenzenesulfonamide (insoluble)

The combination of benzenesulfonyl chloride and sodium hydroxide is referred to as the *Hinsberg reagent*. Production of a base-soluble sulfonamide is a positive test for a primary amine. A secondary amine also will react; but because its product sulfonamide has no hydrogen attached to nitrogen, the sulfonamide is not acidic and so is not soluble in basic solution.

$$C_6H_5\underset{\cdot\cdot}{\overset{CH_3}{\overset{|}{N}}}-H + C_6H_5SO_2Cl \longrightarrow$$

N-Methylaniline — Benzenesulfonyl chloride

$$C_6H_5SO_2-\underset{\cdot\cdot}{\overset{CH_3}{\overset{|}{N}}}-C_6H_5 + HCl \xrightarrow{NaOH} NaCl + H_2O \quad (13\text{-}37)$$

N-Methyl-N-phenylbenzenesulfonamide (insoluble)

As mentioned previously, tertiary amines do not react. The reaction can thus be used not only to classify an unknown amine but also to separate a mixture of a primary, a secondary, and a tertiary amine.

Aromatic Electrophilic substitution

The amino group of aromatic amines is a very strong activating, ortho, para-directing group for electrophilic ring substitution. The amino group is electron-releasing by a very strong positive resonance effect because of its unshared pair of electrons (see Sec. 5-4). For example, if aniline is simply shaken up with bromine water, *both* of the ortho positions and also the para position are immediately brominated to give the solid 2,4,6-tribromoaniline.

$$C_6H_5NH_2 + 3Br_2 \xrightarrow{H_2O} 2,4,6\text{-}Br_3C_6H_2NH_2 + 3HBr \quad (13\text{-}38)$$

Aniline — 2,4,6-Tribromoaniline

Acetylation of the amino group reduces the group's activity to permit monobromination. Subsequent hydrolysis then releases the monobromoaniline.

$$C_6H_5NH_2 \xrightarrow{(CH_3\overset{O}{\overset{\|}{C}})_2O} C_6H_5NH-\overset{O}{\overset{\|}{C}}-CH_3 \xrightarrow{Br_2} p\text{-}BrC_6H_4NH-\overset{O}{\overset{\|}{C}}-CH_3 \quad (13\text{-}39)$$

Aniline — Acetanilide — p-Bromoacetanilide

The ortho isomer is also produced by this reaction.

If aniline is mononitrated with cold nitric acid, about one half of the aniline is oxidized to a complex mixture of products. Electron release from the amino group makes the molecule more easily oxidized. Of the half that is nitrated, almost 40% is *m*-nitroaniline, instead of mostly all *o*- or *p*-nitroaniline as might have been expected. The reason for the large amount of meta isomer is somewhat complex.

Since nitric acid is a proton acid, it reacts with the amino group to form the *anilinium ion*—a substituted ammonium ion with a positive charge. Because of its positive charge, the new group is electron-withdrawing; therefore, it is deactivating and meta-directing. Indeed, the only reason any *o*- and *p*-nitroaniline is produced at all is that the anilinium ion is in equilibrium with a small amount of aniline. The reaction with aniline is much faster; so as the small amount of aniline is consumed, the equilibrium is shifted to produce more aniline. The equilibrium reaction of aniline and anilinium nitrate is

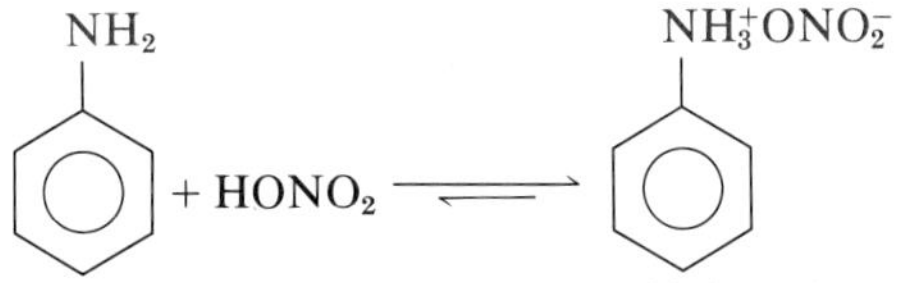

$$C_6H_5NH_2 + HONO_2 \rightleftharpoons C_6H_5NH_3^+ONO_2^- \qquad (13\text{-}40)$$

Aniline → Anilinium nitrate

The two competing reactions are as follows:

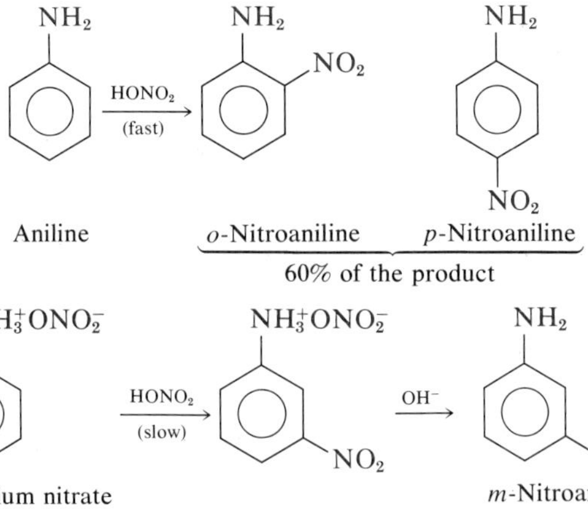

$$C_6H_5NH_2 \xrightarrow[\text{(fast)}]{HONO_2} o\text{-}NO_2C_6H_4NH_2 + p\text{-}NO_2C_6H_4NH_2 \qquad (13\text{-}41)$$

Aniline; *o*-Nitroaniline, *p*-Nitroaniline: 60% of the product

$$C_6H_5NH_3^+ONO_2^- \xrightarrow[\text{(slow)}]{HONO_2} m\text{-}NO_2C_6H_4NH_3^+ONO_2^- \xrightarrow{OH^-} m\text{-}NO_2C_6H_4NH_2 \qquad (13\text{-}42)$$

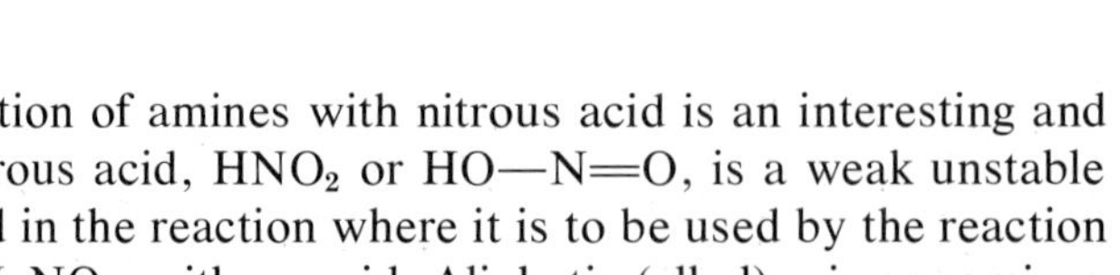

Anilinium nitrate; *m*-Nitroaniline: 40% of the product

Reaction with nitrous acid

The reaction of amines with nitrous acid is an interesting and important one. Nitrous acid, HNO_2 or HO—N=O, is a weak unstable acid. It is produced in the reaction where it is to be used by the reaction of sodium nitrite, $NaNO_2$, with an acid. Aliphatic (alkyl) primary amines react with nitrous acid to produce chiefly an alcohol and nitrogen gas.

$$\underset{\text{Primary amine}}{RNH_2} + \underset{\text{Sodium nitrite}}{NaNO_2} + \underset{\text{Acid}}{H_3O^+} \longrightarrow \underset{\text{Alcohol}}{ROH} + N_2(g) + 2H_2O + Na^+ \qquad (13\text{-}43)$$

An example of the reaction is

$$\underset{\substack{\text{1-Aminobutane}\\(n\text{-Butylamine})}}{CH_3CH_2CH_2CH_2NH_2} + NaNO_2 + H_3O^+ \longrightarrow \underset{\substack{\text{1-Butanol}\\(n\text{-Butyl alcohol})}}{CH_3CH_2CH_2CH_2OH} + N_2(g) + 2H_2O + Na^+ \qquad (13\text{-}44)$$

Aromatic primary amines react to produce an *intermediate diazonium salt* if the temperature is kept below about 10°C. If the aqueous diazonium salt warms up, it produces a phenol and nitrogen.

$$\underset{\substack{\text{Aromatic}\\\text{primary amine}}}{ArNH_2} + NaNO_2 + 2H_3O^+ \longrightarrow \underset{\substack{\text{Diazonium}\\\text{ion}}}{ArN_2^+} + Na^+ + 4H_2O \xrightarrow{\Delta} \underset{\text{Phenol}}{ArOH} + N_2(g) + Na^+ + H_3O^+ + 2H_2O \qquad (13\text{-}45)$$

An example is the following reaction:

$$\underset{\text{Aniline}}{C_6H_5NH_2} + NaNO_2 + 2H_3O^+ \longrightarrow \underset{\substack{\text{Benzenediazonium}\\\text{ion}}}{C_6H_5N_2^+} + Na^+ + 4H_2O \xrightarrow{\Delta} \underset{\text{Phenol}}{C_6H_5OH} + N_2(g) + Na^+ + H_3O^+ + 2H_2O \qquad (13\text{-}46)$$

The reaction of primary amines with nitrous acid to produce nitrogen can be used to distinguish primary amines from secondary amines. It is also the basis for the Van Slyke determination of primary amines. The nitrogen gas evolved is collected and measured to give a quantitative indication of primary amino groups.

Secondary amines produce *N*-nitrosoamines.

$$\underset{\text{Secondary amine}}{R_2NH} + NaNO_2 + H_3O^+ \longrightarrow \underset{N\text{-Nitrosoamine}}{R_2N{-}N{=}O} + Na^+ + 2H_2O \qquad (13\text{-}47)$$

An example is

$$\underset{\text{Diethylamine}}{(C_2H_5)_2NH} + NaNO_2 + H_3O^+ \longrightarrow \underset{N\text{-Nitrosodiethylamine}}{(C_2H_5)_2N{-}N{=}O} + Na^+ + 2H_2O \qquad (13\text{-}48)$$

Aliphatic tertiary amines do not react readily, but *aromatic* tertiary amines react to produce *nitrosation* of the benzene ring in the *para* position. Nitrosation is introduction of the nitroso group, NO, onto the ring.

$$C_6H_5N(CH_3)_2 + NaNO_2 + H_3O^+ \longrightarrow p\text{-}ON{-}C_6H_4{-}N(CH_3)_2 + Na^+ + 2H_2O \qquad (13\text{-}49)$$

N,N-Dimethylaniline (an aromatic tertiary amine) — *p*-Nitroso-*N,N*-dimethylaniline

Like primary amines, primary amides will react with nitrous acid to give off nitrogen. The product is a carboxylic acid.

$$\underset{\text{Amide}}{R{-}\overset{O}{\overset{\|}{C}}{-}NH_2} + HONO \longrightarrow \underset{\text{Acid}}{R{-}COOH} + N_2(g) \qquad (13\text{-}50)$$

For those who would like to rationalize the reactions of the primary and secondary amines, the likely intermediate steps of the mechanism are given in the following series of equations. Note that RNH_2 at the beginning of Eq. (13-53) represents either a primary or a secondary amine; in a secondary amine, one of the H's is replaced by R′. Note also that an *N*-nitrosoamine is the final product formed from a secondary amine; the remaining steps occur only when the original compound was a primary amine.

$$\underset{\text{Sodium nitrite}}{2Na^+ON{=}O^-} + 2H_3O^+ \longrightarrow 2Na^+ + 2H_2O + \underset{\text{Nitrous acid}}{2HO{-}N{=}O} \qquad (13\text{-}51)$$

$$2HO{-}N{=}O \rightleftharpoons H_2O + \underset{\text{Dinitrogen trioxide}}{O{=}N{-}O{-}N{=}O} \qquad (13\text{-}52)$$

$$\underset{\text{Primary or secondary amine}}{R\ddot{N}H_2} + O{=}N{-}O{-}N{=}O \longrightarrow {}^-O{-}N{=}O + R{-}\overset{H}{\underset{H}{N^+}}{-}N{=}O \qquad (13\text{-}53)$$

$$R{-}\overset{H}{\underset{H}{N^+}}{-}N{=}O + H_2O \longrightarrow \underset{\text{An }N\text{-nitrosoamine (final product from secondary amine)}}{R{-}\underset{H}{N}{-}N{=}O} + H_3O^+ \longrightarrow$$

$$R{-}\underset{H}{N}{-}N{=}\overset{+}{O}H + H_2O \longrightarrow \underset{\text{A diazoic acid}}{R{-}N{=}N{-}OH} + H_3O^+ \longrightarrow$$

$$R{-}\ddot{N}{=}\ddot{N}{-}\underset{H}{\overset{+}{O}}H + H_2O \xrightarrow{-2H_2O} \underbrace{R{-}\ddot{N}{=}N{:}^+ \longleftrightarrow R{-}\overset{+}{N}{\equiv}N{:}}_{\text{A diazonium ion (relatively stable if aromatic)}} \longrightarrow$$

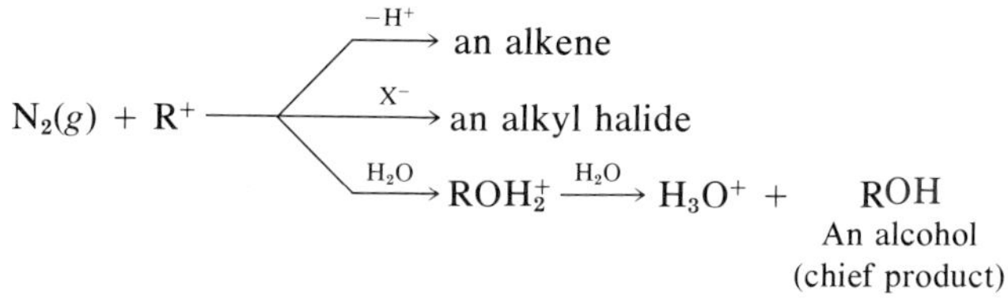

$$N_2(g) + R^+ \longrightarrow \begin{cases} \xrightarrow{-H^+} \text{an alkene} \\ \xrightarrow{X^-} \text{an alkyl halide} \\ \xrightarrow{H_2O} ROH_2^+ \xrightarrow{H_2O} H_3O^+ + \underset{\substack{\text{An alcohol} \\ \text{(chief product)}}}{ROH} \end{cases} \tag{13-54}$$

The intermediate *aromatic* diazonium ions are further stabilized because the positive charge can be delocalized by resonance structures involving the aromatic ring.

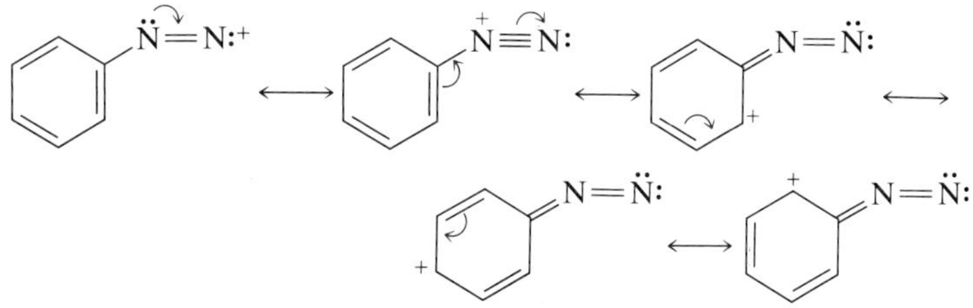

$$\text{(13-55)}$$

Resonance forms of the benzenediazonium ion

The aromatic diazonium salts have two important reactions: substitution for nitrogen and coupling.

Substitution reactions of diazonium salts. The substitution reactions of diazonium salts may be regarded as a type of nucleophilic substitution for nitrogen (N_2) on an aromatic ring. The positive character of the diazonium ion, with its resonance structures involving the benzene ring, makes the ion an electrophile which can react with a nucleophile. Because free nitrogen is a very weak nucleophile, it is a very good leaving group; that is, it is readily displaced by stronger nucleophiles. The reaction is the best general laboratory method for introducing F, I, CN, or OH on an aromatic ring. The mechanism for substitution by iodide ion is given by

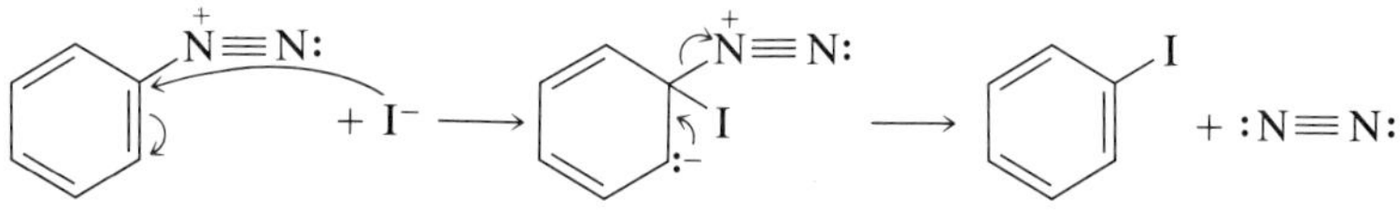

$$\text{Benzenediazonium ion} + I^- \longrightarrow \text{intermediate} \longrightarrow \text{Iodobenzene} + :N{\equiv}N: \text{ (Nitrogen)} \tag{13-56}$$

Table 13-3 gives the general reactions for the substitution of diazonium salts. Ar represents an aryl group.

TABLE 13-3 Substitution reactions of diazonium salts

Aromatic diazonium ion	Reacted with	Yields
ArN_2^+	KI	$ArI + N_2(g)$
	CuCl/HCl	$ArCl + N_2(g)$
	CuBr/HBr	$ArBr + N_2(g)$
	$CuCN/H_2SO_4$	$ArCN + N_2(g)$
	HBF_4	$ArF + N_2(g)$
	HOH	$ArOH + N_2(g)$
	H_3PO_2	$ArH + N_2(g)$

Some specific examples are

$$p\text{-Toluidine } (CH_3C_6H_4NH_2) \xrightarrow{NaNO_2,\ H_3O^+} p\text{-Toluenediazonium salt } (CH_3C_6H_4N_2^+) \xrightarrow{HBF_4} CH_3C_6H_4N_2^+BF_4^- \xrightarrow{\Delta} p\text{-Fluorotoluene } (CH_3C_6H_4F) + N_2(g) + BF_3 \tag{13-57}$$

CH_3 / NH_2 — *p*-Toluidine; CH_3 / N_2^+ — *p*-Toluene-diazonium salt; CH_3 / $N_2^+BF_4^-$; CH_3 / F — *p*-Fluoro-toluene

$$\text{2-Bromo-4-methylaniline} \xrightarrow{NaNO_2,\ H_3O^+} CH_3C_6H_3(Br)N_2^+ \xrightarrow{H_3PO_2,\ \Delta,\ H_2O} m\text{-Bromotoluene } (CH_3C_6H_4Br) + N_2(g) + H_3PO_3 \tag{13-58}$$

CH_3 / Br / NH_2 — 2-Bromo-4-methyl-aniline; CH_3 / Br / N_2^+; CH_3 / Br — *m*-Bromo-toluene

Coupling reactions of diazonium salts. Although the diazonium ion is a rather weak electrophile, it can substitute on a benzene ring which is activated by bearing a stronger activating ortho,para-directing group such as the amino group or the hydroxyl group. The special electrophilic substitution reaction is termed *coupling*. Because of steric (spatial) effects, the orientation is chiefly para with very little ortho isomer.

$$C_6H_5{-}N_2^+ + C_6H_5{-}N(CH_3)_2 \longrightarrow$$

Benzenedi-azonium ion; *N*,*N*-Dimethyl-aniline

$$C_6H_5{-}N{=}N{-}C_6H_4{-}N(CH_3)_2 \tag{13-59}$$

p-Dimethylaminoazobenzene or
p-phenylazo-*N*,*N*-dimethylaniline
(butter yellow)

The *azo compounds* which result from the coupling reaction are colored; they constitute the important *azo dyes*. The example shown, butter yellow, is an oil-soluble yellow dye which was used to color butter and margarine until about 1937. At that time, it was discovered that it could cause cancer. The acid-base indicator methyl orange is another example. It is produced by diazotizing *p*-aminobenzenesulfonic acid and then coupling it with *N,N*-dimethylaniline.

$$HO-SO_2-C_6H_4-NH_2 \xrightarrow{NaNO_2,\ H_2SO_4} HO-SO_2-C_6H_4-N_2^+ \xrightarrow{C_6H_5N(CH_3)_2}$$

$$HO-SO_2-C_6H_4-N{=}N-C_6H_4-N(CH_3)_2 \underset{-H^+}{\overset{H^+}{\rightleftharpoons}}$$

Methyl orange
(yellow when neutral or basic)

$$HO-SO_2-C_6H_4-NH-N{=}C_6H_4{=}\overset{+}{N}(CH_3)_2 \qquad (13\text{-}60)$$

Methyl orange
(red when acidic)

PROBLEM 13-4 Number each group of amines in order of decreasing basicity; that is, the most basic group will be number 1. (Use analogies with compounds listed in Table 13-2 and reasoning based on the principles discussed.)

(a) ___ *N*-ethylaniline ___ *p*-ethylaniline
___ *N*-ethylaminocyclohexane

(b) ___ $p\text{-}CH_3C(=O)-C_6H_4-NH_2$, ___ $C_6H_5NHC(=O)-CH_3$,
___ $p\text{-}CH_3CH_2O-C_6H_4-NH_2$, ___ $p\text{-}HOCH_2-C_6H_4CH_2NH_2$

Solution to (a)

$C_6H_5-NH-CH_2CH_3$ $\quad$ $p\text{-}CH_3CH_2-C_6H_4-NH_2$ $\quad$ $C_6H_{11}-NH-CH_2CH_3$

3 *N*-ethylaniline 2 *p*-ethylaniline 1 *N*-ethylaminocyclohexane

N-Ethylaminocyclohexane is considerably stronger as a base than the aromatic substituted anilines. By analogy with *p*-toluidine and *N*-

methylaniline, *p*-ethylaniline is stronger than *N*-ethylaniline. The alkyl group in a para position is able to exert a positive inductive effect.

PROBLEM 13-5 Write the structural formulas for the products of the reaction of *p*-toluidine ($p\text{-}CH_3—C_6H_4—NH_2$) with each of the following reagents.

(a) $Br_2(aq)$
(b) $NaNO_2$, H_3O^+, 4–10°C
(c) $NaNO_2$, H_3O^+, 25–30°C
(d) $CH_3\overset{O}{\overset{\|}{C}}—Cl$
(e) product of (b) with CuCN
(f) $C_6H_5SO_2Cl$, NaOH
(g) CH_3I (two equivalents)
(h) HCl(*aq*)
(i) $C_6H_5\overset{O}{\overset{\|}{C}}—Cl$
(j) product of (g) with $NaNO_2$, H_3O^+

Solution to (a)

$$p\text{-}CH_3C_6H_4NH_2 + Br_2(aq) \longrightarrow 2,6\text{-}Br_2\text{-}4\text{-}CH_3C_6H_2NH_2$$

4-Methylaniline (*p*-Toluidine) → 2,6-Dibromo-4-methylaniline

The amino group is one of the strongest of the ortho,para-directing groups; a methyl group is one of the weakest such groups. The amino group is such a strong activating group that both available ortho positions are brominated without the need of a catalyst. (Refer to Chap. 5.)

13-6 Color in organic compounds

The color we discern in any object is a result of absorption by the object of light in the visible part of the spectrum. What we see as white light is actually a mixture of all of the wavelengths of light in the visible range of electromagnetic radiation. If the object absorbs any of the light in the visible part of the spectrum, we will see the object as that color associated with the remaining wavelengths of light. The remaining wavelengths will be either reflected or transmitted to our eyes.

The chief colors from longest to shortest wavelength (or lowest to highest frequency, since frequency equals velocity of light divided by the wavelength, $\nu = c/\lambda$) are red, orange, yellow, green, blue, and violet. That is shown more exactly in Fig. 13-4.

All compounds absorb light in the ultraviolet part of the spectrum. If the compound has a pi electron system, the absorption occurs at somewhat longer wavelength. If the pi electron structure becomes extensive, the absorption shifts to still longer wavelength until violet, the shortest wavelength of visible light, is being absorbed. When violet light is absorbed, the eye actually discerns the *complementary color* yellow, which

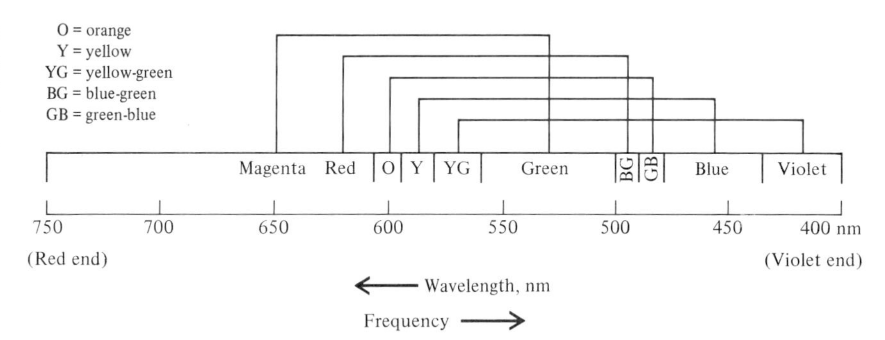

FIGURE 13-4 **The visible spectrum, showing wavelengths of color which are complementary by subtraction.**

is being reflected and/or transmitted to the eye. Therefore, the colored compounds of simplest structure are yellow. As the pi electron system becomes more extensive with more conjugation and opportunities for resonance, the color absorbed is of still longer wavelength, say, blue, and the color seen then is orange. If green is absorbed, red is seen; if yellow is absorbed, violet is the color observed. Therefore, color deepens in the order yellow < orange < red < violet < blue < green < black.

Color-bearing groups

O. N. Witt observed in 1876 that certain functional groups cause a compound to be colored. He termed those groups *chromophores*. Some of them were found to be more effective at producing color than others. Only one unit of a strong chromophore would suffice to cause color, whereas two or more weaker chromophores in series might be required to cause color. Some chromophores are listed in Table 13-4. Note that all of

TABLE 13-4 **Chromophores**

Group	Formula	Relative strength
Carbonyl	$>C{=}O$	Weak
Ethenyl	$>C{=}C<$	Weak
p-Quinoid	$={C_6H_4}=$ (p-quinoid ring)	Strong
Polycyclic aromatic	(fused aromatic rings)$_{n=4\text{ or more}}$	Intermediate
Thiocarbonyl	$>C{=}S$	Intermediate
Nitro	$-NO_2$	Strong
Nitroso	$-N{=}O$	Strong
Azo	$-N{=}N-$	Strong

them contain pi bonds. They are still more effective in conjugated combination with either more of the same chromophore or with different chromophores.

Color-intensifying groups

Witt also observed that the presence of certain groups on a benzene ring bearing a chromophore would not produce color of itself but would intensify the color being caused by the chromophore. He called such assisting groups *auxochromes*. They are the same electron-releasing groups which we recognize today as strongly activating ortho,para-directing groups for aromatic electrophilic substitution. They are effective in the same order: $R_2N— > RHN— > H_2N— > HO— > RO—$.

Bury observed that the possibility of definite resonance for a good delocalization of pi electrons was always involved in production of color. Calvin further observed that one or more of the electronic forms contributing to resonance should be ionic for color production. In the formulas for methyl orange, Eq. (13-60), note how the color deepens from yellow to red as the structure changes from azo to the *p*-benzoquinoid chromophore.

Dyes

Not all colored compounds necessarily make good dyes. A *dye* is a colored substance which can be made to adhere to the fibers of a fabric in such a way as to become an integral part of the fabric, as opposed to a mere surface coating. If a dye is resistant to fading as a result of either washing or exposure to light, it is said to be *fast*. If a dye is not so resistant, it is said to be *fugitive*. A need exists for both types. We like the dyes in our carpets, clothing, and drapes to be fast, but we appreciate fugitive dyes in our children's crayons.

For many years the only available dyes were natural products in the roots, leaves, and barks of various plants and shrubs. Indigo blue, which occurs as a glucoside in the leaf of the indigo plant, was known to the Egyptians as early as 2000 B.C. Indigo blue is still used (for blue denims), but since 1880 it has been made synthetically. Another ancient dye of structure similar to indigo blue but not of plant origin is Tyrian purple. In ancient days it was obtained from a small mollusk (*Murex brandaris*) found on the small Mediterranean island of Tyre. It was so rare and expensive that it was reserved for royalty. Alizarin, another ancient dye, is found in the root of the madder plant. It was used to dye cotton and linen red. The latter two are no longer important dyes because today a great many synthetic dyes of every hue and shade of the rainbow are produced and used. The structural formulas of indigo blue, Tyrian purple, and alizarin are shown in Fig. 13-5.

The first synthetic dye was accidentally discovered by William Henry Perkin in 1856, when Perkin was an 18-year-old student and assistant in A. W. von Hofmann's laboratory in England's Royal College of Chemis-

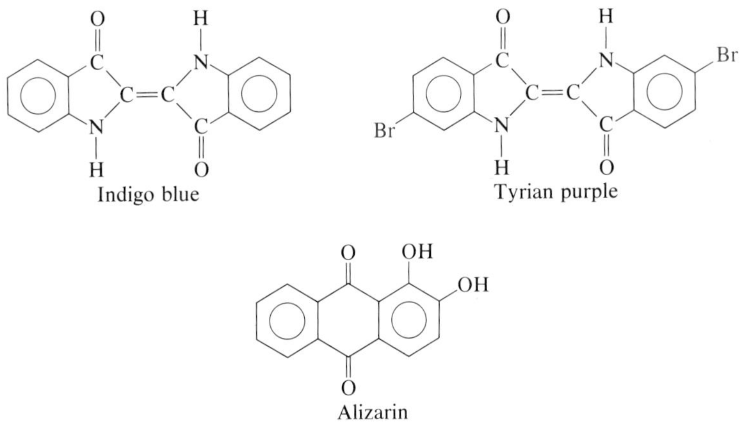

FIGURE 13-5
Three naturally occurring dyes.

Indigo blue

Tyrian purple

Alizarin

try. Perkin was attempting to synthesize quinine by dichromate oxidation of a derivative of aniline. The procedure could never have produced quinine, but it did produce a tarry material out of which Perkin extracted a compound which would dye fabrics a shade of purple. He called his synthetic dye aniline purple or mauve, obtained a patent on it, and built a plant to produce it the following year. Soon many other synthetic dyes were discovered and produced. It was the beginning of the dye industry in England. Hofmann went back to Germany, where a large dye industry also became established.

Direct dyes. *Direct dyes* are those which do not require any other agent to dye material. A dye may be direct for silk or wool but not for cotton; if it is also direct for cotton, it is called a *substantive dye.*

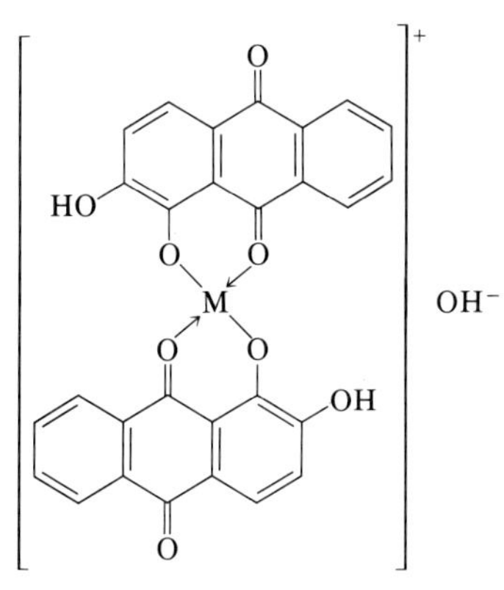

FIGURE 13-6
Lake of the dye alizarin. Suggested structure of the lake produced by the coordination of two molecules of alizarin with a metal ion (M) as a mordant. If the metal ion, M^{3+}, is Al^{3+}, the color is red; if Cr^{+3}, maroon; if Fe^{3+}, purple.

Mordant dyes. Many dyes require a mordant in order to adhere to the material. A *mordant* is generally a metal ion, such as iron, chromium, or aluminum, which coordinates or chelates with certain functional groups on the dye. That forms an insoluble pigment, called a *lake,* within the fibers of the fabric. The fabric is dipped successively in a solution of a salt of the metal and then a solution of the dye, so that the lake is precipitated within the fibers (Fig. 13-6). The exact color produced by a given dye may vary with the mordant used, which indicates that the metal ion may modify the total resonance effect of the molecule.

Azo dyes. The azo dyes are produced by the coupling reactions of diazonium salts. Water solubility is conferred on them by the introduction of sulfonic acid groups to the benzene ring bearing the amino group before it is diazotized. Fast, insoluble azo dyes may be generated right within the fabric by dipping the fabric first in a solution of the diazonium salt and then in a solution of the coupling reagent. Such azo dyes are called *developed dyes.*

Vat dyes. *Vat dyes* are applied in a soluble, almost colorless, reduced form. They are then air-oxidized in the presence of light to an insoluble, colored form. One of the oldest vat dyes is the blue dye indigo.

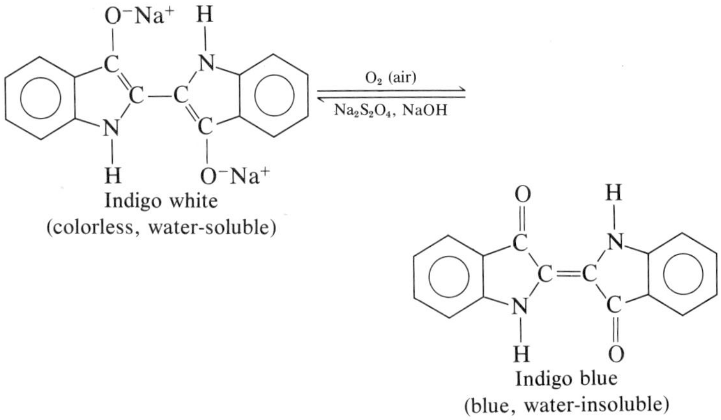

O^-Na^+ H

$\xrightleftharpoons[Na_2S_2O_4,\ NaOH]{O_2\ (air)}$

H O^-Na^+

Indigo white
(colorless, water-soluble)

Indigo blue
(blue, water-insoluble)

(13-61)

The vat dyes are quite resistant to fading by light, since light catalyzes their oxidation to the colored form.

Phthalein dyes. The phthalein dyes are produced by a type of Friedel-Crafts acylation reaction of phenols with phthalic anhydride. The reaction with phenol itself produces phenolphthalein, which is not a dye but is a common acid-base indicator. Phenolphthalein as it is first produced is colorless.

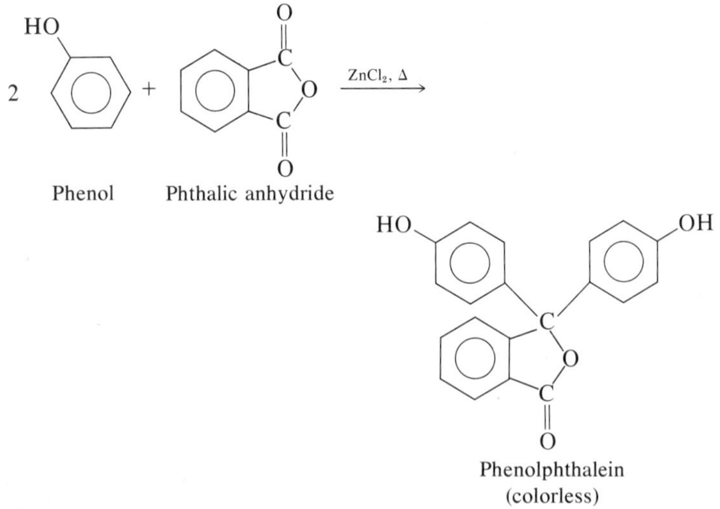

2 Phenol + Phthalic anhydride $\xrightarrow{ZnCl_2,\ \Delta}$

Phenolphthalein
(colorless)

(13-62)

Reaction with 2 mol of sodium hydroxide breaks the lactone ring and also produces a phenoxide ion for which resonance is possible. Thus a red color is produced.

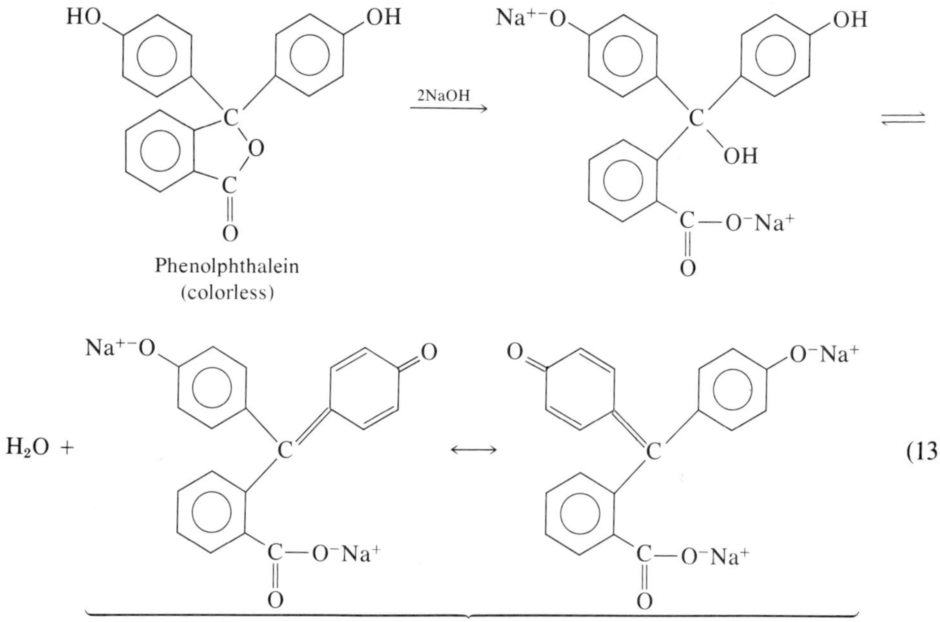

HO OH C O C O

Phenolphthalein (colorless)

$\xrightarrow{2NaOH}$

$Na^{+-}O$ OH C OH $C—O^{-}Na^{+}$ O

⇌

H_2O +

$Na^{+-}O$ O C $C—O^{-}Na^{+}$ O

⟷

O $O^{-}Na^{+}$ C $C—O^{-}Na^{+}$ O

(13-63)

Phenolphthalein (red)

If a third mole of NaOH is added, the trisodium salt of phenolphthalein is formed. It is colorless.

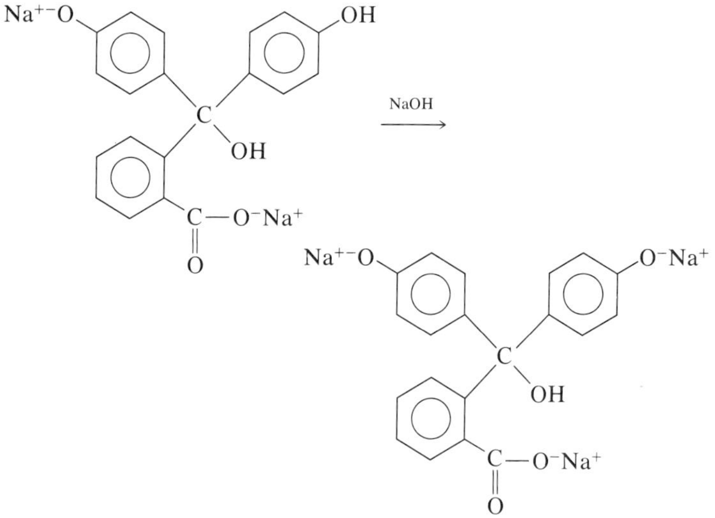

$Na^{+-}O$ OH C OH $C—O^{-}Na^{+}$ O

$\xrightarrow{NaOH}$

$Na^{+-}O$ $O^{-}Na^{+}$ C OH $C—O^{-}Na^{+}$ O

(13-64)

Trisodium salt (colorless)

Phenolphthalein stimulates motor nerve activity of the lower intestinal tract, and so it is used in the formulation of a number of laxatives.

If resorcinol is reacted with phthalic anhydride, the product is similar to phenolphthalein; but in addition a diphenyl ether linkage is formed by a loss of water between two of the resorcinol rings. The product, when made alkaline, is a very strong fluorescent greenish-yellow dye called *fluorescein*.

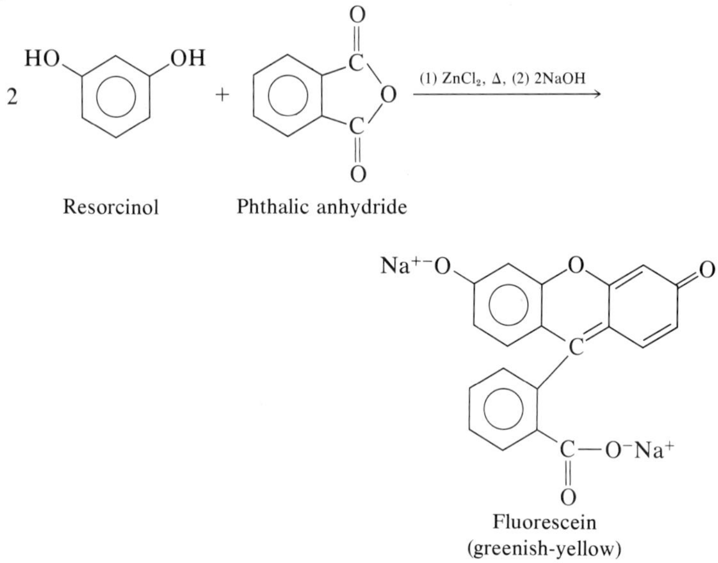

2 Resorcinol + Phthalic anhydride $\xrightarrow{(1)\ ZnCl_2,\ \Delta,\ (2)\ 2NaOH}$ Fluorescein (greenish-yellow) (13-65)

If fluorescein is brominated to the tetrabromo derivative, the red dye *eosin* is produced. Partial bromination and mercuration produces a red dye with germicidal properties known as *Mercurochrome*. The structures of those compounds are shown in Fig. 13-7.

Certified colors. Dyes which have been found safe to use in foods and cosmetics are so certified. Some dyes which at one time were certified have since been found to cause cancer or other disorders, so their use has had to be discontinued. Safe red dyes have been especially difficult to find. The much-used red dye No. 2 has been banned. The extensive use of dyes in food may well be both unnecessary and harmful.

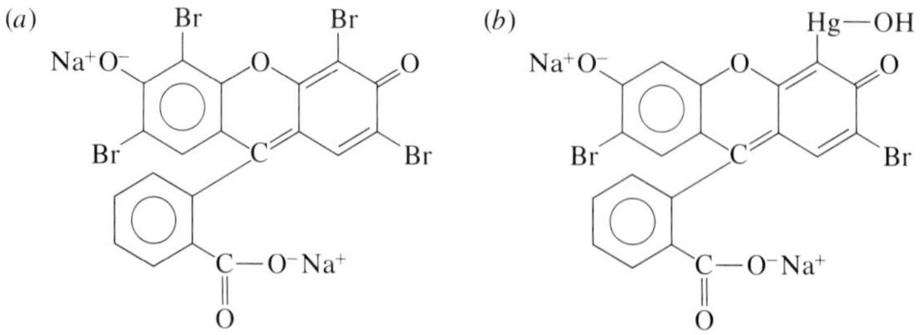

FIGURE 13-7
The structures of (*a*) eosin (red) and (*b*) Mercurochrome.

13-7 Alkaloids

The alkaloids constitute a class of natural products which have important applications in medicine. *Alkaloids* may be defined as basic nitrogenous compounds of botanical origin which have a marked physiological reaction. They may cause such physiological effects as raising the blood pressure, relaxing smooth muscle, stimulating nerves, dilating the pupils of the eyes, causing the blood to clot, stimulating production of urine, relieving pain, and producing sleep. Most of them are complicated molecules consisting of nitrogen heterocyclic rings, but a few have a relatively simple structure.

One of the alkaloids of simple structure is coniine, which occurs in the hemlock; it was the poison in the hemlock tea which Socrates was forced to drink. Coniine is the *n*-propyl derivative of the heterocyclic compound piperidine. It can be readily synthesized from 2-methylpyridine, which is available from coal tar.

$$\text{2-Methylpyridine} + \underset{\text{Acetaldehyde}}{HCCH_3 \ (C{=}O)} \xrightarrow[-H_2O]{OH^-,\ \Delta} \text{(pyridin-2-yl)}{-}CH{=}CHCH_3 \xrightarrow{H_2/Ni,\ \Delta} \underset{\text{(Coniine)}}{\underset{\text{2-}n\text{-Propylpiperidine}}{\text{(piperidin-2-yl, N–H)}{-}CH_2CH_2CH_3}} \quad (13\text{-}66)$$

The reaction of 2-methylpyridine with acetaldehyde is actually a type of aldol condensation. The base removes one of the hydrogens from the methyl group to form a resonance-stabilized carbanion, which adds to the carbonyl group of the aldehyde. The addition is followed by spontaneous dehydration.

Indole alkaloids

The indole alkaloids incorporate the *indole ring system*—a condensed ring system consisting of a benzene ring joined to a pyrrole ring. Representative of the type are the alkaloids found in ergot, a fungus which infects grains—especially rye. In times past, eating bread made from ergot-infected grain caused spontaneous abortion in women. Today, ergonovine, extracted from ergot, is used in medicine to facilitate uterine contraction and prevent excessive loss of blood in difficult childbirth.

Experimental work with ergonovine led to the hydrolysis of the ergonovine amide group to produce lysergic acid. When lysergic acid was converted to the diethylamide, it was found to be the most powerful hallucinogenic substance yet known. As little as 0.5 to 1.0 mg per kilogram of body weight affects a person. Lysergic acid diethylamide is commonly referred to as LSD.

Figure 13-8 shows the structures of indole, ergonovine, and LSD.

FIGURE 13-8
The structures of indole and two indole alkaloids.

$HOCH_2-C(H)(CH_3)-N(H)-C(=O)-$ $N-CH_3$ $(C_2H_5)_2NC(=O)-$ $N-CH_3$

Indole   Ergonovine   LSD

Opium alkaloids

The opium alkaloids are found in a species of poppy. One of the better known ones is morphine, which has long been used to relieve pain. Codeine and heroin are two of the common derivatives of morphine; codeine is the monomethyl ether derivative, and heroin is the diacetyl derivative. Codeine and heroin have been used in cough syrups. All such syrups are addictive, but heroin syrups are especially so. Two synthetic pain relievers are methadone and meperidine (Demerol). Formulas of those compounds are shown in Fig. 13-9.

Purine alkaloids

The purine alkaloids contain the *purine ring system,* which consists of a pyrimidine ring fused to an imidazole ring.

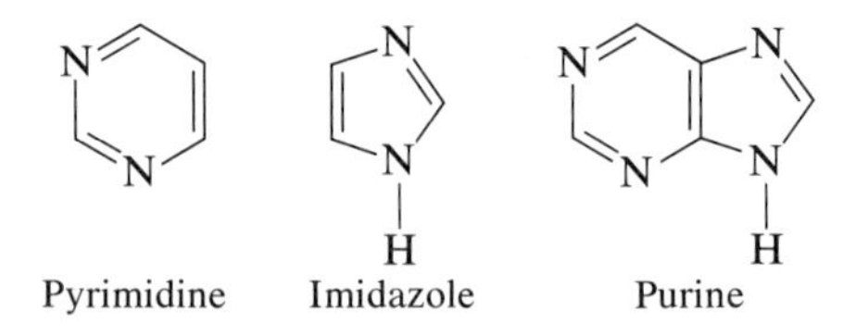

Pyrimidine   Imidazole   Purine

Theobromine and caffeine are two familiar examples of purine alkaloids. Theobromine is found in cacao beans and in chocolate. It stimulates the heart and kidneys. Caffeine is found in coffee beans (1.2%) and in tea leaves (1.9%). It stimulates the heart, kidneys, and nerves. Caffeine is added to all of the cola drinks and to some other drinks as well. The amount added is about that in a cup of tea or coffee. Many people feel that they require such a stimulant to function well. Caffeine is not considered addictive, but it is habituating. It is the active ingredient in No-Doz, and it is frequently used to counteract the unwanted depressant action of certain medications.

A related purine compound of interest is uric acid, which occurs in small amounts in the urine. Individuals suffering from gout have larger than normal amounts of uric acid in their bodies. Uric acid crystals may then precipitate in the joints, especially the joints of the great toes, to produce the characteristic painful symptoms of gout. Water-conserving creatures such as birds and reptiles excrete, as their chief nitrogenous waste, uric acid rather than urea, which requires water to dissolve and carry it out.

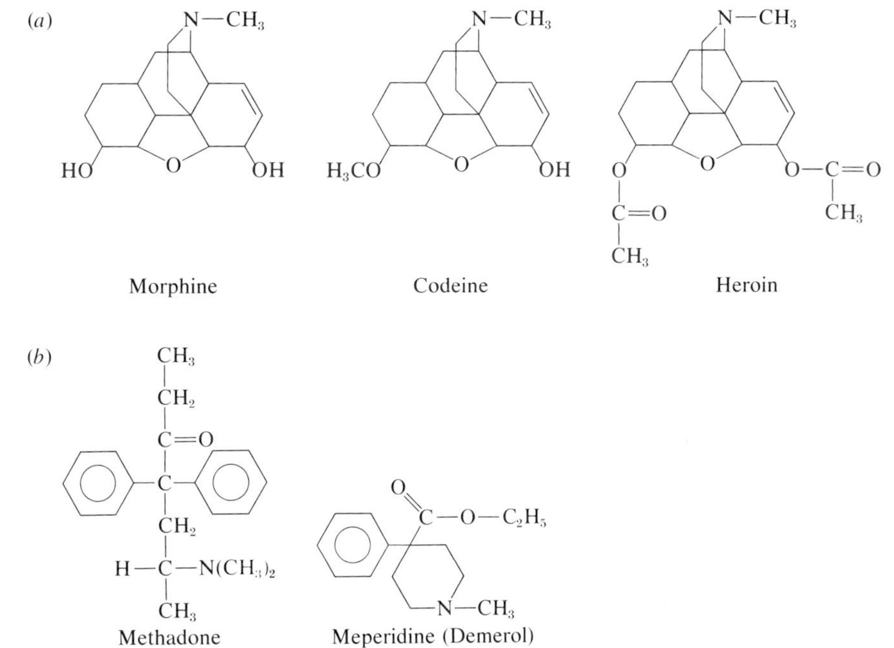

FIGURE 13-9
(*a*) Three opium alkaloids; (*b*) two synthetic pain relievers.

A well-known drug for prevention of motion sickness called Dramamine is a purine derivative. See Fig. 13-10 for the structures of purine derivatives.

Pyridine and quinoline alkaloids Nicotine is another alkaloid of simple structure; it contains a *pyridine ring*. It occurs to the extent of 2 to 7% in tobacco. It is very poisonous and

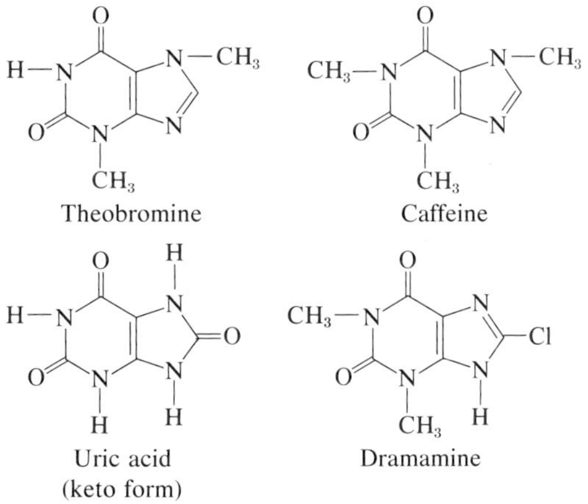

Theobromine

Caffeine

Uric acid
(keto form)

Dramamine

FIGURE 13-10
Four purine alkaloids.

FIGURE 13-11
Nicotine, a pyridine alkaloid, and quinine, a quinoline alkaloid.

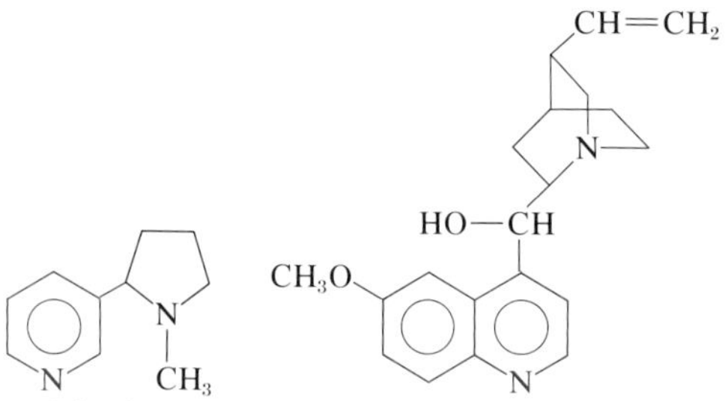

Nicotine

Quinine

has been used in insecticides, particularly those for killing aphids on roses. Even in the small amounts encountered by the users of tobacco, it produces such symptoms as nausea, indigestion, and higher blood pressure. Also, it is addictive.

Quinine, an alkaloid from the bark of the cinchona tree, has long been valued for the treatment of malaria. It is an example of a *quinoline alkaloid;* see Fig. 13-11.

Summary

The *amines* are the *alkyl* and *aryl derivatives of ammonia.* They are classified as *primary, secondary,* or *tertiary* depending on whether they have one, two, or three organic groups substituted for the hydrogen atoms of ammonia. The primary and secondary amines have hydrogen attached to nitrogen, so they, like ammonia, can exhibit hydrogen bonding. As a consequence, their boiling points are somewhat higher than those of alkanes of similar molecular weight but much lower than those of the corresponding alcohols.

A simple *aliphatic amine* is named by naming the alkyl group or groups attached to the nitrogen atom and adding the ending *-amine.* In the IUPAC system, the amino group ($-NH_2$) may be treated as a substituent and named as an amino group on the major structure. An *aryl amine* is named as a derivative of the parent member of the family, *aniline.* The *acyl derivatives* of ammonia ($R-\overset{\overset{\displaystyle O}{\|}}{C}-NH_2$) are called *amides.* The *diacyl derivatives* of ammonia ($R-\overset{\overset{\displaystyle O}{\|}}{C}-\overset{\overset{\displaystyle H}{|}}{N}-\overset{\overset{\displaystyle O}{\|}}{C}-R$), are called *imides,* and the *sulfonyl derivatives* of ammonia ($ArSO_2NH_2$) are called *sulfonamides.* A *secondary* or *tertiary aryl amine* is named by specifying the substituents on nitrogen; the name is preceded by *N* to denote that the attachment is to nitrogen. The same system is used for naming *nitrogen-substituted amides.*

An amine may be prepared by *reduction* of an organic compound possessing nitrogen in a higher oxidation state or by *reductive amination* of an aldehyde or ketone. *Ammonolysis* of an alkyl halide produces a fairly good yield of primary amine if an excess of ammonia is used. Otherwise, a mixture of primary amine, secondary amine, tertiary amine, and quaternary ammonium salt is produced.

If a quaternary ammonium salt is reacted with silver hydroxide, a *quaternary ammonium base,* which is a very strong base, is formed. Higher homologs of quaternary bases have a water-soluble anion and an oil-soluble cation. They have a detergent action with some germicidal activity, and they are referred to as *invert soaps*. Choline and acetylcholine are quaternary salts of importance in human physiology.

Heating an amide with NaOH and Br_2 produces Na_2CO_3, NaBr, and a primary amine. The reaction is called the *Hofmann degradation reaction*.

The amines react *as bases* with water, just as ammonia does, to produce *alkylammonium hydroxides*. The unshared pair of electrons on the nitrogen atom of aniline tends to participate in the resonance hybrid of the aromatic structure through pi overlap with the pi orbital of the ring. That makes the unshared pair of electrons much less available for reaction as a base. Aniline is about a million times weaker as a base than ammonia.

Amines react *with acids* to form *alkylammonium salts*. The reaction is of great utility in the chemical separation of amines from neutral and acidic compounds which are not soluble in acid solution. Amines react *with acid anhydrides* or *acid halides* to form *amides*. Tertiary amines do not react with those reagents.

Primary and secondary amines react *with sulfonyl chlorides* to give N-*substituted sulfonamides*. Tertiary amines do not react. The sulfonamides of primary amines are slightly acidic and are soluble in aqueous NaOH. The combination of benzenesulfonyl chloride and sodium hydroxide is referred to as the *Hinsberg reagent*. Production of a base-soluble sulfonamide is a positive test for a primary amine. A secondary amine reacts to form a sulfonamide which is not soluble in aqueous NaOH, and tertiary amines do not react.

The amino group of aromatic amines is a very strong activating *ortho,para-directing* group for electrophilic ring substitution. However, nitration of aniline produces much *m*-nitroaniline because of the formation of the *anilinium ion,* $C_6H_5NH_3^+$, with the proton from the nitric acid. The ammonium ion, with its positive charge, is a meta-directing group.

Aliphatic primary amines react with *nitrous acid* to produce chiefly an alcohol and nitrogen. *Aromatic primary amines* react with nitrous acid to produce intermediate resonance-stabilized *diazonium salts* if the temperature is kept below about 10°C. If the aqueous diazonium salt warms up, a phenol and nitrogen are produced. *Secondary amines* react with nitrous acid to produce *N*-nitrosoamines. *Aliphatic tertiary amines* do not react readily with nitrous acid, but aromatic tertiary amines react to produce *nitrosation* of the benzene ring. *Primary amides* also react with nitrous acid to give off nitrogen, and the product is a carboxylic acid.

Aromatic diazonium ions undergo a *nucleophilic substitution* for nitrogen. The reaction is the best general laboratory method for introducing

F, I, CN, or OH on an aromatic ring. A diazonium ion can substitute as an electrophile on a benzene ring which is activated by having activating ortho,para-directing groups. The electrophilic substitution reaction is called *coupling*.

The *color* we discern in any object is a result of absorption by the object of light in the visible part of the spectrum. If the object absorbs any of the light in the visible part of the spectrum, we see the object in the color associated with the remaining wavelengths of light, which are reflected or transmitted to our eyes.

Chromophores are certain functional groups that cause a compound to be colored. They generally contain pi bonds. *Auxochromes* are electron-releasing groups which do not produce color themselves but do intensify the color caused by a chromophore. They are effective in the order $R_2N— > RHN— > H_2N— > HO— > RO—$. *Resonance* is always involved in the production of color, and one or more of the contributing forms is generally ionic.

A *dye* is a colored substance which can be made to adhere to the fibers of a fabric in such a way as to become an integral part of the fibers. A *fast dye* is one which is resistant to fading by washing or exposure to light; a *fugitive dye* is one which is not so resistant. *Direct dyes* do not require any other agent to dye material. *Mordant dyes* require a mordant in order to adhere to the material. A *mordant* is generally a metal ion which coordinates with certain functional groups on the dye to form an insoluble pigment called a *lake*. *Azo dyes* are produced by the coupling reactions of diazonium salts; the popular indicator methyl orange belongs to the class. *Vat dyes* are those which are applied in a soluble, almost colorless, reduced form. They are then air-oxidized in the presence of light to an insoluble colored form. Indigo is an example. *Phthalein dyes* are produced by a type of Friedel-Crafts acylation reaction of phenols with phthalic anhydride. The indicator phenolphthalein belongs to the class, and so does Mercurochrome. *Certified colors* are dyes which have been certified as safe to use in foods and cosmetics.

Alkaloids are *basic nitrogenous compounds* of botanical origin which have marked physiological effects. *Indole alkaloids* incorporate the indole ring system. *Opium alkaloids* are found in a species of poppy. *Purine alkaloids* contain the purine ring system. *Pyridine* and *quinoline alkaloids* contain the pyridine ring or the quinoline ring system, respectively.

Key terms

Alkaloid: a basic nitrogenous compound of botanical origin which has a marked physiological effect.

Alkylammonium hydroxide: a compound produced by the reaction of a basic amine with water; the general formula is $R—NH_3^+OH^-$.

Alkylammonium salt: a compound produced by the reaction of an amine with an acid; the cation of such a salt has the general formula $R—NH_3^+$.

Amide: the acyl derivative of ammonia, $R—\underset{\underset{O}{\|}}{C}—NH_2$.

Amine: an alkyl or aryl derivative of ammonia; primary amines have the general formula RNH_2 or $ArNH_2$.

Anilinium ion: the ion $C_6H_5NH_3^+$ produced by reaction of aniline with an acid.

Auxochrome: a group on a benzene ring bearing a chromophore which intensifies the color caused by the chromophore. Auxochromes are electron-releasing and strongly activating ortho,para-directing groups.

Azo dye: a dye produced by the coupling reaction of a diazonium salt.

Certified color: a dye which has been certified as safe to use in foods and cosmetics.

Chromophore: a functional group which causes a compound to be colored; all chromophores contain pi bonds.

Complementary color: the color discerned by our eyes. It comprises the wavelengths of light reflected or transmitted by an object, and it is complementary to the wavelengths of light absorbed by the object.

Coupling reaction: substitution of a diazonium ion on a benzene ring which is activated by bearing the stronger activating ortho,para-directing groups. It produces colored azo compounds.

Developed dye: a fast, insoluble dye generated within a fabric by dipping the fabric in a solution of a diazonium salt and then in a solution of the coupling reagent.

Diazonium ion: the ion $R—N_2^+$ or $Ar—N_2^+$. It is a resonance hybrid of $R—\ddot{N}{=}N:^+$ and $R—\overset{+}{N}{\equiv}N:$ forms, and it is relatively stable if aromatic.

Direct dye: a dye which does not require an auxiliary agent to dye material.

Dye: a colored substance which can be made to adhere to the fibers of a fabric so as to become an integral part of the fibers.

Fast dye: a dye which is resistant to fading as a result of washing or exposure to light.

Fugitive dye: a dye which is not resistant to fading.

Hinsberg reagent: benzenesulfonyl chloride and sodium hydroxide. It reacts with a primary amine to produce a base-soluble sulfonamide and with a secondary amine to produce a base-insoluble sulfonamide. It does not react with a tertiary amine.

Hofmann degradation: reaction of an amide with NaOH and Br_2 to produce a primary amine with one less carbon atom.

Imide: a diacyl derivative of ammonia.

Indole alkaloid: an alkaloid incorporating the indole ring system, a condensed ring system of a benzene ring joined to a pyrrole ring.

Invert soaps: the higher homologs of quaternary ammonium bases; they have a water-soluble anion and an oil-soluble cation.

K_b: the basicity constant of an amine; it is given by the expression

$$K_b = \frac{[RNH_3^+][OH^-]}{[RNH_2]}$$

The larger the value the stronger the amine as a base.

Lake: an insoluble pigment, within the fibers of a fabric, formed by coordination of a mordant with certain functional groups on a dye.

Lecithins: a group of phospholipids important in the construction of cell membranes. Choline, a quaternary ammonium salt, is a molecular part of the lecithins.

Mordant dye: a dye which requires a mordant to adhere to the material; a mordant is a metal ion which coordinates with certain functional groups in the dye to form an insoluble pigment (a lake) within the fibers of the fabric.

***N,N*-Disubstituted amide:** an amide in which both hydrogens on nitrogen are substituted by other groups. It is produced by the reaction of a secondary amine with an acid anhydride or acid halide.

***N*-Nitrosoamines:** a compound of the general formula $R_2—N—N=O$. It is produced by the reaction of a secondary amine with nitrous acid.

***N*-Substituted amide:** an amide in which one of the hydrogens on nitrogen is substituted by another group. It is produced by the reaction of a primary amine with an acid anhydride or an acid halide.

***N*-Substituted sulfonamide:** a sulfonamide in which one or both of the hydrogens on nitrogen is substituted by another group. It is produced by the reaction of a primary or secondary amine with benzene sulfonyl chloride.

Nitrosation: introduction of the nitroso group, —N=O, onto the benzene ring. It may be accomplished by reaction of an aromatic tertiary amine with nitrous acid.

Opium alkaloid: an alkaloid found in a species of poppy. The opium alkaloids include morphine, codeine, and heroin.

Phthalein dye: a dye produced by a type of Friedel-Crafts acylation reaction of a phenol with phthalic anhydride.

pK_b: the negative logarithm of K_b, the basicity constant. The smaller the value of pK_b the stronger the base.

Purine alkaloid: an alkaloid containing the purine ring system, a pyrimidine ring fused to an imidazole ring.

Pyridine alkaloid: an alkaloid containing a pyridine ring.

Quaternary ammonium base: a base of the general formula $R_4N^+OH^-$. It is produced by reaction of a quaternary ammonium salt with silver hydroxide.

Quaternary ammonium salt: a salt of the general formula $R_4N^+X^-$. It can be produced by ammonolysis of an alkyl halide under proper conditions.

Quinoline alkaloid: an alkaloid containing a quinoline ring.

Reductive amination: an excellent method of preparing a primary amine from an aldehyde or ketone.

Substantive dye: a direct dye which is effective for cotton as well as for silk or wool.

Sulfonamide: a sulfonyl derivative of ammonia.

Vat dye: a dye which is applied in a soluble, almost colorless form and then air-oxidized in the presence of light to an insoluble, colored form.

Additional problems

13-6 Name each of the following compounds. Classify the amino function as primary, secondary, or tertiary.

(a) $CH_3C(=O)—N(C_2H_5)_2$ (c) $CH_3CH_2CH(NH_2)—CH_2CH_3$ (e) 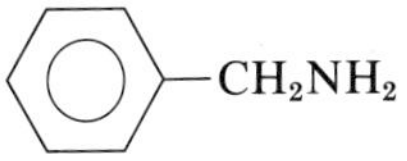 $C_6H_5—CH_2NH_2$

(b) 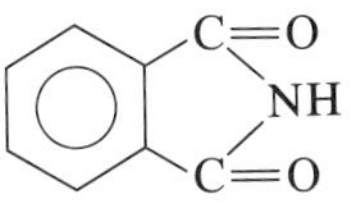 $C_6H_4(C{=}O)_2NH$ (d) $CH_3CH(NH_2)—CO_2H$

13-7 Draw the structural formula of each of the following compounds.

(a) *N*-methylaniline
(b) (*S*)-2-aminopropanoic acid
(c) *N*-methylbenzamide
(d) 2-amino-1-phenylpropane
(e) allylbenzylphenylamine
(f) *p*-aminobenzenesulfonamide

13-8 Write equations for the following reactions:

(a) *n*-butylamine and benzoyl chloride
(b) dimethylamine and water
(c) *n*-propylamine and HNO_2
(d) triethylamine and HCl
(e) *N*-methylaniline and benzenesulfonyl chloride

13-9 Indicate a *simple chemical test* or reagent which will distinguish between the following pairs of compounds. Indicate which member of the pair will give the positive test or greater reaction.

(a) 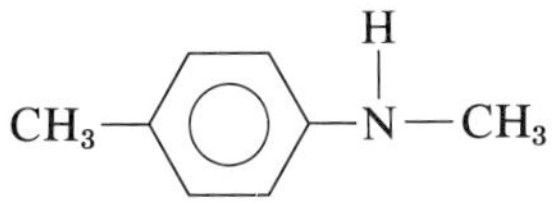 $CH_3—C_6H_4—N(H)—CH_3$ and $C_6H_5—N(CH_3)_2$

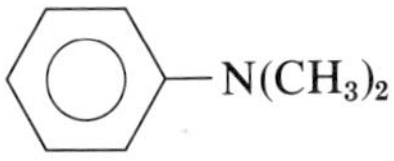

(b) $CH_3CH_2CH_2CH_2NH_2$ and $(CH_3CH_2)_2NH$

(c) 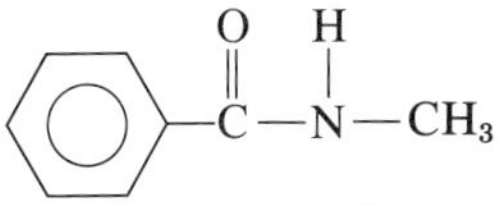 $C_6H_5—C(=O)—N(H)—CH_3$ and $C_6H_5—SO_2—N(H)—CH_3$

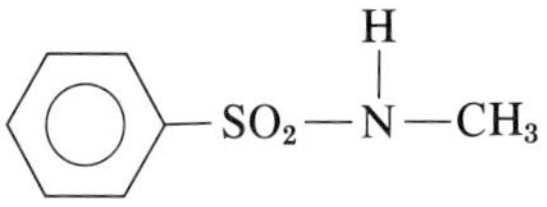

(d) 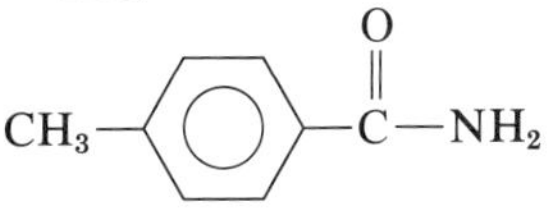 $CH_3—C_6H_4—C(=O)—NH_2$ and $C_6H_5—C(=O)—N(H)—CH_3$

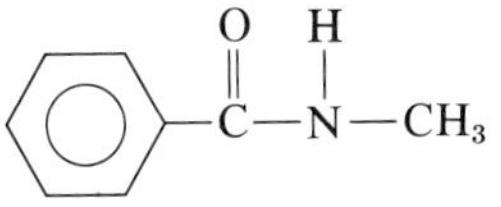

(e) 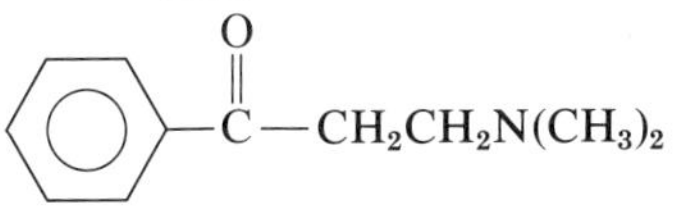 $C_6H_5—C(=O)—CH_2CH_2N(CH_3)_2$ and $C_6H_5—CH_2CH_2C(=O)—N(CH_3)_2$

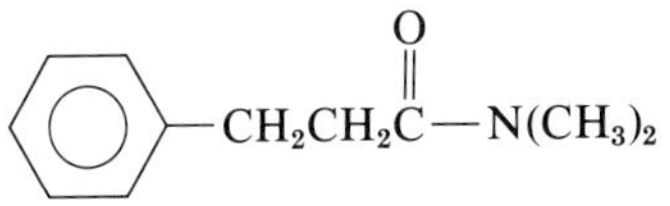

(f) $C_6H_5CH_2OH$ and $C_6H_5CH_2NH_2$

(g) 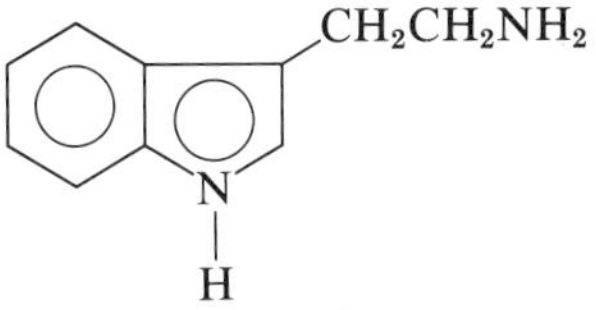 indole(N–H)–$CH_2CH_2NH_2$ and indole(N–H)–$CH_2CH_2N(CH_3)_2$

tryptamine — *N,N*-Dimethyltryptamine

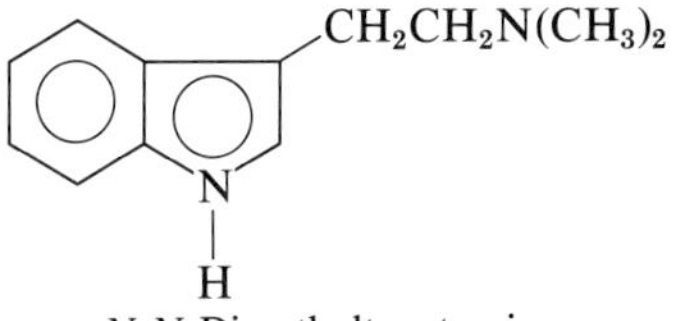

13-10 Write equations for the following reactions.

(a) *p*-toluidine ($p\text{-}CH_3—C_6H_4—NH_2$) and cold HNO_2
(b) aniline and acetic anhydride
(c) *o*-toluidine and bromine water

(d) 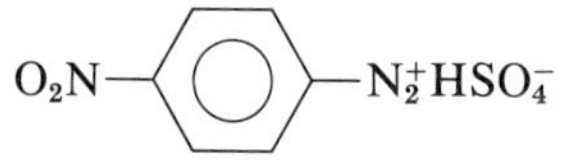 $O_2N—C_6H_4—N_2^+HSO_4^-$ and CuCN

(e) 2 (1-naphthylamine-4-sulfonic acid: NH_2, SO_2OH) + $^+N_2—C_6H_4—C_6H_4—N_2^+$ ⟶ congo red

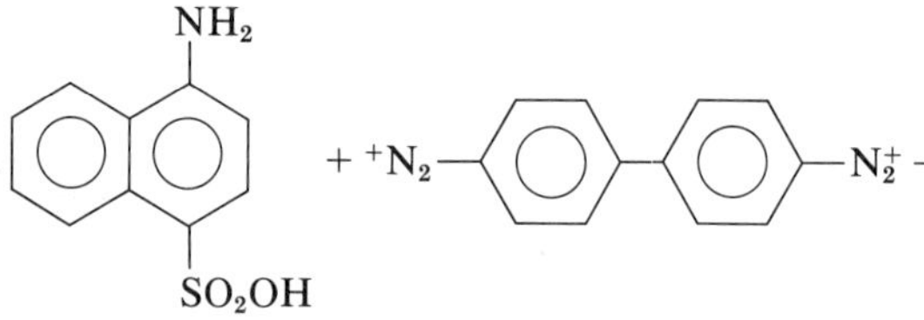

TABLE 13-5
Toxic plant compounds

Common name of plant	Scientific name of plant	Toxic ingredients
Aconite	*Aconitum* spp.	Aconicotine
Black henbane	*Hyoscyamus niger*	Atropine
Black locust	*Robinia pseudoacacia*	i Robin ii Robinin
Death camass	*Zygadenus*, several species	i Zygadenine ii Veratrine
Dutchman's-breeches, bleeding heart	*Dicentra* spp.	Protopine
Elderberry	*Sambucus canadensis*, *S. nigra*	Sambunigrin
False hellebore	*Veratrum viride*, *V. californicum*	i Veratridine ii Veratrine
Foxglove	*Digitalis purpurea*	Digitalis
Indian tobacco	*Lobelia inflata*	Lobeline
Larkspur, delphinium, or staggerweed	*Delphinium* spp.	i Delphinine ii Delphinoidine iii Ajacine
Lupine	*Lupinus* spp.	Lupinine
Mountain laurel, rhododendron	*Kalmia latifolia*, *Rhododendron* spp.	Andromedotoxin or rhodotoxin
Poison hemlock	*Conium maculatum*	Coniine
Wisteria	*Wistaria* spp.	Wistarin
Water hemlock	*Cicuta* spp.	Cicutoxin
Woody nightshade	*Solanum dulcamara*	Solanine
Yellow jasmine	*Gelsemium sempervirens*	i Gelsemine ii Gelseminine
Yew	*Taxus* spp.	Taxine
Ergot	*Claviceps* spp.	i Ergocryptine ii Ergocornine iii Ergocristine
Mistletoe	*Phoradendron flavescens*	β-Phenylethylamine
Mushrooms	*Amanita phalloides*	Phalloidine
Nettle	*Urtica* spp.	i Acetylcholine ii Histamine iii 5-Hydroxytryptamine
Poison ivy, poison oak, poison sumac	*Toxicodendron* spp.	3-*n*-Pentadecylcatechol

13-11 Number the following compounds in order of decreasing basicity; that is, the most basic compound will be number 1.

___ $CH_3-C_6H_4-NH_2$ ___ $C_6H_5-CH_2NH_2$ ___ NH_3

___ $C_6H_5-N^+(CH_3)_3OH^-$ ___ $CH_3O-C_6H_4-N(CH_3)_2$

13-12 Write equations showing how 1-phenyl-2-aminopropane (Benzedrine) could be prepared beginning with benzene and toluene and alcohols of four or fewer carbons. (*Hint:* Use the reductive amination reaction with a ketone produced from an alcohol which is prepared by a Grignard reaction.)

13-13 If tea contains 1.9% caffeine and coffee only 1.2% caffeine, explain why tea drinkers ingest less of the alkaloid in their brew.

13-14 Table 13-5 lists a number of toxic compounds obtained from plants. For each compound, give (1) the structural formula, (2) the alternate chemical name, if any, (3) known antidotes, and (4) medical or pharmacological uses. When alternate chemical names are neither found nor available, give the chemical family to which the substance belongs, e.g., alkaloid, acid, amine, glycoside, phenol. The *Merck Index* and other reference books are your best sources of information.

14
AMINO ACIDS AND PROTEINS

Proteins constitute the third and most complex of the food classes. (The other two classes are fats and carbohydrates.) *Structural substances* such as tendons, cartilage, hair, horns, nails, and feathers are proteins, as are the *enzymes* which catalyze biochemical reactions.

14-1 Nature of proteins

The proteins may be considered to be *polyamide polymers* of more than twenty naturally occurring *α-amino acids* in almost every possible variation of sequence and combination (Fig. 14-1). It was Emil Fischer, the great carbohydrate chemist, who also discovered that hydrolysis of proteins yields chiefly α-amino acids. The great variety of proteins possible from a small number of amino acids is understandable when we consider the great number of words that have been made, and may yet be made, from the 26 letters of the alphabet.

The amino acids are carboxylic acids with an amino group on the carbon atom adjacent to the carboxyl group, $RCH(NH_2)COOH$. Some proteins, upon hydrolysis, also yield a *prosthetic* or *conjugate group* of some sort in addition to the α-amino acids. They are called *conjugate proteins*. Examples are *lipoproteins,* in which the conjugate group is a fat; *glycoproteins* with a carbohydrate for the conjugate group; *nucleoproteins* with a nucleic acid for the conjugate group; and *chromoproteins* with a color-bearing group like the heme of the hemoglobin of red blood cells.

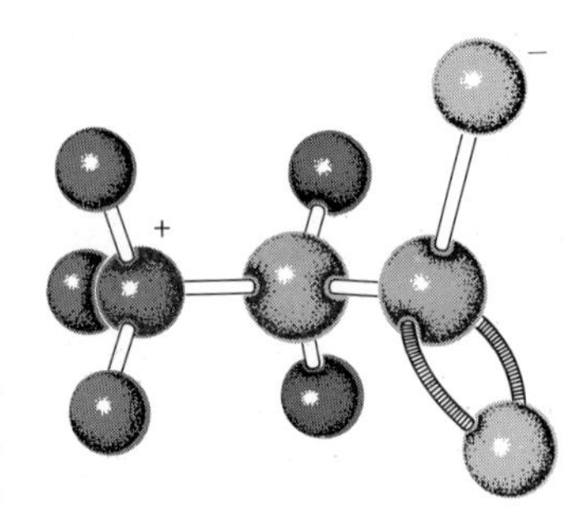

FIGURE 14-1
A ball-and-spring model of glycine, the simplest α-amino acid.

FIGURE 14-2 Configurations of representative compounds of (*a*) the L family and (*b*) the D family. The naturally occurring sugars belong to the D family, but the naturally occurring α-amino acids belong to the L family.

(*a*)

```
    H—C=O
      |
 HO—C—H
      |
     CH2OH
```
L-(−)-Glyceraldehyde

```
    H—C=O
      |
 HO—C—H
      |
  H—C—OH
      |
 HO—C—H
      |
 HO—C—H
      |
     CH2OH
```
L-(−)-Glucose

```
        O
       //
      C
      | \
      |  OH
 H2N—C—H
      |
      R
```
L-α-Amino acid

(*b*)

```
  H—C=O
     |
  H—C—OH
     |
    CH2OH
```
D-(+)-Glyceraldehyde

```
   H—C=O
      |
   H—C—OH
      |
 HO—C—H
      |
   H—C—OH
      |
   H—C—OH
      |
     CH2OH
```
D-(+)-Glucose

```
       O
      //
     C
     | \
     |  OH
  H—C—NH2
     |
     R
```
D-α-Amino acid

14-2 α-Amino acids

Whereas the common sugars and their derivatives belong to the D series, the common α-amino acids belong to the L series (Fig. 14-2).

Eight of the α-amino acids cannot be produced in sufficient quantity by the human body. They must be included in the diet, and so they are called *essential amino acids*. The others are called nonessential amino acids. It must be stressed that the word "nonessential" applies *only* to the dietary requirement.

Structure

All the α-amino acids are similar in structure and may be represented by the general structural formula

```
        O
       //
      C                         R
      | \                       |     O
      |  OH                     |   //
 H2N—C—H        or     H2N—C—C
      |                         |    \
      R                         H     OH
```
L-α-Amino acid

Table 14-1 lists the 20 common L-α-amino acids. They differ only in the nature of the R group. The essential amino acids are marked with an asterisk.

In examining Table 14-1, note that proline differs from the general structure. Its side chain is part of a *pyrrolidine ring* containing the amino group. It is a secondary amine rather than a primary amine as all of the other amino acids are.

TABLE 14-1
L-α-Amino acids

Structural formula	Common name	Abbreviation
	Nonpolar R Groups	
$H_2N-C(H)(H)-COOH$	Glycine	Gly
$H_2N-C(CH_3)(H)-COOH$	Alanine	Ala
$H_2N-C(CH(H_3C)(CH_3))(H)-COOH$	Valine*	Val
$H_2N-C(CH_2-CHCH_3(CH_3))(H)-COOH$	Leucine*	Leu
$H_2N-C(CH(CH_3)CH_2CH_3)(H)-COOH$	Isoleucine*	Ile
$H_2N-C(CH_2C_6H_5)(H)-COOH$	Phenylalanine*	Phe
$HN-C(H)-COOH$, ring: $HN-H_2C-C(H_2)-CH_2-C$	Proline	Pro
	Polar but Neutral R Groups	
$H_2N-C(CH_2OH)(H)-COOH$	Serine	Ser
$H_2N-C(HC(CH_3)-OH)(H)-COOH$	Threonine*	Thr

TABLE 14-1 (Continued)

Structural formula	Common name	Abbreviation
$H_2N-C(H)(CH_2-CH_2-S-CH_3)-COOH$	Methionine*	Met
$H_2N-C(H)(CH_2SH)-COOH$	Cysteine	Cys
$H_2N-C(H)(CH_2\text{-indolyl (N-H)})-COOH$	Tryptophan*	Trp
$H_2N-C(H)(CH_2-C(=O)-NH_2)-COOH$	Asparagine	Asn
$H_2N-C(H)(CH_2CH_2C(=O)-NH_2)-COOH$	Glutamine	Gln
	Acidic or Basic R Groups	
$H_2N-C(H)(CH_2COOH)-COOH$	Aspartic acid	Asp
$H_2N-C(H)(CH_2CH_2COOH)-COOH$	Glutamic Acid	Glu
$H_2N-C(H)(CH_2-C_6H_4-OH)-COOH$	Tyrosine	Tyr
$H_2N-C(H)((CH_2)_4NH_2)-COOH$	Lysine*	Lys

(*table continued on next page*)

TABLE 14-1 (Continued)

Structural formula	Common name	Abbreviation
$H_2N-\underset{H}{\overset{(CH_2)_3NH-\overset{NH}{\overset{\Vert}{C}}-NH_2}{C}}-COOH$	Arginine	Arg
(imidazol-4-yl, N–H)$-CH_2$ attached to $H_2N-\underset{H}{C}-COOH$	Histidine	His

The older literature may indicate more than twenty naturally occurring amino acids. Derivatives of the 20 amino acids listed in Table 14-1 are sometimes found in proteins. Examples are hydroxyproline and hydroxylysine. They can, however, be formed by enzyme-catalyzed air-oxidation *after* the proline and lysine have been incorporated into the protein. Also, two cysteine molecules may be oxidized to produce cystine, which is sometimes listed as an amino acid.

$$2H_2N-\underset{CH_2-SH}{\overset{COOH}{C}}-H \overset{(O)}{\rightleftharpoons} H_2N-\underset{CH_2S}{\overset{COOH}{C}}-H \quad H_2N-\underset{S-CH_2}{\overset{COOH}{C}}-H + (2H) \tag{14-1}$$

Cysteine — Cystine

Essential amino acids. Note that the eight essential amino acids are valine, leucine, isoleucine, phenylalanine, threonine, methionine, tryptophan, and lysine. Most proteins in the human diet contain more than an adequate amount of them, and animal experiments suggest that they may be toxic in excessive amounts. The U.S. Food and Drug Administration has therefore forbidden their addition to food products as supplements.

The three classes of common L-α-amino acids. The amino acids are grouped into three classes based on the nature of the R groups. The first class consists of the amino acids with *nonpolar* R groups. It includes the first seven amino acids in Table 14-1: glycine, alanine, valine, leucine, isoleucine, phenylalanine, and proline. The second class includes amino acids that have *polar but neutral* R groups. The next seven amino acids in Table 14-1 are in that class: serine, threonine, methionine, cysteine, tryptophan, asparagine, and glutamine. The last six amino acids contain *acidic or basic* R groups. Those acids—aspartic acid, glutamic acid, tyrosine, lysine, arginine, and histidine—comprise the third class.

Asparagine and glutamine are the *monoamides* of aspartic acid and glutamic acid, respectively. Examine the formulas in Table 14-1 to satisfy yourself that that is so.

Glycine. Glycine, the first amino acid listed in Table 14-1, is α-aminoacetic acid. It has hydrogen in place of an R group, and so it does not have a chiral center. For that reason, it is not optically active as all the other α-amino acids are.

Glycine constitutes the major component in *fibroin* (the protein of silk), in *keratin* (the protein of hair), and in *collagen* (the protein of bones and connective tissue) (Fig. 14-3). The gelatin used in emulsions for production of photographic films and plates is obtained by extracting collagen from by-products of meat packing plants. Gelatin which may not be of

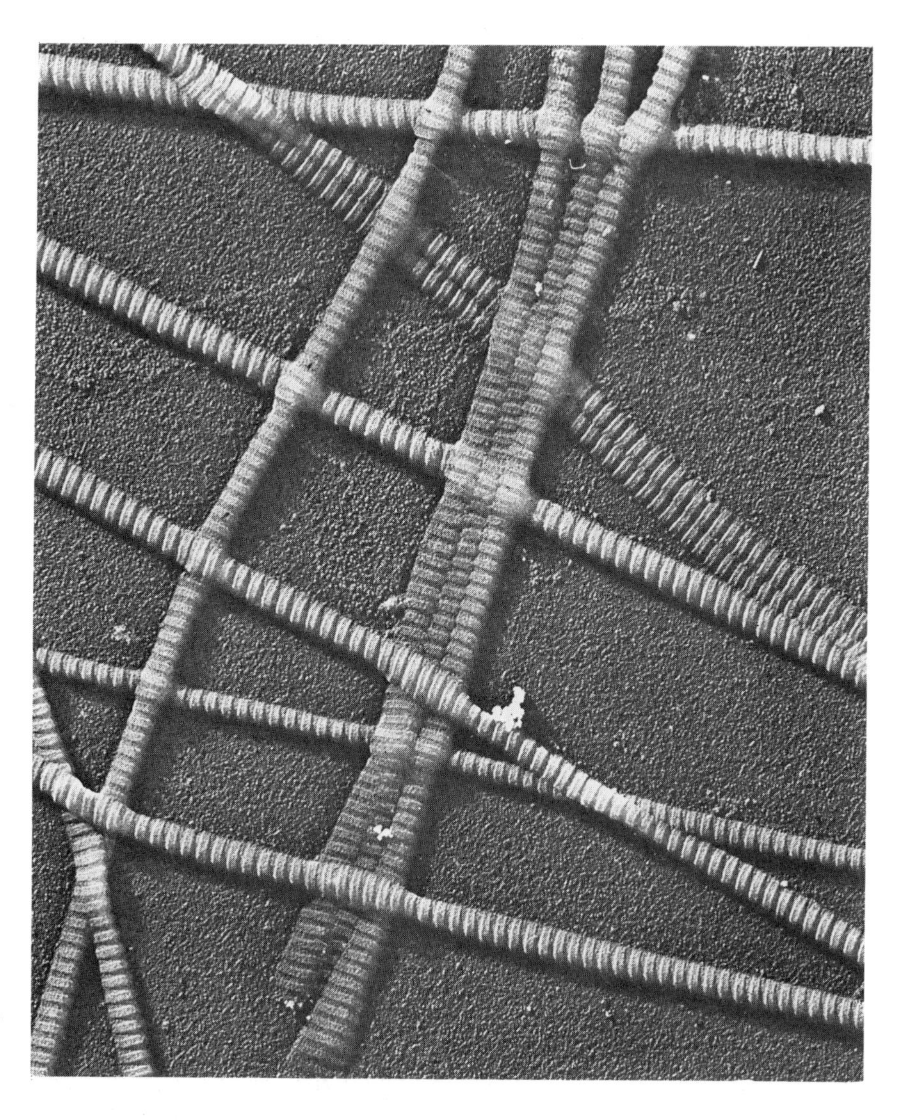

FIGURE 14-3
Collagen fibrils. ×60,000.
(*Courtesy of Dr. Jerome Gross*)

sufficiently high quality for that use may be converted to nutritious gelatin desserts by the addition of sugar, flavoring, and coloring. Collagen that has become too hydrolyzed for good gelatin may be used for making marshmallows. Even the collagen of poorest quality can be used—for making glue.

Glutamic acid. The monosodium salt of glutamic acid has long been used as a meat flavoring agent and flavor enhancer. It is known commercially as MSG or Accent. The formula is

$$\begin{array}{c} CH_2CH_2COONa \\ | \\ H_2N-C-COOH \\ | \\ H \end{array}$$

Monosodium glutamate

Chinese food in the United States often has a lot of MSG in it. MSG was once used in baby food. It is no longer so used because dosages no more than five times greater than that a human infant would receive in baby food were found to produce brain damage in infant mice, rats, rabbits, and monkeys.

Phenylalanine. Phenylalanine is of special interest because of a hereditary disease called *phenylketonuria,* or PKU. The disease is caused by a defective gene resulting in the absence of an enzyme needed for normal metabolism of phenylalanine to tyrosine. In the alternate metabolic pathway, phenylalanine is converted into phenylpyruvic acid, $C_6H_5CH_2\overset{\overset{O}{\|}}{C}-COOH$. The phenylpyruvic acid builds up in the blood until eventually it passes through the kidneys into the urine. That produces the condition called phenylketonuria.

If PKU is not discovered shortly after birth, the phenylketonuric infant will become permanently mentally retarded because part of the brain will fail to develop properly. A simple urine test (black color with $FeCl_3$) or, better, a blood test taken 4 or 5 days after birth will detect PKU. If the test is positive, the infant can be given a special diet especially low in phenylalanine until the critical period of rapid brain growth and development is past.

Properties of the amino acids

Since they have both amino groups and carboxyl groups, the amino acids are at once both basic and acidic; that is, they are *amphoteric.* Moreover, the two functional groups tend to react with each other to form intramolecular salts or dipolar ions called *zwitterions.*

$$\underset{\text{Amino acid}}{\begin{array}{c} R \\ | \\ H_2N-C-COOH \\ | \\ H \end{array}} \rightleftharpoons \underset{\text{Amino acid as a zwitterion}}{\begin{array}{c} R \\ | \\ {}^{+}H_3N-C-COO^{-} \\ | \\ H \end{array}} \qquad (14\text{-}2)$$

That accounts for their relatively high melting points (over 200°C) and water solubility. Amino acids with one amino group and one carboxyl group react with water to give a solution with a pH of about 6. Ions are in equilibrium as follows:

$$\underset{\text{A cation}}{\overset{+}{H_3N}-\underset{H}{\overset{R}{C}}-COOH} + OH^- \underset{-H_2O}{\overset{+H_2O}{\rightleftharpoons}} \underset{\text{A zwitterion}}{\overset{+}{H_3N}-\underset{H}{\overset{R}{C}}-COO^-} \underset{-H_2O}{\overset{+H_2O}{\rightleftharpoons}} \underset{\text{An anion}}{H_2N-\underset{H}{\overset{R}{C}}-COO^-} + H_3O^+ \qquad (14\text{-}3)$$

Those equations indicate that, in order for the pH to be about 6 (slightly acidic), there must be a higher concentration of H_3O^+ ions than OH^- ions at equilibrium and an attendant higher concentration of the organic anion. To maximize production of the zwitterion, a small amount of an aqueous proton acid must be added to repress the reaction to the right and increase the reaction to the left. The resulting pH is called the *isoelectric point* because the positive and negative charges on the zwitterion are then in balance. In an electric field the tendency for migration to either electrode is at a minimum. The exact isoelectric point is characteristic for each amino acid. The amino acids with one amino group and one carboxyl group have isoelectric points in the range 5.02 to 6.53.

Two of the naturally occurring amino acids (aspartic acid and glutamic acid) have two carboxyl groups for one amino group. Such amino acids are more acidic and have isoelectric points in the range 2.87 to 3.22. Three amino acids (lysine, arginine, and histidine) have two amino groups for one carboxyl group. Such basic amino acids have isoelectric points in the range 7.58 to 10.76. It follows that amino acids are more soluble in either acid or base and are least soluble at their isoelectric points.

$$\underset{\substack{\text{Anion}\\\text{(more soluble)}}}{H_2N-\underset{H}{\overset{R}{C}}-COO^-} \underset{OH^-}{\overset{H^+}{\rightleftharpoons}} \underset{\substack{\text{Zwitterion}\\\text{(less soluble)}}}{\overset{+}{H_3N}-\underset{H}{\overset{R}{C}}-COO^-} \underset{OH^-}{\overset{H^+}{\rightleftharpoons}} \underset{\substack{\text{Cation}\\\text{(more soluble)}}}{\overset{+}{H_3N}-\underset{H}{\overset{R}{C}}-COOH} \qquad (14\text{-}4)$$

The isoelectric point is of importance in isolating an amino acid because the amino acid can best be extracted or precipitated out of an aqueous solution when the pH is adjusted to the isoelectric point of the amino acid.

Synthesis of α-amino acids

Many methods have been developed to synthesize the various α-amino acids. One of the more general ones is the reaction of an α-halogen acid with excess ammonia:

$$\mathrm{Br{-}\overset{\displaystyle R}{\underset{\displaystyle H}{C}}{-}COOH} + \underset{\text{(Excess)}}{\mathrm{NH_3}} \xrightarrow{(S_N2)} \mathrm{H_2N{-}\overset{\displaystyle R}{\underset{\displaystyle H}{C}}{-}COOH} + \mathrm{NH_4Br} \qquad (14\text{-}5)$$

Another method is the reaction of an aldehyde with ammonia, hydrogen cyanide, and heat followed by acid hydrolysis. It is called the *Strecker synthesis*.

$$\mathrm{R{-}\overset{\displaystyle H}{C}{=}O} \xrightarrow{NH_3,\ HCN,\ \Delta} \mathrm{H_2N{-}\overset{\displaystyle R}{\underset{\displaystyle H}{C}}{-}CN} \xrightarrow{H_2O,\ H^+,\ \Delta} \mathrm{H_2N{-}\overset{\displaystyle R}{\underset{\displaystyle H}{C}}{-}COOH} \qquad (14\text{-}6)$$

In either synthesis, the product amino acid is obtained as a racemic mixture. Some specific examples are

$$\underset{\alpha\text{-Bromopropionic acid}}{\mathrm{Br{-}\overset{\displaystyle CH_3}{CH}{-}COOH}} + \mathrm{NH_3} \xrightarrow{(S_N2)} \underset{\text{Racemic alanine}}{\mathrm{H_2N{-}\overset{\displaystyle CH_3}{CH}{-}COOH}} \qquad (14\text{-}7)$$

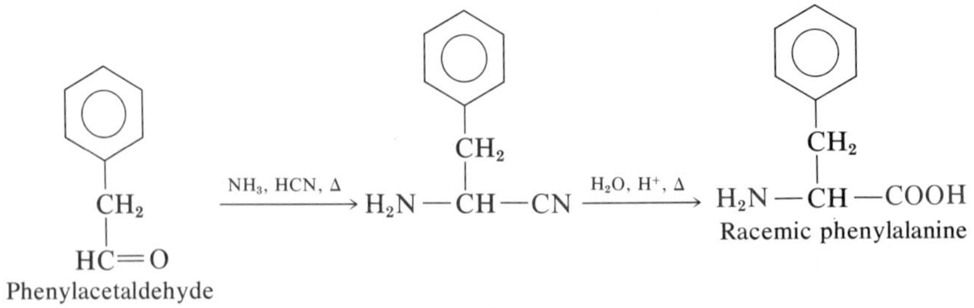

$$\underset{\text{Phenylacetaldehyde}}{\mathrm{C_6H_5{-}CH_2{-}HC{=}O}} \xrightarrow{NH_3,\ HCN,\ \Delta} \mathrm{H_2N{-}\overset{\displaystyle CH_2C_6H_5}{CH}{-}CN} \xrightarrow{H_2O,\ H^+,\ \Delta} \underset{\text{Racemic phenylalanine}}{\mathrm{H_2N{-}\overset{\displaystyle CH_2C_6H_5}{CH}{-}COOH}} \qquad (14\text{-}8)$$

PROBLEM 14-1 Write equations for the following reactions:

(a) alanine with nitrous acid
(b) glutamic acid with 1 equivalent of $NaHCO_3$
(c) leucine with hydrochloric acid
(d) tyrosine with 2 equivalents of acetic anhydride

Solution to (a)

$$\underset{\substack{\text{Alanine}\\ \text{(A primary amide)}}}{\mathrm{H_2N{-}\overset{\displaystyle CH_3}{\underset{\displaystyle H}{C}}{-}COOH}} + \mathrm{NaNO_2} + \mathrm{H_3O^+} \longrightarrow \underset{\substack{\text{2-Hydroxypropanoic acid}\\ \text{(Lactic acid)}}}{\mathrm{HO{-}\overset{\displaystyle CH_3}{\underset{\displaystyle H}{C}}{-}COOH}} + \mathrm{N_2}(g) + 2\mathrm{H_2O} + \mathrm{Na^+}$$

PROBLEM 14-2 If the pH of its solution is 6.00, indicate whether each of the following amino acids will be present as a zwitterion, an anion, or a cation:

(a) aspartic acid
(b) lysine
(c) leucine
(d) histidine
(e) proline
(f) threonine

Solution to (a) The isoelectric point for aspartic acid, which has two carboxyl groups, is in the range of 2.87 to 3.22. That indicates a considerable amount of H_3O^+ must be added to suppress the reaction to the right in Eq. (14-3) to the point at which the zwitterion is present. At the higher pH of 6, the aspartic acid would be present as an anion. Both carboxyl groups will be ionized, and the amino group will be ionized with a positive charge to offset one of the negative charges.

$$^{-}O(O{=})C-CH_2-CH(^{+}NH_3)-C(=O)O^{-}$$

PROBLEM 14-3 Write equations for the preparation of the following amino acids from the starting materials indicated: (a) valine from 2-methylpropanal (isobutyraldehyde), (b) aspartic acid from α-chlorosuccinic acid.

Solution to (a)

$$\underset{\text{2-Methylpropanal}}{H_3C-CH(CH_3)-CH{=}O} \xrightarrow{NH_3,\ HCN,\ \Delta} H_2N-CH(CH(CH_3)_2)-CN \xrightarrow{H_2O,\ H^+,\ \Delta} \underset{\text{Valine}}{H_2N-CH(CH(CH_3)_2)-COOH}$$

14-3 Peptides and the peptide bond

At the beginning of the chapter it was mentioned that Emil Fischer discovered that hydrolysis of proteins yields α-amino acids. The acids are considered to be connected to each other through *amide linkages* to form *polyamides*. The special amide linkage of the carboxyl group of one α-amino acid to the amino group of another α-amino acid, with the resultant elimination of a molecule of water, is called a *peptide bond*.

$$\underset{\alpha\text{-Amino acid}}{H_2N-CHR-C(=O)OH} + \underset{\alpha\text{-Amino acid}}{H_2N-CHR'-C(=O)OH} \longrightarrow H_2O + \underset{\text{Dipeptide}}{H_2N-CHR-\overbrace{C(=O)-NH}^{\text{Peptide bond}}-CHR'-C(=O)OH} \quad (14\text{-}9)$$

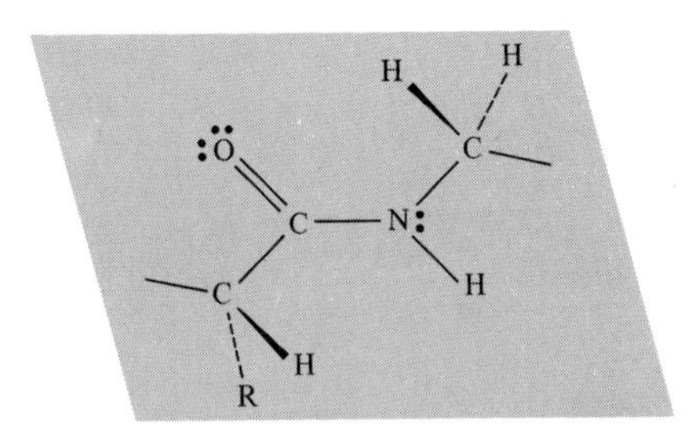

FIGURE 14-4
The planar geometry of the peptide bond.

The linkage cannot actually be made directly as shown; the equation is intended only to show the relation between the free amino acids and the peptide.

Geometry of the peptide bond

In the late 1930s, by use of x-ray crystallographic studies, Linus Pauling and his collaborators discovered that the peptide bond is *planar*. The four atoms of the peptide bond and the two carbon atoms joined to the peptide bond lie in the same plane. The bond angles about the carbonyl carbon atom and about the amide nitrogen atom are approximately 120°. See Fig. 14-4.

The expected orbital hybridization on nitrogen is sp^3 with 109.5° bond angles, and not 120° bond angles. To account for the actual observed geometry, Pauling proposed that the peptide bond is a resonance hybrid of two important contributing structures (Fig. 14-5). In such a hybrid the carbon-nitrogen bond has partial double bond character, so the six-atom group is planar.

There are actually two possible planar arrangements of the atoms of the peptide bond: a cis and a trans (Fig. 14-6). The trans form is observed most frequently. The cis configuration is perhaps slightly less favorable because the close proximity of the substituted carbon atoms in that form causes steric hindrance (spatial crowding).

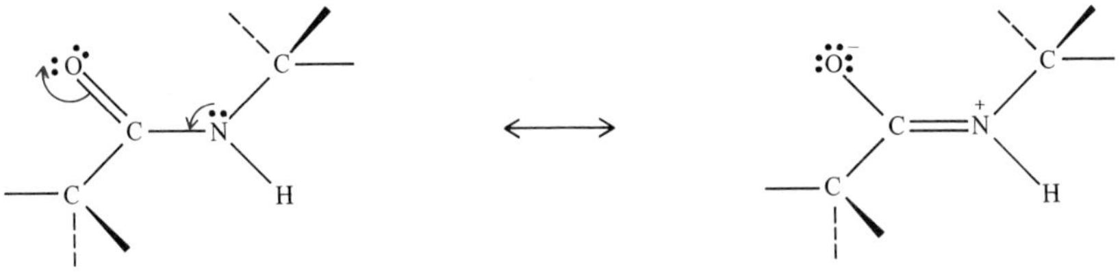

FIGURE 14-5
The resonance structures of the peptide bond.

Cis

Trans

FIGURE 14-6
The two possible planar arrangements of the atoms in the peptide bond.

Peptides Small polymers of amino acids are called *peptides*. A peptide constructed from two amino acids is called a *dipeptide;* one from three amino acids is a *tripeptide;* and one from four amino acids is a *tetrapeptide*. A peptide constructed from ten or more amino acids is usually called a *polypeptide*. The term "peptide" is generally reserved for naturally occurring polyamides of molecular weights less than 10,000. *Proteins* are polyamides of molecular weight over 10,000.

Peptides are named by combining the names (or abbreviations) of the individual amino acids beginning with the amino acid on the end with the free amino group (N-*terminal amino acid*), and ending with the amino acid on the other end possessing the free carboxyl group (C-*terminal amino acid*). An example of a tripeptide is valylhistidylleucine, which can be obtained by hydrolysis of hemoglobin.

```
                                   HN⌒N (imidazole ring)
                                     |                 CH3CHCH3
      CH3CHCH3   H   CH2  O                               |
          |      |    |   ||                             CH2
H2N — C — C — N — C — C — N — C — C — OH
          |     ||         |         |     |    ||
          H     O          H         H     H    O
(N-Terminal end)                          (C-Terminal end)
```

Valylhistidylleucine

By convention, peptides are generally written with the free terminal H_2N— group to the left and proceeding to the right toward the free terminal —COOH. A more abbreviated representation uses the standard abbreviations of the amino acids, each connected by an arrow.

$$\text{Val} \longrightarrow \text{His} \longrightarrow \text{Leu}$$

The tail of the arrow indicates the amino acid contributing the carboxyl group to the peptide bond, and the head of the arrow indicates the amino acid contributing the amino group to the peptide bond.

Some peptides are biologically important in their own right. Oxytocin is a hormone which stimulates uterine contraction. Vasopressin is involved with water balance in the body and can be used as a diuretic. Loss of ability to produce the hormone leads to diabetes insipidus. See Fig. 14-7.

FIGURE 14-7 Two peptide hormones. The amino acids shown in color are the only differences in the two molecules. The carboxyl terminus in each molecule is glycinamide.

```
            NH2
    ↙ Tyr ↖ /
Ile          Cys —— S
 ↓                  |
Gln          Cys —— S
  ↘ Asn ↗   ↘ Pro ⟶ Leu ⟶ Gly —— NH2
```

Bovine oxytocin

```
            NH2
    ↙ Tyr ↖ /
Phe          Cys —— S
 ↓                  |
Gln          Cys —— S
  ↘ Asn ↗   ↘ Pro ⟶ Arg ⟶ Gly —— NH2
```

Bovine vasopressin

Amino acid sequence

A dipeptide constructed from two given amino acids can exist in two possible sequences of the amino acids. For example, a dipeptide constructed from serine and tyrosine could be seryltyrosine or tyrosylserine. Likewise a tripeptide, such as the one constructed from valine, histidine, and leucine, can have the sequence varied to produce six isomeric tripeptides.

$$\begin{array}{ll} \text{Val} \longrightarrow \text{His} \longrightarrow \text{Leu} & \text{Leu} \longrightarrow \text{His} \longrightarrow \text{Val} \\ \text{His} \longrightarrow \text{Val} \longrightarrow \text{Leu} & \text{Leu} \longrightarrow \text{Val} \longrightarrow \text{His} \\ \text{His} \longrightarrow \text{Leu} \longrightarrow \text{Val} & \text{Val} \longrightarrow \text{Leu} \longrightarrow \text{His} \end{array}$$

The number of possible isomers of a polypeptide containing one each of 20 different amino acids runs to $20 \times 19 \times 18 \times \cdots \times 2 \times 1$, or about 2×10^{18}. The number of possible arrangements of larger polypeptides and proteins becomes stupendous.

It is truly remarkable that, when there are so many possible combinations of amino acids, cells produce identical protein molecules generation after generation. The cell's ability to produce the same proteins as its parent produced is vital to life; for if a protein molecule is to fulfill its physiological function, its amino acids must be ordered in a particular, definite sequence. The substitution of even one amino acid in a large protein molecule can completely alter the molecule's biological function. That is what has happened in the genetic disease *sickle-cell anemia*. The hemoglobin of the afflicted individual contains valine where it should have glutamic acid. The substitution occurs in two identical polypeptide chains of the four chains comprising the hemoglobin molecule. Each chain contains 146 amino acids. The seemingly insignificant alteration causes a drastic change in the properties of the molecule: it causes the red blood cells to assume a sickle shape instead of the normal disk shape (Fig. 14-8). The carrying of oxygen is thereby impaired.

$$\begin{array}{lcccccccc} \text{Normal hemoglobin:} & \text{Val} \rightarrow & \text{His} \rightarrow & \text{Leu} \rightarrow & \text{Thr} \rightarrow & \text{Pro} \rightarrow & \text{Glu} \rightarrow & \text{Lys} \rightarrow \cdots \\ & 1 & 2 & 3 & 4 & 5 & 6 & 7 \\ \text{Sickle-cell hemoglobin:} & \text{Val} \rightarrow & \text{His} \rightarrow & \text{Leu} \rightarrow & \text{Thr} \rightarrow & \text{Pro} \rightarrow & \text{Val} \rightarrow & \text{Lys} \rightarrow \cdots \end{array}$$

In determining the amino acid sequence of a peptide, one might first determine which of the amino acids is *N*-terminal. To do so, the peptide is

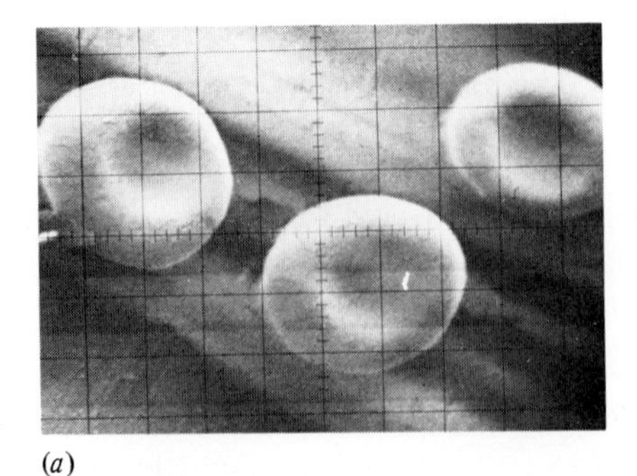

(*a*)

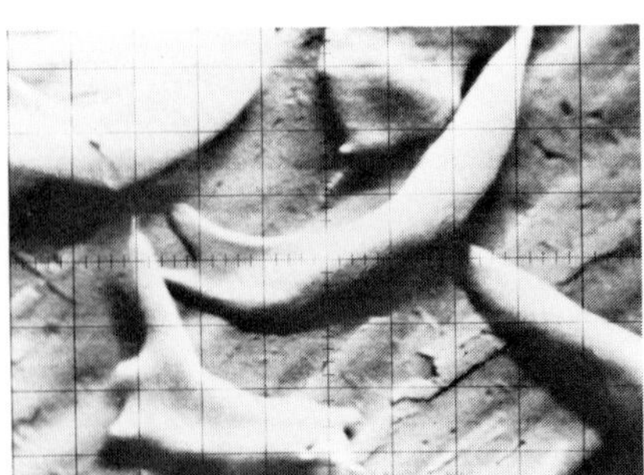

(*b*)

FIGURE 14-8
Scanning electron micrographs of (*a*) normal human red blood cells and (*b*) sickle cells. Note difference in shape which affects the oxygen-carrying ability of the cells. (*Philips Electronics Instruments*)

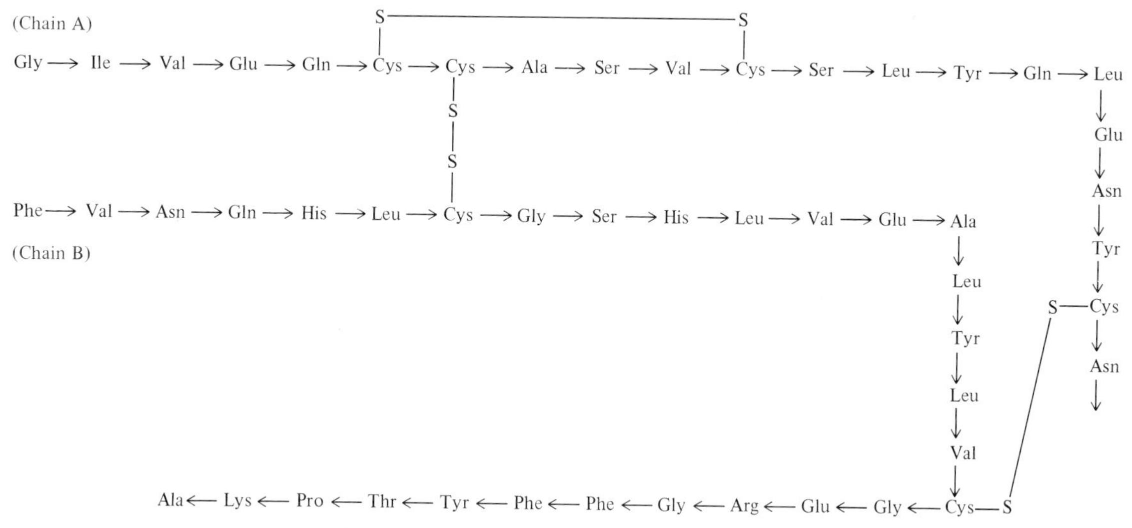

FIGURE 14-9 The amino acid sequence of bovine insulin.

treated with a reagent that combines specifically with free amino groups. One way of proceeding is to treat the peptide with 2,4-dinitrofluorobenzene. Reaction takes place by an aromatic nucleophilic substitution to produce a yellow 2,4-dinitrophenyl derivative. Upon hydrolysis, that yields a 2,4-dinitrophenyl derivative of the *N*-terminal amino acid, which can then be characterized.

The *C*-terminal amino acid may be identified by reduction with lithium borohydride ($LiBH_4$), which reduces the free carboxyl group to a primary alcohol group. Subsequent hydrolysis and chromatography permits identification of the resulting amino alcohol and hence of the amino acid from which it came.

A number of enzymatic and chemical methods have been developed to degrade a peptide partially so that smaller peptides are formed without total hydrolysis into amino acids. The sequence of amino acids in the smaller peptides, especially tripeptides, is determined. Then the sequence of amino acids in the original polypeptide can be reconstructed by the overlapping of amino acids in the small peptide fragments.

One of the first peptides to have its amino acid sequence determined was insulin, which has 51 amino acids (Fig. 14-9). For the achievement Frederick Sanger, of Cambridge, England, was awarded a Nobel prize in 1958. Since that time the primary structures of many polypeptides and proteins have been determined. Gerald Edelman and his coworkers at Rockefeller University determined the amino acid sequence of the four polypeptide chains of a gamma globulin molecule which contains 1320 amino acids with 19,996 atoms and a molecular weight of 150,000.

14-4 Structure of proteins

The different types of proteins result from differences not only in the primary structure but also in the secondary and tertiary structures.

Primary structure

The *primary structure* of a protein is the amino acid content and sequence. Amino acid content is determined by hydrolysis followed by amino acid analysis of the hydrolysate. That is now accomplished by use of an automatic amino acid analyzer developed by Stanford Moore and William H. Stein (Rockefeller University, shared Nobel Prize in Chemistry, 1972). The instrument involves ion-exchange column chromatography.

Separation of a mixture of substances based on their differences in adsorption by an adsorbent is called *chromatography*. In column chromatography, a solution of a mixture is passed through a column containing an adsorbent. The components of the mixture are adsorbed at different rates as the solution passes through the column. The separation is completed by the differential removal of the components from the column by use of different solvents. One version of the amino acid analyzer can use

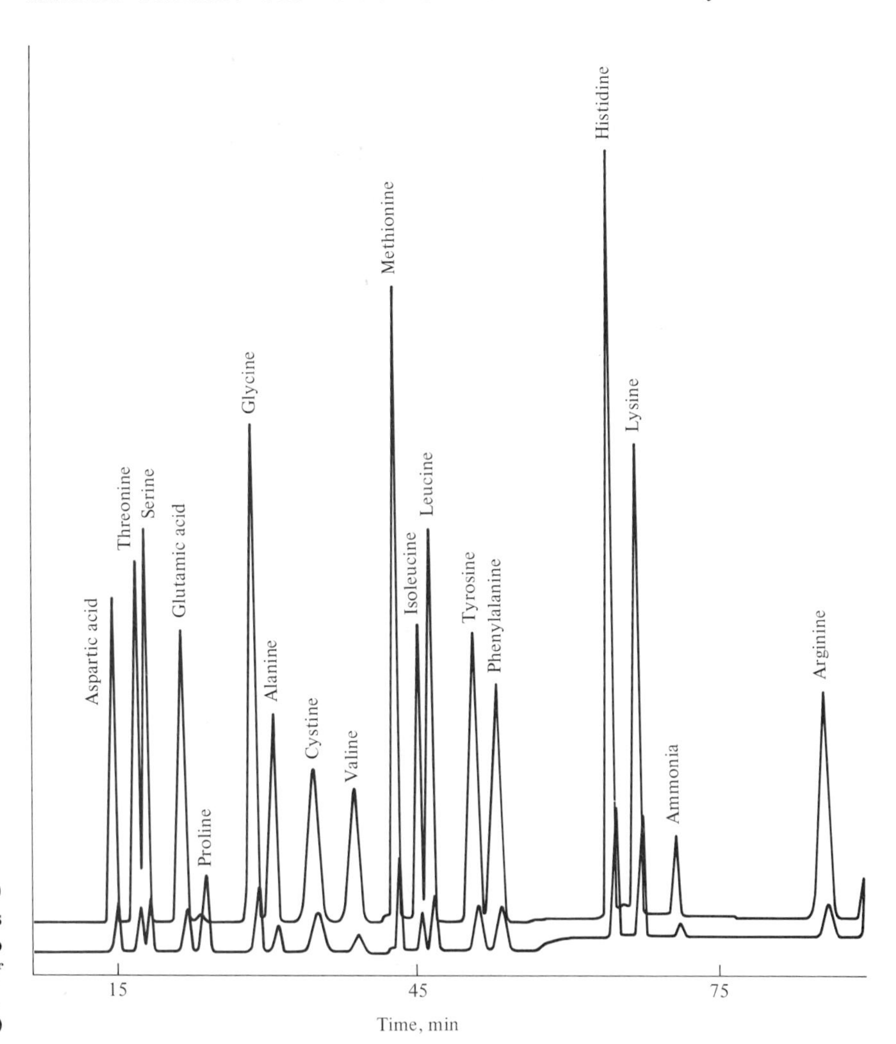

FIGURE 14-10
The record of amino acids produced by a modern amino acid analyzer. (*Courtesy of Beckman Instruments, Inc., Spinco Division.*)

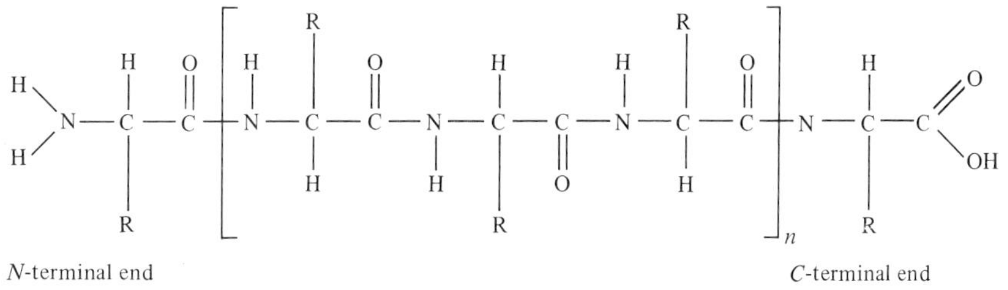

N-terminal end

C-terminal end

(*a*)

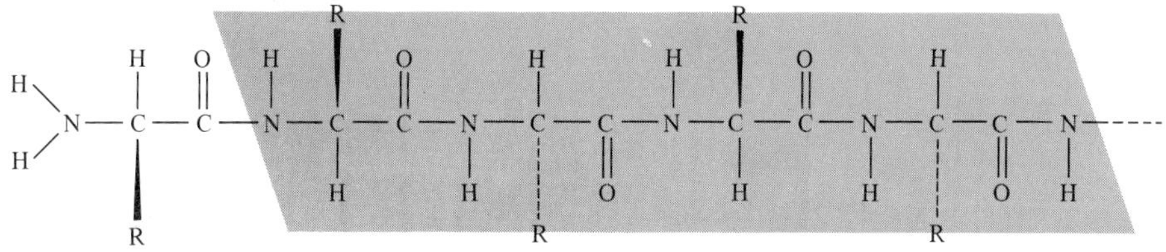

(*b*)

FIGURE 14-11 The primary structure of a protein. (*a*) The primary structure is the protein's backbone consisting of the peptide bonds in proper sequence. (*b*) The peptide bonds are located in a plane with the R groups above and below the plane and trans relative to each other. (*Adapted by permission from H. L. Helmprecht and L. T. Friedman,* **Basic Chemistry for the Life Sciences,** *McGraw-Hill Book Company, New York, 1977, p. 386.*)

samples as small as 1 mg and can perform as many as 72 consecutive analyses without supervision. A single analysis may be completed in 2 h with an accuracy of 3%. (See Fig. 14-10.)

After the amino acid content has been found, the sequence of the amino acids in the molecule must be determined. The methods of doing so were discussed in Sec. 14-3. The primary structure of a protein is shown in Fig. 14-11.

Secondary structure

Because of the possibility of free rotation about their single bonds, protein molecules could assume an almost infinite variety of shapes. They do not, however. It is known from a variety of physical measurements that each protein occurs in nature in a single particular three-dimensional conformation. *The secondary structure of a protein is the fixed configuration of the polypeptide chain.* It involves not only the manner in which the peptide chain is folded and bent but also the hydrogen bonds which stabilize or hold the special structure together.

The two principal secondary structures for proteins have been designated alpha (α) and beta (β). The *alpha structure* was first postulated by Linus Pauling (Nobel Prize for Chemistry, 1954) and Robert Corey as being a *helix* like a spring coiled about an imaginary cylinder. Although the helix could conceivably be either left- or right-handed, the L configuration of the natural amino acids favors the right-handed helix. The helix

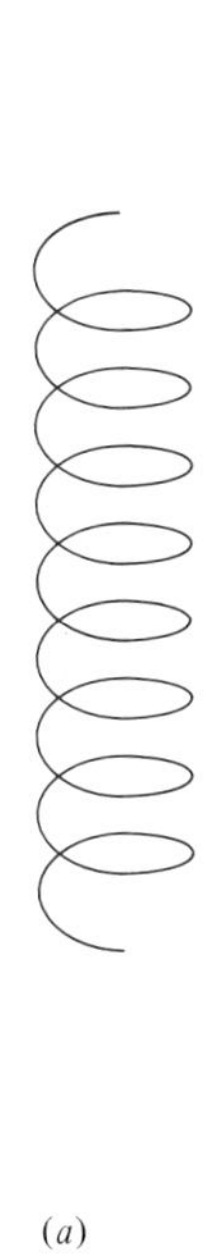

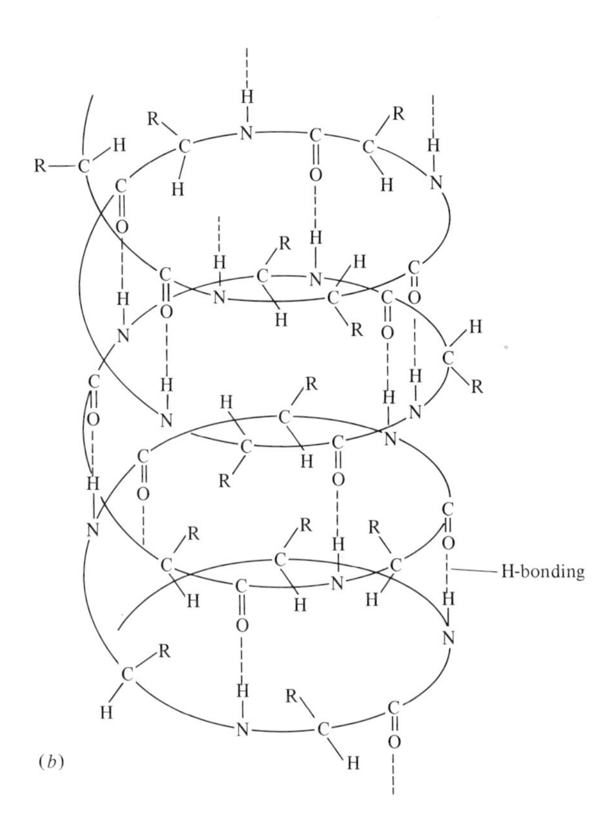

FIGURE 14-12
(*a*) The alpha helix. (*b*) A section of an alpha helix structure of a protein. Hydrogen bonds stabilize the alpha helix, and the R groups protrude out of the helix. (*Adapted by permission from H. L. Helmprecht and L. T. Friedman,* Basic Chemistry for the Life Sciences, *McGraw-Hill Book Company, New York, 1977, p. 387.*)

is held together by hydrogen bond formation between the amide hydrogen of one peptide bond and the carboxyl oxygen which is located on the next turn of the helix. X-ray data indicate that each turn of the helix consists of about four amino acid units with the side chains of the units projecting outward from the coiled peptide backbone. See Fig. 14-12.

The *beta structure* consists of polypeptide chains bonded to *adjacent* polypeptide chains with *interpolypeptide hydrogen bonds* (a form of polymeric cross-linking) in such a fashion as to resemble a pleated sheet (Fig. 14-13). Silk fibroin and β-keratin (hair) have this structure.

Tertiary structure

The *tertiary structure* is the unique three-dimensional shape which results from the precise folding and bending of a polypeptide chain or the alpha helix into which the chain may be formed. That specific shape is essential to the proper biochemical functioning of proteins. Many proteins have a very compact, almost spherical shape with their nonpolar side chains directed toward the interior of the molecule and their polar side chains projecting outward from the molecule toward the aqueous environment. The nonpolar side chains are, of course, insoluble in water; they are termed *hydrophobic* ("water-fearing") *parts*. The polar, water-soluble side chains are called *hydrophilic* ("water-loving") *parts*.

The linkages which hold the protein molecule in its tertiary shape are

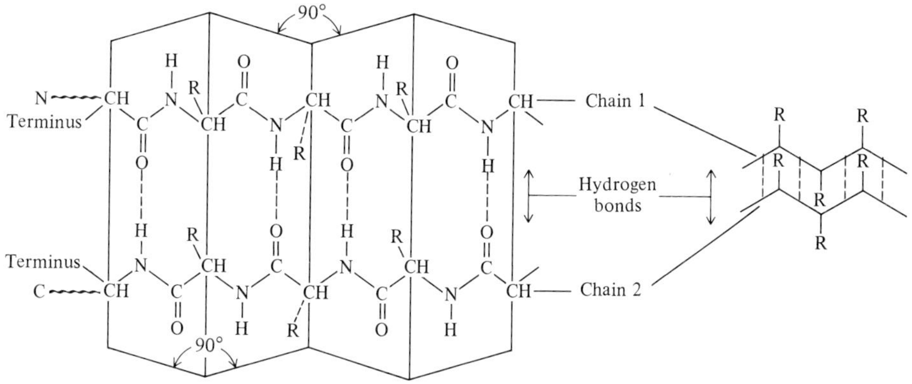

FIGURE 14-13 The beta or pleated-sheet arrangement of the polypeptide chain. All atoms in the face of the pleat are coplanar. (*Adapted by permission from H. L. Helmprecht and L. T. Friedman,* **Basic Chemistry for the Life Sciences,** *McGraw-Hill Book Company, New York, 1977, p. 387.*)

salt linkages, hydrogen bonds, disulfide linkages, and hydrophobic interactions. Figure 14-14 shows the four types.

Salt linkages result from salt formation between the amino groups of the basic amino acids and the side-chain carboxyl groups of the acidic amino acids. The resultant forces of electrostatic attraction, or ionic bonds, are quite strong. An example is shown in Fig. 14-15.

Hydrogen bonding can take place whenever a side chain having a hydrogen donor group ($>$N—H or —OH) can be in proximity to a side chain having a hydrogen acceptor group such as

$$-\ddot{\mathrm{O}}\mathrm{H} \quad \text{or} \quad >\mathrm{C}{=}\ddot{\mathrm{O}}\!: \quad \text{or} \quad -\underset{\cdot\cdot}{\overset{\mathrm{H}}{\overset{|}{\mathrm{N}}}}-$$

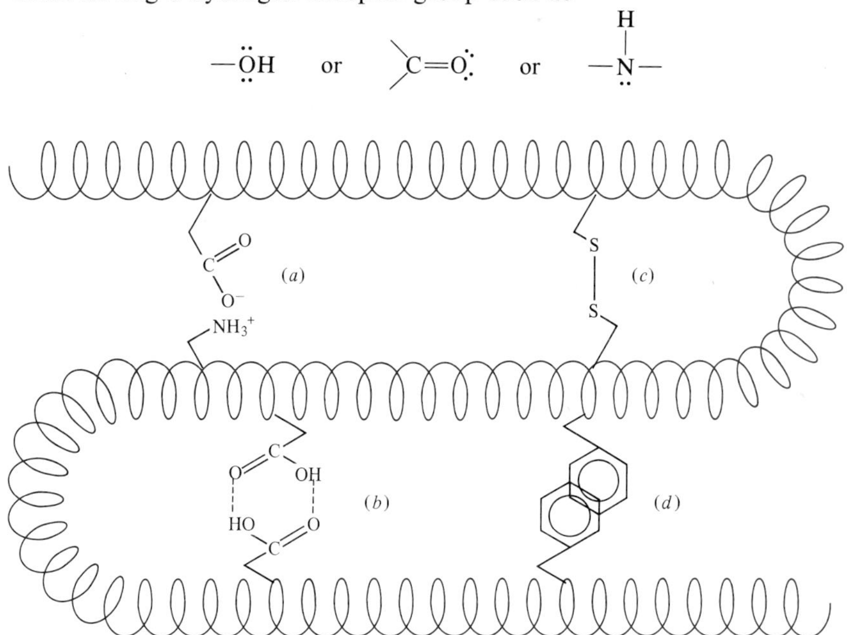

FIGURE 14-14 The tertiary structure of a polypeptide chain showing the four types of bonds that maintain the proper conformation: (*a*) salt linkages, or ionic bonds, (*b*) hydrogen bonds, (*c*) disulfide linkages, (*d*) hydrophobic interactions.

O
(CH2)2C
O−
H3N+(CH2)4
Glutamic acid
Lysine

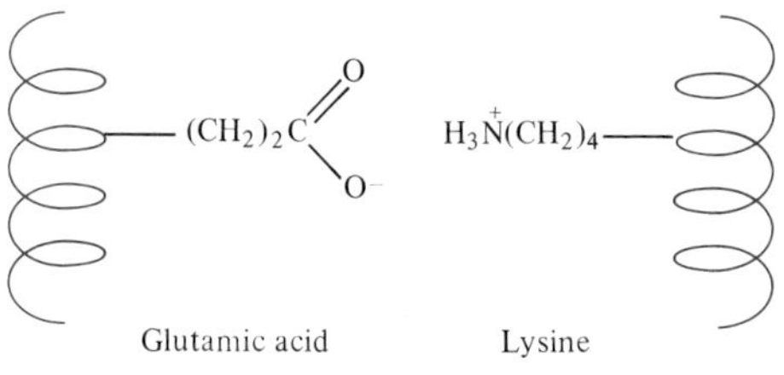

FIGURE 14-15
A salt linkage in a protein molecule.

H
C—OH----O=C—CH2
δ+
δ−
CH3
HO
Threonine
Aspartic acid

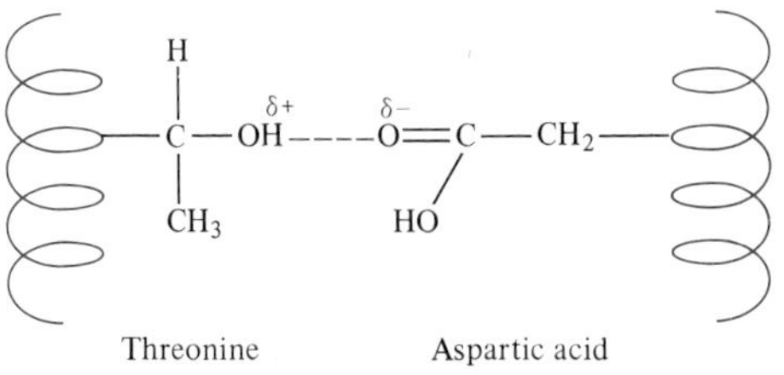

FIGURE 14-16
Hydrogen bonding in a protein molecule.

Figure 14-16 shows an example of hydrogen bonding.

Disulfide linkages are formed by the relatively easy oxidation of sulfhydryl (—SH) groups of two cysteine side chains. That is illustrated in Fig. 14-17.

Hair protein gets a great deal of its body and shape from disulfide bonds, and the chemistry of the permanent wave involves those bonds. First a reducing agent is applied to the hair to break the disulfide bonds by converting them to sulfhydryl groups. That undoes and relaxes the old configuration of the hair, which is then arranged into curls. Reoxidation either by air or by oxidizing agent reforms the disulfide bonds, which then serve to hold the hair in its new curled form.

Hydrophobic interactions are the result of the polar nature of water, the usual solvent in biological reactions, and the nonpolar (hydrophobic) groups on certain amino acids. You will recall that the R groups on the first seven amino acids in Table 14-1 are nonpolar. Such nonpolar groups in an aqueous medium have no attraction for the molecules of water, so they tend to be pushed together by the water molecules surrounding them. An example of hydrophobic interaction of phenylalanine is shown in Fig. 14-14*d*.

Quaternary structure

Some proteins consist of more than one polypeptide chain with no covalent bonds between the chains. *Quaternary structure* refers to the manner

FIGURE 14-17 A disulfide linkage in a protein molecule.

CH2SH
HSCH2
oxidation (−2H)
reduction (+2H)
CH2—S—S—CH2

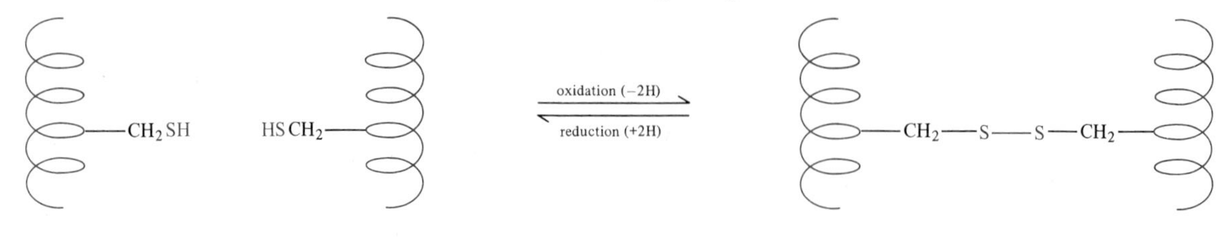

in which separate, noncovalently linked polypeptide chains are arranged in space to form molecular assemblies. Proteins of molecular weight greater than 30,000 to 50,000 are likely to be made up of more than one chain. Hemoglobin, with a molecular weight of about 65,000, consists of four polypeptide chains. Its quaternary structure is the special arrangement of the four chains.

Classes of proteins

Depending on their conformations (both their secondary and tertiary structure), proteins may be classified as fibrous or globular. *Fibrous proteins* are stringy, tough, and generally insoluble. They are composed of long, rodlike chains linked together by the four types of cross-linkages to form stable, insoluble substances. The three major classes of fibrous proteins are the *silks*, the *keratins*, which include hair, wool, skin, claws, horn, scales, and feathers, and the *collagens*, which include tendons, hides, and bones. *Globular proteins* are somewhat spherical in shape. They tend to be soluble in water and salt solutions. *Globulins* and *enzymes* belong to the globular class.

Collagen, a fibrous protein. Collagen is the major extracellular structural protein in connective tissue and bone; it constitutes more than a third of total body protein. It is distinctive in that almost one-third of all its amino acids are glycine and almost one-quarter are proline and hydroxyproline.

```
HO—HC——CH₂
    |    |
   H₂C   CH—COOH
     \  /
      N
      |
      H
```

Hydroxyproline

Both proline and hydroxyproline have the α-amino group incorporated into a five-membered ring which restrains rotation about the α-carbon atom. As a consequence of that restraint, the hydroxyproline and proline molecules cannot rotate properly to acquire the conformation necessary for the formation of an α helix.

Indeed, x-ray crystallographic studies show that collagen, unlike the silks and keratins, does not contain the α helix form. Instead, collagen has three polypeptide chains interwoven like the strands of a rope to form an extended triple helix. The amino and carbonyl groups are extended perpendicular to the axis of the polypeptide chain, and they are bonded from one chain to another. Parallel interchain hydrogen bonding makes the collagen triple helix particularly strong and rigid. Fibers are formed when many such triple helix units line up alongside each other and are cross-linked by covalent bonds. Tendons contain parallel bundles of collagen molecules.

Myoglobin, a globular protein. The primary, secondary, and tertiary structure of myoglobin was described by J. C. Kendrew and M. F. Perutz

(*a*)

10
H_2N—Val→Leu→Ser→Glu→Gly→Glu→Trp→Gln→Leu→Val→Leu→His→Val→Tyr→Ala→Lys→Val→
20 30
Glu→Ala→Asp→Val→Ala→Gly→His→Gly→Gln→Asp→Ile→Leu→Ile→Arg→Leu→Phe→Lys→
40 50
Ser→His→Pro→Glu→Thr→Leu→Glu→Lys→Phe→Asp→Arg→Phe→Lys→His→Leu→Lys→Thr→
60
Glu→Ala→Glu→Met→Lys→Ala→Ser→Glu→Asp→Leu→Lys→Gly→His→His→Glu→Ala→Glu→
70 80
Leu→Thr→Ala→Leu→Gly→Ala→Ile→Leu→Lys→Lys→Gly→His→His→Glu→Ala→Glu→
90 100
Leu→Lys→Pro→Leu→Ala→Gln→Ser→His→Ala→Thr→Lys→His→Lys→Ile→Pro→Ile→Lys→
110
Tyr→Leu→Glu→Phe→Ile→Ser→Glu→Ala→Ile→Ile→His→Val→Leu→His→Ser→Arg→His→
120 130
Pro→Gly→Asn→Phe→Gly→Ala→Asp→Ala→Gln→Gly→Ala→Met→Asn→Lys→Ala→Leu→Glu→
140 150
Leu→Phe→Arg→Lys→Asp→Ile→Ala→Ala→Lys→Tyr→Lys→Glu→Leu→Gly→Tyr→Gln→Gly—COOH

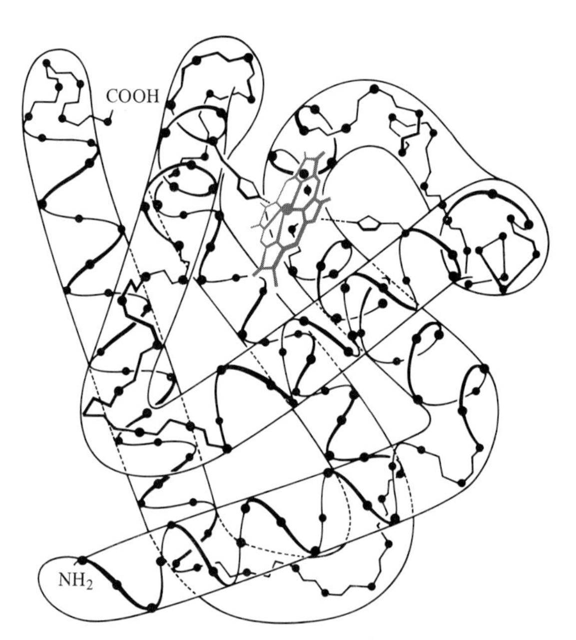

(*b*)

FIGURE 14-18 **(*a*) The primary structure of myoglobin showing the sequence of amino acids in the protein. (*b*) Representation of the structure of myoglobin showing its secondary and tertiary structure. [*Adapted by permission from R. E. Dickerson in Hans Neurath (ed)*: The Proteins, *Vol. II, 2d Ed., Academic Press, Inc., New York, 1964, p. 634.*]**

(Nobel Prize winners, 1963) with the aid of x-ray diffraction studies. Myoglobin is a protein consisting of 153 amino acid units arranged in a single chain. One of its functions is to store oxygen in muscle cells. It is especially abundant in marine diving mammals such as seals, whales, and porpoises; it enables the mammals to remain submerged for long periods of time.

The secondary structure of myoglobin involves coiling about 70% of the single chain into an α helix; the tertiary structure results from nonuniform folding of the helix to form a compact stable structure. Myoglobin is a conjugated protein with a heme moiety attached to the peptide chain. Figure 14-18 shows (*a*) the primary structure of myoglobin and (*b*) aspects of the secondary and tertiary structure as deduced from x-ray diffraction studies.

14-5 Protein denaturation agents

Denaturation is any change in which the unique three-dimensional configuration of the protein molecule is altered without any cleavage of the peptide bonds. Such disruption of the secondary and tertiary structures is accompanied by profound changes in the physical and biochemical properties of the molecule. Denaturation may be reversible (an example is the denaturation of hair in the chemical permanent-waving process) or irreversible. An irreversible denaturation occurs whenever the elastic limit of the α helix is exceeded and the shape is then irreversibly altered. A number of reagents and conditions denature proteins.

Heat and radiation

Heat and ultraviolet radiation provide kinetic energy to cause excessive vibration of the atoms. As a result, hydrogen bonds and salt linkages are broken. The special arrangement of polar and nonpolar groups is destroyed, and *coagulation* occurs. An example is the cooking of an egg. Heat and ultraviolet radiation destroy bacteria by denaturing some of their vital enzymes. Since denatured proteins are easier for enzymes to work on, we cook our food to make its proteins more digestible.

Organic solvents

Organic solvents also may denature proteins. When ethyl alcohol is used as a disinfectant, it denatures the protein of any bacteria present. Reagents such as alcohol are capable of forming *intermolecular* hydrogen bonds with proteins and thereby disrupting the *intramolecular* hydrogen bonds of the molecule.

Acids and bases

Acids and bases may disrupt salt linkages and thereby cause coagulation. Of course, prolonged contact will also cleave the peptide bonds of the primary structure, so that the reaction may go beyond denaturation.

Salts of heavy-metal ions

Salts of heavy-metal ions such as Hg^{2+}, Ag^+, and Pb^{2+} form strong bonds with the carboxylate anions of the acidic amino acids. As a result, they disrupt salt linkages and cause the protein to precipitate as an insoluble salt. That makes some heavy-metal salts suitable for use as antiseptics. Silver nitrate is used to prevent gonorrhea infections in the eyes of newborn infants. Mercuric chloride is used as a disinfectant. Also, organic mercurials like Mercurochrome and Merthiolate are used as germicides. Such agents cannot be taken internally because they precipitate the proteins of all the cells which they contact. Indeed, milk and egg white serve as antidotes for heavy-metal poisoning because their proteins combine with the heavy-metal ions to form insoluble solids. An emetic is then used to cause removal of the insoluble solid from the stomach before digestion of the protein again liberates the poisonous metal ions.

Alkaloid reagents

Alkaloid reagents may be used to denature proteins; the common ones are tannic acid and picric acid. They are called alkaloid reagents because they were initially used in the investigation of the structure of such alkaloids as morphine, cocaine, and quinine. They function by the picrate or tannate anions combining with the positively charged and substituted ammonium groups to form slightly soluble salts. They thereby disrupt salt bridges with resultant loss of three-dimensional structure.

14-6 Functions of proteins

Proteins may be classified according to function, and such a classification includes six major groups.

Structural proteins are those which constitute the body structures. More than half of the total protein of a mammalian body is collagen found in skin, cartilage, and bone.

Contractile proteins include myosin and actin, which are found in skeletal muscle.

Enzymes are organic catalysts for biochemical reactions. They occur in plants and animals. Examples are ribonuclease, emulsin, and glucase.

Many *hormones* are proteins; insulin is one example.

Antibodies are produced by the body to destroy any foreign materials (antigens) released into the body by an infectious agent. The gamma globulins are examples of antibodies.

Among the most important of the *blood proteins* are the albumins, hemoglobin, and fibrinogen.

14-7 Protein synthesis

Although the chemical synthesis of a polypeptide is a tedious and time-consuming endeavor, ribonuclease, which contains over 100 amino acids, has been successfully synthesized. An amide group is easily formed by the reaction of an active acid group, such as an acyl halide, with an amino group. The big problem is that amino acids contain other reactive groups which can react at the same time. For example, if one were simply to react the acyl chloride of alanine with the amino group of glycine, some of the dipeptide alanylglycine might be formed. However, products of side reactions would predominate, since the acyl chloride of alanine would react with the amino group of another molecule of the alanine derivative as well as with the amino group of glycine. If any amino acids which have reactive groups in their side chains were employed, they also would be involved. For that reason it is necessary to cover or protect all reactive sites where reaction is not desired.

Use of protected amino acids

A variety of protecting and activating groups have been used in peptide synthesis. The protecting groups for the carboxyl end are usually esters. They are easier to hydrolyze than amides, and so they can be more readily removed without hydrolyzing the peptide. For protecting the amino group, special types of amides are formed. They can be removed under conditions which do not cleave peptide bonds or esters. Also, any reactive sites in the side chains of the amino acids must be protected.

Solid-phase protein synthesis

It is now possible to buy protected amino acids, which can easily be reacted to form a dipeptide. After the peptide bond is formed, however, a whole series of extractions, washings, and crystallizations is required to purify the protected dipeptide. Those tedious processes must be repeated for each subsequent addition of an amino acid to the growing chain. In recent years, a solid-phase protein synthesis in which these processes may be programmed and automated has been developed.

In solid-phase peptide synthesis, the *C*-terminal amino acid is attached through an ester linkage to about 15% of the units of a polymer formed as an insoluble bead of a polystyrene derivative. The bead of polymer, with the *C*-terminal amino acid now bonded to it, is suspended in a solvent, and an activated amino acid (with its amino group protected) is added. After reaction at the surface of the bead to form a dipeptide on about 15% of the units of the polymer, the excess reagent is removed, the polymer is

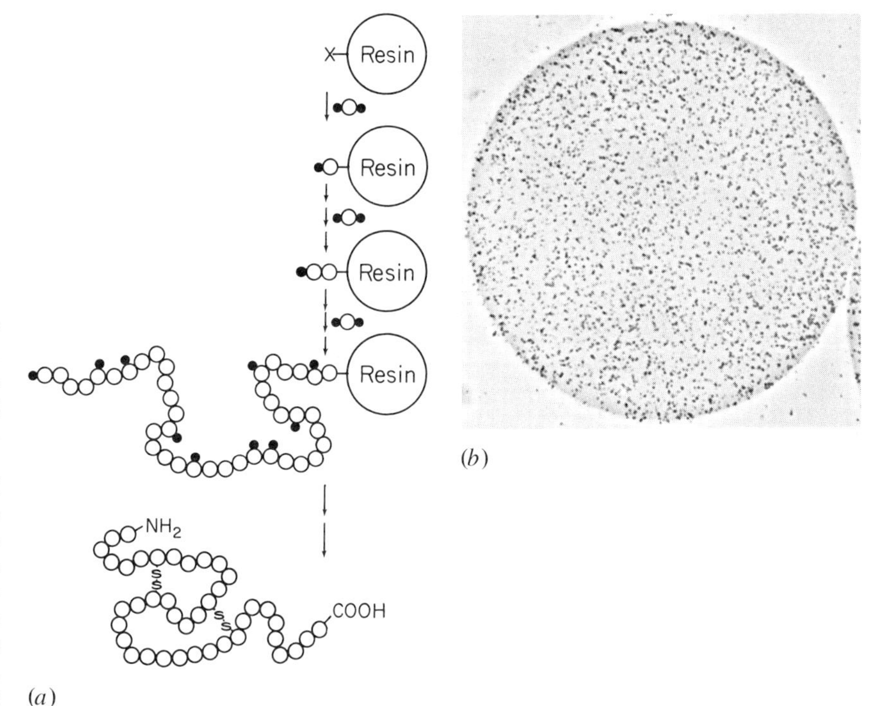

FIGURE 14-19
Solid-phase protein synthesis. (*a*) Schematic view: small open circles represent amino acid residues; solid circles represent protecting groups. (*b*) Thin section of a sphere of resin, the dots are radioactively labeled peptides attached to the resin. (*Courtesy of R. B. Merrifield and the 1970 Intra-Science Research Award Lecture,* Intra-Science Chemistry Reports, *5:183, 1971.*)

washed with pure solvent, and the solution is removed. The protecting group on the end amino group is then removed, and a third activated and protected amino acid is added. The process is repeated over and over.

Many polypeptides have now been synthesized by using this technique. The structural formula for the polymeric bead with the *C*-terminal amino acid bonded to it is shown.

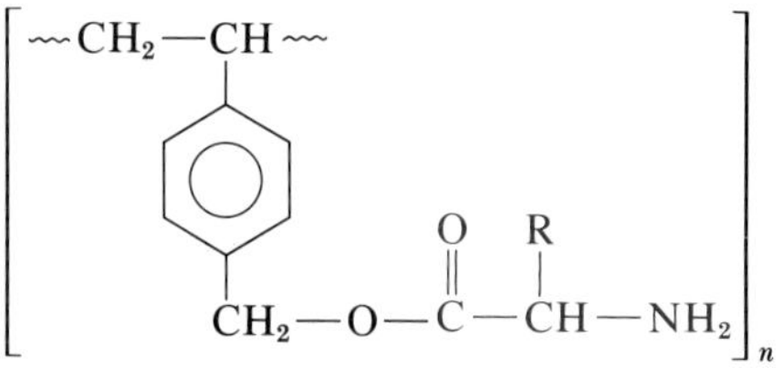

$$\left[\begin{array}{l} \sim CH_2-CH\sim \\ \qquad\quad | \\ \qquad \bigcirc \\ \qquad\quad | \\ \qquad CH_2-O-\overset{\overset{O}{\|}}{C}-\overset{\overset{R}{|}}{CH}-NH_2 \end{array} \right]_n$$

In 1969, Bruce Merrifield, of Rockefeller University, published results of an automated synthesis of the enzyme ribonuclease (Fig. 14-19). In the first step of his synthesis, the *C*-terminal residue, valine, was attached to the polystyrene bead through a reactive chloromethyl group on the polystyrene. The amino group of the valine had been previously protected with a *tert*-butoxycarbonyl (*t*-BOC) group.

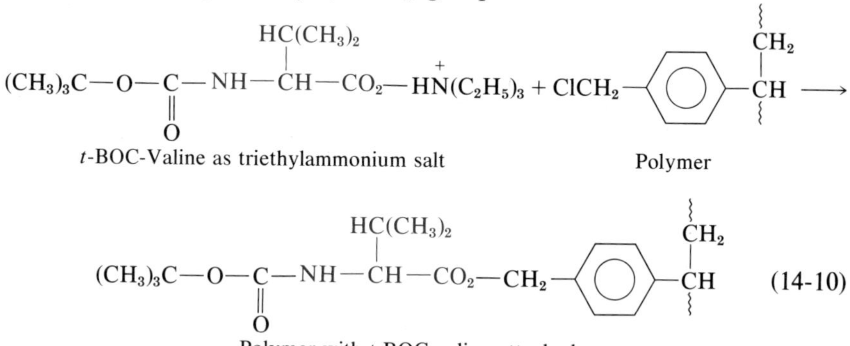

$$(CH_3)_3C-O-\underset{\underset{O}{\|}}{C}-NH-\overset{\overset{HC(CH_3)_2}{|}}{CH}-CO_2-H\overset{+}{N}(C_2H_5)_3 + ClCH_2-\bigcirc-\overset{\overset{\wr}{CH_2}}{\underset{\wr}{CH}} \longrightarrow$$

t-BOC-Valine as triethylammonium salt — Polymer

$$(CH_3)_3C-O-\underset{\underset{O}{\|}}{C}-NH-\overset{\overset{HC(CH_3)_2}{|}}{CH}-CO_2-CH_2-\bigcirc-\overset{\overset{\wr}{CH_2}}{\underset{\wr}{CH}} \qquad (14\text{-}10)$$

Polymer with *t*-BOC-valine attached

The *t*-BOC group was then removed from the amino group of valine.

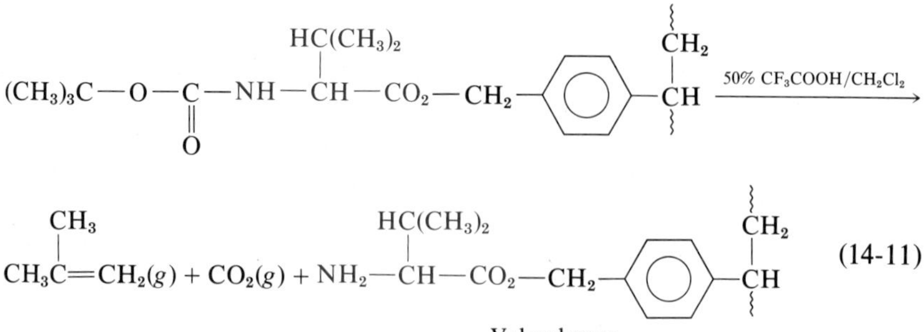

$$(CH_3)_3C-O-\underset{\underset{O}{\|}}{C}-NH-\overset{\overset{HC(CH_3)_2}{|}}{CH}-CO_2-CH_2-\bigcirc-\overset{\overset{\wr}{CH_2}}{\underset{\wr}{CH}} \xrightarrow{50\%\ CF_3COOH/CH_2Cl_2}$$

$$CH_3\overset{\overset{CH_3}{|}}{C}=CH_2(g) + CO_2(g) + NH_2-\overset{\overset{HC(CH_3)_2}{|}}{CH}-CO_2-CH_2-\bigcirc-\overset{\overset{\wr}{CH_2}}{\underset{\wr}{CH}} \qquad (14\text{-}11)$$

Val-polymer

The next protected amino acid, serine, was then added. The —OH group of serine was protected by a benzyl group forming an ether linkage.

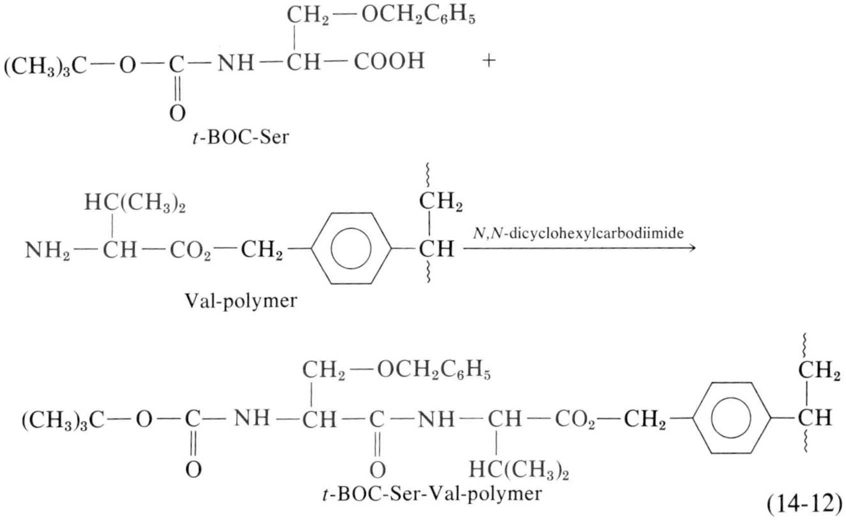

$$\underset{\substack{\| \\ O}}{(CH_3)_3C{-}O{-}C}{-}NH{-}\overset{\substack{CH_2{-}OCH_2C_6H_5 \\ |}}{CH}{-}COOH \quad +$$

t-BOC-Ser

$$NH_2{-}\overset{\substack{HC(CH_3)_2 \\ |}}{CH}{-}CO_2{-}CH_2{-}C_6H_4{-}\overset{\substack{\wr \\ CH_2 \\ |}}{\underset{\wr}{CH}} \xrightarrow{N,N\text{-dicyclohexylcarbodiimide}}$$

Val-polymer

$$\underset{\substack{\| \\ O}}{(CH_3)_3C{-}O{-}C}{-}NH{-}\overset{\substack{CH_2{-}OCH_2C_6H_5 \\ |}}{CH}{-}\underset{\substack{\| \\ O}}{C}{-}NH{-}\underset{\substack{| \\ HC(CH_3)_2}}{CH}{-}CO_2{-}CH_2{-}C_6H_4{-}\overset{\substack{\wr \\ CH_2 \\ |}}{\underset{\wr}{CH}} \qquad (14\text{-}12)$$

t-BOC-Ser-Val-polymer

Each of the other 123 amino acid residues was added one at a time in proper sequence. By automation, the 369 reactions involving 11,931 steps were completed over a period of a few weeks of continuous operation. The final cleavage of the protein from the polystyrene bead was accomplished with HF in a mixture of trifluoroacetic acid (CF_3COOH) and anisole ($C_6H_5OCH_3$).

The resulting synthetic ribonuclease compared well with the natural enzyme in both amino acid analysis and enzymatic activity toward RNA. Solid-phase syntheses have also been reported for cytochrome c, human growth hormone, and ferredoxin.

Summary

Proteins are *polyamide polymers* of more than twenty naturally occurring *α-amino acids* in almost every possible variation of sequence and combination. *Conjugate proteins* are those which, upon hydrolysis, yield a *prosthetic* or *conjugate group* in addition to the α-amino acids. The α-amino acids making up proteins generally belong to the L series.

Essential amino acids are a group of eight α-amino acids which cannot be produced in sufficient quantity by the human body. They must be included in the diet.

With the exception of proline, the 20 common L-α-amino acids differ only in the nature of the R group. Proline has a *pyrrolidine ring* involving the amino group, so that it is a secondary amine. Seven of the L-α-amino acids contain nonpolar R groups; seven of them contain polar, but neutral, R groups; and six of them contain acidic or basic R groups.

The carboxyl group and the amino group of an amino acid tend to react with each other to form a *dipolar ion* or *intramolecular salt* called a *zwitterion*. The *isoelectric point* is the pH at which the positive and negative charges on an amino acid, a peptide, or a protein are in balance, so that in an electric field there is no tendency to migrate to either electrode. Amino acids are least soluble at their isoelectric points. They are *amphoteric* and are more soluble in acid or in base than they are in water.

Two of the general methods for *synthesizing amino acids* are (a) *ammonolysis* of an α-halogen acid with excess ammonia and (b) reaction of an aldehyde or ketone with NH_3, HCN, and heat followed by hydrolysis of the cyanide group to a carboxyl group (*Strecker synthesis*).

The *peptide bond* is the special amide linkage of the carboxyl group of one α-amino acid to the amino group of another α-amino acid, with resultant loss of a molecule of water. The four atoms comprising the peptide bond and the two carbon atoms joined to the peptide bond lie in the same plane. The bond angles about the carbonyl carbon atom and the amide nitrogen atom are approximately 120°. The peptide bond is a *resonance hybrid* of two species. It can exist in both cis and trans forms. The *trans form* with the carbonyl oxygen and the hydrogen on nitrogen on opposite sides is observed most frequently.

Peptides are small polymers of amino acids. A *dipeptide* is derived from two amino acids; a *tripeptide* is derived from three amino acids; and a *polypeptide* is derived from ten or more amino acids. Polyamides with molecular weights over 10,000 are usually referred to as *proteins* rather than polypeptides.

Peptides are named by combining the names (or abbreviations) of the individual amino acids, beginning with the amino acid on the end with the free amino group and ending with the amino acid on the end with the free carboxyl group. Peptides are generally written with the free terminal $—NH_2$ group to the left and proceeding to the right toward the free terminal —COOH. A more abbreviated representation uses the standard abbreviations of the amino acids connected by arrows. The tail of an arrow indicates the amino acid that contributes the carboxyl group to the peptide bond, and the head of the arrow denotes the amino acid that contributes the amino group.

A dipeptide has two possible sequences of amino acids; a tripeptide has six; and a polypeptide containing each of 20 different amino acids has about 2×10^{18}. Determination of the amino acid sequence involves identification of the N-*terminal amino acid,* the C-*terminal amino acid,* and the sequence of all of the amino acids in between. First in the determination process the peptide is partially hydrolyzed into smaller peptides whose sequence is then determined. Next the sequence of amino acids in the original polypeptide is reconstructed by the overlapping of amino acids in the small peptide fragments.

The *primary structure* of a protein is the amino acid content and sequence. The *secondary structure* of a protein is the manner in which the peptide chain is folded or bent and stabilized in that structure with hydrogen bonds. The two principal secondary structures are designated alpha and beta. The α structure is a helix held together by hydrogen bond formation between the amide hydrogen of one peptide bond and the carboxyl oxygen which is located on the next turn of the helix. The β structure consists of the polypeptide chains bonded to adjacent polypeptide chains with interpolypeptide hydrogen bonds in such a manner as to resemble a pleated sheet.

The *tertiary structure* is the unique three-dimensional shape which results from the precise folding and bending of a polypeptide chain or of the α helix into which it may be formed. The linkages which hold the protein molecule in its tertiary shape are *salt linkages, hydrogen bonds, disulfide linkages,* and *hydrophobic interactions.*

The *quaternary structure* of a protein refers to the manner in which separate, noncovalently linked polypeptide chains are arranged in space to form a molecular assembly. Proteins of molecular weight greater than 30,000 to 50,000 are likely to be made of more than one chain and to have a quaternary structure.

Structurally, proteins may be classified as fibrous or globular. *Fibrous proteins* are stringy, tough, and generally insoluble. They have a lot of β structure. Collagen, the major extracellular structural protein in connective tissue and bone, is a fibrous protein. *Globular proteins* are somewhat spherical in shape. They generally have the α-helix structure. Myoglobin is an example of a globular protein.

Denaturation is any change in which the unique three-dimensional configuration of the protein is altered without any cleavage of the peptide bonds. It may be reversible or irreversible. Denaturation may be effected by heat or radiation, organic solvents, acids or bases, salts of heavy metal ions, and alkaloid reagents.

Proteins may also be classified according to *function* as *structural proteins, contractile proteins, enzymes, hormones, antibodies,* and *blood proteins.*

The formation of peptide bonds for *synthesis of a polypeptide* is complicated by the fact that amino acids contain reactive groups other than the amino group. Those groups can react with the acyl chloride used. For that reason, *protecting groups* have to be used in peptide synthesis. In recent years, a *solid-phase protein synthesis* in which the processes have been programmed and automated has been developed. The enzyme ribonuclease and other large peptides have been synthesized by that technique.

Key terms

L-α-Amino acid: one of the naturally occurring amino acids which, in combination, constitute proteins. The general formula is

$$\begin{array}{c} \mathrm{R} \\ | \\ \mathrm{H_2N-C-COOH} \\ | \\ \mathrm{H} \end{array}$$

α Helix: one of the two principal secondary structures for proteins. It is a right-handed helix held together by hydrogen bonds between the amide hydrogen of one peptide bond and the carboxyl oxygen located on the next turn of the helix.

Amino acid sequence: the order of the amino acids in a polyamide polymer.

Amphoteric: having properties of both acids and bases. Amino acids are amphoteric.

Antibody: a protein produced by the body to destroy a foreign material (antigen) released into the body by an infectious agent.

β Structure: one of the two principal secondary structures of proteins. Polypeptide chains are bonded to adjacent polypeptide chains with interpolypeptide hydrogen bonds in such a fashion as to resemble a pleated sheet.

Blood protein: one of the proteins in the blood, which include the albumins, hemoglobin, and fibrinogen.

***C*-Terminal amino acid:** in a peptide, the α-amino acid on the end with the free carboxyl group. Naming of a peptide ends with that amino acid.

Chromatography: separation of a mixture of substances based on their differences in adsorption by an adsorbent.

Chromoprotein: a conjugate protein in which the conjugate group is a color-bearing group.

Collagen: the protein of bones and connective tissue.

Conjugate group: a group present in a conjugate protein in addition to the α-amino acids. Examples of conjugate groups are fats, carbohydrates, nucleic acids, and color-bearing groups. Also called **prosthetic groups.**

Conjugate protein: a protein which, upon hydrolysis, yields a prosthetic or conjugate group in addition to α-amino acids.

Contractile protein: a protein, such as myosin and actin, found in skeletal muscle.

Denaturation: any change in which the unique three-dimensional configuration of the protein molecule is altered without any cleavage of the peptide bonds. It can be effected by heat, radiation, organic solvents, acids or bases, salts of heavy-metal ions, and alkaloid reagents.

Disulfide linkage: one of the types of linkage holding a protein molecule in its tertiary shape. It results from the oxidation of sulfhydryl groups (—SH) of two cysteine side chains.

Enzyme: a globular protein which catalyzes biochemical reactions in plants and animals.

Essential amino acid: one of eight amino acids which cannot be produced in sufficient quantity by the human body and must therefore be included in the diet.

Fibroin: the protein of silk.

Fibrous protein: stringy, tough, and generally insoluble protein composed of long, rodlike chains cross-linked. The fibrous proteins include silks, keratins, and collagens.

Globular protein: a somewhat spherical protein which tends to be soluble in water and salt solutions. The globular proteins include globulins and enzymes.

Glycoprotein: a conjugate protein in which the conjugate group is a carbohydrate.

Hormone: a substance which has a regulatory function in a plant or animal. Some proteins are hormones.

Hydrophilic: "water-loving." Polar, water-soluble parts of a molecule are said to be hydrophilic.

Hydrophobic: "water-fearing." Nonpolar, water-insoluble parts of a molecule are said to be hydrophobic.

Hydrophobic interaction: one of the types of linkage holding a protein molecule in its tertiary shape. It results from the repulsion of nonpolar groups by water, so that the nonpolar groups tend to be pushed together by surrounding water molecules.

Isoelectric point: the pH at which the positive and negative charges on the zwitterions are in balance so that, in an electric field, the tendency for migration to either electrode is at a minimum.

Keratin: the protein of hair.

Lipoprotein: a conjugate protein in which the conjugate group is a fat.

***N*-Terminal amino acid:** in a peptide, the α-amino acid on the end with the free amino group. Naming of a peptide begins with that amino acid.

Nucleoprotein: a conjugate protein in which the conjugate group is a nucleic acid.

Peptide: a small polyamide polymer of amino acids.

Peptide bond: the amide linkage of the carboxyl group of one α-amino acid to the amino group of another α-amino acid. One molecule of water is eliminated.

Phenylketonuria: a genetic disease in which the enzyme needed for normal metabolism of phenylalanine is absent. It results in permanent mental retardation unless it is detected in early infancy and treated with a low-phenylalanine diet.

Polypeptide: a peptide formed from 10 or more amino acids. Generally, the term is used only for polyamide polymers of molecular weights less than 10,000.

Primary structure: the amino acid content and sequence of a protein.

Prosthetic group: a group present in a conjugate protein in addition to α-amino acids. Also called **conjugate group.**

Protecting group: a group, such as an ester or amide, used in peptide synthesis to cover the reactive sites on an amino acid at which reaction is not desired.

Protein: a polyamide polymer of more than twenty naturally occurring L-α-amino acids. Its molecular weight is greater than 10,000. Protein is the third and most important of the food classes.

Pyrrolidine ring: a five-atom, saturated, heterocyclic ring containing a nitrogen atom,

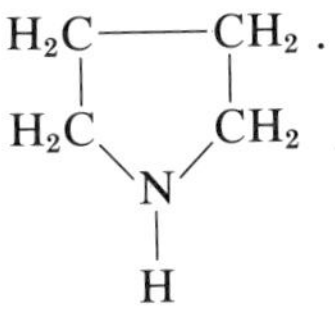

```
H2C——CH2 .
 |    |
H2C   CH2
   \ /
    N
    |
    H
```

Quaternary structure: the manner in which separate, noncovalently linked polypeptide chains are arranged in space to form molecular assemblies. Proteins of molecular weight greater than 30,000 to 50,000 are likely to have a quaternary structure.

Salt linkage: one of the types of linkage holding a protein molecule in its tertiary shape. It results from salt formation between the amino groups of the basic amino acids and the side-chain carboxyl groups of the acidic amino acids.

Secondary structure: the fixed configuration of the polypeptide chain in a protein. It involves the manner in which the peptide chain is folded and bent and the hydrogen bonds which hold the special structure together. The principal secondary structures are the α helix and the β structure.

Sickle-cell anemia: a genetic disease in which two of the four polypeptide chains of hemoglobin have a valine where they should have a glutamic acid. The result of that slight difference is a drastic change in the properties of the hemoglobin molecule and the red blood cells.

Solid-phase protein synthesis: a programmed and automated method of protein synthesis begun with a *C*-terminal amino acid attached through an ester linkage to an insoluble bead of a polystyrene derivative. Through a sequence of reactions, amino acids are added, one at a time, in the desired sequence.

Strecker synthesis: the reaction of an aldehyde or a ketone with NH_3, HCN, and heat followed by hydrolysis. It is a general method for synthesizing α-amino acids.

Structural protein: one of the proteins which constitute the body structure.

Tertiary structure: the unique three-dimensional shape which results from the precise folding and bending of a polypeptide chain or of the α helix into which it may be formed.

Zwitterion: an intramolecular salt or dipolar ion. The two functional groups of amino acids react with each other to form zwitterions.

Additional problems

14-4 Amino acids from hydrolysis of proteins are designated as L-amino acids. What then is the structural relation between L-serine and L-glyceraldehyde? Between L-phenylalanine and D-glucose?

14-5 One of the 20 common amino acids contains more than one chiral center. How many stereoisomers exist for it? Draw Fischer configurational formulas for the stereoisomers.

14-6 Lysine can react with itself under certain conditions to form a cyclic amide. Write the structural formula for the cyclic amide.

14-7 Many α-amino acids react with nitrous acid to give a quantitative volume of nitrogen gas. That is called the *Van Slyke method of analysis*. Which of the 20 common amino acids cannot be analyzed by the Van Slyke method?

14-8 Write equations for the preparation of the following amino acids from the starting materials indicated: (a) leucine from 3-methyl-1-butanol (isoamyl alcohol), (b) valine from 3-methylbutanoic acid (isovaleric acid).

14-9 When concentrated nitric acid contacts your fingers or your hand, the skin becomes yellow. The reaction involves the nitration of phenylalanine, and it is known as the *xanthoproteic reaction*. Write the chemical equation for the reaction.

14-10 Would an aqueous solution of lysine be acidic, basic, or neutral?

14-11 How does one repress the ionization of an α-amino acid? In an electrolysis experiment, do α-amino acids migrate at the isoelectric point?

14-12 An electric current is passed through an aqueous solution at a pH of 6.0. The solution contains the following amino acids with isoelectric points as indicated: Ala (6), Glu (3.2), and Arg (10.7). Describe the migration of those amino acids to the electrodes.

14-13 Which would you predict to have the higher isoelectric point, bovine vasopressin or bovine oxytocin?

14-14 Write equations for the following reactions:

(a) valine with aqueous sodium hydroxide
(b) tryptophan with hydrochloric acid
(c) phenylalanine with nitrous acid
(d) proline with nitrous acid
(e) methionine with acetic anhydride
(f) asparagine with boiling sodium hydroxide solution
(g) lysine and excess benzoyl chloride in the presence of aqueous NaOH
(h) tyrosine warmed with nitric acid.

14-15 Write the structural formula for a tetrapeptide composed of the following amino acids in the order named: isoleucine, lysine, glutamic acid, methionine. Follow the form given on page 429. Give the abbreviated form also.

14-16 Define and give an example of each of the following:

(a) zwitterion
(b) peptide
(c) α-amino acid
(d) conjugated protein
(e) isoelectric point
(f) essential amino acid
(g) primary structure
(h) denaturation

14-17 Distinguish between a fibrous protein and a globular protein.

14-18 A nonapeptide on complete hydrolysis yielded five different amino acids in the following relative quantities: one serine, one glycine, two arginine, two phenylalanine, and three proline. The *N*-terminal amino acid was identified as arginine. Partial hydrolysis liberated the following small peptides:

$$\begin{array}{ll} \text{Ser} \longrightarrow \text{Pro} \longrightarrow \text{Phe} & \text{Pro} \longrightarrow \text{Phe} \longrightarrow \text{Arg} \\ \text{Phe} \longrightarrow \text{Ser} \longrightarrow \text{Pro} & \text{Pro} \longrightarrow \text{Gly} \\ \text{Gly} \longrightarrow \text{Phe} \longrightarrow \text{Ser} & \text{Arg} \longrightarrow \text{Pro} \end{array}$$

Deduce and write out the structure of this nonapeptide.

15

NUCLEIC ACIDS

You have now been introduced to two important classes of *biopolymers:* the polysaccharides and the proteins. The *nucleic acids* constitute a third class of biopolymers. They also are very important; for they are the chemical compounds of the genes. Nucleic acids are polymers with molecular weights ranging from about 25,000 to 1 million or even 1 billion. One type of nucleic acid is *deoxyribonucleic acid* (DNA). It is found mainly in the *nucleus* of the cell as the essential component of the chromosomes. Each chromosome consists of several thousand genes, the units of heredity. The other type of nucleic acid is *ribonucleic acid* (RNA). It is found mainly in the *cytoplasm* of the cell, and it participates in protein synthesis.

15-1 Composition

Upon hydrolysis, all nucleic acids yield three main components: (1) *ribose* or *deoxyribose* (five-carbon sugars), (2) *phosphoric acid,* and (3) a mixture of *heterocylic amine bases* called *purines* and *pyrimidines*. That is shown in Table 15-1. Note two points of difference between DNA and RNA. (1) DNA contains *2-deoxyribose,* whereas RNA contains *ribose*. The term *deoxy* indicates that the structure is the same as that of ribose, but without

TABLE 15-1
Products of hydrolysis of nucleic acids

Nucleic acids	Five-carbon sugar	Phosphoric acid	Heterocyclic amine bases
DNA $\xrightarrow[H_3O^+]{\text{hydrolysis}}$	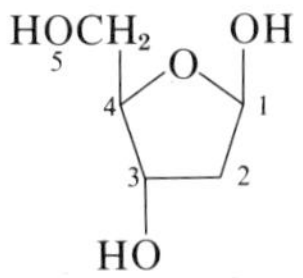 $HOCH_2$ (5), ring O, C-1 OH, C-4, C-3 HO, C-2 — 2-Deoxyribose	+ H_3PO_4	+ adenine, guanine, cytosine, thymine
RNA $\xrightarrow[H_3O^+]{\text{hydrolysis}}$	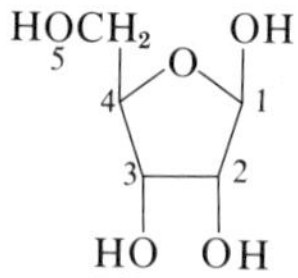 $HOCH_2$ (5), ring O, C-1 OH, C-4, C-3 HO, C-2 OH — Ribose	+ H_3PO_4	+ adenine, guanine, cytosine, uracil

the oxygen atom at position 2. (2) Although both DNA and RNA contain adenine, guanine, and cytosine, DNA contains *thymine,* whereas RNA contains *uracil.*

Following are the structures of the two purines, adenine and guanine, found in both DNA and RNA.

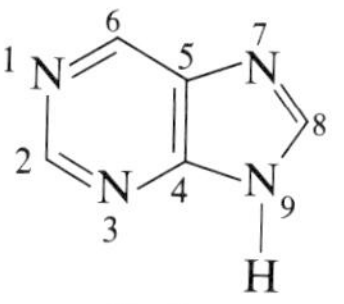

Purine

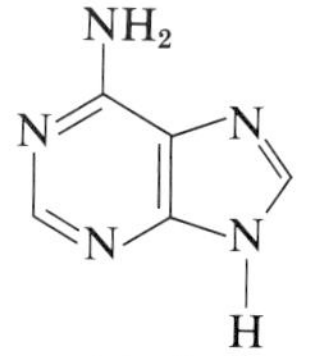

Adenine
(6-Aminopurine)

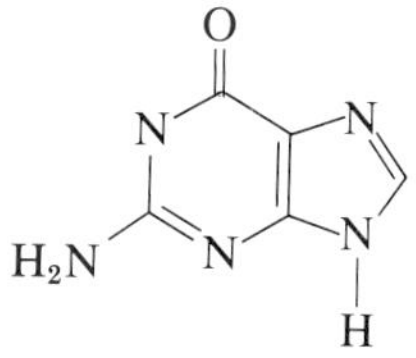

Guanine
(2-Amino-6-oxypurine)

The purine and pyrimidine ring systems are heterocyclic *aromatic systems* containing nitrogen (first introduced in Sec. 5.6). The nitrogen atoms have unshared pairs of electrons which make them basic; that is, they are cyclic amines.

The pyrimidines found in the nucleic acids have the following structures. Note that thymine differs from uracil in having a methyl group.

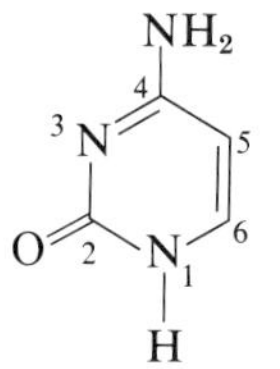

Cytosine
(4-Amino-2-oxypyrimidine)

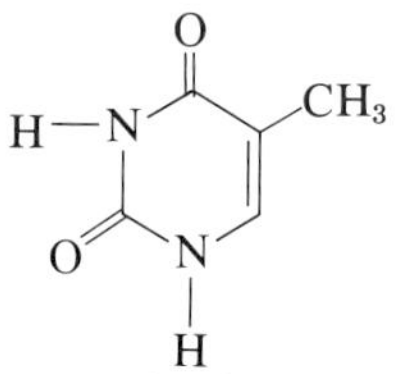

Thymine
(5-Methyl-2,4-dioxypyrimidine)

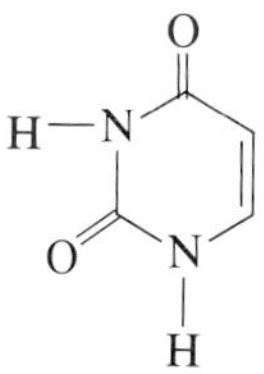

Uracil
(2,4-Dioxypyrimidine)

Nucleosides

A *nucleoside* is a substituted five-carbon sugar (riboside). It is structurally analogous to a glycoside (Sec. 12-7), except that a nitrogen atom of one of the heterocyclic nitrogen bases is bonded at C-1′ instead of an oxygen atom. (In nucleoside nomenclature, the carbon atoms of the pentose are designated by primed numbers to distinguish them from the carbon atoms of the purine or pyrimidine.) The purine or pyrmidine base is bonded to C-1′ of either ribose or 2′-deoxyribose through C-9 of the purine or C-1 of the pyrimidine. The name of the nucleoside incorporates the name of the nitrogen base; it uses the suffix *-idine* if the base is a pyrimidine and *-osine* if the base is a purine.

The four common deoxyribonucleosides present in DNA are *2′-deoxyadenosine, 2′-deoxyguanosine, 2′-deoxycytidine,* and *2′-deoxythymidine*. (The prefix *deoxy-* was not always used with deoxythymidine of DNA because of former confusion over the structure.) The four common ribonucleosides found in RNA are adenosine, guanosine, cytidine, and uridine. The pentose sugar is connected with a *β-glycosidic linkage*. Examples of nucleosides are

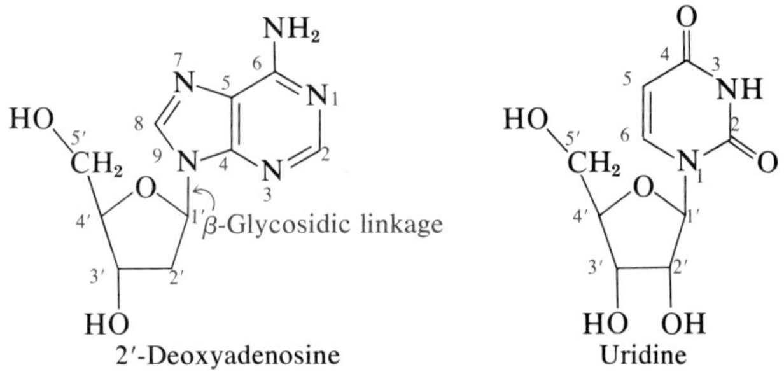

β-Glycosidic linkage

2′-Deoxyadenosine

Uridine

The reason why acidic hydrolysis of the ribosidic linkage of nucleic acids occurs is that C-1′ of a riboside has *both* an oxygen atom and a nitrogen atom attached to it. Because the two heteroatoms are so electronegative, the chemical behavior of the riboside approaches that of an acetal: stable toward bases but hydrolyzed by acids.

Nucleotides

Phosphate esters of the nucleosides, which usually form through C-5′ of the pentose, are referred to as *nucleotides*. Being esters, they are subject to alkaline hydrolysis, which cleaves them to phosphate salts and the nucleosides. (The nucleoside does not cleave further because its acetal-like linkage is stable to base, although not to acid.)

$$\text{Phosphoric acid} + \text{nucleoside} \longrightarrow \text{nucleotide} \qquad (15\text{-}1)$$

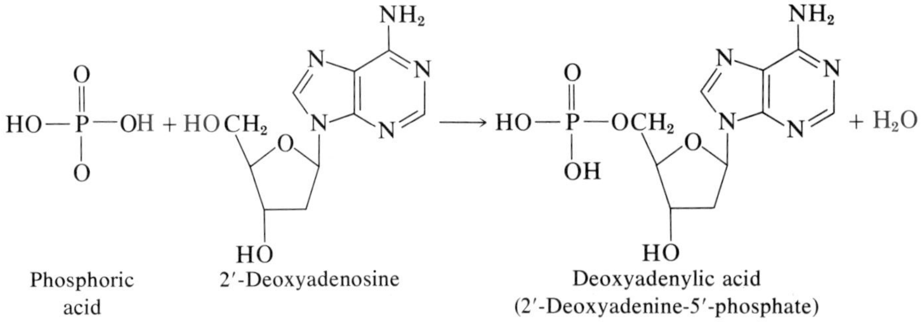

$$\text{Phosphoric acid} + \text{2'-Deoxyadenosine} \longrightarrow \text{Deoxyadenylic acid} + H_2O \qquad (15\text{-}2)$$

Phosphoric acid

2′-Deoxyadenosine

Deoxyadenylic acid (2′-Deoxyadenine-5′-phosphate)

The name of a nucleotide incorporates the name of the nitrogen base, the suffix *-ylic*, and *acid* as a separate word. The names of the nucleotides

TABLE 15-2 Nucleotides

Deoxyribonucleotides	Ribonucleotides
Deoxyadenylic acid	Adenylic acid
Deoxyguanylic acid	Guanylic acid
Deoxycytidylic acid	Cytidylic acid
Deoxythymidylic acid	Uridylic acid

are given in Table 15-2. The structural formulas of two nucleotides are

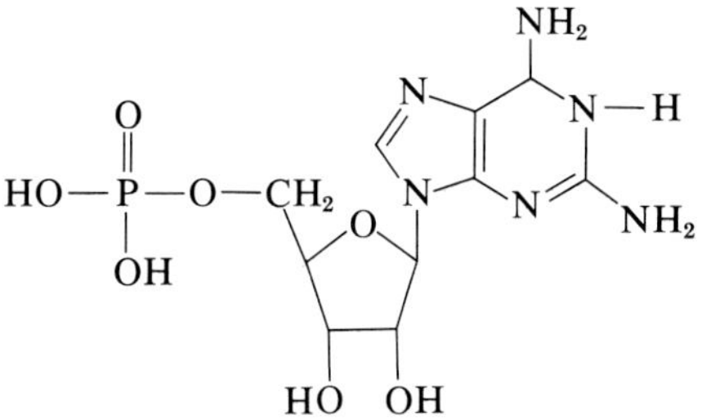

Guanylic acid
(Guanosine-5′-phosphate)

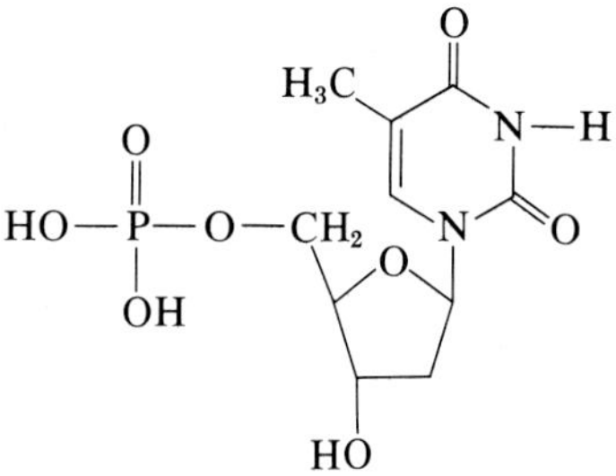

Deoxythymidylic acid
(2′-Deoxythymidine-5′-phosphate)

PROBLEM 15-1 Write structural formulas for the following compounds.

(a) 2′-deoxyguanosine
(b) 2′-deoxycytidine
(c) 2′-deoxythymidine
(d) adenosine
(e) guanosine
(f) cytidine

Solution to (a)

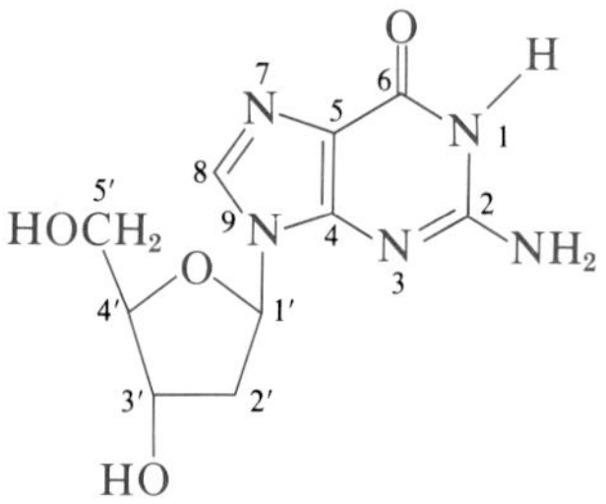

2′-Deoxyguanosine

PROBLEM 15-2 Write structural formulas for the following compounds.

(a) deoxyadenylic acid
(b) deoxyguanylic acid
(c) deoxycytidylic acid
(d) adenylic acid
(e) cytidylic acid
(f) uridylic acid

Solution to (a)

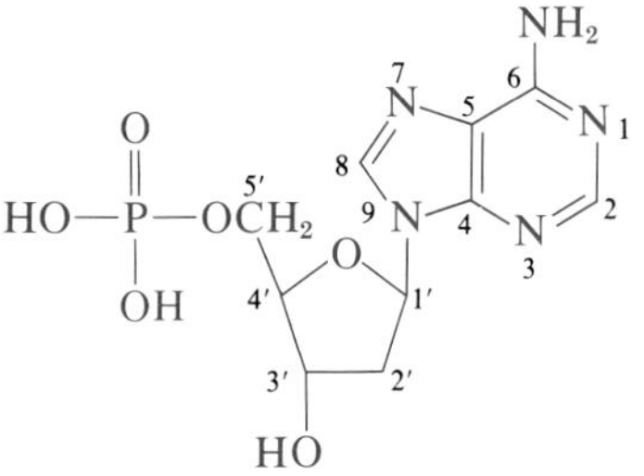

Deoxyadenylic acid
(2′-Deoxyadenine-5′-phosphate)

PROBLEM 15-3 Upon hydrolysis, a nucleotide gave the base uracil. Was the nucleotide from DNA? Give the reason for your answer.

PROBLEM 15-4 Write the equation for the hydrolysis of uridylic acid by using aqueous sodium hydroxide.

15-2 DNA: structure and functions

The nucleic acids, both DNA and RNA, are *polynucleotides* which are formed by the joining of nucleotide units through a *phosphate ester linkage* with the C-3′ hydroxy group of each pentose. Since each phosphoric acid moiety still retains a free hydroxy group in the polymer, the polymer is acidic; hence the name nucleic acids.

The acidic sites of the polymer can combine with ions such as magnesium and with the basic side chains of proteins to form *nucleoproteins* (Fig. 15-1). The genetic material DNA is thus protected by a protein coat. However, only the nucleic acid part of the nucleoprotein determines the genetic characteristics.

The formation of a phosphate ester linkage between deoxythymidylic acid and deoxyadenylic acid to form a *dinucleotide* is given as an example.

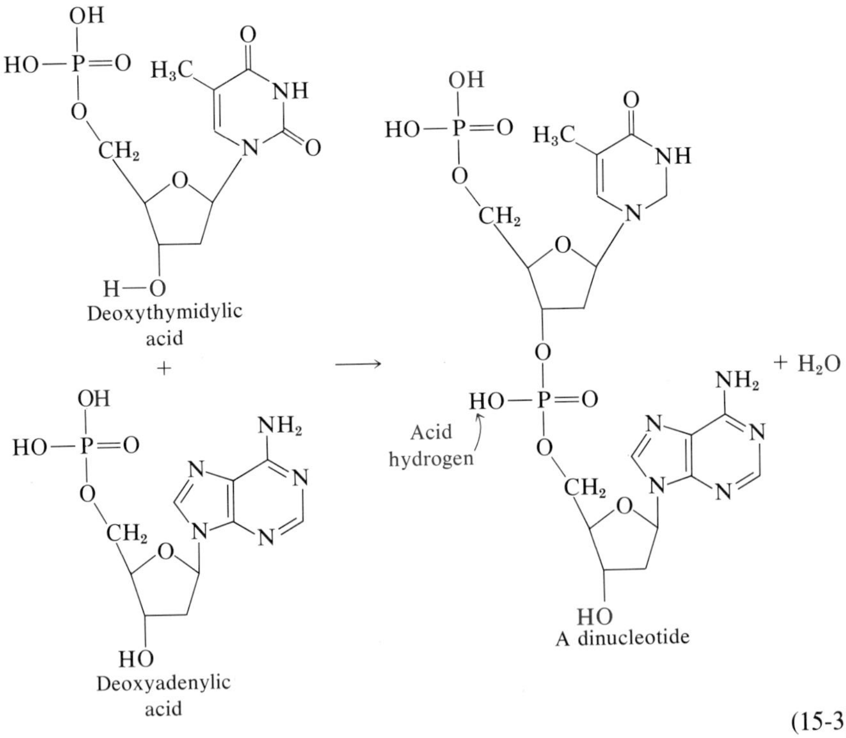

Deoxythymidylic acid + Deoxyadenylic acid ⟶ A dinucleotide (Acid hydrogen) + H_2O

(15-3)

The hydroxide groups shown in color are sites for further phosphate ester formation with other nucleotides. Many additional nucleotides could then be added in similar fashion to form a polynucleotide as a very long chain.

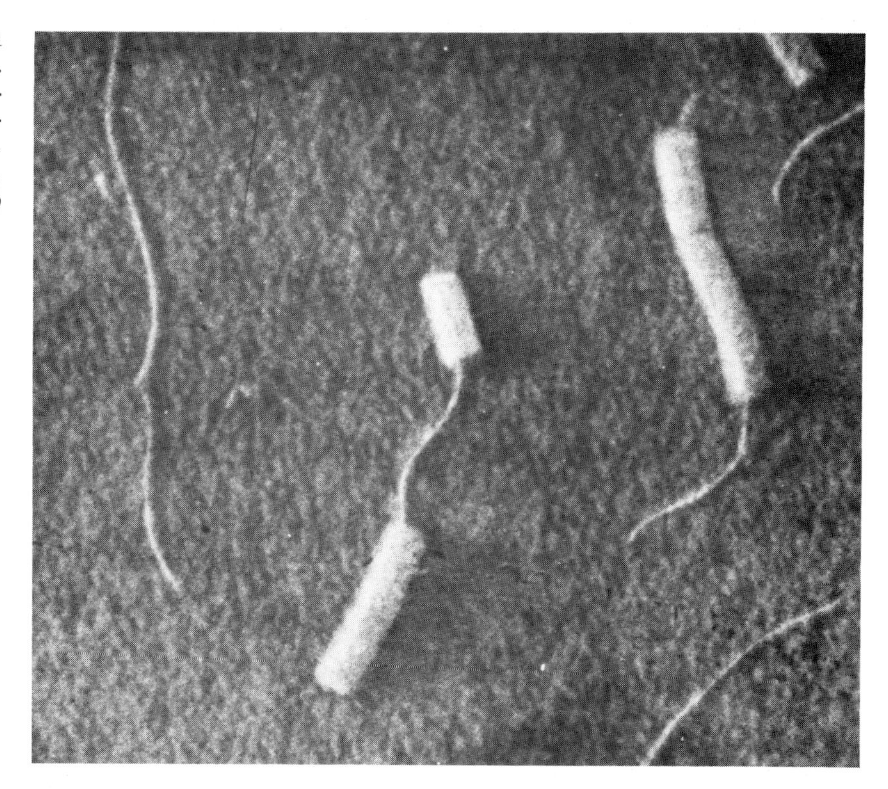

FIGURE 15-1 **Electron micrograph of RNA. Note that both the inner nucleic acid and the outer protein coat are visible.** (*M. K. Corbett and* **Virology,** *22:539–543, 1964*)

Primary structure

The primary structure of DNA is the *nucleotide content* and *sequence* in the polynucleotide polymer. DNA consists of the four deoxyribonucleotides; their various sequences are very important. A short segment of a DNA molecule is shown in Fig. 15-2 as an example of the primary structure of the polymer.

Since the nucleic acids contain so many atoms, the following abbreviations are frequently used in describing the structure of nucleic acids.

Adenine = A
Guanine = G
Cytosine = C
Thymine = T
Uracil = U

Phosphoric acid = (P)
Ribose = ⬠
Deoxyribose = ⬠ D

The abbreviations facilitate depicting the sugar phosphate backbone with projecting purine and pyrimidine bases. An example is shown in Fig. 15-3.

Secondary structure

The many different kinds of DNA are due to the great number of variations possible for the sequencing of the four heterocyclic bases in such a large polymer. However, hydrolysis and analysis of some DNA led to the

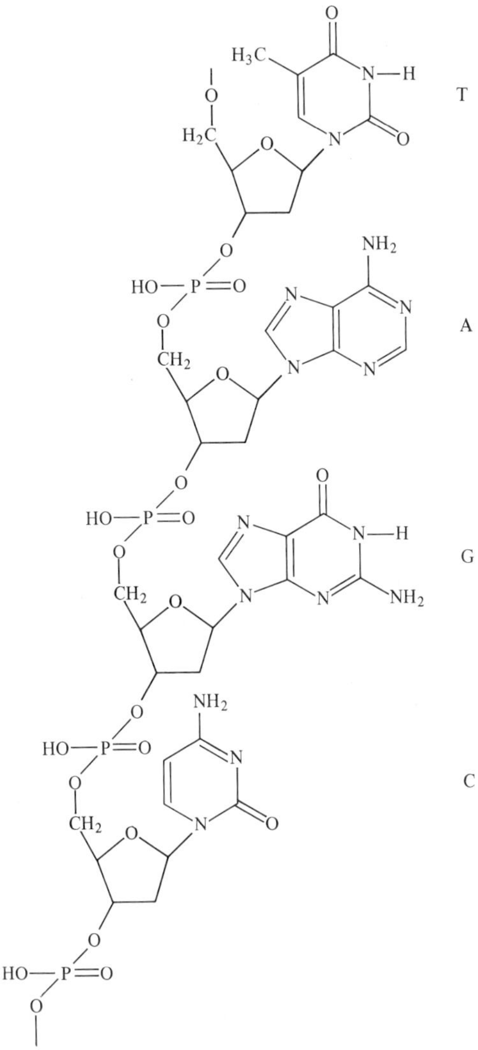

FIGURE 15-2
A segment of a DNA molecule showing the primary structure.

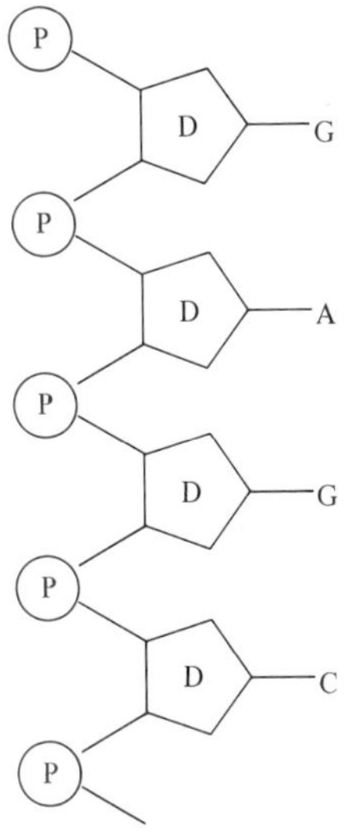

FIGURE 15-3
An abbreviated representation of a segment of a DNA molecule.

discovery that the number of adenine (A) molecules is nearly equal to the number of thymine (T) molecules obtained, and the number of guanine (G) molecules is nearly equal to the number of cytosine (C) molecules obtained. That is, A/T = 1 and G/C = 1. It was the 1:1 ratio, along with the use of x-ray crystallographic data obtained by Rosalind Franklin and Maurice Wilkins, that led J. D. Watson and F. H. C. Crick to suggest, in 1953, that DNA consists of two identical right-handed helical polymeric strands wrapped about each other to form a *double helix*. A further suggestion was that the two strands are held together by a pairing of bases such that hydrogen bonding is most effective. That specifically requires that adenine (A) pair with thymine (T) to produce two hydrogen bonds

FIGURE 15-4
Hydrogen bonding between strands of a DNA molecule.

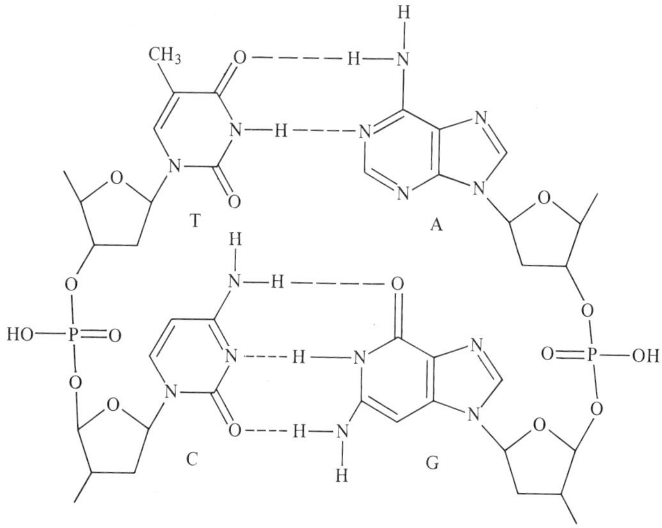

and that guanine (G) pair with cytosine (C) to produce three hydrogen bonds as illustrated in Fig. 15-4.

That particular pairing of the bases produces the greatest number of hydrogen bonds, with maximum closeness of approach for strong bonds. Each pair has at least one of the stronger hydrogen-oxygen hydrogen bonds. You may wish to attempt other pairings to assure yourself that the arrangement is optimal. See Fig. 15-5 for a representation of a segment of the double helix of DNA.

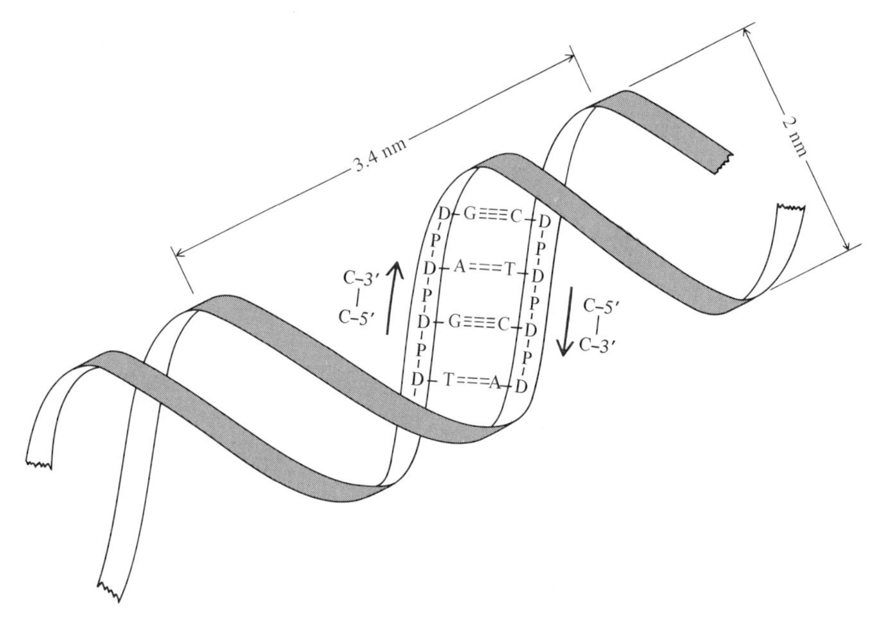

FIGURE 15-5
A segment of the double helix of DNA.

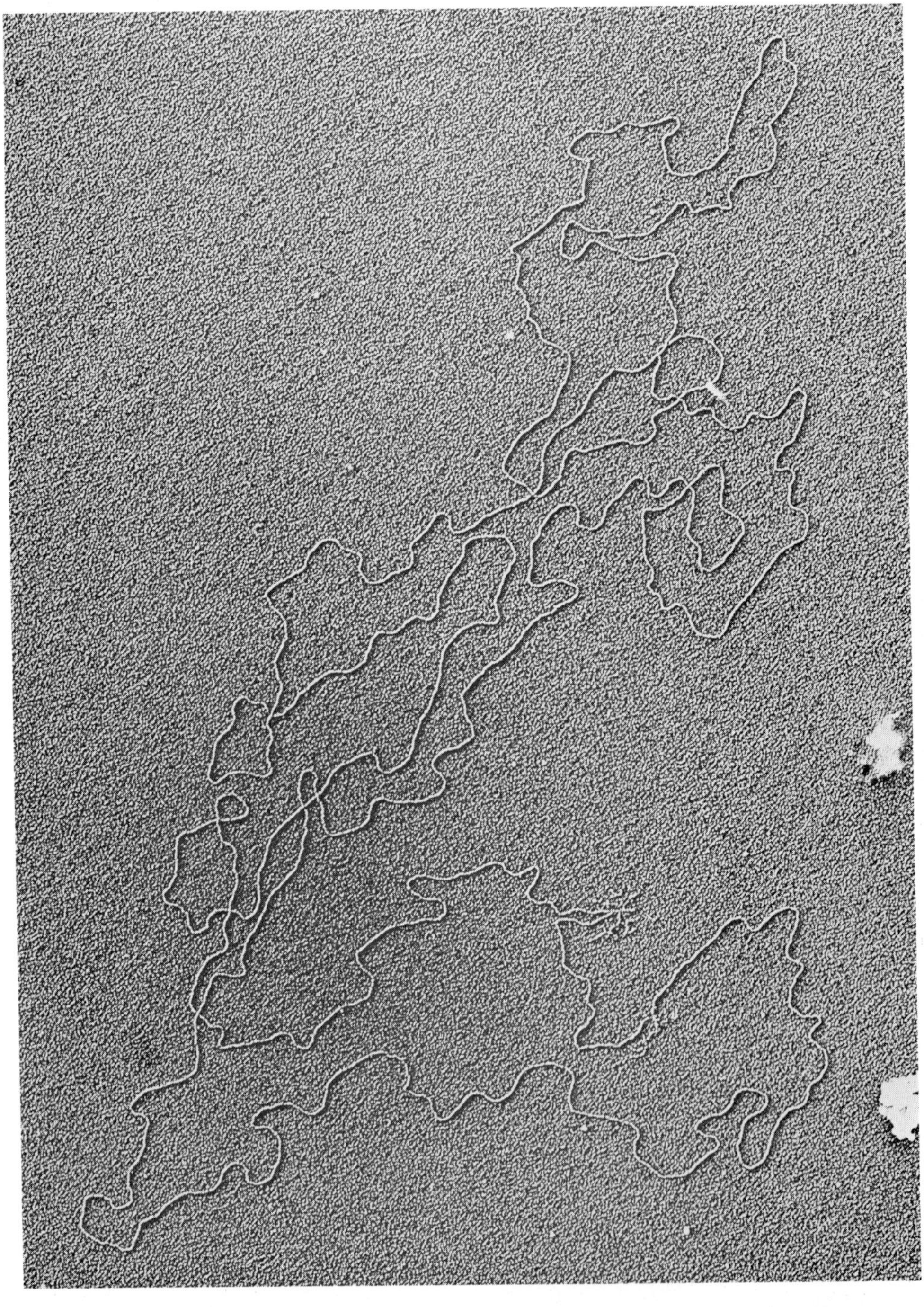

FIGURE 15-6 Electron micrograph of a DNA molecule (from a T4 bacteriophage). This molecule is linear in the virus but was artificially circularized before this micrograph was prepared. (*Courtesy of Charles A. Thomas*)

The *secondary structure* of DNA consists of two helical polynucleotides which coil around a common axis. The two chains run in opposite directions in an antiparallel arrangement. One chain goes up from tail to head, and the other one comes down from tail to head. The purine and pyrimidine bases project *inside* the double helix; the bases from one strand are paired to the bases from the other strand to permit optimum hydrogen bonding (A-T and C-G). Figure 15-6 is an electron micrograph of a DNA molecule.

PROBLEM 15-5 Why doesn't adenine pair with guanine or thymine pair with cytosine in the DNA molecule? Explain your answer with structural formulas.

PROBLEM 15-6 Which base pair, G-C or A-T, would you expect to exhibit the stronger hydrogen bonding? Why?

Functions

The two biochemical functions of DNA are (1) *transmission of genetic information* to succeeding generations (offspring) and (2) *control of protein synthesis*. The transmission of genetic information is accomplished in the nucleus of the cell by a process of *replication* (reproduction) of DNA. Replication occurs during mitosis (cell division). It is illustrated in Fig. 15-7. The first step is the breaking of hydrogen bonds between the bases in the double helix. The two strands of the double helix can then separate into single strands in the presence of a pool of nucleotides. The bases along the

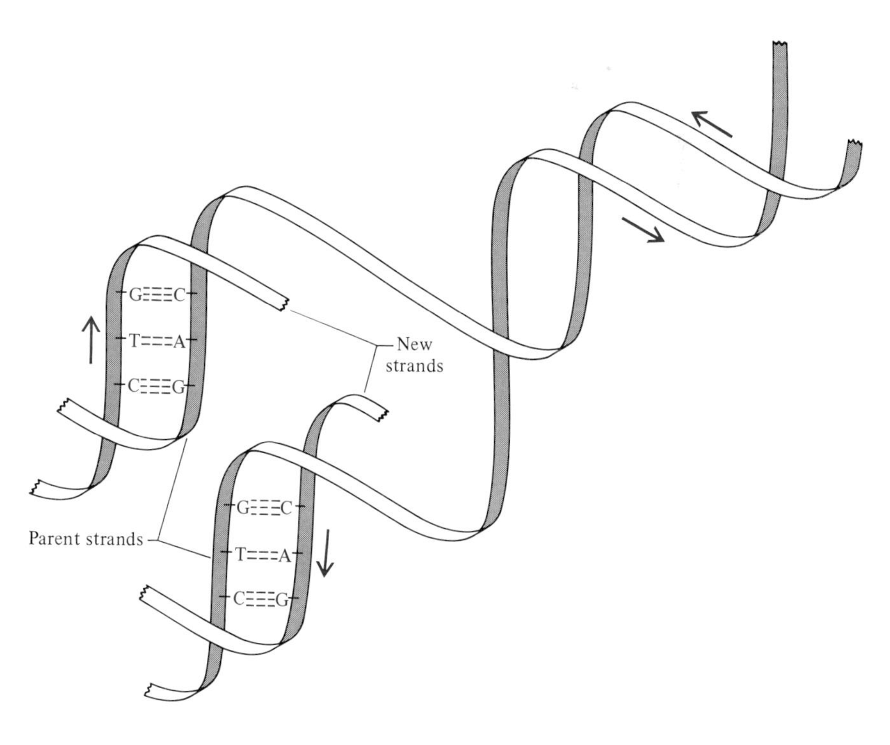

FIGURE 15-7
DNA replication.

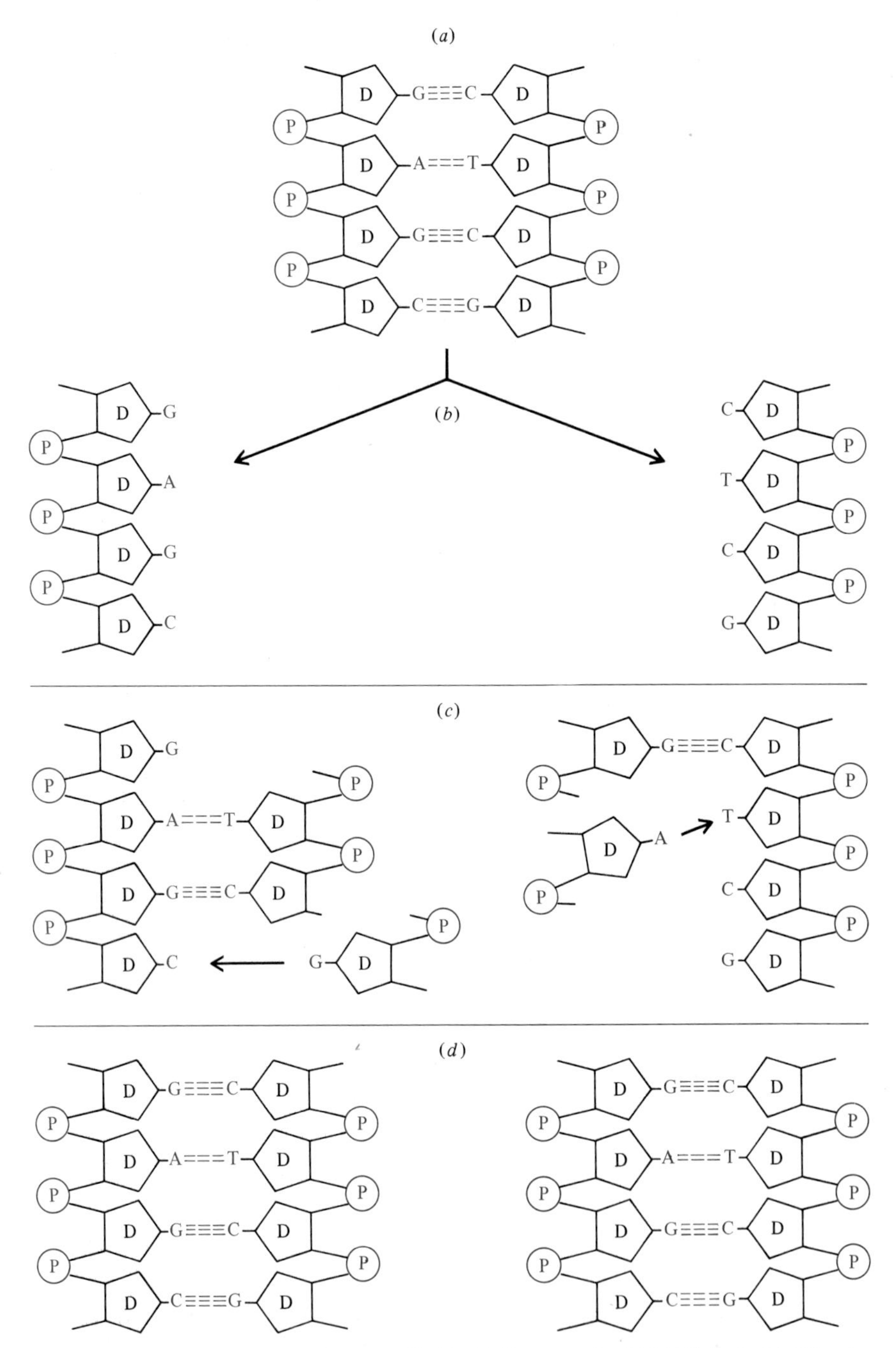

FIGURE 15-8 Replication mechanism of DNA. (*a*) Two strands of DNA linked with H bonds; a segment of a double helix. (*b*) The two strands of the double helix of DNA separate into single strands. (*c*) Individual nucleotides from a pool begin to pair up their bases with the bases on each single strand. (*d*) The new nucleotides are assembled and bonded through phosphate linkages to form a new strand around each of the old strands. Two double helices which are identical with the original double helix are obtained.

newly formed single strands are free to form new hydrogen bonds with the nucleotides surrounding them, pairing adenine to thymine and cytosine to guanine. The new nucleotides become linked in a helical strand through phosphate ester linkages. Thus a new strand is formed around each of the old strands. Two double helices identical with the original double helix are obtained. DNA carries the genetic information, which can thus be transferred to every cell as cell division occurs. See Fig. 15-8.

The second major function of DNA, control of protein synthesis, is accomplished by passing the necessary information to RNA, which then directs the assembly of proteins.

15-3 RNA: structure and functions

Unlike DNA, RNA is generally single-stranded and does not have the almost 1:1 ratio of the heterocyclic bases. The single strands may be folded. There are three types of RNA which have a part in protein synthesis.

Ribosomal RNA

Ribosomal RNA (rRNA) is found in the cytoplasm of the cell, where it occurs in combination with proteins in granules called *ribosomes*. The ribosomes typically consist of 60% rRNA and 40% protein. They are the sites at which protein synthesis occurs, but the details are still obscure.

Messenger RNA

Messenger RNA (mRNA) is involved in obtaining the specific directions from DNA for the synthesis of protein molecules. The molecular weight of mRNA varies considerably, depending on the number of amino acids in the protein which the message from DNA describes, but it is about 1 million.

Transfer RNA

Transfer RNA (tRNA) is also sometimes called *soluble RNA* because it is more soluble in aqueous solutions than other types of RNA are. Its molecular weight is only about 25,000. There is a specific tRNA for each different amino acid. Indeed, the function of tRNA is to pick up and carry a specific amino acid molecule to a ribosomal site for protein synthesis (Fig. 15-9). Only one molecule of an amino acid can become attached to and be carried by a tRNA molecule at a time.

All three types of RNA are synthesized under the control of DNA.

15-4 Natural synthesis of proteins

In order to transcribe a message for the purpose of producing proteins, the DNA molecule partially uncoils to expose a series of bases (purines and pyrimidines) in much the same manner as occurs during DNA replication. An mRNA molecule then forms along a single DNA strand by assembling nucleotides which are complementary to the exposed bases.

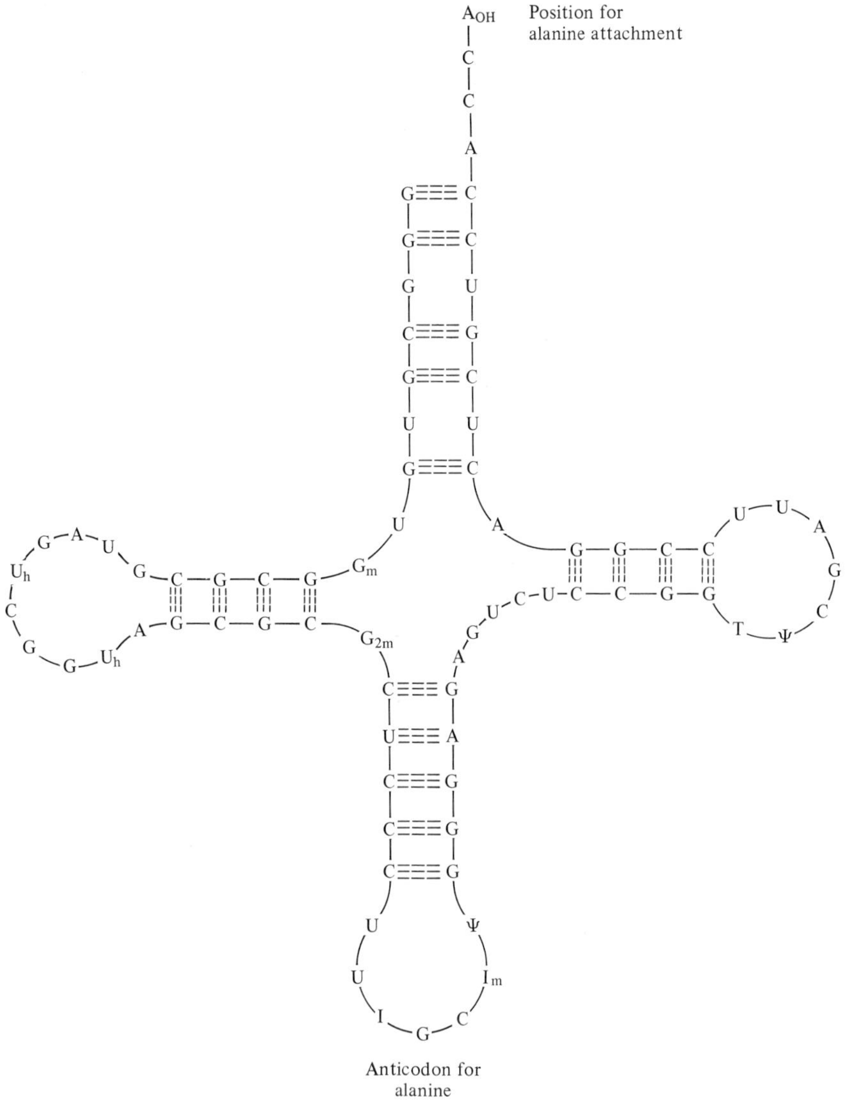

FIGURE 15-9
The base sequence in an alanine tRNA molecule from yeast. The molecule is a single chain; but as it folds back on itself, hydrogen bonds are formed in some regions. A, G, C, and U represent the four common bases in RNA. The other symbols represent less common bases found in RNA.

Position for alanine attachment

Anticodon for alanine

However, RNA uses uracil rather than thymine to pair with adenine. The sequence of bases in the mRNA is determined by the DNA strand the mRNA is formed on. It is that sequence of bases which determines the order in which amino acids are eventually linked together to form a protein.

After the mRNA is formed, it leaves the nucleus of the cell and enters the cytoplasm. There as many as six or eight ribosomes at a time become attached to the mRNA strand. The structure is termed a *polysome*, from polyribosome. The arrangement seems to cause the appropriate tRNA with its attached amino acid molecule to now become temporarily at-

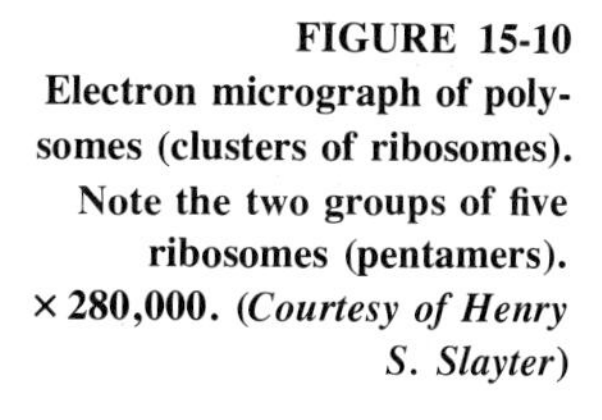

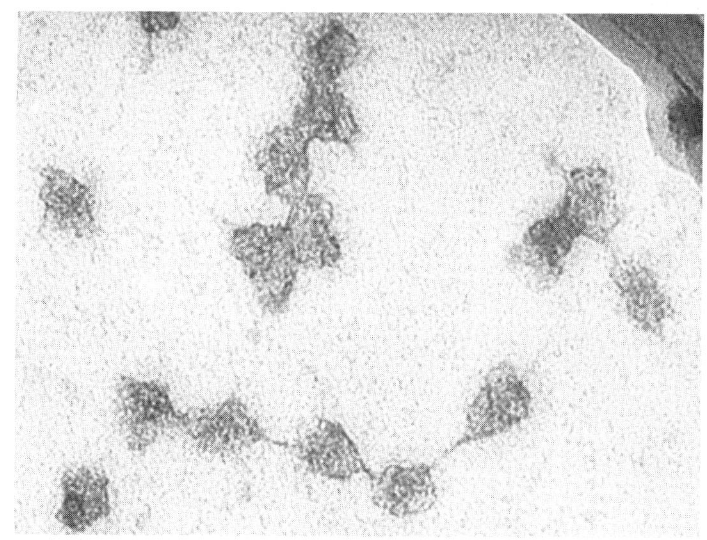

FIGURE 15-10 **Electron micrograph of polysomes (clusters of ribosomes). Note the two groups of five ribosomes (pentamers). × 280,000.** (*Courtesy of Henry S. Slayter*)

tached to the ribosome (Fig. 15-10). The amino acid of the initial tRNA molecule becomes the *N*-terminal end of a protein molecule. A second tRNA molecule is then brought up and attached to the same ribosome; its amino acid molecule becomes linked to the amino acid brought up, and now released, by the initial tRNA molecule. This initial molecule of tRNA now leaves the polysome while the second molecule of tRNA still remaining holds the newly formed dipeptide. A third molecule of tRNA with its attached amino acid then comes in. It attaches to the same ribosome while it delivers its amino acid to the growing peptide chain, and the preceding molecule of tRNA leaves. The process of linking amino acid molecules continues until the protein chain is complete. The mRNA supplies the information concerning the order in which tRNA molecules should arrive. The attached ribosomes seem to move along the mRNA strand as the protein chain becomes longer. Finally, current speculation is that they come off when they reach the end of the mRNA strand.

Codons and anticodons

The process of building an mRNA strand on the basis of the pattern dictated by the separated strand of DNA is called *transcription*. The sequence of bases in the original DNA is believed to constitute a code. Intensive research and study has indicated that the code is made up of "three-letter words." The cracking of the genetic code was the joint accomplishment of several noted scientists; among them are M. Nirenberg, H. Khorana, P. Leder, and S. Ochoa. A set of three bases constitutes one "word" and is called a *codon*. Sequences of codons appear all along an mRNA strand.

After attachment of the mRNA to ribosomes, the codons on the strand are matched by *anticodons* which occur on the tRNA molecules. Only a

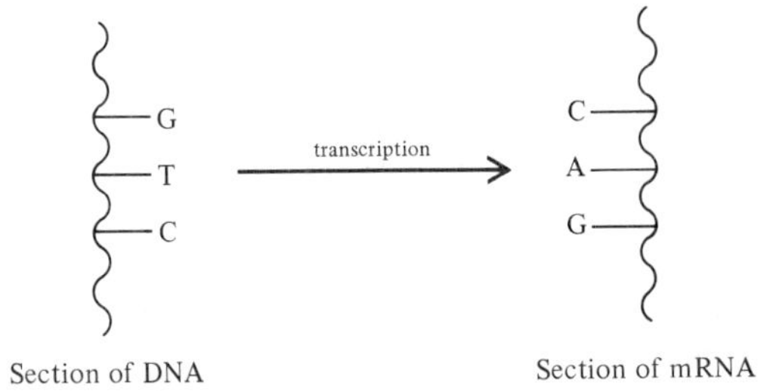

FIGURE 15-11
The code carried by a section of DNA is transcribed to the correct codon on a section of mRNA. The mRNA is shown after it has uncoiled from the DNA.

G
T
C
transcription
C
A
G

Section of DNA   Section of mRNA

tRNA molecule with the proper anticodon in its chain will become attached at the polysome. In that way the codon of the mRNA dictates which particular tRNA molecule can arrive and deposit its amino acid molecule. For example, if a section of DNA carries a sequence of the three bases guanine (G), thymine (T), and cytosine (C), it, when transcribed to mRNA, becomes the codon CAG (Fig. 15-11).

The mRNA then migrates and becomes attached to the ribosomes. There it is matched, through correct hydrogen bonding, by a tRNA molecule with the base sequence GUC, which is the anticodon (Fig. 15-12). [Recall that RNA uses uracil (U) rather than thymine (T).] This particular codon (CAG) on the mRNA matches with the anticodon (GUC) on the tRNA that carries the amino acid glutamine. In this example, a glutamine molecule has been designated as taking a particular place in a protein molecule. The process by which the coded information on the mRNA is used to specifically program the sequence of the amino acids in the synthesis of proteins at the ribosomes is called *translation*.

As the result of much research a large number of codons have been determined. They are given in Table 15-3. Note that more than one codon exists for most of the amino acids. Glutamine, for example, would also be transported to the site of protein synthesis by a tRNA molecule recognizing the codon CAA. A code like that, which is not specific, is said to be *degenerate*. The reason is not well understood. It has been speculated that one codon may be used for a given amino acid in constructing a certain kind of protein and that another codon may be used for the same amino acid in the synthesis of a different kind of protein. Also, it has been suggested that the particular codon utilized is influenced by the amino acid's near neighbors in the protein chain.

Certain codons are referred to as *nonsense codons* because they can-

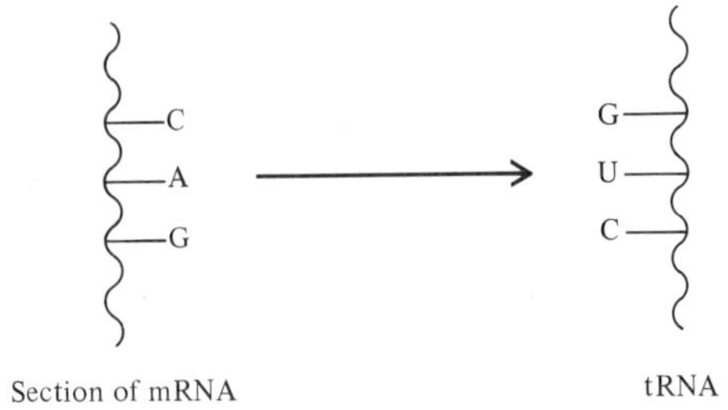

C
A
G
G
U
C

Section of mRNA   tRNA

FIGURE 15-12
A codon (a section of mRNA) is matched with the correct anticodon (a tRNA molecule).

TABLE 15-3
Summary of mRNA codons

Amino acid	mRNA codons					
Alanine	GCU	GCC	GCA			
Arginine	CGU	CGC	CGA	CGG	AGA	GCG
Asparagine	AAU	AAC				
Aspartic acid	GAU	GAC				
Cysteine	UGU	UGC				
Glutamic acid	GAA	GAG				
Glutamine	CAA	CAG				
Glycine	GGU	GGC	GGA	GGG		
Histidine	CAU	CAC				
Isoleucine	AUU	AUC				
Leucine	UUG	CUC	CUG			
Lysine	AAA	AAG				
Methionine	AUG					
Phenylalanine	UUU	UUC				
Proline	CCU	CCC	CCA	CCG		
Serine	UCU	UCC	UCA	UCG	AGU	AGC
Threonine	ACU	ACC	ACA	ACG		
Tryptophan	UGG					
Tyrosine	UAU	UAC				
Valine	GUU	GUC	GUA	GUG		

not be matched by any of the anticodons borne by the tRNA molecules. For that reason the sequence of protein synthesis comes to an end, and such codons are in effect *chain-terminating codons*. Examples are UAA, UAG, and UGA.

Although the codon determinations listed in Table 15-3 were made on the intestinal bacillus, *Escherichia coli,* they have been found to be the same in other organisms. It now appears that all species may use the same code to direct protein synthesis.

Figure 15-13 summarizes the processes of protein synthesis.

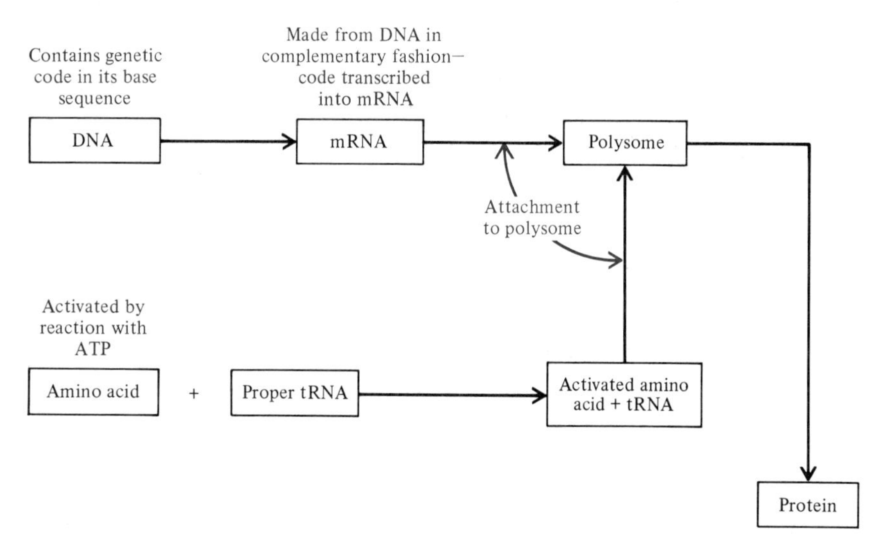

FIGURE 15-13
A summary of protein synthesis.

PROBLEM 15-7 Which amino acid would be designated to take a position in a protein by each of the following codons in the mRNA molecule? (a) GAA, (b) GUA, (c) AUG, (d) AGU, (e) CCC.

Solution to (a) According to Table 15-3, the mRNA codon GAA designates glutamic acid.

PROBLEM 15-8 Which anticodon would match each of the following mRNA codons? (a) UGG, (b) UUU, (c) UGU, (d) GAG, (e) ACU.

Solution to (a) The anticodon on tRNA which matches the mRNA codon UGG is ACC.

PROBLEM 15-9 Which amino acid would be designated to take a position in a protein by each of the following sequences in the *DNA* molecule? (a) CTT, (b) CAT, (c) AAA, (d) TAC, (e) ACC.

Solution to (a) The sequence CTT in the DNA molecule translates to the codon GAA on mRNA. That is one of the codons for glutamic acid.

15-5 Mutations

A *mutation* is a genetic change; it can be caused by the substitution of one heterocyclic base for another in DNA. The result is an altered genetic code and so an abnormal protein. Mutations may be hereditary; they may be caused by the environment; or they may be intentionally brought about by genetic engineering.

Hereditary mutations

In Chap. 14 the genetic disease sickle-cell anemia was discussed. It is caused by the substitution of valine for glutamic acid in a chain of about 300 amino acids. The formation of the abnormal protein, which has a devastating effect, can be interpreted as the result of an inborn error in the genetic code of the person's DNA. Other examples of genetic diseases caused by erroneous coding in the DNA are phenylketonuria, or PKU (Chap. 14), retina blastoma, and even color blindness. There are a great number of genetic diseases.

Mutations caused by environment

Chemicals of various sorts in the environment may produce mutations. For example, nitrous acid will induce a mutation in the tobacco mosaic virus. As you learned in Chap. 13, nitrous acid converts amino groups to hydroxy groups. It can alter adenine and cytosine, as the following equations show.

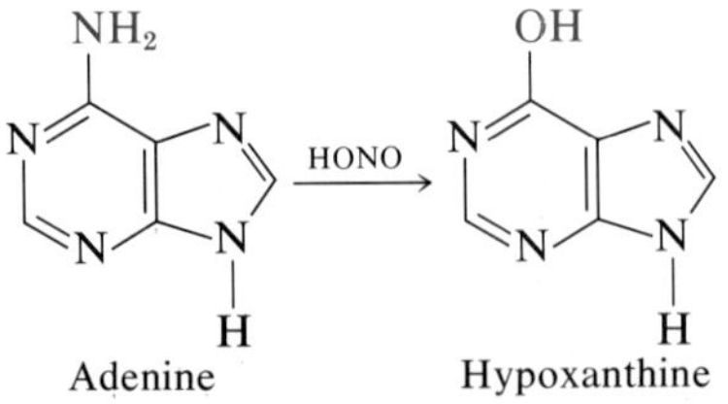

$$\text{Adenine (}NH_2\text{ at C-6; N–H)} \xrightarrow{HONO} \text{Hypoxanthine (OH at C-6; N–H)} \qquad (15\text{-}4)$$

Adenine Hypoxanthine

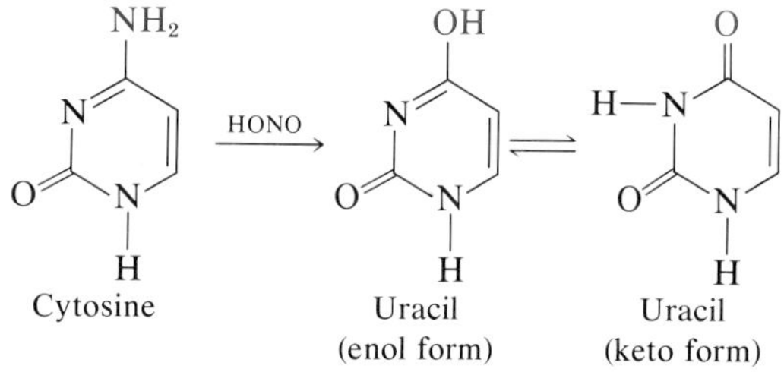

Cytosine $\xrightarrow{HONO}$ Uracil (enol form) $\rightleftharpoons$ Uracil (keto form) (15-5)

These conversions of adenine and cytosine have a considerable effect on base pairing. Hypoxanthine pairs with cytosine rather than thymine; the formation of uracil results in a pairing with adenine rather than with guanine. That, of course, alters the codon and results in substitution of other amino acids in the protein chain. The result is an altered protein.

Sodium nitrite is used in frankfurters, bologna, Vienna sausage, and luncheon meats of all sorts to prevent botulism and to impart a fresh red color to the meat. When the meat is ingested, the sodium nitrite can be converted to nitrous acid in the stomach. Nitrous acid is unstable, and it would have to reach the germinal cells of the body to produce genetic effects. But any nitrous acid produced can also react with any secondary amines present (such as the amino acid proline) to produce *N*-nitrosoamines. Some *N*-nitrosoamines have been demonstrated to cause cancer.

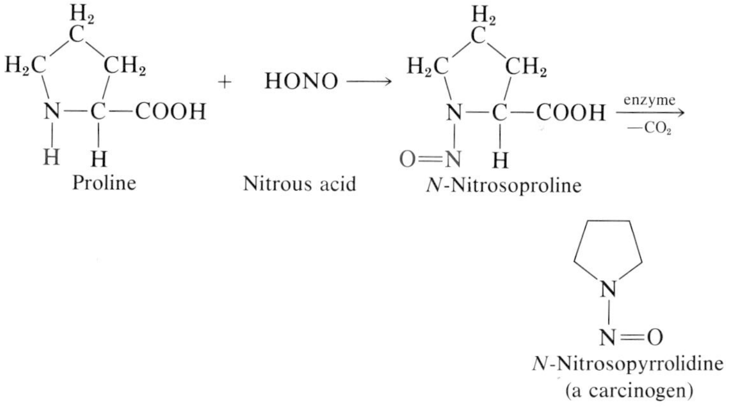

Proline + HONO (Nitrous acid) $\longrightarrow$ *N*-Nitrosoproline $\xrightarrow[-CO_2]{enzyme}$ *N*-Nitrosopyrrolidine (a carcinogen) (15-6)

Caffeine, an alkaloid and stimulant present in coffee, tea, and cola drinks (Chap. 13), may possibly cause mutations. Research has already shown it to cause mutations in *Escherichia coli* and certain fungi and abnormalities in the chromosomes of fruit flies. However, mutagenic effects on mammals have not been demonstrated. Since caffeine is a purine, like adenine and guanine, the possibility exists that it could be substituted for those bases in the DNA chain and thus alter the genetic code.

High-energy radiation, such as x-rays and atomic radiation, is known to produce mutations. Since such radiation is capable of breaking covalent bonds, the structures of the bases in DNA could be altered. That, in turn, would alter the genetic code, and mutations would result.

Alkylating agents such as dimethyl sulfate and methyl iodide have been shown to be mutagenic for lower-life forms. They probably act by alkylating the bases of DNA and thus affecting base pairing. Some anticancer drugs are alkylating agents, and there may be many others which may cause mutations by affecting the DNA molecule in some way. More research regarding the safe usage of chemicals is needed.

Genetic engineering

The development of *genetic engineering* is now possible. Sometime in the future, defective human genes responsible for birth defects may be removed from the living embryo and normal genes substituted. *Cloning* is a technique which has already been applied to simpler plants and animals. In cloning, a few tissue cells are cultivated and treated to stimulate them to differentiate and grow to produce a whole organism. The ultimate of the technique would be to duplicate copies of any human being simply by culturing a few of his or her cells. Genetic engineering also may enable scientists to arrange an embryo's genes so that an individual with certain desired characteristics could be produced. Human life could be lengthened by removing or by breeding out inherited weaknesses. Genetic engineering may make possible the eventual overcoming of immunity reactions so there would be no limit to the number and kinds of organ transplants. Even limb replacement could occur. On the other hand, by increasing the body's immunity, it may be possible to prevent the development of cancer or to destroy cancer once it has started.

Summary

The *nucleic acids* are an important class of *biopolymers*. They constitute the chemical compounds of the *genes*, and they are involved in the natural *synthesis of proteins*. *Deoxyribonucleic acid* (*DNA*) is found mainly in the nucleus of the cell as an essential part of the chromosomes. *Ribonucleic acid* (*RNA*) is found throughout the cell, but mainly in the cytoplasm. It participates in the synthesis of proteins.

Upon hydrolysis, all nucleic acids yield (a) *ribose* or *deoxyribose*, (b) *phosphoric acid*, and (c) a mixture of *purines* and *pyrimidines*. DNA contains *2-deoxyribose*, whereas RNA contains *ribose*. Both DNA and RNA contain adenine, guanine, and cytosine; but DNA contains *thymine*, whereas RNA contains *uracil*.

The combination of a heterocyclic base with a pentose sugar is called a *nucleoside*. A purine or a pyrimidine base is bonded to C-1′ of either ribose or 2′-deoxyribose through C-9 of a purine base or C-1 of a pyrimidine base. The name of a nucleoside incorporates the name of the nitrogen base and the suffix *-idine*, for a pyrimidine base, or *-osine*, for a purine base.

Nucleotides are phosphate esters of the nucleosides which usually form through C-5′ of the pentose. The name of a nucleotide incorporates the name of its nitrogen base and the suffix *-ylic* followed by the word *acid*.

The nucleic acids are *polynucleotides* in which the nucleotide units are bonded through phosphate ester linkages with the C-3′ hydroxy group of each pentose. Each unit of phosphoric acid still retains a free hydroxy group in the polymer, so it is acidic. The acidic sites combine with the basic side chains of proteins to form *nucleoproteins,* so that the genetic material DNA is protected by a coat of protein.

The *primary structure* of DNA is the *nucleotide content* and *sequence* in the polynucleotide polymer. The many different kinds of DNA are due to the great number of variations possible for the sequencing of the four heterocyclic bases in such a large polymer. The almost *1:1 ratio* of purines to pyrimidines, along with crystallographic data, led Watson and Crick to suggest that DNA consists of two identical right-handed helical polymeric strands wrapped about each other to form a double helix. The two strands of the DNA double helix are held together by pairing adenine with thymine to produce two hydrogen bonds and guanine with cytosine to produce three hydrogen bonds. The *secondary structure* of DNA consists of *two helical polynucleotides* which coil around a common axis with the two chains running in opposite directions.

The two *biochemical functions* of DNA are *transmission of genetic information* to succeeding generations and *control of protein synthesis*. The transmission of genetic information is accomplished in the nucleus of the cell by *replication,* which occurs during cell division. In replication, the hydrogen bonds between the bases in the double helix are broken, so that the helix may separate into two single strands in a pool of nucleotides. The bases along a newly formed single strand then form new hydrogen bonds with the nucleotides surrounding them, pairing A to T and C to G. A new strand is thus formed around each of the old strands, and two double helices identical with the original double helix are obtained.

RNA is generally *single-stranded,* and it does not have the almost 1:1 ratio of the heterocyclic bases. The single strands may be folded. There are three types of RNA. *Ribosomal RNA (rRNA)* occurs in the cytoplasm of the cell in combination with proteins in granules called *ribosomes*. The ribosomes are the sites of protein synthesis. *Messenger RNA (mRNA)* is involved in obtaining the specific directions from DNA for the synthesis of protein molecules. *Transfer RNA (tRNA)* is more soluble and is of much lower molecular weight than the other two types of RNA. The function of tRNA is to pick up and carry a specific amino acid molecule to a ribosomal site for protein synthesis. There is a specific tRNA for each amino acid.

To transcribe a message for the purpose of producing proteins, the DNA molecule partially uncoils to expose its bases in much the same manner as in replication. An mRNA molecule then forms along a single DNA strand by assembling nucleotides which are complementary to the DNA molecule's exposed bases. The sequence of bases in the mRNA is thus determined by the DNA strand on which it is formed. Each set of three bases in the sequence constitutes a *codon.* After attachment of the mRNA to ribosomes, the codons on the strand are matched by *anticodons* which occur on the tRNA molecules. In that way, the codon of the mRNA dictates which particular tRNA molecule can arrive and deposit its amino acid for the building of a specific protein. *Translation* is the process by which the coded information on the mRNA is used to program the specific sequence of the amino acids in the synthesis of proteins. More than one codon exists for most of the amino acids, so the code is said to be *degenerate*. Nonsense or *chain-terminating codons* cannot be matched by any of the anticodons borne by the tRNA molecules, and that causes the sequence of protein synthesis to end.

A *mutation* is a genetic change which can be caused by the substitution of one heterocyclic base for another in DNA; the result is an altered genetic code and an abnormal protein. Mutations may be *hereditary,* or they may be caused by *chemicals* in the environment or by high-energy *radiation*.

Genetic engineering may be used in the future to substitute for defective genes or to alter genes to produce more desired characteristics.

Key terms

Anticodon: a set of three bases on a tRNA molecule; it constitutes the code for a specific amino acid. The anticodon is the complement of the codon on mRNA for the particular amino acid.

Chain-terminating codon: a codon on mRNA which cannot be matched by any of the anticodons borne by the tRNA molecules and thus brings the sequence of protein synthesis to an end. Also called **nonsense codon.**

Cloning: a technique in which a few tissue cells are cultivated and treated to stimulate them to differentiate and grow to produce the whole organism.

Codon: a set of three bases on an mRNA molecule; it constitutes the code for a specific amino acid.

Degenerate code: a code in which there is more than one codon for each item coded (in this case, an amino acid).

Deoxyribonucleic acid (DNA): the nucleic acid found in the nuclei of cells; it is the essential component of the chromosomes. It contains the pentose 2-deoxyribose, phosphoric acid, and the bases adenine, guanine, cytosine, and thymine. It exists as a double helix.

Double helix: the secondary structure of DNA. Two helical polynucleotides coil around a common axis. The two chains, joined by hydrogen bonds, run in opposite directions in an antiparallel arrangement.

Genes: the units of heredity; each chromosome consists of several thousand genes.

Genetic engineering: designing, constructing, or managing the genes by substitution or alteration to correct or improve them in some way. Controversy over possible hazards will result in controls.

Messenger RNA (mRNA): one of the three types of RNA. It obtains specific directions from DNA, in the process called transcription, for the synthesis of protein molecules.

Mutation: a genetic change which can be caused by the substitution of one heterocyclic base for another in DNA. The result is an altered genetic code and therefore an abnormal protein.

Nitrogen base: One of the heterocyclic purine and pyrimidine bases which occur in the nucleic acids.

Nonsense codon: one of the codons on mRNA which cannot be matched by any of the anticodons borne by the tRNA molecules and are therefore chain-terminating codons.

Nucleic acid: a polymer of nucleotides with a molecular weight of from 25,000 to about 1 million. It contains three main components: ribose or deoxyribose, phosphoric acid, and a mixture of heterocyclic amine bases (purines and pyrimidines).

Nucleoprotein: a polynucleotide polymer linked to the basic side chains of proteins through the acidic sites on the phosphoric acid groups of the polymer.

Nucleoside: a compound formed by the combination of one of the purine or pyrimidine bases with ribose or 2-deoxyribose; the base and the pentose are bonded through a β-glycosidic linkage.

Nucleotide: a phosphate ester of a nucleoside.

Polynucleotide: a biopolymer formed by the joining of nucleotides through a phosphate ester linkage with the C-3′ hydroxy group of each pentose. The nucleic acids are polynucleotides.

Polysome: a structure consisting of an mRNA molecule and as many as six or eight ribosomes.

Primary structure: the nucleotide content and sequence of a polynucleotide polymer (DNA or RNA).

Purine base: one of the two purine-derivative heterocyclic amines found in DNA and RNA: adenine (6-aminopurine) and guanine (2-amino-6-oxypurine).

Pyrimidine base: one of the three pyrimidine-derivative heterocyclic amines found in nucleic acids. Cytosine occurs in both DNA and RNA; thymine occurs only in DNA; uracil occurs only in RNA.

Replication: reproduction of the DNA molecule in the nucleus of the cell. The two strands of the double helix separate into single strands which form new hydrogen bonds with the nucleotides surrounding them. Two double helices identical with the original helix are thereby produced.

Ribonucleic acid (RNA): the nucleic acid that is found primarily in the cytoplasm of cells and which participates in protein synthesis. It contains the pentose ribose, phosphoric acid, and the bases adenine, guanine, cytosine, and uracil. It is single-stranded.

Ribosomal RNA (rRNA): one of the three types of RNA. It is found in the cytoplasm of the cell in combination with proteins in granules called ribosomes.

Ribosome: one of the granules in the cytoplasm of the cell typically consisting of 60% rRNA and 40% protein. It is a site at which protein synthesis occurs.

Secondary structure: the arrangement in space of a nucleic acid molecule.

The secondary structure of DNA consists of two identical right-handed helical polymeric strands wrapped about each other to form a double helix. The secondary structure of RNA is a single strand, which may be folded.

Transcription: the process by which a DNA molecule partially uncoils and an mRNA molecule then forms along a single DNA strand. The pattern of the mRNA molecule is dictated by the DNA molecule.

Transfer RNA (tRNA): one of the three types of RNA. A specific tRNA exists for each different amino acid; its function is to pick up and carry a specific amino acid molecule to a ribosomal site for protein synthesis.

Translation: the process by which the coded information on mRNA is used to program the specific sequence of the amino acids in the synthesis of proteins at the ribosomes.

Additional problems

15-10 How does deoxyribose differ from ribose?

15-11 How does a nucleotide differ from a nucleoside?

15-12 Draw the structure for a nucleotide found only in DNA.

15-13 Draw the structure for a nucleotide found only in RNA.

15-14 Write the equation for the hydrolysis of deoxythymidylic acid by using aqueous sodium hydroxide. Now write the equation for the hydrolysis of deoxythymidylic acid by using hydrochloric acid.

15-15 Cyclic AMP (adenosine-3′,5′-cyclic monophosphate) is a regulator of metabolic and physiological activity. Draw the structural formula for the compound. (A single phosphate group is esterified with both the 3′- and 5′-hydroxy groups of adenosine.)

15-16 How does RNA differ from DNA? In what ways are the two similar?

15-17 Describe the replication of DNA. How does it resemble the transcription of a message on DNA to a codon on mRNA? In what ways does it differ from transcription?

15-18 The drug 6-mercaptopurine is used in the chemotherapy of leukemia. Draw its structural formula. (The mercapto group is —SH.)

15-19 The codons for glutamic acid are GAA and GAG. What single mononucleotide substitution in each one of those codons would change the codon so it would code for valine? (You will recall that a substitution of valine for glutamic acid in part of the hemoglobin molecule causes sickle-cell anemia.)

15-20 For what amino acid does UAU code? What would be the result of a nucleotide substitution resulting in the codon UAA?

16

ENZYMES

In Chap. 14, you learned that enzymes are largely protein in nature. They have the very important function of serving as catalysts in biochemical reactions. An *enzyme* is a complex organic or organometallic catalyst produced by living cells. You may remember that a *catalyst* is any substance that alters the rate of a chemical reaction without being consumed in the reaction.

16-1 Discovery of enzymes

Since ancient times people have known that yeast is capable of fermenting sugars to produce alcohol and carbon dioxide. Louis Pasteur was among the first to study alcoholic fermentation systematically. He observed that glucose, if stored in sterile and sealed containers, remains unchanged indefinitely. Addition of yeast, however, causes the glucose to be fermented to alcohol and carbon dioxide. Pasteur incorrectly ascribed the catalytic activity of the yeast to intact microorganisms, which he termed *ferment*. In 1878 Willy Kuhne proposed that such organic catalytic substances be called *enzymes* (from the Greek *en,* "in," and *zyme,* "yeast").

In 1897 Edward Buchner (Nobel Prize, 1907) prepared a cell-free extract containing enzymes by grinding yeast cells with sand and then filtering the resultant material. The extract was found to be just as capable of promoting the fermentation of sugars as the original yeast cells were. In 1926, James Sumner (Nobel Prize, 1946) was successful in isolating a pure, crystalline enzyme, urease. It was found to be proteinlike in nature. Since then many other enzymes have been isolated.

16-2 Function and nature of enzymes

The chemical reactions of organic molecules which take place in living things include the usual reactions which may be carried out in the laboratory. Examples are oxidation-reduction, hydrolysis, esterification, formation of amides, and polymerization. But when the reactions are performed in the laboratory, many of them require such rigorous conditions as the use of strong acids, strong bases, or oxidizing agents or high temperatures and pressures. Biochemical reactions, in contrast, must occur at physiological conditions of pH 7 and a temperature about 37°C for mammals. They occur very rapidly and efficiently under those mild conditions because of the complex organic catalysts—enzymes—produced by living cells.

Unlike most chemical catalysts, an enzyme is very specific; it may catalyze only one type of reaction of one particular compound. Hydrochloric acid, on the other hand, will catalyze the hydrolysis of sucrose to glucose and fructose in a reasonable time, if aided by heating. It will also catalyze the hydrolysis of other acetals and of esters and amides as well. But if the enzyme sucrase is added to sucrose, the hydrolysis takes place rapidly at room temperature. And, unlike hydrochloric acid, sucrase will catalyze only the hydrolysis of sucrose. It will not hydrolyze esters and amides or even the other disaccharides.

16-3 Nomenclature of enzymes

The substance upon which an enzyme acts is the *substrate*. In the past enzymes have most often been named by adding the suffix *-ase* to the root of the name of the substrate. Accordingly, the enzyme urease acts upon urea, maltase on maltose, and sucrase on sucrose. Enzymes have also been named after the *products* which are formed as a result of their action. The enzyme sucrase is also known as invertase because it catalyzes the production of invert sugar (a 1:1 mixture of D-glucose and D-fructose). Many enzymes have been given names indicating the *types of reaction* they catalyze. Examples are pyruvic *carboxylase,* lactic *dehydrogenase,* and xanthine *oxidase*. Some of the enzymes which have been known for a longer time, particularly the digestive enzymes, still retain their old common names; examples are ptyalin, pepsin, and rennin.

Over a thousand enzymes have now been identified. In 1961 the International Union of Biochemistry suggested that enzymes be classified according to the general type of reaction that they catalyze. They would then be named on the basis of the precise chemical name of the substrate and the general type of chemical reaction. However, because such names are often ponderous, they have not been enthusiastically accepted for some enzymes. Sucrase, for example, is α-glycopyrano-β-fructofuranohydrolase.

16-4 Enzyme structure

Some enzymes, like pepsin and ribonuclease, are *simple proteins;* others are *conjugated proteins* with both a polypeptide moiety and a nonprotein moiety. Enzymes which are conjugated proteins may be referred to as *holoenzymes*. The polypeptide moiety is called an *apoenzyme,* and the nonprotein moiety (which can often be separated from the apoenzyme) is called the *coenzyme*. The apoenzyme is inactive without the coenzyme.

Metal ions such as Mg^{2+}, Mn^{2+}, and Zn^{2+} are sometimes found bonded to either the apoenzyme or the coenzyme portions. They serve as *enzyme activators*. They probably act by forming a coordination complex between the enzyme and the substrate to facilitate electron shifts. Some examples of enzymes requiring a metal-ion activator are alkaline phosphatase (zinc), carboxypeptidase (zinc), cytochrome c (iron), glycol dehydrase (cobalt), pyruvate oxidase (manganese), and xanthine oxidase (iron and molybdenum). *Activation* is any process in which the inactive protein is converted into an active enzyme; it includes union with a coenzyme or activation by a metal ion.

Proenzymes or *zymogens* are the inactive forms in which some of the simple protein enzymes are first secreted. Such proenzymes are subsequently converted into the active enzymes by other enzymes which remove an inhibitory peptide from the proenzyme molecule. The proenzyme trypsinogen loses a hexapeptide to become the active digestive enzyme trypsin.

Several of the coenzymes are directly related to certain vitamins, especially the B vitamins. *Vitamins* are organic compounds that cannot be synthesized by an organism, and yet very small amounts of them are essential for the maintenance of normal health and growth. They therefore must be included in the diet. Some of the coenzymes involving vitamins will be considered in more detail in Chap. 17.

PROBLEM 16-1 List the function of each of the following enzymes in the human body, i.e., the substrate and the products of the reaction.

(a) ribonuclease
(b) epimerase
(c) racemase
(d) amylase
(e) sucrase
(f) lipase
(g) protease
(h) peroxidase
(i) dehydrogenase

The name gives a clue to either the substrate or the type of reaction. You need not be specific for every name, e.g., racemase.

Solution to (a) Ribonuclease is the enzyme which catalyzes the hydrolysis of nucleic acids (the substrate). The products are pyrimidines, purines, ribose or deoxyribose, and phosphoric acid.

PROBLEM 16-2 Define each of the following terms. Illustrate with appropriate examples:

(a) enzyme
(b) substrate
(c) proenzyme
(d) holoenzyme
(e) prosthetic group
(f) enzyme activator
(g) apoenzyme
(h) coenzyme
(i) vitamin

Solution to (a) An enzyme is a complex organic or organometallic catalyst, largely protein in nature, produced by living cells. An example is ptyalin in the saliva, which hydrolyzes amylase of starches. Another example is sucrase, which hydrolyzes sucrose.

16-5 Enzyme specificity

In catalyzing biochemical reactions, enzymes have some degree of specificity. The different types of specificity are absolute specificity, absolute group specificity, relative group or linkage specificity, and stereochemical specificity.

Absolute

An enzyme with *absolute specificity* catalyzes one and only one reaction. A common example is the hydrolysis of urea by urease.

$$H_2N-\overset{\overset{\displaystyle O}{\|}}{C}-NH_2 + H_2O \xrightarrow{\text{urease}} 2NH_3 + CO_2 \qquad (16\text{-}1)$$

Urease is so specific that it will not catalyze the hydrolysis of any other compound, even a closely related derivative of urea.

Absolute group

An enzyme with *absolute group specificity* catalyzes the reactions of one particular functional group in any compound. Carboxypeptidase is an example. It attacks the peptide linkages of any amino acid moieties containing free carboxyl groups and cleaves them from the rest of the chain. That is, it cleaves *C*-terminal amino acids.

Relative group

Enzymes with *relative group specificity* have still less specificity because they may catalyze a similar reaction (say hydrolysis) of more than one kind of group. For example, an enzyme with relative group specificity might cleave both an ester and an amide linkage. The enzyme chymotrypsin catalyzes the hydrolysis not only of peptide bonds but also of a variety of carboxyl derivatives including amides, alkyl and aryl esters, and anhydrides.

Stereochemical

Enzymes which catalyze the reactions of certain stereoisomers but do not catalyze the reactions of their enantiomers are said to have *stereochemical specificity*. Enzymes in the other specificity categories also may exhibit stereochemical specificity. Lactic acid dehydrogenase catalyzes the oxidation of the L-lactic acid found in muscle cells but not the D-lactic acid found in certain microorganisms. Arginase catalyzes the hydrolysis of L-arginine but will not act on the D isomer.

PROBLEM 16-3 What type of specificity has the enzyme sucrase?

16-6 Theory of enzyme action

In 1888 Svante Arrhenius proposed a theory which applies to all catalytic reactions. He suggested that a catalyst combines with a reactant to form *an intermediate complex*. The intermediate is more reactive than the ini-

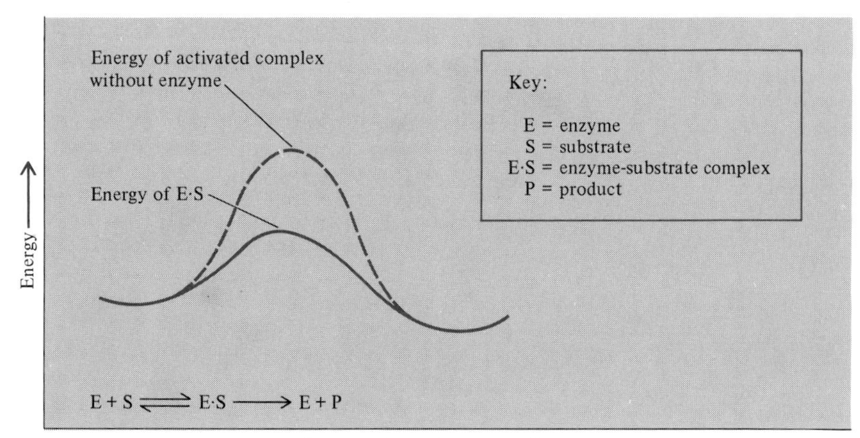

FIGURE 16-1
Progress-of-reaction diagram for a biochemical reaction. The enzyme lowers the activation energy.

tial uncombined complex because it affords a pathway requiring less energy. In effect, the activation energy of the reaction is lowered, and that results in an increased rate of reaction. (You may wish to review Sec. 1-3.) The intermediate complex can then react to form products which disengage from the catalyst.

Applied to enzymes, if E represents the enzyme, S the substrate, E · S the complex, and P the products, the following equation applies.

$$E + S \rightleftharpoons E \cdot S \longrightarrow E + P \qquad (16\text{-}2)$$

The energy relations of such a reaction are shown graphically in Fig. 16-1.

The enzyme, which is usually larger than the substrate, appears to have a relatively small region on its surface which exactly fits a complementary portion on the substrate. The region on the enzyme is called the *active site;* it is where bonding to form the enzyme-substrate complex (E · S) occurs, and it is also the place of product formation. The possible nature of an active site is shown in Fig. 16-2.

The fact that enzymes are deactivated by denaturation (Sec. 14-5) is indicative of the importance of the secondary and tertiary structures of the protein enzyme in maintaining a very precise three-dimensional arrangement for the active site. The concept of complementary structures of enzyme and substrate, which enable the enzyme and substrate to fit together to form the initial complex, is called the *lock-and-key theory*.

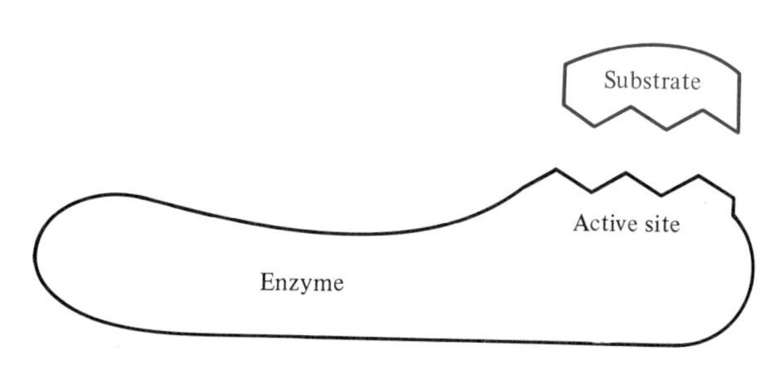

FIGURE 16-2
A schematic representation of a substrate and the active site of an enzyme. The structures are complementary, so they can fit together as a key fits a lock.

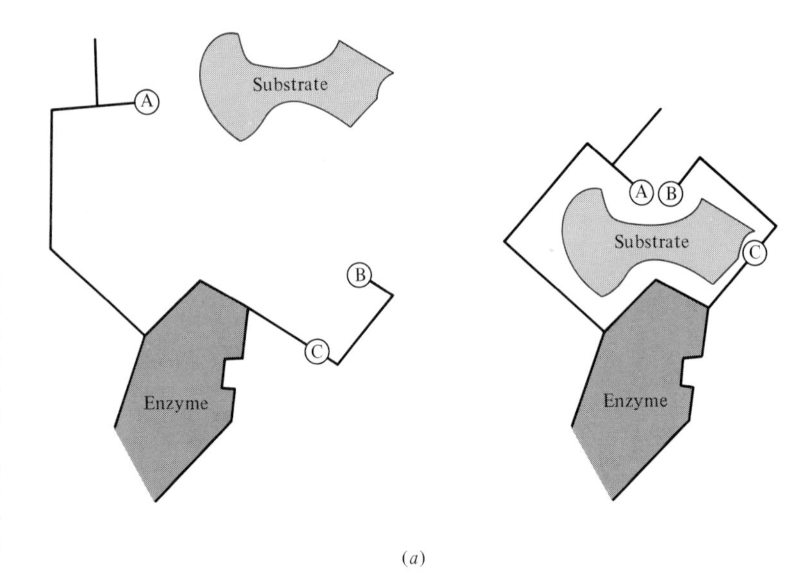

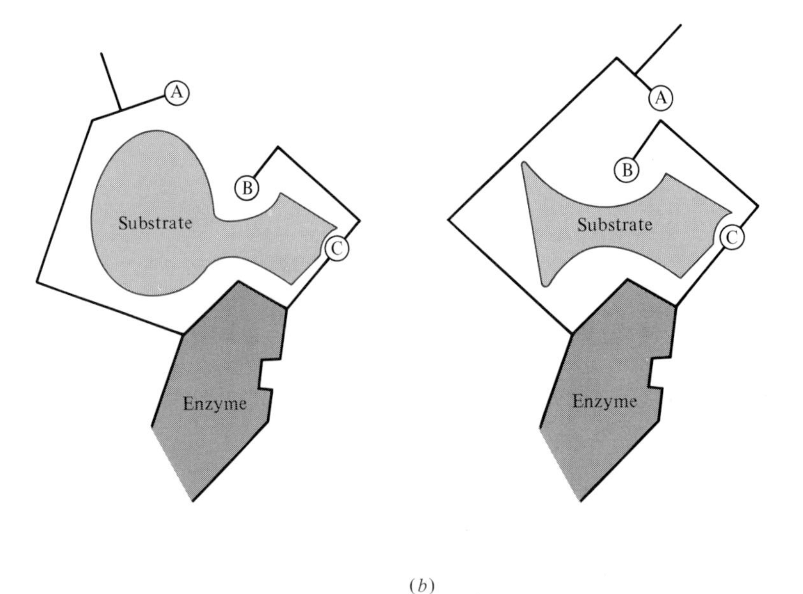

FIGURE 16-3
A representation of a flexible active site of an enzyme. (*a*) Binding to the substrate with contact group C induces proper alignment of catalytic groups A and B so that reaction takes place. (*b*) Representation showing that compounds that are actually too large or too small may be bound to the active site, but fail to produce proper alignment of catalytic groups A and B, so no reaction takes place. (*Adapted from D. E. Koshland, Jr.*, Science, 142: *1539 (1963); Copyright 1963 by the American Association for the Advancement of Science*)

The lock-and-key theory is modified by Koshland's *induced-fit theory*. In 1963, D. E. Koshland, Jr., proposed that an enzyme changes its conformation as it reacts with a substrate molecule in forming an activated complex. After completion of the catalyzed reaction, the enzyme again acquires its original conformation. The substrate causes a change to occur in the structure of the enzyme molecule so that the molecule adapts to fit the substrate. This extension of the lock-and-key theory explains why some compounds which are not properly shaped to bind to the active site nevertheless do bind and are catalyzed to react. Also, it explains why some substrates seem to bind to the enzyme and yet do not react. Koshland explains that the active site of an enzyme consists of *contact groups,* which cause the specificity of the substrate, and also *catalytic groups,* which cause the catalytic action. The active center is thus a center of variable conformation which can be induced to fit more than one structurally related compound, but only the intended substrate will also have its catalytic groups in proper alignment for catalysis. See Fig. 16-3.

16-7 Factors in enzyme activity

Because enzymes are catalysts, they are affected by any factors which in general affect catalysts. Since they are also proteins, they are affected by any factors which affect proteins.

The effect of temperature

Although the rate of most chemical reactions increases with increasing temperature, the rate of enzyme reactions falls off rapidly with temperatures above or below optimum temperature for that particular enzyme (37°C for most enzymes). Heating disrupts the secondary and tertiary structures of the enzyme. That causes a disorientation of the reactive site so that the site is not accessible to the substrate. Still higher temperatures may irreversibly denature the protein of the enzyme and thereby permanently deactivate it. The rate of enzyme-catalyzed reactions is nearly zero at 0°C or at 100°C (Fig. 16-4).

Objects are sterilized by heating them in steam or boiling water to denature the enzymes of any bacteria which may be on them. When we refrigerate or freeze food or biological specimens, enzyme activity is slowed down so that the food or specimens are preserved. Blanching fresh vegetables and fruits in hot water prior to home canning diminishes or destroys enzyme activity.

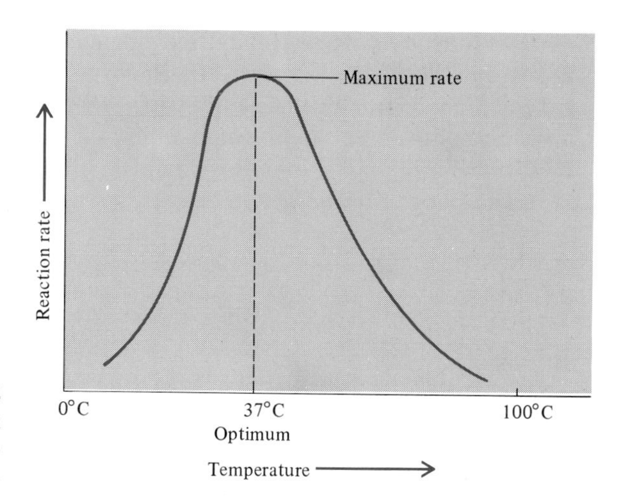

FIGURE 16-4
The effect of temperature on the rate of an enzymatic reaction.

The effect of pH

Enzymes are sensitive to pH changes in their environment. Extreme values of pH cause denaturation. Any change in pH alters the degree of ionization of the basic and acidic groups on both enzyme and substrate. That may adversely affect the formation of the enzyme-substrate complex. With the exception of the enzymes in gastric juice, most body enzymes have an optimum pH between 6 and 8.

The effect of chemicals

Chemical reagents may serve as enzyme inhibitors. Any reagent which denatures proteins will inhibit any enzyme reaction. That is referred to as *nonspecific inhibition*. Some reagents are *specific inhibitors*. They will inhibit only single enzymes or groups of related enzymes. Most poisons act as specific inhibitors. Inhibitors can be effective in extremely small amounts. For example, as low a concentration of cyanide as $4 \times 10^{-6}\ M$ can inhibit enzymatic activity by 85%.

Inhibitors exert their influence in various ways depending on the particular enzyme and inhibitor. A *competitive inhibitor* is structurally related to a particular substrate so that it competes with that substrate for the active site of the enzyme. Therefore, only the first reaction, complex formation, takes place; no products are formed in a second reaction.

$$\mathrm{E + I \rightleftharpoons EI} \qquad (16\text{-}3)$$

where I represents the inhibitor. If inhibitor molecules are present in overwhelming number, they may completely block the active sites on all of the enzyme molecules to effectively stop the reaction. However, since formation of the enzyme-inhibitor complex is reversible, increased substrate concentration permits displacement of the inhibitor from the active site. (See p. 482 for an example.)

Noncompetitive inhibitors form strong covalent bonds with an enzyme and cannot be displaced by the addition of excess substrate. Reactive groups such as $—COO^-$, $—SH$, $—OH$, or $—NH_3^+$ constitute part of the active sites of many enzymes. Any reagent that can combine with one or more of those groups will inhibit the enzyme. The heavy-metal ions, Ag^+, Hg^{2+}, and Pb^{2+}, have strong affinities for $—COO^-$ and $—SH$ groups. Iodoacetic acid combines readily with $—SH$ groups. The so-called *nerve gases* are phosphate esters which inhibit the enzyme acetylcholinesterase by forming an enzyme-inhibitor complex with a hydroxyl group of serine at the active site of the enzyme.

$$\text{Enzyme—OH} + \text{F—}\overset{\overset{\displaystyle O}{\|}}{\text{P}}\text{(OR)}_2 \longrightarrow \text{enzyme—O—}\overset{\overset{\displaystyle O}{\|}}{\text{P}}\text{(OR)}_2 + \text{HF} \qquad (16\text{-}4)$$

Acetylcholinesterase catalyzes the hydrolysis of acetylcholine, a quaternary ammonium salt which is present in an inactive, protein-bound form in nerve cells. The free salt is released when the cell is stimulated by a nerve impulse. A chain reaction along the nerve fiber ensues, and each adjacent nerve cell is stimulated in turn to liberate its own acetylcholine. Once the nerve impulse is so passed, the acetylcholine must be immediately deactivated to leave the cell free to transmit the next impulse. The deactivation is accomplished by the hydrolysis of acetylcholine to choline and acetic acid by the aid of the enzyme acetylcholinesterase.

$$\underset{\text{Acetylcholine}}{CH_3-\overset{\overset{\displaystyle O}{\|}}{C}-OCH_2CH_2\overset{+}{N}(CH_3)_3} + HOH \xrightarrow{\text{acetylcholinesterase}} \underset{\text{Acetic acid}}{CH_3\overset{\overset{\displaystyle O}{\|}}{C}-OH} + \underset{\text{Choline}}{HOCH_2CH_2\overset{+}{N}(CH_3)_3} \qquad (16\text{-}5)$$

If the enzyme is inhibited and inactivation does not occur, then acetylcholine collecting in the cell results in overstimulation of the nerves and the muscles and glands they control. Movements of the entire body become uncoordinated, and tremors, muscular spasms, convulsions, and eventually paralysis and death occur.

The organic phosphate poisons were developed for a weapon in World War II in Germany, and much research has since been done on them. Diisopropylfluorophosphate (DFP) is an example of one of those very toxic agents. Parathion and malathion are related compounds that have been developed as effective, but nonpersistent, insecticides. Parathion is also toxic to mammals, and great care must be exercised when it is used. The structural formulas of those compounds are shown in Fig. 16-5.

Chemotherapy. Chemotherapy is the use of a chemical to destroy or control an invading microorganism without seriously harming the host. Chemicals suitable for chemotherapy are not plentiful. They are difficult to find because the biochemical reactions of all living organisms are so similar that what will harm one organism will usually harm another. An effective chemical may function as an inhibitor of a critical enzyme in the cells of the invading organism.

Antimetabolites serve as competitive inhibitors for a significant meta-

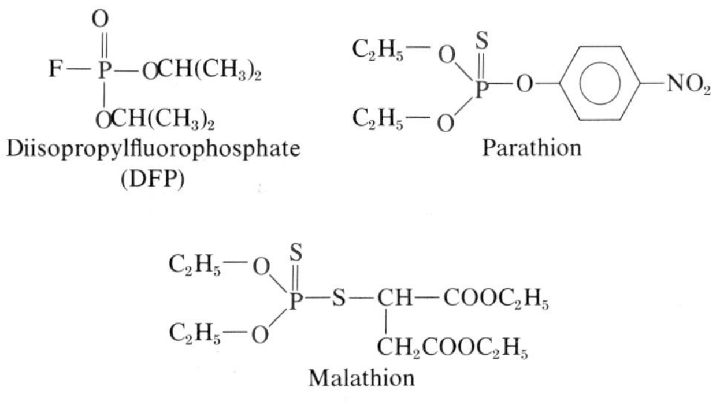

$$F-\overset{\overset{\displaystyle O}{\|}}{\underset{\underset{\displaystyle OCH(CH_3)_2}{|}}{P}}-OCH(CH_3)_2$$

Diisopropylfluorophosphate (DFP)

$$(C_2H_5-O)_2\overset{\overset{\displaystyle S}{\|}}{P}-O-C_6H_4-NO_2$$

Parathion

$$(C_2H_5-O)_2\overset{\overset{\displaystyle S}{\|}}{P}-S-\underset{\underset{\displaystyle CH_2COOC_2H_5}{|}}{CH}-COOC_2H_5$$

Malathion

FIGURE 16-5
Organic phosphate poisons.

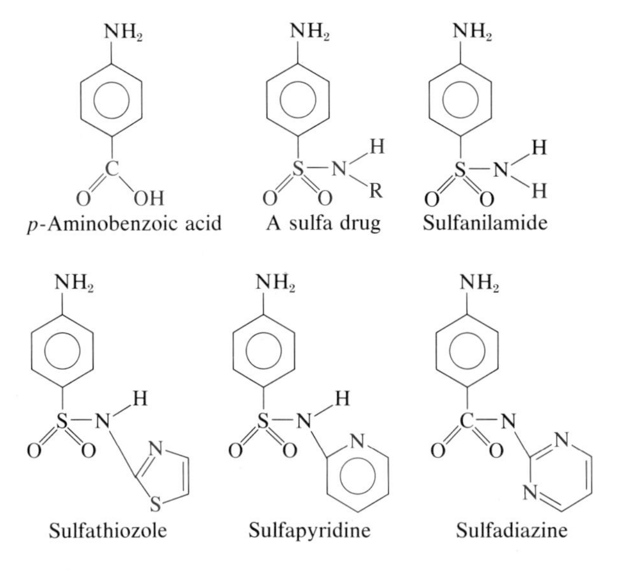

FIGURE 16-6 The relations of the sulfa drugs to *p*-aminobenzoic acid and to each other.

bolic reaction of the invading organism. The *sulfa drugs,* which were discovered and developed in the 1930s as antibacterial agents, are such antimetabolites. They have a structural similarity to *p*-aminobenzoic acid (Fig. 16-6), and so they serve as competitive inhibitors for the enzyme acting on *p*-aminobenzoic acid as a substrate. The enzymatic reaction normally incorporates *p*-aminobenzoic acid into folic acid, a coenzyme needed for body growth and repair. By the action of the sulfa drug, the bacteria are prevented from growing and reproducing, so the body is able to overcome them. Although humans also need folic acid, they obtain it through their diet, so no such human enzyme system is affected. However, allergic effects are sometimes a problem with sulfa drugs.

Antibiotics. An antibiotic is a product of a microorganism which is toxic to another microorganism because it effectively inhibits essential enzymes. It may or may not function as an antimetabolite. Some of the antibiotics have been characterized and synthesized in the laboratory. The oldest and still one of the most widely used antibiotics is penicillin. It was discovered accidentally by Alexander Fleming in 1929, demonstrated to be an effective antibiotic in 1938 by Ernst Chain and Howard Florey, and introduced into medical practice in 1941. It is thought to function by inhibition of an enzyme needed for synthesis of essential components of bacterial cell walls. Several naturally occurring penicillins have been isolated, but pencillin G is the most common (Fig. 16-7). Some strains of bacteria become resistant to penicillin, and some people are allergic to it. In those cases other antibiotics, such as streptomycin, Aureomycin, tetracycline, and Chloromycetin, are employed.

16-8 Enzyme applications

New biochemical methods for the analysis of enzymes developed in 1975 by George F. Sensabaugh and E. F. Blake of the forensic science group,

FIGURE 16-7
The structures of four penicillins.

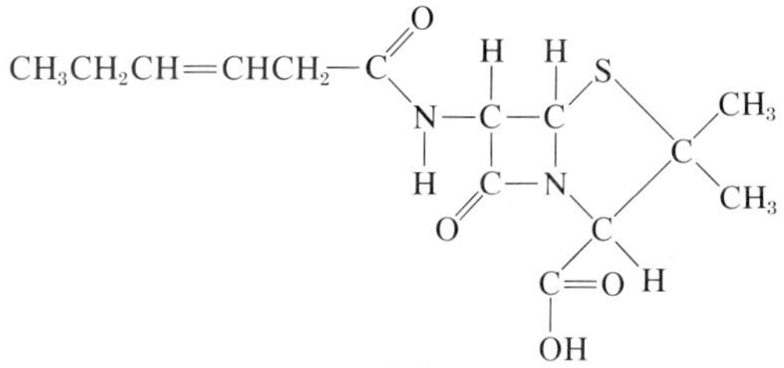

$CH_3CH_2CH{=}CHCH_2{-}C(=O){-}$

Penicillin F

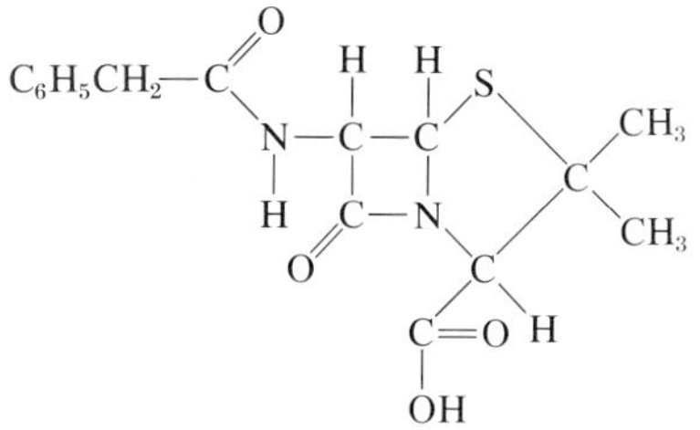

$C_6H_5CH_2{-}C(=O){-}$

Penicillin G

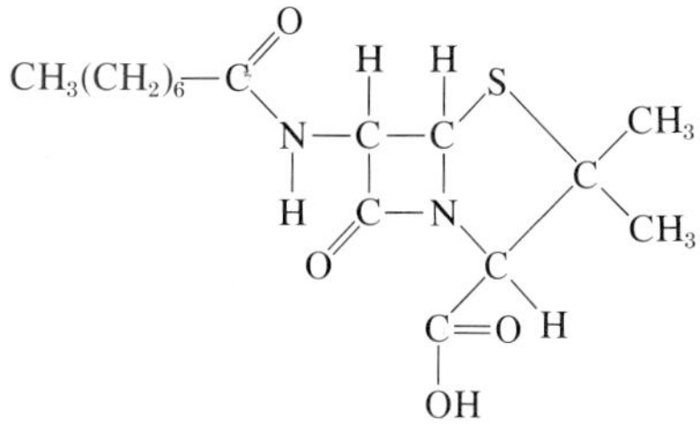

$CH_3(CH_2)_6{-}C(=O){-}$

Penicillin K

$HO{-}C_6H_4{-}CH_2{-}C(=O){-}$

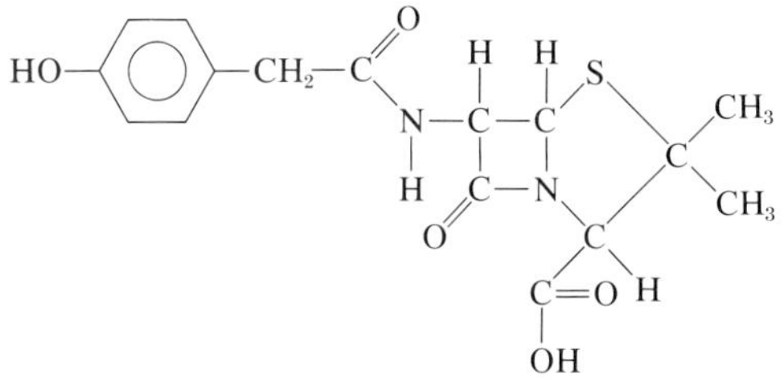

Penicillin X

University of California, Berkeley, will identify semen and sperm sources for evidence in cases of rape. Sensabaugh and Blake found three different genetic variants or patterns in the enzyme diaphorase found in sperm. They also extended their genetic profile to other enzymes found in other tissues as well as in sperm and semen; the others include esterase D, amylase, and glutathione reductase. That makes it possible to distinguish semen from two individuals with a reliability of about 95%.

Enzymes are frequently involved in medical diagnoses. The presence of disease may be indicated by the presence of a specific enzyme in a patient's blood or urine. Damage to tissues, as in the liver, heart, or kidney, allows tissue enzymes to get into the blood. Liver disease can restrict the normal elimination of some enzymes so that concentration of the enzymes in the blood is increased. In hepatitis, damage to the liver causes an increase in the blood of glutamic-oxalacetic transaminase and glutamic-pyruvic transaminase. After myocardial infarction (heart attack), the blood level of creatine phosphokinase and glutamic-oxaloacetic transaminase increases after 1 or 2 days. An increased blood level of amylase may indicate pancreatic or kidney disease. Cancer of the prostate may be indicated by an increased blood level of acid phosphatase.

TABLE 16-1
Genetic disorders and the deficient enzymes

Disease	Deficient enzyme
Acatalasia	Catalase (in red blood cells)
Albinism	Tyrosinase
Alkaptonuria	Homogentisic acid oxidase
Galactosemia	Galactose 1-phosphate uridyl transferase
Glycogen storage disease (several types)	One of the following:
	α-Amylase
	Debranching enzyme
	Glucose 1-phosphatase
	Liver phosphorylase
	Muscle phosphofructokinase
	Muscle phosphorylase
Goiter	Iodotyrosine dehalogenase
Gout	Hypoxanthine-guanine phosphoribosyl transferase
Histidinemia	Histidase
Hypophosphatasia	Alkaline phosphatase
Lesch-Nyhan syndrome	Hypoxanthine-guanine phosphoribosyl transferase
Maple syrup urine disease	α-Keto acid decarboxylase
McArdle's syndrome	Muscle phosphorylase
Metachromatic leukodystrophy	Sphingolipid sulfatase
Methemoglobinemia	NADPH-methemoglobin reductase and NADH-methemoglobin reductase
Niemann-Pick disease	Sphingomyelin-hydrolyzing enzyme
Phenylketonuria	Phenylalanine hydroxylase
Tay-Sachs disease	Ganglioside-degrading enzyme
Tyrosinemia	Hydroxyphenylpyruvate oxidase
Von Gierke's disease	Glucose 6-phosphatase

Reprinted with permission from Robert C. Bohinski, *Modern Concepts in Biochemistry*, 2d ed., Allyn and Bacon, Boston, 1976.

16-9 Enzymatic disorders

Errors in the genetic code of a person's DNA may result in the synthesis of an abnormal enzyme which is unable to catalyze reaction of its usual substrate. Diseases and abnormalities result from such enzymatic disorders. Phenylketonuria has already been mentioned; in it there is a failure in the enzymatic oxidation of phenylalanine to tyrosine.

There are many other such disorders. Galactosemia is a disease resulting from lack of the enzyme required to convert galactose to glucose. It may cause damage to the central nervous system, mental retardation, and even death. It can be controlled by a diet free of galactose. Albinism, a disease in which the person has no pigment in the hair, skin, and eyes, is caused by a lack of the enzyme tyrosinase. Table 16-1 is a partial list of genetic disorders.

Summary

Enzymes are complex organic or organometallic molecules, largely *protein* in nature, which have the important function of serving as *catalysts* in biochemical reactions. Most enzymes function best at physiological conditions of pH 7 and a temperature of 37°C. Unlike chemical catalysts, enzymes are generally very specific in their action; an enzyme may catalyze only one type of reaction of one particular compound.

In the past, an enzyme was most often named by adding the suffix *-ase* to the root of the name of the *substrate*. Many enzymes have been given names indicating the *types of reaction* they catalyze. Some enzymes, particularly the digestive enzymes, retain their old common names. In 1961 the International Union of Biochemistry suggested that enzymes be named on the basis of the *precise chemical name of the substrate* and the *general type of chemical reaction catalyzed*.

Holoenzymes are enzymes which are *conjugated proteins*. The polypeptide part is called an *apoenzyme*, and the nonprotein part is called the *coenzyme*. The apoenzyme is inactive without the coenzyme. *Enzyme activators* are metal ions which may be bonded to either part of a holoenzyme and activate the holoenzyme by forming a coordination complex between the enzyme and the substrate. *Proenzymes*, or *zymogens*, are the inactive forms in which some of the simple (nonconjugated) protein enzymes are first secreted. A proenzyme is subsequently converted into the active enzyme by another enzyme which removes an inhibitory peptide from the proenzyme molecule. Several of the coenzymes are directly related to certain vitamins, especially the B vitamins.

An enzyme with *absolute specificity* will catalyze one and only one reaction. An enzyme with *absolute group specificity* will catalyze the reactions of one particular functional group in a compound. An enzyme with *relative group specificity* may catalyze similar reactions of more than one kind of group, as in the hydrolysis of either an ester or an amide linkage. An enzyme which catalyzes the reactions of certain stereoisomers, but not the reactions of their enantiomers, has *stereochemical specificity*.

An enzyme combines with a substrate to form an *activated complex* which affords a pathway that requires less energy for the formation of products. The *active site* of an enzyme is a relatively small region on the surface of an enzyme which exactly fits a complementary portion on the substrate (*lock-and-key theory*) to form the *enzyme-substrate complex* and subsequently the product. Koshland's *induced-fit theory* suggests that the substrate induces a structural change in the enzyme molecule to form the required activated complex. After completion of the catalyzed reaction, the enzyme resumes its original structure. The active center of an enzyme is a flexible region that can be induced to fit several structurally similar compounds, but only the proper substrate is also capable of correct alignment with the catalytic groups to produce reaction.

The rate of enzyme reactions falls off rapidly with temperatures above or below optimum temperature. Higher temperatures may *irreversibly denature* the protein of the enzyme and thereby permanently deactivate it. Extreme values of pH cause denaturation, but any change in pH alters the degree of ionization of the basic and acidic groups on both enzyme and substrate. That may adversely affect the formation of the enzyme complex. Most human enzymes have an optimum pH between 6 and 8.

Chemical reagents may serve as *enzyme inhibitors*. A *nonspecific inhibitor* is any reagent which denatures protein. A *specific inhibitor* inhibits only a single enzyme or a group of related enzymes. Most poisons act as specific inhibitors. A *competitive inhibitor* is structurally related to a particular substrate so that it competes with that substrate for the active site of the enzyme. Only complex formation with the inhibitor takes place, and no product is formed. However, since formation of the enzyme-inhibitor complex is reversible, increased substrate concentration permits displacement of the inhibitor from the active site and subsequent completion of the enzyme reaction. A *noncompetitive inhibitor* forms strong covalent bonds with an enzyme and cannot be displaced by the addition of excess substrate. Such an inhibitor may react with groups which constitute part of the active sites of many enzymes.

Chemotherapy is the use of a chemical to destroy or control an invading microorganism without seriously harming the host. A chemical of that type may function as an inhibitor of a critical enzyme in the cells of the invading organism. An *antimetabolite* is a competitive inhibitor of a significant metabolic reaction of the invading organism. An *antibiotic* is a product of a microorganism which is toxic to another microorganism by inhibition of an essential enzyme. Some, but not all, antibiotics may function as antimetabolites.

An error in the genetic code of a person's DNA may result in the synthesis of an abnormal enzyme which is unable to catalyze reaction of the usual substrate. Diseases and abnormalities result from such *enzymatic disorders*.

Key terms

Absolute group specificity: limitation of enzyme catalysis to the reactions of one particular functional group in any compound.

Absolute specificity: limitation of enzyme catalysis to one and only one reaction.

Activated complex: an intermediate formed by a catalyst and a reactant. It affords a reaction path requiring less energy and, in effect, lowers the activation energy of the reaction.

Activation: any process in which the inactive protein is converted into an active enzyme; it includes union with a coenzyme or activation by a metal ion.

Active site: a relatively small region on the surface of an enzyme which exactly fits a complementary portion on the substrate.

Antibiotic: a product of a microorganism which is toxic to other microorganisms because it effectively inhibits essential enzymes.

Antimetabolite: a compound which serves as a competitive inhibitor for a significant metabolic reaction of an invading organism.

Apoenzyme: the polypeptide moiety of a holoenzyme; it is inactive without the coenzyme.

Catalytic group: according to the induced-fit theory, the part of the active site of an enzyme responsible for catalysis.

Chemotherapy: use of a chemical to destroy or control an invading microorganism without seriously harming the host.

Coenzyme: the nonprotein moiety of a holoenzyme.

Competitive inhibitor: a compound structurally related to a particular substrate so that it competes with that substrate for the active site of the enzyme.

Contact group: according to the induced-fit theory, the part of the active site of an enzyme responsible for substrate specificity.

Enzymatic disorder: a disease or abnormality that results from an abnormal enzyme being unable to catalyze the reaction of the usual substrate. It results from an error in the genetic code of the person's DNA.

Enzyme: a complex organic or organometallic catalyst, largely protein in nature, produced by living cells.

Enzyme activator: a metal ion, such as Mg^{2+}, Mn^{2+}, or Zn^{2+}, which probably activates the enzyme by forming a coordination complex between enzyme and substrate to facilitate electron shifts.

Enzyme-substrate complex: an activated complex formed by an enzyme and its substrate.

Holoenzyme: an enzyme which is a conjugated protein; it consists of an apoenzyme and a coenzyme.

Induced-fit theory: a theory that suggests that the substrate induces a structural change in the enzyme molecule to form the required activated complex.

Lock-and-key theory: the concept of complementary structures of enzyme and substrate that enable the two to fit together to form the initial complex.

Nerve gas: an organic phosphate ester which inhibits the enzyme cholinesterase by forming an enzyme-inhibitor complex at the active site of the enzyme.

Noncompetitive inhibitor: a compound which forms strong covalent bonds with an enzyme and cannot be displaced by the addition of excess substrate.

Nonspecific inhibition: inhibition of enzyme reactions by any reagent which denatures proteins.

Proenzyme: the inactive form in which a simple protein enzyme may be first secreted. It is converted into the active enzyme by another enzyme which removes an inhibitory peptide from the proenzyme molecule. Also called a **zymogen.**

Relative group specificity: limitation of enzyme catalysis to a similar reaction (for example, hydrolysis) of more than one kind of group.

Simple enzyme: an enzyme which is a simple protein rather than a conjugated protein.

Specific inhibitor: a chemical reagent which will inhibit only a single enzyme or a group of related enzymes.

Stereochemical specificity: limitation of enzyme catalysis to the reactions of certain stereoisomers but not the reactions of their enantiomers.

Substrate: the substance upon which an enzyme acts.

Vitamin: an organic compound that cannot be synthesized by the organism yet, in very small amount, is essential for the maintenance of normal health and growth of the organism and so must be included in the diet.

Zymogen: the inactive form in which a simple protein enzyme may be first secreted. Also called a **proenzyme.**

Additional problems

16-4 Briefly, what are the contributions of Pasteur, Buchner, and Sumner to our understanding of enzymes?

16-5 What is the difference between an apoenzyme and a coenzyme?

16-6 Name the various types of specificity exhibited by enzymes. Give an example of each type.

16-7 How do enzymes differ from the usual chemical catalysts?

16-8 Name the factors which influence the rate of an enzyme-catalyzed reaction. Briefly indicate their effects.

16-9 Remembering that enzymes are essentially proteins, explain why they become inactive above and below the optimum temperature and optimum pH.

16-10 Name the types of enzyme inhibitors.

16-11 What effect has increasing the concentration of the substrate upon (a) the action of a competitive inhibitor and (b) the action of a noncompetitive inhibitor? Explain.

16-12 Mercuric chloride and silver nitrate have been used as bacteriacides. Explain how they function.

16-13 Define, explain, or give an example of each of the following:

(a) chemotherapy
(b) antimetabolite
(c) sulfa drug
(d) antibiotic
(e) active site
(f) inhibitor
(g) induced-fit theory
(h) enzyme activator

16-14 In urban areas of high traffic density, the concentration of lead particles in the atmosphere may become dangerously high. The lead comes from the tetraethyl lead added to some gasolines. Toxicologists have tested the degree of assimilation of lead by humans in the area by analyzing samples of hair for lead content. Why would keratin, the protein of hair, tend to accumulate lead?

16-15 Explain each of the following in terms of inhibiting enzyme action. (a) pasteurization, (b) sterilization with steam, (c) freezing of food or specimens, (d) using alcohol as a disinfectant.

16-16 Consult Table 16-1. What is the substrate for the deficient enzyme in albinism? What is the substrate for the deficient enzyme in tyrosinemia?

METABOLISM

In this chapter we will be concerned with some of the organic chemistry involved in using the food we eat. Our food must be converted into the energy and materials required for the movement, growth, and maintenance of our bodies.

17-1 Energy needs of the body

Every movement that we make requires energy. Speaking, singing, or listening requires energy. The transmittal and processing of light, sound, odor, and contact sensations require energy, as do the thinking processes. We need energy to breathe, to swallow, to digest our food, to excrete our wastes, and indeed to move each and every muscle. We need energy to replace worn out or damaged cells and to build new tissues. Our bodies are actually complicated chemical factories and power plants. The processes involved are referred to as *metabolic processes,* or metabolism.

Definition of metabolism

Metabolism is the total process in which food is taken into the organism and subjected to many complex changes. The process involves release of energy and the production of smaller molecules, which may then be built into the more complex molecules of new cells and tissues. The degradation of absorbed food into smaller molecules is an *exergonic,* or energy-releasing, process called *catabolism.* The biosynthesis of complex molecules from simpler ones, an *endergonic,* or energy-absorbing, process, is called *anabolism.* Any compound (except enzymes) involved in a metabolic reaction is a *metabolite.* The metabolism which supports human life depends upon plant life, as does that of all of the animal world. Plant life, in turn, depends upon photosynthesis.

Photosynthesis

Photosynthesis is the overall process in which plants utilize water, inorganic salts and nitrogenous compounds from the soil, carbon dioxide from the atmosphere, and radiant energy from the sun to synthesize carbohydrates, fats, and proteins. The simplified summary equation is

$$6CO_2 + 6H_2O + 686\text{ kcal} \xrightarrow{\text{sunlight}} \underset{\text{Glucose}}{C_6H_{12}O_6} + 6O_2 \qquad (17\text{-}1)$$

Observe that glucose is the organic product indicated in this summary equation. Glucose and the carbohydrates based on it constitute over half of the food of human beings, and they are important metabolites in the animal kingdom. The utilization of glucose will now be considered in more detail.

Respiration

Carbohydrates based on D-glucose comprise one of the chief sources of cell energy in biological systems. The release of energy requires the oxidation of glucose to yield, ultimately, carbon dioxide and water. The oxidation is a complex one, but the total process is called *respiration*. It can be represented by the summary equation

$$C_6H_{12}O_6 + 6\ O_2 \longrightarrow 6CO_2 + 6H_2O + 686\ \text{kcal} \qquad (17\text{-}2)$$

Note that the summary equation for respiration is exactly the reverse of the summary equation for photosynthesis. The details of the two reactions may, however, differ widely. The energy relations of photosynthesis and respiration are diagramed in Fig. 17-1.

The oxygen for respiration comes from the air via the lungs into the body. Living organisms conserve 33 to 40% of the 686 kcal of energy released from glucose. They do so by a series of stepwise reactions that liberate small amounts of utilizable energy stored in the high-energy phosphate bonds of *adenosine triphosphate* (ATP). The remainder of the energy is used to drive reactions that yield ATP and other products (*coupled reactions*) and to supply heat to the body for maintaining proper temperature. The stepwise release of energy may be contrasted with the single energy-liberating reaction of a laboratory oxidation of glucose to carbon dioxide and water. Such a sudden large release of energy would burn out a living cell.

Coupling of reactions is common in biochemical processes. Key endergonic biochemical reactions, which require energy, are coupled with exergonic reactions, which give off energy, such as occur in the stepwise catabolism of glucose. The energy residing in the ATP so produced may

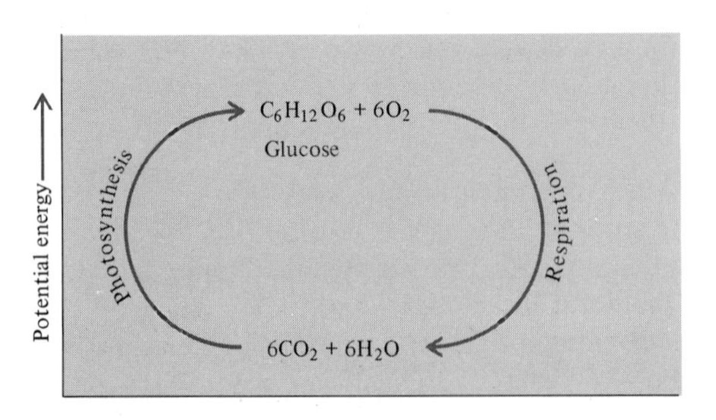

FIGURE 17-1 Cyclical energy relations of photosynthesis and respiration.

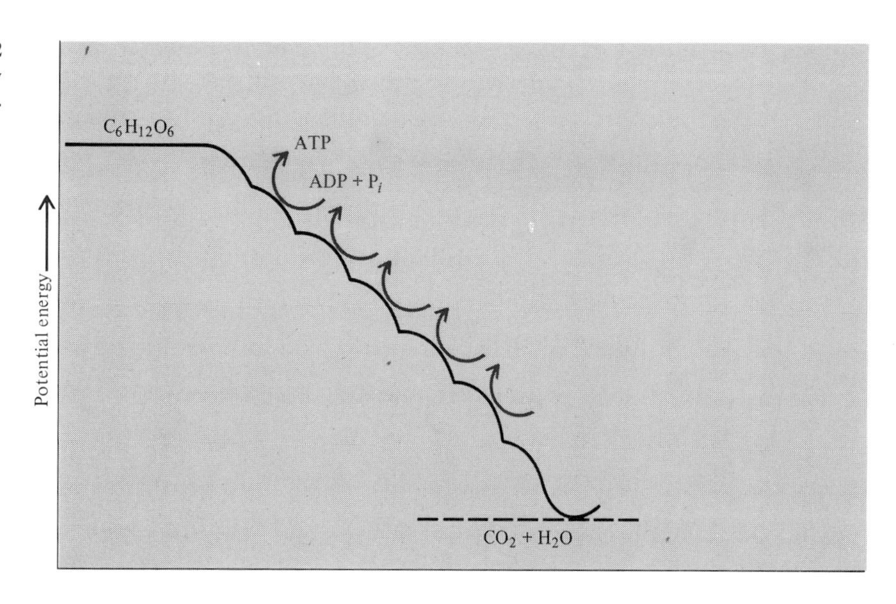

FIGURE 17-2 **Stepwise production of energy by catabolism of glucose coupled to production of ATP. P_i represents inorganic phosphate.**

then be released to drive an endergonic reaction. For example, the formation of glucose-6-phosphate from glucose and phosphate is an endergonic reaction requiring about 3.3 kcal. If the reaction is coupled with the exergonic reaction of ATP to ADP (adenosine diphosphate) in which 7.3 kcal of energy is released, the coupled combination can occur with a net release of 4.0 kcal of energy. The coupled combination of the two reactions could be written as follows:

$$\text{Glucose} + \text{ATP} \xrightarrow{\text{hexokinase}} \text{glucose-6-phosphate} + \text{ADP} \qquad (17\text{-}3)$$

In biochemistry, however, such biochemical reactions are commonly written by using a coupled reaction notation, as follows:

$$\text{Glucose} \xrightarrow[\text{ATP} \curvearrowright \text{ADP}]{\text{hexokinase}} \text{glucose-6-phosphate} \qquad (17\text{-}4)$$

The stepwise oxidation of glucose as it may occur in the cells of the body is diagramed in Fig. 17-2.

Role of adenosine triphosphate (ATP)

Adenosine triphosphate (ATP) is a nucleoside derivative which is built up from adenosine diphosphate (ADP), phosphoric acid, and energy. The breakdown of glucose (catabolism) is coupled with the synthesis of ATP, which results in converting part of the energy to ATP. The energy of ATP can then be used to supply energy for some required biochemical reaction. The new phosphorus-oxygen bond which is formed in producing

ATP is considered a *high-energy bond*. It is sometimes designated by a squiggle, ~.

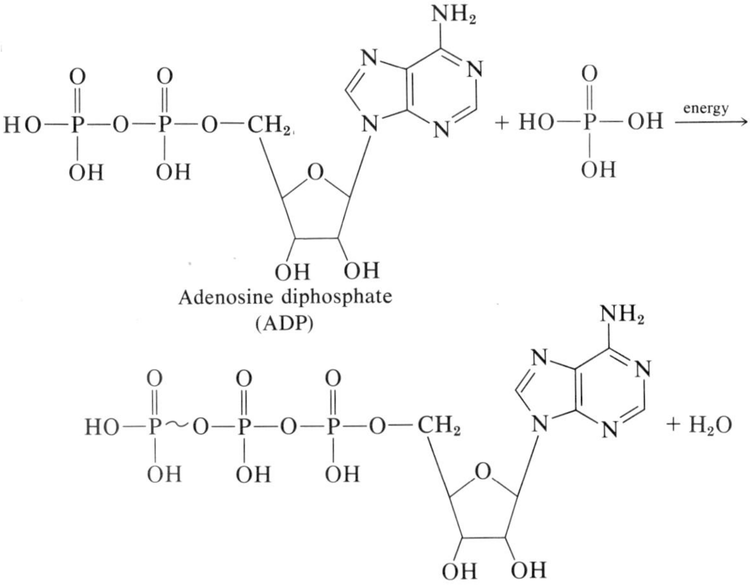

$$HO\text{—}\overset{\overset{O}{\|}}{\underset{\underset{OH}{|}}{P}}\text{—}O\text{—}\overset{\overset{O}{\|}}{\underset{\underset{OH}{|}}{P}}\text{—}O\text{—}CH_2\text{(ribose, OH, OH)—adenine}(NH_2) + HO\text{—}\overset{\overset{O}{\|}}{\underset{\underset{OH}{|}}{P}}\text{—}OH \xrightarrow{\text{energy}}$$

Adenosine diphosphate
(ADP)

$$HO\text{—}\overset{\overset{O}{\|}}{\underset{\underset{OH}{|}}{P}}\sim O\text{—}\overset{\overset{O}{\|}}{\underset{\underset{OH}{|}}{P}}\text{—}O\text{—}\overset{\overset{O}{\|}}{\underset{\underset{OH}{|}}{P}}\text{—}O\text{—}CH_2\text{(ribose, OH, OH)—adenine}(NH_2) + H_2O \qquad (17\text{-}5)$$

Adenosine triphosphate
(ATP)

The equation may be written in the following abbreviated form:

$$ADP + P_i + \text{energy} \longrightarrow ATP + H_2O \qquad (17\text{-}6)$$

where P_i stands for inorganic phosphate. Actually, at physiological pH inorganic phosphate is present not as phosphoric acid, but as some combination of the ions $H_2PO_4^-$ and HPO_4^{2-}.

Reaction (17-6) is called *oxidative phosphorylation*. When the energy is required, the last group can be hydrolyzed off to release energy in the reverse reaction.

$$ATP + H_2O \longrightarrow ADP + P_i + \text{energy} \qquad (17\text{-}7)$$

The energy relations are diagramed in Fig. 17-3.

PROBLEM 17-1 Define each of the following terms:

(a) metabolism
(b) catabolism
(c) anabolism
(d) metabolite
(e) photosynthesis
(f) respiration
(g) ATP
(h) ADP
(i) high-energy bond
(j) coupled reactions

17-2 Overview of metabolism

The three chief classes of foods are carbohydrates, fats, and proteins. The metabolism of each of the classes will be considered in some detail. Fig-

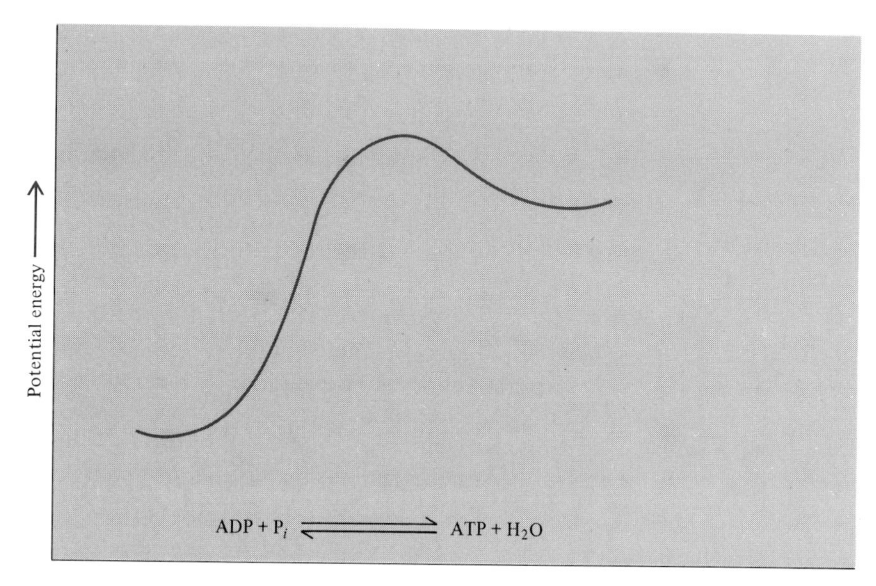

FIGURE 17-3
Energy relation of ADP and ATP.

ure 17-4 is an overview of the interrelations of the three classes as they are involved in metabolism. Note that amino acids from the catabolism of proteins can furnish pyruvic acid, one of the metabolites from the catabolism of carbohydrates. Fatty acids, from the catabolism of fats, can be

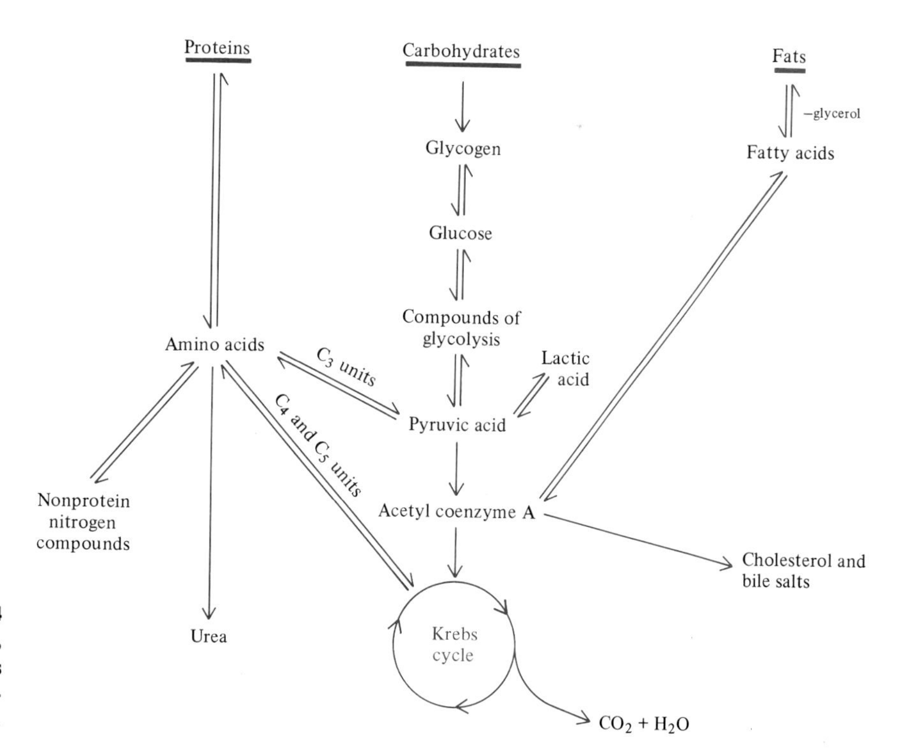

FIGURE 17-4
The interrelation of proteins, carbohydrates, and fats as they are involved in metabolism.

converted to acetyl coenzyme A, a carbohydrate metabolite. The reverse reactions occur to some extent, so that some amino acids, proteins, fatty acids, and fats can be synthesized (anabolism) from metabolites of carbohydrate catabolism. Also, glycogen, the storage form of glucose, can be synthesized from metabolites coming from fats or proteins as well as from carbohydrates. Metabolites from all foods may enter the Krebs cycle to produce carbon dioxide, water, and energy.

17-3 Metabolism of carbohydrates

Most carbohydrates are ingested in the form of starch or sucrose. Digestion begins in the mouth, where hydrolysis of starch to *dextrins* (shorter-chain polysaccharides) is catalyzed by the enzyme pytalin in saliva. In the small intestine, further hydrolysis is catalyzed by the α-amylase, pancreatic amylase, to produce the disaccharide maltose. Also in the small intestine, maltase, an α-glucosidase, catalyzes the hydrolysis of maltose to glucose, sucrase catalyzes the hydrolysis of sucrose to glucose and fructose, and lactase effects the hydrolysis of lactose to galactose and glucose. The monosaccharides glucose, fructose, and galactose are moved by *active transport* through the villi of the small intestine into the bloodstream, where they are carried by the portal vein to the liver. Active transport is the energy-requiring process of moving molecules across a membrane from a region of lower concentration to one of greater concentration. Such a direction of movement is exactly opposite to that of diffusion by osmosis. Active transport is accomplished in the membrane of the cell wall by special proteins embedded in the membrane.

In the liver galactose and fructose are enzymatically converted to glucose. Glucose is the only sugar that circulates in the blood, and so it is referred to as *blood sugar*.

Blood sugar level

Under normal circumstances the blood sugar level remains constant at about 80 mg per 100 ml of blood, but with a range of 70 to 100 mg per 100 ml of blood. Glucose in the bloodstream may be transported to the tissues, or it may be converted into glycogen to be stored in the liver. Glycogen serves as a reservoir for maintenance of the blood sugar level.

Glucose which is transported to the tissues may be oxidized to carbon dioxide and water or to other metabolites needed for biosynthesis. It may be converted into fat, or it may be converted to muscle glycogen to serve as a source of readily available energy for the muscle.

A condition resulting from a lower-than-normal blood sugar level is called *hypoglycemia,* and that resulting from a higher-than-normal level is called *hyperglycemia*. If the blood sugar level reaches as high as 160 to 170 mg per 100 ml of blood, the *renal threshold* is reached and excess glucose is excreted in the urine (*glucosuria*). Extreme hypoglycemia, which is most commonly due to the presence of excessive amounts of the hormone insulin, can cause unconsciousness, lowered blood pressure, and even death. Such loss of consciousness is possibly caused by the lack of glucose in brain tissue, which requires sugar as a source of energy.

The synthesis and breakdown of glycogen maintains the blood sugar level, and several hormones are involved in the control of that level. *Insulin,* which is secreted by the pancreas, acts to lower the blood sugar level by stimulating the formation of glycogen from glucose (*glycogenesis*). If the pancreas does not secrete enough insulin, diabetes may result. All other hormones involved act to raise the concentration of blood sugar. Adrenaline (epinephrine) activates the enzyme phosphorylase, which is involved in the breakdown of glycogen to glucose (*glycogenolysis*).

Glucose catabolism

Glucose catabolism may be considered in three stages. The first stage takes place in the cytoplasm of the cell, where glucose is broken down to pyruvic acid with net formation of 2 mol of ATP per mole of glucose catabolized. That stage is called *glycolysis*. It furnishes just 5% of the total energy realized from catabolism of glucose. Stage two takes place in the mitochondria (cellular bodies located in cytoplasm). In the second stage, pyruvic acid is converted to the C_2 unit, acetyl coenzyme A, which is further broken down in the *citric acid* (*Krebs*) *cycle*. The third stage accompanies the second stage and also occurs in the mitochondria. The third stage is termed the *electron transport system,* or *respiratory chain*. It results in the reduction of oxygen to water in the cells. Figure 17-5 is an overview of glucose catabolism.

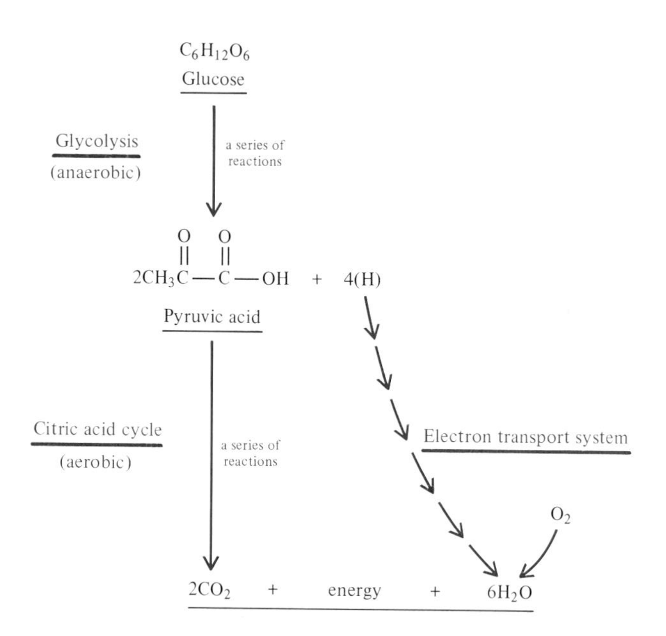

FIGURE 17-5
Summary diagram of glucose catabolism.

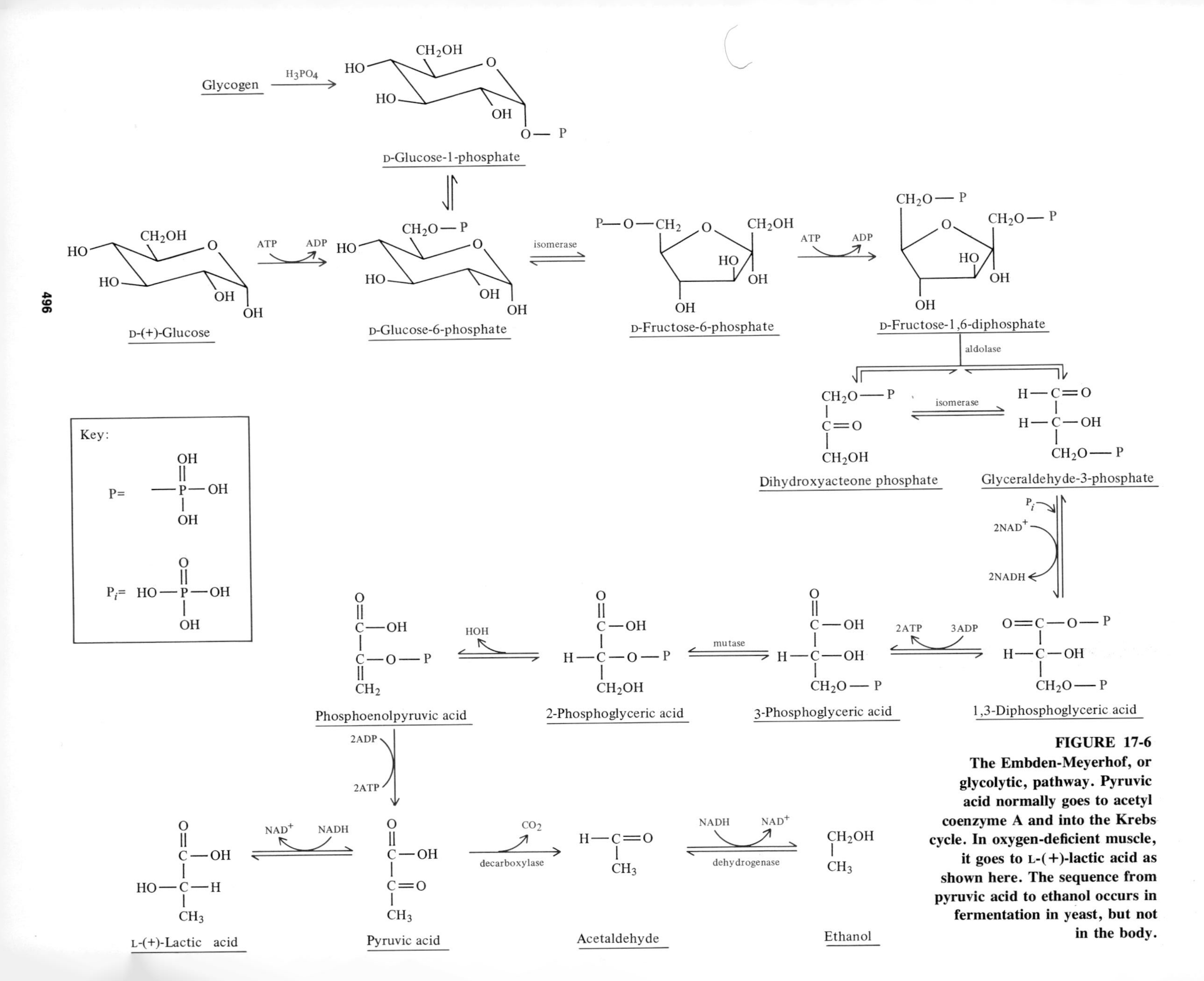

FIGURE 17-6
The Embden-Meyerhof, or glycolytic, pathway. Pyruvic acid normally goes to acetyl coenzyme A and into the Krebs cycle. In oxygen-deficient muscle, it goes to L-(+)-lactic acid as shown here. The sequence from pyruvic acid to ethanol occurs in fermentation in yeast, but not in the body.

Glycolysis. This first stage of glucose catabolism is accomplished without oxygen (*anaerobic*). The sequence of reactions is known as the *Embden-Meyerhof* or *glycolytic pathway*. For every mole of glucose degraded, 2 mol of ATP is initially consumed and 4 mol of ATP is ultimately produced—a net gain of 2 mol of ATP. However, if glycogen is the source of glucose, expenditure of only 1 mol of ATP is necessary, so the net gain is 3 mol of ATP. The reactions are detailed in Fig. 17-6.

The NAD^+ which serves as the *oxidizing agent,* or *electron acceptor,* is nicotinamide adenine dinucleotide. The conjugated enzyme, glyceraldehyde 3-phosphate dehydrogenase, contains NAD^+ as a coenzyme. In the process of oxidizing the aldehyde to a carboxylic acid, the coenzyme is reduced to NADH, which then carries $H:^-$ into the electron transport system. The structures of NAD^+ and NADH are given in Fig. 17-7. Note

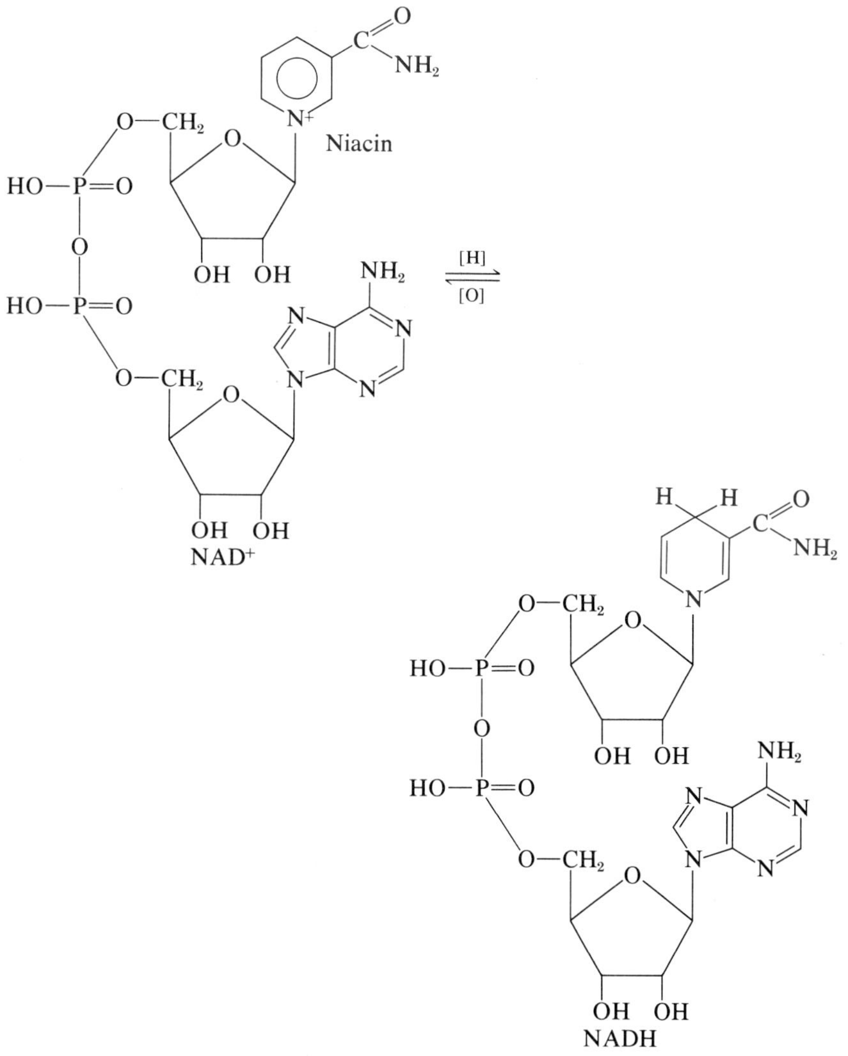

FIGURE 17-7
The structures of NAD^+ and NADH. Note that the reduction of NAD^+ to NADH takes place in the pyridine ring of the nicotinamide moiety.

that the reactive part of NAD^+ is the vitamin *nicotinamide,* or *niacin,* one of the B group of vitamins. A deficiency of that vitamin in humans causes the disease pellagra.

Up to the formation of pyruvic acid, glucose metabolism is the same whether the process takes place in muscle tissue, in the liver, or in the fermentation of sugar to make alcohol. However, in muscle tissue an *anaerobic process* may take place under conditions of oxygen deficiency caused by overexertion. In that process, pyruvic acid is reduced to L-(+)-lactic acid by the reducing agent NADH, which is thereby oxidized back to NAD^+. In that way a small amount of NAD^+ can catalyze the metabolism of a large amount of glucose in the muscle by being oxidized and reduced many times. Overproduction of lactic acid in that way results in accumulation of lactic acid, which causes muscle fatigue. In the fermentation process, pyruvic acid is decarboxylated to ethanal (acetaldehyde), which is then reduced by NADH to ethanol.

Formation of acetyl coenzyme A; the citric acid cycle. In the second stage of glucose catabolism, pyruvic acid is broken down to carbon dioxide and acetyl coenzyme A. Carbon dioxide so produced is quantitatively the chief waste product of the body; an adult produces about one kilogram

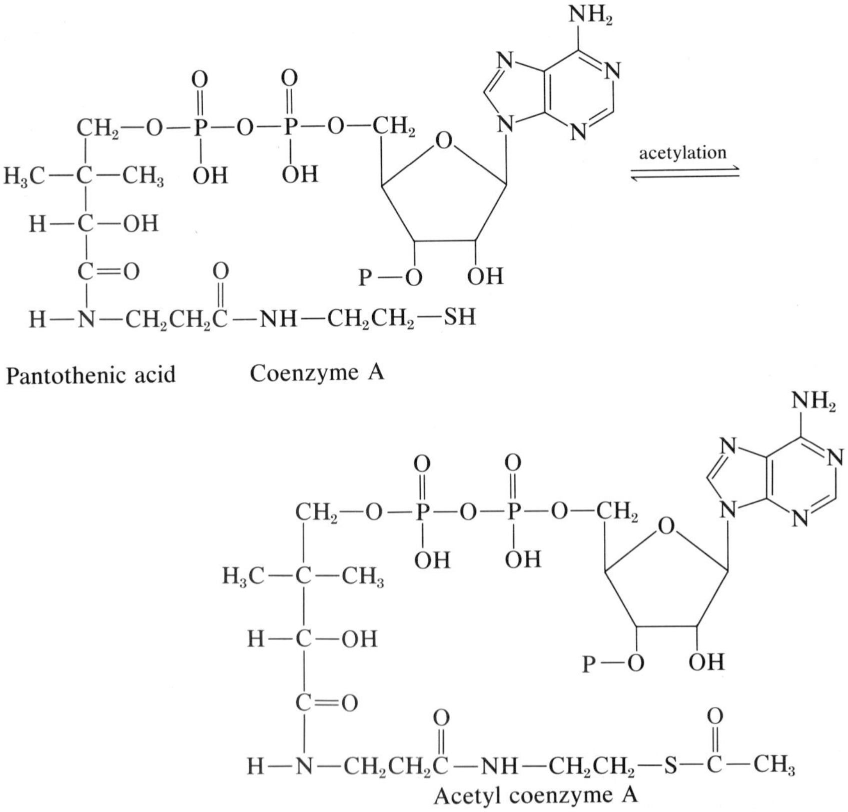

CH_2—O—P(=O)(OH)—O—P(=O)(OH)—O—CH_2 (ribose–adenine; P—O, OH)
H_3C—C—CH_3
H—C—OH
C=O
H—N—CH_2CH_2C(=O)—NH—CH_2CH_2—SH

Pantothenic acid Coenzyme A

acetylation

CH_2—O—P(=O)(OH)—O—P(=O)(OH)—O—CH_2 (ribose–adenine; P—O, OH)
H_3C—C—CH_3
H—C—OH
C=O
H—N—CH_2CH_2C(=O)—NH—CH_2CH_2—S—C(=O)—CH_3

Acetyl coenzyme A

FIGURE 17-8
The structures of coenzyme A and acetyl coenzyme A.

per day. The acetyl coenzyme A which is so produced is also the starting point for the synthesis of fat in the body. The reaction of pyruvic acid is

$$\underset{\text{Pyruvic acid}}{\mathrm{CH_3{-}C({=}O){-}COOH}} + NAD^+ + \underset{\text{Coenzyme A}}{\mathrm{CoASH}} \xrightarrow[\text{dehydrogenase}]{\text{pyruvic}} \underset{\text{Acetyl coenzyme A}}{\mathrm{CH_3C({=}O){-}S{-}CoA}} + NADH + CO_2 + H^+ \qquad (17\text{-}8)$$

The structural formula for coenzyme A is shown in Fig. 17-8. Note that, besides the nucleotide adenylic acid, the molecule contains pantothenic acid, which is part of the vitamin B group.

The enzyme complex for the reaction in Eq. (17-8) requires the vitamins lipoic acid and thiamin (vitamin B_1) as well as NAD^+ and FAD. *FAD* is the abbreviation for *flavin adenine dinucleotide*. It is a coenzyme consisting of a molecule of vitamin B_2 bonded to adenine dinucleotide. The structural formula of FAD is shown in Fig. 17-9.

In aerobic organisms (those that require oxygen to live), most of the pyruvic acid is oxidized in a number of enzymatic reactions to carbon dioxide and water. The oxidation of pyruvic acid liberates about 93% of the energy stored in glucose. Much of the energy is not liberated as heat; instead it is in chemical form in the high-energy phosphate bonds of ATP.

The scheme for this complex series of reactions was first proposed in 1937 by Sir Hans Krebs (Nobel Prize for Medicine, 1953). The sequence of reactions is called the *Krebs cycle*, the *citric acid cycle*, or the *tricar-*

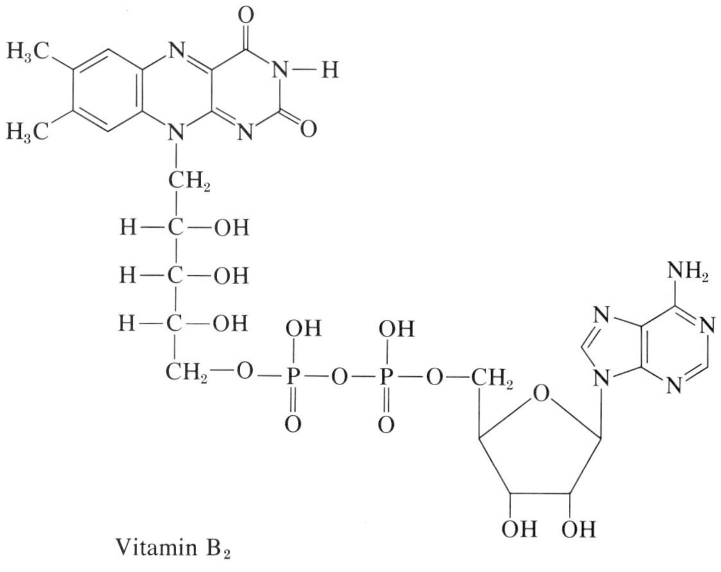

Vitamin B_2

FIGURE 17-9
The structure of FAD, flavin adenine dinucleotide.

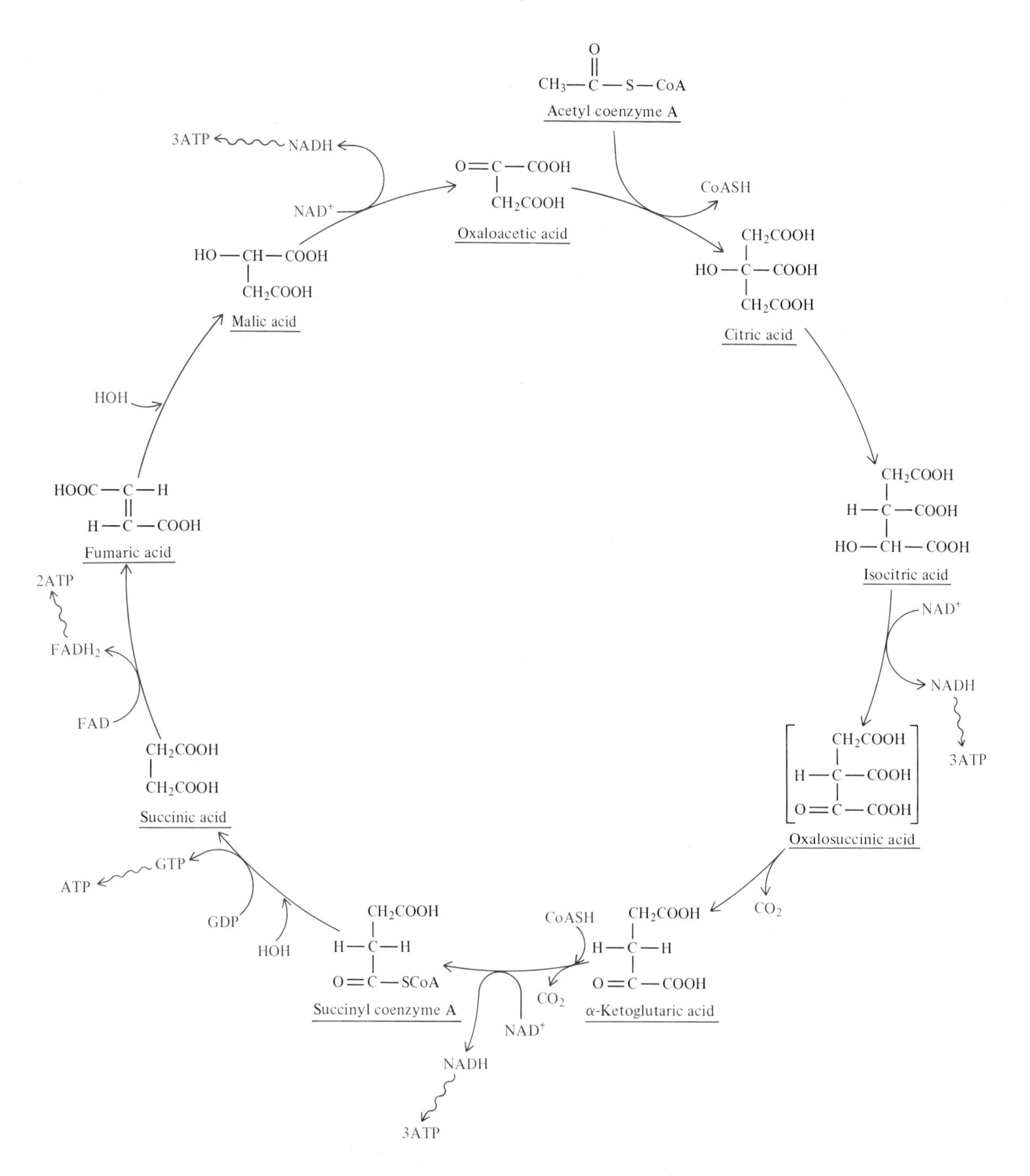

FIGURE 17-10 The citric acid, or Krebs, cycle.

boxylic acid cycle. It is also involved in the metabolism of lipids and proteins. Pyruvic acid itself is not an intermediate in the Krebs cycle; it is first subjected to oxidative decarboxylation to form acetyl coenzyme A. The acetyl coenzyme A likewise is not a true intermediate in the Krebs cycle. It enters the cycle by condensing with a four-carbon dicarboxylic acid, oxaloacetic acid, to yield citric acid (for which the cycle was named) and thereby regenerates coenzyme A. The citric acid cycle may be said to burn the acetate of acetyl CoA.

The compounds and reactions involved in the Krebs cycle are shown in Fig. 17-10. The reaction of guanosine diphosphate (GDP) to form guanosine triphosphate (GTP) is coupled with the cleavage of succinyl CoA.

Respiratory chain, or electron transport system. The chain of chemical events resulting in the reduction of oxygen to water in the cells is called the *respiratory chain,* and the enzymes involved in it are called the *respiratory enzymes*. Cellular respiration is the chief source of the reactions whereby most of the ATP is produced. The respiratory chain facilitates the transfer of chemical energy from foods and oxygen to ATP without the leakage away of energy as unneeded heat.

Reduction of oxygen to water requires electrons. The electron transport occurring in the chain is coupled to the synthesis of ATP from ADP and inorganic phosphate. The electrons are transported as in a bucket brigade, and so the respiratory chain is also known as the *electron transport system*.

All the respiratory enzymes are organized together in tiny particles that occur by the thousands on the inner and outer surfaces of the subcellular organelles called *mitochondria*. Hundreds of mitochondria are present in each cell; they are the primary sites of ATP synthesis. See Fig. 17-11 for electron micrographs of a typical cell and a typical mitochondrion.

The respiratory chain is a complicated thing, but it involves enzymatic oxidation of some intermediate metabolite, say MH_2 (M stands for *metabolite*). The citric acid cycle is the chief supplier of such intermediates, which can give up the parts of elemental hydrogen (H^+ and $H{:}^-$) to the respiratory chain. For example, an enzyme may have nicotinamide adenine dinucleotide (NAD^+) as its coenzyme. The reaction could be written:

$$MH_2 + NAD^+ \longrightarrow NAD{:}H + H^+ + M \qquad (17\text{-}9)$$

The same reaction may be represented as coupled oxidation-reduction reactions:

$$\begin{array}{ccc} MH_2 & \searrow\ \swarrow & NAD^+ \\ M & \swarrow\ \searrow & NAD{:}H + H^+ \end{array} \qquad (17\text{-}10)$$

The NAD^+ is regenerated from NADH by coupling of this oxidation with the reduction of FAD to $FADH_2$.

$$\begin{array}{ccccc} MH_2 & \searrow\ \swarrow & NAD^+ & \nwarrow\ \nearrow & FADH_2 \\ M & \swarrow\ \searrow & NAD{:}H + H^+ & \nearrow\ \swarrow & FAD \end{array} \qquad (17\text{-}11)$$

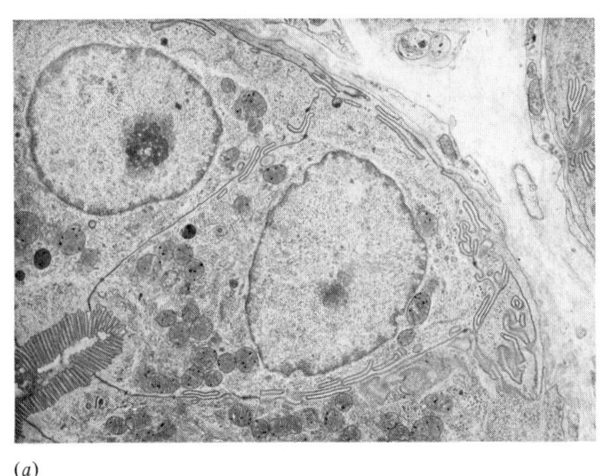

(*a*)

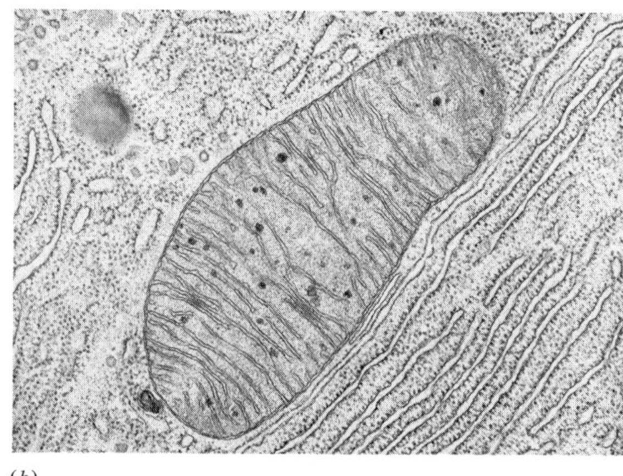

(*b*)

FIGURE 17-11 (*a*) A typical cell; (*b*) a typical mitochondrion. (*Courtesy of Keith R. Porter*)

Figure 17-12 shows a part of the respiratory chain. The *cytochromes* indicated in the diagram are iron porphyrins, and porphyrins are substituted porphins. The porphin ring structure is shown in Fig. 17-13. The atom of iron replaces the amine hydrogens and is bonded between the nitrogen atoms. Heme of hemoglobin and chlorophyll also are substituted porphins, but the metal ion in chlorophyll is magnesium instead of iron.

The reactions of the respiratory chain do not occur unless there is a need to replenish supplies of ATP. If exercise or other events use up ATP, leaving ADP and inorganic phosphate, then the long chain of reactions is automatically activated to put ADP and inorganic phosphate together again. That self-regulating mechanism for regeneration is an example of *homeostasis*.

The process whereby ATP is synthesized as a result of the operation of the respiratory chain is called *respiratory chain oxidative phosphorylation*. The details of the mechanism are still incompletely known. However, the energy required for the production of ATP results from the transport of one pair of electrons from one carrier to the next. Transport of a pair of electrons from NADH to oxygen along the chain releases about 52 kcal. If all that energy were released at once, much of it would be dissipated as heat and could damage the cell. The adverse effect is prevented by the action of the respiratory chain, which delivers the energy in

FIGURE 17-12 A part of the respiratory chain.

MH_2 / M ⇄ NAD^+ / $NADH + H^+$ (ADP + P_i → ATP) ⇄ FADH / FAD ⇄ Quinone / Hydroquinone (coenzyme Q_{10}) ⇄ 2Fe(II) / 2Fe(III) cytochrome b ($2H^+$) ⇄ 2Fe(III) / 2Fe(II) cytochrome c (ADP + P_i → ATP) ⇄ 2Fe(II) / 2Fe(III) cytochrome a ⇄ 2Fe(III) / 2Fe(II) cytochrome a_3 (ADP + P_i → ATP) ⇄ HOH / $\frac{1}{2}O_2$ ($2H^+$)

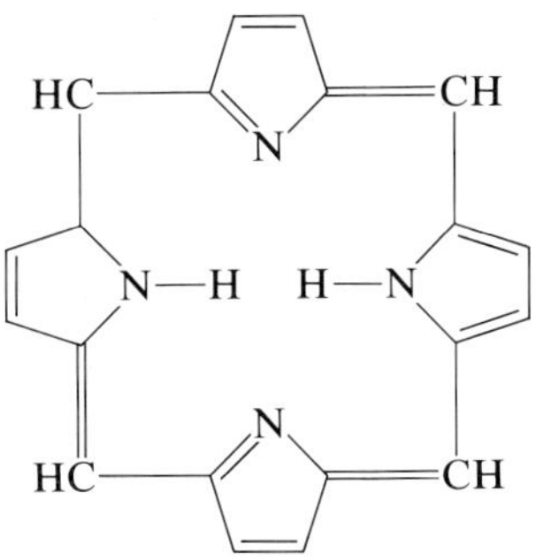

FIGURE 17-13
The porphin ring structure.

small increments for phosphorylating ADP. Almost half of the energy released in the electron transport process is conserved in the formation of high-energy phosphate bonds.

PROBLEM 17-2 Define each of the following terms:

(a) blood sugar level
(b) renal threshold
(c) hypoglycemia
(d) hyperglycemia
(e) glucosuria
(f) glycogenesis
(g) glycogenolysis
(h) respiratory chain
(i) mitochondria
(j) coenzyme A
(k) glycolysis
(l) citric acid cycle

PROBLEM 17-3 In a period of fasting, would the glycogen in the liver increase or decrease? Why or why not?

PROBLEM 17-4 How does the body maintain a fairly constant blood sugar level even though glucose may be ingested only at meals?

PROBLEM 17-5 Too much insulin (hyperinsulinism) results in hypoglycemia and leads quickly to "insulin shock," whereas a low level of insulin for short periods is not so serious. Why is that so?

17-4 Metabolism of fats

The fats which are ingested by higher animals pass through the stomach relatively unchanged by the acidic environment. When they reach the alkaline environment of the small intestine, however, they are hydrolyzed with the aid of enzymes called *lipases* (enzymes that work on lipids or fats). The products of the hydrolysis are fatty acids, monoglycerides, diglycerides, and glycerol. Those products are transported through the intestinal walls into the bloodstream—some directly and some by way of the lymphatic system.

Depot fat

Once in the bloodstream, the fatty acids are reesterified with glycerol. The resulting triglycerides become complexed with blood proteins and transported either to storage sites for deposition as *depot fat* or to body organs, mainly the liver, for metabolism. The depot fat can also be transported to the liver for metabolism when required. An excess of depot fat results in much adipose tissue and the overweight condition known as *obesity*. The chief cause of obesity is consumption of food in excess of the body's needs.

Energy from fats

In the liver, triglycerides are hydrolyzed to fatty acids and ultimately oxidized to carbon dioxide and water. Fats are high-energy foods, and more energy is released per weight of fat in that oxidation than in the oxidation of the same weight of glucose. The oxidation of palmitic acid is an example.

$$CH_3(CH_2)_{14}COOH + 23\ O_2 \longrightarrow 16CO_2 + 16H_2O + 2338\ \text{kcal} \quad (17\text{-}12)$$

You will recall from Eq. (17-2) that the complete oxidation of 1 mol of glucose yields 686 kcal. Since the molecular weight of glucose is 180, 1 g of glucose yields 3.81 kcal:

$$\frac{686\ \text{kcal}}{180\ \text{g}} = 3.81\ \text{kcal/g} \quad (17\text{-}13)$$

Palmitic acid has a molecular weight of 256. Its energy yield per gram is

$$\frac{2338\ \text{kcal}}{256\ \text{g}} = 9.13\ \text{kcal/g} \quad (17\text{-}14)$$

Generally, a value of 4 kcal/g of carbohydrate is used, since polysaccharides are formed by splitting out one molecule of water at each linkage (and thus the molecular weight is 18 units less for each monosaccharide unit). A value of 9 kcal/g of fat is generally used because actual fats incorporate a glycerol moiety. Also, some fats are unsaturated.

Fatty acid cycle

The fatty acids are both broken down and synthesized in units of two carbon atoms, so the fatty acids incorporated in the fats of higher plants and animals contain even numbers of carbon atoms. In fatty acid metabolism, the acid is first converted to its thioester with coenzyme A. (A thioester is an ester containing an atom of sulfur instead of oxygen.) The thioester is then oxidized to an unsaturated acid.

$$CH_3(CH_2)_{12}CH_2CH_2COOH \xrightarrow{\text{CoASH, ATP}} CH_3(CH_2)_{12}CH_2CH_2\overset{\overset{\displaystyle O}{\|}}{C}{-}SCoA \xrightarrow{[O]}$$

$$CH_3(CH_2)_{12}\overset{\overset{\displaystyle H}{|}}{C}{=}\overset{\overset{\displaystyle H}{|}}{C}{-}\overset{\overset{\displaystyle O}{\|}}{C}{-}SCoA \quad (17\text{-}15)$$

Riboflavin (vitamin B_2) is involved in a coenzyme for the catalysis of the second part of the reaction. (Its structure is shown in color in Fig. 17-9.)

In the next step, water adds to the carbon-carbon double bond to produce a β-hydroxy thioester. It, in turn, is oxidized by NAD^+ to the β-keto thioester:

$$CH_3(CH_2)_{12}\overset{H}{\overset{|}{C}}=\overset{H}{\overset{|}{C}}-\overset{O}{\overset{\|}{C}}-SCoA \xrightarrow{H_2O} CH_3(CH_2)_{12}\overset{OH}{\overset{|}{C}}HCH_2\overset{O}{\overset{\|}{C}}-SCoA \xrightarrow{NAD^+}$$

$$CH_3(CH_2)_{12}\overset{O}{\overset{\|}{C}}CH_2\overset{O}{\overset{\|}{C}}-SCoA \qquad (17\text{-}16)$$

Finally, the β-keto thioester is cleaved to give the thioester of acetic acid (acetyl CoA) and the CoA thioester of a fatty acid with two fewer carbon atoms than the original acid. The cleavage reaction is a reverse Claisen condensation reaction.

$$CH_3(CH_2)_{12}\overset{O}{\overset{\|}{C}}CH_2\overset{O}{\overset{\|}{C}}-SCoA + HSCoA \longrightarrow$$

$$CH_3(CH_2)_{12}\overset{O}{\overset{\|}{C}}-SCoA + CH_3-\overset{O}{\overset{\|}{C}}-SCoA \qquad (17\text{-}17)$$

Repetition of these metabolic steps results in the chopping off of two carbon atoms at each cycle, until eventually the fatty acid is completely converted to acetyl CoA. The only ATP required is that to make the CoA thioester out of the original fatty acid.

The product acetyl CoA may now enter the Krebs cycle, where it may become oxidized to carbon dioxide and water with release of energy. Alternatively, it could be used to synthesize any special fats needed by the body or be converted to depot fat if in excess of energy needs. The metabolic reactions involving the fatty acids are sometimes referred to as the *fatty acid cycle,* or *fatty acid spiral.*

PROBLEM 17-6 Define each of the following terms: (a) depot fat, (b) fatty acid cycle, (c) thioester, (d) adipose tissue.

PROBLEM 17-7 Explain how glucose could be metabolized to form depot fat.

PROBLEM 17-8 How can body energy be obtained from depot fat?

PROBLEM 17-9 What are the functions of adipose tissue?

PROBLEM 17-10 What justification might there be for eating more fat in a very cold climate?

PROBLEM 17-11 Beginning with lauric acid, write equations for all the reactions that take place as the acid is catabolyzed through the fatty acid cycle to acetic acid units.

17-5 Metabolism of proteins

The first enzyme to act on proteins after the proteins have been ingested is *pepsin* in the stomach. Pepsin is one of the *proteolytic enzymes,* which are highly specific and hydrolyze only certain types of peptide linkages. Pepsin hydrolyzes the bonds on either side of phenylalanine and tyrosine moieties, and it also hydrolyzes leucine–glutamic acid linkages. In the small intestine, lysine and arginine peptide carbonyls are hydrolyzed by trypsin and the remaining phenylalanine and tyrosine peptide carbonyls are cleaved by chymotrypsin. Also in the small intestine, carboxypeptidases A and B from the pancreas continuously split off *C*-terminal amino acid residues. Aminopeptidase from the intestinal mucosa acts in the same way on *N*-terminal residues. Eventually all of the proteins are broken down into a mixture of amino acids, which are then absorbed into the bloodstream.

The amino acid pool

Unlike carbohydrates and fats, amino acids have no particular storage form or storage place, but the small quantity of free amino acids in circulation is referred to as the *amino acid pool.* Those amino acids may be used for synthesis of body proteins or of nonprotein nitrogen compounds, converted to other amino acids, metabolized for release of energy, or converted into glycogen or fat. Those relations with the amino acid pool are shown in Fig. 17-14. The body is said to be maintaining *nitrogen balance* when the amount of nitrogen excreted (mostly as urea) is equal to the amount of nitrogen taken in as ingested proteins.

Metabolic reactions of amino acids

The metabolic reactions of amino acids include oxidative deamination, transamination, decarboxylation, and alteration of carbon chains.

Oxidative deamination. The oxidative deamination reaction results in the replacement of the amino group by oxygen. It is catalyzed by amino acid

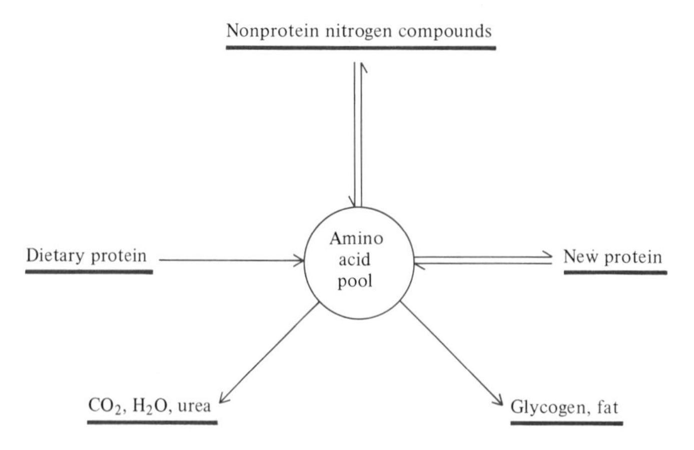

FIGURE 17-14 The amino acid pool. Examples of nonprotein nitrogen compounds are nucleic acids, heme, and coenzymes.

oxidases, which in mammals are found in the liver and kidneys. An example of the reaction is

$$\underset{\text{L-Alanine}}{CH_3\underset{\displaystyle NH_2}{\underset{|}{C}}HCOOH} + H_2O \xrightleftharpoons[\text{(Amino acid oxidase)}]{FAD \quad FADH_2} \underset{\text{Pyruvic acid}}{CH_3\underset{\displaystyle O}{\underset{\|}{C}}-COOH} + NH_3 \qquad (17\text{-}18)$$

The reduced coenzyme is reoxidized by molecular oxygen to form hydrogen peroxide, which is then decomposed by catalase.

$$FADH_2 + O_2 \longrightarrow FAD + H_2O_2 \qquad (17\text{-}19)$$

$$H_2O_2 \xrightarrow{\text{catalase}} H_2O + \tfrac{1}{2}O_2 \qquad (17\text{-}20)$$

In another example, involving glutamic acid, the coenzyme NAD^+ is used.

$$\underset{\text{L-Glutamic acid}}{HOOC-CH_2CH_2\underset{\displaystyle NH_2}{\underset{|}{C}}HCOOH} + H_2O \xrightleftharpoons{NAD^+ \quad NADH} \underset{\alpha\text{-Ketoglutaric acid}}{HOOC-CH_2CH_2\underset{\displaystyle O}{\underset{\|}{C}}-COOH} + NH_3 \qquad (17\text{-}21)$$

The ammonia liberated can be converted to urea for excretion in the urine. The average human excretes about 30 g of urea per day.

Transamination. In transamination, an amino group is transferred from one amino acid and gives rise to another amino acid. An example is the reaction of L-alanine with α-ketoglutaric acid.

$$\underset{\text{L-Alanine}}{CH_3\underset{\displaystyle NH_2}{\underset{|}{C}}HCOOH} + \underset{\alpha\text{-Ketoglutaric acid}}{HOOC-CH_2CH_2\underset{\displaystyle O}{\underset{\|}{C}}-COOH} \xrightleftharpoons{\text{pyridoxal phosphate transaminase}} \underset{\text{Pyruvic acid}}{CH_3\underset{\displaystyle O}{\underset{\|}{C}}-COOH} + \underset{\text{L-Glutamic acid}}{HOOC-CH_2CH_2\overset{\displaystyle H}{\overset{|}{\underset{\displaystyle NH_2}{\underset{|}{C}}}}-COOH} \qquad (17\text{-}22)$$

The fact that these reactions are reversible links the metabolism of protein with that of carbohydrates and fats. Most of the amino acids form metabolites involved in carbohydrate metabolism. Thus the amino acids can be converted to glucose or glycogen or oxidized in the citric acid cycle to CO_2, H_2O, and energy. Other amino acids, such as leucine and lysine, can give rise to acetyl CoA.

The enzyme pyridoxal phosphate transaminase makes use of another of the B vitamins, *pyridoxol* or *vitamin B_6*.

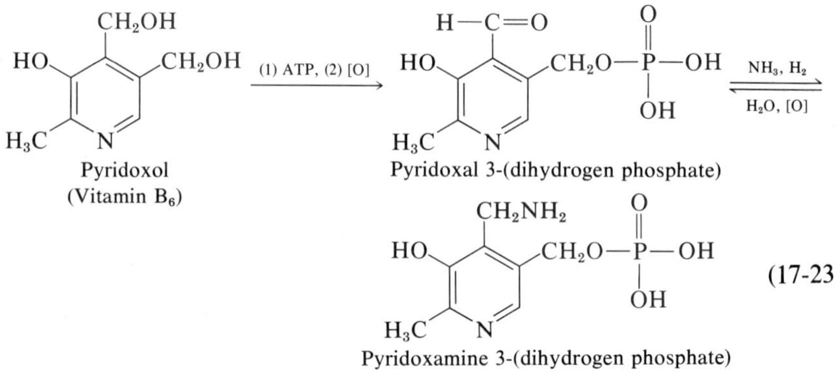

CH_2OH HO CH_2OH H_3C N
Pyridoxol
(Vitamin B_6)

$\xrightarrow{(1)\ ATP,\ (2)\ [O]}$

H—C=O HO $CH_2O—P(=O)(OH)—OH$ H_3C N
Pyridoxal 3-(dihydrogen phosphate)

$\underset{H_2O,\ [O]}{\overset{NH_3,\ H_2}{\rightleftharpoons}}$

CH_2NH_2 HO $CH_2O—P(=O)(OH)—OH$ H_3C N
Pyridoxamine 3-(dihydrogen phosphate)

(17-23)

Pyridoxal and pyridoxamine are interconvertible as shown. They function as coenzymes for the transaminase involved in the interconversion of α-amino and α-keto acids, as well as a number of other transformations involving amino acids.

Pyridoxine is a generic term that includes pyridoxol, pyridoxol phosphate, pyridoxal phosphate, and pyridoxamine phosphate. The daily human requirement for pyridoxine is estimated to be 1.5 mg, but no deficiency diseases have been demonstrated. Nutritional experiments have indicated that a deficiency of pyridoxine results in some disturbance of protein metabolism.

Decarboxylation of amino acids. Pyridoxal phosphate is also the necessary coenzyme for several amino acid decarboxylases. In the decarboxylation reaction, carbon dioxide is eliminated from the amino acid to form a primary amine. The necessary enzymes are found mostly in microorganisms, but they are present to some extent in the body. *Cadaverine* and *putrescine* (Table 13-1) are formed respectively by decarboxylation of lysine and ornithine. They have very putrid odors, and they are the toxins causing *ptomaine poisoning*. Some of the amines so produced in the body also may have pronounced physiological effects. Decarboxylation of histidine produces histamine:

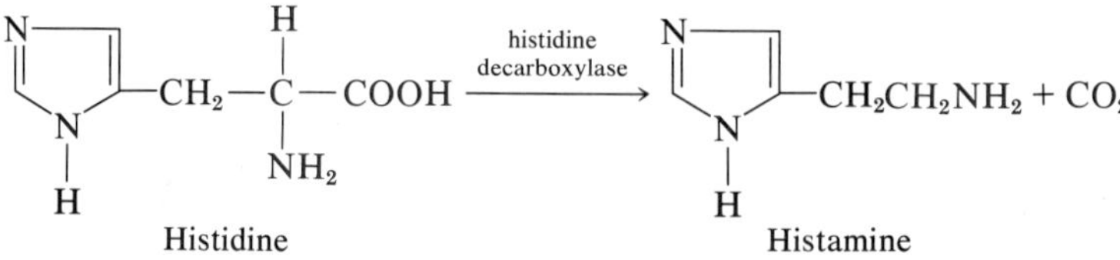

$$\text{(imidazolyl)}-CH_2-\underset{NH_2}{\overset{H}{C}}-COOH \xrightarrow[\text{decarboxylase}]{\text{histidine}} \text{(imidazolyl)}-CH_2CH_2NH_2 + CO_2 \qquad (17\text{-}24)$$

Histidine Histamine

Histamine causes a decrease in blood pressure; it may stimulate gastric secretions; and in excessive amounts it is related to allergic reactions to foreign proteins. Drugs called antihistamines may be prescribed to alleviate the latter condition.

Alteration of the carbon chain of an amino acid. One example of the alteration of the carbon chain of an amino acid, the oxidation of phenylalanine to tyrosine, has already been mentioned (Chap. 14, p. 424). In the normal metabolic pathway, phenylalanine is oxidized to tyrosine. Consequently, phenylalanine is considered an essential amino acid, but tyrosine is not.

$$\underset{\text{L-Phenylalanine}}{C_6H_5\text{—}CH_2\underset{\underset{NH_2}{|}}{CH}\text{—}COOH} \xrightarrow{\text{phenylalanine hydroxylase}} \underset{\text{L-Tyrosine}}{HO\text{—}C_6H_4\text{—}CH_2\underset{\underset{NH_2}{|}}{CH}\text{—}COOH} \quad (17\text{-}25)$$

Phenylalanine hydroxylase is the enzyme lacking in people afflicted with the genetic disease phenylketonuria (PKU). For them tyrosine is an essential amino acid.

17-6 Food and nutrition

Body cells are kept supplied with their needed metabolites by the ingestion and digestion of food. The proper amount and balance of kinds of food are very important for good health and growth. A person could be suffering from malnutrition and an unsatisfied appetite despite a diet of sufficient calories if the diet lacked essential fatty acids, vitamins, and a balance of amino acids. Indeed, such a person could even become obese in trying to quell the appetite and furnish the body with necessary nutrients from inadequate foods.

Grains are not a complete protein, since they are lacking in some of the essential amino acids. Neither are beans and lentils complete proteins for the same reason, but together they complement each other and supply an adequate protein. Fresh fruits and vegetables furnish needed vitamins and minerals. Of course, meat and dairy products have long been recognized as high-protein foods. An old German proverb expresses the importance of a proper diet: "Man ist was Man isst" (one is what one eats).

Summary

Metabolism is the total process in which food is taken into the body and subjected to many complex changes with release of energy and the production of smaller molecules, which may then be built into the more complex molecules of new cells and tissues. *Catabolism* is the degradation of absorbed food into smaller molecules with release of energy. *Anabolism* is the biosynthesis of complex molecules from simpler ones—an endergonic process. A *metabolite* is any compound, except enzymes, involved in a metabolic reaction.

Photosynthesis is the overall process in which plants utilize water, inorganic salts and nitrogenous compounds from the soil, carbon dioxide from the atmosphere, and radiant energy from the sun to synthesize carbohydrates, fats, and proteins. *Respiration* is the total process by which glucose is oxidized in the cells of the body in a stepwise fashion to produce carbon dioxide, water, and energy. Key *endergonic* biochemical reactions, which require energy, are *coupled* with *exergonic* reactions, which give off energy.

Adenosine triphosphate (ATP) is a nucleoside derivative which is built up from adenosine diphosphate (ADP), phosphoric acid, and energy. The breakdown of glucose is coupled with the synthesis of ATP, which results in converting part of the energy to ATP. The new phosphorus-oxygen bond which is formed in producing ATP is a *high-energy bond*. *Oxidative phosphorylation* consists of the coupled reactions whereby ADP is combined with inorganic phosphate to produce ATP.

Amino acids from catabolism of proteins can furnish pyruvic acid, one of the metabolites from catabolism of carbohydrates. Fatty acids from catabolism of fats can be converted to acetyl coenzyme A, a carbohydrate metabolite. Some amino acids, proteins, fatty acids, and fats can be synthesized from metabolites of carbohydrate catabolism. Also, glycogen can be synthesized from metabolites coming from fats or proteins, as well as from carbohydrates.

Most carbohydrates are ingested in the form of starch or sucrose. Digestion begins in the mouth and extends into the small intestine with the aid of enzymes which convert all the carbohydrates to monosaccharides. In the liver, those monosaccharides are enzymatically converted to glucose, which is the only sugar that circulates in the blood. Glucose is referred to as *blood sugar*. Glucose which is transported to the tissues may be oxidized to carbon dioxide and water or other metabolites needed for biosynthesis. It may be converted into fat, or it may be converted to muscle glycogen. Some glucose is converted to liver glycogen and stored in the liver as a reservoir for maintenance of the blood sugar.

Hypoglycemia is a condition resulting from a lower-than-normal blood sugar level. *Hyperglycemia* is a condition resulting from a higher-than-normal blood sugar level. *Insulin*, which is secreted by the pancreas, acts to lower the blood sugar level by stimulating the formation of glycogen from glucose (*glycogenesis*). If the pancreas does not secrete enough insulin, *diabetes* may result. *Adrenaline* (epinephrine) activates the enzyme phosphorylase, which is involved in the breakdown of glycogen to produce glucose (*glycogenolysis*).

Glycolysis is the first stage of *glucose catabolism*. It is anaerobic and takes place in the *cytoplasm* of the cell. There glucose is broken down to pyruvic acid with net formation of 2 mol of ATP per mole of glucose catabolized. The sequence of reactions is known as the *Embden-Meyerhof* or *glycolytic pathway*. *Nicotinamide adenine dinucleotide* (NAD^+), which serves as an electron acceptor in the glycolytic sequence, incorporates niacin or nicotinamide (one of the B vitamins) in its structure.

Stage two of glucose catabolism occurs in the *mitochondria* of the cells. *Pyruvic acid* is converted to *acetyl coenzyme A,* which is further broken down in the *citric acid* (*Krebs*) *cycle.* In aerobic organisms, most of the pyruvic acid is oxidized in a series of enzymatic reactions to carbon dioxide and water. About 93% of the energy of glucose is liberated in the oxidation of pyruvic acid, but it is mostly in the chemical form of high-energy phosphate bonds of ATP.

The *respiratory chain* is the chain of chemical events resulting in the reduction of oxygen to water in the cells. It is the chief source of the reactions whereby most of the ATP is produced. The respiratory chain is also known as the *electron transport system.*

The *fats* which are ingested by higher animals are mostly hydrolyzed with the aid of *lipases* in the small intestine to fatty acids, monoglycerides, diglycerides, and glycerol. Those products are transported through the intestinal walls into the bloodstream, most of them by way of the lymphatic system. In the bloodstream, the fatty acids are reesterified with glycerol. The resulting triglycerides are complexed with blood proteins and transported to storage sites for deposition as *depot fat* or to body organs for metabolism. An excess of depot fat results in much adipose tissue and the overweight condition known as *obesity.* Fats are high-energy foods. About 9 kcal is produced per gram of fat catabolized, compared with about 4 kcal per gram of carbohydrate. The fatty acids are both broken down and synthesized in units of two carbon atoms. The metabolic reactions involving the fatty acids are referred to as the *fatty acid cycle.*

Proteins are hydrolyzed by various *proteolytic enzymes* in the stomach and small intestine. Those enzymes are highly specific. The amino acids which result from hydrolysis of proteins are absorbed into the bloodstream. There is no particular storage form or place for amino acids, but the small quantity of amino acids in circulation is referred to as the *amino acid pool.* The amino acids may be used for synthesis of body proteins or of nonprotein nitrogen compounds, converted to other amino acids, metabolized for release of energy, or converted into glycogen or fat. The metabolic reactions of amino acids include (a) *oxidative deamination,* (b) *transamination,* (c) *decarboxylation,* and (d) *alteration of the carbon chain.* The ammonia liberated in metabolic reactions of amino acids can be converted to urea for excretion in the urine.

Body cells are kept supplied with the metabolites they use by the ingestion and digestion of food. The proper amount and balance of kinds of food are very important for good health and growth.

Key terms

Acetyl coenzyme A: a thioester of acetic acid and coenzyme A. It is an important intermediate in the metabolism of carbohydrates and fats.

Active transport: the energy-requiring process of moving molecules across a membrane from a region of lower concentration to one of greater concentration—a direction opposite that of osmotic diffusion. It is accomplished in the membrane of the cell wall by special proteins embedded in the membrane.

ADP: adenosine diphosphate, a nucleoside derivative. It can add inorganic phosphate in a high-energy bond to yield ATP.

Amino acid pool: the small quantity of free amino acids in circulation in the blood.

Anabolism: the biosynthesis of complex molecules from simpler ones; it is an endergonic process.

ATP: adenosine triphosphate, a nucleoside derivative built up from ADP, inorganic phosphate, and energy. The new phosphorus-oxygen bond formed in producing ATP is a high-energy bond that is used to store energy for biochemical reactions.

Blood sugar: glucose, the only sugar that circulates in the blood.

Catabolism: the degradation of absorbed food into smaller molecules; it is an exergonic process.

Cellular respiration: the use of oxygen in metabolic reactions in the cell. It is accomplished by respiratory enzymes in the respiratory, or electron transport, chain.

Citric acid cycle: the complex series of reactions in which the acetate of acetyl CoA is converted to carbon dioxide, which releases energy that is stored in ATP. Also called the **Krebs cycle.**

Coenzyme A: a coenzyme containing the nucleotide adenylic acid and the vitamin pantothenic acid (one of the B vitamins). It is of great importance in the metabolism of carbohydrates and fats.

Coupled reactions: key endergonic biochemical reactions that are coupled with exergonic reactions.

Cytochrome: one of the iron porphyrin enzymes, which are important respiratory enzymes.

Decarboxylation: elimination of carbon dioxide from an amino acid to form a primary amine.

Depot fat: triglycerides stored in adipose tissue.

Electron transport system: the chain of chemical events resulting in the reduction of oxygen to water in the mitochondria; it produces most of the ATP in the cell. Also called the **respiratory chain.**

Embden-Meyerhof pathway: the sequence of reactions by which glycolysis occurs. Also known as the **glycolytic pathway.**

FAD: flavin adenine dinucleotide, a coenzyme consisting of a molecule of vitamin B_2 bonded to adenine dinucleotide. It serves as an oxidizing agent or electron acceptor in a number of metabolic reactions in which it is reduced to $FADH_2$.

Fatty acid cycle: the sequence of metabolic reactions involving the fatty acids. Also called the **fatty acid spiral.**

Glucosuria: a condition in which excess glucose is excreted in the urine; it occurs when the blood sugar level exceeds the renal threshold.

Glycogenesis: formation of glycogen from glucose; it is stimulated by the hormone insulin.

Glycogenolysis: the breakdown of glycogen to glucose; it is catalyzed by the enzyme phosphorylase, which is activated by the hormone adrenaline.

Glycolysis: the first stage of glucose catabolism, which occurs in the cytoplasm. Glucose is broken down to pyruvic acid with net formation of 2 mol of ATP per mole of glucose catabolized. The sequence of reactions is known as the Embden-Meyerhof or glycolytic pathway.

Glycolytic pathway: the sequence of reactions by which glycolysis occurs. Also known as the **Embden-Meyerhof pathway.**

High-energy bond: the phosphorus-oxygen bond formed in producing ATP from ADP and inorganic phosphate.

Homeostasis: the response of an organism to stimuli that disturb its various balances, whereby a series of reactions which restores the unbalanced system of the organism to normal is started. Called **feedback** in engineering.

Hyperglycemia: a condition resulting from a higher-than-normal blood sugar level.

Hypoglycemia: a condition resulting from a lower-than-normal blood sugar level.

Krebs cycle: the complex series of reactions in which the acetate of acetyl CoA is converted to carbon dioxide and thereby releases energy which is stored in ATP. Also called the **citric acid cycle.**

Lipase: one of the enzymes in the small intestine which catalyze hydrolysis of lipids (fats).

Metabolism: the total process in which food is taken into the organism and subjected to many complex changes. The process involves release of energy and smaller molecules, which may then be built into the more complex molecules of new cells and tissues.

Metabolite: any compound, except an enzyme, involved in a metabolic reaction.

Mitochondria: subcellular organelles. All of the respiratory enzymes are organized together in tiny particles on the inner and outer surfaces of the mitochondria.

NAD^+: nicotinamide adenine dinucleotide, a coenzyme which serves as an oxidizing agent or electron acceptor in a number of metabolic reactions in which it is reduced to NADH. Its reactive portion is the vitamin nicotinamide (niacin).

Nitrogen balance: a condition in which the amount of nitrogen excreted is equal to the amount of nitrogen taken in as ingested proteins.

Oxidative deamination: replacement of the amino group of an amino acid by oxygen. It is catalyzed by amino acid oxidases which occur in the liver and kidneys of mammals.

Oxidative phosphorylation: the reaction in which inorganic phosphate and ADP are joined in a high-energy bond to produce ATP.

Photosynthesis: the overall process in which plants utilize water, inorganic salts and nitrogenous compounds from the soil, carbon dioxide from the atmosphere, and radiant energy from the sun to synthesize carbohydrates, fats, and proteins.

Porphin: an aromatic super-ring system of conjugated double bonds consisting of four pyrrole nuclei joined together by =CH— groups from the 2 position of each pyrrole ring to the 5 position of the adjoining pyrrole ring.

Porphyrin: a substituted porphin. The cytochromes, important respiratory enzymes, are iron porphyrins.

Proteolytic enzyme: a highly specific enzyme which hydrolyzes a certain type of peptide linkage.

Pyridoxol: vitamin B_6. It is involved as a coenzyme in transamination reactions.

Renal threshold: a blood sugar level of 160 to 170 mg per 100 ml of blood or higher. At that level, the kidneys can no longer process all of the glucose, and excess glucose is excreted in the urine.

Respiration: the total process by which glucose is oxidized in the cells of the body in a stepwise fashion to produce carbon dioxide, water, and energy.

Respiratory chain: the chain of chemical events resulting in the reduction of oxygen to water in the mitochondria of cells. It produces most of the ATP in the cell. Also called the **electron transport system.**

Respiratory enzymes: the enzymes involved in the respiratory chain.

Transamination: the process of removal of the α-amino group of an amino acid and transfer of the group to a molecule of an α-keto acid. The original amino acid thereby becomes an α-keto acid, and the original α-keto acid becomes an α-amino acid. The process is accomplished by transaminases.

Additional problems

17-12 Write the equations for the enzymatic decarboxylation of (a) tryptophan, (b) lysine, (c) arginine.

17-13 Some food faddists in affluent countries have become obsessed with eating an abundance of protein foods. What happens to proteins which are consumed in excess of the body's needs for energy and for building new protein and necessary nonprotein nitrogenous compounds?

17-14 Why is it necessary to eat a balanced diet? What may be the result of a diet which is not properly balanced for essential nutrients?

17-15 What is the difference in both structure and energy content between ADP and ATP?

17-16 In general, how are the B vitamins utilized?

17-17 List the function of each of the following: (a) aldolase, (b) insulin, (c) lipase, (d) cholinesterase.

17-18 List at least one enzyme found in each of the items listed below. For each enzyme, give its substrate(s) and the end product(s).

(a) saliva
(b) gastric juice
(c) pancreatic juice
(d) intestinal juice
(e) pancreas
(f) liver
(g) papaya (look up *papain* in the dictionary)
(h) blood
(i) kidneys

INDEX

Index to Some Applications of Organic Chemistry